Springer Collected Works in Mathematics

For further volumes:
http://www.springer.com/series/11104

Brown University, 1990s

Peter D. Lax

Selected Papers II

Editors

Peter Sarnak · Andrew J. Majda

Reprint of the 2005 Edition

 Springer

Author
Peter D. Lax
Courant Institute
New York, NY
USA

Editors
Peter Sarnak
Andrew Majda
Princeton University, Princeton, NJ
USA

ISSN 2194–9875
ISBN 978-1-4614-9431-7   (Softcover)
    978-0-387-22926-3   (Hardcover)
DOI 10.1007/978-0-387-28149-0
Springer New York Heidelberg Dordrecht London

Library of Congress Control Number: 2012954381

Printed on acid-free paper

Springer is part of Springer Science+Business Media (www.springer.com)

Contents

Part VII. Functional Analysis

Part VIII. Analysis

Part IX. Algebra

PETER D. LAX
List of Publications

1944

[1] Proof of a conjecture of P. Erdös on the derivative of a polynomial, *Bull. Amer. Math. Soc.* **50**, 509–513.

1948

[2] The quotient of exponential polynomials, *Duke Math. J.* **15**, 967–970.

1950

[3] Partial Diffential Equations, Lecture Notes, NYU, IMS (1950–51).

1951

[4] A remark on the method of orthogonal projections, *Comm. Pure Appl. Math.* **4**, 457–464.

1952

[5] On the existence of Green's function, *Proc. Amer. Math. Soc.* **3**, 526–531.

1953

[6] Nonlinear hyperbolic equations, *Comm. Pure Appl. Math.* **6**, 231–258.

1954

[7] Weak solutions of nonlinear equations and their numerical computation, *Comm. Pure Appl. Math.* **7**, 159–194.

[8] Symmetrizable linear transformations, *Comm. Pure Appl. Math.* **7**, 633–648.

[9] (with A. Milgram) Parabolic equations, *Ann. Math. Studies 33* (Princeton) 167–190.

[10] The initial value problem for nonlinear hyperbolic equations, *Ann. Math. Studies 33* (Princeton) 211–229.

1955

[11] Reciprocal extremal problems in function theory, *Comm. Pure Appl. Math.* **8**, 437–454.

[12] On Cauchy's problem for hyperbolic equations and the differentiability of solutions of elliptic equations, *Comm. Pure Appl. Math.* **8**, 615–633.

[13] (with R. Courant) Cauchy's problem for hyperbolic differential equations, *Ann. Mat. Pura Appl.* **40**, 161–166.

1956

[14] A stability theorem for solutions of abstract differential equations, and its application to the study of local behavior of solutions of elliptic equations, *Comm. Pure Appl. Math.* **9**, 747–766.

[15] (with R.D. Richtmyer) Survey of the stability of linear finite difference equations, *Comm. Pure Appl. Math.* **9**, 267-293.

[16] (with R. Courant) The propagation of discontinuities in wave motion, *Proc. Nat. Acad. Sci.* **42**, 872–876.

1957

[17] A Phragmen–Lindelöf theorem in harmonic analysis and its application to some questions in the theory of elliptic equations, *Comm. Pure Appl. Math.* **10**, 361–389.

[18] Hyperbolic systems of conservation laws, II, *Comm. Pure Appl. Math.* **10**, 537–566.

[19] Remarks on the preceding paper, *Comm. Pure Appl. Math.* **10**, 617–622.

[20] Asymptotic solutions of oscillatory initial value problems, *Duke Math. J.* **24**, 627–646.

1958

[21] Differential equations, difference equations and matrix theory, *Comm. Pure Appl. Math.* **11**, 175–194.

1959

[22] Translation invariant spaces, *Acta Math.* **101**, 163-178.

1960

[23] (with B. Wendroff) Systems of conservation laws, *Comm. Pure Appl. Math.* **13**, 217–237.

[24] (with R.S. Phillips) Local boundary conditions for dissipative symmetric linear differential operators, *Comm. Pure Appl. Math.* **13**, 427–455.

1961

[25] Translation invariant spaces, Proc. Int. Symp. on Linear Spaces, Israeli Acad. of Sciences and Humanities, Jerusalem (1960) (Pergamon), 299–307.

[26] On the stability of difference approximations to solutions of hyperbolic equations with variable coefficients, *Comm. Pure Appl. Math.* **14**, 497–520.

1962

[27] (with R.S. Phillips) The wave equation in exterior domains, *Bull. Amer. Math. Soc.* **68**, 47–79.

[28] A procedure for obtaining upper bounds for the eigenvalues of a Hermitian symmetric operator, *Studies in Mathematical Analysis and Related Topics* (Stanford Univ. Press), 199–201.

1963

[29] On the regularity of spectral densities, *Teoriia Veroiatnosteii i ee Prim.* **8**, 337–340.

[30] An inequality for functions of exponential type, *Comm. Pure Appl. Math.* **16**, 241–246.

[31] (With C.S. Morawetz and R. Phillips) Exponential decay of solutions of the wave equation in the exterior of a star-shaped obstacle, *Comm. Pure Appl. Math.* **16**, 477–486.

[32] Survey of stability of difference schemes for solving initial value problems for hyperbolic equations, *Symp. Appl. Math.* **15**, 251–258.

1964

[33] Development of singularities of solutions of nonlinear hyperbolic partial differential equations, *J. Math. Phys.* **5**, 611–613.

[34] (with B. Wendroff) Difference schemes for hyperbolic equations with high order of accuracy, *Comm. Pure Appl. Math.* **17**, 381–398.

[35] (with R.S. Phillips) Scattering theory, *Amer. Math. Soc. Bull.* **70**, 130–142.

1965

[36] (with J.F. Adams and R.S. Phillips) On matrices whose real linear combinations are nonsingular, *Proc. AMS* **16**, 318–322; correction, *ibid.* **17** (1966) 945–947.

[37] Numerical solution of partial differential equations, *Amer. Math. Monthly* **72** II, 74–84.

[38] (with K.O. Friedrichs) Boundary value problems for first order operators, *Comm. Pure Appl. Math.* **18**, 355–388.

1966

[39] (with R.S. Phillips) Analytic properties of the Schrödinger scattering matrix, in *Perturbation Theory and Its Application in Quantum Mechanics*, Proc., Madison, 1965 (John Wiley & Sons), 243–253.

[40] Scattering theory; remarks on the energy theory, and on scattering theory for a geometrical optics model, in *Proc. Conf. on Dispersion Theory*, Cambridge, MA, 1966, M 11, pp. 36–39, 40–42.

[41] (with L. Nirenberg) On stability for difference schemes; a sharp form of Garding's inequality, *Comm. Pure Appl. Math.* **19**, 473–492.

[42] (with J.P. Auffray) Aspects mathématiques de la mécanique de phase, *Acad. Sci. Paris, Compt. Rend.* (B) **263**, 1355–1357.

1967

[43] (with J. Glimm) Decay of solutions of systems of hyperbolic conservation laws, *Bull. AMS* **73**, 105.

[44] Hyperbolic difference equations: a review of the Courant–Friedrichs–Lewy paper in the light of recent developments, *IBM J. Res. Develop.* II(2), 235–238.

[45] (with R.S. Phillips) *Scattering Theory* (Academic Press).

[46] (with R.S. Phillips) The acoustic equation with an indefinite energy form and the Schrödinger equation, *J. Func. Anal.* **1**, 37–83.

[47] (with R.S. Phillips) Scattering theory for transport phenomena, in *Functional Analysis*, Proc. of Conf. at Univ. of Calif., Irvine, 1966, ed. Gelbaum (Thompson Book Co.), 119–130.

[48] (with K.O. Friedrichs) On symmetrizable differential operators, Symp. on Pure Math., Singular Integrals, *AMS* **10**, 128–137.

1968

[49] Integrals of nonlinear equations of evolution and solitary waves, *Comm. Pure Appl. Math.* **21**, 467–490.

1969

[50] Nonlinear partial differential equations and computing, *SIAM Rev.* **11**, 7–19.

[51] (with R.S. Phillips) Decaying modes for the wave equation in the exterior of an obstacle, *Comm. Pure Appl. Math.* **22**, 737–787.

[52] Toeplitz operators, in *Lecture Series in Differential Equations 2*, ed. A.K. Aziz (Van Nostrand), 257–282.

1970

[53] (with J. Glimm) Decay of solutions of systems of nonlinear hyperbolic conservation laws, *Memoirs of the AMS* **101**.

[54] (with R.S. Phillips) The Paley–Wiener theorem for the Radon transform, *Comm. Pure Appl. Math.* **23**, 409–424; Errata *ibid.* **24** (1971) 279.

1971

[55] Nonlinear partial differential equations of evolution, *Proc. Int. Conf. of Mathematicians*, Nice, September 1970 (Gauthier-Villars), 831–840.

[56] (with R.S. Phillips) Scattering theory, *Rocky Mountain J. Math.* **1**, 173–223.

[57] (with H. Brezis, W. Rosenkrantz, and B. Singer) On a degenerate elliptic parabolic equation occurring in the theory of probability, *Comm. Pure Appl. Math.* **24**, 410–415; appendix by P.D. Lax.

[58] Shock waves and entropy, *Contributions to Functional Analysis*, ed. E.H. Zarantonello (Academic Press), 603–634.

[59] Approximation of measure preserving transformations, *Comm. Pure Appl. Math.* **24**, 133–135.

[60] (with K.O. Friedrichs) Systems of conservation equations with a convex extension, *Proc. Nat. Acad. Sci.* **68**, 1686–1688.

[61] (with R.S. Phillips) A logarithmic bound on the location of the poles of the scattering matrix, *Arch. Rational Mech. Anal.* **40**, 268–280.

1972

[62] (with R.S. Phillips) On the scattering frequencies of the Laplace operator for exterior domains, *Comm. Pure Appl. Math.* **25**, 85–101.

[63] The formation and decay of shock waves, *Amer. Math. Monthly* **79**, 227–241.

[64] (with R.S. Phillips) Scattering theory for the acoustic equation in an even number space dimensions, *Ind. U. Math. J.* **22**, 101–134.

[65] Exponential modes of the linear Boltzmann equation, in *The Boltzmann Equation*, ed. F.A. Grunbaum, NYU, CIMS, 111–123.

1973

[66] Hyperbolic systems of conservation laws and the mathematical theory of shock waves, Conf. Board of the Mathematical Sciences, Regional Conf. Series in Appl. Math. (*SIAM*), 11.

[67] The differentiability of Pólya's function, *Adv. Math.* **10**, 456–465.

[68] (with R.S. Phillips) Scattering theory for dissipative hyperbolic systems, *J. Func. Anal.* **14**, 172–235.

1974

[69] Invariant functionals of nonlinear equations of evolution, in *Nonlinear Waves*, eds. S. Leibovich and R. Seebass (Cornell Univ. Press), 291–310.

[70] Applied mathematics and computing, *Symp. on Appl. Math.* (AMS) **20**, 57–66.

[71] Periodic solutions of the KdV equations, in *Nonlinear Wave Motion*, ed. A.C. Newell, Lectures in Appl. Math. **15**, 85–96.

1975

[72] Periodic solutions of the KdV equation, *Comm. Pure Appl. Math.* **28**, 141–188.

[73] Almost periodic behavior of nonlinear waves, *Adv. Math.* **16**, 368–379.

1976

[74] Almost periodic solutions of the KdV equation, *SIAM Rev.* **18**, 351–375.

[75] (with A. Harten, J.M. Hyman, and B. Keyfitz) On finite-difference approximations and entropy conditions for shocks, *Comm. Pure Appl. Math.* **29**, 297–322.

[76] On the factorization of matrix-valued functions, *Comm. Pure Appl. Math.* **29**, 683–688.

[77] (with R.S. Phillips) *Scattering Theory for Automorphic Functions*, Ann. Math. Studies 87 (Princeton Univ. Press and Univ. of Tokyo Press).

[78] (with S. Burstein and A. Lax) *Calculus with Applications and Computing*, Undergrad. Texts in Math. 1 (Springer).

1977

[79] The bomb, Sputnik, computers and European mathematicians, the Bicentenial Tribute to American Mathematics, San Antonio, 1976 (Math. Assoc. of America), 129–135.

[80] (with R.S. Phillips) The scattering of sound waves by an obstacle, *Comm. Pure Appl. Math.* **30**, 195–233.

1978

[81] (with M.S. Mock) The computation of discontinuous solutions of linear hyperbolic equations, *Comm. Pure Appl. Math.* **31**, 423–430.

[82] (with R.S. Phillips) An example of Huygens' principle, *Comm. Pure Appl. Math.* **31**, 415–421.

[83] (with R.S. Phillips) The time delay operator and a related trace formula, *Topics in Functional Analysis*, Adv. Math. Suppl. Studies, 3, eds. I.T. Gohberg and M. Kac (Academic Press), 197–215.

[84] Accuracy and resolution in the computation of solutions of linear and nonlinear equations, *Recent Adv. in Numer. Anal.*, Proc. of Symp., Madison (Academic Press), 107–117.

[85] (with A. Lax) On sums of squares, *Linear Algebra and Its Appl.* **20**, 71–75.

[86] (with R.S. Phillips) Scattering theory for domains with nonsmooth boundaries, *Arch. Rational Mech. Anal.* **68**, 93–98.

[87] Chemical kinetics, in *Lectures on Combustion Theory*, eds. S. Burstein, P.D. Lax, and G.A. Sod, NYU, COO-3077-153, 122–136.

1979

[88] (with R.S. Phillips) Translation representations for the solution of the non-Euclidean wave equation, *Comm. Pure Appl. Math.* **32**, 617–667.

[89] (with C.D. Levermore) The zero dispersion limit for the Korteweg–de Vries equation, *Proc. Natl. Acad. Sci.* **76**, 3602–3606.

[90] Recent methods for computing discontinuous solutions – a review, in *Computing Methods in Applied Sciences and Engineering*, 1977, II; 3rd Int. Symp. IRIA, eds. R. Glowinski and D.L. Lions, Springer Lecture Notes in Physics, **91**, 3–12.

1980

[91] (with R.S. Phillips) Scattering theory for automorphic functions (AMS), *Bull.* **2**, 161–195.

1981

[92] On the notion of hyperbolicity, *Comm. Pure Appl. Math.* **33**, 395–397.

[93] (with A. Harten) A random choice finite difference scheme for hyperbolic conservation laws, *SIAM J. Numer. Anal.* **18**, 289–315.

[94] (with R.S. Phillips) The translation representation theorem, *Integral Equations and Operator Theory* **4**, 416–421.

[95] (with R.S. Phillips) Translation representations for the solution of the non-Euclidean wave equation, II, *Comm. Pure Appl. Math.* **34**, 347–358.

[96] Applied mathematics 1945 to 1975, *Amer. Math. Heritage, Algebra & Appl. Math.*, 95–100.

[97] *Mathematical Analysis and Applications*, Part B, essays dedicated to Laurent Schwartz, ed. L. Nachbin (Academic Press) Advances in Math. Suppl. Studies, 7B, 483–487.

1982

[98] (with R.S. Phillips) The asymptotic distribution of lattice points in Euclidean and non-Euclidean spaces, *J. Func. Anal.* **46**, 280–350.

[99] The multiplicity of eigenvalues, *AMS Bull.*, 213–214.

[100] (with R.S. Phillips) A local Paley–Wiener theorem or the Radon transform of $L2$ functions in a non-Euclidean setting, *Comm. Pure Appl. Math.* **35**, 531–554.

1983

[101] Problems solved and unsolved concerning linear and nonlinear partial differential equations, *Proc. Int. Cong. Mathematicians*, 1 (North-Holland), 119–137.

[102] (with A. Harten and B. van Leer) On upstream differencing and Godunovo-type schemes for hyperbolic conservation laws, *SIAM Rev.* **25**, 35–61.

[103] (with C.D. Levermore) The small dispersion limit of the Korteweg–de Vries equation, *Comm. Pure Appl. Math.* **36**, I 253–290, II 571–593, III 809–930.

1984

[104] On a class of high resolution total-variation-stable finite difference schemes, Ami Harten with Appendix by Peter D. Lax, *SIAM* **21**, 1–23.

[105] (with R.S. Phillips) Translation representations for automorphic solutions of the wave equation in non-Euclidean spaces, I, II, III. *Comm. Pure Appl. Math.* **37**, I 303–328, II 779–813, III **38** (1985), 179–207.

[106] *Shock Waves, Increase of Entropy and Loss of Information, Seminar on Non-linear Partial Differential Equations*, ed. S.S. Chern (Math. Sci. Res. Inst. Publ.), 129–171.

1985

[107] *Large Scale Computing in Science, Engineering and Mathematics*, Rome.

[108] (with R.J. Leveque and C.S. Peskin) Solution of a two-dimensional cochlea model using transform techniques, *SIAM J. Appl. Math.* **45**, 450–464.

[109] (with R. Phillips) Translation representations for automorphic solutions of the wave equation in non-Euclidean spaces; the case of finite volume, *Trans. AMS* **289**, 715–735.

[110] (with P. Constantin and A. Majda) A simple one-dimensional model for the three-dimensional vorticity equation, *Comm. Pure Appl. Math.* **38**, 715–724.

1986

[111] On dispersive difference schemes, 1985, Kruskal Symposium, *Physica* **18D**, 250–255.

[112] Mathematics and computing, *J. Stat. Phys.* **43**, 749–756.

[113] (with A. Jameson) Conditions for the construction of multipoint total variation diminishing difference schemes, *Appl. Numer. Math.* **2**, 335–345.

[114] Mathematics and its applications, *Math. Intelligencer* **8**, 14–17.

[115] Hyperbolic systems of conservation laws in several space variables, *Current Topics in Partial Differential Equations*, papers dedicated to Segeru Mizohata (Tokyo Press), 327–341.

1987

[116] *The Soul of Mathematics*, Studies in Mathematics and Its Applications 16, Patterns and Waves (North-Holland).

1988

[117] Oscillatory solutions of partial differential and difference equations, Mathematics Applied to Science (Academic Press), 155–170.

[118] (with R.J. Leveque and C. Peskin) Solution of a two-dimensional cochlea model with fluid viscosity, *SIAM J. Appl. Math.* **48**, 191–213.

[119] The flowering of applied mathematics in America, AMS Centennial Celebration Proc., 455–466; *SIAM Rev.* **31**, 65–75.

[120] (with J. Goodman) On dispersive difference schemes I, *Comm. Pure Appl. Math.* **41**, 591–613.

1989

[121] Science and computing, *Proc. IEEE* **77**.

[122] Writing mathematics well, Leonard Gillman review, *Amer. Math. Monthly* **96**, 380–381.

[123] From cardinals to chaos: Reflections on the life and legacy of Stanislaw Ulam, reviewed in *Phys. Today* **42**, 69–72; *Bull. AMS* **22** (1990) 304–310, *St. Petersburg Math. J.* **4** (1993) 629–632.

[124] Deterministic turbulence, *Symmetry in Nature*, Volume in Honor of Luigi A. Radicati di Brozolo II (Scuola Normale Superiore), 485–490.

1990

[125] Remembering John von Neumann, *Proc. Symp. Pure Math.* 50.

[126] The ergodic character of sequences of pedal triangles, *Amer. Math. Monthly* 97, 377–381.

1991

[127] Deterministic analogues of turbulence, *Comm. Pure Appl. Math.* 44, 1047–1055.

[128] (with T. Hou) Dispersive approximations in fluid dynamics, *Comm. Pure Appl. Math.* 44.

1992

[129] (with R.S. Phillips) Translation representation for automorphic solutions of the wave equation in non-Euclidean spaces. IV. *Comm. Pure Appl. Math.* 45 (1992), no. 2, 179–201.

1993

[130] (with C.D. Levermore and S. Venakides) The generation and propagation of oscillations in dispersive IVP's and their limiting behavior, in *Important Developments in Soliton Theory 1980–1990*, eds. T. Fokas and V.E. Zakharov (Springer-Verlag).

[131] The existence of eigenvalues of integral operators, in Honor of C. Foias, ed. R. Temam, *Ind. Univ. Math. J.* 42, 889–991.

1994

[132] Trace formulas for the Schrödinger operator, *Comm. Pure Appl. Math.* 47, 503–512.

[133] Cornelius Lanczos and the Hungarian phenomenon in science and mathematics, *Proc. Lanczos Centennary Conf.* (N.C. State University Press).

1995

[134] Computational fluid dynamics at the Courant Institute 1-5, *Computational Fluid Dynamics Review*, eds. M. Hafez and K. Oshima (John Wiley & Sons).

[135] A short path to the shortest path, *Amer. Math. Monthly* 102, 158–159.

1996

[136] Outline of a theory of the KdV equation, Lecture Notes in Mathematics, *Recent Mathematical Methods in Nonlinear Wave Propagation* (Springer) 1640, 70–102.

[137] (with Xu-Dong Liu) Positive schemes for solving multidimensional hyperbolic conservation laws, *Comp. Fluid Dynamics J.* 5, 133–156.

[138] *The Old Days: A Century of Mathematical Meetings* (AMS), 281–283.

[139] (with A. Harten, C.D. Levermore, and W.J. Morokoff) Convex entropies and hyperbolicity for general Euler equations, *SIAM J. Numer. Anal.*

1997

[140] *Linear Algebra*, Pure and Applied Math. Series (Wiley-Interscience).

1998

[141] (with Xu-Dong Liu) Solution of the two-dimensional Riemann problem of gas dynamics by positive schemes, *SIAM J. Sci. Comput.* **19**, 319–340.

[142] Jean Leray and Partial Differential Equations, Introduction to Volume II, *Selected Papers of Jean Leray* (Springer-Verlag).

[143] On the discriminant of real symmetric matrices, *Comm. Pure Appl. Math.* **51**, 1387–1396.

[144] The beginning of applied mathematics after the Second World War, *Quart. Appl. Math.* **56**, 607–615.

1999

[145] A mathematician who lived for mathematics, Book Review, *Review, Phys. Today*, 69–70.

[146] The mathematical heritage of Otto Toeplitz, in *Otto Toeplitz*, Bonner Mathematische Schriften, 319, 85–100.

[147] Mathematics and computing, in *Useful Knowledge*, ed. A.G. Bearn (Amer. Philos. Soc.), 23–44.

[148] Change of variables in multiple integrals, *Amer. Math. Monthly* **105**, 497–501.

2000

[149] Mathematics and computing, in IMU, Mathematics: Frontiers and Perspectives (AMS), 417–432.

2001

[150] Change of variables in multiple integrals II, *Amer. Math. Monthly* **108**, 115–119.

[151] On the accuracy of Glimm's scheme, *Math. Appl. Anal.*, **7**, 473–478.

[152] The Radon transform and translation representation, *J. Evol. Equ.* **1**, 311–323.

2002

[153] *Functional Analysis*, Pure and Applied Mathematics Series (Wiley-Interscience).

[154] Jaques-Louis Lions, International Scientist (SMAI Journal MATAPLI), to appear.

[155] Richard Courant, *National Academy of Sciences*, Biographical Memoirs, **82**.

[156] Jürgen Moser, *Ergod. Th. & Dynam. Sys.* **22**, 1337–1342.

2003

[157] *John von Neumann: The Early Years, the Years at Los Alamos and the Road to Computing*, to appear.

[158] (with G. Francsics) A fundamental domain for the Picard modular group in C^2, *ESI* **1273**, 1–18.

PART V

SCATTERING THEORY IN EUCLIDEAN SPACE

THE WAVE EQUATION IN EXTERIOR DOMAINS

BY PETER D. LAX[1] AND RALPH S. PHILLIPS[2]

Communicated by Lipman Bers, October 26, 1961

This note deals with solutions of the wave equation in three dimensions in the exterior of a finite number of smooth obstacles, on whose boundaries the solution is subject to boundary conditions of the form $u=0$ or $u_n=\sigma u$, σ a non-negative function. We shall show that every such solution of finite energy propagates eventually out to infinity and behaves asymptotically like a free space solution.

THEOREM I. *For every nonzero solution there exists a positive constant d less than the total energy of the solution such that given any bounded domain, there is a time at which the energy contained in the exterior of this domain exceeds d.*

SKETCH OF PROOF. Suppose the theorem is false for some u; then given any positive ϵ there exists a bounded domain such that the energy contained in its exterior is less than ϵ for all time. Assume that u is a smooth function;[3] applying the law of conservation of energy to u_t and using standard estimates for u_{xx} in terms of Δu, we see that the square integral of the second partial derivatives of u over the exterior of the obstacles is uniformly bounded for all time. Let $\{t_n\}$ be an arbitrary sequence of numbers; using the Rellich selection theorem[4] and a diagonal process we can choose a subsequence such that the first partial derivatives of u form a Cauchy sequence in the square integral sense over any bounded domain in x-space. Since we have supposed that the energy contained outside bounded domains is uniformly small for all t, it follows that $\{u(t_n)\}$ is a Cauchy sequence in the energy norm over the whole exterior. This shows that $u(t)$ is a vector-valued almost periodic function of t. By the main theorem of a.p. functions $u(t)$ is a superposition of exponentials:

$$u \approx \sum a_j(x)e^{i\omega_j t}.$$

From the mean value expression for the coefficient a_j it follows that it is a solution of the reduced wave equation $\Delta a + \omega_j^2 a = 0$. But according to a theorem of Rellich,[5] the reduced wave equation has for $\omega_j \neq 0$

[1] Sloan Fellow.
[2] Sponsored by the National Science Foundation, contract NSF-G16434.
[3] u can be made smooth by mollifying it with respect to t.
[4] F. Rellich, Göttinger Nachrichten, 1930.
[5] F. Rellich, Jber. Deutsch. Math. Verein. 53 (1943), 57–65.

47

no nonzero solution which vanishes near infinity; so all Fourier coefficients with $w_j \neq 0$ vanish. This shows that u is a time independent harmonic function; since u satisfies the imposed boundary conditions it follows that $u = 0$.

Another application of the selection principle yields this

COROLLARY. *Given any bounded domain D and any ϵ, there exist arbitrarily large values of t for which the energy contained in D at time t is less than ϵ.*

We define a *detached solution* of the mixed problem as one which vanishes in a cone $|x| \leqq t - t_0$ for some t_0. According to Huygens' principle, any solution of the wave equation in free space whose Cauchy data at $t = 0$ have compact support becomes detached for t large enough.

THEOREM II. *The Cauchy data of detached solutions are dense in the space of all Cauchy data relative to the energy norm.*

SKETCH OF PROOF. Let ϕ denote Cauchy data which are orthogonal in the energy norm to the Cauchy data of all detached solutions. Let $v(t)$ be the solution of the mixed problem with Cauchy data equal to ϕ at $t = 0$. It is easy to show that the Cauchy data of v at any other time have the same orthogonality property. Denote by w_s a solution of the wave equation *in free space* whose Cauchy data at $t = s$ agree with those of v outside the obstacles. Using the orthogonality of v to the detached solutions constructed above one shows easily that w_s is zero in the backward cone $|x| < s - t - r$, where r is the radius of a sphere around the origin which contains the obstacles. Thus for $t < s - 2r$, w_s is a solution of the mixed problem as well.

Since v and w_s have the same Cauchy data at $t = s$ outside the obstacles, they are equal in the domain of dependence $|x| > |t - s| + r$. Applying the law of conservation of energy to $v - w_s$ we see that the energy of v at $t = 0$ inside the sphere of radius $s - r$ is bounded by the energy of $v - w_s$ at time $t = s - 2r$ inside the sphere of radius $3r$. This is bounded by a constant times the energy of v at time $t = s$ inside the sphere of radius $5r$. But according to the corollary of Theorem I there exist arbitrarily large values of s for which the latter quantity is arbitrarily small. This shows that v has zero energy, q.e.d.

Theorem II shows that any solution with finite energy can be approximated by detached solutions. Thus for any solution the amount of energy contained in a given bounded set tends to zero with increasing t. Previously, Cathleen S. Morawetz[6] has shown for the ex-

---

[6] Comm. Pure Appl. Math. (1961).

terior of a single star-shaped obstacle and for any solution which satisfies the boundary condition $u = 0$ with Cauchy data at $t = 0$ vanishing outside a sphere S, that the energy contained in sphere D is less than[7] const. E/t, where E is the total energy and the value of the constant depends only on the radii of S and D. We wish to point out that such a quantitative result about energy decay can hold only if the obstacle satisfies the following geometric condition: No two boundary points of it form a segment exterior to the obstacle and perpendicular to it at both endpoints. For a narrow high frequency beam directed along such a segment would be reflected back and forth for a length of time proportional to the reciprocal of wave length.

If one regards the presence of the obstacle as a perturbation of free space and applies the usual scattering theory formalism,[8] then it follows from Huygens' principle in free space that the so-called wave operators exist, and it follows further from Theorem II that they are unitary. This proves

THEOREM III. *The perturbed and unperturbed problems are unitarily equivalent; in particular, the spectrum of the wave equation generator is unaffected by the presence of obstacles.*

Added in proof. Recently we have succeeded in deriving Theorem II without making use of Huygens' principle, by using solutions which are superpositions of plane waves of finite width. Thus the results of this note can be extended to a large class of hyperbolic equations.

INSTITUTE OF MATHEMATICAL SCIENCES, NEW YORK UNIVERSITY AND
    STANFORD UNIVERSITY

---

[7] In a more recent paper, Morawetz has shown that energy decays like $1/t^2$.

[8] See S. T. Kuroda, *Perturbation of continuous spectra by unbounded operators.* I, J. Math. Soc. Japan 11 (1959), 247–262; or J. M. Jauch, *Theory of the scattering operators*, Helv. Phys. Acta 31 (1958), 127–158.

COMMUNICATIONS ON PURE AND APPLIED MATHEMATICS, VOL XVI, 477–486 (1963)

Exponential Decay of Solutions of the Wave Equation in the Exterior of a Star-Shaped Obstacle*

P. D. LAX, C. S. MORAWETZ,
AND R. S. PHILLIPS

Introduction

In this paper we study the behavior for large time of solutions of the wave equation in three space dimensions in the exterior of some smooth, bounded reflecting obstacle, assumed to be *star-shaped*. We shall prove that, given an initial disturbance, the bulk of its energy is propagated to infinity. The precise statement of the result is Theorem III; it has as corollary the following:

For a solution of the above exterior problem whose initial values have finite energy and vanish outside of some bounded region, the energy remaining in any bounded region decays exponentially with time.

When the scattering obstacle is a sphere, this result has been deduced by Wilcox [5], by analyzing the explicit expression for the solution obtained by separation of variables. In [3], C. S. Morawetz has proved that energy decays like the inverse of time, and it follows that the solution decays like $1/\sqrt{t}$. The present paper makes essential use of that result.

Theorem III of this paper implies that the scattering matrix associated with the above problem can be continued analytically from the lower half-plane into a horizontal strip of the upper half-plane. The details of this development are explained in [2].

§ 1. Consider pairs ϕ of smooth functions (ϕ_1, ϕ_2) of $x = (x_1, x_2, x_3)$ with compact support. We introduce the energy norm

$$\|\phi\|^2 = \int (|D\phi_1|^2 + \phi_2^2) \, dx,$$

where $|D\phi_1|^2$ is the sum of the squares of the first partial derivatives of ϕ_1. Denote by H_0 the Hilbert space obtained by completion of the set of functions ϕ

---

* This paper represents results obtained under the sponsorship of the Office of Naval Research, contract number Nonr-285(46), Office of Ordinance Research, Department of the Army, contract number DA-ARO-(D)-31-124-G156, and the National Science Foundation, contract number NSF-G16434. Reproduction in full or in part is permitted for any purpose of the United States Government.

477

with respect to this norm. The scalar product associated with the above norm will be indicated by brackets.

Denote by $u(x, t)$ the solution of the wave equation

$$u_{tt} - \Delta u = 0$$

with initial values

$$u(x, 0) = \phi_1, \qquad u_t(x, 0) = \phi_2.$$

Define the operators $U_0(t)$ as mapping the initial Cauchy data $\phi = (\phi_1, \phi_2)$ into $\psi = (\psi_1, \psi_2)$, where ψ_1 and ψ_2 are the Cauchy data at time t, i.e.,

$$\psi_1 = u(x, t), \qquad \psi_2 = u_t(x, t),$$

where u satisfies the wave equation in all of space. According to the classical theory of the initial value problem for the wave equation (see [6], Chapter VI), for each t, $U_0(t)$ is a unitary transformation of H_0 onto itself (conservation of energy), and these transformations form a one-parameter group.

Denote by H the closure in the energy norm of pairs (ϕ_1, ϕ_2) which are zero inside the obstacle. Clearly H is a closed subspace of H_0.

Denote by $u(x, t)$ the solution of the exterior problem of the wave equation:

$$u_{tt} - \Delta u = 0 \text{ in the exterior,}$$

$$u(x, t) = 0 \text{ on the obstacle,}$$

with initial values

$$u(x, 0) = \phi_1, \qquad u_t(x, 0) = \phi_2,$$

ϕ_1, ϕ_2 in H. The operators $U(t)$ are defined as mapping $\phi = (\phi_1, \phi_2)$ into $\psi = (\psi_1, \psi_2)$:

$$\psi_1 = u(x, t), \qquad \psi_2 = u_t(x, t).$$

According to the classical theory of the mixed problem, for each t the operator $U(t)$ is a unitary transformation of H onto itself and these transformations form a one-parameter group.

A solution $u(x, t)$ of the wave equation is called *incoming*[1] if $u(x, t) = 0$ for t negative and $|x| \leqq \rho - t$ (the cross-hatched region in Figure 1); here ρ is some fixed number so chosen that the sphere of radius ρ around the origin contains the obstacle.

Similarly, a solution is called *outgoing* if it is zero for t positive and $|x| \geqq \rho + t$. The set of Cauchy data at $t = 0$ of incoming and outgoing solutions will be denoted by D_- and D_+. Since ρ was chosen so large that the obstacle is contained in the sphere of radius ρ, it follows that D_- and D_+ are contained in H. Clearly, D_- and D_+ form closed subspaces of H.

---

[1] If one merely requires that u be zero inside the sphere of radius ρ for all negative t, it follows by the Holmgren uniqueness theorem that u vanishes in the layer region indicated in Figure 1.

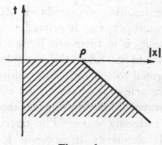

Figure 1

The following proposition plays a crucial role; its proof will be given in Section 2.

THEOREM I. D_+ and D_- are orthogonal

The following further properties of D_+ and D_- will be needed in the sequel; they follow immediately from the definitions:

(i) $U_0(t) = U(t)$ on D_+ for positive t, on D_- for negative t.

(ii) $U(t)$ maps D_+ into itself for t positive, D_- into itself for t negative.

(iii) $U(t)$ maps the orthogonal complement of D_- with respect to H into itself for t positive and the orthogonal complement of D_+ into itself for t negative.

Clearly statement (iii) is the dual of (ii).

Denote by K the set of vectors in H which are orthogonal to both D_+ and D_-, and denote by P_+ and P_- the orthogonal projections onto the orthogonal complements of D_+ and D_-, respectively.

THEOREM II. *For t positive the operators*

$$Z(t) = P_+U(t)P_-$$

map H into K (annihilating D_+ and D_-) and form a one-parameter semigroup over K.

Proof: By (iii), $U(t)P_-\phi$ is orthogonal to D_- for t positive. By Theorem I, projection of $U(t)\phi$ into the orthogonal complement of D_+ leaves it orthogonal to D_-. This shows that $Z(t)$ maps H into K.

To show the semigroup property we observe that for ϕ in K, $P_-\phi = \phi$; so we can write

$$Z(t)\phi = U(t)\phi - \phi_+,$$

where ϕ_+ lies in D_+. It follows from Theorem I and (ii) that for s positive $Z(s)$ annihilates D_+, as well as D_-. Applying $Z(s)$ to both sides of the above relation gives

$$Z(s)Z(t)\phi = Z(s)U(t)\phi = P_+U(s)P_-U(t)\phi.$$

Since, as we saw before, $U(t)\phi$ is orthogonal to D_-, we can, on the extreme right, omit the operator P_-; the resulting expression is $Z(s + t)\phi$ since $U(s)U(t) = U(s + t)$. Thus the semigroup[2] property is proved.

The role of the operator P_- is to filter out signals which are coming in from very far away and which would not reach the obstacle for a very long time. Disturbances which are orthogonal to the incoming solutions would be expected to start interacting with the obstacle immediately; the effect of this interaction is to scatter that portion of the wave which impinges on the obstacle and to convert it after a finite time into an outgoing wave. The restriction that the obstacle be starshaped plays a crucial role here; it prevents long lasting reverberations. This intuitive picture of the physical situation suggests that the norm of the operator $Z(t)$ tends to zero as t tends to infinity; this is indeed so and is the main result of this paper.

THEOREM III. *For the exterior problem of a star-shaped obstacle, there exist positive constants S and k depending only on ρ such that*

$$\|Z(t)\| \le e^{-kt} \quad for \quad t > S.$$

Proof: It suffices to show that for some value of T the norm of $Z(T)$ is less than 1:

$$\|Z(T)\| = e^{-h} < 1.$$

For, we can write any t greater than T in the form

$$t = nT + t_1, \qquad\qquad 0 \le t_1 < T,$$

and hence

$$n > \frac{t}{T} - 1.$$

By the semigroup property,

$$Z(t) = Z(t_1)\,Z^n(T),$$

so

$$\|Z(t)\| \le \|Z(t_1)\|\,\|Z(T)\|^n \le e^{-h(t/T-1)},$$

which is the desired exponential decay.

We proceed to show that for T large enough the norm of $Z(T)$ is less than 1; we shall make use of the following lemma, whose proof is given in Section 2.

LEMMA 1. *If ϕ in H_0 is orthogonal to D_-, then $U_0(t)\phi$ is zero for $|x| < t - \rho$.*

$U_0(t)\phi$ stands for the solution of the wave equation in free space. If such a solution vanishes in the cone $|x| < t - \rho$, then for $t \ge 2\rho$ it vanishes on the boundary of the obstacle and hence it is a solution of the mixed problem as well. In fact this implies:

---

[2] It is shown in [2] that the semigroup $Z(t)$ is closely related to the scattering operator; in particular the point spectrum of its infinitesimal generator coincides with the set of poles of the scattering matrix.

COROLLARY. *If ϕ is orthogonal to D_-, then $U_0(2\rho)\phi$ belongs to D_+.*

Next we define the operator A as

(1.1) $$A = U(2\rho) - U_0(2\rho).$$

LEMMA 2. *The operator A has the following properties:*
(a) *A maps the orthogonal complement of D_- into the orthogonal complement of D_-,*
(b) *for all ψ orthogonal to D_-*

(1.2) $$Z(t)\psi = P_+ A U(t - 4\rho) A\psi,$$

(c) *$A\psi$ is zero for $|x| > 3\rho$,*
(d) *$\|A\| \leq 2$,*
(e) *$\|A\psi\| \leq 2\,\|\psi\|_{5\rho}$,*

where $\|\psi\|_k$ denotes the energy norm over the sphere $|x| \leq k$, i.e., the square root of the energy contained in this sphere:

$$\|\psi\|_k^2 = \int_{|x| \leq k} |D\psi_1|^2 + |\psi_2|^2 \, dx.$$

Proof: Both $U(2\rho)$ and $U_0(2\rho)$ have property (a), therefore so does their difference.

(b) follows from our previous observation in the corollary to Lemma 1 that $U_0(2\rho)$ maps the orthogonal complement of D_- into D_+ and thus its effect is wiped out by P_+.

(c) expresses the fact that the solutions of the pure and the mixed initial value problems agree outside the domain of influence of the boundary.

(d) holds since A is the difference of two isometric operators.

(e) follows from (c) and the known fact that the energy of $U(2\rho)\psi$ and of $U_0(2\rho)\psi$ inside the sphere $|x| < 3\rho$ is bounded by the energy of ψ contained inside the sphere $|x| < 5\rho$.

The key item in the proof is the following inequality contained in [3]:

LEMMA 3.   *There exists a number T_0 such that for all elements χ of H whose support is contained in the sphere $|x| \leq 3\rho$ the inequality*

(1.3) $$\|U(T_0)\chi\|_{5\rho} \leq \tfrac{1}{8}\,\|\chi\|$$

holds.

Take now any element ϕ orthogonal to D_-; according to (c) of Lemma 2, $\chi = A\phi$ satisfies the hypothesis of Lemma 3. Therefore by (1.3) we have, for $T = T_0 + 4\rho$,

(1.4) $$\|U(T - 4\rho)A\phi\|_{5\rho} \leq \tfrac{1}{8}\,\|A\phi\| \leq \tfrac{1}{4}\,\|\phi\|;$$

in the last step we have made use of inequality (d) of Lemma 2. Using inequality (e) of Lemma 2 with $U(t - 4\rho)A\phi$ in place of ψ we get from (1.4) that

$$\|AU(T - 4\rho)A\phi\| \leq \tfrac{1}{2}\,\|\phi\|.$$

Since P_+ does not increase norm, it follows from this last inequality that

$$\|P_+A(T - 4\rho)A\phi\| \leqq \tfrac{1}{2} \|\phi\|$$

for all ϕ orthogonal to D_-. In view of identity (b) of Lemma 2 this shows that

$$\|Z(T)\phi\| \leqq \tfrac{1}{2} \|\phi\|$$

for all such ϕ. This completes the proof of Theorem III.

The corollary stated in the introduction can now be verified by choosing ρ so that $|x| < \rho$ contains the given bounded region and noting that for any ϕ in H, $P_+\phi$ is equal to ϕ inside $|x| < \rho$.

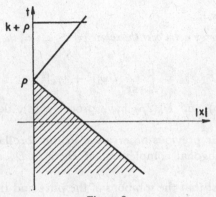

Figure 2

We close this section with two further observations:

1) From the exponential decay of the energy contained in solutions in bounded regions it is not hard to derive the pointwise decay of smooth solutions by using the inequalities of Sobolev and known estimates for the square integrals of higher derivatives of solutions of the wave equations.

2) It can be shown that the constant k in Theorem III does not depend on ρ, although the constant S does.

§ 2. In this section we shall supply proofs for Lemma 1 and Theorem I. Lemma 1 asserts:

If ϕ in H_0 is orthogonal to D_-, then the solution w of the wave equation in free space with initial value ϕ vanishes in the case $|x| < t - \rho$.

In order to prove this we shall construct a class of elements which belong to D_-.

Let κ be some positive constant and denote by v any solution whose data at time $t = \kappa + \rho$ vanish outside the sphere $|x| = \kappa$. We can construct such solutions v by solving a Cauchy problem backward, the initial time being $t = \kappa + \rho$.

According to Huygens' principle, such a solution v is zero at all points whose cones at time $t = \kappa + \rho$ do not pass through the sphere $|x| \leqq \kappa$; in particular, v

will be zero in the cross-hatched portion in Figure 2. This shows that all such solutions v are incoming, i.e., $v(0)$ lies in D_-.

Denote by w the solution in free space of the wave equation with initial value ϕ. Since ϕ was assumed to be orthogonal to D_-, it follows that

$$[v(0), \phi] = [v(0), w(0)] = 0,$$

where the bracket expression denotes the scalar product associated with energy.

By the conservation of energy, the scalar product is time independent. Therefore,

$$[v(t), w(t)] = 0$$

for all t, in particular $t = \kappa + \rho$. But for $t = \kappa + \rho$ the data of v are arbitrary, except that they vanish for $|x| > \kappa$. We conclude from this that at $t = \kappa + \rho$, w_t is zero and w a harmonic function of x inside $|x| < \kappa$. Since κ is varied from 0 to ∞ this shows that w is a time independent harmonic function inside the cone $|x| < t - \rho$. Since the total energy of this harmonic function is finite, it follows that it is a constant. We shall show now that this constant is zero.

LEMMA 4. *Let f be a smooth function with compact support; then*

(2.1)
$$\int_{|\omega|=1} f^2(R\omega) \, d\omega \leqq \frac{\|f\|_1^2}{R},$$

where $\|f\|_1^2 = \int f_x^2 \, dx$.

Proof: Let f be any smooth function with compact support in x-space. Let y be any point; then

$$f(y) = -\int_y^\infty f_r(x) \, dr,$$

where the subscript r denotes differentiation in the radial direction and the integration is with respect to arc length along the radius from y to ∞. By the Schwarz inequality,

$$|f(y)|^2 \leqq \int_y^\infty \frac{dr}{r^2} \cdot \int_y^\infty f_r^2 r^2 \, dr$$

$$\leqq \frac{1}{|y|} \int_0^\infty f_r^2 r^2 \, dr.$$

Put $y = R\omega$, $|\omega| = 1$ and integrate; inequality (2.1) results.

It follows by continuity that (2.1) is valid for all functions which are limits in the norm $\|f\|_1$ of smooth functions with compact support, such as the solutions under discussion of the wave equations with finite energy.

Denote by c the constant value of w inside the cone $|x| \leqq t - \rho$. Applying (2.1) to $f(x) = w(t + \rho, x)$ and $R = t$ we get

$$4\pi c^2 \leqq E/t,$$

11

where E is the total energy of w. Letting t tend to infinity, we conclude that $c = 0$. This completes the proof of Lemma 1.

LEMMA 1′.   *If $U_0(t)\phi$ is zero for $|x| < t - \rho$, then ϕ is orthogonal to D_-.*

Since every ϕ in D_+ satisfies the hypothesis of Lemma 1′, Theorem I follows from Lemma 1′.

To prove Lemma 1′ we reverse the steps leading to the proof of Lemma 1: from the assumption that $U_0(t)\phi = 0$ for $|x| < t - \rho$ it follows that ϕ is orthogonal to all elements of the form

$$(2.2) \qquad\qquad v(0),$$

where v is a solution of the kind considered before. As we saw earlier, all elements of this form lie in D_-. We shall show that they *span D_-*; this would complete the proof of Lemma 1′.

Assume to the contrary that elements of the form (2.2) do not span D_-. Then, since D_- is closed, it follows by the projection theorem that there is a nonzero element ψ in D_- orthogonal to all elements of the form (2.2). Denote by $u(x, t)$ the solution of the wave equation in free space with initial value ψ; we conclude, just as in the proof of Lemma 1, that

$$u(x, t) = 0 \qquad \text{for} \qquad |x| < t - \rho.$$

On the other hand, since ψ belongs to D_-, it follows that

$$u(x, t) = 0 \qquad \text{for} \qquad |x| < \rho - t.$$

Thus $u(x, t)$ vanishes in the double cone $|x| < |t - \rho|$; we now assert that this implies $u = 0$, contrary to our assumption that $\psi \neq 0$.

THEOREM IV.   *Let $u(x, t)$ be a weak solution with finite energy of the wave equation in free space which vanishes inside the double cone $|x| < |t|$. Then u is identically zero.*

Proof: Since u vanishes in the double cone, its Fourier transform $\hat{u}$, given by

$$(2.3) \qquad \hat{u}(x, \lambda) = \int_{-|x|}^{|x|} e^{i\lambda t} u(x, t) \, dt = \int_{-\infty}^{\infty} e^{i\lambda t} u(x, t) \, dt,$$

exists as a distribution and satisfies, in the sense of distributions, the reduced wave equation

$$\Delta \hat{u} + \lambda^2 \hat{u} = 0.$$

Since this is an elliptic equation, every distribution solution of it is smooth; thus $\hat{u}$ is smooth.

Since $\hat{u}$ is a solution of the reduced wave equation, its spherical harmonic coefficients

$$(2.4) \qquad \hat{u}_\alpha(r, \lambda) = \int \hat{u}(r\omega, \lambda) Y_\alpha(\omega) \, d\omega$$

are solutions of Bessel's equations. Since $\hat{u}$ is a smooth function, $\hat{u}_\alpha$ is bounded as r goes to zero and so equals a constant multiple of the regular solution of the Bessel equation:

$$(2.5) \qquad u_\alpha(r, \lambda) = a_\alpha(\lambda)(r\lambda)^{-1/2} J(r\lambda),$$

where J is the Bessel function of order $|\alpha| + \frac{1}{2}$ (which can be expressed as a trigonometric function).

Let E denote the total energy of u; by Lemma 4,

$$(2.6) \qquad \int_{|\omega|=1} u^2(r\omega, t)\, d\omega \leqq \frac{E}{r}.$$

This inequality shows that the integral (2.3) exists for every value of λ and r, $r \neq 0$, in the L_2 sense over the unit sphere. Thus we can substitute (2.3) in (2.4) and interchange the order of the ω and t integrations:

$$(2.7) \qquad \hat{u}_\alpha(r, \lambda) = \int_{-r}^{r} \int_{|\omega|=1} e^{i\lambda t} u(r\omega, t) Y_\alpha(\omega)\, d\omega\, dt.$$

This formula shows that for fixed r, $\hat{u}_\alpha(r, \lambda)$ is an entire analytic function of λ. Furthermore, estimating the right side of (2.7) by the Schwarz inequality and using (2.6) we get

$$(2.8) \qquad |\hat{u}_\alpha(r, \lambda)| \leqq \text{const. } r^{1/2} e^{r|\mathscr{I}m\,\lambda|}.$$

As already noted, $\hat{u}_\alpha(r, \lambda)$ is entire in λ for each $r > 0$; hence it follows from (2.5) that $a_\alpha(\lambda)$ is holomorphic except perhaps at the zeroes of $J(r\lambda)$. Since these zeroes vary with r, we conclude that $a_\alpha(\lambda)$ is holomorphic except perhaps at $\lambda = 0$. Next choose $\delta, \varepsilon > 0$ so that $|J(z)| \geqq \varepsilon$ for $|z| = \delta$. Setting $r = \delta/|\lambda|$ and making use of (2.8) we obtain

$$|a_\alpha(\lambda)| \leqq \frac{\text{const}}{\varepsilon} \left(\frac{\delta}{|\lambda|} \right)^{1/2} \exp\{\delta\} = \text{const} \cdot |\lambda|^{-1/2}.$$

An analytic function which satisfies such a growth condition in the whole complex λ-plane punctured at the origin must be identically zero. Thus it follows that $a_\alpha(\lambda) \equiv 0$. Retracing our steps, we see that $\hat{u}_\alpha(r, \lambda) \equiv 0$, so that $\hat{u}(x, \lambda) \equiv 0$, and finally that $u(x, t) \equiv 0$, which was to be proved.

Slight modifications of this proof yield Theorem IV in an arbitrary number of space dimensions.

Bibliography

[1] Lax, P. D., and Phillips, R. S., *The wave equation in exterior domains*, Bull. Amer. Math. Soc. Vol. 68, 1962, pp. 47–49.

[2] Lax, P. D., and Phillips, R. S., *Spectral theory for the wave equation*, to appear.

[3] Morawetz, C. S., *The decay of solutions of the exterior initial-boundary value problem for the wave equation*, Comm. Pure Appl. Math., Vol. 14, 1961, pp. 561–568.

[4] Morawetz, C. S., *The limiting amplitude principle*, Comm. Pure Appl. Math., Vol. 15, 1962, pp. 349–361.

[5] Wilcox, C., *The initial-boundary value problem for the wave equation in an exterior domain with spherical boundary*, Amer. Math. Soc. Not., Abstract No. 564-20, Vol. 6, 1959.

[6] Courant, R., and Hilbert, D., *Methods of Mathematical Physics*, Vol. 2, Interscience Publishers, New York, 1962.

[7] Lax, P. D., Morawetz, C. S., and Phillips, R. S., *The exponential decay of solutions of the wave equation in the exterior of a star-shaped obstacle*, Bull. Amer. Math. Soc., Vol. 68, 1962, pp. 593–595.

Received April, 1963.

SCATTERING THEORY

BY PETER D. LAX[1] AND RALPH S. PHILLIPS[2]

Communicated September 6, 1963

1. Let H be a Hilbert space, $U(t)$ a group of unitary operators. A closed subspace D_+ of H will be called *outgoing* if it has the following properties:

(i) $U(t)D_+ \subset D_+$ for t positive.
(ii) $\bigcap_{t>0} U(t)D_+ = \{0\}$.
(iii) $\bigcup_{t<0} U(t)D_+$ dense in H.

A prototype of the above situation is when H is $L_2(-\infty, \infty; N)$, i.e., the space of square integrable functions on the whole real axis whose values lie in some accessory Hilbert space N, $U(t)$ is translation by t, and D_+ is $L_2(0, \infty; N)$.

THEOREM 1.[3] *If D_+ is outgoing for the group $U(t)$, then H can be represented isometrically as $L_2(-\infty, \infty; N)$ so that $U(t)$ is translation and D_+ is the space of functions with support on the positive reals. This representation is unique up to isomorphisms of N.*

We shall call this representation an *outgoing translation representation* of the group.

Taking the Fourier transform we obtain an *outgoing spectral representation* of the group $U(t)$, where elements of D_+ are represented as functions in $A_+(N)$, that is the Fourier transform of $L(0, \infty; N)$. According to the Paley-Wiener theorem $A_+(N)$ consists of boundary values of functions with values in N, analytic in the upper half-plane whose square integrals along lines $\text{Im } z = \text{const}$ are uniformly bounded.

An incoming subspace D_- is defined similarly and an analogous representation theorem holds, D_- being represented by functions with support on the negative axis, that is, by $L_2(-\infty, 0; N_-)$. N_- and N are unitarily equivalent and will henceforth be identified. In the application to the wave equation there is a natural identification of N and N_-.

Let D_+ and D_- be outgoing and incoming subspaces respectively for the same unitary group, and suppose that D_+ and D_- are *orthogonal*. To each function $f \in H$ there are associated two functions k_- and k_+, the respective incoming and outgoing translation representa-

[1] Sloan Fellow.

[2] Sponsored by the National Science Foundation, contract NSF-G 16434.

[3] We were informed by Professor Sinai that he has obtained and used a similar theorem.

130

tions of f. The mapping $k_- \to k_+$, denoted by S, is called the *scattering operator* and has the following properties:

(i) S is unitary.

(ii) S commutes with translation.

(iii) S maps $L_2(-\infty, 0, N)$ into $L_2(-\infty, 0, N)$.

Properties (i) and (ii) follow from the fact that S is defined in terms of two different unitary translation representations of the same group. To deduce property (iii), we note that every function in $L_2(-\infty, 0, N)$ of the incoming representation corresponds to an element f of D_-; since we have assumed that D_- is orthogonal to D_+ it follows that the function representing f in the outgoing representation will be orthogonal to $L_2(0, \infty, N)$, i.e., will belong to $L_2(-\infty, 0, N)$, as asserted in (iii).

We take now Fourier transforms and define the operator $\mathcal{S}$ as FSF^{-1}, F denoting the Fourier transformation. Properties (i)–(iii) for S translate into

(i)$'$ $\mathcal{S}$ is unitary.

(ii)$'$ $\mathcal{S}$ commutes with multiplication by scalar functions.

(iii)$'$ $\mathcal{S}$ maps $A(N)$ into $A(N)$

where $A(N)$ is the Fourier transform of $L_2(-\infty, 0, N)$ and thus consists of the boundary values of functions analytic in the lower half-plane.

According to a simple special case of a theorem of Segal and Fourès [13], an operator with properties (i)$'$, (ii)$'$ and (iii)$'$ is multiplication by an operator valued function $\mathcal{S}(z)$, mapping N into N, with the following properties:

THEOREM 2. (a) $\mathcal{S}(z)$ *is analytic in the lower half-plane.*

(b) *The norm of $\mathcal{S}(z)$ is not greater than one for every z.*

(c) $\mathcal{S}(z)$ *is unitary for z real.*

$\mathcal{S}(z)$ is the Heisenberg *scattering matrix*. Extending the terminology of Beurling [1], to the operator case $\mathcal{S}(z)$ is also an *inner factor*.[4]

2. Let D_+ and D_- be as before and denote by P_+ and P_- orthogonal projection onto the orthogonal complements of D_+ and D_- respectively. Consider the one-parameter family of operators $Z(t)$ defined as

$$Z(t) = P_+ U(t) P_-.$$

It follows easily that for positive t, $Z(t)$ annihilates both D_+ and D_-; consider $Z(t)$ for positive values of t and acting on $K = H \ominus D_+ \ominus D_-$.

THEOREM 3. $Z(t)$ *forms a semigroup over K.*

---

[4] See [1], [7] and [2] for the theory of inner factors.

This is very easy to prove directly from the postulated relations of D_+ and D_- to each other and $U(t)$. It also follows from the interpretation of $Z(t)$ in the, say, outgoing translation representation. For, since $H \ominus D_+$ is represented by $L_2(0, \infty, N)$ and D_- by $L_2(0, \infty, N)$, K is represented by

$$K \Leftrightarrow L_2(0, \infty, N) \ominus L_2(0, \infty, N),$$

and *the action of $Z(t)$ consists in shifting to the right followed by restriction to the negative real axis.*

A subspace of functions K which is mapped into itself under such an operation is called a *translation invariant space*. It is not surprising that K and $Z(t)$ can be so represented, since according to a simple generalization of a theorem of Beurling every contraction semigroup $Z(t)$ can be so represented, provided that for every u in K, $\|Z(t)u\|$ tends to zero as t tends to infinity.

In the outgoing spectral representation K is represented as

$$A (N) \ominus \mathcal{S} A(N);$$

$\mathcal{S}$ is called the *inner factor* associated with the translation invariant space K. Again this is no surprise since according to a generalization due to Lax, [6], [7] of a theorem of Beurling, see also Halmos [2], the orthogonal complement of the Fourier transform of every translation invariant space is of the form $\mathcal{S}A$, where $\mathcal{S}$ is an inner factor. The importance of this representation is that the associated inner factor contains almost complete information about the spectrum of $Z(t)$ over K:

THEOREM 4.[5] (a) *Let μ be a complex number with negative real part; μ belongs to the resolvent set of the infinitesimal generator B of $Z(t)$ if and only if the operator*

$$\mathcal{S}(i\bar{\mu})$$

is invertible.

(b) *Let λ be a complex number of absolute value less than one; λ belongs to the resolvent set of $Z(t)$ if and only if*

$$\mathcal{S}\left(\frac{i}{t}\bar{\mu}\right)$$

is invertible for all numbers μ for which $e^{\mu t} = \lambda$, and if the norms of the inverses are uniformly bounded for all such μ.

As corollary we obtain another proof of the well-known result of Phillips, see [11] or [3], that if μ belongs to the spectrum of B then

----

[5] See [10] for the scalar case.

17

$e^{\mu t}$ belongs to the spectrum of $Z(t)$, but in general not conversely.

THEOREM 5. *If for some value of T, $\|Z(T)\| = a < 1$, then $S(z)$ can be continued analytically into the strip* $\operatorname{Im} z \leqq -\log a/T$.

Analogous expressions can be derived for the location of the spectrum of any function of $Z(t)$ and B; from this we deduce

THEOREM 5'. *If for some T and some μ, $Z(T)(B-\mu)^{-1}$ is completely continuous, then $S(z)$ can be continued into the upper half-plane as a meromorphic function.*

The definition of the scattering matrix depends on the choice of a pair of orthogonal incoming and outgoing subspaces. Let us call two outgoing subspaces D_+ and D'_+ *equivalent* if for sufficiently large positive T, $U(-T)D_+$ contains D'_+ and $U(T)D_+$ is contained in D'_+. Following the derivation of Theorem 2 one can easily show that the outgoing spectral representations with respect to D_+ and D'_+ are related by multiplication by an operator valued function $\mathfrak{M}(z)$ which is entire analytic, of exponential growth, and unitary on the real axis. Such a function satisfies the relation

$$\mathfrak{M}^*(\bar{z}) = \mathfrak{M}^{-1}(z)$$

for z real, so by analytic continuation for all z; this shows that $\mathfrak{M}^{-1}$ exists for all z and is of exponential growth.

Suppose that D_+, D_- and D'_+, D'_- are two pairs of orthogonal incoming and outgoing subspaces which are equivalent. Then the associated scattering matrices are related by

$$S' = \mathfrak{M}_+ S \mathfrak{M}_-^{-1}.$$

Since the factors $\mathfrak{M}_+$, $\mathfrak{M}_-$ and their inverses are uniformly bounded in any strip, we conclude from Theorem 4 that the associated semigroups have the same spectrum.

Choose in particular D^a_+ as $U(a)D_+$, and D^a_- as $U(-a)D_-$, a positive. As is easily shown,

$$S^a(z) = e^{2iaz}S(z).$$

We denote the operator $Z(t)$ corresponding to $D^a_\pm$ by $Z_a(t)$.

THEOREM 6. *If f^b is an eigenvector of $Z_b(t)$ with eigenvalue $e^{\mu t}$, then for $a < b$*

$$f^a = P^a_+ f^b$$

is an eigenvector of $Z_a(t)$ with the same eigenvalue.

3. Let H_0 denote the Hilbert space of pairs of functions $f = [f_1, f_2]$ defined in R_n, normed by the energy norm:

$$\|f\|^2 = \int (|\, Df_1\,|^2 + |f_2|^2)\, dx.$$

Define $U_0(t)$ as the operator which relates the Cauchy data at time zero of solutions of the wave equation to their Cauchy data at time t. $U_0(t)$ forms a one-parameter group of unitary operators mapping H_0 onto H_0 (conservation of energy).

Consider a smooth, bounded, reflecting obstacle. Denote by H the subspace of H_0 consisting of pairs of functions which vanish inside the obstacle, and denote by $U(t)$ the operator which relates the initial data to data at time t of solutions of the wave equation defined outside of the obstacle and vanishing on it. $U(t)$ forms a one-parameter group of unitary operators mapping H onto H.

We shall call a solution of the wave equation defined for all values of x and t *outgoing* (*incoming*) if it vanishes inside the cone $|x| < t(|x| < -t)$. We denote by $D_{\pm}^0$ the data at time zero of outgoing (incoming) solutions.

D_{+}^0 and D_{-}^0 are outgoing and incoming subspaces for the group $U_0(t)$ in the sense of §1; the first two properties are obviously satisfied and the third is an easy consequence of Huygens' principle. As shown in [9], D_{+}^0 and D_{-}^0 are orthogonal for n odd; we give here a new proof based on an explicit form for the translation representation.

We start with the representation of functions in terms of their Radon transforms:

$$(3.1) \qquad\qquad f(x) = \int_{|\omega|=1} h(x \cdot \omega, \omega)\, d\omega$$

where $h(s, \omega)$, the Radon transform of f, is a function of s and ω defined for all real s and all vectors ω on S_{n-1} which is even:

$$h(-s, -\omega) = h(s, \omega).$$

A Parseval relation holds:

$$(3.2) \qquad\qquad \|f\|^2 = \|h\|_{-(n-1)/2},$$

where we define

$$\|h\|_{-q}^2 = \int |\, k(s, \omega)\,|^2 ds\, d\omega, \qquad \frac{\partial^q}{\partial s^q} k = h.$$

Corollary.

$$(3.2)' \qquad\qquad \|f\|_1 = \|h\|_{-(n-3)/2}$$

Let h_1 and h_2 denote the Radon transforms of f_1 and f_2 respectively and define h as

$$h = h_1 - \int h_2.$$

It can be verified immediately that the function

$$(3.3) \qquad\qquad u(x, t) = \int h(x \cdot \omega - t, \omega) d\omega$$

is a solution of the wave equation and that its initial data are f_1 and f_2. Furthermore, by (3.2) and (3.2)′,

$$(3.4) \qquad\qquad \|f\| = \|h\|_{-(n-3)/2}$$

Let k be the $(n-3)/2$ fold integral of h; regarding k as a function of s whose values lie in the Hilbert space $N = L_2(S_{n-1})$, we conclude from (3.3), (3.4) that $f \to k$ is a translation representation for $U_0(t)$. We claim that for n odd this representation is both incoming and outgoing.

It follows from (3.3) that if $h(s)$ vanishes for negative (positive) values of s, then $u(x, t)$ vanishes in the forward (backward) cone $|x| < t$ ($|x| < -t$). Conversely:

Theorem 7. *If $u(x, t)$ vanishes in the forward (backward) cone then h vanishes on the negative (positive) axis.*

Sketch of proof. If u vanishes in the forward cone, all its space derivatives vanish on the positive t axis:

$$0 = (D_x^j u)(0, t) = \int \omega^j h^{|j|}(-t, \omega) d\omega.$$

Multiply this by any smooth test function $\phi(t)$ whose support lies on the positive t-axis, integrate with respect to t and perform $|j|$ integrations by parts:

$$(3.5) \qquad\qquad 0 = \int \omega^j h(-t, \omega) \phi^{|j|}(t) d\omega dt.$$

From (3.5) and the fact that h has finite $(3-n)/2$ norm it follows by an approximation procedure that for every smooth test function $\chi(t)$ with compact support on the positive t axis and every multi index j

$$\int\int h(-t, \omega)\omega^j\chi(t)\partial\omega dt = 0.$$

But this implies that $h(s)$ vanishes for s negative; then so does k for n odd.

REMARK. In the proof we only used the fact that $u(x, t)$ has a zero of infinite order on the positive t-axis; thus we have shown that this condition implies that u vanishes in the forward cone—a new proof for a special case of a theorem of Fritz John.

COROLLARY. *For n odd, D_+ and D_- are orthogonal.*

For the group $U_0(t)$ we have found a representation with D_+^0 and D_-^0 as outgoing and incoming subspaces. The associated scattering operator is the identity. We turn now to the group $U(t)$ and take for D_+ and D_- the initial data of solutions which vanish in $|x| < \rho + t$ for $t > 0$ and $|x| < \rho - t$ for $t < 0$ respectively;[6] we claim that these subspaces are outgoing and incoming respectively for $U(t)$: Properties (i) and (ii) are immediate while property (iii) is proved in [8]. D_+ and D_- are orthogonal since they are subspaces of D_+^0 and D_-^0. *Thus there exists an associated scattering matrix.* Conversely, we can prove

THEOREM 8. *The scattering matrix uniquely determines the scattering obstacle.*

In [9], Cathleen Morawetz and the authors have shown that for star-shaped obstacles $\|Z(t)\|$ is less than one for t large enough. By Theorem 5 it follows that the associated scattering matrix can be continued analytically into a strip $0 \leq \mathrm{Im}\, z \leq \tau$. For any obstacle we have this result:

THEOREM 9. *For $\mathrm{Re}\,\lambda$ positive $Z(2\rho)(B-\lambda)^{-1}$ is completely continuous.*

By Theorem 5' this implies that $S(z)$ can be continued into the upper half-plane as a meromorphic function. This implies that the zeros of $S(z)$ in the lower half-plane are discrete; furthermore, for each z, $S(z)$ has a closed range whose codimension is finite and equal to the dimension of the nullspace of $S(z)$.

SKETCH OF PROOF.

LEMMA 1. *The operator*

$$M = U(2\rho) - U_0(2\rho)$$

_____

[6] ρ is chosen so large that the sphere $|x| < \rho$ contains the obstacle.

annihilates all f in H which vanish in $|x| < 3\rho$.

LEMMA 2. $U_0(2\rho)$ *maps the complement of* D_- *into* D_+.

Lemma 2 implies that $P_+U_0(2\rho)P_- = 0$, whence for $t \geqq 2\rho$

$$(3.6) \qquad Z(t) = P_+U(t)P_- = P_+MU(t - 2\rho)P_-.$$

So

$$Z(2\rho)(B - \lambda)^{-1} = e^{2\lambda\rho} \int_{2\rho}^{\infty} Z(t)e^{-\lambda t}dt$$

$$(3.7)$$

$$= e^{2\lambda\rho}P_+M \int_{2\rho}^{\infty} U(t - 2\rho)e^{-\lambda t}dt P_- = P_+M(A - \lambda)^{-1}P_-,$$

where A denotes the infinitesimal generator of $U(t)$. It is easy to show that $(A - \lambda)^{-1}$ raises by one the degree of differentiability; since by Lemma 1 the value of Mf does not depend on the values of f outside the sphere $|x| \leqq 3\rho$, it follows by Rellich's compactness criterion that $M(A - \lambda)^{-1}$ is a completely continuous operator.

For the pure initial value problem for hyperbolic equations with variable coefficients it is known that the sharp propagation of signals is along characteristic rays. This *generalized Huygens principle* can be reformulated as follows:

Let G_1 and G_2 be two closed sets in R_n with the property that no characteristic ray starting at time zero in G_1 passes through G_2 at time t. Let P_1 and P_2 be operators such that the range of P_1 consists of functions which vanish outside of G_1, while P_2 annihilates all functions whose support lies outside G_2. Then

$$P_2U_0(t)P_1,$$

is completely continuous.

We believe that this principle also holds for the mixed problem as well (for general hyperbolic equations with variable coefficients), provided that rays are interpreted as reflected rays. A demonstration of this for the interior problem for convex domains has been given by Povsner and Suharevskiĭ [12].

We say that an obstacle has property L if there exists a number l such that any ray starting in the sphere $|x| \leqq 3\rho$ leaves the sphere $|x| \leqq 3\rho$ after time l.

REMARK. Star-shaped obstacles have property L.

Assuming the generalized Huygens principle to hold we assert:

THEOREM 10. $Z(t)$ *is eventually compact if and only if the obstacle has property* L.

PROOF. The identity (see [9])

$$(3.6)' \qquad\qquad Z(t + 4\rho) = P_+ M U(t) M P_-$$

follows similarly as (3.6). Take both G_1 and G_2 as the sphere $|x| \leq 3\rho$; the operator MP_- has the property required of P_1 while M has the property required of P_2. So by the generalized Huygens principle $Z(l+4\rho)$ is completely continuous.

The necessity of property L follows from known properties of propagation of high frequency signals along rays.

Theorem 10 implies that $Z(t)$ has a standard discrete spectrum. There can be no eigenvalue of absolute value one since this would correspond to a solution of the wave equation which is a purely imaginary exponential in time, and according to a theorem of Rellich there are no such solutions with finite energy. Thus the spectral radius of $Z(t)$ is less than one; by the Gelfand formula we conclude that $\|Z(t)\|$ decays exponentially. Thus Theorem 10 gives another proof of the result of [9].

Similar reasoning gives the following result: let f be any element of K, $\sum a_j f_j$ its formal Fourier expansion in terms of the eigenfunctions of $Z(t)$; then

$$\sum a_j e^{\mu_j t} f_j$$

is an asymptotic expansion for $Z(t)f$.

Next we wish to characterize the eigenvalues and eigenfunctions of the generator B of $Z(t)$. For this purpose we say that a solution of the reduced wave equation

$$(3.8) \qquad\qquad \Delta u - \mu^2 u = 0$$

in the exterior domain is *outgoing* if the free space solution of the wave equation with initial data $f = [u, -\mu u]$, in symbols $U_0(t)f$, vanishes for $|x| < t - \rho$ for all $t > \rho$. This notion is equivalent with the Sommerfeld definition of outgoing when μ is imaginary. Moreover for arbitrary μ in the case $n = 3$ such a solution of the reduced wave equation can be represented as

$$u(x) = \frac{1}{4\pi} \int_\Gamma \left(u \frac{\partial u}{\partial n} - v \frac{\partial u}{\partial n} \right) dS_y$$

where $v = e^{\mu r}/r$, $r = |x - y|$, and Γ is any smooth surface containing the obstacle but not containing x. The converse is also true.

THEOREM 11. *μ is an eigenvalue of the generator of $Z(t)$ if and only if there exists an outgoing solution of the reduced wave equation* (3.8) *satisfying the boundary conditions.*

SKETCH OF PROOF. We consider $Z_a(t) = P_+^a U(t) P_-^a$ as $a \to \infty$; in the limit this is simply $U(t)$. According to Theorem 4 the eigenvalues μ of the generator of $Z_a(t)$ are simply related to the zeros of the scattering operator; thus they are independent of $a \geq \rho$. The eigenfunctions depend upon "a" but according to Theorem 6 in a rather trivial fashion. In fact for $b > a \geq \rho$,

$$(3.9) \qquad\qquad f^a = P_+^a f^b.$$

Since P_+^a does not alter the data inside the sphere $|x| < a$, it follows that $f^a(x) = f^b(x)$ for $|x| < a$. This shows that the limit

$$\lim_{a \to \infty} f^a(x) \equiv f(x)$$

exists.

Each f_a satisfies

$$Z_a(t) f_a = e^{-\mu t} f_a.$$

Since $Z_a(t)f = U(t)f$ for $|x| < a$, f_a is a solution of the reduced wave equation there. So for $|x| < a$, f_a is of the form

$$f_a = (u_a, -\mu u_a),$$

u_a a solution of the reduced wave equation

$$\Delta u_a - \mu^2 u_a = 0$$

which is zero on the obstacle. Since $u_a(x) = u_b(x)$ for $|x| < a$, the limit

$$\lim_{a \to \infty} u_a = u$$

exists. u is in the exterior a solution of the reduced wave equation and is zero on the obstacle. The data $f = [u, \mu u]$ can be thought of as a generalized eigenfunction of $U(t)$; not only does it not lie in H, but it blows up exponentially in $|x|$. f^ρ is orthogonal to D_- and so is f^a for all $a \geq \rho$ by (3.9). As a consequence the free space solution $U_0(t)f$ vanishes in $|x| < t - \rho$ for $t > \rho$ so that u is outgoing.

Conversely if u is an outgoing solution of the reduced wave equation (3.8) satisfying the boundary conditions, then $e^{-\mu t}$ is an eigenvalue of $Z(t)$. To prove this one shows that the free space translation representation of $f = [u, -\mu u]$ is of the form

$$h(s, \omega) = \begin{cases} 0, & s < -\rho, \\ n(\omega) e^{\mu s}, & s > \rho. \end{cases}$$

Setting

$$h_a(s, \omega) = \begin{cases} h(s, \omega), & s < a, \\ 0, & s > a, \end{cases}$$

one proves that h_a is the free space translation representation of the eigenfunction of $Z_a(t)$ corresponding to the eigenvalue $e^{-\mu t}$.

The above ideas can also be employed to obtain an explicit description of the incoming and outgoing spectral representations of $U(t)$ from which we will in turn be able to obtain an explicit formula for the scattering operator $S(z)$. We shall denote by $\hat{f}_0$, $\hat{f}_-$, and $\hat{f}_+$ the free space, incoming, and outgoing spectral representations respectively of a given initial data f.

These spectral representations are given by scalar products of f with certain improper eigenfunctions of $U_0(t)$, respectively $U(t)$. We shall show that these improper eigenfunctions consist of exponentials plus certain incoming and outgoing solutions. We recall that the free space spectral representation for $U_0(t)$ is simultaneously incoming and outgoing. Thus D^0_- and D^0_+ map onto $A_-(N)$ and $A_+(N)$ respectively, while D_- and D_+ map onto $e^{-ipz}A_-(N)$ and $e^{ipz}A_+(N)$ respectively. We shall limit our considerations to the case $n = 3$.

THEOREM 12 (SPECTRAL REPRESENTATION FOR $U_0(t)$).

(3.10) $\hat{f}_0(z, \omega) = (f, \phi_0(\cdot, z, \omega))$

where $(\ ,\ )$ denotes the H_0 inner product and

$$4\pi^{3/2}\phi_0(x, z, \omega) = [e^{-izx\cdot\omega}, iz\,e^{-izx\cdot\omega}].$$

The main tool employed in the derivation of (3.10) is the Fourier transform.

THEOREM 13 (INCOMING AND OUTGOING SPECTRAL REPRESENTATIONS FOR $U(t)$). Let $v_+(v_-)$ be the outgoing (incoming) solution of the reduced wave equation

$$\Delta v + z^2 v = 0$$

satisfying $v + e^{-izx\cdot\omega} = 0$ on the obstacle. Set

$$4\pi^{3/2}\psi_\pm(x, z, \omega) = [v_\pm(x, z, \omega), izv_\pm(x, z, \omega)]$$

and define

$$\phi_\pm = \phi_0 + \psi_\pm.$$

Then

(3.11) $\hat{f}_\pm(z, \omega) = (f, \phi_\mp(\cdot, z, \omega)),$

where the $(\ ,\ )$ denotes the inner product in H. Note the switch in signs.

25

SKETCH OF PROOF FOR THE INCOMING REPRESENTATION FORMULA. *Step one.* To verify (3.11) for data in D_-. It is required that data in D_- have the same representation as in the free space spectral representation. This in turn requires that $(f, \psi_+) = 0$ for all f in D_-. Now for $f = U_0(-\tau - \rho)w$ where w has support in $|x| < \tau$, it is clear that

$$(f, \psi_+) = (w, U_0(\tau + \rho)\psi_+) = 0$$

since $U_0(\tau + \rho)\psi_+$ vanishes for $|x| < \tau$. It is proved in [9] that linear combinations of such f are dense in D_- and hence (3.11) shares with (3.10) the property of being an isometry in D_-.

Step two. Extend the isometric property of the representation to all translates of D_-. A simple integration by parts shows for any f in D_A that

$$\frac{d}{dt}\hat{f}_-(t) = iz\hat{f}_-(t),$$

where $f(t) = U(t)f$. As a consequence

$$\hat{f}_-(t) = e^{izt}\hat{f}_-.$$

This extends the isometry of the map to all of the translates of D_- and hence to all of H since the translates of D_- are dense in H (see [8]). It also follows that $U(t)$ is represented as multiplication by e^{izt} in this representation.

Step three. The map $f \to \hat{f}_-$ is onto $L_2(-\infty, \infty; N)$. In the case of the free space representation of $U_0(t)$ it is known that the translates of D_- fill out $L_2(-\infty, \infty; N)$ in the representation space. Since D_- and translation are represented by the same objects in both the free space and incoming spectral representations, it follows that the map $f \to \hat{f}_-$ is onto.

THEOREM 14. *The scattering operator is given by*

$$\hat{f}_+(z, \omega) = [\mathcal{S}(z)\hat{f}_-(z, \cdot)](\omega)$$

(3.12)
$$= \hat{f}_-(z, \omega) - 2(2\pi)^{1/2}iz \int_{|\theta|=1} s(-\theta, \omega, z)^* \, \hat{f}_-(z, \theta)d\theta,$$

where

$$\psi_-(r\xi, z, \omega) \sim r^{-1}e^{izr} s(\xi, \omega, z) \ as \ r \to \infty.$$

SKETCH OF PROOF. It suffices to determine the behavior of $\mathcal{S}(z)$ on D_- since S commutes with $U(t)$ and since translates of D_- are dense in H. Now for f in D_-, $\hat{f}_- = \hat{f}_0$ so that $\mathcal{S}(z)\hat{f}_- = \hat{f}_+$ simply becomes

$$\mathcal{S}(z)\hat{f}_- = \hat{f}_- + (f, \psi_-), \qquad\qquad f \in D_-.$$

A straightforward calculation now yields (3.12). This is roughly what one expects from the classical theory and shows in particular that $\mathcal{S}(z)$ differs from the identity on N by an operator with a smooth kernel.

Much of the foregoing can be generalized to solutions of symmetric hyperbolic equations which satisfy conservative boundary conditions on some obstacle, provided that these boundary conditions are elliptic for the spatial part of the operator. That there are conservative boundary conditions which are not elliptic is somewhat surprising.

REFERENCES

1. A. Beurling, *On two problems concerning linear transformations in Hilbert space*, Acta Math. 81 (1949), 239–255.

2. P. R. Halmos, *Shifts on Hilbert spaces*, J. Reine Angew. Math. 108 (1961), 102–112.

3. E. Hille and R. Phillips, *Functional analysis and semi-groups*, Amer. Math. Soc. Colloq. Publ. Vol. 31, Amer. Math. Soc., Providence, R. I., 1957.

4. H. Helson and D. Lowdenslager, *Prediction theory and Fourier series in several variables*, Acta. Math. 99 (1958), 165–202.

5. P. D. Lax, *Remark on the preceding paper of Koosis*, Comm. Pure Appl. Math. 10 (1957), 617–622.

6. ————, *Translation invariant spaces*, Acta Math. 101 (1959), 163–178.

7. ————, Proceedings of the International Symposium on Linear Spaces, Hebrew University, Jerusalem, Pergamon Press, 1961, pp. 299–306.

8. P. D. Lax and R. S. Phillips, *The wave equation in exterior domains*, Bull. Amer. Math. Soc. 68 (1962), 47–49.

9. P. D. Lax, C. S. Morawetz and R. S. Phillips, *The exponential decay of solutions of the wave equation in the exterior of a star-shaped obstacle*, Bull. Amer. Math. Soc. 68 (1962), 593–595.

10. J. W. Moeller, *On the spectra of some translation invariant spaces*, J. Math. Anal. Appl. 4 (1962), 276–296.

11. R. S. Phillips, *Spectral theory for semigroups of linear operators*, Trans. Amer. Math. Soc. 71 (1951), 393–415.

12. A. Ya. Povzner and I. V. Suharevskiĭ, *Discontinuities of the Green's function of mixed problems for the wave equation*, Mat. Sb. (N.S.) 51 (1960), (93) 3–26.

13. Y. Fourès and I. E. Segal, *Causality and analyticity*, Trans. Amer. Math. Soc. 78 (1955), 385–405.

NEW YORK UNIVERSITY AND
    STANFORD UNIVERSITY

COMMUNICATIONS ON PURE AND APPLIED MATHEMATICS, VOL. XXII, 737–787 (1969)

Decaying Modes for the Wave Equation in the Exterior of an Obstacle*

P. D. LAX AND R. S. PHILLIPS

Abstract

In this paper we study the dependence of the set of 'exterior' eigenvalues $\{\lambda_k\}$ of Δ on the geometry of the obstacle $\mathcal{O}$. In particular we show that the *real* eigenvalues, corresponding to purely decaying modes, *depend monotonically on the obstacle $\mathcal{O}$, both for the Dirichlet and Neumann boundary conditions*. From this we deduce, by comparison with spheres—for which the eigenvalues $\{\lambda_k\}$ can be determined as roots of special functions—upper and lower bounds for the density of the real $\{\lambda_k\}$, and upper and lower bounds for λ_1, the rate of decay of the fundamental real decaying mode. We also consider the wave equation with a positive potential and establish an analogous monotonicity theorem for such problems. We obtain a second proof for the above Dirichlet problem in the limit as the potential becomes infinite on $\mathcal{O}$.

Finally we derive an integral equation for the decaying modes; this equation bears strong resemblance to one appearing in the transport theory of mono-energetic neutrons in homogeneous media, and can be used to demonstrate the existence of infinitely many modes.

1. Introduction

We are concerned in this paper with the behavior of solutions of the wave equation in the *exterior* of an obstacle. The differences between the exterior and interior problems are so great that they tend to hide the points of similarity. Our purpose is to bring out certain analogies between these two problems and to this end we first discuss the relevant, but familiar, facts about the interior problem, that is, the behavior of solutions of the wave equation

$$(1.1) \qquad\qquad u_{tt} - \Delta u = 0$$

in some smoothly bounded compact domain $\mathcal{O}$ on whose boundary u is required to satisfy a boundary condition, say

$$(1.2) \qquad\qquad u = 0 \quad \text{or} \quad \partial_n u = 0 \quad \text{on} \quad \partial\mathcal{O}.$$

---

* The first author was partially supported by the U.S. Atomic Energy Commission under Contract AT(30-1)-1480 at the Courant Institute of Mathematical Sciences. The second author was supported partly by the National Science Foundation, grant NSF GP-8857 and the Air Force Office of Scientific Research Contract AF-F44620-68-C-0054. Reproduction in whole or in part is permitted for any purpose of the United States Government.

737

The spectral decomposition of Δ over $\mathcal{O}$ leads to the following representation of the totality of solutions of (1.1), (1.2):

$$(1.3) \qquad\qquad u(x, t) = \sum_{k=1}^{\infty} (a_k e^{i\mu_k t} + b_k e^{-i\mu_k t}) v_k(x) \; ;$$

here $\{\mu_k^2\}$ are the eigenvalues of $-\Delta$ arranged in increasing order with $\mu_k > 0$ and $\{v_k\}$ are the corresponding eigenfunctions:

$$(1.4) \qquad\qquad \Delta v_k + \mu_k^2 v_k = 0, \qquad v_k(x) = 0 \quad \text{on} \quad \partial\mathcal{O} .$$

For the interior problem, $(-\Delta)^{-1}$ is a positive compact operator and therefore the $\{\mu_k\}$ form a sequence of positive numbers tending to ∞; each solution of the wave equation is represented in (1.3) as a superposition of harmonic motions with frequencies μ_k. These frequencies are functionals of the domain $\mathcal{O}$, and much effort has gone into studying the dependence of the set of numbers $\{\mu_k\}$ on the geometrical properties of $\mathcal{O}$. The following results described in Vol. I of [1] are particularly interesting mathematically and significant from the point of view of physics (for more recent results see also [12]):

1. $\mu_k(\mathcal{O})$ depends *monotonically* on $\mathcal{O}$; that is, if $\mathcal{O}_1 \subset \mathcal{O}_2$, then

$$(1.5) \qquad\qquad \mu_k(\mathcal{O}_1) \geqq \mu_k(\mathcal{O}_2) \quad \text{for} \quad \text{all} \quad k .$$

2. The *asymptotic distribution* of the μ_k for large k is

$$(1.6) \qquad\qquad \mu_k \sim 2\pi \left(\frac{k}{\Omega V}\right)^{1/n} ,$$

where n is the dimension of the x-space, Ω the volume of the n-dimensional unit ball and V the volume of $\mathcal{O}$.

3. The *isoperimetric inequality*. Among all domains $\mathcal{O}$ with given volume V, the sphere has the smallest fundamental frequency μ_1.

We turn now to the behavior of solutions of the wave equation in the exterior $\mathcal{G}$ of $\mathcal{O}$, subject to the same boundary condition (1.2). In this case, $-\Delta$ has a continuous spectrum (of infinite multiplicity) extending from 0 to ∞ and one can again express all solutions of the wave equation as a superposition of harmonic motions involving all frequencies. It turns out that such a representation as it stands sheds no light on the asymptotic behavior of $u(x, t)$ for large t with x fixed.

To get some idea of what kind of asymptotic representation to look for, we first recall that the solution to the wave equation in free space of an odd number of dimensions obeys Huyghens' principle; thus for initial data having compact support, say contained in $\{x : |x| < R\}$, the solution will vanish in the cone $\{x : |x| < t - R\}$. If an obstacle is present, this is no longer true. Nevertheless, if the obstacle satisfies certain geometrical conditions described below and if the space dimension is odd, then all such solutions decay exponentially for fixed x as t tends to infinity. In fact, for large t such solutions behave asymptotically as follows:

$$(1.7) \qquad u(x, t) \sim \sum_{k=0}^{\infty} c_k e^{\lambda_k t} w_k(x) ,$$

where the numbers c_k depend on the initial data but the numbers λ_k and the functions w_k are determined solely by the obstacle $\mathcal{O}$ and are in a generalized sense eigenpairs for the operator Δ in the exterior domain. Each λ_k has a negative real part and they have been indexed so that

$$(1.8) \qquad 0 > \mathcal{R}e\, \lambda_1 \geqq \mathcal{R}e\, \lambda_2 \geqq \cdots \to -\infty .$$

We now give a brief resumé of the theoretical basis for (1.7) and (1.8). Let H denote the Hilbert space of all initial data with finite energy, normed by the energy norm

$$(1.9) \qquad ||u|| = \frac{1}{2} \int [|\partial u|^2 + |u_t|^2]\, dx ,$$

integrated over the exterior domain. Let ρ be chosen large enough so that $\mathcal{O}$ lies inside of the ball $\{x : |x| < \rho\}$. We call a solution *outgoing* if it is zero for $|x| < t + \rho$, $t \geqq 0$, and *incoming* if it is zero for $|x| < \rho - t$, $t \leqq 0$. The set of initial data for all outgoing (incoming) solutions we denote by D_+^ρ (D_-^ρ). Let $U(t)$ denote the operator which maps initial data into data at time t; it is clear that the $\{U(t)\}$ form a one-parameter group of unitary (energy conserving) operators on H.

Next we define the operators

$$(1.10) \qquad Z(t) = P_+^\rho U(t) P_-^\rho , \qquad\qquad t \geqq 0 ,$$

where P_+^ρ (P_-^ρ) is the orthogonal projection onto the orthogonal complement of D_+^ρ (D_-^ρ). The effect of the projection P_-^ρ is to remove signals which might be coming in from far away and the effect of P_+^ρ is to remove that part of the signal which has already been converted into an outgoing wave and no longer interacts with the obstacle. Since the data in D_+^ρ and D_-^ρ are zero inside the ball $\{|x| < \rho\}$, we see that, for data f with support in this ball,

$$(1.11) \qquad [Z(t)f](x) = [U(t)f](x)$$

for all $|x| < \rho$; thus for such data the two sides are equal near the obstacle.

In an odd number of space dimensions the subspaces D_+^ρ and D_-^ρ are orthogonal; this and the fact that, for $t > 0$, $U(t)$ maps D_+^ρ into itself implies that the operators $\{Z(t)\}$ annihilate D_+^ρ and D_-^ρ and form a *semi-group* of operators on the

30

subspace

$$(1.12) \qquad K^\rho = H \ominus (D_+^\rho \oplus D_-^\rho).$$

It is a well-known fact in the theory of semi-groups of operators that if $Z(t)$ is a compact operator for some $t > 0$, then for every f in K^ρ one can express $Z(t)f$ asymptotically as

$$(1.13) \qquad Z(t)f \sim \sum c_k e^{\lambda_k t} w_k(x),$$

where the $\{\lambda_k\}$ are the eigenvalues and the $\{w_k\}$ the eigenfunctions of the *infinitesimal generator* B of $\{Z(t)\}$. Since the operators $Z(t)$ are contractions, $\mathcal{R}e \, \lambda_k < 0$; and this combined with (1.10) gives the desired relations (1.6) and (1.7).

The parameter ρ is arbitrary; happily however the eigenvalues $\{\lambda_k\}$ do not depend on ρ, and neither do the eigenfunctions for $|x| < \rho$. In fact the $w_k(x)$ obtained for various values of ρ converge as $\rho \to \infty$ to an eigenfunction of Δ, with λ_k^2 as eigenvalue:

$$(1.14) \qquad \Delta w_k = \lambda_k^2 w_k \quad \text{in} \quad \mathcal{G}.$$

These eigenfunctions behave asymptotically like $|x|^{-1} \exp\{-\lambda_k |x|\}$ for large $|x|$ and therefore lie outside the Hilbert space H. They do however satisfy an outgoing radiation condition.

To connect the eventual compactness of $Z(t)$ with the geometrical properties of $\mathcal{O}$, we introduce the following notation: Consider all rays starting on the sphere of radius ρ which proceed toward the obstacle and are continued according to the law of reflection whenever they impinge on $\mathcal{O}$ until they leave the ball $\{|x| < \rho\}$. We call $\mathcal{O}$ *confining* if there are arbitrarily long rays of this kind; otherwise, $\mathcal{O}$ is called *nonconfining*. Surmising that sharp signals propagate along rays we conjectured (see pages 155–157 of [4]) that $Z(t)$ is eventually compact if and only if $\mathcal{O}$ is nonconfining. Ralston [10] has shown in an important special case of confining obstacles that $Z(t)$ is not compact for any t. In the opposite direction, Ludwig and Morawetz [5] (see also Phillips [8]) have shown that if $\mathcal{O}$ is convex then $Z(t)$ is eventually compact.

In this paper we study the dependence of the set of "exterior" eigenvalues $\{\lambda_k\}$ on the geometry of the obstacle $\mathcal{O}$. In particular, we show that the *real* eigenvalues, corresponding to purely decaying modes, *depend monotonically on the obstacle $\mathcal{O}$, both for the Dirichlet and Neumann boundary conditions.* From this we deduce, by comparison with spheres—for which the eigenvalues $\{\lambda_k\}$ can be determined as roots of special functions—upper and lower bounds for the density of the real $\{\lambda_k\}$ and upper and lower bounds for λ_1, the rate of decay of the fundamental real decaying mode. We also consider the wave equation with a positive potential and establish an analogous monotonicity theorem for such problems. Finally we obtain a second proof for the above Dirichlet problem in the limit as the potential becomes infinite on $\mathcal{O}$.

Our results indicate, roughly speaking, that the real exterior eigenvalues are influenced by bulk properties of the obstacle, while the complex ones are sensitive to surface details. This is merely the first step toward discovering the true relation between the shape of the obstacle and the location of the exterior eigenvalues. We have written our results in an expository style in the frank hope of attracting others to this interesting and important problem.

CONTENTS: Section 2 contains the monotonicity theorem for the transmission coefficients. Section 3 applies these results to derive comparison theorems for the imaginary zeros of the scattering matrix. Section 4 contains the proof of a comparison theorem for the eigenvalues of nonsymmetric operators. Section 5 describes an approximation to the transmission coefficient which is used to determine the asymptotic distribution of real eigenvalues for the sphere. Section 6 contains a comparison theorem for transmission coefficients for potentials in place of obstacles, and Section 7 contains a rigorous proof of the folk theorem that the transmission coefficient for an obstacle with Dirichlet boundary condition is the limit of the transmission coefficients for a sequence of potentials which are zero outside the obstacle and tend to infinity on the obstacle.

In Section 8 we derive an integral equation for the decaying modes and use it to show that there are infinitely many such modes. In Section 9 we point out a strong similarity of this integral equation to one which appears in the transport theory of mono-energetic neutrons in homogeneous media.

In an appendix we give a heuristic derivation of the relation between the transmission coefficient and the scattering matrix, and the relation between the zeros of the scattering matrix and the decaying modes of the exterior problem.

2. The Transmission Coefficient

The eigenvalue problem (1.14) imposes an outgoing radiation condition on the eigenfunction. We found it more convenient to make use of a different characterization of the eigenvalues $\{\lambda_k\}$ of the generator B of $Z(t)$, namely one which involves the *scattering matrix*. As we shall see the "near-symmetry" of the scattering matrix is especially helpful. In this section we merely state this connection; however in the appendix we present a short course on scattering theory in which all of the relations used are derived or at least made plausible.

The correspondence between the scattering matrix and the eigenvalues of $Z(t)$ is easily stated: λ is an eigenvalue of B if and only if $i\bar{\lambda}$ is a zero of the scattering matrix $\mathscr{S}(z)$ and the degree of multiplicity is the same (see [4], Theorem 3.1 of Chapter 3). Moreover, $\mathscr{S}(z)$ is meromorphic having as its poles precisely the points $-i\lambda$ for which $i\bar{\lambda}$ is a zero of $\mathscr{S}(z)$ (see [4], Theorem 5.1 of Chapter 3).

The scattering matrix $\mathscr{S}(z)$ is an operator on $L_2(S_2)$ and can be represented as

$$(2.1) \qquad \mathscr{S}(z) = I + K^{sc}(z),$$

where $K^{sc}(z)$ is an integral operator with kernel

$$(2.2) \qquad K^{sc}(\omega, \theta; z) = \frac{iz}{2\pi} k^{sc}(\omega, -\theta; z) ;$$

here $k^{sc}(\omega, \theta; z)$ is the *transmission coefficient* (see [4], Theorem 5.4 of Chapter 5). We shall be concerned with values of z in the lower half-plane, $\mathscr{I}m \, z \leqq 0$, and for such z the transmission coefficient is determined by the solution of the reduced wave equation:

$$(2.3) \qquad \begin{array}{lll} z^2 v + \Delta v = 0 & \text{in} & \mathscr{G}, \\ v(x, \omega; z) = \exp\{izx \cdot \omega\} & \text{on} & \partial\mathscr{G}. \end{array} \qquad \text{(Dirichlet problem)}$$

It can be shown that the asymptotic behavior of the solution v for large $|x|$ is given by

$$v(r\theta, \omega; z) = \frac{e^{-jzr}}{r} \left[k^{sc}(\omega, \theta; z) + O\left(\frac{1}{r}\right) \right],$$

where θ is a unit vector and $x = r\theta$. Thus, $k^{sc}(\omega, \theta; z)$ represents the asymptotic scattering amplitude created by a plane wave traveling in the ω-direction and reflected by the scattering body in the direction θ. It is known that the transmission coefficient is smooth in ω, θ and analytic in z in the lower half-plane.

We begin by deriving a useful integral representation for the transmission coefficient. Since we shall be concerned with the purely imaginary zeros of $\mathscr{S}(z)$ in the lower half-plane, it is convenient to work with $\sigma = iz$. In what follows, σ will denote a positive real number and, again for notational convenience, we set

$$(2.4) \qquad \begin{array}{l} k(\omega, \theta; \sigma) = k^{sc}(\omega, \theta; -i\sigma), \\[6pt] K(\omega, \theta, \sigma) = \dfrac{\sigma}{2\pi} k(\omega, -\theta; \sigma) = K^{sc}(\omega, \theta, -i\sigma). \end{array}$$

In this case, (2.3) becomes

$$(2.3)_{\mathrm{D}} \qquad \begin{array}{lll} \sigma^2 v - \Delta v = 0 & \text{in} & \mathscr{G}, \\ v(x, \omega; \sigma) = e^{\sigma x \cdot \omega} & \text{on} & \partial\mathscr{G}. \end{array} \qquad \text{(Dirichlet problem)}$$

and the asymptotic behavior of v is given by

$$v(r\theta, \omega; \sigma) = \frac{e^{-\sigma r}}{r} \left[k(\omega, \theta; \sigma) + O\left(\frac{1}{r}\right) \right].$$

The transmission coefficient for the corresponding Neumann problem is defined

33

in terms of the bounded solution v of

$$\sigma^2 v - \Delta v = 0 \qquad \text{in} \qquad \mathscr{G},$$

$(2.3)_\text{N}$

$$\frac{\partial v(x, \omega; \sigma)}{\partial n} = \frac{\partial e^{\sigma x \cdot \omega}}{\partial n} \qquad \text{on} \qquad \partial \mathscr{G},$$

where n denotes the outer normal to $\partial\mathscr{G}$. Again, the asymptotic behavior of v near infinity is given by

$$v(r\theta, \omega; \sigma) = \frac{e^{-\sigma r}}{r} \left[k(\omega, \theta; \sigma) + O\left(\frac{1}{r}\right) \right]$$

THEOREM 2.1.   *If we denote by* q_σ *the bilinear form*

$$q_\sigma(u, w) = \sigma^2 u w + \nabla u \cdot \nabla w,$$

then the transmission coefficient is given by

$$k(\omega, \theta; \sigma) = \frac{\alpha}{4\pi} \int_{\mathcal{O}} q_\sigma(e^{\sigma x \cdot \omega}, e^{\sigma x \cdot \theta}) \, dx$$

(2.5)

$$+ \frac{\alpha}{4\pi} \int_{\mathscr{G}} q_\sigma(v(x, \omega; \sigma), v(x, \theta; \sigma)) \, dx,$$

where $\alpha = 1$ *for the Dirichlet problem and* -1 *for the Neumann problem,* v *is the solution of* (2.3) *and* $\mathcal{O}$ *the interior,* $\mathscr{G}$ *the exterior of the obstacle.*

Proof: Since the proof is essentially the same for both the Dirichlet and the Neumann problem, we shall limit ourselves to the former. Let $\mathscr{G}_R$ denote the domain $\mathscr{G} \cap \{x : |x| < R\}$. Making use of the fact that $\exp\{\sigma x \cdot \omega\}$ and $v(x, \omega; \sigma)$ both satisfy the differential equation

$$L_\sigma u \equiv \sigma^2 u - \Delta u = 0,$$

we obtain from Green's formula the relation

$$0 = \int_{\mathscr{G}_R} [e^{\sigma x \cdot \theta} L_\sigma v(x, \omega; \sigma) - v(x, \omega; \sigma) L_\sigma e^{\sigma x \cdot \theta}] \, dx$$

(2.6)

$$= \int_{\partial\mathscr{G}_R} [e^{\sigma x \cdot \theta} \, \partial_n v(x, \omega; \sigma) - v(x, \omega; \sigma) \, \partial_n e^{\sigma x \cdot \theta}] \, dS.$$

Now, $\partial\mathscr{G}_R$ consists of two parts: $\partial\mathscr{G}$ and $\{x : |x| = R\}$; denote the integrals on the right-hand side of (2.6) over these parts by I_1 and I_2, respectively. On $\partial\mathscr{G}$

34

we have $v(x, \omega; \sigma) = \exp\{\sigma x \cdot \omega\}$ and hence I_1 can be written as

$$I_1 = \int_{\partial \mathscr{G}} [v(x, \theta; \sigma) \, \partial_n v(x, \omega; \sigma) - e^{\sigma x \cdot \omega} \, \partial_n e^{\sigma x \cdot \theta}] \, dS \,.$$

Next we apply the Green's formula

$$\int_{\mathscr{F}} u L_\sigma w \, dx = \int_{\mathscr{F}} q_\sigma(u, w) \, dx - \int_{\partial \mathscr{F}} u \, \partial_n w \, dS$$

to the functions $v(x, \theta; \sigma)$ and $v(x, \omega; \sigma)$ in the exterior $\mathscr{G}$ and to the functions $\exp\{\sigma x \cdot \omega\}$ and $\exp\{\sigma x \cdot \theta\}$ in the interior $\mathcal{O}$. Using the fact that the exterior normal to $\mathscr{G}$ is the interior normal for $\mathcal{O}$, we see that I_1 is equal to the right-hand side of (2.5).

Setting $x = \eta R$ on $\{x : |x| = R\}$ and making use of the asymptotic form of v given in $(2.3)_D$, we get

$$I_2 = \int_{|\eta|=1} \left[e^{\sigma R \eta \cdot \theta} \, \partial_R \left(\frac{e^{-\sigma R}}{R} k(\omega, \eta; \sigma) \right) \right.$$

$$\left. - \frac{e^{-\sigma R}}{R} k(\omega, \eta; \sigma) \, \partial_R \left(\frac{e^{\sigma R \eta \cdot \theta}}{R} \right) \right] R^2 \, d\eta + O\left(\frac{1}{R}\right)$$

$$= -\sigma \int_{|\eta|=1} e^{-\sigma(1 - \eta \cdot \theta)R} (1 + \eta \cdot \theta) k(\omega, \eta; \sigma) \, R \, d\eta + O\left(\frac{1}{R}\right) .$$

The kernel $R \exp\{-\sigma(1 - \eta \cdot \theta)R\}$ peaks at $\eta = \theta$ and it is easy to show that, as R becomes infinite,

$$\sigma \int_{|\eta|=1} R \exp\{-\sigma(1 - \eta \cdot \theta)R\} \, d\eta = 2\pi\sigma \int_{-1}^{1} R \exp\{-\sigma(1 - \tau)R\} \, d\tau \to 2\pi \,.$$

On replacing I_2 in (2.6) by its limiting value $-4\pi k(\omega, \theta; \sigma)$, we obtain (2.5).

COROLLARY 2.2. *For the Dirichlet and Neumann problems* $\alpha k(\omega, \theta; \sigma)$ *is the kernel of a symmetric non-negative Hilbert-Schmidt operator on* $L_2(S_2)$.

Proof: The symmetry of $k(\omega, \theta; \sigma)$, that is,

$$k(\omega, \theta; \sigma) = k(\theta, \omega; \sigma) \,,$$

follows from (2.5) and the symmetry of the form q_σ. Since v is real-valued, so is $k(\omega, \theta; \sigma)$ and, since k is smooth,

$$\iint |k(\omega, \theta; \sigma)|^2 \, d\omega \, d\theta < \infty \,,$$

so that the operator is Hilbert-Schmidt. It remains to show that

$$\alpha(ka, a) \equiv \alpha \iint k(\omega, \theta; \sigma)\, a(\omega)\, a(\theta)\, d\omega\, d\theta \geqq 0$$

for all real-valued $a(\,\cdot\,)$ in $L_2(S_2)$. Setting

$$(2.7) \qquad A(x) = \begin{cases} \displaystyle\int e^{\sigma x \cdot \omega}\, a(\omega)\, d\omega & \text{for } x \text{ in } \mathcal{O}, \\[2ex] \displaystyle\int v(x, \omega; \sigma)\, a(\omega)\, d\omega & \text{for } x \text{ in } \mathcal{G}, \end{cases}$$

it follows from (2.5) that

$$(2.8) \qquad \alpha(ka, a) = \frac{1}{4\pi}\int_{\mathbb{R}_3} [\sigma^2 A^2 + (\nabla A)^2]\, dx \geqq 0,$$

which proves the corollary.

We note that A satisfies the differential equation

$$(2.9) \qquad \sigma^2 A - \Delta A = 0$$

in $\mathcal{O} \cup \mathcal{G}$ and in the case of the Dirichlet problem that A is continuous across $\partial\mathcal{G}$, whereas in the case of the Neumann problem $\partial_n A$ is continuous across $\partial\mathcal{G}$, but A itself is not. In both cases, $A(x)$ tends to zero as $|x| \to \infty$. Since (2.8) is a positive definite quadratic form and since (2.9) is the Euler equation associated with that quadratic form, we have the following *extremal characterization* of (2.8):

COROLLARY 2.3. *In the case of the Dirichlet problem,*

$$(ka, a) = \inf \frac{1}{4\pi}\int_{\mathbb{R}_3} [\sigma^2 B^2 + (\nabla B)^2]\, dx$$

over all smooth functions B with compact support in $\mathbb{R}_3$ which, in $\mathcal{O}$, are equal to $\int \exp\{\sigma x \cdot \omega\} a(\omega)\, d\omega$. In the case of the Neumann problem,

$$-(ka, a) = \inf \frac{1}{4\pi}\int_{\mathbb{R}_3} [\sigma^2 B^2 + (\nabla B)^2]\, dx$$

over all smooth functions B in $\mathcal{O} \cup \mathcal{G}$ which vanish near infinity, and which are equal to $\int \exp\{\sigma x \cdot \omega\}\, a(\omega)\, d\omega$ in $\mathcal{O}$ and have a continuous normal derivative (but need not be continuous) across $\partial\mathcal{G}$.

We come now to one of our main results.

THEOREM 2.4. *Denote by k_1 and k_2 the transmission coefficients for the scattering objects $\mathcal{O}_1$ and $\mathcal{O}_2$, respectively. If $\mathcal{O}_1 \subset \mathcal{O}_2$, then considered as operators on $L_2(S_2)$,*

$$(2.10) \qquad \alpha k_1(\sigma) \leqq \alpha k_2(\sigma) \qquad \text{for all} \quad \sigma > 0 \text{ ;}$$

here $\alpha = 1$ for the Dirichlet problem and $\alpha = -1$ for the Neumann problem.

Proof: In the case of the Dirichlet problem, the result follows directly from Corollary 2.3 since in going from $\mathcal{O}_1$ to $\mathcal{O}_2$ we decrease the class of admissible functions. To prove the corresponding assertion in the Neumann case we set

$$A_i(x) = \begin{cases} \int \exp\{\sigma x \cdot \omega\}\, a(\omega)\, d\omega & \text{for} \quad x \text{ in } \mathcal{O}_i, \\[2mm] \int v_i(x, \omega; \sigma)\, a(\omega)\, d\omega & \text{for} \quad x \text{ in } \mathcal{G}_i, \end{cases} \qquad i = 1, 2,$$

and

$$C(x) = A_2(x) - A_1(x) .$$

Then $C(x)$ is smooth and satisfies the differential equation (2.9) in $\mathcal{O}_1$, $\mathcal{O}_2 - \overline{\mathcal{O}}_1$ and $\mathcal{G}_2$; further, $\partial_n C = 0$ on both sides of $\partial \mathcal{G}_1$, and $\partial_n C$ is continuous across $\partial \mathcal{G}_2$. Writing

$$E(A, \mathcal{G}) \equiv \int_{\mathcal{G}} [\sigma^2 A^2 + (\nabla A)^2]\, dx ,$$

we have

$$(2.11) \quad E(A_2, \mathcal{G}_1) = E(A_1 + C, \mathcal{G}_1) = E(A_1, \mathcal{G}_1) + E(C, \mathcal{G}_1) + Q ,$$

where

$$Q = 2 \int_{\mathcal{G}_1} [\sigma^2 A_1 C + \nabla A_1 \cdot \nabla C]\, dx .$$

Integrating by parts over $\mathcal{G}_1 - \mathcal{G}_2$ and $\mathcal{G}_2$ separately, and using the fact that $\partial_n C = 0$ on $\partial \mathcal{G}_1$ and is continuous across $\partial \mathcal{G}_2$, we get

$$Q = 2 \int_{\mathcal{G}_1} [(\sigma^2 C - \Delta C)A_1]\, dx + \int_{\partial \mathcal{G}_1} A_1\, \partial_n C\, dS + \int_{\partial \mathcal{G}_2} A_1[\partial_n C|_{\text{out}} - \partial_n C|_{\text{in}}]\, dS = 0 .$$

Substituting this into (2.11) and recalling that E is positive, we see that

$$E(A_2, \mathcal{G}_1) = E(A_1, \mathcal{G}_1) + E(C, \mathcal{G}_1) \geqq E(A_1, \mathcal{G}_1) .$$

It follows that

$$4\pi\alpha(k_1 a, a) = E(A_1, \mathbb{R}_3) = E(A_1, \mathcal{O}_1) + E(A_1, \mathcal{G}_1)$$

$$\leqq E(A_1, \mathcal{O}_1) + E(A_2, \mathcal{G}_1) = E(A_2, \mathbb{R}_3)$$

$$= 4\pi\alpha(k_2 a, a) \,.$$

This proves the inequality (2.10).

According to Corollary 2.2, $\alpha k(\sigma)$ is the kernel of a non-negative and symmetric operator; in Section 5 we shall show for a spherical scatterer that $\alpha k(\sigma)$ defines a strictly positive operator. Since every scatterer with a nonempty interior contains a sphere, we conclude from this and the monotonicity theorem:

COROLLARY 2.5. *If the scatterer $\mathcal{O}$ has a nonempty interior, α times the transmission coefficient is the kernel of a strictly positive operator.*

A few concluding remarks about the boundary value problem $(2.3)_D$: The operator $\sigma^2 - \Delta$ satisfies the maximum principle; consequently, the solution v which has positive boundary values and vanishes at infinity is non-negative throughout $\mathcal{G}$. This implies that the transmission coefficient k is a non-negative function.

Further, the function $\exp\{\sigma x \cdot \omega\}$ is itself a solution of the reduced wave equation. On the boundary it equals v and, from the asymptotic description of v for large $|x|$, we see that $\exp\{\sigma x \cdot \omega\}$ is greater than v near infinity. Again by the maximum principle, we conclude that

$$\exp\{\sigma x \cdot \omega\} \geqq v(x, \omega; \sigma)$$

throughout $\mathcal{G}$.

Suppose that the scatterer $\mathcal{O}_2$ contains $\mathcal{O}_1$; then it follows from the foregoing that on the boundary of $\mathcal{G}_2$ we have

$$v_1(x, \omega; \sigma) \leqq \exp\{\sigma x \cdot \omega\} = v_2(x, \omega; \sigma) \,.$$

Since both v_1 and v_2 are solutions of the reduced wave equation in $\mathcal{G}_2$ which vanish at infinity, it follows that $v_1(x, \omega; \sigma) \leqq v_2(x, \omega; \sigma)$ throughout $\mathcal{G}_2$, and so the same is true of the corresponding transmission coefficients:

$$k_1(\omega, \theta; \sigma) \leqq k_2(\omega, \theta; \sigma) \,.$$

Thus, positivity and monotonicity of the transmission coefficients hold not only in the operator sense but also in the pointwise sense for the Dirichlet problem. Of course, as is well known, these two concepts of positivity are quite distinct, neither implying the other; nevertheless there is some interest attached to those kernels which are positive in both senses and it amused us to find that the transmission coefficients belong to this class of kernels.

3. On the Purely Imaginary Zeros of the Scattering Matrix

We recall that a purely decaying mode of $Z(t)$ with eigenvalue $e^{-\sigma t}$ corresponds to a purely imaginary zero of the scattering matrix $\mathcal{S}(z)$ at $z = -i\sigma$ with the same degree of multiplicity. Since, in the notation of the previous section, $\mathcal{S}(-i\sigma) = I + K(\sigma)$, this simply means that the purely decaying modes of $Z(t)$ correspond to those positive values of σ for which -1 is an eigenvalue of $K(\sigma)$; the kernel of $K(\sigma)$ is given by

$$(3.1) \qquad K(\omega, \theta; \sigma) = \frac{\sigma}{2\pi} k(\omega, -\theta; \sigma),$$

where k is the transmission coefficient. Denoting reflection through the origin by W:

$$[Wa](\theta) = a(-\theta),$$

the relation (3.1) can be written in operator form as

$$(3.2) \qquad K(\sigma) = \frac{\sigma}{2\pi} k(\sigma) W.$$

According to Corollary 2.2 and Theorem 2.4, $\alpha k(\sigma)$ is a symmetric strictly positive Hilbert-Schmidt operator on $L_2(S_2)$; $\alpha = 1$ for the Dirichlet problem and -1 for the Neumann problem.

The presence of W complicates the problem since K need not be symmetric. Nevertheless, the following comparison theorem for K is valid.

THEOREM 3.1. (a) *The eigenvalues of $K(\sigma)$ are real.*

(b) *Let $K_1(\sigma)$ and $K_2(\sigma)$ be defined as above for obstacles $\mathcal{O}_1$ and $\mathcal{O}_2$, respectively, and suppose that $\mathcal{O}_1 \subset \mathcal{O}_2$. For fixed $\sigma > 0$, arrange according to size the positive and negative eigenvalues of $K_i(\sigma)$ taking multiplicities into account:*

$$\nu_1^{(i)} \geq \nu_2^{(i)} \geq \cdots > 0 > \cdots \geq \kappa_2^{(i)} \geq \kappa_1^{(i)}, \qquad\qquad i = 1, 2.$$

Then, for all integers n,

$$(3.3) \qquad \nu_n^{(1)} \leq \nu_n^{(2)} \qquad and \qquad \kappa_n^{(1)} \geq \kappa_n^{(2)}.$$

We shall postpone the proof of this theorem in its full generality until the next section. However, when $\mathcal{O}$ is centrally symmetric the discussion is considerably simplified and it is both instructive and useful to have a special proof for this case:

Step 1. If $\mathcal{O}$ is centrally symmetric, then $k(\sigma)$ and W commute:

$$(3.4) \qquad k(\sigma) W = Wk(\sigma),$$

and hence $K(\sigma)$ is symmetric for each $\sigma > 0$.

Proof: In this case the solutions of $(2.3)_D$ and $(2.3)_N$ have the property

$$v(x, -\omega; \sigma) = v(-x, \omega; \sigma),$$

so that

$$k(-\omega, \theta; \sigma) = k(\omega, -\theta; \sigma).$$

Consequently,

$$[kWa](\omega) = \int k(\omega, \theta; \sigma) \, a(-\theta) \, d\theta = \int k(\omega, -\theta; \sigma) \, a(\theta) \, d\theta$$

$$= \int k(-\omega, \theta; \sigma) \, a(\theta) \, d\theta = [Wka](\omega).$$

Moreover, $K^* = (kW)^* = Wk = kW = K$, as asserted.

Step 2. For the Dirichlet problem the positive (negative) eigenspaces of K consist of the even (odd) eigenspaces of k, whereas for the Neumann problem the positive (negative) eigenspaces of K consist of the odd (even) eigenspaces of k.

Proof: The involution W decomposes $L_2(S_2)$ into two orthogonal subspaces H_+ and H_- consisting of the even and odd functions, respectively:

$$Wa = a \quad \text{for all } a \text{ in } H_+ \quad \text{and} \quad Wa = -a \quad \text{for all } a \text{ in } H_-.$$

Since both k and K commute with W, each eigenspace of k and K decomposes into the orthogonal sum of a subspace in H_+ and one in H_-. Further, if $\{v, \lambda\}$ is an eigenpair of k with v even, then $\{v, \sigma\lambda/2\pi\}$ is an eigenpair for K, whereas if v is odd, then $\{v, -\sigma\lambda/2\pi\}$ is an eigenpair for K, and conversely. The assertion of step 2 now follows from the fact that k is strictly positive for the Dirichlet problem and strictly negative for the Neumann problem.

Step 3. To complete the proof of the theorem in the case of centrally symmetric objects it suffices to combine steps 1 and 2 with Theorem 2.4 and the *minimax principle* applied to H_- in the Dirichlet case and H_+ in the Neumann case.

We now have a substantial grip on the problem of determining the purely imaginary zeros of the scattering matrix; we must find positive values of σ for which -1 is an eigenvalue of $K(\sigma)$. It therefore suffices to study the growth of the negative eigenvalues $\{\kappa_n(\sigma)\}$ of $K(\sigma)$ as a function of σ, picking out those values of σ and n for which $\kappa_n(\sigma) = -1$. It is known that $k(\sigma)$ is analytic in σ for real $\sigma \geq 0$; it therefore follows from the relation (3.2) that $K(\sigma)$ converges to zero as $\sigma \to 0+$. Thus the smallest purely imaginary zero of the scattering matrix comes from the smallest root σ_0 of $\kappa_1(\sigma) = -1$ and hence we obtain, as an immediate consequence of Theorems 2.4 and 3.1,

THEOREM 3.2. *If $\mathcal{O}_1 \subset \mathcal{O}_2$, then the smallest purely imaginary zero of $\mathcal{S}_1$ is greater than or equal to the smallest purely imaginary zero of $\mathcal{S}_2$; that is, $\sigma_0^{(1)} \geqq \sigma_0^{(2)}$.*

In general the negative eigenvalues $\{\lambda_n(\sigma)\}$ are not monotone decreasing functions of σ. However, the situation is comparatively simple for star-shaped obstacles.

LEMMA 3.3. *If $\mathcal{O}$ is star-shaped, then the negative eigenvalues of $K(\sigma)$ are monotone decreasing functions of σ.*

Proof: We may as well suppose that $\mathcal{O}$ is star-shaped with respect to the origin. Let $\tau\mathcal{O}$ be the figure obtained from $\mathcal{O}$ by the transformation: $x \to \tau x$; $\mathcal{O}$ being star-shaped means that for $\tau > 1$ we will have $\tau\mathcal{O} \supset \mathcal{O}$. It is easy to verify that the corresponding solution of $(2.3)_D$ (or $(2.3)_N$) for $\tau\mathcal{O}$ and $\mathcal{O}$ are related by

$$v(x, \omega; \sigma, \tau\mathcal{O}) = v(\tau^{-1}x, \omega; \tau\sigma, \mathcal{O}) ,$$

so that

(3.5) $k(\omega, \theta; \sigma, \tau\mathcal{O}) = \tau k(\omega, \theta; \tau\sigma, \mathcal{O}) .$

By the relation (3.2), it follows from (3.5) that

(3.6) · $K(\sigma, \tau\mathcal{O}) = K(\tau\sigma, \mathcal{O}) .$

Since $\mathcal{O} \subset \tau\mathcal{O}$, Theorem 3.1 applies; we have

$$\kappa_n(\sigma, \tau\mathcal{O}) \leqq \kappa_n(\sigma, \mathcal{O}) , \qquad\qquad \tau > 1 ,$$

and combining this with (3.6) we obtain

(3.7) $\kappa_n(\tau\sigma) \leqq \kappa_n(\sigma) , \qquad\qquad \tau > 1 .$

This proves that $\kappa_n(\sigma)$ is a non-increasing function of σ. It is known however for an analytic family of compact operators such as $K(\sigma)$ that the eigenvalues are analytic except for crossing points. Since $\kappa_n(\sigma)$ tends to zero from below as $\sigma \to 0+$, we see that $\kappa_n(\sigma)$ can not remain constant in any interval and therefore must be monotone decreasing.

COROLLARY 3.4. *If $\mathcal{O}$ is star-shaped and if for a given σ, n of the eigenvalues of $K(\sigma)$ are less than or equal to -1, then the scattering matrix has exactly n purely imaginary zeros $\{-i\sigma_k\}$ with $\sigma_k \leqq \sigma$.*

For a general obstacle it seems likely that the negative eigenvalues of $K(\sigma)$ will not be monotone decreasing functions of σ. In this case, the comparison theorem furnishes us with a lower bound for the number of zeros of the scattering matrix in a given interval.

THEOREM 3.5. *If $\mathcal{O} \subset \mathcal{O}_s$, where $\mathcal{O}_s$ is star-shaped, and if $\mathcal{S}_s$ has n purely imaginary zeros $\{-i\sigma_k^s\}$ with $\sigma_k^s \leqq \sigma$, then $\mathcal{S}$ has at least n purely imaginary zeros $\{-i\sigma_k\}$ with $\sigma_k \leqq \sigma$.*

It is clear that a comparison of $\mathcal{O}$ with contained and containing spheres will furnish us with good estimates of the purely imaginary zeros of $\mathcal{S}$. This will be discussed in Section 5.

Let $\mathcal{O}_0$ be an obstacle for which one of the $\kappa_n(\sigma, \mathcal{O}_0)$ has more than one intersection with the line $\kappa = -1$. It is instructive to observe what happens when $\mathcal{O}_0$ is deformed continuously into another obstacle $\mathcal{O}_1$, say a star-shaped one, for which $\kappa_n(\sigma, \mathcal{O}_1)$ is monotonic in σ. We start with Rellich's observation that even when $K(\sigma)$ has multiple eigenvalues at some σ's, the eigenvalues can be continued as regular analytic functions of σ across these points. Let $\alpha_m(\sigma)$ denote the regular branches of these eigenvalues. It is easy to show that if one of the $\kappa_n(\sigma)$ is not monotone and crosses the line $\kappa = -1$ more than once, then one of the analytic branches $\alpha_m(\sigma)$ will also cross the line $\alpha = -1$ more than once. On the other hand, if the $\kappa_n(\sigma)$ are monotonic, so are the $\alpha_m(\sigma)$.

Suppose now we deform $\mathcal{O}_0$ continuously into $\mathcal{O}_1$ through an increasing family $\mathcal{O}_t$ of obstacles, $0 \leqq t \leqq 1$. Denote by $\alpha_m(\sigma, t)$ the analytic branch corresponding to $\mathcal{O}_t$. As t increases we shall eventually reach a t_0 for which $\alpha_m(\sigma, t_0)$ has a local maximum on the line $\alpha = -1$. For $t < t_0$, the curve $\alpha_m(\sigma, t)$ has at least three intersections with $\alpha = -1$, two of which coalesce at $t = t_0$ and disappear for $t > t_0$, becoming in the process a pair of complex zeros of the scattering matrix. This process of annihilation and creation of purely imaginary zeros out of pairs of complex ones may be worth further study.

4. A comparison Theorem for K

We now return to the proof of Theorem 3.1; we shall in fact give two proofs. The first is fairly direct and is accomplished by means of a symmetric operator with the same spectrum as K. The second proof gives a minimax characterization of the eigenvalues of K in terms of K itself in the setting of a Hilbert space with an indefinite metric.

Throughout this section, k will denote a compact strictly positive symmetric operator, W a symmetric involution (that is $W = W^*$ and $W^2 = I$) and

$$(4.1) \qquad\qquad K = kW.$$

Setting A equal to the positive square root of k,

$$(4.2) \qquad\qquad A = k^{1/2},$$

we now define

$$(4.3) \qquad\qquad C = AWA.$$

It is clear that C is compact and symmetric and further that

(4.4) $$KA = AC.$$

Actually K and C are even more closely related.

LEMMA 4.1. *K and C have the same eigenvalues with the same multiplicities. The κ-eigenspace of K is the A-transform of the κ-eigenspace of C.*

Proof: If $\{v, \kappa\}$ is an eigenpair for C, that is, if $Cv = \kappa v$, then by (4.4)

$$KAv = ACv = \kappa Av,$$

so that $\{Av, \kappa\}$ is an eigenpair for K. Conversely, if $Kw = \kappa w$, then since $K = A^2 W$ we see that w lies in the range of A and hence that $v = A^{-1}w$ exists. Again by (4.4)

$$ACv = KAv = \kappa Av,$$

and hence $\{A^{-1}w, \kappa\}$ is an eigenpair for C. Since A is one-to-one, it follows that A maps the eigenspaces of C onto the corresponding eigenspaces of K.

We have now reduced the study of the eigenvalues of K to that of the more tractable compact symmetric operator C. We proceed to investigate C.

LEMMA 4.2. *Order the eigenvalues of C taking account of multiplicities:*

$$\nu_1 \geqq \nu_2 \geqq \cdots > 0 > \cdots \geqq \kappa_2 \geqq \kappa_1 .$$

Then

(4.5) $$\nu_n = \max_{\substack{V \\ V \subset R(A)}} \min_{w \in V} \frac{(Ww, w)}{(A^{-1}w, A^{-1}w)}$$

and

(4.6) $$\kappa_n = \min_{\substack{V \\ V \subset R(A)}} \max_{w \in V} \frac{(Ww, w)}{(A^{-1}w, A^{-1}w)} ,$$

where the maximum and minimum with respect to V are taken over all subspaces V of dimension n.

Proof: According to the familiar minimax theory,

(4.7) $$\nu_n = \max_{V} \min_{v \in V} \frac{(Cv, v)}{(v, v)} ,$$

where the maximum is taken over all n-dimensional subspaces V. Writing $(Cv, v) = (WAv, Av)$ and setting $w = Av$, we get

$$\frac{(Cv, v)}{(v, v)} = \frac{(Ww, w)}{(A^{-1}w, A^{-1}w)} .$$

Now v in V can be written as w in AV. The assertion (4.5) is therefore a simple consequence of (4.7); (4.6) is proved in an analogous way.

We wish to make use of (4.5) and (4.6) to establish a comparison theorem for the eigenvalues of two different operators $K_1 = k_1 W$ and $K_2 = k_2 W$ when $k_1 \leq k_2$. This requires an additional lemma proving in effect that $k_2^{-1} \leq k_1^{-1}$. Again we set $A_i = k_i^{1/2}$.

LEMMA 4.3. *If* $0 \leq k_1 \leq k_2$, *then* $R(A_1) \subset R(A_2)$ *and*

$$\|A_2^{-1}w\| \leq \|A_1^{-1}w\| \quad for \quad all \quad w \quad in \quad R(A_1) .$$

Proof: Assume to begin with that $0 < \delta I \leq k_1 \leq k_2$ and write

$$k(t) = k_1 + (k_2 - k_1)t , \qquad\qquad 0 \leq t \leq 1 .$$

Then $k(t)^{-1}$ is uniformly bounded and

$$\frac{dk(t)^{-1}}{dt} = -k(t)^{-1}[k_2 - k_1]k(t)^{-1}$$

is obviously non-positive. It follows from this that

(4.8) $$k_2^{-1} \leq k_1^{-1} .$$

If we merely assume that $0 \leq k_1 \leq k_2$, we can obtain the desired result by a limiting procedure. To this end define

$$k_i(n) = k_i + n^{-1}I \qquad and \qquad A_i(n) = k_i(n)^{1/2} .$$

Since $A_i(n)$ is invertible, given u in H there will exist a v_n such that

(4.9) $$A_2(n)v_n = A_1(n)u .$$

The inequality (4.8) takes the form

$$\|A_2(n)^{-1}w\|^2 \leq \|A_1(n)^{-1}w\|^2 ,$$

and setting $w = A_1(n)u = A_2(n)v_n$ we get

(4.10) $$\|v_n\| \leq \|u\| .$$

Since the v_n are bounded in norm, a subsequence will converge weakly to some limit v and $\|v\| \leqq \varliminf \|v_n\|$. On the other hand, $A_i(n) = [k_i + n^{-1}]^{1/2}$ tends in norm to A_i. Hence letting n tend to infinity in (4.9), we obtain, as a weak limit on the left and a strong limit on the right,

$$A_2 v = A_1 u \, .$$

This proves that $R(A_1) \subset R(A_2)$ and also by (4.10) that $\|A_2^{-1}w\| \leqq \|A_1^{-1}w\|$ for all w in $R(A_1)$.

We now have all of the ingredients at hand for the proof of our comparison theorem.

THEOREM 4.4. *Suppose k_1 and k_2 are compact strictly positive symmetric operators such that $0 < k_1 \leqq k_2$ and set $K_i = k_i W$. Order the eigenvalues of K_i taking into account multiplicities*:

$$v_1^{(i)} \geqq v_2^{(i)} \geqq \cdots > 0 > \cdots \geqq \kappa_2^{(i)} \geqq \kappa_1^{(i)} \, .$$

Then, for all integers n,

(4.11) $$v_n^{(1)} \leqq v_n^{(2)} \quad and \quad \kappa_n^{(1)} \geqq \kappa_n^{(2)} \, .$$

Proof: For all w in $R(A_1)$ it follows from Lemma 4.3 that

$$\frac{(Ww, w)}{\|A_1^{-1}w\|^2} \leqq \frac{(Ww, w)}{\|A_2^{-1}w\|^2} \qquad \text{if} \qquad (Ww, w) \geqq 0 \, ,$$

and

$$\frac{(Ww, w)}{\|A_1^{-1}w\|^2} \geqq \frac{(Ww, w)}{\|A_2^{-1}w\|^2} \qquad \text{if} \qquad (Ww, w) \leqq 0 \, .$$

The assertion of the theorem is now an immediate consequence of Lemma 4.2.

Before starting the second proof of this theorem we introduce some concepts associated with the theory of indefinite forms (see [7]). A subspace N (or P) is called negative (positive) if $Q(x, x) \leqq 0$ ($\geqq 0$) for all x in N (P), and it is called maximal negative (maximal positive) if it is not the proper subspace of any other negative (positive) subspace. If these inequalities are strict inequalities for all non-zero elements of N (or P), then the subspace is called strictly negative (strictly positive).

The next theorem generalizes the familiar minimax characterization for the eigenvalues of a symmetric operator to operators which are symmetric with respect to an indefinite form Q.

THEOREM 4.5.   *Assume that K is a compact Q-symmetric, strictly Q-positive operator in H. Order the positive and negative eigenvalues taking into account multiplicities:*

$$(4.12) \qquad \nu_1 \geqq \nu_2 \geqq \cdots > 0 > \cdots \geqq \kappa_2 \geqq \kappa_1 .$$

Let V denote an arbitrary subspace of dimension $n-1$ and let N and P denote strictly negative and strictly positive maximal subspaces. Finally set

$$(4.13) \qquad
\begin{aligned}
U(V, P) &= \sup_{\substack{u \in P \\ Q(u, V) = 0}} \frac{Q(Ku, u)}{Q(u, u)} , \\[2ex]
\Lambda(V, N) &= \inf_{\substack{u \in N \\ Q(u, V) = 0}} \frac{Q(Ku, u)}{Q(u, u)} .
\end{aligned}
$$

Then

$$(4.14) \qquad
\begin{aligned}
\nu_n &= \inf_{P} \inf_{V \subset P} U(V, P) , \\[2ex]
\kappa_n &= \sup_{N} \sup_{V \subset N} \Lambda(V, P) .
\end{aligned}
$$

Applying this to

$$Q(x, y) = (Wx, y)$$

and $K = kW$, we obtain a characterization of the eigenvalues of K which shows directly their monotonic dependence on k.

A proof of Theorem 4.5 will appear in [9].

5. Estimates for the Distribution of the Imaginary
Zeros of the Scattering Matrix

THEOREM 5.1.   *Define*

$$(5.1) \qquad k_0(\omega, \theta, \sigma) = \frac{\alpha}{4\pi} (1 + \omega \cdot \theta)\sigma^2 \int_{\mathcal{O}} \exp \{\sigma x \cdot (\omega + \theta)\} \, dx ,$$

where $\alpha = 1$ for the Dirichlet problem and -1 for the Neumann problem. Then

$$(5.2) \qquad \alpha k_0(\sigma) \leqq \alpha k(\sigma) \leqq 3\alpha k_0(\sigma) ;$$

the first inequality holds for all $\sigma > 0$, the second for all sufficiently large σ.

Proof: The lower bound follows directly from Corollary 2.3 by restricting the integration to $\mathcal{O}$. In fact, since $A(x)$ in $\mathcal{O}$ is given by (2.7), we obtain

$$\alpha(ka, a) \geqq \frac{1}{4\pi} \int_{\mathcal{O}} [\sigma^2 A^2 + (\nabla A)^2] \, dx$$

$$= \frac{1}{4\pi} \iiint_{\mathcal{O}} \sigma^2 [1 + \omega \cdot \theta] e^{\sigma x \cdot \omega} e^{\sigma x \cdot \theta} u(\omega) \, a(\theta) \, d\omega \, d\theta \, dx .$$

Carrying out the x integration first, we can rewrite the right side as $\alpha(k_0 a, a)$ and so obtain

(5.3) $\alpha(ka, a) \geqq \alpha(k_0 a, a) .$

To obtain an upper bound we need only choose a particular admissible function B. Denoting this function by B_δ, we have of necessity $B_\delta(x) = A(x)$ in $\mathcal{O}$. We extend B_δ outside of $\mathcal{O}$ by reflection across the boundary of $\mathcal{O}$, followed by multiplication by a smooth function φ, constructed so that $0 \leqq \varphi(x) \leqq 1$ and

$$\varphi(x) = \begin{cases} 1 \text{ in } \mathcal{O} , \\ 0 \text{ at all points whose distance from } \mathcal{O} \text{ is } > \delta . \end{cases}$$

In other words, we set

(5.4) $B_\delta(x) = \begin{cases} A(x) \text{ in } \mathcal{O} , \\ \alpha\varphi(x) A(T(x)) \text{ outside } \mathcal{O} ; \end{cases}$

here $T(x)$ is the image of x under reflection across $\partial\mathcal{O}$. Note that $B_\delta(x)$ vanishes at all points whose distance from $\mathcal{O}$ is greater than δ, that B_δ is continuous across $\partial\mathcal{O}$ for $\alpha = 1$ and that $\partial_n B_\delta$ is continuous across $\partial\mathcal{O}$ for $\alpha = -1$.

Let J denote the Jacobian of T; then J is unitary on $\partial\mathcal{O}$ and hence almost unitary near $\mathcal{O}$. In particular, at all points in the support of φ but outside of $\mathcal{O}$ the norm of J can be estimated by

(5.5) $|J| \leqq 1 + \varepsilon ,$

where ε tends to zero with δ. Now,

$$\nabla B_\delta = \nabla\varphi A + \varphi J \nabla A .$$

If β denotes an upper bound for $|\nabla\varphi|$, then (5.5) and an obvious estimate give

(5.6)
$$(\nabla B_\delta)^2 \leqq \beta^2 A^2 + 2\beta |A| (1 + \varepsilon) |\nabla A| + (1 + \varepsilon)^2 (\nabla A)^2$$

$$\leqq \left(\beta^2 + \frac{\beta^2}{\varepsilon} \right) (A)^2 + (1 + \varepsilon)^3 (\nabla A)^2 .$$

Making use of (5.5) once again and of (5.6), we get

$$\int_{\mathscr{G}} [\sigma^2 B_\delta^2 + (\nabla B_\delta)^2] dx \leqq (1 + \varepsilon) \int_{\mathcal{O}} \left[\left(\sigma^2 + \beta^2 + \frac{\beta^2}{\varepsilon} \right) A^2 + (1 + \varepsilon)^3 (\nabla A)^2 \right] dx .$$

Thus by choosing δ sufficiently small so that $(1 + \varepsilon)^4 \leqq 2$ and σ sufficiently large, we have

$$\int_{\mathscr{G}} [\sigma^2 B_\delta^2 + (\nabla B_\delta)^2] \, dx \leqq 2 \int_{\mathcal{O}} [\sigma^2 A^2 + (\nabla A)^2] \, dx .$$

Since B_δ is admissible for both the Dirichlet and the Neumann problem, Corollary 2.3 shows that

(5.7)
$$\alpha(ka, a) \leqq \frac{1}{4\pi} \int_{\mathbb{R}_3} [\sigma^2 B_\delta^2 + (\nabla B_\delta)^2] \, dx = \frac{1}{4\pi} \left\{ \int_{\mathcal{O}} + \int_{\mathscr{G}} \right\}$$

$$\leqq \frac{3}{4\pi} \int_{\mathcal{O}} [\sigma^2 A^2 + (\nabla A)^2] \, dx = 3\alpha(k_0 a, a) .$$

This establishes the upper bound in (5.4).

We recall that the purely imaginary zeros of the scattering matrix correspond to those values of σ for which -1 is an eigenvalue of $K(\sigma)$. According to the theory sketched in Section 3, for symmetric obstacles this means that plus 1 is an eigenvalue of $(\sigma/2\pi)\alpha k(\sigma)$ with an odd eigenfunction in the case of the Dirichlet problem and an even eigenfunction in the case of the Neumann problem. Denote by $C_{\mathrm{D}}(\sigma)$ and $C_{\mathrm{N}}(\sigma)$ the number of eigenvalues of $(\sigma/2\pi)\alpha k(\sigma)$ greater than 1 which correspond to odd, respectively even, eigenfunctions. Since $(\sigma/2\pi)\alpha k(\sigma)$ depends continuously on σ and vanishes at $\sigma = 0$, we conclude:

THEOREM 5.2a. *For symmetric obstacles the scattering matrix has at least $C_{\mathrm{D}}(\sigma)$ $(C_{\mathrm{N}}(\sigma))$ purely imaginary zeros of absolute value less than σ for Dirichlet (Neumann) boundary conditions.*

Now for star-shaped obstacles we showed in Lemma 3.3 that $(\sigma/2\pi)\alpha k(\sigma)$ increases monotonically with σ. This implies

THEOREM 5.2b. *For symmetric star-shaped obstacles the scattering matrix has exactly $C_{\mathrm{D}}(\sigma)$ $(C_{\mathrm{N}}(\sigma))$ purely imaginary zeros of absolute value less than σ for Dirichlet (Neumann) boundary conditions.*

According to Theorem 5.1, for large σ the operator $(\sigma/2\pi)\alpha k(\sigma)$ is bracketed between $(\sigma/2\pi)\alpha k_0(\sigma)$ and $(3\sigma/2\pi)\alpha k_0(\sigma)$; furthermore, for symmetric obstacles the subspaces of even and odd functions reduce both $k(\sigma)$ and $k_0(\sigma)$. Therefore, again for large σ, it follows by the minimax principle that

$$C_D^0(\sigma) \leqq C_D(\hat{\sigma}) \leqq C_D'(\sigma) ,$$

and

$$C_N^0(\sigma) \leqq C_N(\sigma) \leqq C_N'(\sigma) ;$$

here $C_D^0(\sigma)$ and $C_N^0(\sigma)$ are the numbers of eigenvalues with odd, respectively even, eigenfunctions of $(\sigma/2\pi)\alpha k_0(\sigma)$ whose values are greater than 1, and $C_D'(\sigma)$ and $C_N'(\sigma)$ are the numbers of eigenvalues with odd, respectively even, eigenfunctions of $(\sigma/2\pi)\alpha k_0(\sigma)$ whose values are greater than $\frac{1}{3}$. As will be seen in the example given below, it is reasonable to expect that $C_D'(\sigma)$ will not differ appreciably from $C_D^0(\sigma)$, nor $C_N'(\sigma)$ from $C_N^0(\sigma)$. If this were so, the asymptotic distribution of the purely imaginary zeros of the scattering matrix of a symmetric star-shaped obstacle would be completely determined by the functions $C_D^0(\sigma)$ and $C_N^0(\sigma)$. The kernel $k_0(\omega, \theta; \sigma)$ is given explicitly by the rather simple formula (5.3) and depends *additively on* $\mathcal{O}$. We expected no great difficulty in determining the asymptotic distribution of its eigenvalues, but so far we have succeeded in this only for the case of the sphere, which we now present.

We begin by transforming the quadratic form defining $k_0(\sigma)$ by Green's formula: Using the fact that A satisfies the equation $\sigma^2 A - \Delta A = 0$, we get

$$\alpha(k_0 a, a) = \frac{1}{4\pi} \int_{\mathcal{O}} [\sigma^2 A^2 + (\nabla A)^2] \, dx = \frac{1}{4\pi} \int_{\partial \mathcal{O}} A \, \partial_n A \, dS ;$$

substituting for A as defined in (2.7) we get, after interchanging orders of integration, that

$$(5.8) \qquad \alpha k_0(\omega, \theta; \sigma) = \frac{1}{8\pi} \int_{\partial \mathcal{O}} [e^{\sigma x \cdot \omega} \partial_n e^{\sigma x \cdot \theta} + e^{\sigma x \cdot \theta} \partial_n e^{\sigma x \cdot \omega}] \, dS .$$

Next we expand $\exp\{\sigma x \cdot \omega\}$ into spherical harmonics; we start with the expansion of the exponential function:

$$(5.9) \qquad e^{a\tau} = \sum_{n=0}^{\infty} c_n(a) P_n(\tau) ,$$

where P_n is the n-th Legendre polynomial. The coefficients c_n are given by

$$(5.10) \qquad c_n(a) = \frac{2n+1}{2} \int_{-1}^{1} e^{a\tau} P_n(\tau) \, d\tau .$$

In particular,

$$(5.11) \qquad e^{a\eta \cdot \omega} = \sum c_n(a) P_n(\eta \cdot \omega) .$$

We now take $\mathcal{O}$ to be a spherical ball of radius R about the origin; in this case (5.8) can be rewritten as

$$\alpha k_0(\omega, \theta; \sigma) = \frac{R^2}{8\pi} \frac{d}{dR} \int_{|\eta|=1} e^{\sigma R \eta \cdot \omega} e^{\sigma R \eta \cdot \theta} \, d\eta \; .$$

Substituting the expansion (5.11) and using the orthogonality relations

(5.12) $$\int_{|\eta|=1} P_n(\eta \cdot \omega) P_m(\eta \cdot \theta) \, d\eta = \frac{4\pi}{2n+1} \delta_{m,n} P(\omega \cdot \theta) \; ,$$

we get

(5.13) $$\alpha k_0(\omega, \theta; \sigma) = \frac{1}{2} \sum_{n=0}^{\infty} \frac{R^2}{2n+1} \frac{d}{dR} c_n^2(\sigma R) P_n(\omega \cdot \theta) \; .$$

According to the orthogonality relations (5.12) the operator with kernel $P_n(\omega \cdot \theta)$ annihilates all spherical harmonics of order different from n and is equal to $4\pi(2n+1)^{-1}I$ on the space of n-th order spherical harmonics. This proves

PROPOSITION 5.3.   *The eigenvalues of $(\sigma/2\pi)\alpha k_0(\sigma)$ are*

(5.14) $$\lambda_n = \frac{\sigma}{(2n+1)^2} \frac{d}{dR} c_n^2(\sigma R)$$

with multiplicity $2n+1$. The eigenfunctions in the n-th eigenspace have the parity of n.

Next we estimate the number of those eigenvalues (5.14) which are greater than 1 (or $> \frac{1}{3}$). For this we need to know the asymptotic behavior of the coefficients c_n defined in (5.10). Making use of Rodrigue's formula for P_n, namely

$$P_n(\tau) = \frac{(-1)^n}{2^n n!} \frac{d^n}{d\tau^n} (1 - \tau^2)^n \; ,$$

we obtain, after n integrations by parts, that

$$c_n(a) = \frac{2n+1}{2} \int_{-1}^{1} \frac{a^n}{2^n n!} e^{a\tau} (1 - \tau^2)^n \, d\tau \; .$$

Hence,

$$\frac{d}{da} c_n^2(a) = \left(\frac{2n+1}{2}\right)^2 \left(\frac{a^n}{2^n n!}\right)^2 \int_{-1}^{1}\int_{-1}^{1} \left(\frac{2n}{a} + \tau + t\right) e^{a(\tau+t)} (1 - \tau^2)^n (1 - t^2)^n \, d\tau \, dt \; .$$

Applying the mean value theorem, this can be written as

$$(5.15) \quad \frac{d}{da} c_n^2(a) = \left(\frac{2n+1}{2}\right)^2 \left(\frac{a^n}{2^n n!}\right)^2 \left(\frac{2n}{a} + \bar{\tau} + \bar{l}\right) \left[\int_{-1}^{1} e^{a\tau}(1 - \tau^2)^n \, d\tau\right]^2 .$$

For $a > 0$, the exponential factor $\exp\{a(\tau + t)\}$ will weight the positive values of $(\tau + t)$ more than the negative values, and therefore $\bar{\tau} + \bar{l} > 0$. We set

$$(5.16) \qquad\qquad\qquad \gamma n = a = \sigma R ,$$

substitute the expression (5.15) into (5.14) and apply Stirling's approximation. We then get

$$\lambda_n = f \left\{\int_{-1}^{1} \left[\frac{\gamma e}{2} e^{\gamma \tau}(1 - \tau^2)\right] d\tau\right\}^2 ,$$

where

$$f_n \sim (2\pi n)^{-1/2} \left(\frac{\gamma n}{2}\right)^2 \left(\frac{2}{\gamma} + \bar{\tau} + \bar{l}\right) .$$

We wish to find the value of γ, say γ_n, for which $\lambda_n = c$ where c is either 1 or $\frac{1}{3}$. To this end we study the integrand in the above representation for λ_n. The maximum of the expression in the square brackets for $-1 \leq \tau \leq 1$ is attained at a root of its derivative:

$$\gamma(1 - \tau^2) - 2\tau = 0 ;$$

the root lying in the interval $(-1, 1)$ is

$$\tau_0 = \frac{\sqrt{1 + \gamma^2} - 1}{\gamma} .$$

At this point the value of the square bracket is

$$\exp\{\sqrt{1 + \gamma^2}\} \frac{\sqrt{1 + \gamma^2} - 1}{\gamma} .$$

It is not difficult to show that this is an increasing function of γ. Denote by γ_0 the value for which this expression equals one:

$$\exp\{\sqrt{1 + \gamma_0^2}\} \frac{\sqrt{1 + \gamma_0^2} - 1}{\gamma_0} = 1 .$$

This equation has but one solution and its value is

$$(5.17) \qquad\qquad\qquad \gamma_0 = 0.66274 \cdots .$$

For $\gamma > \gamma_0$ the bracket expression has a maximum greater than one and hence λ_n becomes infinite with n. On the other hand, for $\gamma < \gamma_0$ the bracket expression has a maximum less than one so that the integral tends to zero very rapidly; in fact, even with the increasing factor f_n, the eigenvalue λ_n tends to zero as n becomes infinite. It follows that γ_n tends to γ_0 as $n \to \infty$. Thus for a given σ the number of indices n for which the n-th eigenvalue λ_n is greater than 1 is by (5.16) approximately $\sigma R/\gamma_0$. Since λ_n has multiplicity $2n + 1$, the number of eigenvalues of $(\sigma/2\pi)\alpha k_0(\sigma)$ greater than 1 is

$$\left(\frac{\sigma R}{\gamma_0}\right)^2 .$$

Approximately half of the corresponding eigenfunctions are even, the other half being odd; so the number of each is

$$\frac{1}{2}\left(\frac{\sigma R}{\gamma_0}\right)^2 .$$

We have therefore proved

PROPOSITION 5.4. *Let $C(\sigma)$ denote the number of purely imaginary zeros of the scattering matrix for a sphere of radius R which are less than or equal to σ in absolute value, under either Dirichlet or Neumann boundary conditions. Then*

$$C(\sigma) \sim \frac{1}{2}\left(\frac{\sigma R}{\gamma_0}\right)^2 ,$$

where $\gamma_0 = 0.66274\cdots$.

Remark. The exact values of the purely imaginary zeros of the scattering matrix for a sphere of radius R can of course be computed from the eigenvalues of $(\sigma/2\pi)\alpha k(\sigma)$. These have been determined by Wilcox [13] for the Dirichlet problem; σ_n for the n-th mode occurs at the real zero of $K_{n+1/2}(-\sigma R)$, where $K_{n+1/2}$ is the modified Hankel function. The asymptotic expression for this zero for large n has been found by Olver [6]; it is

$$\sigma_n R \sim \gamma_0 n ,$$

in agreement with our estimate. The exact value for the lowest mode for both the Dirichlet and Neumann problems is easily computed; it is simply $\sigma_1 = 1/R$.

We can now apply the comparison Theorems 3.2 and 3.5 to any obstacle which is bracketed between two spheres.

THEOREM 5.5. *Suppose that the obstacle $\mathcal{O}$ contains a sphere of radius R_1 and is contained in a sphere of radius R_2. Let $C(\sigma)$ denote the number of purely imaginary zeros* *of the scattering matrix for $\mathcal{O}$ under either Dirichlet or Neumann boundary conditions which are less than σ in absolute value. Then,*

$$\liminf_{\sigma\to\infty} \frac{C(\sigma)}{\sigma^2} \geqq \frac{1}{2}\left(\frac{R_1}{\gamma_0}\right)^2,$$

where $\gamma_0 = 0.66274\cdots$. If in addition $\mathcal{O}$ is star-shaped, then

$$\limsup_{\sigma\to\infty} \frac{C(\sigma)}{\sigma^2} \leqq \frac{1}{2}\left(\frac{R_2}{\gamma_0}\right)^2.$$

We surmise that the limit

$$\lim_{\sigma\to\infty} \frac{C(\sigma)}{\sigma^2}$$

exists.

Formula (5.15) shows that λ_n, defined by (5.14), is positive for all n. This proves that the operator $(\sigma/2\pi)\alpha k_0(\sigma)$ is strictly positive for all $\sigma > 0$; but then according to Theorem 5.1 so is $(\sigma/2\pi)\alpha k(\sigma)$. This fact was anticipated in the proof of Corollary 2.5.

6. The Monotonicity Theorem for Scattering by Positive Potentials

In this section we consider scattering by a positive potential and show that the operator whose kernel is the transmission coefficient depends monotonically on the potential. The dynamic equation governing the motion is

$$u_{tt} = \Delta u - qu.$$

As before the transmission coefficient is defined by means of the reduced equation

$$(6.1) \qquad\qquad \sigma^2 u - \Delta u + qu = 0 \; ;$$

we consider solutions of the form

$$(6.2) \qquad\qquad u(x, \omega; \sigma) = e^{\sigma x \cdot \omega} - v(x, \omega; \sigma) \, .$$

The function v satisfies the inhomogeneous equation

$$(6.3) \qquad\qquad -\Delta v + \sigma^2 v + qv = q e^{\sigma x \cdot \omega} \, .$$

It is hard not to deduce from this that if $q(x) \exp\{\sigma x \cdot \omega\}$ tends to zero fast enough as $|x| \to \infty$ (see condition (iii) below), then (6.3) has exactly one solution v which vanishes at infinity; for large $|x|$ this solution has the following asymptotic behavior:

(6.4) $$v(r, \theta; \sigma) = \frac{e^{-\sigma r}}{r} \left[k(\omega, \theta; \sigma) + O\left(\frac{1}{r}\right) \right].$$

We shall assume that q satisfies the following requirements:

(i) $q(x)$ is non-negative and Hölder continuous except at a finite number of points,

(ii) q is locally square integrable,

(iii) q decays exponentially, i.e., there exist positive constants α, C, R such that

$$|q(x)| \leqq Ce^{-\alpha|x|} \qquad \text{for} \qquad |x| > R.$$

In what follows we take $0 < \sigma < \frac{1}{2}\alpha$.

Let G denote the *inverse* of the selfadjoint differential operator $-\Delta + \sigma^2$ acting in $L_2(\mathbb{R}_3)$; since $-\Delta + \sigma^2$ is strictly positive so is its inverse G. Applying G to (6.3) we get

(6.5) $$v = G(q \exp\{\sigma x \cdot \omega\}) - Gqv.$$

As is well known, G is an integral operator with kernel

$$G(x, y; \sigma) = \frac{1}{4\pi} \frac{e^{-\sigma|x-y|}}{|x-y|}.$$

The asymptotic behavior of this kernel for fixed y and large $|x|$ is readily seen to be

(6.6) $$G(r\theta, y; \sigma) = \frac{1}{4\pi} \frac{e^{-\sigma r}}{r} \left[e^{\sigma \theta \cdot y} + O\left(\frac{1}{r}\right) \right].$$

Writing the right side of (6.5) as an integral and substituting into it the above asymptotic expression for $G(x, y; \sigma)$, we obtain an asymptotic description for $v(x, \omega; \sigma)$ for large $|x|$. Finally, recalling the definition (6.4) of k, we have

(6.7)
$$k(\omega, \theta; \sigma) = \lim_{r \to \infty} re^{\sigma r} v(r, \omega; \sigma)$$

$$= \frac{1}{4\pi} \int_{\mathbb{R}_3} e^{\sigma \theta \cdot y} q(y)(e^{\sigma \omega \cdot y} - v(y, \omega; \sigma)) \, dy.$$

Next we solve for v in (6.5); the following lemma shows that this is possible.

LEMMA 6.1. *The operator* $I + Gq$ *is invertible in* $L_2(\mathbb{R}_3)$.

Proof: It is easy to establish that Gq is compact, in fact Hilbert-Schmidt. It follows by the Fredholm alternative that $I + Gq$ is invertible if the only solution of

(6.8)
$$w + Gqw = 0$$

which lies in $L_2(\mathbb{R}_3)$ is $w = 0$. To verify this condition note that (6.8) implies that $w = (Gq)^4 w$ and it can be shown from this by familiar estimates that $w(x)$ is bounded. Thus, qw belongs to $L_2(\mathbb{R}_3)$ and we can take the $L_2(\mathbb{R}_3)$ scalar product of (6.8) with qw; denoting this L_2 scalar product by brackets we get

$$[qw, w] + [qw, Gqw] = 0.$$

We have remarked earlier that G is a strictly positive operator and, by assumption, the potential q is non-negative. It therefore follows from the above relation that $qw = 0$; but it then also follows by (6.8) that $w = 0$. This completes the proof of the lemma.

We can now express v from (6.5) as

$$v = (I + Gq)^{-1} qe(\omega),$$

where $e(\omega)$ stands for the function

$$e(x, \omega) = e^{\sigma x \cdot \omega};$$

notice that $qe(\omega)$ belongs to $L_2(\mathbb{R}_3)$. Substituting this expression for v on the right side of (6.7) and using the above abbreviation, we get the following representation for k:

(6.9)
$$k(\omega, \theta; \sigma) = \frac{1}{4\pi} \int e(y, \theta) q(y) \{e(y, \omega) - v(y, \omega; \sigma)\} \, dy$$

$$= \frac{1}{4\pi} \int e(y, \theta) q(y) \{e(y, \omega) - (I + Gq)^{-1} Gqe(\omega)\} \, dy.$$

This could be put in a more convenient form if it were not for the fact that $e(\omega)$ is not square integrable. To get around this we set

$$e_n(x, \omega) = \begin{cases} e(x, \omega) & \text{for} \quad |x| < n, \\ 0 & \text{elsewhere}, \end{cases}$$

and define

$$k_n(\omega, \theta; \sigma) = \frac{1}{4\pi} \int e_n(y, \theta) q(y) \{e_n(y, \omega) - (I + Gq)^{-1} Gqe_n(\omega)\} \, dy$$

(6.10)
$$= \frac{1}{4\pi} [qe_n(\theta), \{I - (I + Gq)^{-1} Gq\} e_n(\omega)]$$

$$= \frac{1}{4\pi} [qe_n(\theta), (I + Gq)^{-1} e_n(\omega)].$$

Since, by assumption, $q(x)$ decays like $\exp\{-\alpha|x|\}$, it is easy to see for fixed $\sigma < \frac{1}{2}\alpha$ that $k_n(\omega, \theta; \sigma)$ is bounded and converges uniformly to $k(\omega, \theta; \sigma)$ as $n \to \infty$; *a fortiori*, $k_n(\sigma) \to k(\sigma)$ in the operator sense. Finally, let $a(\omega)$ be any real square integrable function on the unit sphere S_2; multiplying (6.10) by $a(\theta)a(\omega)$ and integrating over S_2 in θ and ω, we obtain the following identity after inverting the order of integration:

$$(6.11) \qquad (k_n a, a) = [qA_n, (I + Gq)^{-1}A_n],$$

where $(\,,\,)$ denotes the $L_2(S_2)$ inner product and A_n is the function

$$(6.12) \qquad A_n(x) = \int e_n(x, \omega)\, a(\omega)\, d\omega.$$

Note that both A_n and qA_n belong to $L_2(\mathbb{R}_3)$.

LEMMA 6.2. *For bounded non-negative potentials, the operator*

$$(6.13) \qquad q(I + Gq)^{-1}$$

is (a) *symmetric,* (b) *non-negative and* (c) *an increasing function of q.*

Proof: Suppose for the moment that q is also bounded from below by a positive constant; then the following identity holds:

$$(6.14) \qquad q(I + Gq)^{-1} = (q^{-1} + G)^{-1}.$$

Symmetry and non-negativity follow immediately from this representation of the operator (6.12). To show that the operator (6.13) increases with q we note that q^{-1} decreases and that according to Lemma 4.4 inversion reverses the order relation among positive operators.

To rid ourselves of the restriction that q is bounded from below by a positive constant, we consider the family of potentials $q_\varepsilon = q + \varepsilon$. Then in the operator topology, $q_\varepsilon \to q$, $Gq_\varepsilon \to Gq$ and $(I + Gq_\varepsilon)^{-1} \to (I + Gq)^{-1}$ as $\varepsilon \to 0$. We therefore obtain the properties (a)–(c) in the limit.

LEMMA 6.3. *The operator k on $L_2(S_2)$ whose kernel is $k(\omega, \theta; \sigma)$ is* (a) *symmetric,* (b) *non-negative and* (c) *an increasing function of q.*

Proof: Since $k_n \to k$ as n becomes infinite, it suffices to prove these properties for k_n. If q is bounded and non-negative, these properties follow for k_n from the corresponding properties of the operator $q(I + Gq)^{-1}$ listed in Lemma 6.2 and the formula (6.11). Finally, if q merely satisfies the assumptions (i) — (iii) we set

$$q_m(x) = \begin{cases} q(x) & \text{if} \quad q(x) < m, \\ m & \text{otherwise.} \end{cases}$$

Then, as m becomes infinite, $q_m A_n \to q A_n$ in the $L_2(\mathbb{R}_3)$ norm, $G q_m \to G q$ in the operator topology (even in the Hilbert-Schmidt norm) and hence $(I + G q_m)^{-1} \to (I + G q)^{-1}$ also in the operator topology. In the limit then we obtain the properties (a) — (c) for these general potentials from formula (6.11).

As before the scattering matrix is of the form

$$\mathscr{S}(-i\sigma) = I + K(\sigma) ,$$

where $K(\sigma)$ is the operator on $L_2(S_2)$ whose kernel is

$$K(\omega, \theta; \sigma) = \frac{\sigma}{2\pi} k(\omega, -\theta; \sigma) .$$

Proceeding as in Section 3 we obtain as an immediate corollary of Lemma 6.3

THEOREM 6.4. *Suppose that, for all* x,

$$0 \leqq q_1(x) \leqq q_2(x) ,$$

and that $q_1 \not\equiv 0$. *Then the smallest purely imaginary zero of* $\mathscr{S}_1$ *is in absolute value greater than or equal to the smallest imaginary zero of* $\mathscr{S}_2$.

At the moment we do not have an analogue for Theorem 3.5.

7. Scattering by an Obstacle as the Limit of Potential Scattering: A New Proof of the Monotonicity Theorem

It is part of the folklore in scattering theory that the exterior problem under Dirichlet boundary conditions is the same as scattering by a potential which is infinite on the obstacle and zero elsewhere. In this section we give a precise formulation and a rigorous proof for this proposition; and in the process we obtain an amusing new proof for the monotonicity theorem for the case of scattering by an obstacle for the Dirichlet problem.

Let $\mathcal{O}$ be a smoothly bounded obstacle. We choose a smooth non-negative function $g(x)$ which is everywhere positive inside the obstacle $\mathcal{O}$ and which vanishes identically outside $\mathcal{O}$. For each integer n, we set

$$q_n = n g .$$

Consider a plane wave $\exp\{\sigma x \cdot \omega\}$ and denote by $v_n = v_n(x, \omega; \sigma)$ the resultant signal scattered by the potential q_n; that is, v_n is the solution of equation $(6.3)_n$:

(7.1) $$-\Delta v_n + \sigma^2 v_n + q_n v_n = q_n e^{\sigma x \cdot \omega} .$$

Finally, let $v = v(x, \omega; \sigma)$ denote the corresponding signal scattered by the obstacle $\mathcal{O}$ under Dirichlet boundary conditions; that is, v is the solution of

$$(7.2) \qquad\qquad -\Delta v + \sigma^2 v = 0$$

in the exterior $\mathcal{G}$ of $\mathcal{O}$ with

$$(7.3) \qquad\qquad v = e^{\sigma x \cdot \omega} \qquad \text{on} \qquad \mathcal{O}$$

and vanishing at infinity.

THEOREM 7.1. *For each $\sigma > 0$ the limit*

$$\lim_{n \to \infty} v_n(x, \omega; \sigma) = v(x, \omega; \sigma)$$

exists uniformly for all x on compact subsets of $\mathcal{G}$ and all ω in S_2.

Before presenting the proof of Theorem 7.1 we shall point out the most important implications of this result.

COROLLARY 7.2. *Denote by $k_n(\omega, \theta; \sigma)$ the transmission coefficient associated with the potential q_n and by $k(\omega, \theta; \sigma)$ that associated with the obstacle $\mathcal{O}$ under Dirichlet boundary conditions; then*

$$\lim_{n \to \infty} k_n(\omega, \theta; \sigma) = k(\omega, \theta; \sigma)$$

uniformly in ω and θ.

Proof of Corollary 7.2: The transmission coefficients k_n and k are determined from the asymptotic behavior at infinity of the scattered waves v_n and v; and this asymptotic behavior can in turn be related to values of v_n and v in a compact portion of the exterior domain. To find such a relation we apply Green's formula to the function v_n, respectively v, and the fundamental singularity

$$G(x, y; \sigma) = \frac{1}{4\pi} \frac{e^{-\sigma|x-y|}}{|x-y|}$$

in the exterior of the sphere $\{x : |x| = R\}$ so chosen that the obstacle lies entirely inside the sphere. For $|x| > R$, we obtain

$$(7.4) \qquad v_n(x) = \int_{|y|=R} [v_n(y) \, \partial_\nu G(x, y; \sigma) - G(x, y; \sigma) \, \partial_\nu v_n(y)] \, d_y S,$$

where ∂_ν denotes differentiation with respect to $|y|$.

According to relation (6.6), for y in a compact set and $|x|$ large,

$$G(r\theta, y; \sigma) = \frac{1}{4\pi} \frac{e^{-\sigma r}}{r} \left(e^{\sigma \theta \cdot y} + O\left(\frac{1}{r}\right) \right) ;$$

and similarly

$$\partial_v G(r\theta, y; \sigma) = \frac{1}{4\omega} \frac{e^{-\sigma r}}{r} \left(\frac{\sigma\theta \cdot y}{|y|} e^{\sigma\theta \cdot y} + O\left(\frac{1}{r}\right) \right).$$

Substituting these estimates into (7.4) and recalling the definition (6.4) of k_n, we get

$$k_n(\omega, \theta; \sigma) = \lim_{r \to \infty} re^{\sigma r} v_n(r\theta, \omega; \sigma)$$

(7.5)

$$= \frac{1}{4\pi} \int_{|y|=R} \left[v_n(y, \omega; \sigma) \frac{\sigma\theta \cdot y}{|y|} - \partial_v v_n(y, \omega; \sigma) \right] e^{\sigma\theta \cdot y} \, dS.$$

According to Theorem 7.1, as $n \to \infty$, $v_n(y, \omega; \sigma)$ tends to $v(y, \omega; \sigma)$, uniformly in ω and on compact subsets of $\mathscr{G}$. Since $q_n = 0$ in $\mathscr{G}$, the differential equation satisfied by the functions v_n in $\mathscr{G}$ is simply

$$-\Delta v_n + \sigma^2 v_n = 0.$$

According to standard interior estimates for solutions of elliptic equations, we can conclude that each derivative of v_n converges to the corresponding derivative of v uniformly in ω and on compact subsets of $\mathscr{G}$. In particular,

$$\lim_{n \to \infty} v_n(y, \omega; \sigma) = v(y, \omega; \sigma)$$

and

$$\lim_{n \to \infty} \partial_v v_n(y, \omega; \sigma) = \partial_v v(y, \omega; \sigma),$$

uniformly for all y on $\{y : |y| = R\}$ and all ω. In view of (7.5) this implies that

$$\lim_{n \to \infty} k_n(\omega, \theta; \sigma) = k(\omega, \theta; \sigma)$$

uniformly for all ω and θ, as asserted in Corollary 7.2.

The monotonicity theorem for the exterior Dirichlet problem is a simple consequence of Lemma 6.3 and Corollary 7.2. In fact, let $\mathcal{O}_1$ and $\mathcal{O}_2$ be two obstacles, $\mathcal{O}_1 \subset \mathcal{O}_2$, and denote the corresponding transmission coefficients under Dirichlet boundary conditions by $k_1(\sigma)$ and $k_2(\sigma)$. Let g_1 and g_2 be smooth non-negative auxiliary functions with support $\mathcal{O}_1$ and $\mathcal{O}_2$, respectively; it is clear that they can be chosen so that, for all x,

$$g_1(x) \leqq g_2(x).$$

Finally, denote by $k_{1,n}(\sigma)$ and $k_{2,n}(\sigma)$ the transmission coefficients corresponding to the potentials

$$q_{1,n} = ng_1 \qquad \text{and} \qquad q_{2,n} = ng_2,$$

respectively. Clearly $q_{1,n}(x) \leqq q_{2,n}(x)$ and therefore, according to Lemma 6.3,

$$k_{1,n}(\sigma) \leqq k_{2,n}(\sigma)$$

as operators. Letting n become infinite and applying Corollary 7.2, we deduce that

$$k_1(\sigma) \leqq k_2(\sigma) \;;$$

this is the monotonicity theorem for scattering by obstacles for the Dirichlet problem.

We turn now to the proof of Theorem 7.1.

Proof: Each function v_n satisfies equation (7.1) in $\mathbb{R}_3$:

$$-\Delta v_n + (q_n + \sigma^2)v_n = q_n e^{\sigma x \cdot \omega} ,$$

and vanishes at infinity. We claim that v_n is non-negative throughout $\mathbb{R}_3$; for otherwise, v_n would have a *negative minimum* somewhere and at this point we would have $\Delta v_n \geqq 0$ and hence the left side of (7.1) would be negative, the right side non-negative, which is impossible.

Let x_n denote the point at which v_n attains its maximum; at this point, $\Delta v_n \leqq 0$ and we deduce from (7.1) that

$$v_n(x_n) \leqq \frac{q_n(x_n)}{q_n(x_n) + \sigma^2} \exp\{\sigma x_n \cdot \omega\} .$$

Since q_n vanishes outside of $\mathcal{O}$, we conclude that x_n lies in $\mathcal{O}$ and that

$$v_n(x_n) = \max |v_n(x)| \leqq e^{\sigma R} ,$$

where

$$R = \max_{x \in \mathcal{O}} |x| .$$

This shows that the functions $v_n(x, \omega; \sigma)$ are bounded *uniformly* with respect to n and ω for fixed $\sigma > 0$.

According to (6.2), u_n is related to the scattered signal v_n by

$$u_n = e^{\sigma x \cdot \omega} - v_n ;$$

and it follows from this that the functions $u_n(x, \omega; \sigma)$ *are also uniformly bounded in n and ω over any compact subset of* $\mathbb{R}_3$. Now, u_n satisfies the equation $(6.1)_n$:

(7.6) $$-\Delta u_n + (q_n + \sigma^2)u_n = 0 \quad \text{in} \quad \mathbb{R}_3 .$$

Outside of $\mathcal{O}$ these equations do not depend on n so that

$$-\Delta u_n + \sigma^2 u_n = 0 \quad \text{in} \quad \mathscr{G} .$$

It follows from *interior estimates* for solutions of elliptic equations that the first derivatives of the $u_n(x, \omega; \sigma)$ are uniformly bounded in n and ω in any compact subset of $\mathcal{G}$. Multiplying equation (7.6) by u_n and applying Green's formula in the ball $\{x : |x| < R\}$, chosen so that the obstacle lies in its interior, we get

$$(7.7) \qquad \int_{|x|<R} [(\Delta u_n)^2 + (q_n + \sigma^2)u_n^2] \, dx = \int_{|x|=R} u_n \, \partial_r u_n \, dS.$$

Since the sphere $\{|x| = R\}$ lies in a compact subset of $\mathcal{G}$, we can assert that u_n and $\partial_r u_n$ are uniformly bounded in n and ω on this sphere. It therefore follows that the integrals on the left in (7.7) are also uniformly bounded in n and ω.

Now the potentials q_n are defined as ng where $g(x)$ is positive throughout $\mathcal{O}$, vanishing in $\mathcal{G}$. We can therefore conclude from the boundedness of (7.7) that, for every compact subset K of the interior of $\mathcal{O}$,

$$(7.8) \qquad \lim_{n \to \infty} \int_K u_n^2 \, dx = 0$$

uniformly in ω.

Next we prove

LEMMA 7.3.

$$\lim_{n \to \infty} \int_{\partial \mathcal{O}} u_n^2 \, dS = 0$$

uniformly in ω.

Proof: After a partition of unity it suffices to show for every small boundary patch S on the boundary of $\mathcal{O}$ that

$$(7.9) \qquad \lim_{n \to \infty} \int_S u_n^2 \, dS = 0$$

uniformly in ω. It is convenient to introduce new coordinates in $\mathcal{O}$ near S, i.e., a smooth mapping f of the cylinder $\mathcal{D} \times [0, 1]$ into $\mathcal{O}$, so that $\mathcal{D} \times 0$ is S. The image of $\mathcal{D} \times [\delta, 2\delta]$ is a compact subset K_δ of $\mathcal{O}$ and hence, according to the assertion (7.8),

$$(7.10) \qquad \varepsilon(n, \delta) = \int_{K_\delta} u_n^2 \, dx = \int_\sigma^{2\delta} \int_{\mathcal{D}} w_n^2 \, J \, dy \, ds$$

tends to zero uniformly in ω as $n \to \infty$; here $w_n = u_n \circ f^{-1}$ and J denotes the Jacobian of f.

If we now apply the mean value theorem to (7.10), we see that there is a value $s(n, \delta)$ (here we omit the dependence on ω),

$$(7.11) \qquad \delta < s(n, \delta) < 2\delta,$$

such that

$$(7.12) \qquad \int_{\mathscr{D}} w_n^2(y, s(n, \delta)) \, dy = \frac{\varepsilon(n, \delta)}{j\delta} \, ,$$

where j is the value of J at some point.

Next we make use of the identity

$$(7.13) \qquad \int_0^{s(n,\delta)} \int_{\mathscr{D}} \partial_s w_n^2 \, dy \, ds = \int_{\mathscr{D}} w_n^2(y, s(n, \delta)) \cdot dy - \int_{\mathscr{D}} w_n^2(y, 0) \, dy \, .$$

The integrand of the left side can be written as $2 w_n \, \partial_s w_n$; using the fact established earlier that the functions u_n , and thereby the functions w_n , are uniformly bounded in n and w, we can estimate the left side of (7.13) by the Schwarz inequality and see that it is bounded by

$$\text{const.} \left(\delta \int_{\partial}^{2\partial} \int_{\mathscr{D}} (\partial_s w_n)^2 \, dy \, ds \right)^{1/2} .$$

According to a calculus inequality,

$$\int_0^{2\delta} \int_{\mathscr{D}} (\partial_s w_n)^2 \, dy \, ds \leqq \text{const.} \int_{|x| < R} (\Delta u_n)^2 \, dx \; ;$$

and according to (7.7), the integrals on the right are bounded uniformly in n and ω. Putting these estimates together, we see that the double integral on the left in (7.13) is bounded by

$$\text{const.} \; \delta^{1/2} \, ,$$

the value of the constant being independent of n or ω. The first term on the right in (7.13) is expressed by (7.12); consequently the remaining term in (7.13) can be bounded as follows:

$$\int_{\mathscr{D}} w_n^2(y, 0) \, dy \leqq \text{const.} \; \delta^{1/2} + \text{const.} \; \frac{\varepsilon(n, \delta)}{\delta} \, .$$

We can choose δ first and then n' so that the right side of this expression is less than any preassigned number uniformly in $n \geqq n'$ and ω; this shows that

$$\lim_{n \to \infty} \int_{\mathscr{D}} w_n^2(y, 0) \, dy = 0$$

uniformly in ω; it is clear that (7.9) is equivalent with this and thus the proof of Lemma 7.3 is complete.

Now by (6.2),

$$u_n = e^{\sigma x \cdot \omega} - v_n \, ,$$

and by (7.3),

$$v = e^{\sigma x \cdot \omega} \qquad \text{on} \qquad \partial \mathcal{O}.$$

Hence, Lemma 7.3 can be expressed as

$$(7.14) \qquad \lim_{n \to \infty} \int_{\partial \mathcal{O}} |v - v_n|^2 \, dS = 0$$

uniformly in ω. According to (7.1) and (7.2) both v_n and v are annihilated outside of $\mathcal{O}$ by the operator $-\Delta + \sigma^2$ and hence so is their difference $v_n - v$. We can therefore represent $v_n - v$ at any point x in $\mathcal{G}$ in terms of the value of $v_n - v$ on $\partial \mathcal{O}$ with the aid of the Green's function F for the exterior domain $\mathcal{G}$:

$$(7.15) \qquad v_n(x) - v(x) = \int_{\partial \mathcal{O}} (v_n - v) F(x, y; \sigma) \, d_y S.$$

Expressing $F(x, y; \sigma)$ as the sum of the fundamental singularity and a regular part, we see, with the aid of the maximum principle, that

$$|F(x, y; \sigma)| \leqq \frac{1}{4\pi d} e^{-\sigma d},$$

where d is the distance of x to $\partial \mathcal{O}$. Estimating the right side of (7.15) by the Schwarz inequality and using the above estimate for F as well as (7.14), we conclude that $v_n(x, \omega; \sigma) \to v(x, \omega; \sigma)$, uniformly for all x bounded away from $\partial \mathcal{O}$ and for all ω. This completes the proof of Theorem 7.1.

8. An Integral Equation for the Exponentially Decaying Modes of the Wave Equation

There is an interesting relation between the purely decaying modes of the wave equation and those of the transport equation. In order to bring out this similarity we shall give a different and more direct proof of the fact that there are infinitely many purely decaying modes for the wave equation with the following simple type of potential:

$$(8.1) \qquad q(x) = \begin{cases} c & \text{in} & \mathcal{O}, \\ 0 & \text{outside} & \mathcal{O}, \end{cases}$$

where c is a positive constant. We shall use a direct method to prove that *there are infinitely many purely decaying modes*.

Let $u = e^{-\sigma t} w(x)$, $\sigma > 0$, be a purely decaying mode; w satisfies the relation

$$(8.2) \qquad \sigma^2 w - \Delta w + qw = 0.$$

We introduce the following abbreviations: L_σ is the operator

$$L_\sigma = \sigma^2 - \Delta .$$

e_σ denotes the outgoing fundamental solution of L_σ :

(8.3)
$$e_\sigma = \frac{e^{\sigma|x|}}{|x|} ;$$

it satisfies

(8.4)
$$L_\sigma e_\sigma = \delta .$$

We denote the translates of e_σ by $e_{\sigma,y}$:

$$e_{\sigma,y} = e_\sigma(x - y) ;$$

they satisfy

(8.5)
$$L_\sigma e_{\sigma,y} = \delta(x - y) .$$

Equation (8.2) can be abbreviated as

(8.6)
$$L_\sigma w + qw = 0 .$$

In addition to satisfying the eigenfunction equation (8.6), the purely decaying mode w has to be outgoing at ∞. Among the several equivalent formulations of this property the following is the most convenient to use here:

A function w which satisfies

$$L_\sigma w = 0$$

outside of a bounded domain $\mathcal{O}$ is outgoing if it is a superposition of outgoing fundamental solutions $e_{\sigma,y}$, y inside $\mathcal{O}$.

We shall derive now an integral equation for outgoing solutions w of (8.5). We shall use Green's formula

$$\int_{\mathcal{D}} (v L_\sigma z - z L_\sigma v) \, dx = \int_{\partial \mathcal{D}} \left(z \frac{dv}{dn} - v \frac{dz}{dn} \right) dS .$$

LEMMA 8.1. *Let $\mathcal{D}$ be any domain, p and y any pair of points contained in $\mathcal{D}$; then*

$$\int_{\partial \mathcal{D}} \left(e_{\sigma,p} \frac{de_{\sigma,y}}{dn} - e_{\sigma,y} \frac{de_{\sigma,p}}{dn} \right) dS = 0 .$$

Proof: Using Green's formula we can express the above boundary integral as the volume integral

$$\int_{\mathcal{D}} (e_{\sigma,y} L_\sigma e_{\sigma,p} - e_{\sigma,p} L_\sigma e_{\sigma,y}) \, dx .$$

Using (8.5) we see that the value of this integral is

$$e_{\sigma,y}(p) - e_{\sigma,p}(y)$$

which by definition equals

$$e_\sigma(p - y) - e_\sigma(y - p) .$$

Since $e_\sigma(x)$ is an even function of x, the above quantity is zero. This proves Lemma 8.1 when $p \neq y$; the case $p = y$ can be handled by letting p approach y.

LEMMA 8.2. *Suppose that w is an outgoing solution of $L_\sigma w = 0$ outside $\mathcal{O}$; then, for any domain $\mathcal{D}$ such that $\partial \mathcal{D}$ lies outside $\mathcal{O}$,*

$$\int_{\partial \mathcal{D}} \left(e_{\sigma,p} \frac{dw}{dn} - w \frac{de_{\sigma,p}}{dn} \right) dS = 0 .$$

Proof: Outgoing means that, outside of $\mathcal{O}$, w is a superposition of functions $e_{\sigma,y}$, y in $\mathcal{O}$. Forming a superposition of the integrals asserted to be zero in Lemma 8.1 we obtain Lemma 8.2.

We apply now Green's formula to w and $e_{\sigma,p}$, over any domain containing $\mathcal{O}$. According to Lemma 8.2 the boundary integral is zero; therefore so is the volume integral. Since $L_\sigma e_{\sigma,p} = \delta(x - p)$ and, according to (8.6), $L_\sigma w = -q w$, that volume integral is

$$(8.7) \qquad\qquad w(p) + \int_{\mathcal{D}} e_{\sigma,p} q w \, dx = 0 .$$

Recalling the definitions of $e_{\sigma,p}$ and of q we can rewrite this as

$$(8.8) \qquad\qquad -\frac{1}{c} w(p) = \int_{\mathcal{O}} \frac{e^{\sigma|x-p|}}{|x - p|} w(x) \, dx .$$

Conversely, let w be a function defined in $\mathcal{O}$ which satisfies equation (8.8) for p in $\mathcal{O}$; let us define w outside of $\mathcal{O}$ by (8.8). Clearly, the function w thus extended satisfies

$$L_\sigma w + q w = 0$$

and, being a superposition of function $e_{\sigma,x}$, x in $\mathcal{O}$, is outgoing.

What we have shown so far can be summarized as

THEOREM 8.3. *The wave equation with the potential (8.1) has an exponential mode $e^{-\sigma t}w$ if and only if $-1/c$ is an eigenvalue of $K_\sigma(\mathcal{O})$, an integral operator over $\mathcal{O}$ with kernel*

$$\frac{e^{\sigma|x-p|}}{|x - p|} .$$

We turn now to investigating the eigenvalues of K_σ. Since the kernel of K_σ is only mildly singular, K_σ is a compact operator and has a standard discrete spectrum. Also, the kernel of K_σ is real and symmetric so that the eigenvalues of K_σ are real.

Denote by $\kappa_n(\sigma, \mathcal{O})$ the n-th negative eigenvalue of K_σ when K_σ has n negative eigenvalues. Otherwise set $\kappa_n(\sigma) = 0$.

LEMMA 8.4. $\kappa_n(\sigma)$ *is a continuous function of σ for each n.*

Proof: It follows from the minimax characterization of eigenvalues that

$$|\kappa_n(\sigma) - \kappa_n(\tau)| \leqq \|K_\sigma - K_\tau\| .$$

It can be shown easily that K_σ depends continuously on σ in the norm topology, from which Lemma 8.4 follows.

LEMMA 8.5. $\kappa_n(\sigma) = 0$ *for $\sigma \leqq 0$ for all n.*

Proof: We shall prove that $K_\sigma \geqq 0$ for $\sigma \leqq 0$; clearly this implies the lemma. To prove the positivity of K_σ we note that K_σ is the restriction to $\mathcal{O}$ of convolution with

$$\frac{e^{\sigma|x|}}{|x|} .$$

For $\sigma < 0$ this function has the Fourier transform

$$\frac{1}{\xi^2 + \sigma^2}$$

which is positive. This proves the positive definiteness of K_σ, $\sigma < 0$.

LEMMA 8.6.   *For all n,*

$$\lim_{\sigma \to \infty} \kappa_n(\sigma) = -\infty .$$

Combining the last three lemmas we see that each $\kappa_n(\sigma)$ is a continuous function whose value changes from 0 to $-\infty$ as σ goes from 0 to ∞. Therefore, according to the intermediate value theorem, each equation

$$\kappa_n(\sigma) = -\frac{1}{c}$$

has at least one solution $\sigma_n > 0$. This proves

THEOREM 8.7.   *There are infinitely many decaying modes.*

Proof of Lemma 8.6:   We choose a ball inside $\mathcal{O}$; denote its diameter by D. We select n distinct points $x_1, \cdots, x_n$ on the surface of the ball, so chosen that no two are antipodal. Denote by d the maximum distance of any point x_i to any point x_j or $\bar{x}_j$, $j \neq i$, where $\bar{x}_j$ denotes the antipode of x_j. By construction,

$$d < D .$$

Let B_i be the ball of radius r and center x_i, $\bar{B}_i$ its antipode. For r small enough these balls belong to $\mathcal{O}$ and are disjoint.

Define n functions u_j, $j = 1, \cdots, n$, as follows

$$u_j(x) = \begin{cases} a & \text{in} & B_j , \\ -a & \text{in} & \bar{B}_j , \end{cases}$$

a so chosen that the L_2 norm of u_j is 1. Since the u_j have disjoint supports, they are orthogonal. We claim that for any u belonging to the span of the $\{u_j\}$, for σ large enough and for r small enough,

$$(8.9) \qquad (u, K_\sigma u) \leqq -e^{(D-\varepsilon)\sigma} \|u\| .$$

The assertion of Lemma 8.6 follows from this via the minimax principle.

We have

$$u = \sum a_j u_j ;$$

since the u_j are orthonormal,

$$\|u\|^2 = \sum a_j^2 .$$

Now

$$(8.10) \quad \begin{aligned} (u, K_\sigma n) &= \sum_{i,j} (u_i, K_\sigma u_j)\, a_i\, a_j \\ &= \sum_i (u_i, K_\sigma u_i)\, a_i^2 + \sum_{i \neq j} (u_i, K_\sigma u_j)\, a_i\, a_j , \end{aligned}$$

and by definition,

$$(u_i, K_\sigma u_i) = \iint \frac{e^{\sigma|x-y|}}{|x-y|}\, u_i(x)\, u_i(y)\, dx\, dy .$$

Since the support of u_i consists of two antipodal balls whose minimum distance is $D - 2r$, and since u_i was chosen as *odd*, we see that for σ large enough

$$(8.11) \qquad (u_i, K_\sigma u_i) \leqq -\text{const.}\ e^{\sigma(D-2r)} .$$

On the other hand, the maximum distance of any point x in the support of u_i to any point y in the support of u_j is, for $i \neq j$, less than $d + 2r$. Therefore,

$$(8.12) \qquad |(u_i, K_\sigma u_j)| \leqq \text{const.}\ e^{\sigma(d+2r)} .$$

Substituting (8.11) and (8.12) into (8.10), we see that for $r < \frac{1}{4}(D - d)$ inequality (8.9) holds.

It follows from the minimax principle that the eigenvalues of $K_\sigma(\mathcal{O})$ depend monotonically on $\mathcal{O}$; this implies that for star-shaped $\mathcal{O}$ the number of negative eigenvalues of $K_\sigma(\mathcal{O})$ increases with σ. Using the folk theorem of Section 7 this yields yet another proof of Theorem 3.5.

9. Exponential Modes of the Linear Boltzmann Equation

We shall derive now an integral equation for the exponential modes of the linear Boltzmann equation for the transport of mono-energetic neutrons in homogeneous domains. This turns out to be very similar to the integral equation obtained for the wave equation in the preceding section. Let $f = f(t, x, v)$ be the distribution function of the neutron density, where x is position in space, v velocity. Since the particles are assumed mono-energetic, the velocities all have the same absolute value, say 1.

The transport equation asserts that

$$(9.1) \qquad f_t + v \cdot f_x + \lambda f - c \int_{|v|=1} f \, dv = 0 \,.$$

Here λ is the removal cross section and c the combined scattering and fission cross section; λ and c are constants. The equation asserts that the rate of change of f along a particle path is the sum of two terms: one, proportional to f, is due to the removal of scattered particles from the beam, the other, proportional to $\int f \, dv$, is due to particles scattered into the beam or created by fission; both processes are assumed to be isotropic.

The boundary conditions are that no particles enter $\mathcal{O}$ from the outside; for convex $\mathcal{O}$ this means that

$$(9.2) \qquad f(t, x, v) = 0$$

at all boundary points for those velocities v which point into $\mathcal{O}$.

Let us look at exponential solutions $f = e^{-\mu t} g(x, v)$; we get

$$(9.3) \qquad (-\mu + \lambda) g + v \cdot g_x - c \, w = 0 \,,$$

where $w = w(x)$ stands for

$$(9.4) \qquad w(x) = \int_{|v|=1} g(x, v) \, dv \,.$$

Abbreviating $\mu - \lambda$ by σ we can write (9.3) as

$$\sigma g + \frac{d}{ds} g(x - s v, v) = -c w .$$

Multiplying by $e^{\sigma s}$ and integrating with respect to s from 0 to d, where d is the distance of x to the boundary in the direction v, we get, using the boundary conditions,

$$g(x, v) = c \int_0^d e^{\sigma s} w(x - s v) \, ds .$$

Substituting this into the definition (9.4) of w we obtain

$$(9.5) \qquad w(x) = c \int_{|v|=1} \int_0^d e^{\sigma s} w(x - s v) \, ds \, dv .$$

We introduce instead of s and v a new variable of integration y:

$$y = x - s v .$$

Since $dy = s^2 \, ds \, dv$ and $s = |x - y|$, (9.5) becomes

$$\frac{1}{c} w(x) = \int \frac{e^{\sigma |x - y|}}{|x - y|^2} w(y) \, dy .$$

We can summarize this result as follows:

The Boltzmann equation (9.1), (9.2) *has an exponential mode* $e^{-\mu t} g$ *if and only if* $1/c$ *is an eigenvalue of* $H_\sigma(\mathcal{O})$, $\sigma = \mu - \lambda$, *where* $H_\sigma(\mathcal{O})$ *is the integral operator over* $\mathcal{O}$ *with kernel*

$$\frac{e^{\sigma |x - y|}}{|x - y|^2} .$$

This kernel is very similar to the kernel of the integral operator $K_\sigma(\mathcal{O})$ which appeared in the description of the exponential modes of the wave equation with potential c over $\mathcal{O}$. The same argument as was employed in Section 8, except that now we choose even instead of odd test functions u_j, shows that the n-th positive eigenvalue of H_σ tends to ∞ as $\sigma \to +\infty$, and tends to 0 as $\sigma \to -\infty$. The existence of infinitely many eigenvalues follows as before.

The argument presented above is a crude version of one given by Van Norton in [12] for the existence of infinitely many exponential modes for a ball. All the eigenfunctions found by him had spherical symmetry. Van Norton's result was extended to arbitrary convex $\mathcal{O}$ in an interesting paper of Ukai [11].

Appendix

A Short Course in Scattering Theory

Many of the intuitive ideas associated with scattering theory for the wave equation are quite distinct from those associated with the more familiar quantum mechanical theory. It may therefore be helpful if we sketch in a heuristic way the connection between the spectrum of the infinitesimal generator of the semi-group $\{Z(t)\}$ defined in Section 1 and the zeros of the scattering matrix; rigorous proofs can be found in our book [4].

To begin with, scattering theory deals with the asymptotic behavior of solutions for large values of t, more precisely with the relation of the asymptotic behavior of solutions as $t \to \infty$ and as $t \to -\infty$. For a solution u of the wave equation in the exterior of an obstacle which has finite energy, it can be shown that for large $|t|$ almost all of the energy of u is carried out to infinity along rays; in other words, such solutions can be described asymptotically as follows:

$$\text{(A.1)} \qquad u(x, t) \simeq \begin{cases} f_+\left(t - |x|, \dfrac{x}{|x|}\right) & \text{as} \qquad t \to \infty\,, \\[2em] f_-\left(t + |x|, \dfrac{x}{|x|}\right) & \text{as} \qquad t \to -\infty\,. \end{cases}$$

This asymptotic form shows that, for large $|t|$, u_t and u_r are approximately equal in absolute value and much larger than spacial derivatives of u in directions orthogonal to the radius vector; and it also follows that the energy outside the shell $|t| - c < |x| < |t| + c$ is small for large c. Thus for large $|t|$ and c the total energy of u is approximated by

$$\text{(A.2)} \qquad \int_{|t|-c<|x|<|t|+c} u_t^2 \, dx \simeq \iint t^2 u_t^2 \, dr \, d\theta\,.$$

Motivated by these considerations we study the following limits[1] for s in $\mathbb{R}$ and θ in S_2 :

$$\text{(A.3)} \qquad k_\pm(s, \theta) = \lim_{t \to \pm\infty} t u_t((t + s)\theta, t)\,.$$

The functions k_+ and k_- assigned in this way to a given solution u are called the *outgoing* and *incoming translation representations*, respectively.

---

[1] Limits of this kind were first considered by F. G. Friedländer, [2].

THEOREM A.1   (i) *The limits* (A.3) *exist for a dense set of u.*

(ii) *The representations defined by* (A.3) *are unitary; that is, the* $L_2(\mathbb{R}, S_2)$ *norm of* $k_\pm$ *is equal to the energy norm of u:*

$$\|u(\,\cdot\,, t)\|_E^2 = \frac{1}{2} \int \left[|\partial_x u|^2 + |u_t|^2 \right] dx = \iint |k_\pm(s, \theta)|^2 \, d\theta \, ds \,,$$

and every $k_\pm$ *in* $L_2(\mathbb{R}, S_2)$ *represents some u of finite energy.*

(iii) *Translation of u in t corresponds to translation of k in s:*

$$u(x, t + \tau) \sim k_\pm(s - \tau, \theta) \,.$$

The proof of this theorem can be made to depend on a related result for the wave equation in free space. To this end we associate to a given k in $L_2(\mathbb{R}, S_2)$ the function $v(x, t)$ defined as

(A.4) $$v(x, t) = \frac{1}{2\pi} \int k(x \cdot \theta) - t, \theta) \, d\theta \,.$$

LEMMA A.2.   (a) *The function v satisfies the wave equation in* $\mathbb{R}_3$.

(b) *For smooth k with compact support, the limit*

$$\lim_{t \to \pm \infty} t v_t((t + s)0, t) = k(s, \theta)$$

exists.

(c) $\|v(\,\cdot\,, t)\|_E^2 = \iint |k(s, \theta)|^2 \, d\theta \, ds = \|k\|^2 \,.$

(d) *If* $k(s, \theta) = 0$ *for* $s < \rho$, *then* $v(x, t) = 0$ *for* $|x| < \rho + t$; *similarly, if* $k(s, \theta) = 0$ *for* $s > -\rho$, *then* $v(x, t) = 0$ *for* $|x| < \rho - t$.

(e) *Conversely, if v is a solution of the wave equation in* $\mathbb{R}_3$ *which has finite energy and which vanishes for* $|x| < \rho + t$, *then v can be represented in the form* (A.4) *with* $k(s, \theta) = 0$ *for* $s < \rho$. *There is a similar statement with t replaced by* $-t$.

Proof of Lemma A.2:   Part (a) follows directly by differentiating the right side of (A.4). Part (b) requires a straightforward asymptotic evaluation; the details can be found on pages 107–108, Theorem 2.4, of [4]. Part (c) is simply the Parseval relation for the Radon transform (which, after a Fourier transformation with respect to s, reduces to the Parseval relation for the Fourier transform). Part (d) follows directly from (A.4), part (e) is contained in Theorem 2.3, Chapter IV of [4].

We turn now to the proof of Theorem A.1; for this proof and throughout the appendix, ρ is chosen so that the scatterer is contained in the ball $|x| < \rho$. We consider the set of all *outgoing* solutions, i.e., solutions $u(x, t)$ which vanish in the truncated cone $|x| < \rho + t$. Then, for $t > 0$, u does not interact with the scatterer and hence can be thought of as a solution of the wave equation in free space. According to part (e) of Lemma A.2, such a u can be represented for $t > 0$ in the form (A.4) with a k which is zero for $s < \rho$. If this k turns out to be smooth and has compact support, then according to part (b) of Lemma A.2 the limit (A.3)$_+$ exists. The set of all such nice k is dense in the L_2 sense among all k with support in (ρ, ∞); it therefore follows from part (c) of Lemma A.2 that the set of u corresponding to such nice k form a dense subset of the outgoing solutions in the sense of the energy norm.

Next we make use of Theorem 2.1 from Chapter V of [4], which asserts that translates of outgoing solutions are dense in the energy norm among all solutions. It follows that the translates of a dense subset of the outgoing solutions are also dense among all solutions. We have just shown the existence of the limit (A.3)$_+$ for a dense subset of outgoing solutions; and it can be seen by inspection that if the limit k_+ in (A.3) exists for a given solution u, then it also exists for any t-translate of u, the value of this limit being the translate of k_+. This proves part (i) of Theorem A.1, that is, the existence of the limit (A.3)$_+$ for a dense set of u; in the process we have also proved part (iii).

To prove part (ii) we note that, according to part (d) of Lemma A.2, each k with support in (ρ, ∞) represents an outgoing solution and that, according to part (c), this representation is isometric. Translates of k represent translates of u, by (iii), and since translates of those k with support in (ρ, ∞) are dense in $L_2(\mathbb{R}, S_2)$, this proves part (ii).

DEFINITION. The operator S relating functions k_- and k_+ which represent the same solution u is called the *scattering operator*:

$$S : k_- \to k_+ .$$

According to Theorem A.1, translating u in t corresponds to a translation of both k_- and k_+ by the same amount. This shows that S *commutes with translation;* it follows from this that S is a convolution operator:

$$(A.5) \qquad k_+(s) = S * k_- = \int S(r)k_-(s - r)\, dr .$$

If we use a Fourier transformation on (A.5), denoting the transform of S by $\mathscr{S}$, those of $k_\pm$ by $a_\pm$, and the variable dual to s by z, we then obtain

$$(A.6) \qquad a_+(z) = \mathscr{S}(z)a_-(z) .$$

The temporary suppression of the variable θ means that $k_\pm$ and $a_\pm$ are thought of as vector-valued (that is $L_2(S_2)$-valued) functions and that $\mathscr{S}(z)$ is an operator-valued $(L_2(S_2) \to L_2(S))$ function.

We wish to derive an explicit expression for $\mathscr{S}(z)$; by definition,

$$(A.7) \qquad \mathscr{S}(z) = \int e^{irz} S(r)\, dr\,.$$

In order to compute $\mathscr{S}(z)$ it is convenient to determine how S acts on

$$(A.8) \qquad k_-(s) = e^{-izs} n\,, \qquad\qquad n \text{ in } L_2(S_2);$$

for purposes of exposition we overlook the fact that such k_- are not square integrable. It follows from (A.5) and (A.7) that

$$(A.9) \qquad k_+(s) = e^{-izs} \mathscr{S}(z) n\,.$$

To find an explicit relation between k_- and k_+ in (A.8) and (A.9), we examine the solution u which they both represent. First of all, since both k_- and k_+ are exponential functions of s, it follows from part (iii) of Theorem A.1 that u also depends exponentially on t:

$$(A.10) \qquad u(x, t) = e^{izt} w(x)\,.$$

Clearly, w is a solution of the reduced wave equation

$$(A.11) \qquad z^2 w + \Delta w = 0 \quad \text{in} \quad \mathscr{G}\,,$$

which satisfies the boundary condition

$$(A.12) \qquad w = 0 \text{ on the obstacle}\,.$$

One way of constructing solutions of this type is to solve (A.11) for solutions w_+ and w_- which are outgoing and incoming, respectively, in the sense of Sommerfeld and which are both equal to some given smooth function on the obstacle (see Theorem 4.2 of Chapter V in [4]). Then,

$$(A.13) \qquad w(x) = w_+(x) - w_-(x)$$

satisfies both (A.11) and (A.12); moreover, for large $|x|$,

$$(A.14)_+ \qquad w_+(x) = \frac{e^{-iz|x|}}{|x|}\, \varphi_+\left(\frac{x}{|x|}\right) + O\left(\frac{1}{|x|^2}\right)$$

and

$$(A.14)_- \qquad w_-(x) = \frac{e^{iz|x|}}{|x|}\, \varphi_-\left(\frac{x}{|x|}\right) + O\left(\frac{1}{|x|^2}\right)\,.$$

Next we evaluate the limits (A.3); it follows from (A.13) and (A.14) that these limits exist in some mean, say Cesaro, sense [2]:

$$k_+(s, \theta) = C - \lim_{t \to \infty} tu_t((t + s)\theta, t)$$

$$= C - \lim_{t \to \infty} izte^{izt}w((t + s)\theta)$$

(A.15)$_+$

$$= C - \lim_{t \to \infty} izte^{izt} \left[\frac{e^{-i(t+s)z}}{t + s} \varphi_+(\theta) - \frac{e^{i(t+s)z}}{t + s} \varphi_-(\theta) \right]$$

$$= ize^{-isz}\varphi_+(\theta) .$$

Similarly we obtain, noting that for t large negative $|(t + s)\theta| = -s - t$,

(A.15)$_-$ $$k_-(s, \theta) = ize^{-isz}\varphi_-(-\theta) .$$

Comparing this with (A.8) and (A.9) we see that

(A.16) $$\mathscr{S}(z) : \varphi_-(-\theta) \to \varphi_+(\theta) .$$

In what follows we need an expression for φ_- and φ_+ directly in terms of w; as above this can be obtained from (A.13) and (A.14):

$$\varphi_+(\theta) = C - \lim_{r \to \infty} re^{irz}w(r\theta) ,$$

(A.17)

$$\varphi_-(\theta) = C - \lim_{r \to \infty} re^{-irz}w(r\theta) .$$

Let e denote the function

$$e(x, \omega; z) = e^{izx \cdot \omega} .$$

Clearly, $e(x)$ is a solution of the reduced wave equation; equally clearly it does not vanish on the obstacle nor at infinity. Ignoring these defects for the moment, we evaluate the limits (A.17) and obtain after a brief calculation

$$C - \lim_{r \to \infty} re^{irz}e(r\theta, \omega; z) = \frac{2\pi}{iz} \delta_{-\omega}(\theta) ,$$

(A.18)

$$C - \lim_{r \to \infty} re^{-irz}e(r\theta, \omega; z) = \frac{2\pi}{iz} \delta_{\omega}(\theta) ;$$

here δ_{ω} denotes the δ-function at the point ω. The limits (A.18) exist in the topology of distributions in θ; it is reassuring to note that we can recover

---

[2] The limits do not exist in the ordinary sense because the solution u in question has infinite energy.

$e^{izt}e(x, \omega; z)$ from (A.4) if we set $k(s, \theta) = 2\pi e^{-isz}\delta_{-\omega}$, as required by (A.15), (A.17) and (A.18).

Next let $v(x, \omega; z)$ denote that solution of the reduced wave equation (A.11) in the exterior domain which satisfies the boundary condition

(A.19) $v(x) = e^{izx\cdot\omega}$ on   the   obstacle,

and which is outgoing at infinity, i.e., has the asymptotic form

(A.20) $v(r\theta, \omega; z) = \dfrac{e^{-izr}}{r} k^{sc}(\omega, \theta; z) + O\left(\dfrac{1}{r^2}\right).$

As above, this boundary value problem has a unique solution. The function

(A.21) $w(x) = e(x) - v(x)$

is then a solution of the reduced wave equation in the exterior domain which is zero on the boundary of the obstacle. From (A.18) and (A.20) we get the following values for φ_- and φ_+ associated with the function w defined in (A.21):

$$\varphi_-(\theta) = \frac{2\pi}{iz} \delta_\omega(\theta), \qquad \varphi_+(\theta) = \frac{2\pi}{iz} \delta_{-\omega}(\theta) + k^{sc}(\omega, \theta; z).$$

Hence, using the description (A.16) of $\mathscr{S}(z)$, we have

$$\mathscr{S}(z) : \delta_{-\omega}(\theta) \to \delta_{-\omega}(\theta) + \frac{iz}{2\pi} k^{sc}(\omega, \theta; z).$$

The operator $\mathscr{S}(z)$ is of course completely determined by how it acts on δ-functions; in fact, writing

$$a(\omega) = \int \delta_{-\omega}(\theta) a(-\theta) d\theta,$$

we see that

$$[\mathscr{S}(z)a](\omega) = \int [\mathscr{S}(z)\delta_{-\omega}](\theta) a(-\theta) d\theta = a(\omega) + \frac{iz}{2\pi} \int k^{sc}(\omega, -\theta; z) a(\theta) d\theta,$$

in other words,

(A.22) $\mathscr{S}(z) = I + K^{sc}(z),$

where $K^{sc}(z)$ is an integral operator with kernel

$$K^{sc}(\omega, \theta; z) = \frac{iz}{2\pi} k^{sc}(\omega, -\theta; z).$$

75

Next we establish the connection between the scattering matrix and the semigroup $\{Z(t)\}$ defined in Section 1. Up to this point we have regarded the functions k_+ and k_- as representing solutions of the wave equation in the exterior domain; from now on we shall regard them as representing the *initial data* of these solutions. Also, it will be convenient to regard $k_\pm(s, \theta)$ as vector-valued functions of s with values in $L_2(S_2)$.

We recall that H denotes the space of all initial data with finite energy, and D_+^ρ and D_-^ρ denote the initial data of *outgoing*, respectively *incoming*, solutions, i.e., solutions which vanish for $|x| < \rho + t$, $t > 0$, respectively $|x| < \rho - t$, $t < 0$.

LEMMA A.3. (i) *In the outgoing translation representation D_+^ρ is represented by* $L_2(\rho, \infty)$.

(ii) *In the incoming translation representation D_-^ρ is represented by $L_2(-\infty, -\rho)$.*

(iii) *In the outgoing translation representation D_-^ρ is represented by $SL_2(-\infty, -\rho)$.*

Proof: As in the proof of Theorem A.1, parts (i) and (ii) are restatements of parts (d) and (e) of Lemma A.2; and part (iii) is a direct consequence of the definition of the scattering operator.

THEOREM A.4. *The scattering matrix $\mathscr{S}(z)$ is analytic in the lower half-plane and grows at most like* $\exp\{-2\rho \mathscr{I}m\, z\}$.

Proof: As stated in Section 1, D_-^ρ is orthogonal to D_+^ρ; since the representations we are using are unitary, it follows that the subspaces representing D_-^ρ and D_+^ρ in the outgoing translation representation are also orthogonal. In view of parts (i) and (iii) of Lemma A.3, this means that $SL_2(-\infty, -\rho)$ is orthogonal to $L_2(\rho, \infty)$, which implies that $SL_2(-\infty, -\rho)$ is contained in $L_2(-\infty, \rho)$. Now according to (A.5), $SL_2(-\infty, -\rho)$ is the convolution of $S(r)$ with $L_2(-\infty, -\rho)$; since we have shown that the result of these convolutions always has its support in $(-\infty, \rho)$, it follows that the support of $S(r)$ itself lies in $(-\infty, 2\rho)$; this is known as the *principle of causality*. The Fourier transform of a function with support in $(-\infty, 2\rho)$ is analytic in the lower half-plane and of exponential growth there, as asserted in Theorem A.4.

As in Section 1, the semigroup $\{Z(t)\}$ is defined by

$$Z(t) = P_+^\rho U(t) P_-^\rho\,, \qquad\qquad t \geqq 0\,,$$

where P_-^ρ and P_+^ρ denote the orthogonal projections onto the orthogonal complements of D_-^ρ and D_+^ρ, respectively. The domain of $Z(t)$ is K^ρ:

$$K^\rho = H \ominus (D_+^\rho \oplus D_-^\rho)\,.$$

It follows from Lemma A.3 that the outgoing translation representation of K^ρ is

(A.23) $$L_2(-\infty, \rho) \ominus SL_2(-\infty, -\rho)\,.$$

Since this is a translation representation, $U(t)$ acts as right shift by t units; P_+^ρ, on the other hand, corresponds to multiplication by the characteristic function of $(-\infty, \rho)$. So, in this representation, $Z(t)$ acts, on the space of functions (A.23), as a right shift by t units followed by restriction to $(-\infty, \rho)$.

If λ is an eigenvalue of the infinitesimal generator B of $\{Z(t)\}$ and w is the corresponding eigenfunction, then

$$Z(t)w = e^{\lambda t}w .$$

Denoting the outgoing translation representer of w by h, it is readily seen from the above description of $Z(t)$ that h has to be an exponential function on $(-\infty, \rho)$:

(A.24)
$$h(s) = \begin{cases} e^{-\lambda s}n & \text{on} & (-\infty, \rho) , \\ 0 & \text{on} & (\rho, \infty) , \end{cases}$$

where n is an element of $L_2(S_2)$. According to (A.23), w belongs to K^ρ only if its representer h is orthogonal to $SL_2(-\infty, -\rho)$, i.e.,

$$(Sk, h) = 0$$

for every k in $L_2(-\infty, -\rho)$. By Parseval's formula, this can be rewritten in terms of Fourier transforms as

(A.25)
$$(\mathscr{S}\tilde{k}, \tilde{h}) = 0 .$$

We can calculate the Fourier transform of h explicitly:

$$\tilde{h}(z) = \frac{1}{\sqrt{2\pi}} \frac{e^{(iz-\lambda)\rho}}{iz - \lambda} n .$$

Substituting this into (A.25) we get

(A.26)
$$\int_{-\infty}^{\infty} \mathscr{S}(z)\tilde{k}(z) \cdot n \frac{e^{-(iz+\bar{\lambda})\rho}}{iz + \bar{\lambda}} \, dz = 0$$

for all k in $L_2(-\infty, -\rho)$. Now according to Theorem A.4, $\mathscr{S}(z)$ is analytic in the lower half-plane and grows there at most exponentially. Choosing the support of k to lie sufficiently to the left of $-\rho$, we can make the integrand in (A.26) tend to zero exponentially as $\mathscr{I}m\, z \to -\infty$. We then shift the contour from the real axis to the parallel line $\mathscr{I}m\, z = c$ and let $c \to -\infty$. The integral tends to zero, leaving us with only the residue at $z = i\bar{\lambda}$. Denoting

(A.27)
$$\tilde{k}(i\bar{\lambda}) = m$$

and omitting the exponential factor we get from (A.26)

$$(A.28) \qquad \mathscr{S}(i\bar{\lambda})m \cdot n = 0$$

for all m of the form (A.27). Clearly every m in $L_2(S_2)$ can be represented in this way by taking $k(s) = g(s)m$, g some scalar function; hence (A.28) implies that the range of $\mathscr{S}(i\bar{\lambda})$ is orthogonal to n. Since by (A.22), $\mathscr{S}(z)$ is of the form I plus an integral operator, it follows by the Fredholm alternative that $\mathscr{S}(i\bar{\lambda})$ *annihilates some nonzero function in* $L_2(S_2)$. This completes the identification of the eigenvalues of the infinitesimal generator of $\{Z(t)\}$ with the zeros of the scattering matrix.

Bibliography

[1] Courant, R., and Hilbert, D., *Methods of Mathematical Physics*, Vol. 2, Interscience Publishers, New York, 1962.

[2] Friedländer, F. G., *On the radiation field of pulse solutions of the wave equation, II*, Proc. Royal Soc. London, Vol. 279A, 1964, pp. 386–394.

[3] Kac, M., *Can one hear the shape of a drum?* Amer. Math. Monthly, Vol. 73, No. 4, 1966.

[4] Lax, P. D., and Phillips, R. S., *Scattering Theory*, Academic Press, New York, 1967.

[5] Morawetz, C. S., and Ludwig, D., *The generalized Huyghens principle for reflecting bodies*, Comm. Pure Appl. Math., Vol. 22, 1969, pp. 189–205.

[6] Olver, F. W. J., *The asymptotic expansion of Bessel functions of large order*, Philos. Trans. Royal Soc. London, Ser. A., Vol. 247, 1954, pp. 328–368.

[7] Phillips, R. S., *The extension of dual subspaces invariant under an algebra*, Proc. International Symp. Linear Spaces, Jerusalem, 1960, pp. 366–398.

[8] Phillips, R. S., *A remark on the preceding paper of C. S. Morawetz and D. Ludwig*, Comm. Pure Appl. Math., Vol. 22, 1969, pp. 207–211.

[9] Phillips, R. S., *A minimax principle for the eigenvalues of symmetric operators in a space with an indefinite metric*, J. Functional Anal., to appear.

[10] Ralston, J., *Energy decay problems in scattering theory*, Thesis, Stanford University, 1968.

[11] Ukai, S., *Real eigenvalues of the mono-energetic transport operator for a homogeneous medium*, J. of Nuclear Science and Technology, Vol. 3, No. 7, 1966, pp. 263–266.

[12] Van Norton, R., *On the real spectrum of a mono-energetic neutron transport operator*, Comm. Pure Appl. Math., Vol. 15, 1962, pp. 149–158.

[13] Wilcox, C. H., *The initial-boundary value problem for the wave equation in an exterior domain with spherical boundary*, Amer. Math. Soc. Notices, Vol. 6, 1959, pp. 869–870.

Received January, 1969.

COMMUNICATIONS ON PURE AND APPLIED MATHEMATICS, VOL. XXV, 85–101 (1972)

On the Scattering Frequencies of the Laplace Operator for Exterior Domains*

PETER D. LAX

Courant Institute, N. Y. U.

AND

RALPH S. PHILLIPS

Stanford University

Abstract

In this paper we show that the so-called *scattering frequencies* of the Laplace operator over an exterior domain, subject to Robin or Dirichlet boundary condition, cannot lie in certain portions of the upper half-plane. The excluded sets depend only on the type of boundary condition and the radius of the smallest sphere containing the scattering obstacle.

1. Introduction

Let O be a compact, smoothly bounded set in $\mathbb{R}^3$ called the *obstacle;* we denote by G the complement of O.

The *scattering eigenfunctions* v of the Laplace operator Δ over the exterior domain G are defined as solutions of

$$(1.1) \qquad \Delta v + \zeta^2 v = 0 \quad \text{in} \quad G,$$

subject to Dirichlet, Neumann or Robin boundary conditions on ∂G, and subject to a *radiation condition* at infinity:

For $|x|$ large, $v(x)$ is a superposition of *fundamental solutions*:

$$(1.2) \qquad \frac{e^{-i\zeta|x-y|}}{|x-y|}, \qquad \mathscr{I}m\,\zeta > 0,$$

y restricted to some compact set. The number ζ will be called a *scattering frequency;* it is a complex frequency, with positive imaginary part, so that it

* This work was supported in part by the United States Atomic Energy Commission under Contract AT(30-1)-1480, by the National Science Foundation under Grant GP-8867 and by the United States Air Force under Contract AF-44620-71-C-0037. Reproduction in whole or in part is permitted for any purpose of the United States Government.

85

corresponds to a mode that decays in time as well as oscillates. The quantity ζ^2 is called a *scattering eigenvalue* of $-\Delta$.

Note that the incoming fundamental solution (1.2) grows exponentially as $|x| \to \infty$; therefore so do the scattering eigenfunctions.

The spectrum of Δ over G is absolutely continuous; the scattering eigenfunctions play no role in the spectral resolution of Δ. However, these eigenfunctions, as explained in [2] and [3], enter the asymptotic description for large t of solutions of the wave equation

$$u_{tt} - \Delta u = 0 \,,$$

x in G, t real, u subject to prescribed boundary conditions on ∂G.

In two previous papers [3] and [4] we have studied the distribution of the scattering frequencies of the Laplace operator over the domain G; in this paper we continue this study. We establish the existence of regions of the upper half-plane which are free of scattering frequencies; these regions depend only on the radius ρ of the smallest sphere which contains the scattering obstacle and, in case of the Robin boundary condition

(1.3) $$\frac{\partial v}{\partial n} + \kappa v = 0 \quad \text{on} \quad \partial G \,,$$

on the lower bound κ_0 of the function κ. For the Dirichlet problem, a region empty of scattering frequencies and containing the origin is determined in [7] by an estimate depending on the geometry of the body.

In Section 2 we show that the disk of diameter $1/\rho$ located in the upper half-plane and containing the origin is free of scattering frequencies ζ. The proof relies on the characterization of such ζ values as poles of the scattering matrix, the description of the scattering matrix in terms of the so-called transmission coefficients, and on the holomorphic character and exponential boundedness of the scattering matrix in the lower half-plane.

It is known that for a spherical scatterer of radius ρ with either Dirichlet or Neumann boundary conditions the smallest purely imaginary radiation frequency is i/ρ. Furthermore, approximate solutions for the *Helmholtz resonator* (see Morse and Feshbach [5]), that is a punctured spherical shell with Neumann boundary conditions, indicate that, as the hole in the shell becomes smaller, two frequencies symmetrically positioned with respect to the imaginary axis tend toward the origin. Thus the disk is quite sharp as far as the purely imaginary scattering frequencies are concerned.

Nevertheless the disk can be improved upon for Dirichlet and Robin boundary conditions. In fact, we show in Section 3 that there is an open set in the complex plane containing the interval $(-\pi/\rho, \pi/\rho)$ which is free of scattering frequencies for the Dirichlet case.

The intuitive reason for this can be described as follows:

Suppose on the contrary that there were a sequence O_j of obstacles contained in the ball of radius ρ around the origin, and corresponding to these obstacles a sequence of radiation frequencies ζ_j tending to a point χ in $(-\pi/\rho, \pi/\rho)$. Let us assume that the exterior domains G_j converge; the limit will be an exterior domain G plus possibly a number of interior domains D which get pinched off in the limiting process. The sequence of scattering eigenfunctions u_j may tend either to a scattering eigenfunction of G or to an ordinary eigenfunction of $-\Delta$ for one of the interior domains D, with eigenvalue $\chi^2 < (\pi/\rho)^2$. The first alternative is ruled out by a theorem of Rellich, which states that there are no real scattering frequencies. The second alternative is ruled out also: an ordinary eigenvalue of $-\Delta$ over D must be greater than or equal to the lowest eigenvalue of $-\Delta$ over the ball of radius ρ, because of the monotone dependence of the eigenvalue of $-\Delta$ on the underlying domain (see Courant-Hilbert [1]). Since the lowest eigenvalue of $-\Delta$ over the ball of radius ρ and subject to the Dirichlet condition is $(\pi/\rho)^2$, we have a contradiction.

Of course the reasoning sketched above is only heuristic, since a sequence of exterior domains can behave in a very complicated fashion; nevertheless a proof, presented in Section 3, can be carried out along these lines.

The same reasoning applies as well to the Robin boundary condition (1.3) with positive κ:

$$(1.4) \qquad\qquad \kappa \geqq \kappa_0 > 0 .$$

In this case the role of the theorem on monotonic dependence of the lowest eigenvalue on the underlying interior domain is taken over by

THEOREM 1.1. *Denote by* λ *the lowest eigenvalue of* $-\Delta$ *over* D, *subject to the boundary condition* (1.3). *Denote by* λ_0 *the lowest eigenfunction of* $-\Delta$ *over a ball* B *containing* D, *subject to the boundary condition*

$$(1.5) \qquad\qquad \frac{\partial v_0}{\partial n} + \kappa_0 v_0 = 0 \qquad on \qquad \partial B ,$$

where κ_0 *is a constant as in* (1.4).

Conclusion:

$$(1.6) \qquad\qquad \lambda \geqq \lambda_0 .$$

This result is due to Payne and Weinberger [6]; we conclude this section with a new proof of Theorem 1.1, based on the maximum principle.

Proof: The lowest eigenfunction v_0 of $-\Delta$ in a ball is of the form

$$(1.7) \qquad v_0(x) = \frac{\sin \chi r}{r}, \qquad r = |x| .$$

The boundary condition (1.5) imposed at $r = \rho$ leads to the transcendental equation

$$(1.8) \qquad \tan \chi\rho = \frac{\chi\rho}{1 - \kappa_0\,\rho} .$$

The smallest positive solution χ_0 gives the smallest eigenvalue $\lambda_0 = \chi_0^2$.

The lowest eigenfunction v_0 has the following property:

$$\left| \frac{\partial v_0}{\partial r} \right| < \kappa_0\, v_0$$

at every interior point of the ball B. Then, in particular, for any unit vector n,

$$(1.9) \qquad \left| \frac{\partial v_0}{\partial n} \right| < \kappa_0\, v_0$$

at all points interior to B.

Let v be the fundamental eigenfunction of $-\Delta$ over some domain D contained in a ball B of radius ρ, subject to the boundary condition (1.3); denote by λ the lowest eigenvalue. Since $v_0 > 0$ in B, we can write v as

$$v = v_0\, w .$$

Substituting this into the eigenvalue equation for v, and using the eigenvalue equation which v_0 satisfies, we get the following equation for w:

$$(1.10) \qquad \Delta w + 2\frac{\nabla v_0}{v_0} \cdot \nabla w + (\lambda - \lambda_0)w = 0 .$$

On the boundary of D we have

$$(1.11) \qquad \frac{\partial w}{\partial n} + \left(\frac{1}{v_0}\frac{\partial v_0}{\partial n} + \kappa \right)w = 0 .$$

To prove that $\lambda \geqq \lambda_0$ we assume the contrary; then solutions w of equation (1.10) satisfy the maximum principle. Suppose w takes on positive values; it assumes its maximum at some point y on the boundary of D. At y, relation (1.11) holds; using (1.9) and the fact that $w(y) > 0$ we deduce from (1.11) that

$$\frac{\partial w}{\partial n} + (-\kappa_0 + \kappa)w < 0 .$$

By (1.4), $\kappa_0 \leqq \kappa$; therefore we see that, at y, $\partial w / \partial n < 0$; but then w cannot assume its maximum at y.

The same reasoning shows that w cannot have a negative minimum. These two facts show that $w \equiv 0$; a contradiction which we obtained by denying conclusion (1.6) of Theorem 1.1.

2. A Circular Disk Free of Scattering Frequencies

An operator-valued *inner factor* $\Phi(\zeta)$ is an analytic function of ζ defined for $\mathscr{Im}\, \zeta \leqq 0$, whose values are operators mapping some Hilbert space into itself and which has the following properties:

$$(2.1) \qquad \begin{array}{l} \text{(i)}\ \ |\Phi(\zeta)| \leqq 1 \cdot \quad \text{for} \quad \mathscr{Im}\, \zeta < 0 , \\[1em] \text{(ii)}\ \ \Phi(\zeta)\ \text{is unitary for}\ \zeta\ \text{real} . \end{array}$$

LEMMA 2.1. *Let $\Phi(\zeta)$ be an operator-valued inner factor and suppose that*

$$(2.2) \qquad |\Phi^{-1}(\zeta_0)| \leqq c^{-1}$$

for some $\zeta_0 = \chi_0 + i\eta_0$, $\eta_0 < 0$. Then $\Phi(\zeta)$ can have no nontrivial null vectors for any ζ in the circle of radius $2c\,|\eta_0|\,(1 - c^2)^{-1}$ about the point $(\chi_0\,,\ \eta_0(1 + c^2)(1 - c^2)^{-1})$.

Proof: Suppose that $\Phi(\zeta_1)n = 0$, $\mathscr{Im}\, \zeta < 0$, and choose $|n| = 1$. It follows from (2.1i) that, for arbitrary m of norm one,

$$f(\zeta) \equiv (\Phi(\zeta)n, m)$$

is holomorphic and of absolute value at most equal to 1 in the lower half-plane; in addition, $f(\zeta_1) = 0$. Hence, f can be factored as

$$f(\zeta) = \frac{\zeta - \zeta_1}{\zeta - \bar{\zeta}_1} f_1(\zeta) ,$$

where f_1 is again holomorphic in the lower half-plane. Moreover, since $(\zeta - \zeta_1)(\zeta - \bar{\zeta}_1)^{-1}$ is of absolute value one for real ζ, the maximum principle implies that $|f_1(\zeta)| \leqq 1$ for all ζ in the lower half-plane. Consequently,

$$|f(\zeta)| \leqq \left| \frac{\zeta - \zeta_1}{\zeta - \bar\zeta_1} \right| \qquad \text{for} \qquad \mathscr{I}m\, \zeta < 0 \,.$$

Since this holds for all m, we see that

$$|\Phi(\zeta)\, n| \leqq \left| \frac{\zeta - \zeta_1}{\zeta - \bar\zeta_1} \right| \qquad \text{for} \qquad \mathscr{I}m\, \zeta < 0 \,.$$

According to the hypothesis (2.2),

$$c \leqq |\Phi(\zeta_0)\, n| \leqq \left| \frac{\zeta_0 - \zeta_1}{\zeta_0 - \bar\zeta_1} \right| \,,$$

and it follows from the extreme members of this inequality that ζ_1 lies outside the circle of radius $2c\,|\eta_0|\,(1 - c^2)^{-1}$ about the point $(\chi_0,\, \eta_0(1 + c^2)(1 - c^2)^{-1})$, as asserted.

COROLLARY 2.2. *Suppose that the inner factor* $\Phi(\zeta)$ *is holomorphic for* $\mathscr{I}m\, \zeta \leqq 0$ *and that for some real* χ_0

$$(2.3) \qquad\qquad\qquad |\Phi'(\chi_0)| = \alpha \,.$$

Then $\Phi(\zeta)$ *is one-to-one for all* ζ *in the circle of radius* α^{-1} *about the point* $(\chi_0,\, -\alpha^{-1})$.

Proof: Expanding $\Phi(\zeta)$ about the point $\zeta = \chi_0$, we see that

$$\Phi(\chi_0 + i\eta_0) = \Phi(\chi_0) + i\eta_0\, \Phi'(\chi_0) + O(\eta_0^2) \qquad \text{for} \qquad \eta_0 < 0 \,.$$

Since $\Phi(\chi_0)$ is unitary by (2.1), we have

$$\Phi^*(\chi_0)\Phi(\chi_0 + i\eta_0) = I + i\eta_0\, \Phi^*(\chi_0)\Phi'(\chi_0) + O(\eta_0^2) \,,$$

and making use of (2.3) we obtain

$$|\Phi^{-1}(\chi_0 + i\eta_0)| = |[\Phi^*(\chi_0)\Phi(\chi_0 + i\eta_0)]^{-1}| \leqq [1 - \alpha\,|\eta_0| + O(\eta_0^2)]^{-1} \,.$$

Thus, replacing c in Lemma 2.1 by $[1 - \alpha\,|\eta_0| + O(\eta_0^2)]$, we see that $\Phi(\zeta)$ can have no null vector for ζ in the circle of radius $\alpha^{-1} + O(|\eta_0|)$ and center $(\chi_0,\, -\alpha^{-1} + O(|\eta_0|))$, and passing to the limit as $\eta_0 \to 0$ we obtain the desired result.

We shall apply Corollary 2.2 above to find a circular region free of scattering frequencies of the Laplace operator over G. To accomplish this we need to relate the scattering frequencies of $-\Delta$ to the poles of the acoustic scattering matrix associated with the exterior domain G in question. The relation is the following:

> ζ is a scattering frequency of $-\Delta$ in G if ζ is a pole located in the upper half-plane of the scattering matrix $\mathscr{S}(\zeta)$.

We shall use the following description of the scattering matrix $\mathscr{S}$ (see [2], p. 170).

(A) $\mathscr{S}(\zeta)$ is an operator-valued meromorphic function of ζ, holomorphic in the lower half-plane, and of the form

$$(2.4) \qquad \mathscr{S}(\zeta) = e^{2i\rho\zeta}\,\Phi(\zeta),$$

where Φ is an inner factor.

(B) For $\mathscr{I}m\,\zeta \leqq 0$, $\mathscr{S}$ is of the form

$$(2.5) \qquad \mathscr{S}(\zeta) = I + K(\zeta),$$

where K is an integral operator, mapping functions on the unit sphere S_2 into S_2, whose kernel K is

$$(2.6) \qquad K(\omega, \theta; \zeta) = \frac{-i\zeta}{2\pi}\,k(-\theta, \omega; \zeta).$$

Here k is the *transmission coefficient* associated with the exterior G of an obstacle in the following fashion:

For $\mathscr{I}m\,\zeta < 0$, $v(x) = v(x; \omega, \zeta)$ is the solution of

$$(2.7) \qquad \Delta v + \zeta^2 v = 0 \quad \text{in} \quad G,$$

subject to the boundary condition that $v + \exp\{-i\zeta x \cdot \omega\}$ has zero boundary data on ∂G; the boundary data can be of Dirichlet, Neumann or Robin type.

The solution v is required to vanish at ∞; this implies that v tends to zero exponentially as $|x| \to \infty$, and has there the asymptotic behavior

$$(2.8) \qquad v(r\theta; \omega, \zeta) = \frac{e^{-i\zeta r}}{r}\left[k(\theta, \omega; \zeta) + O\left(\frac{1}{r}\right)\right].$$

The function k appearing in this formula is defined to be the transmission coefficient.

It follows from (2.4) that $\mathcal{S}$ is unitary on the real axis; therefore, it can be continued into the upper half-plane by the Schwarz reflection principle. This shows that the poles of $\mathcal{S}(\zeta)$ in the upper half-plane are reflections across the real axis of those points ζ in the lower half-plane where $\mathcal{S}(\zeta)$ is not invertible. Since $\mathcal{S}$ is the sum of the identity and a (compact) integral operator, $\mathcal{S}(\zeta)$ is noninvertible if and only if it has a nontrivial nullvector.

We shall apply Corollary 2.2 to the inner factor appearing in (2.4). Differentiating (2.4) with respect to ζ and setting $\zeta = 0$, we get

$$(2.9) \qquad\qquad \Phi'(0) = \mathcal{S}'(0) - 2i\rho\mathcal{S}(0) \ .$$

Setting $\zeta = 0$ in (2.5) we see that

$$(2.10) \qquad\qquad \mathcal{S}(0) = I \ .$$

LEMMA 2.3. *For the scattering matrix* $\mathcal{S}$

$$(2.11) \qquad\qquad \mathcal{S}'(0) = icP \ ,$$

where c *is a number satisfying*

$$(2.12) \qquad\qquad 0 \leqq c \leqq 2\rho \ ,$$

and P *is the orthogonal projection of* $L_2(S_2)$ *onto the constant functions.*

Proof: Differentiating (2.5) with respect to ζ, using (2.6) and setting $\zeta = 0$, we see that $\mathcal{S}'(0)$ *is an integral operator with kernel* $-(i/2\pi)k(-\theta, \omega, 0)$, defined by (2.8).

The boundary condition for v when $\zeta = 0$ is independent of the parameter ω; therefore so is $k(\theta, \omega; 0)$. The function v defined by (2.7) is harmonic and hence asymptotic for $|x|$ large,

$$(2.13) \qquad\qquad v(x) \simeq \frac{k_0}{|x|} , \qquad\qquad k_0 = \text{const.} \ ;$$

this shows that $k(\theta, \omega; 0)$ is independent of θ as well as ω. Thus the kernel of the integral operator $\mathcal{S}'(v)$ is a constant so that $\mathcal{S}'(v)$ is a constant times the orthogonal projection P onto the constants, as asserted in (2.11). The value of the constant c in (2.11) is

$$(2.14) \qquad\qquad c = -2k_0 \ ,$$

where k_0 is the constant appearing in (2.13).

Let us consider the Robin boundary condition:

$$(2.15) \qquad \frac{\partial}{\partial n} v + \kappa v + \kappa = 0 \quad \text{on} \quad \partial G,$$

where κ is some positive function defined on the boundary. *We claim that*

$$(2.16) \qquad -1 \leqq v(x) \leqq 0 \quad in \quad G.$$

Proof: Since v tends to zero at ∞ if it has negative values, it will, by the minimum principle, assume its minimum value at some boundary point y. At this point the outward normal derivative of v is at most 0, so it follows from (2.15) that $v(y) \geq -1$. Using the maximum principle we deduce similarly that v cannot assume a positive value; this completes the proof of (2.16). Letting $\kappa \to \infty$, respectively $\to 0$, we conclude that (2.16) holds also in the limiting case of the Dirichlet, respectively Neumann, boundary conditions.

It follows from (2.16) that the harmonic function

$$v(x) + \frac{\rho}{|x|}$$

is non-negative on the sphere $|x| = \rho$; therefore by the minimum principle the same is true for $|x| > \rho$. In particular, it follows that $k_0 + \rho \geqq 0$, where k_0 is the constant in (2.13). It follows likewise from (2.16) that $k_0 \leqq 0$. This shows that $c = -2k_0$ lies between 0 and 2ρ, as asserted in (2.17) and completes the proof of Lemma 2.3.

Substituting (2.10) and (2.11) into (2.9), we get

$$-i\Phi'(0) = cP - 2\rho I = (c - 2\rho)P - 2\rho[I - P].$$

Since c lies between 0 and 2ρ, we obtain

$$(2.17) \qquad |\Phi'(0)| \leqq 2\rho.$$

THEOREM 2.4. *The acoustic scattering matrix, associated with an obstacle contained in a ball of radius ρ and subject to Dirichlet, Neumann or Robin boundary conditions, has no pole in the disk of radius $1/2\rho$ about the point $(0, i/2\rho)$ shown in Figure 1.*

Proof: We apply Corollary 2.2 to the inner factor Φ appearing in the factorization (2.4) of $\mathscr{S}$. Inequality (2.17) shows that (2.3) is satisfied for $\chi_0 = 0$ with $\alpha = 2\rho$; hence, according to the corollary, Φ does not vanish in the indicated circle. It follows that $\mathscr{S}(\zeta)$ is invertible in the same circle, and

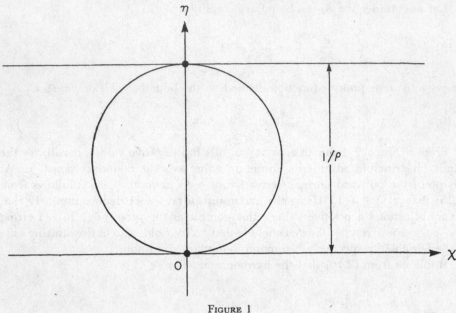

FIGURE 1

therefore $\mathscr{S}(\zeta)$ is pole-free in the reflected image of that circle; this completes the proof of Theorem 2.4.

In view of the relation between poles of $\mathscr{S}$ and scattering frequencies, we have

COROLLARY 2.5. *In the exterior of an obstacle contained in a ball of radius ρ, all scattering frequencies ζ of $-\Delta$ lie outside the disk described in Theorem 2.4.*

3. An Additional Frequency-Free Region

In this section we show that there is an open set of the complex plane, containing the open interval $(-\pi/\rho, \pi/\rho)$, which is free of scattering frequencies of $-\Delta$ for any domain whose complement is contained in $|x| \leqq \rho$, and on whose boundary the Dirichlet condition $v = 0$ is imposed. A similar result holds for the Robin boundary condition

$$(3.1) \qquad \frac{\partial v}{\partial n} + \kappa v = 0 \quad \text{on} \quad \partial G, \qquad \kappa \geqq \kappa_0 > 0,$$

except that now the excluded interval is

$$\left(-\frac{\nu}{\rho}, \frac{\nu}{\rho}\right),$$

where ν is the smallest positive solution of

$$(3.2) \qquad \tan \nu = \frac{\nu}{1 - \rho \kappa_0} \, .$$

We recall that the scattering frequency ζ, $\mathcal{I}m\, \zeta > 0$, and scattering eigenfunctions v are defined as:

$$(3.3) \qquad
\begin{array}{lll}
\text{(i)} & \Delta v + \zeta^2 v = 0 & \text{in} \quad G \, , \\[4pt]
\text{(ii)} & v = 0 \quad \text{on} \quad \partial G & \qquad \text{(Dirichlet condition)} \, , \\[4pt]
\text{(iii)} & v \text{ satisfies the radiation condition (1.2)} \, . &
\end{array}$$

This last property is equivalent with the following: for any surface S containing O but not containing x,

$$(3.4) \qquad v(x) = \int_S \left\{ v(y) \frac{\partial e(x-y)}{\partial n} - e(x-y) \frac{\partial v}{\partial n} \right\} dS \, ,$$

where e is the fundamental solution

$$e(x, \zeta) = \frac{e^{-i\zeta |x|}}{4\pi |x|} \, .$$

Our proof relies on the energy identity, derived by multiplying (3.3i) by $\bar{v}$, and integrating over the truncated domain G_R defined as

$$G_R = G \cap \{ |x| < R \} \, .$$

After one integration by parts we get

$$(3.5) \qquad \int_{G_R} \{ |v_x|^2 - \zeta^2 |v|^2 \} \, dx = \int_{|x|=R} \bar{v} \frac{\partial v}{\partial n} \, dS \, .$$

The imaginary part of the above relation is

$$(3.6) \qquad (\bar{\zeta}^2 - \zeta^2) \int_{G_R} |v|^2 \, dx = \int_{|x|=R} \left\{ \bar{v} \frac{\partial v}{\partial n} - v \frac{\partial \bar{v}}{\partial n} \right\} dS \, .$$

Next we introduce the norm

$$(3.7) \qquad \| v \|^2 = \int_{G_{2\rho}} |v(x)|^2 \, dx$$

89

and normalize all scattering eigenfunctions by $\|v\| = 1$. We denote by V_Z the set $\{v\}$ of all scattering eigenfunctions thus normalized, corresponding to all smooth *obstacles* contained in the ball $\{|x| < \rho\}$ and to any ζ in some compact set Z of the complex plane.

LEMMA 3.1. *The set V_Z is contained in a compact set in the sense of the norm*

$$\|v\|_R^2 = \int\limits_{\rho < |x| < R} |v(x)|^2 \, dx \,,$$

R arbitrary.

Proof: Every v in V_Z satisfies an elliptic equation (3.3) for $|x| > \rho$. Because of the imposed normalization, $\displaystyle\int\limits_{\rho < |x| < 2\rho} |v(x)|^2 \, dx \leqq 1$; elliptic estimates applied in the shell $\rho < |x| < 2\rho$ show that all $v(x)$ and $v_x(x)$ are uniformly bounded on a sphere $\{|x| = d\}$, $\rho < d < 2\rho$. Taking S in (3.4) to be the sphere $\{|x| = d\}$, we get a representation for v from which, using this uniform bound for v, v_x, we can get an estimate for all derivatives of v, uniformly on any compact subset K of $\{|x| > d\}$:

$$(3.8) \qquad\qquad |\partial_x^n v(x)| \leqq c(K, n)$$

for all v in V_Z, all x in K and all $n = 0, 1, \cdots$.

In order to estimate the right side of (3.5), we apply (3.8) for $n = 0, 1$ and take K as $\{|x| = R\}$, $R > 2\rho$; we see that for all v in V_Z,

$$(3.9) \qquad\qquad \int_{G_R} |v_x|^2 \, dx \leqq c(R) \,.$$

Since G_R includes the set $\rho < |x| < R$, Lemma 3.1 follows from Rellich's compactness theorem.

Let $\{G_j\}$ be a sequence of domains whose complements are contained inside the ρ-sphere, ζ_j a sequence of scattering frequencies in Z for Δ over G_j which tend to a *real* value χ. Let v_j denote the corresponding scattering eigenfunctions, normalized by (3.7). According to Lemma 3.1 we can select a subsequence which converges in L_2 locally outside the sphere $|x| = \rho$.

LEMMA 3.2. *Denote by v the local L_2 limit in $|x| > \rho$ of the normalized eigenfunctions v_j; v has the following properties*:

    (i) v *is locally L_2 for $|x| > \rho$*;
    (ii) v *satisfies*

$$(3.10) \qquad\qquad \Delta v + \chi^2 v = 0 \qquad for \qquad |x| > \rho \,;$$

(iii) *for any* d, $\rho < d < |x|$,

$$(3.11) \qquad v(x) = \int_{|v|=d} \left[v(y) \frac{\partial e(x-y)}{\partial n} - e(x-y) \frac{\partial v}{\partial n} \right] dS_y \, ;$$

(iv) *for every* $R > 0$,

$$(3.12) \qquad \int_{|x|=R} \left[\bar{v} \frac{\partial v}{\partial n} - v \frac{\partial \bar{v}}{\partial n} \right] dS = 0 \, .$$

Proof: Part (i) follows from Lemma 3.1. To prove part (ii) we note that each v_j satisfies, for $|x| > \rho$, a second order elliptic equation $\Delta v_j + \zeta_j^2 v_j = 0$. Using standard elliptic estimates, we deduce from the uniform boundedness of

$$\int_{\rho < |x| < R} |v_j|^2 \, dx$$ that all derivatives of v_j are locally uniformly bounded. Using the theorem of Arzela-Ascoli we can select a subsequence such that v_j and Δv_j converge uniformly inside any ball. The limit v satisfies the limiting equation (3.10).

Relation (3.11), where e is the fundamental solution, follows by taking the limit of the analogous integral relation (3.4) satisfied by each v_j. Likewise, we see that (3.12) is the limit of the analogous relation (3.6) for each v_j if we observe that the left side of (3.6) tends to zero when ζ_j approaches a real value χ; here we make use of the uniform boundedness of $\int |v_j|^2 \, dx$ over G_R.

THEOREM 3.3. *There is an open set of the complex ζ-plane containing the interval $(-\pi/\rho, \pi/\rho)$ which is free of scattering frequencies of $-\Delta$ with Dirichlet boundary conditions for all domains exterior to obstacles contained inside the ball $\{|x| < \rho\}$.*

Proof: Assuming the contrary, there exists a sequence of scattering eigenfunctions v_j on domains G_j such that the corresponding frequencies ζ_j tend to a point χ in $(-\pi/\rho, \pi/\rho)$. Normalize v_j as before; then, by Lemma 3.1, a subsequence converges to a limit v and, by Lemma 3.2, v satisfies (3.10)–(3.12) with χ real. It follows from (3.11) that for $|x|$ large v has the asymptotic behavior

$$(3.13) \qquad v(x) = \frac{e^{-i\chi r}}{r} k(\theta) + O\left(\frac{1}{r^2} \right),$$

where $r = |x|$, $\theta = x/r$, and

$$k(\theta) = \int_{|v|=d} e^{i\chi\theta \cdot v} \left\{ i\chi\theta \cdot nv - \frac{\partial v}{\partial n} \right\} dS \, .$$

Furthermore, the asymptotic behavior of the derivatives is obtained by differentiating (3.13); in particular,

$$\frac{\partial v(x)}{\partial r} = -i\chi \frac{e^{-i\chi r}}{r} k(\theta) + O\left(\frac{1}{r^2}\right). \tag{3.14}$$

Substituting the asymptotic description (3.13) and (3.14) into (3.12) we get

$$\int_{|x|=R} |k(\theta)|^2 \, d\theta + O\left(\frac{1}{R}\right) = 0.$$

Letting $R \to \infty$, we conclude that the square integral of $|k|$ is 0; since k is continuous, it follows that $k(\theta) \equiv 0$. Looking again at (3.13) we deduce from this that

$$v(x) = O\left(\frac{1}{r^2}\right) \qquad \text{as} \qquad |x| \to \infty.$$

Such a function in $\mathbb{R}^3$ is square integrable; but according to Rellich, if $\chi \neq 0$, the only square integrable solution of (3.10) in $|x| > \rho$ is identically zero. So we conclude that $rv \equiv 0$ for $|x| > \rho$.

If $\chi = 0$, we have to argue somewhat differently to reach the same conclusion. In this case, v is harmonic; hence its asymptotic behavior is

$$v(x) \sim O\left(\frac{1}{r}\right), \qquad \frac{\partial v}{\partial r} \sim O\left(\frac{1}{r^2}\right). \tag{3.15}$$

Consider now relation (3.5) for v_j :

$$\int_{G_R} \{|v_{j_x}|^2 - \zeta_j^2 |v_j|^2\} \, dx = \int_{|x|=R} \bar{v}_j \frac{\partial v_j}{\partial n} \, dS. \tag{3.16}$$

As $j \to \infty$, the right side of (3.16) tends to

$$\int_{|x|=R} \bar{v} \frac{\partial v}{\partial n} \, dS$$

which, according to (3.15), is $O(1/R)$. Since $\zeta_j \to \chi = 0$, the second term in (3.16) tends to 0 as $j \to \infty$; thus we see that, for large j,

$$\int_{G_R} |(v_j)_x|^2 \, dx \leqq \frac{\text{const.}}{R}.$$

Now $(v_j)_x \to v_x$, uniformly for $2\rho < |x| < \rho$, and we conclude from the above that, as $j \to \infty$,

$$\int_{2\rho < |x| < R} |v_x|^2 \, dx \leqq \frac{\text{const.}}{R} .$$

Letting $R \to \infty$, we get

$$\int_{2\rho < |x|} |v_x|^2 \, dx = 0 ,$$

so that v is constant; because of (3.15), that constant is zero. Again we see that $v \equiv 0$ for $|x| > \rho$.

Having shown that $v \equiv 0$ for $|x| > \rho$, we return to the identity (3.16) choosing $R = \rho + \delta$ with δ small but greater than 0. Since on $\{|x| = \rho + \delta\}$, v_j and $\partial v_j / \partial n$ tend uniformly to v and $\partial v / \partial n$, respectively, both of which are zero, it follows that the right side of (3.16) tends to zero as $j \to \infty$. Likewise, $\zeta_j \to \chi$ and, since $\int_{G_R} (v_j)^2$ is uniformly bounded, (3.16) implies that, as $j \to \infty$,

$$(3.17) \qquad \int_{G_{\rho+\delta}} \{|(v_j)_x|^2 - \chi^2 |v_j|^2\} \, dx \to 0 .$$

Let $\varphi(|x|)$ be a C^∞ function which is equal to 1 for $|x| < \rho + \frac{1}{2}\delta$, and drops off to 0 at $|x| = \rho + \delta$. Define w_j by

$$w_j = \varphi v_j .$$

Since we saw earlier that, as $j \to \infty$, v_j and $(v_j)_x$ tend to zero uniformly in $\rho + \frac{1}{2}\delta < |x| < \rho + \delta$, it follows that, for j large, w_j differs little from v_j and $(w_j)_x$ differs little from $(v_j)_x$. Hence, it follows from (3.17) that, as $j \to \infty$,

$$(3.18) \qquad \int_{G_{\rho+\delta}} \{|(w_j)_x|^2 - \chi^2 |w_j|^2\} \, dx \to 0 .$$

Finally, we remark that the normalization and the fact that $v_j \to 0$ for $\rho + \frac{1}{2}\delta < |x| < 2\rho$, imply that $\int_{G_{\rho+\delta}} |v_j|^2 \, dx \to 1$ and therefore that

$$(3.19) \qquad \lim_{j \to \infty} \int_{G_{\rho+\delta}} |w_j|^2 \, dx = 1 .$$

The domains $G_{\rho+\delta}$ are all contained in the ball $|x| \leqq \rho + \delta$. According to the monotonicity principle, the spectrum of $-\Delta$ over $G_{\rho+\delta}$ lies above the lowest eigenvalue λ_0 of $-\Delta$ over the ball of radius $\rho + \delta$, under Dirichlet boundary conditions. Consequently, for any function w which is zero on the boundary of $G_{\rho+\delta}$, the Rayleigh quotients of w cannot be less than λ_0 :

$$(3.20) \qquad \int_{G_{\rho+\delta}} |w_x|^2 \, dx \geqq \lambda_0 \int_{G_{\rho+\delta}} |w|^2 \, dx \ .$$

The functions w_j are zero on the boundary of $G_{\rho+\delta}$ and hence they satisfy (3.20). If χ^2 were less than λ_0, then (3.18) and (3.19) would contradict (3.20). This proves that $|\chi^2| \geqq \lambda_0$. As is well known, for a ball of radius $(\rho + \delta)$, λ_0 is $(\pi/(\rho + \delta))^2$, and therefore, $|\chi| \geqq \pi/(\rho + \delta)$. Since this holds for all $\delta > 0$, $|\chi|$ must be at least π/ρ, as asserted in Theorem 3.3.

A similar result holds for the Robin boundary condition (3.1). One slight difference is that in this case the left side of (3.5) has to be augmented by the term

$$\int_{\partial G} \kappa \, |v|^2 \, dS \ ;$$

the same term has to be added to the left sides of (3.16), (3.17), (3.18) and (3.20). The second difference is that to conclude as before that $\chi^2 \geqq \lambda_0$, we appeal to Theorem 1.1, instead of the monotonicity theorem. Since the lowest eigenvalue λ_0 of $-\Delta$ over a ball of radius ρ, subject to the boundary condition (1.5), is given by the smallest solution ν of (3.2), we conclude:

THEOREM 3.4. *There is an open set of the complex ζ-plane containing the interval $(-\nu/\rho, \nu/\rho)$, where ν is defined by (3.2), which is free of scattering frequencies of $-\Delta$ under Robin boundary condition (3.1) for all domains exterior to obstacles contained inside the ball $\{|x| < \rho\}$.*

4. Remarks

All of our results carry over for odd-dimensional spaces of dimension $n > 1$. For $n > 3$ and odd, it is easy to see from the form of S (see [2]; p. 170) that $S'(0) = 0$ and it follows from this that Theorem 2.4 remains valid. The analogues of Theorems 3.3 and 3.4 are also true, only the interval needs to be adjusted to the fundamental frequency for the ball of radius ρ in $\mathbb{R}^n$; the proofs are the same.

The even-dimensional case is different. In this case the scattering matrix is not simply related to an inner factor as in (2.4), so it is unlikely that Theorem 2.4 is valid. On the other hand, if interpreted properly, Theorems 3.3 and 3.4 do carry over. Here the essential difference between the even and odd-dimensional problem is that the incoming fundamental solution $e(x, \zeta)$, while single-valued for n odd, is a multi-valued function of ζ for n even. However, if we treat e as single-valued in the ζ-plane, cut along the positive imaginary axis, Theorems 3.3 and 3.4 do remain valid except at the origin for $n = 2$, which is the only case for which $e(x, \zeta)$ does not converge as $\zeta \to 0$ to the fundamental solution for the Laplacian.

Bibliography

[1] Courant, R., and Hilbert, D., *Methods of Mathematical Physics*, Vol. 1, Interscience Publ., New York, 1962.

[2] Lax, P. D., and Phillips, R. S., *Scattering Theory*, Academic Press, New York, 1967.

[3] Lax, P. D. and Phillips, R. S., *Decaying modes for the wave equation in the exterior of an obstacle*, Comm. Pure and Appl. Math., Vol. 22, 6, 1969, pp. 737–787.

[4] Lax, P. D., and Phillips, R. S., *A logarithmic bound on the location of the poles of the scattering matrix*, Archive Rat. Mech. and Analysis, Vol. 40, 1971, pp. 268–280.

[5] Morse, P. M., and Feshbach, H., *Methods of Theoretical Physics*, McGraw-Hill, New York, 1953.

[6] Payne, L. E., and Weinberger, H. F., *Lower bounds for vibration frequencies of elastically supported membranes and plates*, J. Soc. Industrial and Appl. Math., Vol. 5, 1957, pp. 171–182.

[7] Morawetz, C. S., *On the modes of decay for the wave equation in the exterior of a reflecting body*, to appear.

Received August, 1971.

27, 31, 35, 51, 71

The Lax and Phillips approach to scattering theory was originally developed by them for the Euclidean wave equation in the exterior of a compact obstacle Ω in $\mathbb{R}^n$. The papers included in this volume were critical ones in the development of the theory. In [27] it is shown that any solution of the wave equation as above has some of its energy propagate out to infinity, where it behaves like a solution to the free space wave equation. A number of the ingredients that constitute their abstract functional-analytic setup for scattering theory, such as the incoming and outgoing solutions, are already present in this paper. The paper [31], written jointly with C. Morawetz contains most of the key notions for the general setup of their scattering theory axioms. Specifically, the semigroup $Z(t)$ and its infinitesimal generator B, whose eigenvalues in $\mathbb{C}$ correspond to the poles of the scattering operator, are introduced. The main result in [31] is the exponential decay of solutions to the wave equation when Ω is star-shaped (in dimension 3). The proof relies on the work of Morawetz [Mo]. The paper [35] gives the setup and techniques for what is known today as the Lax–Phillips time-dependent scattering theory. A comprehensive treatment of the theory is given in their well-known book *Scattering Theory* [L-P]. The revised edition of this book [L-P2] contains an up-to-date epilogue with a wealth of information about recent developments stemming from the book. In [35] one finds far-reaching insights and conjectures about the relation between the geometry of an obstacle Ω and especially the motion of rays in $\mathbb{R}^n \backslash \Omega$ and the distribution of scattering poles. Ω is said to be nontrapping if there is a ball C containing Ω and a $t_0 > 0$ such that any ray entering B and obeying the laws of linear motion and reflection in the complement of Ω exits B within time t_0. A basic conjecture put forth in [35] is that the semigroup $Z(t)$ is eventually compact (which implies that the imaginary parts of the poles tend to infinity) iff Ω is nontrapping. If Ω is trapping, this was established by Ralston [R]. In the other direction Ludwig and Morawetz [L-M] established the conjecture if Ω is convex. Further results were obtained by Morawetz–Ralston–Strauss [M-R-S], and the complete solution (that is, if Ω is nontrapping, then $Z(t)$ is eventually compact) was given by Melrose [Me1]. His work makes use of the modern machinery of propagation of singularities for such wave equations as well as the corresponding microlocal analysis.

Paper [51] initiates the finer study of the distribution of the poles of the scattering operator in analogy with the well-studied Weyl law for eigenvalues of the Laplacian on a compact domain. Unlike the latter case, the poles are not restricted to lie on a line, so that counting them asymptotically or studying their possible monotonicity properties under changes of the obstacle are problematic even in their formulation. It is therefore quite surprising that in [51] Lax and Phillips establish a monotonicity property (in terms of increasing Ω) for the purely imaginary poles of the scattering operator. Using this they give asymptotic order of magnitude estimates for the number of such purely imaginary poles for a general Ω. More recently there have been a number of works concerning bounds on the number of poles in a disk of radius r as $r \to \infty$. We mention in particular the works of Melrose [Me2] and Zworski [Z]. Another striking result is that of Ikawa [I], who gives a precise description of the poles near the real axis for the case that Ω consists of two disjoint convex bodies containing

25

an unstable periodic ray bouncing between them. A closer analogue to Weyl's law in this exterior scattering problem concerns the winding number of the phase of the scattering operator along the real axis. Majda and Ralston [M-R] (see also the references on page 281 of the revised edition of [L-P]) establish an exact analogue of Weyl's law for this winding number.

The Lax–Phillips scattering theory as well as the set of problems and ideas that come out of it are still very active areas of research today.

References

[I]   Ikawa, M. On the poles of the scattering matrix for two strictly convex obstacles. *J. Math. Kyoto Univ.* **23**, 127–194 (1983).

[L-M]   Ludwig, D.; Morawetz, C. The generalized Huygens principle for reflecting bodies. *CPAM* **22**, 189–205 (1969).

[L-P]   Lax, P.; Phillips, R. Scattering Theory revised edition, A.P. 1989.

[M-R]   Majda, A.; Ralston, J. An analogue of Weyl's formula for unbounded domains. *Duke Math. J.* **45**, 183–196 (1978).

[Mo]   Morawetz, C. The decay of solutions of the exterior initial-boundary value problem for the wave equation. *CPAM* **14**, 561–568 (1961).

[M-R-S] Morawetz, C.; Ralston, J.; Strauss, W. Decay of solutions of the wave equation outside nontrapping obstacles. *CPAM* **30**, 447–508 (1977).

[Me1]   Melrose, R. Singularities and energy decay in acoustical scattering. *D. M. J.* **46**, 43–59 (1979).

[Me2]   Melrose, R. Polynomial bound on the number of scattering poles. *J. F. A.* **53**, 287–303 (1983).

[R]   Ralston, J. Solutions of the wave equation with localized energy *CPAM* **22**, 807–823 (1969).

[Z]   Zworski, M. Sharp polynomial bounds on the number of scattering poles for radical potentials. *J. Funct. Anal.* **82** (1989) 370–403.

P. Sarnak

26

PART VI

SCATTERING THEORY FOR AUTOMORPHIC FUNCTIONS

Translation Representations for the Solution of the Non-Euclidean Wave Equation*

PETER D. LAX

Courant Institute

AND

RALPH S. PHILLIPS

Stanford University

0. Introduction

In 1972, Pavlov and Faddeev [10] discovered an elegant connection between harmonic analysis for functions automorphic with respect to a discrete subgroup of $SL(2, R)$ and the Lax–Phillips scattering theory as applied to the non-Euclidean wave equation; we have elaborated their idea in our monograph [8]. Recently, M. A. Semenov–Tian–Shansky [11] has treated harmonic analysis and scattering theory for Riemannian symmetric spaces of negative curvature also along the lines of the Lax–Phillips theory. We have thought it worthwhile in the present paper to go over the same material as Semenov–Tian–Shansky, but in much greater detail, for the special case of non-Euclidean three-space. The main difference between the two approaches derives from the fact that Semenov–Tian–Shansky obtains the translation representations from the asymptotic behavior of solutions to an intrinsic hyperbolic system of partial differential equations, which reduces to the non-Euclidean wave equation for symmetric spaces of rank one. In our work we obtain the translation representations from the Radon transform; this leads to many more detailed properties of these representations.

We start with the Cauchy problem for the non-Euclidean wave equation on $\Pi = \{w = (x, y, z); \; z > 0\}$:

$$(0.1) \qquad \begin{aligned} u_{tt} &= Lu \equiv z^3 \partial z^{-1} \partial u + u, \\ u(w, 0) &= f_1, \qquad u_t(w, 0) = f_2. \end{aligned}$$

Let $U(t)$ denote the solution operator relating initial date $F = (f_1, f_2)$ to data

* The work of both authors was supported in part by the National Science Foundation; the first author under Grant No. NSF MCS-76-07039 and the second under Grant No. NSF MCS-77-04908. Reproduction in whole or in part is permitted for any purpose of the United States Government.

Communications on Pure and Applied Mathematics, Vol. XXXII, 617–667 (1979)
0010-3640/79/0032-0617$01.00

at time $t : \{u(t), u_t(t)\}$. Since energy is conserved, $U(t)$ defines a one-parameter family of unitary operators on the Hilbert space $\mathcal{H}$ of all initial data F, normed by the energy form

$$(0.2) \qquad \|F\|_E^2 = -(f_1, Lf_1) + (f_2, f_2) ;$$

here the inner product on the right is the L_2 inner product with the non-Euclidean volume element dw:

$$(0.3) \qquad (f, g) = \int_\Pi f(w)\overline{g(w)} \, dw .$$

A solution u of (0.1) is called *outgoing* from the point j if it vanishes in the forward cone with vertex j:

$$(0.4)_+ \qquad u(w, t) = 0 \text{ for non-Euclidean dist.} (w, j) < t, \qquad t \geqq 0 ,$$

and u is called *incoming* from j if

$$(0.4)_- \qquad u(w, t) = 0 \text{ for non-Euclidean dist.} (w, j) < -t, \qquad t \leqq 0 .$$

The set of initial data for all incoming (outgoing) solutions form the *incoming* (*outgoing*) subspace of $\mathcal{H}$ is denoted by $\mathcal{D}_-(\mathcal{D}_+)$. We prove that the subspaces $\mathcal{D}_-$ and $\mathcal{D}_+$ have the following properties:

$$(0.5) \qquad \begin{array}{ll} \text{(i)} & U(t)\mathcal{D}_\pm \subset \mathcal{D}_\pm \quad \text{for} \quad t \gtrless 0, \\ \text{(ii)} & \bigcap U(t)\mathcal{D}_- = 0, \qquad \bigcap U(t)\mathcal{D}_+ = 0, \\ \text{(iii)} & \overline{\bigcup U(t)\mathcal{D}_-} = \mathcal{H}, \qquad \overline{\bigcup U(t)\mathcal{D}_+} = \mathcal{H}, \\ \text{(iv)} & \mathcal{D}_-^\perp \perp \mathcal{D}_+^\perp . \end{array}$$

Our main tool in proving these properties is the *Radon transform* introduced by Gelfand *et al.* [1] and studied in great detail by Helgason in a series of publications. The Radon transform is defined as

$$(0.6) \qquad \hat{f}(s, \beta) = \int_{(s, \beta)} f(w) \, dS ,$$

where the domain of integration is the horosphere (s, β), that is the sphere tangent to the x, y-plane at β whose non-Euclidean distance from j is s. Setting

$$(0.7)_\pm \qquad R_\pm F = \text{const.} (\partial_s^2 e^s \hat{f}_1 \mp \partial_s e^s \hat{f}_2)$$

and denoting the reflection $s \to -s$ by $^\vee$, we show in Theorems 3.4 and 3.7 that

$$(0.8) \qquad\qquad T_1 = R_+ \quad \text{and} \quad T_2 = R_-^\vee$$

are translation representations of U, that is, T is a unitary mapping of $\mathcal{H}$ onto $L_2(\mathbb{R}, N)$, N some auxiliary Hilbert space, with the intertwining property

$$(0.9) \qquad\qquad TU(t) = U_t T,$$

where U_t denotes right translation by t units. One obtains a *spectral representation* from a translation representation by Fourier transformation.

The *scattering operator* S relates the above translation representations:

$$(0.10) \qquad\qquad S : T_1 F \to T_2 F.$$

The spectral representer of S is called the *scattering matrix*.

In Theorem 3.7 we prove that

$$(0.11) \qquad R_\pm(\mathcal{D}_\pm) = L_2(\mathbb{R}_+, N) \quad \text{and} \quad R_\pm(\mathcal{D}_\mp) \supset L_2(\mathbb{R}_-, N).$$

$\mathcal{D}_-$ and $\mathcal{D}_+$ have in common an infinite-dimensional subspace and Theorem 3.7 also furnishes us with an explicit characterization of $R_\pm(\mathcal{D}_- \cap \mathcal{D}_+)$. The situation is quite different in the Euclidean space where the incoming and outgoing subspaces are orthogonal; the non-Euclidean analogue of this property is (0.5. iv): $\mathcal{D}_-^\perp \perp \mathcal{D}_+^\perp$, established in Lemma 4.1.

As a by-product of our analysis we prove the following Paley–Wiener result (Theorem 3.14): A function f in $L_2(\Pi)$ vanishes at all points outside of a given horosphere (or sphere) Σ if and only if $\partial_s e^s \hat{f}(s, \beta)$ vanishes on all horospheres lying outside of Σ. In Theorem 3.15 we derive a similar result for data in $\mathcal{H}$. These results are in contrast to the Euclidean case where the vanishing of $\hat{f}$ on all hyperplanes not intersecting a given sphere is necessary but not sufficient for f to be equal to 0 outside the sphere. Our result is a somewhat surprising sharpening of a theorem of Helgason in [5].

For completeness we have included a brief description of the geometry of non-Euclidean space in Section 1. In Section 2 the Euclidean analogues of the results quoted above are derived to give the reader a better perspective on the problem; this material can be found in one form or another in [6], Chapter 4. Finally, an explicit representation for the scattering matrix is derived in Section 4.

1. Non-Euclidean Space

The three standard models of non-Euclidean space Π are the upper half-space

$(1.1)_i$ $$(x, y, z), \qquad z > 0,$$

the unit ball

$(1.1)_{ii}$ $$X^2 + Y^2 + Z^2 < 1,$$

and the semi-hyperboloid

$(1.1)_{iii}$ $$\tau^2 - (\xi^2 + \eta^2 + \zeta^2) = 1, \qquad \tau > 0.$$

We shall refer to these as models (i), (ii) and (iii). In this paper we shall be using the first two of these models. The models are entirely equivalent to each other; the mappings relating (i) and (iii) are

(1.2)
$$(\tau, \xi, \eta, \zeta) = \left(\frac{x^2 + y^2 + z^2 + 1}{2z}, \frac{x}{z}, \frac{y}{z}, \frac{x^2 + y^2 + z^2 - 1}{2z} \right)$$

and

$(1.2)'$
$$(x, y, z) = \left(\frac{\xi}{\tau - \zeta}, \frac{\eta}{\tau - \zeta}, \frac{1}{\tau - \zeta} \right).$$

The mappings relating (i) and (ii) are conformal maps between the x, z upper half-plane and the unit disk in the X, Z-plane mapping $(0, z)$ on $(0, Z)$, combined with rotation around the z-(respectively Z-) axes.

The Riemannian metric in (i) is

$(1.3)_i$
$$ds^2 = \frac{dx^2 + dx^2 + dz^2}{z^2},$$

in (ii)

$(1.3)_{ii}$
$$dS^2 = 4 \frac{dX^2 + dY^2 + dZ^2}{(1 - (X^2 + Y^2 + Z^2))^2}.$$

Note that (1.3) is *conformal* with the underlying Euclidean metric.

In model (iii) we endow the hyperboloid with the Lorentz metric of

4-space in which the hyperboloid is imbedded:

$$(1.3)_{iii} \qquad d\sigma^2 = -d\tau^2 + d\xi^2 + d\eta^2 + d\zeta^2 .$$

The geodesics in both model (i) and model (ii) are circular arcs perpendicular to the boundary. The length of each geodestic is ∞; for this reason the points of the boundary of Π—the plane $z = 0$, including the point at ∞, in model (i), the sphere $X^2 + Y^2 + Z^2 = 1$ in model (ii)—are called points at infinity of Π. We denote this set by B, and its points by β.

Π equipped with the metric (1.3) is a space of constant negative curvature, Π possesses a six-parameter group of isometries called non-Euclidean motions and denoted by Γ. These are most easily described for (iii) as the group of Lorentz transformations, i.e., linear maps of τ, ξ, η, ζ-space into itself which preserves the Lorentz form $\tau^2 - (\xi^2 + \eta^2 + \zeta^2)$, as well as preserving the direction of time. It is well known that the Lorentz group is isomorphic to SL$(2, C)$ modulo $\pm I$. The isometries in model (i) can be described elegantly[1] with the aid of the 2×2 matrices $\binom{ab}{cd}$ of SL$(2, C)$,

$$(1.4) \qquad ad - bc = 1 ,$$

in the following fashion: to each point (x, y, z) of Π we assign a *quaternion w with three components*,

$$(1.5) \qquad w = x + yi + zj ,$$

where $i^2 = j^2 = -1$, $ij = -ji$. It is not hard to verify that the mapping defined by

$$(1.6) \qquad \tau w = w' = (aw + b)(cw + d)^{-1}$$

maps Π onto Π, with inverse τ^{-1} given by

$$(1.6)' \qquad w = (dw' - b)(-cw' + a)^{-1} ,$$

and that the assignment $\binom{ab}{cd} \to \tau$ is an isomorphism. Note that these mappings restricted to the boundary B of Π are the classical Möbius transformations of the complex $(x + iy)$-plane onto itself.

It it easy to verify using (1.6) that the group of motions Γ is *symmetric* in the sense that, given any two points w_1 and w_2 of Π, there is a motion τ such that $\tau w_1 = w_2$, $\tau w_2 = w_1$; this τ is uniquely determined, being reflection across the geodesics through the midpoint w_m on the geodesic arc connecting w_1 and w_2. The motion τ is an involution, i.e., $\tau^2 = I$.

The subgroup of Γ keeping a point w_0 fixed is isomorphic to the group of

---

[1] We thank Professor Werner Fenchel for acquainting us with this description.

rotations $SO(3, \mathbb{R})$; this is most easily seen in model (ii), taking w_0 to be the origin. This subgroup acts transitively on the boundary. It follows that, given two boundary points β_1 and β_2, and two interior points w_1 and w_2, there exists a motion τ such that $\tau\beta_1 = \beta_2$, $\tau w_1 = w_2$.

The isometric character of the mapping (1.6) is easily verified.

The mappings τ of Π induce mappings of functions u defined on Π; we define u^τ by

$$(1.7) \qquad u^\tau(w) = u(\tau^{-1}w).$$

Since τ is isometric,

$$(1.7)' \qquad \int_\Pi |u^\tau(w)|^2 \, dw = \int_\Pi |u(w)|^2 \, dw,$$

where dw is the non-Euclidean volume element, which in model (i) is

$$(1.8)_\mathrm{i} \qquad dw = \frac{dx \, dy \, dz}{z^3}.$$

The quantity $(1.7)'$ is called the invariant L_2 norm.

Denote the differential of the mapping $\tau w = w'$ by $\partial w'/\partial w$; then

$$\delta w' = \frac{\partial w'}{\partial w} \, \delta w.$$

It follows from the isometry of τ with respect to the metric $(1.3)_\mathrm{i}$ that $|\delta w'|/z' = |\delta w|/z$; therefore,

$$(1.9) \qquad \frac{\partial w'}{\partial w} = \frac{z'}{z} \, U,$$

where U is an orthogonal transformation.

Let u be any C_0^∞ function, u^τ defined by (1.7), i.e., $u^\tau(w') = u(w)$. Denote the gradient by

$$u_w = (u_x, u_y, u_z).$$

Then

$$u_w = u^\tau(w')_w = u^\tau(w')_{w'} \cdot \frac{\partial w'}{\partial w}.$$

Hence, by (1.9),

$$|u_w|z = |u_{w'}^\tau|z'.$$

Since $dw = dw'$ by definition $(1.8)_i$, we get

$$\int |u_w|^2 z^2 \, dw = \int |u_{w'}^\tau|^2 z'^2 \, dw'.$$

The quantity above,

(1.10)
$$\int |u_w|^2 \frac{dx \, dy \, dz}{z},$$

is called the *invariant Dirichlet integral*. Its first variation is also invariant:

$$-\int \left(\partial \frac{1}{z} \partial u\right) v \, dx \, dy \, dz = -\int z^3 \left(\partial \frac{1}{z} \partial u\right) v \, dw.$$

This shows that

(1.11)
$$z^3 \partial \frac{1}{z} \partial = L_0,$$

called the Laplace–Beltrami operator, is *invariant* under non-Euclidean motions.

Note that L_0, being defined as the first variation of the Dirichlet integral, is symmetric with respect to the invariant L_2 norm.

Next we define a non-Euclidean equivalent to the Euclidean notion of plane. A plane with normal β through a point w in Euclidean space can be thought of as the limit of spheres of radius R through w as the center of the sphere tends to ∞ in the direction β. Similarly we define a *horosphere* through a point w in non-Euclidean space as a limit of spheres through w whose centers tend to the point β at ∞. Since geodesics from the (non-Euclidean) center of a sphere are perpendicular to the sphere, it follows that all geodesics from β are perpendicular to all horospheres through β. From this we deduce that horospheres through β are spheres tangent to the boundary of π at β; see Figure 1.1 for a geometric proof, keeping in mind that Euclidean and non-Euclidean angles are equal, since the two metrics are conformal.

In model (i) the point at infinity on the β-plane is an exceptional and an exceptionally simple point at infinity for Π; we denote it by β_∞. The horospheres through β_∞ are the planes $z = \text{const.}$, and the geodesic lines through β_∞ are parallel to the z-axis.

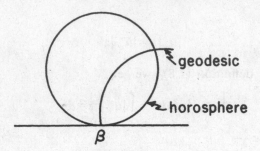

Figure 1. Geometric proof.

Since horospheres are defined in terms of the metric, it is clear that non-Euclidean motion carries one horosphere into another. It is often convenient to specialize to horospheres through β_∞.

In the Euclidean case a convenient parametrization of planes is their normal β and their distance s from the origin. Similarly we parametrize horospheres by the point β at ∞ through which they pass, and their distance s from a reference point which in model (ii) is taken as the origin and in model (i) as $(0, 0, 1)$. We denote this horosphere as $\Sigma(s, \beta)$.

Note that, in contrast to the Euclidean case, a horosphere divides Π into an *inside* and *outside*.

For a point w in Euclidean space and unit vector β the scalar product $w \cdot \beta$ can be interpreted geometrically, aside from sign, as the distance of the origin from the plane through w with normal β. The analogous concept in non-Euclidean space is:

(1.12) $\quad$ $\langle w, \beta \rangle = \pm$ distance of j from horosphere through w and β, where $j = (0, 0, 1)$ in model (i), $j = (0, 0, 0)$ in model (ii); the plus sign holds if j lies outside the horosphere and the minus sign holds if j lies inside the horosphere.

Under isometry $\langle w, \beta \rangle$ changes somewhat like its Euclidean counterpart:

(1.13) $$\langle \tau w, \tau \beta \rangle = \langle w, \beta \rangle + \langle \tau j, \tau \beta \rangle .$$

This is obvious for τ which keeps j fixed. In general, we factor $\tau = \mu \sigma$, where σ keeps β fixed and μ keeps j fixed, and verify (1.13) for σ in place of τ; this can be done by inspection in model (i) if β is chosen as β_∞.

It is easy to see in model (i) that

(1.14) $$\langle w, \beta_\infty \rangle = \log z ,$$

since non-Euclidean distance along the geodesics through β_∞ are the logarithms of the Euclidean distance. For arbitrary β we deduce, using (1.13), that in model (i)

$$(1.15) \qquad \langle w, \beta \rangle = \log \frac{1+|\beta|^2}{|w-\beta|^2}\, z\,,$$

where $|\,|$ denotes Euclidean distance. The analogous formula in model (ii) is

$$(1.16) \qquad \langle w, \beta \rangle = \log \frac{1-|w|^2}{1-2w\cdot\beta+|w|^2}\,,$$

where $|\,|$ and $\cdot$ denote the Euclidean distance and scalar product.

In model (ii) the points at infinity are represented as the unit sphere; we endow B with the standard surface measure of the unit sphere. The mapping relating model (ii) to model (i) carries this into the area element

$$(1.17) \qquad d\beta = \frac{dx\,dy}{(1+|\beta|^2)^2}\,, \qquad |\beta|^2 = x^2 + y^2\,.$$

The element $d\beta$ is invariant under all motions τ that keep j fixed; under arbitrary τ,

$$(1.18) \qquad d(\tau\beta) = \exp\{2\langle\tau^{-1}j, \beta\rangle\}\, d\beta\,.$$

This may be verified with the aid of (1.13).

2. The Radon Transform in Euclidean Space

We start with the Fourier transform of a function $f(w)$ in $\mathbb{R}^3$:

$$(2.1) \qquad \tilde{f}(\xi) = \frac{1}{(2\pi)^{3/2}} \int e^{-iw\cdot\xi} f(w)\, dw\,.$$

It is convenient to express ξ as $\lambda\beta$, where $\lambda = |\xi|$ and β is a unit vector; then we break up the integration in (2.1) by first integrating over the planes $w\cdot\beta = s$ and then integrating with respect to s. We get

$$(2.2) \qquad \tilde{f}(\lambda\beta) = \frac{1}{(2\pi)^{3/2}} \int e^{-is\lambda} \hat{f}(s, \beta)\, ds\,,$$

where

$$(2.3) \qquad \hat{f}(s, \beta) = \int_{w \cdot \beta = s} f(w) \, ds \; ;$$

$\hat{f}$ is called the *Radon transform* of f. Thus $\tilde{f}$ is the Fourier transform of $\hat{f}$ with respect to s. From this we deduce that

$$(2.4) \qquad \int \tilde{f}(\lambda \beta) e^{i\lambda s} \lambda^2 \, d\lambda = \frac{-1}{(2\pi)^{1/2}} \partial_s^2 \hat{f}(s, \beta) \, .$$

Fourier inversion of (2.1) gives

$$(2.1)' \qquad f(w) = \frac{1}{(2\pi)^{3/2}} \int \tilde{f}(\xi) e^{iw \cdot \xi} \, d\xi \, .$$

We set $\xi = \lambda \beta$, $d\xi = \lambda^2 \, d\lambda \, d\beta$; extending the range of λ integration from $\mathbb{R}_+$ to $\mathbb{R}$ and using (2.4) with $s = w \cdot \beta$, we get

$$(2.5) \qquad f(w) = \frac{-1}{8\pi^2} \int \partial_s^2 \hat{f}(w \cdot \beta, \beta) \, d\beta \, ,$$

the inversion formula for the Radon transform. Note that the integration in (2.5) takes place over planes that pass through w.

It follows from (2.2) that $\lambda \tilde{f}(\lambda \beta)$ is the Fourier transform in s of $\partial_s \hat{f}(s, \beta)$; thus, by Plancherel's formula,

$$4\pi^2 \int_{-\infty}^{\infty} \lambda^2 |\tilde{f}(\lambda \beta)|^2 \, d\lambda = \int |\partial_s \hat{f}(s, \beta)|^2 \, ds \, .$$

Substituting this into the Plancherel relation between $\tilde{f}$ and f:

$$\int_{S^2} \int_0^{\infty} |\tilde{f}(\lambda \beta)|^2 \lambda^2 \, d\lambda \, d\beta = \int |f(w)|^2 \, dw \, ,$$

we get

$$(2.6) \qquad \frac{1}{8\pi^2} \int |\partial_s \hat{f}(s, \beta)|^2 \, ds \, d\beta = \int |f(w)|^2 \, dw \, ,$$

the Plancherel relation for the Radon transform. From (2.6) we deduce

$$(2.6)' \qquad \frac{1}{8\pi^2} \int \partial_s \hat{f} \, \partial_s \overline{\hat{g}} \, ds \, d\beta = \int f \overline{g} \, dw \, .$$

We state now the main properties of the Radon transform.

THEOREM 2.1.   (i) *For every f, $\hat{f}$ is even:*

$$(2.7) \qquad \hat{f}(-s, -\beta) = \hat{f}(s, \beta) \, .$$

   (ii) *The Plancherel relation (2.6) holds.*
   (iii) *Every even function $k(s, \beta)$ for which $\partial_s k$ is in L_2 is the Radon transform of some f in L_2.*
   (iv) *For any Euclidean motion τ define*

$$(2.8) \qquad f^\tau(w) = f(\tau^{-1} w) \, .$$

Then

$$(2.9) \qquad \hat{f}^\tau(s, \beta) = \hat{f}(s - \tau(0) \cdot \beta, \tau^{-1} \beta) \, .$$

   (v) *For f in C_0^1,*

$$(2.10) \qquad \widehat{\partial_w f} = \beta \, \partial_s \hat{f}, \qquad\qquad \partial_w = gradient \, .$$

   (vi) *For every f in C_0^2,*

$$(2.11) \qquad \widehat{\Delta f} = \partial_s^2 \hat{f}, \qquad\qquad \Delta = Laplace \ operator \, .$$

   (vii) *The inverse of the Radon transform is described by (2.5).*

Proof: Property (i) follows directly from (2.3). We have already proved (ii) and (vii). Property (iv) is obvious for translations and rotations; therefore by composition it holds for all τ. To prove (iii) we observe from relation (2.2) that every even function in C_0^∞ is the Radon transform of some square integrable function f; from this and (ii) we deduce (iii).

Property (v) follows from (2.9) applied to translations, and (vi) follows from (v).

We turn now to the wave equation

$$(2.12) \qquad \partial_t^2 u - \Delta u = 0 \, .$$

It is easy to prove that the *energy*, denoted by $|u|_E^2$ and defined by

(2.13)
$$|u|_E^2 = |u_t|^2 - (u, \Delta u)$$
$$= \int |u_t|^2 + |u_w|^2 \, dw \,,$$

is independent of time.

We denote the Cauchy data of u at time t by $F(t)$:

$$F(t) = \{u(t), u_t(t)\} \,.$$

Conservation of energy shows that $F(t)$ is determined by $F(0)$; we denote the operator relating $F(0)$ to $F(t)$ by $U(t)$. Conservation of energy also shows that $U(t)$ is unitary in the energy norm. We denote by $\mathcal{H}$ the completion in the energy norm of C_0^∞ data. It is known (cf. [6], page 95) that the first component of every F in $\mathcal{H}$ can be identified with an element of L_2^{loc}.

The wave equation can be solved with the aid of the Radon transform. Taking the transform of (2.12) and using (2.11) we get

$$\partial_t^2 \hat{u} - \partial_s^2 \hat{u} = 0 \,,$$

which can be factored as

(2.14)
$$(\partial_t + \partial_s)(\partial_t \hat{u} - \partial_s \hat{u}) = 0 \,.$$

This shows that $\partial_t \hat{u} - \partial_s \hat{u}$ is a function of $s - t$ alone, and suggests the

DEFINITION. *Let $F = \{f_1, f_2\}$ be data in $\mathcal{H}$; the operator R_+ on $\mathcal{H}$ is defined by*

(2.15)+
$$R_+ F = c(\partial_s^2 \hat{f}_1 - \partial_s \hat{f}_2) \,, \qquad\qquad c = -(2^{3/2} \pi)^{-1} \,.$$

R_+ *is called the outgoing translation representation.*

THEOREM 2.2. R_+ *gives a unitary representation of $\mathcal{H}$ for the group of operators $U(t)$:*

  (i) *Denote $R_+ F$ by $k_+(s, \beta)$, F in $\mathcal{H}$; then*

(2.16)
$$R_+ U(t)F = k_+(s - t, \beta) \,.$$

  (ii) *For every F in $\mathcal{H}$,*

(2.17)
$$|F|_E^2 = |R_+ F|^2 \,,$$

where || *on the right is the* L_2 *norm;*

(iii) R_+ *maps* $\mathscr{H}$ *onto* L_2.

Proof: (i) Let u be the solution of (2.12) with initial value F. Apply ∂_s to (2.14); using the definition (2.15)$_+$ of R_+ we can express the resulting relation as

$$(\partial_t + \partial_s) R_+ F(t) = 0.$$

This shows that $R_+ F(t)$ is a function of $s - t$ alone. Since $F(t) = U(t)F$, (2.16) follows.

(ii) As remarked in (2.7), $\hat{f}$ is an even function of s, β. It follows that $\partial_s^2 \hat{f}_1$ is even and $\partial_s \hat{f}_2$ odd; thus the two components are orthogonal to each other in the L_2 norm. Hence, in view of definition (2.15), we have

$$(2.18) \qquad |R_+ F|^2 = c^2 (|\partial_s^2 \hat{f}_1|^2 + |\partial_s \hat{f}_2|^2).$$

By (2.6), $c^2 |\partial_s \hat{f}_2|^2 = |f_2|^2$; using (2.6)' with $g = \Delta f$, as well as (2.11), we get

$$-(f_1, \Delta f_1) = -c^2 (\partial_s \hat{f}_1, \partial_s \widehat{\Delta f_1}) = -c^2 (\partial_s \hat{f}_1, \partial_s^3 \hat{f}_1) = c^2 (\partial_s^2 \hat{f}_1, \partial_s^2 \hat{f}_1).$$

Definition (2.13) and the last two relations yield

$$|F|_E^2 = |f_2|^2 - (f_1; \Delta f_1) = c^2 (|\partial_s \hat{f}_2|^2 + |\partial_s^2 \hat{f}_1|^2).$$

Combining this with (2.18) we deduce (2.16).

(iii) now follows from (2.17), combined with part (iii) of Theorem 2.1, applied separately to f_1 and f_2.

The incoming translation representer k_- of f is defined analogously, as $c(\partial_s^2 \hat{f}_1 + \partial_s \hat{f}_2)$. It follows from the evenness property (2.7) that

$$k_-(s, \beta) = k_+(-s, -\beta).$$

The data f_1, f_2 are easily expressed in terms of their representer k_+:

$$(2.19) \qquad f_1(w) = |c| \int k_+(w \cdot \beta, \beta) \, d\beta.$$

To see this substitute the definition (2.15)$_+$ of k_+ into (2.19); since $\partial_s \hat{f}_2$ is an odd function, its contribution to the integral in (2.19) is zero, while that of $\partial_s^2 \hat{f}_1$ is, according to (2.5), equal to $f_1(w)$.

According to part (i) of Theorem 2.2, $R_+\{u(t), u_t(t)\} = k_+(s-t, \beta)$. Substituting this in (2.19) we obtain

$$(2.20) \qquad u(w, t) = |c| \int k_+(w \cdot \beta - t, \beta) \, d\beta \; ;$$

this is a representation of solutions of the wave equation as a superposition of plane waves.

THEOREM 2.3. *Let $\{f_1, f_2\}$ be initial data in $\mathcal{H}$ that are equal to 0 for $|w| > a$. Then the solution u of the wave equation with these initial data is zero at $w = 0$ for $|t| > a$.*

Proof: Let $|s|$ be greater than a; then every point w of the plane $w \cdot \beta = s$ lies in $|w| > a$, where f_1 and f_2 are equal to 0. It follows from (2.3) that $\hat{f}_1$ and $\hat{f}_2$ are both equal to 0 for $|s| > a$; therefore, by (2.15)$_+$ we conclude that

$$k_+(s, \beta) = 0 \quad \text{for} \quad |s| > a \, .$$

Consequently, for $w = 0$, $|t| > a$, the integrand on the right side of (2.20) is equal to 0 for all β; this proves the theorem.

Theorem 2.3 shows that the value of $u(0, t)$ does not depend on values of the initial data inside the ball $|w| < t$; this property of solutions of the wave equation is called *Huygens' principle*.

THEOREM 2.4. *Let k be any L_2 function; define u by (2.20).*
(i) *$u(w, t) \equiv 0$ if and only if $k \equiv 0$.*
(ii) *$u(w, t) = 0$ in the forward cone $|w| < t$ if and only if $k(s, \beta) = 0$ for $s < 0$.*
(iii) *$u(w, t) = 0$ in the backward cone $|w| < -t$ if and only if $k(s, \beta) = 0$ for $s > 0$.*
(iv) *$u(w, t) = 0$ for $|w| < a - |t|$ if and only if*

$$(2.21) \qquad k(s, \beta) = \sum p_m(s) Y_m(\beta) \quad \text{for} \quad |s| < a,$$

where Y_m is a spherical harmonic of degree m, and $p_m(s)$ a polynomial of degree less than m.

Proof: The "if" parts of Theorem 2.4 are obviously true. Let $\phi(t)$ be a C_0^∞ test function; multiply (2.20) by $\phi(t)$ and integrate; interchanging the order of integration and introducing $w \cdot \beta - t = s$ as a new variable, we get

$$(2.22) \qquad \int \phi(t) u(w, t) \, dt = |c| \int \int \phi(w \cdot \beta - s) k(s, \beta) \, ds \, d\beta \, .$$

In case (i) the left side is 0 for all w. Differentiating the right side of (2.22) m

times with respect to w and setting $w = 0$ gives

(2.23)
$$0 = \int \int \beta^N (\partial_s^n \phi(-s)) k(s, \beta) \, ds \, d\beta ,$$

where N is any multi-index of order $|N| = n$. Fundtions of the form $\partial_s^n \phi$, $\phi \in C_0^\infty$, are dense in $L_2(\mathbb{R})$, since their orthogonal complement, which would consist of L_2 polynomials, is 0. So it follows from (2.23) that k is orthogonal to all functions of the form $\psi(s) Y_n$, ψ in L_2. Since these functions span L_2, $k \equiv 0$ as asserted.

In case (ii) the left side is equal to 0 for $|w| < \delta$ if the support of ϕ is confined to $t > \delta > 0$. Then (2.23) holds for all such ϕ. Since functions of the form $\partial_s^n \phi(-s)$, $\phi \in C_0^\infty$, $\phi(t) = 0$ for $t > \delta$, are dense in $L_2(\mathbb{R}_-)$, it follows as before that $k(s, \beta) = 0$ for $s < 0$. This proves (ii); the proof of (iii) is analogous.

In case (iv), (2.23) holds for all ϕ supported in $|s| < a$. In this case functions of the form $\partial_s^m \phi$ do not span $L_2(-a, a)$; their orthogonal complement consists precisely of all polynomials of order less than m. It follows then from (2.23) that

$$\int \beta^N k(s, \beta) \, d\beta$$

are polynomials of degree less than $n = |N|$. This completes the proof of Theorem 2.4.

A solution $u(w, t)$ of the wave equation is called *outgoing* if it vanishes in the forward cone $|w| < t$, *incoming* if it vanishes in the backward cone $|w| < -t$. The initial data of outgoing, respectively incoming, solutions with finite energy are called the outgoing and incoming subspaces of the space of data with finite energy, and are denoted by $\mathcal{D}_+$ and $\mathcal{D}_-$, respectively. In this language, parts (ii) and (iii) of Theorem 2.4 can be restated as follows:

THEOREM 2.5. *In the outgoing translation representation, $\mathcal{D}_+$ is represented by $L_2(\mathbb{R}_+)$, $\mathcal{D}_-$ by $L_2(\mathbb{R}_-)$.*

Finally we turn to the question of characterizing the Radon transform of functions that vanish outside a ball. The answer is contained in

THEOREM 2.6. *Let f be an L_2 function which is equal to 0 for $|w| > r$. Denote $\hat{f}$ by h. Then*
  (i) $h(s, \beta) = 0$ *for* $|s| > r$,
  (ii) *for every non-negative integer n,*

(2.24)
$$\int s^n h(s, \beta) \, ds$$

is a polynomial in β of degree at most n.

Conversely, if h is an even function such that $\partial_s h$ is in L_2 and if h satisfies (i) and (ii), then $h = \hat{f}$ where f is in L_2 and equal to 0 for $|w| > r$.

Proof: We have already shown in the course of proving Theorem 2.3 that if $f(w) = 0$ for $|w| > r$, then $\hat{f}(s, \beta) = 0$ for $|s| > r$. To derive property (ii) consider

$$(2.25) \qquad \int (w \cdot \beta)^n f(w) \, dw,$$

clearly a polynomial in β of degree at most n. Performing the integration first on the plane $w \cdot \beta = s$ and then integrating with respect to s, we see that $(2.25) = (2.24)$.

To prove the converse, we assume that $\hat{f}$ satisfies (i) and (ii) and deduce from this that $f(w) = 0$ for $w > r$ by showing that

$$(f, g) = 0$$

for all g that vanish for $|w| < r + \varepsilon$. By (2.6)',

$$(2.26) \qquad (f, g) = c^2 \int \partial_s \hat{f} \, \partial_s \hat{g} \, ds \, d\beta .$$

Define

$$u(w, t) = c^2 \int \partial_s \hat{g}(w \cdot \beta - t, \beta) \, d\beta ;$$

clearly u is a solution of the wave equation, and since $\partial_s \hat{g}(s, \beta)$ is odd, $u(w, 0) = 0$; by (2.5), $u_t(w, 0) = -g(w)$. Since $g(w) = 0$ for $|w| < r + \varepsilon$, it follows by the classical uniqueness theorem for solutions of the wave equation that $u(w, t) = 0$ for $|w| < r + \varepsilon - |t|$. Then part (iv) of Theorem 2.4 is applicable, with $k = c^2 \partial_s \hat{g}$. In view of (2.21),

$$c^2 \partial_s \hat{g} = \sum p_m(s) Y_m(\beta) ,$$

p_m being a polynomial in s of degree less than m, Y_m a spherical harmonic of order m. Substituting this into (2.26) and using (2.24) we get

$$(f, g) = \sum \int \partial_s \hat{f} p_m(s) Y_m(\beta) \, ds \, d\beta$$

$$= \sum \int q_m(\beta) Y_m(\beta) \, d\beta ,$$

where $q_m(\beta)$ is a polynomial of degree less than m; therefore, $\int q_m Y^m \, d\beta = 0$ and Theorem 2.6 is proved.

Theorem 2.6 is the Paley–Wiener theorem for the Radon transform; the above proof is from [7]. A slightly different version of the result is contained in [9]; another version is given in [2].

An entirely analogous result holds for data; see page 113 of [6]:

THEOREM 2.7. *Let* $F = \{f_1, f_2\}$ *be data with finite energy, and suppose that*
(i) $Lf_1 = 0$, $f_2 = 0$ *for* $|w| > r$.
Denote R_+F *by* k. *Then,*
(ii) $k(s, \beta) = 0$ *for* $|s| > r$,
(iii) *for every non-negative integer* n,

$$\int s^n k(s, \beta)\, ds$$

is a polynomial in β *of degree at most* n. *Conversely, if* k *is in* L_2 *and satisfies* (ii) *and* (iii), *then* $k = R_+F$ *for some* F *of finite energy, satisfying* (i).

3. The Radon Transform in Non-Euclidean Space

Helgason in [3] defines the Fourier transform of a function f on Π in analogy with the Euclidean case, but replacing $w \cdot \beta$ by $\langle w, \beta \rangle$ as defined in (1.12):

$$(3.1) \qquad \tilde{f}(\lambda, \beta) = \frac{1}{(2\pi^3)^{1/2}} \int_{\Pi} f(w) \exp\{(1 - i\lambda)\langle w, \beta \rangle\}\, dw.$$

Just as the Fourier transform in Euclidean space gives a spectral representation of the Laplace operator, so (3.1) gives a spectral representation of the non-Euclidean Laplacean L_0 defined in (1.11). To see this we note that in the Euclidean case each exponential $\exp\{-iw \cdot \xi\}$ is an eigenfunction of Δ:

$$(3.2) \qquad \Delta \exp\{-iw \cdot \xi\} = -|\xi|^2 \exp\{-iw \cdot \xi\}.$$

From this we deduce, using the self-adjointness of Δ, that for every f in C_0^∞

$$(3.3) \qquad \Delta f = (\Delta f, \exp) = (f, \Delta \exp) = -|\xi|^2 \tilde{f}.$$

A similar equation is satisfied by the non-Euclidean exponentials $e = e(w, \beta, \lambda) = \exp\{(1 - i\lambda)\langle w, \beta \rangle\}$:

$$(3.4) \qquad L_0 e = -(1 + \lambda^2)e,$$

where L_0 is given by (1.11). Since L_0 is invariant under τ, and since (1.13) implies

$$e(\tau w, \tau\beta) = \text{const.} \; e(w, \beta),$$

it follows that it suffices to verify (3.4) for a single value of β. We choose $\beta = \beta_\infty$ in model (i); by (1.14), $e(w, \beta_\infty) = z^{(1-i\lambda)}$, and hence (3.4) can be verified by a simple calculation.

Formula (3.4) suggests that we replace L_0 by L defined as

$$(3.5) \qquad\qquad L = L_0 + 1 = z^3 \, \partial z^{-1} \, \partial + 1 \; ;$$

L satisfies

$$(3.4)' \qquad\qquad Le = -\lambda^2 e.$$

Since L is selfadjoint, we get for any f in C_0^∞

$$(3.6) \qquad\qquad \widetilde{Lf} = (Lf, e) = (f, Le) = -\lambda^2 \tilde{f},$$

where $(\;,\;)$ denotes the invariant L_2 scalar product. This shows that the Fourier transform (3.1) furnishes a spectral representation for L.

The inverse of the Fourier transform is quite analogous to the Euclidean case:

$$(3.7) \qquad f(w) = \frac{1}{(2\pi^3)^{1/2}} \int\limits_{R_+} \int\limits_{B} \tilde{f}(\lambda, \beta) \exp\{(1 + i\lambda)\langle w, \beta\rangle\}\lambda^2 \, d\lambda \, d\beta,$$

where $d\beta$ is defined in (1.17). Note that only positive values of λ are used; of course, an entirely analogous inversion formula can be based on $\lambda < 0$. For verification we refer to [3].

Denote by T the operator defined in (3.1) linking f to $\tilde{f}$, and by S the inverse operator defined in (3.7). Since the kernel of S is the conjugate transpose of that of T, S and T are adjoint to each other:

$$(3.8) \qquad\qquad (Tf, g) = (f, Sg)$$

for all C_0^∞ f and g. Setting $g = \tilde{f}$ we see using (3.7) that $Sg = S\tilde{f} = f$. Substituting this into (3.8) we get

$$(3.9) \qquad\qquad |\tilde{f}|^2 = |f|^2,$$

where the norm on the right is $L_2(\Pi)$, on the left $L_2(\mathbb{R}_+ \times B)$ with respect to the measure $\lambda^2 \, d\lambda \, d\beta$; (3.9) is the *Plancherel relation*.

If in the definition (3.1) of $\tilde{f}$ we first integrate over $\langle w, \beta \rangle = s$ and then integrate with respect to s, we get

$$(3.10) \qquad \tilde{f}(\lambda, \beta) = \frac{1}{(2\pi^3)^{1/2}} \int \exp\{(1 - i\lambda)s\} \hat{f}(s, \beta) \, ds ,$$

where

$$(3.11) \qquad \hat{f}(s, \beta) = \int_{\langle w, \beta \rangle = s} f(w) \, dS ,$$

dS being the non-Euclidean surface element on the horosphere $\langle w, \beta \rangle = s$. The function $\hat{f}$ is called the *Radon transform* of f.

Formula (3.10) shows that $\tilde{f}$ is the Fourier transform of $e^s \hat{f}$ with respect to s; from (3.10) we deduce that

$$(3.12) \qquad \int_{-\infty}^{\infty} \tilde{f}(\lambda, \beta) e^{i\lambda s} \lambda^2 \, d\lambda = -\sqrt{2/\pi} \, \partial_s^2 e^s \hat{f}(s, \beta) .$$

Substituting (3.12) into (3.7), after extending the λ integration to all of $\mathbb{R}$, we get

$$(3.13) \qquad \begin{aligned} f(w) &= \frac{-1}{2\pi^2} \int_B \exp\{\langle w, \beta \rangle\} \, \partial_s^2 e^s \hat{f}(s, \beta)\big|_{s = \langle w, \beta \rangle} \, d\beta \\ &= -\frac{1}{2\pi^2} \int_B \exp\{2\langle w, \beta \rangle\} \\ &\quad \times [\hat{f}(\langle w, \beta \rangle, \beta) + 2\partial_s \hat{f}(\langle w, \beta \rangle, \beta) + \partial_s^2 \hat{f}(\langle w, \beta \rangle, \beta)] \, d\beta . \end{aligned}$$

This is the inversion formula for the Radon transform. Note that integration is over all horospheres that pass through w.

Just as in the Euclidean case, we deduce from (3.10) that $\pi\lambda\tilde{f}(\lambda, \beta)$ is the Fourier transform in s of $\partial_s e^s \hat{f}(s, \beta)$; thus, by Plancherel's formula,

$$\pi^2 \int_{-\infty}^{\infty} \lambda^2 |\tilde{f}(\lambda, \beta)|^2 \, d\lambda = \int |\partial_s e^s \hat{f}(s, \beta)|^2 \, ds .$$

Integrating this with respect to $d\beta$ and using Plancherel's relation between $\tilde{f}$ and f given by (3.9), we get

$$(3.14) \qquad \frac{1}{2\pi^2} \int \int |\partial_s e^s \hat{f}|^2 \, ds \, d\beta = \int_\pi |f(w)|^2 \, dw ,$$

the Plancherel relation for the Radon transform. By polarization we obtain, from (3.14),

(3.14)′
$$\frac{1}{2\pi^2} \int \int \partial_s e^s \hat{f} \, \partial_s e^s \bar{\hat{g}} \, ds \, d\beta = \int f \bar{g} \, dw \, .$$

We now state the main properties of the Radon transform.

THEOREM 3.1. *All functions f in $C_0^\infty(\Pi)$ have the following properties:*
(i) *The Plancherel relation (3.14) holds.*
(ii) *For any motion τ define*

(3.15)
$$f^\tau(w) = f(\tau^{-1} w) \, ;$$

then

(3.15)′
$$\widehat{f^\tau}(s, \beta) = \hat{f}(s - \langle \tau j, \beta \rangle, \tau^{-1} \beta) \, .$$

(iii) *For any $a > 0$,*

(3.16)
$$e^a \int \hat{f}(a, \beta) \, d\beta = e^{-a} \int \hat{f}(-a, \beta) \, d\beta \, .$$

(iv)

(3.17)
$$\widehat{Lf} = e^{-s} \, \partial_s^2 e^s \hat{f} \, ,$$

where L is defined in (3.5).
(v) *The inverse of the Radon transform is given by (3.13).*

Proof: We have already established (i) and (v). To prove (ii) we perform a change of variable $w = \tau w'$ in (3.11):

$$\widehat{f^\tau}(s, \beta) = \int_{\langle w, \beta \rangle = s} f^\tau(w) \, dS = \int_{\langle w, \beta \rangle = s} f(\tau^{-1} w) \, d_w S$$

$$= \int_{\langle \tau w', \beta \rangle = s} f(w') \, d_{w'} S \, .$$

Using (1.13), with $\tau^{-1} \beta$ in place of β, we have

$$\langle \tau w', \beta \rangle = \langle w', \tau^{-1} \beta \rangle + \langle \tau j, \beta \rangle \, .$$

So the domain of integration above is

$$\langle w', \tau^{-1} \beta \rangle = s - \langle \tau j, \beta \rangle \, .$$

Substituting this in the above integral yields (3.15)'.

We note in particular that if we specialize τ to be a rotation around j, that is to a τ for which $\tau j = j$, then $\langle \tau j, \beta \rangle = \langle j, \beta \rangle = 0$ and (3.15)' gives

$$(3.18) \qquad \widehat{f^\tau}(s, \beta) = \hat{f}(s, \tau^{-1}\beta).$$

If f is rotationally symmetric, that is if $f^\tau = f$, then we conclude that $\hat{f}(s, \beta) = \hat{f}(s, \tau^{-1}\beta)$ for all such τ. Since any point of B can be moved to any other by a rotation about j, it follows that f is independent of β.

We begin the proof of (iii) by setting

$$(3.19) \qquad f_{av}(w) = \int f^\tau(w) \, d\tau,$$

where the integration is over the rotation group keeping j fixed. Clearly, f_{av} is rotationally symmetric so that its Radon transform is independent of β. Hence if we take the Radon transform of (3.19), making use of (3.18), we obtain

$$(3.19)' \qquad \hat{f}_{av}(s) = \int \hat{f}(s, \tau^{-1}\beta) \, d\tau.$$

Obviously,

$$(3.19)'' \qquad \int \hat{f}(s, \tau^{-1}\beta) \, d\tau = \int \hat{f}(s, \beta) \, d\beta.$$

Thus in order to prove (3.16) it suffices to show for a rotationally symmetric function f that

$$(3.20) \qquad e^s \hat{f}(s) = e^{-s} \hat{f}(-s).$$

To this end we consider two horospheres $\Sigma_+ = \Sigma(s, \beta)$ and $\Sigma_- = \Sigma(-s, \beta)$. Denote by $A_+(r)$ and $A_-(r)$ the surface area of that part of Σ_+ and Σ_- whose distance to j does not exceed r. We claim that, for all r,

$$(3.21) \qquad A_-(r) = e^{2s} A_+(r).$$

Since f is rotationally symmetric around j,

$$f(w) = g(r), \qquad\qquad r = \text{distance of } j \text{ to } w,$$

and thus

$$\hat{f}(s) = \int_{\Sigma_+} f(w)\, dS = \int g(r)\, dA_+(r),$$

$$\hat{f}(-s) = \int_{\Sigma_-} f(w)\, dS = \int g(r)\, dA_-(r).$$

From this and (3.21) we deduce (3.20).

The proof of (3.21) is a calculation best performed in model (ii). It suffices to verify that

$$\frac{d}{dr} A_- = e^{2s} \frac{d}{dr} A_+ ;$$

we leave the calculation to the reader.

To prove (iv) we apply (3.12) with Lf in place of f, and make use of (3.6):

$$\partial_s^2 e^s \widehat{Lf} = -\sqrt{\tfrac{1}{2}\pi} \int \widetilde{Lf} e^{i\lambda s} \lambda^2\, d\lambda$$

$$= \sqrt{\tfrac{1}{2}\pi} \int \tilde{f} e^{i\lambda s} \lambda^4\, d\lambda = \partial_s^4 e^s \hat{f}.$$

It follows from this that $e^s \widehat{Lf} = \partial_s^2 e^s \hat{f} + a + bs$. Since, for f with compact support, $\hat{f}$, and likewise $\widehat{Lf}$, have compact support in s (see the proof of Theorem 3.7), $a = b = 0$ and (3.17) follows.

In the course of proving Theorem 3.1 we have also proved:

COROLLARY 3.2. *Let f be a C_0^∞ function in Π with rotational symmetry, that is $f(\tau w) = f(w)$ when $\tau j = j$. Then $\hat{f}(s, \beta)$ is independent of β and (3.20) holds.*

A more complete treatment of the Radon transform in the setting of model (iii) can be found in [1], Chapter V; see also [5]. It should be pointed out that condition (3.16) is the non-Euclidean analogue of the evenness of the Radon transform in the Euclidean case (see Theorem 2.1, (i)). Helgason [4], page 473, has shown that this property characterizes the range of the Radon transform.

We turn now to the *non-Euclidean wave equation*:

$$(3.22) \qquad\qquad u_{tt} - Lu = 0,$$

where L is defined in (3.5).

The following facts are furnished by the classical theory of hyperbolic equations:

(i) The initial value problem for (3.22),

$$u(w, 0) = f_1(w), \qquad u_t(w, 0) = f_2(w),$$

has, for arbitrary C_0^∞ functions f_1 and f_2 defined in Π, a unique C^∞ solution $u(w, t)$ in $\Pi \times \mathbb{R}$.

(ii) Signals propagate with speed at most 1, i.e., the value of $u(w_0, t)$ does not depend on values of f_1 and f_2 at point w whose distance from w_0 exceeds $|t|$.

(iii) The non-Euclidean energy

$$(3.23) \qquad |u|_E^2 = |u|_L^2 + |u_t|^2,$$

where

$$(3.24) \qquad |u|_L^2 = -(u, Lu),$$

is independent of t.

Yet another property of solutions of (3.22), Huygens' principle, will be stated later as Theorem 3.7.

We claim that $|u|_L^2$ is non-negative[2] for u in $C_0^2(\Pi)$. In fact, it follows from the definition of L and an integration by parts, with $dw = dx\, dy\, dz/z^3$, or directly from (1.10) that

$$-(u, Lu) = -\int \left[uz^3 \partial \frac{1}{z} \partial u + u^2 \right] dw$$

$$= \int \left[\frac{u_x^2 + u_y^2 + u_z^2}{z} - \frac{u^2}{z^3} \right] dx\, dy\, dz.$$

Another integration by parts shows that for u in C_0^∞

$$\int \left(\frac{u_z^2}{z} - \frac{u^2}{z^3} \right) dz = \int z \left[\partial_z \left(\frac{u}{z} \right) \right]^2 dz,$$

so that we can write

$$(3.24)' \qquad |u|_L^2 = \int z^4 (v_x^2 + v_y^2 + v_z^2)\, dw,$$

---

[2] We shall show in a subsequent publication that, in contrast, energy over the fundamental domain with respect to some discrete subgroups can be negative.

where

(3.24)″ $$v = u/z.$$

Even more is true.

THEOREM 3.3. *For any compact subset K of Π there is a constant c_K such that*

(3.25) $$|u|_K^2 \equiv \int_K u^2 \, dw \leqq c_K |u|_L^2$$

for all u in $C_0^2(\Pi)$.

Proof: We define a new norm $|u|_{L,K}$ by

$$|u|_{L,K}^2 \equiv |u|_L^2 + |u|_K^2$$

and work in the function space obtained by completion of C_0^2 with respect to this norm. It follows easily from the Rellich compactness theorem that $|u|_K^2$ is compact with respect to $|u|_{L,K}^2$. Therefore, $|u|_K^2$ can be written as

(3.25)′ $$|u|_K^2 = (u, T_K u)_{L,K},$$

where T_K is a compact selfadjoint operator. We claim that all eigenvalues λ of T_K are less than 1. For if

$$T_K u = \lambda u,$$

then by (3.25)′, $|u|_K^2 = \lambda |u|_{L,K}^2$. Substituting this into the definition of $|u|_{L,K}$ we get

$$(1 - \lambda)|u|_{L,K}^2 = |u|_L^2.$$

Since, by (3.24)′, $|u|_L^2 \geqq 0$, it follows that $\lambda \leqq 1$. If λ were equal to 1, $|u|_L$ would be 0. To show that this cannot happen, we recall that $|u|_L$ is invariant under all motions. Therefore, we can apply (3.24)′, (3.24)″ to u^τ and obtain

$$|u|_L^2 = |u^\tau|_L^2 = \int z^4 (\partial_w v^\tau)^2 \, dw,$$

where

$$v^\tau = u^\tau/z.$$

Thus $|u|_L = 0$ implies that $v^\tau = $ const. and so $u^\tau = $ const. z; but this can hold for all τ only if $u = 0$.

Since T_K is compact, its eigenvalues accumulate only at 0. Denote by λ the largest of them; as shown above, $\lambda < 1$. Since T_K is selfadjoint,

$$|u|_K^2 = (u, T_K u)_{L,K} \leqq \lambda |u|_{L,K}^2.$$

Substituting $|u|_{L,K}^2 = |u|_L^2 + |u|_K^2$ gives (3.25), with $c_K = \lambda/(1-\lambda)$.

We shall denote by $\mathcal{H}$ the completion in the energy norm (3.23) of all $C_0^2(\Pi)$ data. It follows from Theorem 3.3 that for all data $F = \{f_1, f_2\}$ in $\mathcal{H}$

$$|f_1|_K \leqq c_K |F|_E.$$

Let u be a C_0^2 solution of (3.17). We denote by $F(t)$ its data $\{u(t), u_t(t)\}$ at time t. It follows from conservation of energy and positivity of energy that if $F(0) = 0$, $F(t) = 0$ for all t. We denote by $U(t)$ the operator relating $F(0)$ to $F(t)$; since the wave equation (3.22) is hyperbolic, there exist solutions with arbitrary initial data. Thus the $U(t)$ are unitary operators with respect to the energy norm, and they form a one-parameter group. We denote the generator of this group by A, i.e.,

$$AF = \frac{d}{dt} U(t)F|_{t=0}.$$

In view of (3.22),

$$A = \begin{pmatrix} 0 & 1 \\ L & 0 \end{pmatrix}.$$

As in the Euclidean case, the non-Euclidean wave equation (3.22) can be solved by the Radon transform; taking the transform of (3.22) and using (3.17) we get

$$\hat{u}_{tt} - e^{-s} \partial_s^2 e^s \hat{u} = 0.$$

Defining m as $e^s \hat{u}$ we can rewrite the above in the form

$$\partial_t^2 m - \partial_s^2 m = 0,$$

which can be factored as

(3.26) $$(\partial_t \pm \partial_s)(\partial_t m \mp \partial_s m) = 0.$$

It follows that

$$m_t \mp m_s$$

is a function of $s \mp t$ alone.

Denote the initial data of u by

$$u(w, 0) = f_1, \qquad u_t(w, 0) = f_2.$$

We define the *incoming* and *outgoing translation representers* k_- and k_+ of $F = \{f_1, f_2\}$ by

$$(3.27) \qquad \begin{aligned} k_+(s, \beta) &= c(\partial_s^2 e^s \hat{f}_1 - \partial_s e^s \hat{f}_2), \\ k_-(s, \beta) &= c(\partial_s^2 e^s \hat{f}_1 + \partial_s e^s \hat{f}_2), \end{aligned}$$

where $c = -(\sqrt{2}\, \pi)^{-1}$. We denote by $R_\pm$ the operators

$$(3.27)' \qquad R_+ F = k_+, \qquad R_- F = k_-.$$

THEOREM 3.4. (i) *For every* C_0^∞ *data* F,

$$(3.28) \qquad \begin{aligned} R_+ U(t)F &= k_+(s - t, \beta), \\ R_- U(t)F &= k_-(s + t, \beta), \end{aligned}$$

where $k_\pm$ *are defined in* (3.27), (3.27)'.

(ii) *For every* F *in* C_0^∞,

$$(3.29) \qquad |F|_E^2 = |R_+ F|^2 = |R_- F|^2,$$

where the norm on the right is the L_2 *norm with respect to the measure* $ds\, d\beta$, *as defined in* (1.17).

(iii) R_+ *and* R_- *each map* C_0^∞ *onto a dense subset of* $L_2(\mathbb{R} \times B)$.

Note. It follows that R_+ and R_- can be extended by completion to all of $\mathcal{H}$, and give the unitary translation representations of $\mathcal{H}$ for the group $U(t)$. The relation of these representations is furnished by the scattering operator which was described in the introduction.

Proof: Let u be the solution of (3.22) with initial data F; then (3.26) is satisfied. Applying ∂_s to (3.26) we obtain, in the notation (3.27),

$$(\partial_t \pm \partial_s) R_\pm F(t) = 0$$

which shows that $R_+F(t)$ is a function of $s-t$, $R_-F(t)$ of $s+t$. Since $F(t) = U(t)F$, (3.28) follows.

We turn to (ii); by definition,

$$|R_\pm F|^2 = c^2 \left\{ \int |\partial_s^2 e^s \hat{f}_1|^2 \mp 2\,\mathfrak{Re} \int \partial_s^2 e^s \hat{f}_1 \, \partial_s e^s \bar{\hat{f}}_2 + \int |\partial_s e^s \hat{f}_2|^2 \right\}.$$

To prove (3.29) we have to show two things:

(3.30)
$$|F|_E^2 = c^2 \left\{ \int |\partial_s^2 e^s \hat{f}_1|^2 + \int |\partial_s e^s \bar{\hat{f}}_2|^2 \right\}$$

and

(3.30)′
$$\int \partial_s^2 e^s \hat{f}_1 \, \partial_s e^s \bar{\hat{f}}_2 = 0.$$

By definition (3.23),

$$|F|_E^2 = -(f_1, Lf_1) + |f_2|^2.$$

We shall verify (3.30) by showing separately that

$$c^2 \int |\partial_s e^s \hat{f}_2|^2 = |f_2|^2 \quad \text{and} \quad c^2 \int |\partial_s^2 e^s f_1|^2 = -(f_1, Lf_1).$$

The first equality is just the Plancherel relation (3.14) with f_2 in place of f. The second follows from the bilinear Plancherel relation (3.14)′ upon setting $f = f_1$; $g = Lf_1$; for, using (3.17), we get

$$(f_1, \overline{Lf_1}) = c^2 \int \partial_s(e^s \hat{f}_1) \, \partial_s(e^s \overline{\widehat{Lf_1}}) \, ds \, d\beta$$

$$= c^2 \int \partial_s(e^s \hat{f}_1) \, \partial_s^3(e^s \bar{\hat{f}}_1) \, ds \, d\beta$$

$$= -c^2 \int |\partial_s^2(e^s \hat{f}_1)|^2 \, ds \, d\beta.$$

This completes the proof of (3.30).

Relation (3.30)′ is easily proved when both f_1 and f_2 have rotational symmetry. For then, by Corollary 3.2, $\partial_s^2 e^s f_1$ is an even, $\partial_s e^s f_2$ is an odd function of s, and both are independent of β. Next we consider the case that f_1 is rotationally symmetric but f_2 is not. Then only $\hat{f}_2$ depends on β. Recall the definition of f_{av} in (3.19):

$$f_{av}(w) = \int f_2^\tau(w) \, d\tau,$$

where the integration is over the rotation group keeping j fixed; it follows from (3.19)′ and (3.19)″ that

$$\hat{f}_{av}(s) = \int \hat{f}_2(s, \beta)\, d\beta .$$

We can therefore write (3.30)′ as

$$\int \int \partial_s^2(e^s \hat{f}_1)\, \partial_s(e^s \bar{\hat{f}}_2)\, d\beta\, ds = \int \partial_s^2(e^s \hat{f}_1)\, \partial_s(e^s \bar{\hat{f}}_{av})\, ds$$

which is zero since both f_1 and f_{av} are rotationally symmetric.

Next we take the case when f_2 is arbitrary and f_1 has rotational symmetry around some point other than j; i.e., we take $f_1 = h^\tau$, where h is symmetric around j and τ is arbitrary. By (3.15),

$$\widehat{h^\tau} = \hat{h}(s - \langle \tau j, \beta \rangle) .$$

Setting this into (3.30) we obtain

$$\int \partial_s^2 e^s \hat{h}(s - \langle \tau j, \beta \rangle)\, \partial_s e^s \bar{\hat{f}}_2(s, \beta)\, ds\, d\beta .$$

Introducing $t = s - \langle \tau j, \beta \rangle$ as a new variable of integration gives

$$\int \partial_t^2 \exp\{t + \langle \tau j, \beta \rangle\} \hat{h}(t)\, \partial_t \exp\{t + \langle \tau j, \beta \rangle\} \bar{\hat{f}}_2(t + \langle \tau j, \beta \rangle, \beta)\, dt\, d\beta .$$

We introduce $\gamma = \tau^{-1}\beta$ as a new variable of integration. Relation (1.23), with τ^{-1} in place of τ, yields

$$d(\tau^{-1}\beta) = \exp\{2\langle \tau j, \beta \rangle\}\, d\beta .$$

This enables us to rewrite the above integral as

$$\int \partial_t^2 e^t \hat{h}(t)\, \partial_t e^t \bar{\hat{h}}_2(t + \langle \tau j, \tau\gamma \rangle, \tau\gamma)\, d\gamma\, dt .$$

We use now (1.13), with $w = \tau^{-1} j$, $\beta = \gamma$:

$$0 = \langle j, \tau\gamma \rangle = \langle \tau^{-1} j, \gamma \rangle + \langle \tau j, \tau\gamma \rangle .$$

From this relation and (3.15)′ we get

$$\hat{f}_2(t + \langle \tau j, \tau\gamma \rangle, \tau\gamma) = \hat{g}(t, \gamma) ,$$

where $g = f_2^{\tau^{-1}}$. Therefore, the above integral can be written as

$$\int \partial_t^2 e^t \hat{h}(t) \, \partial_t e^t \bar{\hat{g}}(t, \gamma) \, d\gamma \, dt,$$

an integral previously shown to be zero.

To complete the proof of part (ii) of Theorem 3.4 we have to show that every f_1 can be approximated by linear combinations of functions symmetric about some point. A function $h(z, w)$ which is rotationally symmetric around z is simply a function of the non-Euclidean distance from z:

$$(3.31) \qquad\qquad h(z, w) = g(\text{dist.}\,(w, z)).$$

Obviously $h(z, w)$ is symmetric in z, w.

Given any $\varepsilon > 0$ we can choose g in $C_0^\infty(\mathbb{R})$ so that

   (i) $g(t) \geqq 0$ for all t,

   (ii) $g(t) = 0$ for $t \geqq \varepsilon$,

   (iii) $\displaystyle\int g(\text{dist.}\,(w, j)) \, dw = 1.$

The foregoing implies that

   (i)′ $h(z, w) \geqq 0$ for all z, w,

   (ii)′ $h(z, w) = 0$ when dist. $(z, w) \geqq \varepsilon$,

   (iii)′ $\displaystyle\int h(z, w) \, dz = 1$ for all w.

Given any function f in Π, continuous and of compact support, we define

$$f_\varepsilon(w) = \int h(z, w) f(z) \, dz.$$

It follows from a standard result in analysis that $f_\varepsilon(w) \to f(w)$ uniformly as $\varepsilon \to 0$. Since f_ε is a superposition of symmetric functions, we are done. This completes the proof of part (ii).

We turn now to part (iii) of Theorem 3.4; since $\mathscr{H}$ is the closure in the energy norm of C_0^∞, and since, by part (ii), R_+ is isometric from the energy norm into the L_2 norm, it suffices to show that there exists a set of $h(s, \beta)$ in the range of R_+ that is dense in L_2. Since in model (ii) the L_2 space of functions is spanned by functions of the form $h_n(s) Y_n(\beta)$, Y_n some spherical harmonic, it suffices to display, for each Y_n, a set of $h(s)$ which are dense in L_2 and for which $h(s) Y_n(\beta) = R_+ f$. For this we shall use

LEMMA 3.5. *If, in model* (ii), *f is of the form*

$$(3.32) \qquad\qquad f(w) = g Y_n(\theta),$$

where g depends only on $|w|$, $\theta = w/|w|$ and Y_n is a spherical harmonic, then the

Radon and Fourier transforms are likewise of the form

(3.32)′
$$\hat{f}(s, \beta) = k(s) Y_n(\beta),$$
$$\bar{f}(\lambda, \beta) = h(\lambda) Y_n(\beta).$$

Sketch of proof: It is well known that a linear mapping of the space of functions on the sphere S^2 into itself which commutes with rotations has each spherical harmonic of order n, $n = 0, 1, \ldots$, as eigenfunction. Let ρ be a rotation around j, i.e., $\rho j = j$; according to relations (3.15), (3.15)′ with ρ in place of τ,

$$\widehat{f^\rho}(s, \beta) = \hat{f}(s, \rho^{-1}\beta).$$

Consider functions f of the form

$$f(w) = gp(\theta), \qquad\qquad \theta = w/|w|,$$

and g a fixed function of $|w|$. It follows from the above relation that for any weight function $m(s)$ the mapping

$$p \rightarrow \int \hat{f}(s, \beta) \, dm(s)$$

commutes with rotations ρ and therefore maps spherical harmonics into spherical harmonics. Relations (3.32), (3.32)′ follow.

Consider now f of the form (3.32); then $\hat{f}$ is of the form (3.32)′. Using formula (3.27)′ on the data $F = (0, f)$ we see that $R_+ F$ is of the form

$$R_+ F = k(s) Y_n(\beta).$$

According to part (i) of Theorem 3.4, $k(s-t) Y_n(\beta)$ also belongs to the range of R_+. We claim that, for an appropriate choice of the function g appearing in (3.32), the translates of k span $L_2(\mathbb{R})$. To see this we recall that translates of an L_2 function k span L_2 if and only if the Fourier transform $\bar{k}$ of k does not vanish on a set of positive measure. According to formula (3.10), the Fourier transform of $e^s \hat{f}$ with respect to s is the Fourier transform $\bar{f}$ of f. Thus, for f of form (3.32),

$$\pi i \lambda \bar{f}(\lambda, \beta) = \bar{k}(\lambda) Y_n(\beta).$$

Now choose g to have compact support; it follows from the definition (3.1) of the Fourier transform that, for f of compact support, $\bar{f}(\lambda, \beta)$ is entire analytic as a function of λ. Such a function cannot vanish on a set of positive measure unless it vanishes for all λ. We show now how to choose g so that $\bar{f} \neq 0$, in particular so that

$$\tilde{f}(-\tfrac{1}{2}i) \neq 0 .$$

By definition (3.1), with $f = gY_n$,

(3.33) $\qquad \tilde{f}(-\tfrac{1}{2}i, \beta) = \dfrac{1}{(2\pi^3)^{1/2}} \displaystyle\int g(|w|) \exp\{\tfrac{1}{2}\langle w, \beta\rangle\} Y_n(\theta)\, dw .$

Next we use formula (1.16) for $\langle w, \beta\rangle$, and the well-known generating function for the Legendre polynomials:

(3.34) $\quad \exp\{\tfrac{1}{2}\langle w,\beta\rangle\} = \left(\dfrac{1-|w|^2}{1-2w\cdot\beta^+|w|^2}\right)^{1/2} = (1-|w|^2)^{1/2} \displaystyle\sum |w|^k P_k(\theta\cdot\beta) ,$

where P_k is the k-th Legendre polynomial. It is well known that, for fixed β, $P_k(\theta\cdot\beta)$ is a spherical harmonic of order k; moreover,

(3.35) $\qquad \displaystyle\int_{S^2} P_k(\theta\cdot\beta) Y_n(\theta)\, d\theta = \begin{cases} 0 & \text{for } k \neq n , \\ Y_n(\beta) & \text{for } k = n . \end{cases}$

Now we express the integration in (3.33) by polar coordinates:

$$dw = dr A(r)\, d\theta ,$$

where r is the non-Euclidean distance from j, and $A(r)$ the non-Euclidean area of the sphere of radius r. Substituting (3.34) into (3.33) and using (3.35) we get

$$\tilde{f}(-\tfrac{1}{2}i, \beta) = \int r^n (1-r^2)^{1/2} g(r) A(r)\, dr\, Y_n(\beta) .$$

All we have to do is to choose g so that g has compact support and the above integral is not equal to 0. This proves part (iii) and so completes the proof of Theorem 3.4.

THEOREM 3.6. *Denote* $R_+ F = k_+$; *then*

(3.36) $\qquad f_1(w) = |c| \displaystyle\int \exp\{\langle w,\beta\rangle\} k_+(\langle w, \beta\rangle, \beta)\, d\beta .$

A similar formula can be given for the second component f_2 *of* F.

Proof: Substituting the definition of k_+ in (3.27) into the integral, we get two terms; the first,

$$-\frac{1}{2\pi^2} \int \exp\{\langle w, \beta \rangle\} \partial_s^2 e^s \hat{f}_1|_{s=\langle w, \beta \rangle} \, d\beta \, ,$$

is, according to (3.13), $f_1(w)$; thus to prove (3.36) we have to verify that the second term is zero; that is,

$$(3.37) \qquad\qquad \int \exp\{\langle w, \beta \rangle\} \partial_s e^s \hat{f}_2|_{s=\langle w, \beta \rangle} \, d\beta = 0 \, .$$

Now it follows from (3.16) that

$$p(s) \equiv \int e^s \hat{f}_2(s, \beta) \, d\beta$$

is an even function of s; hence $p'(s)$ is odd and therefore vanishes at $s = 0$:

$$(3.38) \qquad\qquad p'(0) = \int \partial_s e^s \hat{f}_2|_{s=0} \, d\beta = 0 \, .$$

Consequently, (3.37) holds for $w = j$.

One could argue that all of our concepts are invariant under the non-Euclidean motions and that, under a different parametrization, (3.38) implies (3.37) for arbitrary w. We shall make this argument explicit. For given w let τ denote the motion interchanging j and w:

$$(3.39) \qquad\qquad \tau = \tau^{-1}, \quad \tau w = j, \quad \tau j = w \, .$$

By (3.15)',

$$(3.40) \qquad\qquad \hat{f}_2^{\tau}(s, \beta) = \hat{f}_2(s - \langle \tau j, \beta \rangle, \tau^{-1}\beta) \, ;$$

by (1.13),

$$(3.40)' \qquad\qquad \langle w, \beta \rangle = -\langle w, \tau\beta \rangle \, ,$$

and by (1.18)

$$(3.40)'' \qquad\qquad d(\tau\beta) = \exp\{2\langle \tau^{-1}j, \beta \rangle\} \, d\beta \, .$$

Replacing f_2 in (3.38) by f_2^{τ} and making use of the relations (3.39)–(3.40)'', we get

$$0 = \int \partial_s e^s \hat{f}_2(s + \langle w, \tau\beta \rangle, \tau\beta)|_{s=0} \exp\{2\langle w, \tau\beta \rangle\} \, d(\tau\beta) \, .$$

Finally, making the substitutions

$$\tau\beta \to \gamma \quad \text{and} \quad s \to s - \langle w, \gamma \rangle,$$

we get (3.37) with β replaced by γ. This completes the proof of Theorem 3.6.

Let $U(t)$ be the solution operator of the non-Euclidean wave equation, F some element of $\mathcal{H}$ and $R_+F = k_+$. Then, according to (3.28), $R_+U(t)F = k_+(s-t, \beta)$. Setting this into the inversion formula (3.36) we obtain the following formula for the value of the solution of the wave equation $u(w, t)$ with initial data F:

$$(3.41)_+ \qquad u(w, t) = |c| \int \exp\{\langle w, \beta \rangle\} k_+(\langle w, \beta \rangle - t, \beta) \, d\beta.$$

Similarly, we have

$$(3.41)_- \qquad u(w, t) = |c| \int \exp\{\langle w, \beta \rangle\} k_-(\langle w, \beta \rangle + t, \beta) \, d\beta.$$

We show now that, conversely, if k_+ is any C^∞ function, $(3.41)_+$ defines a C^∞ solution of the wave equation (3.22). To see this let us abbreviate by q the function

$$(3.42) \qquad q(w) = \exp\{\langle w, \beta \rangle\}.$$

According to (3.4)′, $q^{1-i\lambda}$ satisfies

$$Lq^{1-i\lambda} = -\lambda^2 q^{1-i\lambda},$$

which implies that

$$(3.42)' \qquad Lq = 0 \quad \text{and} \quad L(q \log q) = 0.$$

It is easy to deduce from these two relations that

$$(3.42)'' \qquad q_x^2 + q_y^2 + q_z^2 = q^2/z^2.$$

Using these relations we can verify that u satisfies the wave equation by directly differentiating $(3.41)_+$. The same holds for $(3.41)_-$.

Note that if $k(s, \beta)$ is a distribution in $\mathbb{R} \times B$, then $(3.41)_+$ defines u as a distribution that satifies the wave equation. The same holds for $(3.41)_-$. The analogous representation for solutions of the Euclidean wave equation was given in (2.20). Just as in the Euclidean case, Huygens' principle is an easy consequence of $(3.41)_+$:

THEOREM 3.7. *Let F be initial data in $\mathscr{H}$ whose components are zero at all w whose distance from j exceeds a. Then the solution u of the wave equation (3.17) with initial data F is zero at $w = j$ for $|t| > a$.*

Proof: All points of the horosphere $\langle w, \beta \rangle = s$ have distance at least $|s|$ from j. Therefore, if $F = 0$ at points whose distance from j exceeds a, it follows from the definition (3.11) of the Radon transform that both $\hat{f}_1$ and $\hat{f}_2$ are equal to 0 for $|s| > a$; therefore, so is $R_+ f = k_+$. The solution u of the wave equation with initial value F is given by $(3.41)_+$; setting $w = j$, we get

$$u(j, t) = |c| \int k_+(-t, \beta) \, d\beta;$$

clearly the integrand equals 0 for $|t| > a$.

We now draw some further consequences of formula $(3.41)_+$ which are analogues of Theorem 2.4; again it is convenient to work with model (ii).

THEOREM 3.8. *Let k_+ be any distribution on $\mathbb{R} \times S^2$, and define u by $(3.41)_+$.*

(i) $u(w, t) = 0$ for all w, t if and only if the n-th spherical harmonic components of k_+ are of the form

$$(3.43) \qquad k_n(s) = \int k_+ Y_n(\beta) \, d\beta = \sum k_{m,n} e^{-ms}, \qquad 0 < m \leqq n.$$

(ii) $u(w, t) = 0$ in the forward cone: dist. $(w, j) < t$ if and only if (3.43) is satisfied for $s < 0$.

(iii) $u(w, t) = 0$ in the backward cone: dist. $(w, j) < -t$ if and only if (3.43) is satisfied for $s > 0$.

(iv) $u(w, t) = 0$ for dist. $(w, j) < a - |t|$ if and only if (3.43) is satisfied for $|s| < a$.

Proof: (i) Let ϕ be any C_0^∞ function; we form

$$(3.44) \qquad v(w) = \int u(w, t) \phi(t) \, dt.$$

Substituting for u as defined in (3.41) we get, after the change of the variable $s = \langle w, \beta \rangle - t$ in the integration, that

$$(3.45) \qquad \begin{aligned} v(w) &= \int \int \exp\{\langle w, \beta \rangle\} k(\langle w, \beta \rangle - t, \beta) \phi(t) \, d\beta \, dt \\ &= \int \exp\{\langle w, \beta \rangle\} k(s, \beta) \phi(\langle w, \beta \rangle - s) \, ds \, d\beta. \end{aligned}$$

This formula shows that $v(w)$ is a C^∞ function. If $u(w, t) \equiv 0$, then $v(w) \equiv 0$, and in particular all its partial derivatives with respect to w at j are zero. We compute now these derivatives by differentiating (3.45) and setting $w = j$. Let $N = (N_1, \cdots, N_n)$ be a multi-index. Using the abbreviation $\partial_w^N = \Pi \, \partial_{w_i}^N$, we form

$$(3.46) \qquad \partial_w^N v|_{w=j} = \int \partial_w^N [\exp\{\langle w, \beta \rangle\} \phi(\langle w, \beta \rangle - s)]_{w=j} k \, ds \, d\beta .$$

Clearly,

$$(3.47) \qquad \partial_w^N [\exp\{\langle w, \beta \rangle\} \phi(\langle w, \beta \rangle - s)]_{w=j} = Q_N(\partial_s, \beta) \phi ,$$

where Q_N is a differential operator of order $|N|$ whose coefficients depend on β.

The leading coefficient of Q_N is easily computed; it is $2^{|N|} \beta^N$, where we have used the fact that $\partial \langle w, \beta \rangle / \partial w_l|_{w=j} = 2\beta_l$, a fact easily deduced from formula (1.16) for $\langle w, \beta \rangle$. So we set

$$(3.48) \qquad Q_N(s, \beta) = \beta^N P_N .$$

We claim that all the coefficients of P_N are independent of β. To see this we set (3.48) into (3.47) and (3.47) into (3.46); we get

$$(3.49) \qquad \partial_w^N v(w)|_{w=j} = \int k(s, \beta) \beta^N P_N(\partial_s, \beta) \phi \, ds \, d\beta .$$

We take now any rotation around j, i.e., a motion for which $\rho j = j$. According to (1.13), with ρ for τ and $\rho^{-1}\beta$ for β,

$$(3.50) \qquad \langle \rho w, \beta \rangle = \langle w, \rho^{-1}\beta \rangle ,$$

since $\langle j, \beta \rangle = 0$. We deduce from (3.45) with the aid of (3.50) and introducing $\rho^{-1}\beta = \gamma$ that

$$v(\rho w) = \int \exp\{\langle \rho w, \beta \rangle\} \phi(\langle \rho w, \beta \rangle - s) k(s, \beta) \, ds \, d\beta$$

$$= \int \exp\{\langle w, \gamma \rangle\} \phi(\langle w, \gamma \rangle - s) k(s, \rho\gamma) \, ds \, d\gamma .$$

Analogously to (3.49) we get after differentiation

$$(3.51) \qquad \partial_w^N v(\rho w)|_{w=j} = \int k(s, \rho\gamma) \gamma^N P_N(\partial_s, \gamma) \, ds \, d\gamma .$$

Next we compute $\partial_w^N v(\rho w)$ by using the chain rule. We note that in model (ii), with $j = 0$, non-Euclidean rotations ρ are Euclidean rotations:

$$(\rho w)_k = \sum \rho_{kl} w_l .$$

Consequently,

$$\partial_{w_l} v(\rho w)|_{w=j} = \sum \rho_{kl} \frac{\partial}{\partial w_k} v(w)\Big|_{w=j}$$

and

$$\partial_w^N v(\rho w)|_{w=j} = \prod_l \left(\sum_k \rho_{kl} \frac{\partial}{\partial w_k} \right)^{N_l} v(w)\Big|_{w=j} .$$

Using formula (3.49) for derivatives of order $|N|$ of $v(w)$ at $w = j$ we get

$$\partial_w^N v(\rho w)|_{w=j} = \int k(s, \beta)(\rho^T \beta)^N P_N(\partial_s, \beta)\phi \, ds \, d\beta ,$$

where ρ^T is the transpose of ρ. Since ρ is a rotation, $\rho^T = \rho^{-1}$, and hence the above can be rewritten, with $\rho^{-1}\beta = \gamma$ as a new integration variable, in the form

$$\int k(s, \rho\gamma)\gamma^N P_N(\partial_s, \rho\gamma)\phi \, ds \, d\gamma .$$

Comparing this with (3.51) we conclude since k is arbitrary that

$$P_N(\partial_s, \rho\gamma) = P_N(\partial_s, \gamma) ,$$

i.e., that P_N has constant coefficients. We make use now of the fact that $v(w) \equiv 0$ to conclude from (3.49) that

$$\int k(s, \beta) Y_n(\beta) P_n(\partial_s)\phi \, d\beta \, ds = 0$$

for any spherical harmonic Y_n of order n, since $Y_n(\beta)$ is a sum of terms of the form β^N for $|N| = n$. In the notation of (3.43) we can, after carrying out the β integration, write this as

$$\int k_n(s) P_n(\partial_s)\phi(-s) \, ds = 0$$

for all ϕ in C_0^∞. Denoting the adjoint of P by P^*, it follows that $P_n^*(\partial_s)k_n = 0$. All solutions of this ordinary differential equation are of the form

$$(3.52) \qquad\qquad k_n(s) = \sum_{m=1}^n k_{m,n} \exp\{\lambda_m s\} ,$$

where $P_n^*(\lambda_n) = 0$; since P_n is of order n, there are n roots $\lambda_1, \cdots, \lambda_n$. We claim that $\lambda_m = -m$; to see this we substitute

$$(3.53) \qquad\qquad k_+(s, \beta) = e^{-ms} Y_n(\beta), \qquad\qquad m = 1, \cdots, n,$$

into $(3.41)_+$:

$$u(w, t) = \int \exp\{\langle w, \beta \rangle\} \exp\{(-\langle w, \beta \rangle + t)m\} Y_n(\beta) \, d\beta$$

$$= e^{tm} \int \exp\{\langle w, \beta \rangle(1 - m)\} Y_n(\beta) \, d\beta$$

$$= e^{tm} \int \left(\frac{1 - 2w \cdot \beta + |w|^2}{1 - |w|^2} \right)^{m-1} Y_n(\beta) \, d\beta,$$

where in the last step we have used formula (1.16) for $\langle w, \beta \rangle$. Clearly for $m = 1, \cdots, n$ the first factor is a polynomial in β of degree less than n, and thus is orthogonal to Y_n. This proves that, for k of the form (3.53), $u(w, t) \equiv 0$. Since we have shown that all k for which this is true are of the form (3.52), it follows that the values $-1, \cdots, -n$ belong to the set $\{\lambda_m\}$. Since this set consists of at most n numbers, it follows that $\lambda_m = -m$, $m = 1, \cdots, n$, as asserted in part (i) of Theorem 3.8.

The other parts of Theorem 3.8 are proved in the same way. To show the sufficiency of condition (3.43) we note that the value of $u(w, t)$ as given by $(3.41)_+$ for (w, t) in the forward cone only depends on values of $k_+(s, \beta)$ on $\mathbb{R}_-$, while the values of u in the backward cone only depend on the values of k_+ on $\mathbb{R}_+$; the value of u for dist. $(w, j) < a - |t|$ only depends on values of k_+ in $(-a, a)$.

To show the necessity of condition (3.43) we argue as before, except that in case (ii) we restrict the support of ϕ to $\mathbb{R}_+$, in case (iii) to $\mathbb{R}_-$, and in case (iv) to $(-a, a)$. This completes the proof.

We denote as before by $U(t)$ the solution operator of the wave equation (3.22). We define the *outgoing* and *incoming* subspace $\mathcal{D}_+$ and $\mathcal{D}_-$ of $\mathcal{H}$ as follows:

$\mathcal{D}_+$ consists of all data F in $\mathcal{H}$ for which $U(t)F$ vanishes in the forward cone:

$$\text{dist. } (w, j) < t.$$

$\mathcal{D}_-$ consists of all data F in $\mathcal{H}$ for which $U(t)F$ vanishes in the backward cone:

$$\text{dist. } (w, j) < -t.$$

$\mathscr{C}_a$ consists of all data in $\mathscr{H}$ that are equal to 0 inside the ball:

$$\text{dist.}\,(w, j) < a .$$

THEOREM 3.9.   (i) $R_+\mathscr{D}_+ = L_2(\mathbb{R}_+)$, the space of L_2 functions $k_+(s, \beta) = 0$ for $s < 0$.

(ii) $R_+\mathscr{D}_-$ consists of all functions k_+ in L_2 that satisfy (3.43) for $s > 0$.

(iii) $R_+\mathscr{C}_a$ consists of all functions k_+ in $L_2(\mathbb{R})$ which satisfy condition (3.43) for $-a < s < a$.

Similar results hold for R_-:

(i)_

$$R_-\mathscr{D}_- = L_2(\mathbb{R}_+) .$$

(ii)_ $R_-\mathscr{D}_+$ consists of all functions k_+ in $L_2(\mathbb{R})$ which satisfy (3.43) for $s > 0$.

(iii)_ $R_-\mathscr{C}_a$ consists of all functions k_+ in $L_2(\mathbb{R})$ which satisfy condition (3.43) for $-a < s < a$.

Proof: We have shown that if F is in $\mathscr{H}$, $U(t)F$ is given by formula $(3.41)_+$, with $k_+ = R_+F$. Applying part (ii) of Theorem 3.8 we see that F is in $\mathscr{D}_+$ if and only if each spherical harmonic component k_n of $k_+ = R_+F$ is of spherical form (3.43) for $s < 0$. The exponentials appearing in (3.43) tend to ∞ as $s \to -\infty$ and hence they are not in $L_2(\mathbb{R})$. On the other hand, according to Theorem 3.4, (ii) and (iii), for F in $\mathscr{H}$, $R_+F = k_+$ belongs to L_2 and so does each harmonic component k_n. From this we conclude that each k_n, and thereby k_+ itself, is zero for $s < 0$. This proves $(i)_+$. Part $(ii)_+$ is proved in the same way, except that the exponentials in (3.43) belong to L_2 on $\mathbb{R}_+$ and so are not excluded. To prove part $(iii)_+$ we observed that, by a classical uniqueness theorem for the wave equation, if data F vanish when $\text{dist.}\,(w, j) < a$, then $U(t)F = 0$ when $\text{dist.}\,(w, j) < a - |t|$. We complete the proof by appealing to part (iv) of Theorem 3.8.

Parts $(i)_-$, $(ii)_-$, and $(iii)_-$ can be proved similarly, based on the analogue of Theorem 3.8 concerning solutions u of the wave equation of the form $(3.41)_-$. This completes the proof of Theorem 3.9, an analogue of Theorem 2.4.

We define now

(3.54) $\mathscr{D}_+^a = U(a)\mathscr{D}_+ , \qquad \mathscr{D}_-^a = U(-a)\mathscr{D}_- .$

Since R_+ and R_- are translation representations, see part (i) of Theorem 3.4, part $(t)_\pm$, $(ii)_\pm$ of Theorem 3.9 yield

COROLLARY 3.10.   (i)$_+^a$ $R_+\mathscr{D}_+^a$ consists of all L_2 functions $k_+(s, \beta)$ that are zero for $s < a$.

(ii)$_+^a$ $R_+\mathscr{D}_-^a$ consists of all L_2 functions k_+ that satisfy (3.43) for $s > -a$.
Similarly, (i)$_-^a$ $R_-\mathscr{D}_-^a$ consists of all L_2 functions k_- that are zero for $s < a$.

(ii)$_-^a$ $R_-\mathscr{D}_+^a$ consists of all L_2 functions k_- that satisfy (3.43) for $s > -a$.

LEMMA 3.11.   The closure of $\mathscr{D}_+^a + \mathscr{D}_-^a$ is $\mathscr{C}_a$.

Proof: It follows from the definition of $\mathscr{D}_+$ that every F in $U(a)\mathscr{D}_+$ is zero in the ball dist. $(w, j) < a$; this shows that $\mathscr{D}_+^a \subset \mathscr{C}_a$. One can see similarly that $\mathscr{D}_-^a \subset \mathscr{C}_a$; it then follows that $\mathscr{D}_+^a + \mathscr{D}_-^a \subset \mathscr{C}_a$.

To prove that $\mathscr{D}_+^a + \mathscr{D}_-^a$ is dense in $\mathscr{C}_a$, take any F in $\mathscr{C}_a$. For any positive integer N, denote by $F^{(N)}$ the projection of F obtained by removing from F all spherical harmonic components of order greater than N. It is clear that $F^{(N)}$ vanishes in the ball dist. $(w, j) < a$ so that $F^{(N)}$ also belongs to $\mathscr{C}_a$. We shall show that $F^{(N)}$ belongs to $\mathscr{D}_+^a + \mathscr{D}_-^a$. By Lemma 3.5, $k = R_+ F^{(N)}$ likewise has no spherical harmonic components of order exceeding N. So, by part (iii) of Theorem 3.9, k satisfies

$$(3.55) \qquad k(s, \beta) = \sum_{0 < m \leq n \leq N} k_{m,n} e^{-ms} Y_n(\beta) \quad \text{for} \quad -a < s < a.$$

We define k_+ and k_- as follows:

$$k_-(s, \beta) = \begin{cases} k(s, \beta) & \text{for} \quad s < a, \\ \sum k_{m,n} e^{-ms} Y_n(\beta) & \text{for} \quad a < s, \end{cases}$$

$$k_+(s, \beta) = \begin{cases} 0 & \text{for} \quad s < a, \\ k(s, \beta) - \sum k_{m,n} e^{-ms} Y_n(\beta) & \text{for} \quad a < s. \end{cases}$$

It follows from (3.55) and part (ii)$_-^a$ of Corollary 3.10 that $k_- \subset R_-\mathscr{D}_-^a$, and from (i)$_+^a$ that $k_+ \subset R_+\mathscr{D}_+^a$. Since, clearly $k_- + k_+ = k$ and $k = R_+ f^{(N)}$, we have accomplished the decomposition of $f^{(N)}$ into a sum of an element of $\mathscr{D}_-^a$ and $\mathscr{D}_+^a$. Since $|f - f^{(N)}|_E \to 0$ as $N \to \infty$, the lemma is proved.

THEOREM 3.12.   The following assertions are equivalent for data G in $\mathscr{H}$:
   (i) $G \perp \mathscr{C}_a$,
   (ii) $G \perp \mathscr{D}_+^a$ and $G \perp \mathscr{D}_-^a$,
   (iii) $Lg_1 = 0$ and $g_2 = 0$ outside the ball $B(j, a)$ of non-Euclidean radius a about j,
   (iv) $\partial_s^2 e^s \hat{g}_1 = 0 = \partial_s e^s \hat{g}_2$ for $s > a$.

Proof: Lemma 3.11 implies that the orthogonal complement of $\mathscr{C}_a$ is the intersection of the orthogonal complements of $\mathscr{D}_+^a$ and $\mathscr{D}_-^a$. Both are easily determined. Suppose $G \perp \mathscr{C}_a$, that is,

$$(G, F)_E = 0 \quad \text{for all} \quad F \quad \text{in} \quad \mathscr{C}_a.$$

Then, by definition (3.23), (3.24) of the energy inner product, we have

$$(3.56) \qquad\qquad (G, F)_E = -(Lg_1, f_1) + (g_2, f_2)$$

for all f_1, f_2 in C_0^∞. $G \perp \mathscr{C}_a$ implies that this is zero for all F in C_0^∞ which vanish in the ball $B(j, a)$. Since such F are dense in $\mathscr{C}_a$, we conclude that $G \perp \mathscr{C}_a$ if and only if (iii) holds. On the other hand, according to parts (i)$_+^a$ and (i)$_-^a$ of Corollary 3.10, G is orthogonal to $\mathscr{D}_+^a$ (respectively, $\mathscr{D}_-^a$) if and only if $R_+ G = 0$ $(R_- G = 0)$ for $s > a$. Recalling (3.27):

$$R_\pm G = c(\partial_s^2 e^s \hat{g}_1 \mp \partial_s e^s \hat{g}_2),$$

we see from Corollary 3.10, (i)$_\pm$ that G is orthogonal to both $\mathscr{D}_+^a$ and $\mathscr{D}_-^a$ if and only if assertion (iv) holds. This concludes the proof of Theorem 3.12.

If in particular we take $g_1 \equiv 0$ and set $g_2 = g$, we conclude from the above that the following theorem holds.

THEOREM 3.13. *Let g be in* $L_2(\Pi)$. *Then* $g = 0$ *outside the ball* $B(j, a)$ *if and only if* $\partial_s e^s \hat{g} = 0$ *for* $s > a$.

An analogous characterization can of course be given for $L_2(\Pi)$ functions that vanish outside any given ball. A weaker version of Theorem 3.13 can be found in Helgason [5], page 128.

Remark. This result is a non-Euclidean analogue of the Paley–Wiener Theorem 2.6, with these differences: necessary conditions for g to be supported in the ball $B(j, a)$ are for $\hat{g}$ to be equal to 0 for $|s| > a$ and for

$$(3.57) \qquad\qquad \int e^{-ms} \hat{g}(s, \beta) \, ds$$

to be polynomial in β of degree at most m. The first of these conditions follows from the observation that the horosphere $\langle w, \beta \rangle = s$ does not intersect the ball $B(j, a)$ when $|s| > a$. The second follows if we insert formula (1.16) for $\langle w, \beta \rangle$ into the definition (3.11) of $\hat{g}$. We get

$$\int e^{-ms}\hat{g}(s, \beta)\, ds = \int_{-\infty}^{\infty} \int_{\langle w, \beta \rangle = s} \exp\{-m\langle w, \beta \rangle\} g(w)\, dS\, ds$$

$$= \int \left(\frac{1 - 2w \cdot \beta + |w|^2}{1 - |w|^2} \right)^m g(w)\, dw,$$

which is indeed a polynomial in β of degree at most m. What is remarkable in Theorem 3.13 is that, in contrast to the Euclidean case, only a part of these necessary conditions already suffice to make g vanish outside the ball $B(j, a)$. In particular, $\hat{g}$ is not required to vanish for $s < -a$, nor is $\hat{g}$ required to satisfy (3.57); rather, these properties are consequences of the fact that $\partial_s e^s \hat{g}$ vanishes for $s > a$.

Since horospheres are limits of spheres one would conjecture

THEOREM 3.14. *Let g be in $L_2(\Pi)$; then $g = 0$ outside a horosphere Σ if and only if*

(3.58) $$\partial_s e^s \hat{g} = 0$$

on almost all horospheres that lie outside of Σ.

Remark. We were unable to deduce this from Theorem 3.13; an independent proof is given below.

Proof: Since the necessity argument is obvious from the definition of the Radon transform, we treat only the sufficiency.

It is convenient to deal with model (i) and to specialize Σ to a horosphere associated with β_∞, say $z = 2a$. In this proof we shall reparameterize the horospheres exterior to Σ by $\beta \in \mathbb{R}^2$ and the Euclidean radius r of the horosphere. In terms of the non-Euclidean distance s of the horosphere from $j = (0, 0, 1)$, it is easy to see from (1.15) that

(3.59) $$e^s = \frac{1 + |\beta|^2}{2r}.$$

Denoting the Radon transform of f in terms of the above parameters by $\hat{f}(r, \beta)$ and using spherical coordinates about the center of such a horosphere, we can write

$$\overset{\blacktriangle}{f}(r, \beta) = \int_0^{2\pi} \int_0^{\pi} f(\beta_1 + r \sin \theta \cos \phi, \beta_2 + r \sin \theta \sin \phi, r(1 - \cos \theta))$$

(3.60)

$$\times \frac{\sin \theta}{(1 - \cos \theta)^2} \, d\theta \, d\phi.$$

Making use of (3.59), condition (3.58) becomes

(3.58)' $$\partial_r \left(\frac{1}{r} f(r, \beta) \right) = 0 \quad \text{for} \quad r < a.$$

Our strategy is to reformulate this condition as a Volterra integral equation from which the conclusion of the theorem will follow.

Let us suppose to begin with that f belongs to $C_0^1(\Pi)$. We denote the partial Fourier transform in the x, y direction by $\bar{f}(\xi, z)$:

(3.61) $$\bar{f}(\xi, z) = \frac{1}{2\pi} \int_{\mathbf{R}^2} \exp\{-i(x\xi_1 + y\xi_2)\} f(x, y, z) \, dx \, dy.$$

We now derive an expression for the β-Fourier transform of $\overset{\blacktriangle}{f}$ in terms of $\bar{f}$:

$$\frac{1}{2\pi} \int_{\mathbf{R}^2} \exp\{-i\beta \cdot \xi\} \overset{\blacktriangle}{f}(r, \beta) \, d\beta$$

$$= \int_0^{2\pi} \int_0^{\pi} \exp\{iv(r, \theta, \phi) \cdot \xi\} \bar{f}(\xi, r(1 - \cos \theta)) \frac{\sin \theta}{(1 - \cos \theta)^2} \, d\theta \, d\phi,$$

where $v(r, \theta, \phi) = (r \sin \theta \cos \phi, r \sin \theta \sin \phi)$. Interchanging the order of integration and performing the ϕ-integration we get

$$\frac{1}{2\pi} \int \exp\{-i\beta \cdot \xi\} \overset{\blacktriangle}{f}(r, \beta) \, d\beta$$

(3.62)

$$= 2\pi \int_0^{\pi} J_0(r|\xi| \sin \theta) \bar{f}(\xi, r(1 - \cos \theta)) \frac{\sin \theta}{(1 - \cos \theta)^2} \, d\theta,$$

where

$$J_0(u) = \frac{1}{2\pi} \int_0^{2\pi} \exp\{iu \cos\phi\} \, d\phi$$

is the Bessel function of order zero. A change of the variable of integration from θ to $z = r(1 - \cos\theta)$ allows us to write (3.62) as

$$(3.62)' \qquad \frac{1}{(2\pi)^2} \int \exp\{-i\beta \cdot \xi\} \hat{f}(r, \beta) \, d\beta = r \int_0^{2r} J_0(|\xi|(2zr - z^2)^{1/2}) \tilde{f}(\xi, z) \frac{dz}{z^2}.$$

The corresponding expression for $r^{1/2} \partial_r (1/r) \hat{f}(r, \beta))$, obtained by differentiating (3.62)′, is then

$$(3.63)$$
$$\frac{1}{(2\pi)^2} \int \exp\{-i\beta \cdot \xi\} r^{1/2} \partial_r \left(\frac{1}{r} \hat{f}(r, \beta)\right) d\beta$$
$$= \frac{\tilde{f}(\xi, 2r)}{4r^{3/2}} - \int_0^{2r} J_1(|\xi|(2zr - z^2)^{1/2}) \frac{|\xi|(rz)^{1/2}}{(2zr - z^2)^{1/2}} \frac{\tilde{f}(\xi, z)}{z^{3/2}} \, dz,$$

where $J_1 = J_0'$ is the Bessel function of order one. We note that the kernel of this integral operator, that is

$$J_1(|\xi|(2zr - z^2)^{1/2}) \frac{|\xi|(rz)^{1/2}}{(2zr - z^2)^{1/2}},$$

is continuous on $0 \le z \le 2r < 2a$.

The familiar Fourier transform Plancherel relation gives

$$(3.64) \qquad \int_\Pi |f(w)|^2 \, dw = \int_0^\infty \int_{\mathbb{R}^2} \left|\frac{\tilde{f}(\xi, z)}{z^{3/2}}\right|^2 \, d\xi \, dz.$$

Using (1.17) and (3.59) the Radon transform Plancherel relation (3.14) can be rewritten in terms of the (r, β) variables as

$$(3.64)' \qquad \int_\Pi |f(w)|^2 \, dw = \frac{1}{8\pi^2} \int_{\mathbb{R}^2} \int_0^\infty \left|r^{1/2} \partial_r \left(\frac{1}{r} \hat{f}(r, \beta)\right)\right|^2 \, dr \, d\beta_1 \, d\beta_2.$$

Given f in $L_2(\Pi)$, by approximating f by $C_0^1(\Pi)$ functions, it follows that, for almost all ξ, (i) $\bar{f}(\xi, z)/z^{3/2}$ belongs to $L_2(\mathbb{R})$ and (ii) the relation (3.63) holds for almost all r. In particular, if f satisfies (3.58)' for $r < a$, then for almost all ξ the function $\bar{f}(\xi, \cdot) \in L_2(\mathbb{R})$ satisfies a homogeneous Volterra integral equation of the first kind with smooth kernel for $r < a$. Since such an integral equation has no non-trivial solution, we conclude that, for almost all ξ, $\bar{f}(\xi, z) = 0$ for almost all $z < 2a$ and it follows that $f(w) = 0$ for almost all w with $z < 2a$. This completes the proof of Theorem 3.14.

We note that Theorem 3.13 follows from Theorem 3.14. The analogous generalization of Theorem 3.12 is

THEOREM 3.15. *Suppose* $F = \{f_1, f_2\}$ *belongs to* $\mathcal{H}$. *Then*

$$(3.65) \qquad\qquad Lf_1 = 0 \quad and \quad f_2 = 0$$

outside the horosphere Σ *if and only if both* R_+F *and* R_-F *vanish in all horospheres lying outside of* Σ.

Proof: We see from (3.27) that both R_+F and R_-F vanish in all horospheres lying outside of Σ if and only if

$$(3.66) \qquad\qquad \partial_s^2 e^s \hat{f}_1 = 0 = \partial_s e^s \hat{f}_2$$

in all such horospheres. Hence the assertion of the theorem for the second component of F, that is for f_2, follows directly from the previous theorem.

Remark. Set $G = \{f_1, 0\}$. If G belongs to $D(A)$, then $AG = \{0, Lf\} \in \mathcal{H}$. According to (3.17),

$$\widehat{Lf_1} = e^{-s} \partial_s^2 \hat{f}_1$$

and hence as above the assertion of the theorem for the first component would follow from the previous theorem.

If f_1 is not smooth, we can achieve the same end by mollifying. Choose ϕ in $C_0^\infty(-1, 1)$ to be even with $\phi \geq 0$, $\int \phi = 1$ and set

$$\phi_\varepsilon(t) = \frac{1}{\varepsilon} \phi(t/\varepsilon), \qquad G_\varepsilon = \int \phi_\varepsilon(t) U(t) G \, dt.$$

Then G_ε belongs to $D(A)$, converges to G in $\mathcal{H}$ as $\varepsilon \to 0$ and

$$(3.67) \qquad R_\pm G_\varepsilon = \int \phi_\varepsilon(t) k(s \mp t, \beta) \, dt = \int \phi_\varepsilon(|s - t|) k(t, \beta) \, dt,$$

where by (3.27)

$$k = R_\pm G = c\partial_s^2 e^s \hat{f}_1.$$

Working with model (i) and specializing Σ to the horosphere $z = 2a$, we can be quite explicit. Assuming (3.66) and making use of (3.59), we see that $k(s, \beta)$ vanishes for all $s > \log[(1+|\beta|^2)/2a]$. It therefore follows from (3.67) that $R_\pm G_\varepsilon$ vanishes for all $s > \log[1+|\beta|^2)/2a] + \varepsilon$, in other words on all horospheres lying in the set $z < 2ae^{-\varepsilon}$. By the above remark we conclude that

$$Lg_{\varepsilon,1} = 0 \quad \text{for} \quad z < 2ae^{-\varepsilon}.$$

Taking the limit as $\varepsilon \to 0$, we see from Theorem 3.3 that $g_{\varepsilon,1} \to f_1$ locally in L_2 and it follows by elliptic theory that $Lf_1 = 0$ for $z < 2a$.

Conversely, suppose that $Lf_1 = 0$ for $z < 2a$. Then, by (3.56), $(G, H)_E = 0$ for all smooth H with compact support in $z < 2a$. Taking into account the unit speed of propagation we deduce similarly that for $|t| < \varepsilon$

(3.68) $$(U(t)G, H)_E = (G, U(-t)H)_E = 0$$

for all smooth H with compact support in $z < 2ae^{-\varepsilon}$. Multiplying (3.68) by $\phi_\varepsilon(t)$ and integrating with respect to t, we see that

$$0 = (G_\varepsilon, H)_E = -(g_{\varepsilon,1}, Lh_1) + (g_{\varepsilon,2}, h_2)$$

for all such H. It follows that $Lg_{\varepsilon,1} = 0$ for all $z < 2ae^{-\varepsilon}$ and, since G_ε belongs to $D(A)$, we conclude from the above remark that $\partial_s^2 e^s \hat{g}_{\varepsilon,1} = 0$ on all horospheres lying in the set $z < 2ae^{-\varepsilon}$. Since

$$c\partial_s^2 e^s \hat{g}_{\varepsilon,1} = R_\pm(g_{\varepsilon,1}, 0)$$

and since $\{g_{\varepsilon,1}, 0\}$ converges to G in the energy norm, we see that $R_\pm\{g_{\varepsilon,1}, 0\} \to R_\pm G = c\,\partial_s^2 e^s \hat{f}_1$ in the L_2 norm and hence that $\partial_s^2 e^s \hat{f}_1$ vanishes on all horospheres lying outside of Σ. This concludes the proof of Theorem 3.15.

4. The Scattering Operator

A scattering operator S for the group U is defined by means of two translations representations T_1 and T_2 as the mapping

(4.1) $$S : T_1 F \to T_2 F.$$

In the case of the non-Euclidean wave equation on Π the obvious candidates for T_1 and T_2 are R_+ and R_-; we shall in fact use R_+ and a slightly modified R_-.

We shall require of a translation representation T that it maps $\mathcal{H}$ unitarily onto $L_2(\mathbb{R}, N)$, where $N = L_2(B)$, and that

$$(4.2) \qquad TU(t)F = U_t TF,$$

where U_t denotes translation to the *right* by t units; that is, $[U_t k](s) = k(s - t)$. According to Theorem 3.4, R_+ is a translation representation of U in the above sense whereas R_- fails only because it translates to the left instead of the right. This defect is easily remedied by reflection; we set

$$(4.3) \qquad T_1 = R_+, \qquad T_2 = R_-^{\vee},$$

where $k^{\vee}(s) = k(-s)$. For $F \in \mathcal{H}$, we write

$$(4.4) \qquad k_i(s, \beta) = T_i F.$$

The corresponding *spectral representations* $\mathcal{T}_1$ and $\mathcal{T}_2$ are obtained by Fourier transformation Φ with respect to s from T_1 and T_2:

$$(4.5) \qquad \mathcal{T}_i = \Phi^{-1} T_i;$$

that is,

$$(4.5)' \qquad \tilde{k}_i(\sigma, \beta) = \mathcal{T}_i F = \frac{1}{\sqrt{2\pi}} \int e^{i\sigma s} k_i(s, \beta) \, ds.$$

In the spectral representation the action of $U(t)$ is multiplication by $e^{i\sigma t}$. The spectral representation of S is

$$(4.6) \qquad \mathcal{S} = \Phi^{-1} S \Phi,$$

called the *scattering matrix*.

It is easy to show (see [6], Chapter 2) that S has the properties
  (i) S is unitary on $L_2(\mathbb{R}, N)$,
  (ii) S commutes with translation.
The corresponding properties of $\mathcal{S}$ are:
  (i)' $\mathcal{S}$ is unitary in $L_2(\mathbb{R}, N)$,
  (ii)' $\mathcal{S}$ commutes with scalar multipliers.

It follows from this that $\mathscr{S}$ is a multiplicative operator, that is

$$(4.7) \qquad \tilde{k}_2(\sigma) \equiv \mathscr{S}(\sigma)\tilde{k}_1(\sigma),$$

and that $\mathscr{S}(\sigma)$ is unitary on N for each σ.

If in addition S has the *causal* property:

(iii) $SL_2(\mathbb{R}_-) \subset L_2(\mathbb{R}_-)$,

then obviously

(iii)' $\mathscr{S}\mathscr{H}_2 \subset \mathscr{H}_2$,

where $\mathscr{H}_2$ denotes the Hardy class $\Phi^{-1}L_2(\mathbb{R}_-)$. In this case it can be shown that $\mathscr{S}(\sigma)$ is the boundary value of an operator-valued function $\mathscr{S}(z)$ which is holmorphic in the lower half-plane, $\mathscr{I}m\, z < 0$, where $\mathscr{S}(z)$ is a contraction on N.

LEMMA 4.1.   *Denote by $\mathscr{D}'_\pm$ the orthogonal complement of $\mathscr{D}_\mp$. Then*

$$(4.8) \qquad \mathscr{D}'_\pm \subset \mathscr{D}_\pm$$

and

$$(4.9) \qquad \mathscr{D}'_- \perp \mathscr{D}'_+ .$$

Proof: According to Theorem 3.9,

$$R_+\mathscr{D}_+ = L_2(\mathbb{R}_+) \quad \text{and} \quad R_+\mathscr{D}_- \supset L_2(\mathbb{R}_-) .$$

It follows from the left relation that

$$(4.10) \qquad R_+\mathscr{D}'_- = L_2(\mathbb{R}_-)$$

and this together with the right relation shows that $\mathscr{D}'_- \subset \mathscr{D}_-$. The "plus" part of (4.8) is proved similarly. Relation (4.9) is now immediate since

$$\mathscr{D}'_- \subset \mathscr{D}_- \quad \text{and} \quad \mathscr{D}_- \perp \mathscr{D}^\perp_- = \mathscr{D}'_+ .$$

COROLLARY 4.2.   *S has property* (iii).

Proof: In view of (4.10), $L_2(\mathbb{R}_-) = T_1(\mathscr{D}'_-)$. Likewise, $R_-(\mathscr{D}'_+) = L_2(\mathbb{R}_-)$ so that

$$(4.10)' \qquad L_2(\mathbb{R}_+) = L_2(\mathbb{R}_-)^\vee = R_-(\mathscr{D}'_+)^\vee = T_2(\mathscr{D}'_+) .$$

By definition, S takes $L_2(\mathbb{R}_-) = T_1(\mathscr{D}'_-)$ into $T_2(\mathscr{D}'_-)$. According to (4.9), $T_2(\mathscr{D}'_-)$ is orthogonal to $T_2(\mathscr{D}'_+) = L_2(\mathbb{R}_+)$ and hence is contained in $L_2(\mathbb{R}_-)$ as desired.

LEMMA 4.3. *Set*

$$(4.11) \qquad \mathcal{K} = \mathcal{H} \ominus (\mathcal{D}'_- \oplus \mathcal{D}'_+) .$$

Then

$$(4.12) \qquad \mathcal{K} = \mathcal{D}_- \cap \mathcal{D}_+ .$$

Proof: It is evident that $(\mathcal{D}'_\mp)^\perp = \mathcal{D}_\pm$ and hence

$$(\mathcal{D}'_- \oplus \mathcal{D}'_+)^\perp \subset (\mathcal{D}'_\mp)^\perp = \mathcal{D}_\pm .$$

This proves that

$$(4.13) \qquad (\mathcal{D}'_- \oplus \mathcal{D}'_+)^\perp \subset \mathcal{D}_- \cap \mathcal{D}_+ .$$

On the other hand,

$$\mathcal{D}'_\mp = \mathcal{D}_\pm^\perp \subset (\mathcal{D}_- \cap \mathcal{D}_+)^\perp$$

and hence

$$\mathcal{D}'_- \oplus \mathcal{D}'_+ \subset (\mathcal{D}_- \cap \mathcal{D}_+)^\perp ,$$

or by duality

$$(4.13)' \qquad (\mathcal{D}'_- \oplus \mathcal{D}'_+)^\perp \supset \mathcal{D}_- \cap \mathcal{D}_+ .$$

The relations (4.13) and (4.13)' imply the assertion of the lemma.

It is proved in [6], page 53, that $\mathcal{S}$ is completely determined, up to right multiplication by a unitary operator on N, by the properties (i)'–(iii)' and the range of $\mathcal{S}$ on $\mathcal{H}_2$. This indeterminacy is in this case further restricted by the fact that $\mathcal{S}(\sigma)$ commutes with rotations about j. As in Lemma 3.5, this implies that $\mathcal{S}$ takes $\bar{k}(\sigma) Y_n(\beta)$ into a function of the same form. Moreover, since the rotations act transitively on the sperical harmonics of a given order,

$$(4.14) \qquad \text{the right multiplier is a scalar of absolute value one on each such subspace.}$$

The range of $\mathcal{S}$ in $\mathcal{H}_2$ is precisely $\mathcal{T}_2(\mathcal{D}'_-)$ and

$$\mathcal{H}_2 \ominus \mathcal{T}_2(\mathcal{D}'_-) = (L_2(\mathbb{R}) \ominus \mathcal{T}_2(\mathcal{D}'_+)) \ominus \mathcal{T}_2(\mathcal{D}'_-)$$
$$= \mathcal{T}_2(\mathcal{K}) = \mathcal{T}_2(\mathcal{D}_- \cap \mathcal{D}_+) ,$$

by Lemma 4.3. In terms of model (ii),

$$T_2(\mathcal{D}_- \cap \mathcal{D}_+) = R_-(\mathcal{D}_- \cap \mathcal{D}_+)^\vee$$

is, according to Theorem 3.9, the set of all functions $k(s, \beta)$ in L_2 whose n-th spherical harmonic components are of the form

(4.15)
$$\left(\sum_{m=1}^n c_m e_m(s) \right) Y_n(\beta),$$

where

(4.16)
$$e_m(s) = \begin{cases} 0 & \text{for } s > 0, \\ e^{ms} & \text{for } s < 0. \end{cases}$$

Note that

(4.16)'
$$\Phi^{-1} e_m(s) = \frac{-i}{\sqrt{2\pi}} \frac{1}{\sigma - im}.$$

Since $\mathcal{S}$ is an inner factor in the sense of Beurling, the obvious candidate for $\mathcal{S}$ is

(4.17)
$$\mathcal{S}_0(\sigma) = \sum_{n=0}^\infty \oplus B_n(\sigma) Q_n,$$

where

$$B_n(\sigma) = \varepsilon_n \prod_{m=1}^n \frac{\sigma + im}{\sigma + im}, \qquad |\varepsilon_n| = 1,$$

is the Blaschke product with the same poles as the $\Phi^{-1} e_m$, $m = 1, 2, \cdots, n$, and Q_n is the orthogonal projection of N onto the spherical harmonics of order n. We can express Q_n as an integral operator with kernel

$$\frac{2n+1}{4\pi} P_n(\beta \cdot \theta),$$

where P_n is the Legendre polynomial of order n and β, θ are unit vectors.

THEOREM 4.4. *The scattering matrix $\mathcal{S}$ mapping $\mathcal{T}_1 F \to \mathcal{T}_2 F$ is given by* (4.17).

Proof: Since $|B_n(\sigma)| \equiv 1$ for real σ and since the Q_n span N, it is clear that $\mathscr{S}_0(\sigma)$ is unitary on N for each real σ. $\mathscr{S}_0(\sigma)$ obviously commutes with scalar functions of σ; $B_n(\sigma)$, being holomorphic and bounded in the lower half-plane, maps the Hardy class into itself and the same is true of $\mathscr{S}_0$. Thus $\mathscr{S}_0$ satisfies properties (i)′–(iii)′. Moreover, $\mathscr{S}_0(\sigma)$ commutes with rotations and has the maximum amount of indeterminacy permitted by (4.14).

It remains only to show that the range of $\mathscr{S}_0$ and $\mathscr{H}_2$ is $\mathscr{T}_2(\mathscr{D}'_-)$. To this end we prove:

(I) $\mathscr{S}_0(\mathscr{H}_2) \subset \mathscr{T}_2(\mathscr{D}'_-)$. Here it suffices to show that $\mathscr{S}_0(\mathscr{H}_2) \perp \Phi^{-1} e_m Y_n$ for $0 < m \le n$. Now for arbitrary $\bar{k} \in \mathscr{H}_2$, we have by (4.16)′

$$(4.18) \qquad (\mathscr{S}_0 \bar{k}, \Phi^{-1} e_m Y_n) = \frac{1}{\sqrt{2\pi}} \int \prod_{p=1}^{n} \frac{\sigma + ip}{\sigma - ip} \bar{k}_n(\sigma) \frac{i}{\sigma + im} \cdot d\sigma,$$

where

$$\bar{k}_n(\sigma) = \int_{S^2} \bar{k}(\sigma, \beta) Y_n(\beta) \, d\beta$$

also belongs to the Hardy class. Since

$$\frac{1}{\sigma - im} \prod_{p \ne m} \frac{\sigma + ip}{\sigma - ip}$$

obviously belongs to the Hardy class, the usual contour integration argument shows that the integral in (4.18) vanishes, as desired.

(II) $\mathscr{S}_0 \mathscr{H}_2 \supset \mathscr{T}_2(\mathscr{D}'_-)$. Denoting the conjugate Hardy class by $\bar{\mathscr{H}}_2$, it follows from properties (i)′–(iii)′ that $\mathscr{S}_0 \mathscr{H}_2 \supset \bar{\mathscr{H}}_2$. Given $\bar{k} \in \mathscr{H}_2$ we now compute the $\mathscr{H}_2$ component of $\mathscr{S}_0 \bar{k}$; i.e., the projection of $\Phi \mathscr{S}_0 \bar{k}$ into $L_2(\mathbb{R}_-)$.

$$\Phi \mathscr{S}_0 \bar{k} = \sum_n \frac{1}{\sqrt{2\pi}} Y_n(\beta) \int e^{-i\sigma s} \prod_1^n \frac{\sigma + ip}{\sigma - ip} \bar{k}_n(\sigma) \, d\sigma.$$

Since $\bar{k}_n$ belongs to the conjugate Hardy class (and hence is holomorphic in the upper half-plane), the usual contour integration argument shows that, for $s < 0$, $\Phi \mathscr{S}_0 k$ is a sum of the form (4.15). Thus

$$\mathscr{S}_0(\bar{\mathscr{H}}_2) \subset \bar{\mathscr{H}}_2 \oplus \mathscr{T}_2(\mathscr{K})$$

and since $\mathscr{S}_0$ is unitary and $\mathscr{H}_2 = \bar{\mathscr{H}}_2^\perp$, we conclude that $\mathscr{S}_0(\mathscr{H}_2) \supset (\bar{\mathscr{H}}_2 \oplus \mathscr{T}_2(\mathscr{K}))^\perp = \mathscr{T}_2(\mathscr{D}'_-)$. Together (I) and (II) prove that $\mathscr{S}_0(\mathscr{H}_2) = \mathscr{T}_2(\mathscr{D}'_-)$. Consequently, for a suitable choice of constants $\{\varepsilon_n\}$, $\mathscr{S}_0 = \mathscr{S}$. This completes the proof of Theorem 4.4.

Bibliography

[1] Gel'fand, I. M., Graev, M. I., and Vilenkin Ya. N., *Generalized Functions*, Vol. 5, Academic Press, New York, 1966.

[2] Helgason, S., *The Radon transform on Euclidean spaces, compact two-point homogeneous spaces, and Grassman manifolds*, Acta Math. 113, 1965, pp. 153–180.

[3] Helgason, S., *Lie Groups and Symmetric Spaces*, Battelle Rencontres, Benjamin Pub. Co., New York, 1967.

[4] Helgason, S., *The surjectivity of invariant differential operators in symmetric spaces*, I, Ann. Math. 98, 1973, pp. 451–479.

[5] Helgason, S., *Harmonic analysis on homogeneous spaces*, Proc Symposia in Pure Math., 26, Amer. Math. Soc. 1973, pp. 101–146.

[6] Lax, P. D., and Phillips, R. S., *Scattering Theory*, Academic Press, New York, 1967.

[7] Lax, P. D., and Phillips, R. S., *The Paley–Wiener theorem for the Radon transform*, Comm. Pure Appl. Math. 23, 1970, pp. 409–424.

[8] Lax, P. D., and Phillips, R. S., *Scattering Theory for Automorphic Functions*, Annals of Math. Studies, No. 87, Princeton Univ. Press, 1976.

[9] Ludwig, D., *The Radon transform on Euclidean space*, Comm. Pure Appl. Math. 19, 1966, pp. 49–81.

[10] Pavlov, B. S., and Faddeev, L. D., *Scattering theory and automorphic functions*, Seminar of the Steklov Mathematical Institute of Leningrad, 27, 1972, pp. 161–193.

[11] Semenov-Tian-Shansky, M. A., *Harmonic Analysis on Riemannian symmetric spaces of negative curvature and scattering theory*, Math. U.S.S.R. Izv. 10, 1976, pp. 535–563.

Received October, 1978.

BULLETIN (New Series) OF THE
AMERICAN MATHEMATICAL SOCIETY
Volume 2, Number 2, March 1980

SCATTERING THEORY FOR AUTOMORPHIC FUNCTIONS

BY PETER D. LAX AND RALPH S. PHILLIPS[1]

ABSTRACT. This paper is an expository account of our 1976 monograph [6] on *Scattering theory for automorphic functions*. Several improvements have been incorporated: a more direct proof of the meromorphic character of the Eisenstein series, an explicit formula for the translation representations and a simpler derivation of the spectral representations. Our hyperbolic approach to the Selberg trace formula is also included.

1. Introduction. In 1972 Faddeev and Pavlov [2] discovered a revealing connection between the harmonic analysis of functions automorphic with respect to a discrete subgroup of SL(2, *R*) and the Lax-Phillips scattering theory as applied to the non-Euclidean wave equation. Their work is based on the spectral theory for the Laplace-Beltrami operator previously developed by Faddeev [1] using elliptic arguments. In our 1976 monograph [6] we redid the Faddeev-Pavlov paper entirely within the framework of our theory, basing our development on the non-Euclidean wave equation. We obtained new treatments for (i) the spectral theory of the Laplace-Beltrami operator over noncompact domains of finite area; (ii) the meromorphic character of the Eisenstein series over the whole complex plane; and (iii) a new form of the Selberg trace formula.

In this paper we sketch a revised version of our monograph including a more direct proof of the meromorphic character of the Eisenstein series, an explicit formula for the translation representations and a simpler derivation of the spectral representations.

The harmonic analysis of automorphic functions has been extensively studied; references to the pertinent parts of this theory are contained in our monograph. We recall that the Poincafe plane Π, that is the upper half plane

$$w = x + iy, \quad y > 0, \tag{1.1}$$

serves as a model for a non-Euclidean geometry in which the motions are given by the group *G* of fractional linear transformations:

$$w \to \frac{aw + b}{cw + d} \tag{1.2}$$

where *a*, *b*, *c*, *d* are real and

$$ad - bc = 1; \tag{1.3}$$

Received by the editors August 1, 1979.

AMS (MOS) subject classifications (1970). Primary

[1]The work of both authors was supported in part by the National Science Foundation, the first author under Grant No. MCS-76-07039 and the second under Grant No. MCS-77-04908 A 01.

261

G is isomorphic with $SL(2, R)/ \pm I$. The Riemannian metric

$$\frac{dx^2 + dy^2}{y^2} \qquad (1.4)$$

is invariant under this group of motions. The invariant L_2 form is

$$\iint u^2 \frac{dx\, dy}{y^2}. \qquad (1.5)$$

The invariant Dirichlet form is

$$\iint \left(u_x^2 + u_y^2\right) dx\, dy. \qquad (1.6)$$

The corresponding Laplace-Beltrami operator

$$L_0 = y^2 \Delta = y^2 \left(\partial_x^2 + \partial_y^2\right) \qquad (1.7)$$

is then clearly invariant. It turns out that the operator L defined as

$$L = L_0 + 1/4, \qquad (1.7)'$$

also invariant, has more useful analytic properties, as will be seen in what follows.

A subgroup Γ of G is called *discrete* if the identity is not a limit point of Γ. A *fundamental domain* F for a discrete subgroup Γ is a subdomain of Π such that every point of Π can be carried into a point of $\bar{F}$ by a transformation in Γ and no point of F is carried into another point of F by such a transformation. A function f defined on Π is called *automorphic* with respect to Γ if

$$f(\gamma w) = f(w) \qquad (1.8)$$

for all γ in Γ. Because of (1.8) an automorphic function is completely determined by its values on $\bar{F}$. If f is continuous then (1.8) imposes a relation between values of f at those pairs of boundary points of F which can be mapped into each other by some γ in Γ. If f is C^1, then (1.8) imposes a similar relation on the first derivatives of f at such pairs of boundary points.

The Laplace-Beltrami operator, being invariant, maps automorphic functions into automorphic functions. Alternatively we can consider automorphic functions as being defined on F: In this case we introduce the space $L_2(F)$ of functions on F square integrable with respect to the invariant measure and define L as the selfadjoint extension of the differential operator defined by $(1.7)'$, subject to the above mentioned boundary conditions.

We shall study the spectral properties of L by means of the non-Euclidean wave equation

$$u_{tt} = y^2 \Delta u + u/4 = Lu. \qquad (1.9)$$

This turns out to be a convenient analytic tool. As Semenov-Tian-Shansky [11] has recently shown, this equation also has an intrinsic meaning in the Lie algebra framework for problems of this kind. Indeed many of the classical concepts previously introduced in the study of these problems (see Kubota [4]) appear in a natural way in the scattering theory setting for the non-Euclidean wave equation. In particular the translation representations for the wave equation are closely related to the Radon transform associated with symmetric spaces as well as the theta series employed by Kubota in the study of Eisenstein series.

Perhaps the simplest discrete subgroup of G is the modular group, consisting of the fractional linear transformations with a, b, c, d integers satisfying (1.3). A convenient fundamental domain for this subgroup is the geodesic triangle

$$F: -1/2 < x < 1/2, \quad y > \sqrt{1 - x^2} \ . \tag{1.10}$$

For the sake of simplicity, most of the detailed discussion in this paper will be carried out for this spectral subgroup.

We conclude this introduction with a brief description of our abstract scattering theory and its application to the classical wave equation in Euclidean space. For details we refer to our book [5]. The theory concerns a one-parameter group of unitary operators $U(t)$ acting on a Hilbert space $\mathcal{H}$ for which there exist certain subspaces $\mathcal{D}_-$ and $\mathcal{D}_+$, called *incoming* and *outgoing* with the following properties:

(i) $U(t)\mathcal{D}_- \subset \mathcal{D}_-$ for $t < 0$ and $U(t)\mathcal{D}_+ \subset \mathcal{D}_+$ for $t > 0$,

(ii) $\bigcap U(t)\mathcal{D}_- = \{0\} = \bigcap U(t)\mathcal{D}_+$,

(iii) $\overline{\bigcup U(t)\mathcal{D}_-} = \mathcal{H} = \overline{\bigcup U(t)\mathcal{D}_+}$. $\tag{1.11}$

An example of an outgoing subspace is given by the following: $\mathcal{H} = L_2(\mathbf{R}, \mathcal{N})$, $\mathcal{N}$ some auxiliary Hilbert space, $U(t)$ translation to the right by an amount t, and $\mathcal{D}_+ = L_2(\mathbf{R}_+, \mathcal{N})$. It turns out that this class of examples exhausts all possibilities since every Hilbert space $\mathcal{H}$ with a unitary group for which there is an outgoing subspace can be represented as above, essentially uniquely; this is called the *outgoing translation representation* of $\mathcal{H}$. Corresponding to an incoming subspace there is an incoming translation representation in which $\mathcal{D}_-$ maps onto $L_2(\mathbf{R}_-, \mathcal{N})$.

In the applications of the theory we are about to describe, t represents time and the group $U(t)$ describes the propagation of waves. The significance of incoming and outgoing subspaces for wave propagation is that they furnish an asymptotic description of signals in the remote past and distant future respectively. The *scattering operator* S relates the incoming translation representer of a given data to its outgoing representer. Thus S gives a description of the scattering process, i.e. it relates directly the asymptotic behavior of waves for large negative and positive times.

S maps $L_2(\mathbf{R}, \mathcal{N})$ onto itself; since it relates two translation representations it is unitary and commutes with translation. Therefore S is convolution with an operator valued distribution ($\mathcal{N} \to \mathcal{N}$) called the *scattering function* and denoted as $S(t)$ (cf. [7]).

This approach is particularly fruitful when $\mathcal{D}_-$ and $\mathcal{D}_+$ are orthogonal to each other; in this case S is a so-called *causal operator*, in the sense that the scattering function $S(t) = 0$ for $t > 0$. In this case the Fourier transform of $S(t)$, called the *scattering matrix* and denoted as $\mathcal{S}(z)$, is analytic in the lower half plane. It turns out that many important properties of the scattering process can be related to the behavior of the meromorphic continuation of $S(z)$ into the upper half plane.

This connection can be studied most directly through the one-parameter semigroup of operators $Z(t)$, carved out of the group $U(t)$ as follows:

$$Z(t) = P_+ U(t) P_-, \qquad t > 0, \tag{1.12}$$

where P_+ and P_- are the orthogonal projections that remove the components of data that belong to $\mathcal{D}_+$ and $\mathcal{D}_-$, respectively. It is easy to show that $Z(t)$, $t > 0$, annihilates $\mathcal{D}_+$ and $\mathcal{D}_-$ and acts as a semigroup of operators on $\mathcal{K} = \mathcal{H} \ominus (\mathcal{D}_- \oplus \mathcal{D}_+)$. The resolvent of the infinitesimal generator of Z turns out to be intimately related to the scattering matrix.

Before returning to the automorphic case we would like to indicate how these ideas are used to study wave propagation in Euclidean space in the presence of an obstacle Θ. In the exterior $\mathcal{E}$ of the obstacle the signal satisfies the Euclidean wave equation:

$$u_{tt} = \Delta u. \tag{1.13}$$

The interaction of the wave with the obstacle is described as a boundary condition on u; if the interaction conserves the energy contained in the wave, the boundary condition makes Δ a selfadjoint operator on $L_2(\mathcal{E})$. Typical of such boundary conditions are $u = 0$ (or $\partial_n u = 0$) on $\partial\Theta$.

The underlying Hilbert space is the space of initial data $\{u, u_t\}$ defined on $\mathcal{E}$ and normed by the energy norm:

$$E_{\mathcal{E}}(u) = -(u, \Delta u)_{\mathcal{E}} + (u_t, u_t)_{\mathcal{E}}, \tag{1.14}$$

where $(\ ,\ )_{\mathcal{E}}$ denotes the L_2 inner product over $\mathcal{E}$. Since Δ is selfadjoint, $E_{\mathcal{E}}(u)$ is independent of t for solutions u of (1.14) satisfying the boundary conditions. This shows that the operator $U(t)$ relating initial data at time 0 to data at time t is an isometry; since the initial-boundary value problem is reversible in time and can be solved for a dense set of data, the operators $U(t)$ are actually unitary and form a group.

We call a solution u *outgoing* if $u(x, t) = 0$ at all points x whose distance from the boundary is less than t; an incoming solution is described similarly with t replaced by $-t$. $\mathcal{D}_-$ and $\mathcal{D}_+$ are defined as the initial data of incoming and outgoing solutions, respectively. Properties (i) and (ii) can be immediately verified; property (iii) lies considerably deeper and says essentially that all incoming signals eventually become outgoing signals after a complicated process of reflection from the boundary.

Roughly speaking, the same picture can be painted of automorphic waves over noncompact fundamental domains. Signals coming in from or going out to infinity can be defined analogously. Again one of the main results to be proved is that all incoming signals are eventually turned into outgoing signals. The analysis is complicated by the possible presence of standing waves due to a rather rich point spectrum, and of exponentially increasing solutions due to the indefiniteness of the energy form. A treatment of the corresponding free space problem can be found in [8].

2. The non-Euclidean wave equation. The first step in our analysis consists in finding a suitable Hilbert space setting for the initial value problem associated with the non-Euclidean wave equation:

$$u_{tt} = Lu,$$
$$u(x, 0) = f_1(x), \qquad u_t(x, 0) = f_2(x). \tag{2.1}$$

For a more complete treatment we refer the reader to [6, § 5].

In terms of the $L_2(F)$ inner product $(\, , \,)_F$, the *energy form* for the wave equation is

$$E_F(u) = - (u, Lu)_F + (u_t, u_t)_F. \tag{2.2}$$

Since L is selfadjoint in $L_2(F)$, $E_F(u)$ is independent of t for automorphic solutions of (2.1). In general E_F is indefinite for automorphic data; this can be seen if we bring (2.2) into a more symmetric form by an integration by parts:

$$E_F(u) = \iint_F \left\{ |u_x|^2 + |u_y|^2 - \frac{|u|^2}{4y^2} + \frac{|u_t|^2}{y^2} \right\} dx\, dy. \tag{2.2}'$$

The indefiniteness introduces complications analogous to those produced by bound states in the theory of scattering for quantum mechanics (see [5, Chapter 6]).

It is useful at this point to write the energy form in a way which depends on the shape of the fundamental domain F. In the case of the modular group with fundamental domain described in (1.10) we split F up into two parts:

$$F_0 = F \cap \{y < a\}, \tag{2.3}_0$$

which is compact, and a neighborhood of infinity

$$F_1 = F \cap \{Y > a\}; \tag{2.3}_1$$

here a is restricted to *be* > 1. Writing (2.2)' as the sum of integrals over F_0 and F_1 an integrating by parts over F_1, the energy form becomes

$$
\begin{aligned}
E_F(u) = &\iint_{F_0} \left\{ |u_x|^2 + |u_y|^2 - \frac{|u|^2}{4y^2} + \frac{|u_t|^2}{y^2} \right\} dx\, dy \\
&+ \iint_{F_1} \left\{ |u_x|^2 + y \left| \partial_y \frac{u}{\sqrt{y}} \right|^2 + \frac{|u_t|^2}{y^2} \right\} dx\, dy \\
&- \frac{1}{2a} \int_{-1/2}^{1/2} |u(x, a)|^2\, dx.
\end{aligned}
\tag{2.2}''
$$

From now on we omit the subscript F except where it is needed for clarity.

Next we introduce a new form which is very close to (2.2)'' but which has the advantage of being positive definite:

$$G(u) = E(u) + 2K(u), \tag{2.4}$$

where

$$K(u) = \iint_{F_0} \frac{|u|^2}{y^2}\, dx\, dy. \tag{2.5}$$

Notice that for data with support in F_1 the E and G forms are the same. It can be shown by simple estimates that G is positive definite and equivalent with

$$G'(u) = \iint_{F_0} \left\{ |u_x|^2 + |u_y|^2 + \frac{|u|^2}{y^2} \right\} dx\, dy$$

$$+ \iint_{F_1} \left\{ |u_x|^2 + y \left| \partial_y \frac{u}{\sqrt{y}} \right|^2 \right\} dx\, dy + \iint_{F} \frac{|u_t|^2}{y^2} dx\, dy. \quad (2.6)$$

By Rellich's theorem, K is compact with respect to G'.

We denote by $\mathcal{H}_G$ the completion in the G-norm of the space of C^∞ automorphic data with compact support in F. Since the K-form is compact with respect to the G-form it follows that the E and G-forms are equivalent on any closed subspace of $\mathcal{H}_G$ on which E is positive.

In order to treat u and u_t on an equal footing we rewrite the wave equation in component form as

$$u_t = v, \qquad v_t = Lu; \quad (2.7)$$

or in matrix notation as

$$V_t = AV \quad (2.7)'$$

where

$$V = \begin{pmatrix} u \\ v \end{pmatrix} \quad \text{and} \quad A = \begin{pmatrix} 0 & I \\ L & 0 \end{pmatrix}. \quad (2.7)''$$

We take as $D(A)$, the domain of A, the set of data (u, v) for which both (u, v) and (v, Lu) belongs to $\mathcal{H}_G$, Lu being defined in the weak sense. It can be shown that A generates on $\mathcal{H}_G$ a group of bounded operators $U(t)$ which grow exponentially in the G-norm but are unitary with respect to the indefinite energy form E. Since (2.1) is hyperbolic, signals carried by solutions of (2.1) propagate with non-Euclidean speed ≤ 1.

Of particular interest to us are certain subspaces of $\mathcal{H}_G$ which we call incoming and outgoing spaces and denote by $\mathcal{D}_-$ and $\mathcal{D}_+$, respectively. For the modular group and F of the form (1.10) these subspaces are the initial data for solutions of (2.1) with support in F_1 and which are independent of x for $t < 0$ ($t > 0$) in the case of $\mathcal{D}_-$ ($\mathcal{D}_+$); see §6 of [6]. A solution of (2.1) which is independent of x satisfies the differential equation

$$u_{tt} = y^2 \Delta u + u/4. \quad (2.8)$$

The change of variables

$$s = \log y, \qquad v = u/\sqrt{y}, \quad (2.9)$$

transforms (2.8) into the classsical wave equation:

$$v_{tt} = v_{ss}, \quad (2.10)$$

whose general solution is

$$v = l(s + t) + r(s - t). \quad (2.11)$$

The first term on the right corresponds to a wave traveling to the left, the second term to a wave traveling to the right. The corresponding incoming and outgoing solutions of (2.1) are of the form

$$u(w, t) = y^{1/2}\varphi(ye^t), \tag{2.12}_-$$

$$u(w, t) = y^{1/2}\varphi(ye^{-t}); \tag{2.12}_+$$

here φ is chosen to be C^∞ and vanishing for $y < a$.

We define the incoming and outgoing subspaces $\mathcal{D}_\pm$ as the closure in $\mathcal{H}_G$ of the initial data of the above respective incoming and outgoing solutions:

$$\mathcal{D}_- = \text{closure}\{y^{1/2}\varphi(y), y^{3/2}\varphi'(y)\},$$

$$\mathcal{D}_+ = \text{closure}\{y^{1/2}\varphi(y), -y^{3/2}\varphi'(y)\}, \tag{2.13}$$

φ in C^∞, zero for $y < a$, $\varphi' = \partial_y\varphi$.

It is clear from this that $\mathcal{D}_\pm$ have the properties, see (1.11):

$$\text{(i)} \quad U(t)\mathcal{D}_\pm \subset \mathcal{D}_\pm \quad \text{for } t \gtrless 0,$$

$$\text{(ii)} \quad \bigcap U(t)\mathcal{D}_- = \{0\} = \bigcap U(t)\mathcal{D}_+. \tag{2.14}$$

The change of variables (2.9) maps the energy form E defined by (2.2)' into the classical energy form $\int\{|v_s|^2 + |v_t|^2\} \, ds$ for (2.10). In these variables a trivial calculation shows that

$$\mathcal{D}_- \text{ and } \mathcal{D}_+ \text{ are } E\text{-orthogonal.} \tag{2.14}''$$

Since data in $\mathcal{D}_-$ and $\mathcal{D}_+$ are zero in F_0, it follows that $\mathcal{D}_-$ and $\mathcal{D}_+$ are also G-orthogonal.

We denote the zero Fourier coefficient of f with respect to x by $f^{(0)}$:

$$f^{(0)}(y) = \int_{-1/2}^{1/2} f(x, y) \, dx, \quad y > a. \tag{2.15}$$

Suppose the data $f = (f_1, f_2)$ in $\mathcal{H}_G$ is E-orthogonal to $\mathcal{D}_-$; it is easy to show by using the definition (2.13)$_-$ of $\mathcal{D}_-$ and the E-scalar product derived from (2.2)'' that the components of $f^{(0)}$ satisfy

$$f_2^{(0)} = -y^{3/2}\partial_y\left(\frac{f_1^{(0)}}{\sqrt{y}}\right) \quad \text{for } y > a. \tag{2.16}_-$$

Similarly if it is E-orthogonal to $\mathcal{D}_+$ then

$$f_2^{(0)} = y^{3/2}\partial_y\left(\frac{f_1^{(0)}}{\sqrt{y}}\right) \quad \text{for } y > a. \tag{2.16}_+$$

If f in $\mathcal{H}_G$ is orthogonal to both $\mathcal{D}_-$ and $\mathcal{D}_+$ then (2.16) implies that the zero Fourier coefficient of f is of the form

$$f^{(0)} = (c\sqrt{y}, 0) \quad \text{for } y > a. \tag{2.17}$$

We denote the set of all such data by $\mathcal{K}$:

$$\mathcal{K} = \mathcal{H}_G \ominus (\mathcal{D}_- \oplus \mathcal{D}_+). \tag{2.18}$$

The next theorem is the basic technical tool in this application of our scattering theory.

THEOREM 2.1. *For every* λ *in the resolvent set of* A, *the resolvent operator* $(\lambda I - A)^{-1}$ *maps the unit ball in* $\mathcal{K}$ *into a compact subset of* $\mathcal{K}_G$.

We give a new proof of this theorem in the appendix.

COROLLARY 2.2. A *has at most a denumerable set of eigenfunctions in* $\mathcal{K}$ *and the corresponding eigenvalues are discrete.*

3. The associated semigroup of operators. In this section the group of operators U and the incoming and outgoing subspaces are combined to form a semigroup of operators which plays an important role in our theory.

To this end we introduce the orthogonal projections P_- and P_+ which remove the $\mathcal{D}_-$ and $\mathcal{D}_+$ components, respectively. Again since the G and E-forms are the same for $y > a$, $P_\pm$ is orthogonal with respect to both forms. Notice also that since $\mathcal{D}_-$ and $\mathcal{D}_+$ are orthogonal,

$$P = P_- + P_+ \tag{3.1}$$

is the orthogonal projection on $\mathcal{K}$. Using the relations (2.16) and (2.17) it is easy to verify for any f in $\mathcal{K}_G$ that $g = Pf$ can be decribed as follows:

$$\left. \begin{array}{l} g = f \quad \text{for } y < a, \\[2mm] g_1(w) = f_1(w) - f_1^{(0)}(y) + \left(\dfrac{y}{a}\right)^{1/2} f_1^{(0)}(a) \\[2mm] g_2(w) = f_2(w) - f_2^{(0)}(y) \end{array} \right\} \quad \text{for } y > a. \tag{3.2}$$

We now set

$$Z(t) = P_+ U(t) P_- \quad \text{for } t > 0. \tag{3.3}$$

THEOREM 3.1. *The operators* $Z(t)$, $t > 0$, *form a strongly continuous semi-group of operators on* $\mathcal{K}$. $Z(t)$ *annihilates both* $\mathcal{D}_-$ *and* $\mathcal{D}_+$.

A proof of this theorem can be based entirely on the properties (2.14)$_i$ and (2.14)' (see Theorem 2.7 of [6]).

We denote the infinitesimal generator of $Z(t)$ by B. We have noted earlier that U grows at most exponentially:

$$\|U(t)\| < Ce^{\omega|t|}. \tag{3.4}_U$$

Obviously Z satisfies an analogous inequality:

$$\|Z(t)\| < Ce^{\omega t}. \tag{3.4}_Z$$

Consequently for f in $\mathcal{K}$ and $\text{Re } \lambda > \omega$ we can write

$$(\lambda I - B)^{-1} f = \int_0^\infty e^{-\lambda t} Z(t) f \, dt$$

$$= P_+ \int_0^\infty e^{-\lambda t} U(t) f \, dt = P_+ (\lambda I - A)^{-1} f. \tag{3.5}$$

Combining this with Theorem 2.1 we can now state

THEOREM 3.2 *The resolvent of* B *is compact on* $\mathcal{K}$.

COROLLARY 3.3. *The resolvent of* B *is meromorphic in the entire complex plane and* B *has a pure point spectrum.*

Since by (2.14) $U(t)\mathfrak{D}_+ \subset \mathfrak{D}_+$ for $t > 0$, we see that

$$P_+ U(t)(I - P_+) = 0. \tag{3.6}$$

Hence for f orthogonal to $\mathfrak{D}_-$ we have

$$Z(t)P_+ f = P_+ U(t)P_+ f = P_+ U(t)f.$$

If in addition f belongs to the domain of A, then we can differentiate the above relation at $t = 0$ and obtain

$$BP_+ f = P_+ Af, \tag{3.7}$$

a relation we shall use in §4.

4. The Eisenstein series and its analytic continuation. We are now ready to construct the generalized eigenfunctions of A associated with the continuous part of its spectrum. It is readily verified that for any complex z the data

$$h(w) = \left\{ y^{1/2+iz}, izy^{1/2+iz} \right\} \tag{4.1}$$

locally satisfies

$$Ah = izh. \tag{4.2}$$

Since A is invariant, the function $h^\gamma(w) = h(\gamma w)$ satisfies

$$Ah^\gamma = izh^\gamma \tag{4.2}'$$

for every γ in G. Since h is by definition independent of x, it is automorphic with respect to the subgroup Γ_∞ consisting of $\gamma: w \to w + n$, n integer; that is

$$h(\gamma w) = h(w), \qquad \gamma \in \Gamma_\infty. \tag{4.3}$$

In order to construct an automorphic function out of h for the entire modular group Γ we have to sum $h(\gamma w)$ over all right cosets of Γ modulo Γ_∞. This sum is the *Eisenstein series*:

$$e(w, z) = \sum_{\gamma \in \Gamma_\infty \backslash \Gamma} h(\gamma w). \tag{4.4}$$

It is well known that this series converges for $\operatorname{Im} z < -1/2$; except for the first term $h(w)$, the series converges in the G-norm for such z. Since $(A - iz)$ annihilates each term we conclude that

$$(A - iz)e = 0 \tag{4.5}$$

in the sense of distributions for $\operatorname{Im} z < -1/2$. It follows from elliptic regularity theory that e is C^∞ and satisfies (4.5) pointwise.

Recall the definition (2.7)″ of A as $\left(\begin{smallmatrix} 0 & I \\ L & 0 \end{smallmatrix}\right)$; it follows from this and (4.5) that the first component e_1 of e satisfies the equation

$$Le_1 = -z^2 e_1. \tag{4.6}$$

Integrating this with respect to x gives the following ordinary differential equation for the zero Fourier coefficient of e_1:

$$\left(y^2 \partial_y^2 + 1/4\right)e_1^{(0)} = -z^2 e_1^{(0)}. \tag{4.6}'$$

The solutions to this equation are superpositions of y^κ, where κ satisfies

$$\kappa(\kappa - 1) + 1/4 = -z^2,$$

that is $\kappa = 1/2 \pm iz$. Since by (4.5), $e_2 = ize_1$ we deduce for $\mathrm{Im}\, z < -1/2$ that the zero Fourier coefficient of e defined by (4.4) is of the form

$$e^{(0)}(y, z) = \left\{ y^{1/2+iz}, izy^{1/2+iz} \right\} + s(z)\left\{ y^{1/2-iz}, izy^{1/2-iz} \right\}, \qquad (4.7)$$

where $s(z)$ is some function of z. A straightforward calculation, using (4.4), shows that

$$s(z) = \frac{\Gamma(1/2)\Gamma(iz)\zeta(2iz)}{\Gamma(1/2 + iz)\zeta(1 + 2iz)}, \qquad (4.7)'$$

ζ being the Riemann ζ-function.

Recalling the definition of P described in (3.2), it is clear that Pe has finite G-norm and hence belongs to $\mathcal{K}$. For our purposes it is more convenient to truncate e in a somewhat different fashion but still bring it into $\mathcal{K}$. We choose

$$1 < a_0 < b < c < a \qquad (4.8)$$

and ξ in $C^\infty(\mathbf{R})$ so that

$$\xi(y) = \begin{cases} 0 & \text{for } y > c, \\ 1 & \text{for } y < b. \end{cases} \qquad (4.8)'$$

We now define f equal to e except for the zero Fourier coefficient in the region $y > b$ where we set

$$f^{(0)}(y) = \left\{ y^{1/2}\xi y^{iz}, y^{3/2}(\xi y^{iz})' \right\} + s(z)\left\{ y^{1/2-iz}, izy^{1/2-iz} \right\}. \qquad (4.9)$$

It follows from the above discussion that f belongs to $\mathcal{K}_G$ and from (2.16)$_-$ that f is orthogonal to $\mathfrak{D}_-$. Comparing (4.7) and (4.9) shows that

$$f = e + \left\{ y^{1/2}(\xi - 1)y^{iz}, y^{3/2}((\xi - 1)y^{iz})' \right\}; \qquad (4.9)'$$

clearly f belongs to $D(A)$ when $\mathrm{Im}\, z < -1/2$. It follows then from (4.5) that

$$Af = izf + k, \qquad (4.10)$$

where

$$k = k^{(0)} = \left\{ y^{1/2}\xi' y^{1+iz}, y^{3/2}(\xi' y^{1+iz})' \right\}. \qquad (4.11)$$

Recalling the definition of $\mathfrak{D}_\pm$, it is clear that k is orthogonal to $\mathfrak{D}_-$ and $\mathfrak{D}_+$ and so lies in $\mathcal{K}$; one sees by inspection that $k(z)$ is an entire function of z.

Next we operate on (4.10) by P_+; since k lies in $\mathcal{K}$, $P_+ k = k$ and hence we get

$$P_+ Af = izP_+ f + k. \qquad (4.12)$$

Set $g = P_+ f$. Since f is orthogonal to $\mathfrak{D}_-$ and belongs to $D(A)$, the relation (3.7) is applicable. Using (4.12) we obtain

$$Bg = izg + k. \qquad (4.13)$$

For iz in the resolvent set of B, the relation (4.13) can be rewritten as

$$g = -(izI - B)^{-1}k. \qquad (4.14)$$

We conclude from this and Corollary 3.3 that

LEMMA 4.1. $g(z)$ *can be extended to be meromorphic in the whole complex plane, having poles at most at the points* $-i$ *times the spectrum of* B, *that is* $-i\sigma(B)$.

We can now prove

THEOREM 4.2. *The Eisenstein series* e *can be extended to be meromorphic in the whole complex plane with poles at most at the points* $-i\sigma(B)$; *the analytic continuation of* e *is an eigenfunction of* A.

REMARK. The analytic continuation of $Pe(z)$ is in the $\mathcal{H}_G$ topology, that of the zero Fourier coefficient of e is pointwise.

PROOF. By construction $e(z) = g(z)$ for Im $z < -1/2$ except for the zero Fourier coefficient when $y > b$ and there $e^{(0)}(z)$ is given by (4.7) which holds for $y > a_0$. We see from (3.2) that the action of P_+ effects only the zero Fourier coefficient of data and this only for $y > a$. Hence it follows from (4.9)' and (4.8) that the zero Fourier coefficient of $g(z)$ is also given by (4.7) for $a_0 < y < b$. Thus for Im $z < -1/2$, $e(w, z)$ can be reconstructed from $g(w, z)$. Now according to Lemma 4.1, $g(z)$ has an analytic continuation into $-i$ times the resolvent set of B, in symbols $-i\rho(B)$. Since the first term in (4.7) obviously extends analytically into $-i\rho(B)$ so does the second; in particular $s(z)$ can be continued analytically into this region. Finally we note that this implies that the relation (4.5) can also be continued analytically at least in the sense of distributions. It then follows by elliptic regularity theory that the so extended function satisfies (4.5) in the pointwise sense. This completes the proof of Theorem 4.2.

From formula (4.7) we conclude

COROLLARY 4.3. *The poles of* $s(z)$ *occur at most at the points* $-i\sigma(B)$.

5. The translation representations. So far we have managed to avoid the complications which result from the energy form being indefinite; in fact E has played only a minor role in our discussion. We must now confront this problem.

By definition $G = E + 2K$; since K is compact with respect to G, E is positive on a subspace of finite codimension. It is clear from the expression (2.2) that E is nonnegative on the E-orthogonal complement of the positive eigenspaces of L. Denote the positive eigenfunctions and eigenvalues of L by $(q_j, \lambda_j^2), j = 1, \ldots, m$ with $\lambda_j > 0$:

$$Lq_j = \lambda_j^2 q_j. \tag{5.1}$$

We set

$$f_j^{\pm} = \{ q_j, \pm\lambda_j q_j \}. \tag{5.2}$$

It is easy to verify that $f_j^{\pm}$ belongs to $\mathcal{H}_G$ and that

$$Af_j^{\pm} = \pm\lambda_j f_j^{\pm}; \tag{5.1}'$$

$$E(f_j^+, f_k^+) = 0 = E(f_j^-, f_k^-) \quad \text{for all } j, k; \tag{5.3}$$

and

$$E(f_j^+, f_k^-) = \begin{cases} 0 & \text{for } j \neq k, \\ -\lambda_j^2 & \text{for } j = k. \end{cases} \tag{5.3}'$$

We now set

$$\mathcal{P}_+ = \text{span of} \{f_j^+\}, \qquad \mathcal{P}_- = \text{span of} \{f_j^-\},$$

$$\mathcal{P} = \mathcal{P}_+ + \mathcal{P}_- \tag{5.4}$$

and denote the E-orthogonal complement of $\mathcal{P}$ in $\mathcal{H}_G$ by $\mathcal{H}_G'$. E is nonnegative on $\mathcal{H}_G'$, positive if A has no null vectors.

We denote by Q' the E-orthogonal projection of $\mathcal{H}_G$ onto $\mathcal{H}_G'$; using formulas (5.3) and (5.3)' we can write

$$Q'f = f + \sum a_j^+ f_j^+ + \sum a_j^- f_j^- \tag{5.5}$$

where

$$a_j^+ = E(f, f_j^-)/\lambda_j^2, \, a_j^- = E(f, f_j^+)/\lambda_j^2. \tag{5.5}'$$

If A has null vectors, they will span a finite dimensional subspace $\mathcal{J}$ and on the quotient space

$$\mathcal{H}_E' = \mathcal{H}_G'/\mathcal{J}$$

E is positive definite and equivalent to G. Clearly $\mathcal{H}_E'$ is an invariant space for the operator $U(t)$.

It is easy to show that

$$\mathcal{D}_- \perp \mathcal{P}_+, \quad \mathcal{D}_+ \perp \mathcal{P}_- \quad \text{and} \quad \mathcal{P} \cap \mathcal{D}_- = \{0\} = \mathcal{P} \cap \mathcal{D}_+, \tag{5.6}$$

but it is not true in general that $\mathcal{D}_- \perp \mathcal{P}_-$ or that $\mathcal{D}_+ \perp \mathcal{P}_+$; thus the subspaces $\mathcal{D}_-$ and $\mathcal{D}_+$ do *not* in general lie in $\mathcal{H}_E'$. The next best thing is to replace $\mathcal{D}_\pm$ in our considerations by their E-orthogonal projections $\mathcal{D}_\pm'$ in $\mathcal{H}_E'$; that is by

$$\mathcal{D}_\pm' = Q'\mathcal{D}_\pm. \tag{5.6}'$$

Strictly speaking one has to mod out $\mathcal{J}$ in the right member of (5.6)'; however since $\mathcal{D}_\pm \cap \mathcal{J} = \{0\}$ this causes no confusion.

The projection operator Q' commutes with $U(t)$. It follows from this and properties (2.14) (i) and (ii) that

(i) $U(t)\mathcal{D}_\pm' \subset \mathcal{D}_\pm' \quad \text{for } t \gtrless 0,$

(ii) $\bigcap U(t)\mathcal{D}_-' = \{0\} = \bigcap U(t)\mathcal{D}_+'.$ $\tag{5.7}$

It can be shown that $\mathcal{D}_\pm'$ also satisfy

(iii) $\overline{\bigcup U(t)\mathcal{D}_-'} = \mathcal{H}_c' = \overline{\bigcup U(t)\mathcal{D}_+'},$

where $\mathcal{H}_c'$ is the E-orthogonal complement of the eigenfunctions of A in $\mathcal{H}_E'$. For the proof of this property we refer to Corollary 6.14 of our monograph [6]; the main ingredient of the proof is Theorem 2.1.

According to the translation representation theorem (see Chapter 2 of [6]), any subspace satisfying (i)$_-$–(iii)$_-$ (or (i)$_+$–(iii)$_+$) of (5.7) can be used to construct a translation representation of $\mathcal{K}_c'$, i.e. a unitary map T_- (or T_+) of $\mathcal{K}_c'$ onto a vector valued space of functions of the form $L_2(\mathbf{R}, \mathfrak{N})$, $\mathfrak{N}$ being an auxiliary Hilbert space, with the properties:

$$T_\pm : f \in \mathcal{K}_c' \to k_\pm \in L_2(\mathbf{R}, \mathfrak{N}), \qquad (5.8)$$

$$(\alpha) \quad E(f) = \int \|k_\pm\|_{\mathfrak{N}}^2 \, ds,$$

$$(\beta) \quad U(t)f \to k_\pm(s - t),$$

$$(\gamma) \quad T_\pm \text{ maps } \mathcal{D}_\pm' \text{ onto } L_2(\mathbf{R}_\pm, \mathfrak{N}).$$

We need not rely on this general existence theorem for we can, as we shall show below, obtain explicit formulas for these primed translation representations. We treat only the $\mathcal{D}_+'$-representation but it will be clear that the $\mathcal{D}_-'$-representation can be treated analogously. We begin by constructing a translation representation for the elements of $U(t)\mathcal{D}_+$. Given automorphic data $f = (f_1, f_2)$ defined on all Π with locally finite G-norm, denote, as in (2.15), by $f_i^{(0)}$ the zero Fourier coefficients of f_i, $i = 1, 2$:

$$f_i^{(0)}(y) = \int_{-1/2}^{1/2} f_i(x, y) \, dx, \qquad 0 < y < \infty. \qquad (5.9)$$

We define the translation representer of f to be

$$T_+ f = \frac{1}{\sqrt{2}} \left[\partial_s \left(e^{-s/2} f_1^{(0)}(e^s) \right) - e^{-s/2} f_2^{(0)}(e^s) \right], \quad -\infty < s < \infty. \quad (5.10)_+$$

It is easy to see that T_+ is a linear mapping, continuous from $\mathcal{K}_G$ into $L_2^{\text{loc}}(\mathbf{R})$.

LEMMA 5.1. T_+ *commutes with translations*:

$$T_+ U(t)f = T(t)T_+ f; \qquad (5.11)$$

here $T(t)$ denotes translation to the right by t units.

PROOF. Let $u(x, y, t)$ be the solution to the non-Euclidean wave equation with automorphic initial data f:

$$u_{tt} = Lu; \quad u(0) = f_1 \quad \text{and} \quad u_t(0) = f_2.$$

Then for each time t, u will be automorphic and in particular it will be periodic in x of period 1 for *all* $y > 0$. The zero Fourier coefficient $u^{(0)}(y, t)$ will therefore satisfy the equation

$$u_{tt}^{(0)} = y^2 u_{yy}^{(0)} + u^{(0)}/4 \quad \text{for all } y > 0. \qquad (5.12)$$

As we have seen in §2, the change of variables

$$s = \log y, \qquad v = u^{(0)}/\sqrt{y} \qquad (5.13)$$

transforms (5.12) into the classical wave equation:

$$v_{tt} = v_{ss}; \qquad (5.14)$$

the initial data goes over into

$$v(0) = e^{-s/2} f_1^{(0)}(e^s) \quad \text{and} \quad v_t(0) = e^{-s/2} f_2^{(0)}(e^s). \qquad (5.15)$$

Setting

$$k(s, t) = v_s - v_t, \tag{5.16}$$

it follows from (5.10) and (5.15) that

$$k(s, 0) = \sqrt{2}\, T_+\, f. \tag{5.17}_0$$

Since the solution data at time t is $U(t)f$, we have similarly

$$k(s, t) = \sqrt{2}\, T_+\, U(t)f. \tag{5.17}_t$$

Combining (5.14) and (5.16) we see that

$$\partial_t k + \partial_s k = 0$$

and hence that

$$k(s, t) = k(s - t, 0). \tag{5.18}$$

Finally on combining (5.18) with (5.17) we obtain the assertion of the lemma.

Note that in the proof of Lemma 5.1 we used only the fact that f was periodic in x with period 1. We therefore have

COROLLARY 5.2 *Let f be data periodic in x with period 1 and with locally finite G-norm. Define T_+ as before by (5.9) and (5.10)$_+$. Then Lemma 5.1 holds for f.*

Next we consider data d_0 defined as in (2.13):

$$d_0 = \{y^{1/2}\varphi(y), -y^{3/2}\varphi'(y)\} \tag{5.19}$$

where φ lies in $C_0^\infty(\mathbf{R})$ and vanishes for $y < a$. Since d_0 is periodic in x, the corresponding data d obtained by summing its γ-translates over the right cosets $\Gamma_\infty \setminus \Gamma$ is automorphic:

$$d(w) = \sum_{\Gamma_\infty \setminus \Gamma} d_0(\gamma w). \tag{5.20}$$

Since $\varphi = 0$ for $y < a$, $d_0(w) = 0$ for $y < a$; so $d_0(\gamma w) = 0$ for w in F and $\gamma \notin \Gamma_\infty$ and therefore $d = d_0$ on the fundamental domain F; hence (5.19) shows that d belongs to $\mathcal{D}_+$. As we showed in §2, the solution of the non-Euclidean wave equation with initial data d_0 of the form (5.19) is given by

$$u_0(w, t) = y^{1/2}\varphi(ye^{-t}). \tag{5.21}$$

Clearly $u_0(w, t)$ is periodic in x. The automorphic solution of the non-Euclidean wave equation with initial data d given by (5.20) is

$$u(w, t) = \sum_{\Gamma_\infty \setminus \Gamma} u_0(\gamma w, t). \tag{5.20}_t$$

The reasoning used above shows that $u(w, t) = u_0(w, t)$ for w in F and $t > 0$.

LEMMA 5.3. *For d_0 and d given by (5.19) and (5.20), respectively,*
(a) $T_+ d_0 = T_+ d$;
(b) $E_F(d) = \displaystyle\int_{-\infty}^{\infty} |T_+ d|^2\, ds.$

PROOF. Since $u = u_0$ in F for $t \geq 0$, we have

$$T_+(U(t)d) = T_+(U(t)d_0) \quad \text{for } t \geq 0 \text{ and } s > \log a.$$

Using this relation and (5.11) of Lemma 5.1 and its corollary, we conclude that

$$T(t)T_+d = T(t)T_+d_0 \quad \text{for } t \geq 0 \text{ and } s > \log a.$$

Since this holds for all $t \geq 0$, it follows that it must hold for all s. This proves part (a).

Substituting (5.19) into the definition (5.10)$_+$ of T_+ we obtain

$$T_+d_0 = \sqrt{2}\, e^s \varphi'(e^s). \tag{5.22}$$

The proof of part (b) is now immediate since by definition (2.2)″

$$E_F(d) = \iint_F \left\{ y|\partial_y\varphi|^2 + \frac{|y^{3/2}\varphi'(y)|^2}{y^2} \right\} dx\, dy$$

$$= 2\int_0^\infty y|\varphi'|^2\, dy = 2\int_{-\infty}^\infty e^{2s}|\varphi'(e^s)|^2\, ds \tag{5.23}$$

which by (5.22)

$$= \int_{-\infty}^\infty |T_+d_0|^2\, ds.$$

Combining this with the assertion of part (a) gives part (b). This completes the proof of the lemma.

Combining Lemma 5.3 with Lemma 5.1 and its corollary, we deduce the following

COROLLARY 5.4. (a) $T_+(U(t)d_0) = T_+(U(t)d)$;

(b) $E_F(U(t)d) = \int_{-\infty}^\infty |T_+d|^2\, ds.$

We are now ready to construct the $\mathcal{D}'_+$-translation representaion. Recall that $\mathcal{D}'_+ = Q'\mathcal{D}_+$; that is the elements d'_+ of $\mathcal{D}'_+$ are obtained by projecting the elements d_+ of $\mathcal{D}_+$ into $\mathcal{K}'_E$:

$$d'_+ = d_+ + p_- + p_+,$$

where $p_\pm$ belongs to $\mathcal{P}_\pm$. According to (5.6), $\mathcal{D}_+$ is orthogonal to $\mathcal{P}_-$ and it follows from this and the relations (5.5) and (5.5)′ that $p_+ = 0$:

$$d'_+ = d_+ + p_-. \tag{5.24}$$

Using (5.3) and (5.3)′ we deduce that

$$E_F(d_+) = E_F(d'_+). \tag{5.25}$$

This shows that Q' defines an isometric map on $\mathcal{D}_+$ to $\mathcal{D}'_+$. It follows that the translation representer of d'_+ can be obtained directly by means of the operator T_+ as we now prove:

LEMMA 5.5. $T_+d'_+ = T_+d_+.$

PROOF. Since d_+ and d'_+ differ only by an element of $\mathcal{P}_-$ and since $\mathcal{P}_-$ is spanned by the $\{f_j^-\}$ of (5.2), it suffices to show that

$$T_+ f_j^- = 0. \tag{5.26}$$

We note that $U(t)f_j^- = e^{-\lambda_j t}f_j^-$. Using this fact and Lemma 5.1 we conclude that

$$T_+ f_j^- = \text{const } e^{\lambda_j s}. \tag{5.27}$$

As noted in (5.6), $\mathcal{P}_-$ is orthogonal to $\mathcal{D}_+$; we now show that $T_+ f = 0$ for $s > \log a$ for any element f in $\mathcal{H}_G$ which is orthogonal to $\mathcal{D}_+$. To see this take d in $\mathcal{D}_+$ of the form (5.20); the restriction of d to F is given by (5.19). Hence we have

$$0 = E_F(f, d) = \iint_F \left\{ y\left(\partial_y \frac{f_1}{\sqrt{y}}\right)\overline{\varphi'} - \frac{f_2 y^{3/2}\overline{\varphi'}}{y^2}\right\} dx \, dy.$$

Performing that x integration, we can replace f_1 and f_2 by their zero Fourier coefficients $f_1^{(0)}$ and $f_2^{(0)}$; changing the other variable of integration y into $s = \log y$, the above equation becomes

$$0 = \int_{-\infty}^{\infty} \left\{\partial_s\left(e^{-s/2}\hat{f}_1\right) - e^{-s/2}\hat{f}_2\right\}\partial_s \overline{\varphi(e^s)} \, ds.$$

Recalling the definition $(5.10)_+$ of T_+ we see that this relation says that $T_+ f$ is L_2-orthogonal to $\partial_s\varphi(e^s)$. Since such functions are dense in $L_2(\mathbf{R}_+ + \log a)$, we can conclude that $T_+ f = 0$ for $s > \log a$. Comparing this with (5.27) we see that $T_+ f_j^- \equiv 0$.

LEMMA 5.6. (a) *Let p be an eigenfunction of L with negative eigenvalue $-\nu^2$. Denote by $g_\pm = \{p, \pm i\nu p\}$ the corresponding eigenfunctions of A. Then*

$$T_+(g_+) = 0.$$

(b) *Let f be an arbitrary automorphic data in $\mathcal{H}'_E$ and denote by f'_c the projection of f into $\mathcal{H}'_c$. Then*

$$T_+ f = T_+ f'_c. \tag{5.28}$$

PROOF. The proof of (a) is entirely analogous to that of (5.26) since $g_\pm$ belongs to the point spectrum of A in $\mathcal{H}'_E$ and because of this can easily be shown to be orthogonal to $\mathcal{D}'_\pm$ and hence to be orthogonal to $\mathcal{D}_\pm$. Part (b) follows from part (a) since $f - f'_c$ can be expanded as a convergent series in eigenfunctions of the kind considered in part (a). Recalling that T_+ is continuous from $\mathcal{H}_G$ to L_2^{loc} and that the G and E norms are equivalent on $\mathcal{H}'_E$, assertion (b) now follows.

THEOREM 5.7. T_+ *is the $\mathcal{D}'_+$-translation representation of $\mathcal{H}'_c$.*

PROOF. We shall verify properties (5.8) where auxiliary Hilbert space $\mathcal{N}$ is C: By Lemma 5.1, T_+ is a translation representation and hence satisfies property (β). By part (b) of Lemma 5.3, T_+ is isometric on $\mathcal{D}_+$. Formula (5.22) shows that T_+ maps $\mathcal{D}_+$ onto $L_2(R + \log a, \mathbf{C})$. Since by Lemma 5.5, $T_+ d'_+ = T_+ d_+$ and by (5.25) $E_F(d'_+) = E_F(d_+)$, it follows that T_+ maps $\mathcal{D}'_+$

isometrically onto $L_2(\mathbf{R} + \log a, \mathbf{C})$. This proves property ($\gamma$). Since by property (iii) of (5.7) the translates of $\mathcal{D}'_+$ by $U(t)$ span $\mathcal{H}'_c$ and since the isometry persists under such translations, it follows that T_+ is unitary and hence is an outgoing translation representation of $\mathcal{H}'_c$ with respect to $\mathcal{D}'_+$.

The analogue of Theorem 5.7 holds for the $\mathcal{D}'_-$-translation representation of $\mathcal{H}'_c$ with T_+ replaced by

$$T_- f = -\frac{1}{\sqrt{2}} \left[\partial_s \left(e^{s/2} f_1^{(0)}(e^{-s}) \right) - e^{s/2} f_2^{(0)}(e^{-s}) \right]. \tag{5.10}_-$$

Formulas $(5.10)_\pm$ are the explicit translation representation formulas alluded to in the introduction. We recall that for the Euclidean wave equation, the translation representation is given in terms of the Radon transform of the data defined by integrals along all straight lines in the plane. The integrals in (5.8) are over a one-parameter set of curves in F that are the images of the horocycles through infinity under the projection of the Poincaré plane Π onto F.

The inversion of T_+ can be accomplished in three steps. Let $k(s) = T_+ f$ and define φ by

$$\varphi(y) = \int_{-\infty}^{\log y} k(s)\, ds. \tag{5.29}$$

(i) Set

$$g_0(w) = \left\{ y^{1/2}\varphi(y), -y^{3/2}\varphi'(y) \right\}. \tag{5.30}$$

(ii) Define

$$g(w) = \sum_{\Gamma_\infty \backslash \Gamma} g_0(\gamma w). \tag{5.31}$$

(iii) Project g into $\mathcal{H}'_G$:

$$f = Q'g. \tag{5.32}$$

PROOF. Suppose f belongs to $\mathcal{D}'_+$. Then by definition there exists a d_+ in $\mathcal{D}_+$ such that $Q'd_+ = d'_+ = f$. It is clear from (5.19), (5.20) and (5.22) that g, defined as above, is this d_+; this proves (5.32) when $f \in \mathcal{D}'_+$. Next suppose that $f = U(t)d_+$; then, since Q' commutes with $U(t)$, $f = Q'U(t)d_+$. Comparing the relations (5.19) through (5.20), with (5.29) to (5.31) we conclude that $g = U(t)d_+$; the relation (5.32) therefore holds in this case as well. Finally since $\cup\, U(t)\mathcal{D}'_+$ is dense in $\mathcal{H}'_c$, it follows by continuity that (5.32) holds for all f in $\mathcal{H}'_c$.

Using Lemma 5.6, we see that formula (5.32) can be used to project any f in $\mathcal{H}'_E$ into $\mathcal{H}'_c$.

We conclude this section by defining the *scattering operator* S' as the mapping of the $\mathcal{D}'_-$-representation onto the $\mathcal{D}'_+$-representation:

$$S': T_- f \to T_+ f \quad \text{for } f \text{ in } \mathcal{H}'_c. \tag{5.33}$$

6. The spectral representation. A representation in which the action of $U(t)$ is multiplication by $\exp(i\sigma t)$ is called a *spectral representation*. It is clear that the Fourier transform of a translation representation satisfies this criterion. In this section we shall study the spectral representation of U, restricted to $\mathcal{H}'_c$, so obtained from the $\mathcal{D}'_+$ and $\mathcal{D}'_-$ translation representation treated in §5:

$$(\mathcal{T}_\pm f)(\sigma) = \frac{1}{\sqrt{2\pi}} \int_{-\infty}^{\infty} e^{i\sigma s}(T_+ f)(s)\, ds. \tag{6.1}$$

We now prove

THEOREM 6.1. *The $\mathcal{D}'_\pm$-spectral representation can be expressed as*

$$\mathcal{T}_\pm f(\sigma) = \frac{1}{\sqrt{2\pi}} E_F\left(f,\, \overline{e_\pm(\sigma)}\right), \tag{6.2}$$

where $e_\pm(z)$, $z = \sigma + i\tau$, are analytic continuations of the Eisenstein series generated by

$$h_\pm(w, z) = \frac{\pm 1}{\sqrt{2}\, iz}\left\{ y^{1/2\pm iz}, -izy^{1/2\pm iz} \right\}. \tag{6.3}$$

REMARK. It should be noted that h_+ and h_- are closely related to but somewhat different than h in (4.1). In particular

$$h_-(z) = -h(-z)/\sqrt{2}\, iz;$$

since the Eisenstein series e corresponding to h converges for $\operatorname{Im} z < -1/2$, the Eisenstein series $e_-(z)$ converges for $\operatorname{Im} z > 1/2$, and the zero Fourier coefficient of e_- is related to that of e by

$$e_-^{(0)}(z) = \frac{-1}{\sqrt{2}\, iz}\, e^{(0)}(-z). \tag{6.4}_-$$

An explicit formula for $e^{(0)}$ is given in (4.7). On the other hand $h_+(z)$ is essentially the same as $(\sqrt{2}\, iz)^{-1}h(z)$ except that its second component has the opposite sign. This sign change has no effect on the convergence of the corresponding Eisenstein series $e_+(z)$ nor on its analytic extension. The analogue of (4.7) is now

$$\sqrt{2}\, iz\, e_+^{(0)}(z) = \left\{ y^{1/2+iz}, -izy^{1/2+iz} \right\} + s(z)\left\{ y^{1/2-iz}, -izy^{1/2-iz} \right\}. \tag{6.4}_+$$

PROOF. We shall treat only the $\mathcal{D}'_+$-representation. Since $\mathcal{T}_+$ is the composite of the two unitary maps T_+ and the Fourier transform, it is itself unitary. Thus $\mathcal{T}_+$ is continuous; so it suffices to prove (6.2) for a dense subset of $\mathcal{H}'_c$; for our purpose this will be the set of data f for which $T_+ f$ is smooth with compact support, and for which

$$\int T_+ f(s)\, ds = 0.$$

We shall derive bounds for such f by inverting T_+; see the end of §5. Set

$$\varphi(y) = \frac{1}{\sqrt{2}} \int_{-\infty}^{\log y} T_+ f(s)\, ds$$

and

$$g_0(y) = \left\{ y^{1/2}\varphi(y), -y^{3/2}\varphi'(y) \right\}.$$

Then $g_0(y)$ has compact support in y; as a consequence only a finite number of images $\gamma^{-1}F$ of F lie in that portion of the support of g_0 which is contained in the strip

$$S = \{-1/2 < x < 1/2, y > 0\}.$$

Thus the sum

$$g(w) = \sum_{\Gamma_\infty \backslash \Gamma} g_0(\gamma w) \tag{6.5}$$

has only a finite number of nonvanishing terms on F and is of compact support in F. g is automorphic and, as shown at the end of §5,

$$f = Q'g, \tag{6.6}$$

and by (5.26)

$$T_+ f = T_+ g. \tag{6.6}'$$

By definition (5.5) of Q',

$$f - g \in \mathcal{P}. \tag{6.7}$$

N.b. It is easy to show that $f - g$ actually belongs to $\mathcal{P}_-$, but the weaker result (6.7) suffices, here.

Substituting (6.6) into (6.1) we get

$$\mathcal{T}_+ f = \frac{1}{\sqrt{2\pi}} \int_{-\infty}^{\infty} e^{ias}(T_+ g)(s) \, ds. \tag{6.8}$$

Recalling formula $(5.10)_+$ defining T_+ and making the change of variable: $s \to \log y$, the relation (6.8) becomes

$$\mathcal{T}_+ f = \frac{1}{2\sqrt{\pi}} \int_0^\infty \left[\int_{-1/2}^{1/2} y^{ia} \left\{ y \partial_y \left(\frac{g_1(x,y)}{\sqrt{y}} \right) - \frac{g_2(x,y)}{\sqrt{y}} \right\} dx \right] \frac{dy}{y}. \tag{6.8}'$$

The remainder of the proof of Theorem 6.1 splits into three parts:

Step 1. Continue $(6.8)'$ analytically into the half plane Im $z < -1/2$ and then rewrite it as

$$\mathcal{T}_+ f = \frac{1}{\sqrt{2\pi}} E_S(g, \overline{h_+(z)}); \tag{6.9}$$

here E_S denotes the invariant form of the energy integral given in $(2.2)'$ but integrated over the strip S instead of F.

Step 2. Replace the integral over the strip by a sum of integrals over all the $\gamma^{-1}F$'s contained in S. Using the fact that g is automorphic this gives

$$\mathcal{T}_+ f = \frac{1}{\sqrt{2\pi}} \sum_{\Gamma_\infty \backslash \Gamma} E_{\gamma^{-1}F}(g, \overline{h_+(z)})$$

$$= \frac{1}{\sqrt{2\pi}} \sum_{\Gamma_\infty \backslash \Gamma} E_F(g, \overline{h_+(\gamma w, z)})$$

$$= \frac{1}{\sqrt{2\pi}} E_F\left(g, \sum_{\Gamma_\infty \backslash \Gamma} \overline{h_+(\gamma w, z)} \right) = E_F(g, \overline{e_+(z)}). \tag{6.10}$$

Note that the sum defining the Eisenstein series converges for Im $z < -1/2$

and that $e_+(z)$ can be continued analytically back to real z.

Step 3. Show for the eigenfunctions $\{f_j^\pm\}$ that

$$E_F(f_j^\pm, \bar{e}_+(\sigma)) = 0. \tag{6.11}$$

Since the $f_j^\pm$ span $\mathcal{P}$ and since by (6.7) $f - g \in \mathcal{P}$, the relation (6.2) is an immediate consequence of (6.10) and (6.11).

PROOF OF STEP 1. Since $T_+ g(s)$ has compact support, it is clear that $\mathfrak{T}_+ f$ as defined by (6.8), and hence (6.8)′, is entire in σ and can be continued analytically into the half-plane Im $z < -1/2$, $z = \sigma + i\tau$. Notice that the support for the individual terms in (6.8)′ are bounded away from $y = \infty$ but not from $y = 0$. It follows that each of these terms is integrable for large y; for small y the factor y^{iz} decays fast enough if Im $z < -1/2$, and we now make use of this fact to show that (6.8)′ exists as a double integral.

As noted before, all but a finite number of the terms in (6.5) vanish on F. Thus $g(w)$ is bounded on F and since g is automorphic it is also bounded on S. It follows that for Im $z < -1/2$ the second term on the right in (6.8)′ is integrable near $y = 0$. The first term can be written as

$$y\partial_y\left(\frac{g_1}{\sqrt{y}}\right) = \sqrt{y}\,\partial_y g_1 - \frac{g_1}{2\sqrt{y}}\,;$$

the same reasoning as above shows that the contribution of the second term on the right in this expression to (6.8)′ is also integrable for small y. Thus it only remains to consider the integrability of

$$y^{iz-1/2}\partial_y g_1. \tag{6.12}$$

Recall that by (6.5)

$$g = \sum_{\Gamma_\infty \backslash \Gamma} g_0(\gamma w),$$

where g_0 is a smooth function of y alone with compact support, say in the interval (m, M). Hence

$$g_0(\gamma w) = g_0(v)$$

where

$$u + iv = \gamma w = \frac{aw + b}{cw + d} \quad \text{and} \quad v = \frac{y}{|cw + d|^2}.$$

Now

$$\frac{\partial g_0(\gamma w)}{\partial y} = \frac{\partial g_0(v)}{\partial v} \cdot \frac{\partial v}{\partial y} \quad \text{and} \quad \left|\frac{\partial v}{\partial y}\right| < \frac{1}{|cw + d|^2} = \frac{v}{y}.$$

It follows that

$$\left|\frac{\partial g_0(\gamma w)}{\partial y}\right| < \text{const}\,\frac{M}{y} \quad \text{for all } \gamma.$$

Since the number of terms in (6.5) which do not vanish on any transform γF of F is finite and independent of γ, we conclude that

$$\left|\frac{\partial g_1}{\partial y}\right| < \text{const }\frac{M}{y}.$$

This estimate is sufficient to assure the integrability of (6.12) over the strip S when $\text{Im } z < -1/2$.

We may now proceed formally to bring (6.8)' into the desired invariant form. We write

$$\iint_S y^{iz}\partial_y\left(\frac{g_1}{\sqrt{y}}\right)dx\,dy = \iint_S \left\{y^{iz-1/2}\partial_y g_1 - \frac{1}{2}y^{iz-3/2}g_1\right\}dx\,dy; \quad (6.13)$$

splitting the first term, this can be rewritten as

$$= \iint_S \left\{\frac{iz+1/2}{iz}\,y^{iz-1/2}\partial_y g_1 - \frac{1}{2iz}\,y^{iz-1/2}\partial_y g_1 - \frac{1}{2}\,y^{iz-3/2}g_1\right\}dx\,dy;$$

and integrating the middle term by parts gives

$$\int_0^\infty \frac{1}{2iz}\,y^{iz-1/2}\partial_y g_1\,dy = \frac{1}{2iz}\,y^{iz-1/2}g_1\big|_0^\infty - \int_0^\infty \frac{iz-1/2}{2iz}\,y^{iz-3/2}g_1\,dy.$$

The integrated term vanishes; substituting the remaining term back into (6.13) we get

$$\iint_S y^{iz}\partial_y\left(\frac{g_1}{\sqrt{y}}\right)dx\,dy = \frac{1}{iz}\,\iint_S \left\{\partial_y g_1 \cdot \partial_y y^{iz+1/2} - \frac{g_1 \cdot y^{iz+1/2}}{4y^2}\right\}dx\,dy.$$

Note that this integration by parts is the reverse of the one that brought (2.2)' into the form (2.2)". The right member of the above equation is in invariant form as can be seen from (1.5) and (1.6). Substituting this back in (6.8)' and making use of the definition of h_+ as given in (6.3) we obtain (6.9).

PROOF OF STEP 2. In order to justify the interchange in the order of summation and integration occurring in (6.10), it suffices to recall the fact, noted in §4, that the Eisenstein series converges in $\mathcal{H}_G$ when $\text{Im } z < -1/2$, except for the first term $h_+(w)$. However $h_+(w)$ is locally in $\mathcal{H}_G$ and since g is of compact support in F, even this term causes no difficulty.

According to Theorem 4.2, the Eisenstein function $e_+(z)$ is meromorphic in the *local* $\mathcal{H}_G$ topology. Hence if we again make use of the fact that g has compact support in F, we see that $E_F(g, \overline{e_+(z)})$ is meromorphic and can be continued analytically back to the real axis. It follows that

$$\mathcal{T}_+ f(\sigma) = \frac{1}{\sqrt{2\pi}}\,E_F\big(g,\,\overline{e_+(\sigma)}\big). \quad (6.9)'$$

PROOF OF STEP 3. This proof is based on the observation that both $f_j^\pm$ and $e_+(\sigma)$ are eigenfunctions of the generator A and that the standard proof for the orthogonality of eigenfunctions of a skew-Hermitian partial differential operator can be applied in this case even though one of the eigenfunctions is a generalized one. As defined in (5.1)

$$f_j^\pm = \{q_j, \pm\lambda_j q_j\},$$

where q_j is an eigenfunction of $L: Lq_j = \lambda_j^2 q_j$, belonging to $L_2(F)$. The zero

Fourier coefficient of q_j satisfies the ordinary differential equation

$$y^2 \frac{d^2}{dy^2} q_j^{(0)} + \frac{1}{4} q_j^{(0)} = \lambda_j^2 q_j^{(0)}$$

and hence is of the form

$$q_j^{(0)} = cy^{1/2+\lambda_j} + dy^{1/2-\lambda_j} \quad \text{for } y > a.$$

Since q_j is square integrable on F, the coefficient c must be zero. On the other hand $e_+(\sigma)$ is in $\mathcal{K}_G$ except for its zero Fourier coefficient which by (4.7) is of the form

$$\frac{1}{\sqrt{2}\, iz} \left[\left\{ y^{1/2+i\sigma}, -i\sigma y^{1/2+i\sigma} \right\} + s(\sigma) \left\{ y^{1/2-i\sigma}, -i\sigma y^{1/2-i\sigma} \right\} \right].$$

It follows that the E_F inner product of $f_j^\pm$ and $\overline{e_+(\sigma)}$ is integrable and that integration by parts is permissible. This gives

$$i\sigma E_F\left(f_j^\pm,\ \overline{e_+(\sigma)}\right) = E_F\left(f_j^\pm,\ \overline{Ae_+(\sigma)}\right)$$

$$= -E_F\left(Af_j^\pm,\ \overline{e_+(\sigma)}\right) = \mp\lambda_j E_F\left(f_j^\pm,\ \overline{e_+(\sigma)}\right) (6.14)$$

from which (6.11) follows. This completes the proof of Theorem 6.1.

In analogy with Lemma 5.6 we have

LEMMA 6.2. *Let f be an arbitrary automorphic data in $\mathcal{K}_E'$ and denote by f_c' the projection of f into $\mathcal{K}_c'$. Then*

$$E_F\left(f,\ \overline{e_+(\sigma)}\right) = E_F\left(f_c',\ \overline{e_+(\sigma)}\right). \tag{6.15}$$

Since f differs from f_c' by a sum of genuine eigenvectors of A, this follows by essentially the same argument as that used in (6.14).

REMARK 6.3. It is easy to see that A has only a discrete set of purely imaginary point eigenvalues. In fact since the zero Fourier coefficient of an associated eigenfunction is of the form

$$c\left\{ y^{1/2+i\sigma}, i\sigma y^{1/2+i\sigma} \right\} + d\left\{ y^{1/2-i\sigma}, i\sigma y^{1/2-i\sigma} \right\}$$

and since a point eigenfunction is by definition of finite G-norm, both c and d must be zero. It then follows by (2.17) that any such eigenfunction belongs to the subspace $\mathcal{K}$; the assertion is therefore an immediate consequence of Corollary 2.2.

We conclude this section by obtaining the spectral representer for the scattering operator S' defined in (5.33). To this end we prove

LEMMA 6.4.

$$e_+(\sigma) = -s(\sigma)e_-(\sigma). \tag{6.16}$$

PROOF. It is readily verified that $h_\pm$ satisfy locally

$$Ah_\pm = -izh_\pm.$$

It follows, along the lines indicated at the beginning of §4, that $e_\pm$ also satisfy locally

$$Ae_\pm = -ize_\pm.$$

It follows from (6.4)$_+$ that

$$\sqrt{2}\, ize_+^{(0)}(z) = \{y^{1/2+iz}, -izy^{1/2+iz}\} + s(z)\{y^{1/2-iz}, -izy^{1/2-iz}\}$$

while according to (6.4)$_-$ and (4.7)

$$-\sqrt{2}\, ize_-^{(0)}(z) = \{y^{1/2-iz}, -izy^{1/2-iz}\} + s(-z)\{y^{1/2+iz}, -izy^{1/2+iz}\}.$$

Therefore

$$f = e_+(z) + s(z)e_-(z)$$

satisfies

$$f^{(0)}(z) = \text{const}\,\{y^{1/2+iz}, -izy^{1/2+iz}\}.$$

Thus for Im $z > 0$, $f^{(0)}(z)$ belongs to $\mathcal{H}_G$. According to Lemma 4.1, $e_\pm(z) - e_\pm^{(0)}(z)$, belong to $\mathcal{H}_G$ except when z lies in $-i\sigma(B)$; it follows that for all such z with Im $z > 0$, $f(z)$ belongs to $\mathcal{H}_G$. Since f is a linear combination of e_+ and e_-, it satisfies

$$Af = -izf.$$

Now A has only a finite set of nonimaginary eigenvalues. These are the values $\{\pm\lambda_j\}$ appearing in (5.1)′ which correspond to the positive eigenpairs of L. In fact, as explained in §5, in the quotient space $\mathcal{H}_E' = \mathcal{H}_G'/\mathcal{J}$, where $\mathcal{J}$ is the null space of A, E is positive definite and equivalent to G. Since A restricted to $\mathcal{H}_E'$ is skew-Hermitian, it has no nonimaginary eigenvalues on $\mathcal{H}_E'$. Thus for Im $z > 0$ and z not in $\{i\lambda_j\}$, it follows from the above that $f(z) \equiv 0$. By analytic continuation this holds for real z as well, as asserted in Lemma 6.4.

THEOREM 6.5. *The spectral representer of the scattering operator*:

$$S': \mathcal{T}_- f \to \mathcal{T}_+ f \tag{6.17}$$

is the multiplicative operator:

$$S'(\sigma) = -s(\sigma). \tag{6.18}$$

PROOF. It follows from Theorem 6.1 and Lemma 6.4 that

$$\mathcal{T}_+ f = E_F(f, \overline{e_+(\sigma)}) = -s(\sigma)E_F(f, \overline{e_-(\sigma)}) = -s(\sigma)\mathcal{T}_- f,$$

as desired.

In the case of the modular group, the scattering operator S' can be expressed in terms of the Riemann ζ-function and the gamma function as in (4.7)′. The poles of S' are the nontrivial zeroes of $\zeta(1 + 2iz)$. Thus if the Riemann hypothesis is true these poles will lie on the line Im $z = 1/4$. A further connection between our theory and the ζ-function can be expressed in terms of the semigroup of operators Z'' defined in analogy with Z of §3 using $\mathcal{D}_\pm'' = \mathcal{D}_\pm \cap \mathcal{H}_E'$ instead of $\mathcal{D}_\pm$. Pavlov and Faddeev [2] have formulated the Riemann hypothesis in terms of properties of Z'': The Riemann hypothesis is true if and only if $Z''(t)$ decays exponentially. Specifically, in [6] we have shown: The Riemann hypothesis is true if

$$\limsup t^{-1}\log\|Z''(t)f\|_E < -1/4 \tag{6.19}$$

for a set of f dense in $\mathcal{H}_c'$. The relation (6.19) is an assertion about the rate at which energy is propagated to infinity for those solutions of the automorphic wave equation which contain no standing waves, i.e. are orthogonal to point eigenvectors. This latter condition prevents the application of hyperbolic techniques for proving local energy decay via nonstandard energy identities.

Another possible approach is to use in (6.19) vectors f obtained by application of $Z''(t)$ to $\mathcal{D}''_-$; this is discussed in an appendix to §7 in [6] entitled "How not to prove the Riemann hypothesis".

Hilbert is reputed to have envisaged a proof of the Riemann hypothesis by constructing a skew-hermitian operator on a Hilbert space whose eigenvalues are of the form $\kappa - 1/2$, κ being the nontrivial zeroes of ζ. By an analogue of Corollary 4.3 the infinitesimal generator B'' of Z'' has as its spectrum the values $\{(\kappa - 1)/2\}$. It follows that

$$2B'' + 1/2 \tag{6.20}$$

is the kind of operator on $\mathcal{H}'_c$ that was envisaged by Hilbert; all that is missing is an appropriate inner product on $\mathcal{H}'_c$ which makes (6.20) skew-hermitian.

7. The point spectrum. In addition to the continuous spectrum analysed in §§4, 5 and 6, the operator A has a substantial point spectrum. As noted in Remark 6.3, the point spectrum is discrete. In the case of the modular group, analysis of the point spectrum is made particularly easy by exploiting the bilateral symmetry of the fundamental domain F described in (1.10), that is the mapping $x \to -x$ which takes F into itself. Since the operator L is invariant under the reflection: $x \to -x$, the domain of L can be reduced to the direct sum of even and odd functions. The boundary conditions induced by a function being automorphic become Dirichlet boundary conditions on half of F for odd functions and Neumann boundary conditions on half of F for even functions. It is not hard to show that under Dirichlet boundary conditions on half of F, the proper eigenfunctions form a complete orthonormal system. Thus the point eigenvalues corresponding to odd eigenfunctions of L over F accumulate at infinity. It can also be shown that the number of proper eigenvalues corresponding to even eigenfunctions over F is also infinite. The asymptotic distribution of the proper eigenvalues of L can be related to the asymptotic behavior of the scattering operator, see Theorem 8.6, p. 205 of [6].

Finally we note that if $-\nu^2$ is an eigenvalue of L with eigenfunction p:

$$Lp = -\nu^2 p, \tag{7.1}$$

then

$$g_\pm = \{p, \pm i\nu p\}$$

are eigenvectors of A with eigenvalues $\pm i\nu$:

$$Ag_\pm = \pm i\nu g_\pm. \tag{7.2}$$

8. The Selberg trace formula. The hyperbolic theory developed here lends itself to two distinct versions of the formula which Selberg derived using the concept of trace for operators of trace class. On the one hand the trace of such an operator is equal to the sum of its eigenvalues; on the other hand when the operator is an integral operator on L_2 with a smooth kernel, it is equal to the integral of the kernel along the diagonal. For Hermitian trace class operators this is standard; the general theorem is due to Lidskii (see [3]). In our first version of the trace formula we need only the Hermitian case, but in our second version we require the general theorem.

We now give a simple illustration of how the trace formula is used. Let $\varphi(t)$ be a real valued smooth even function which decays reasonably fast as $|t| \to \infty$. Given a positive real number p, let H be the space of p-periodic functions endowed with the $L_2(0, p)$ topology. Finally define $K = K_\varphi$ as the convolution operator:

$$(K_\varphi f)(x) = \int_{-\infty}^{\infty} \varphi(t)f(x - t)\, dt = \varphi * f. \tag{8.1}$$

Clearly K_φ maps $L_2(0, p)$ into itself and is Hermitian. When written as

$$K_\varphi f(x) = \sum_n \int_{np}^{(n+1)p} \varphi(x - t)f(t)\, dt = \int_0^p \sum_n \varphi(x - t + np)f(t)\, dt,$$

we see that its kernel:

$$\sum \varphi(x - t + np)$$

is smooth; therefore K_φ is of trace class and its trace is equal to

$$p\sum \varphi(np).$$

The eigenfunctions of K_φ are $\exp(2\pi imt/p)$, $m \in \mathbf{Z}$; the eigenvalues are

$$\lambda_m = \int_{-\infty}^{\infty} \varphi(t)\exp\left(\frac{2\pi i}{p}mt\right) dt = \hat{\varphi}\left(\frac{2\pi m}{p}\right),$$

where $\hat{\varphi}$ denotes the Fourier transform of φ. The trace formula asserts that

$$p\sum \varphi(np) = \sum \hat{\varphi}\left(\frac{2\pi m}{p}\right); \tag{8.2}$$

this is the Poisson summation formula.

Returning to the problem at hand, let $U(t)$ be the solution operator for the automorphic wave equation, defined as in §2, and let A denote the infinitesimal generator of the group U. As we saw earlier, the domain of A can be reduced to two E-orthogonal subspaces, one spanned by the proper eigenvectors of A and the other spanned in some sense by the generalized eigenvectors corresponding to the continuous spectrum of A. We denote the restrictions of U to these subspaces by U_p and U_c, respectively; clearly

$$U(t) = U_p(t) + U_c(t). \tag{8.3}$$

Next we denote by $\mathfrak{K}_1$ the space of initial data of the form $\{0, f\}$. Note that E is positive definite on $\mathfrak{K}_1$, in fact $\mathfrak{K}_1 = L_2(F)$. Suppose that $u(t)$ is a solution of the automorphic wave equation with initial data in $\mathfrak{K}_1$, that is $u(0) = 0$. Then $-u(-t)$ will have the same initial data as $u(t)$, which shows that $u(t)$ is an odd function of t. It follows that $U(t) + U(-t)$ maps $\mathfrak{K}_1$ into $\mathfrak{K}_1$.

Let φ be a real valued function which is even and decreases exponentially as $|t| \to \infty$. We define $K = K_\varphi$ by

$$K_\varphi = \int_{-\infty}^{\infty} U_p(t)\varphi(t)\, dt \tag{8.4}$$

restricted to $\mathcal{K}_1$. Clearly K_φ maps $\mathcal{K}_1$ into $\mathcal{K}_1$ and is Hermitian.

If $-\nu_j^2$ is a negative eigenvalue of L with eigenfunction p_j, then $\{0, p_j\}$ is an eigenvector of K_φ with eigenvalue

$$\kappa_j = \int_{-\infty}^{\infty} e^{i\nu_j t} \varphi(t)\, dt = \hat{\varphi}(\nu_j); \qquad (8.5)$$

as before $\hat{\varphi}$ denotes the Fourier transform of φ. As we noted earlier, L also has a finite number of positive eigenvalues λ_j^2, $j = 1, \ldots, m$ (see (5.1) and (5.2)); these give rise to eigenvalues $\hat{\varphi}(i\lambda_j)$ of K_φ. Note that since $\varphi(t)$ decreases exponentially, $\hat{\varphi}$ is defined for complex arguments.

The estimates on the point eigenvalues of $\mathcal{A}$ referred to in §7 show that the number of ν_j that do not exceed σ is $O(\sigma^2)$. It follows from this that the sum $\Sigma |\kappa_j|$, where κ_j is given by (8.5), converges provided that $\hat{\varphi}(\tau)$ tends to zero sufficiently fast as $|\tau| \to \infty$. To this end we further impose some mild assumptions on φ and a few of its derivatives. Then K_φ is of trace class and

$$\operatorname{tr} K_\varphi = \sum \hat{\varphi}(\nu_j) + \sum \hat{\varphi}(i\lambda_j) + \eta \hat{\varphi}(0); \qquad (8.6)$$

here η is the multiplicity of the zero eigenvalue of L.

Next in order to evaluate the trace of K_φ in another way, we express K_φ as an integral operator. Using (8.3) and (8.4) we write

$$K_\varphi = \int [U(t) - U_c(t)] \varphi(t)\, dt. \qquad (8.7)$$

We can express U_c in terms of the operational calculus as

$$U_c(t) = \exp A_c t,$$

where A_c is the absolutely continuous part of A. Now in §6 we have explicity determined the spectral representation of A_c. The generalized eigenfunctions of A_c are given by the extended Eisenstein series $e(w, \sigma)$. Since A_c acts as multiplication by $i\sigma$ in the spectral representation, any function $\Phi(-iA_c)$ of $-iA_c$ is an integral operator with kernel

$$\int \Phi(\sigma) e(w, \sigma) e(w', \sigma)\, d\sigma.$$

In particular

$$\int U_c(t) \varphi(t)\, dt = \int \varphi(t) \exp(t, A_c)\, dt = \hat{\varphi}(-iA_c)$$

is an integral operator with kernel

$$\int \hat{\varphi}(\sigma) e(w, \sigma) e(w', \sigma)\, d\sigma. \qquad (8.8)$$

Next we show how to write $U(t)$ as an integral operator. Denote by $U_0(t)$ the solution operator for the non-Euclidean wave operator over all of Π. Since L is invariant under the non-Euclidean motions, so is $U_0(t)$; in particular if f_0 is automorphic in Π, so is $U_0(t)f_0$. Thus if we denote the restriction of f_0 to F by

$$f = \chi_F f_0,$$

then

$$U(t)f = \chi_F U_0(t)f_0.$$

Since

$$f_0 = \sum_\Gamma T_\gamma f,$$

where

$$(T_\gamma f)(w) = f(\gamma w),$$

we conclude that

$$U(t)f = \chi_F \sum_\Gamma U_0 T_\gamma f. \tag{8.9}$$

The non-Euclidean wave equation in Π can be solved explicitly: For initial data

$$u(w, 0) = 0, \qquad u_t(w, 0) = f(w),$$

we have

$$u_t(w, t) = \frac{\operatorname{sgn} t}{\sqrt{8}\,\pi} \; \partial_t \int_{r<|t|} \frac{f(w')}{\sqrt{\cosh t - \cosh r}} \; dw'$$

where r is the non-Euclidean distance from w to w' and dw' is the non-Euclidean area element. Using this formula we can represent the restriction of U_0 to $\mathcal{H}_1$ as an integral operator with a distribution kernel R, called the Riemann function:

$$R(w, w', t) = \frac{\operatorname{sgn} t}{\sqrt{8}\,\pi} \; \partial_t \frac{1}{\sqrt{\cosh t - \cosh r}}. \tag{8.10}$$

Substituting this into (8.9) we see that the kernel of $\int U(t)\varphi(t) \, dt$ is

$$\frac{1}{\sqrt{8}\,\pi} \int \varphi(t)\operatorname{sgn} t \, \partial_t \sum_\Gamma \frac{1}{\sqrt{\cosh t - \cosh r_\gamma}} \; dt. \tag{8.11}$$

Following Selberg we separate the sum in this integral into four parts corresponding to the identity, the hyperbolic, the elliptic and the parabolic elements of Γ. Within each part we group the terms into conjugacy classes of Γ. Accordingly, (8.11) is broken into the sum of four kernels, three of which are of trace class. The trace of the part coming from the identity is

$$-\frac{\operatorname{area}(F)}{4\pi} \int \frac{\varphi'(t)}{\sinh t/2} \; dt. \tag{8.12}_{\mathrm{id}}$$

For the modular group area $(F) = \pi/3$. The trace of the part coming from the hyperbolic elements is the weighted sum

$$\sum_\tau \sum_{k \neq 0} \frac{l}{2|\sinh(kl/2)|} \varphi(kl) \tag{8.12}_{\mathrm{hyp}}$$

where the first sum is over all inequivalent primitive hyperbolic elements τ of Γ, and $l = l(\tau)$ is defined by

$$2 \cosh l = \operatorname{tr} \tau. \tag{8.13}$$

The trace of the part coming from the elliptic elements is a sum of weighted integrals:

$$\sum_\tau \sum_1^{m-1} \int \varphi(t) \frac{\cosh t/2}{\sinh^2 t/2 + \sin^2 k\pi/m} \, dt \qquad (8.12)_{\text{ell}}$$

where the first sum is over all inequivalent primitive elliptic elements τ and $m = m(\tau)$ is the order of τ, i.e. the smallest m such that $\tau^m = I$. Alternatively

$$2 \cos \pi/m = \operatorname{tr} \tau.$$

The part corresponding to the parabolic elements is not of trace class, but its difference from $\hat{\varphi}(-iA_c)$ is. Using (8.8) the trace of the difference can be expressed as a weighted integral of $\hat{\varphi}$; for the modular group this difference is the sum of two terms:

$$\frac{1}{4} \hat{\varphi}(0) - \log 2\varphi(0) - \frac{1}{2\pi} \int \hat{\varphi}(\sigma) \frac{d}{d\sigma} \log \Gamma(1 + i\sigma) \, d\sigma \qquad (8.12)_{\text{par}}$$

and

$$\frac{1}{4} \hat{\varphi}(0) + \frac{1}{4\pi} \int \hat{\varphi}(\sigma) \frac{1}{i} \frac{d}{d\sigma} \log s(\sigma) \, d\sigma; \qquad (8.12)_{\text{cont}}$$

here $s(\sigma)$ is the scattering function appearing in formulas (4.7)′ and (6.18). For arbitrary discrete subgroups (with noncompact fundamental domain of finite area) $s(\sigma)$ has to be replaced by the determinant of the so-called scattering matrix, and the three terms in $(8.12)_{\text{par}}$ have to be multiplied by the number of cusps, i.e. points at infinity, of the fundamental domain (which is the same as the number of inequivalent primitive parabolic elements in Γ). The trace formula says that

$$(8.6) = (8.12)_{\text{id}} + (8.12)_{\text{hyp}} + (8.12)_{\text{ell}} + (8.12)_{\text{par}} + (8.12)_{\text{cont}}.$$

For further details we refer to §9 of [6] and to Kubota [4].

It should be remarked that a similar trace formula has been derived by Selberg for discrete subgroups whose fundamental domain F is compact. The formula and its derivation are simpler in this case since there is no continuous spectrum and since Γ contains no parabolic elements; see [9] for an illuminating exposition and a geometrical interpretation of the trace formula.

We remark finally that when the procedure described above is applied to the trivial case of the classical wave equation for functions of x and t which are periodic in x, the operator (8.4) reduces to the operator (8.1) and, as remarked earlier, the trace formula becomes the Poisson summation formula.

We turn now to a description, even sketchier than the foregoing, of our other approach to the Selberg trace formula; this is described more fully in an appendix to §9 of [6]. It is based on the notion of trace for non-Hermitian operators.

Let $Z(t)$ denote the semigroup of operators

$$Z(t) = P_+ U(t) P_-, \qquad (8.15)$$

where, as in §3, $P_\pm$ are the orthogonal projections onto $\mathfrak{D}_\pm^\perp$ and $U(t)$ is as before; we take as the domain of $Z(t)$ all of the original Hilbert space $\mathcal{H}_G$. As so defined $Z(t)$ does not converge strongly to the identity as $t \to 0$; however the restriction to $\mathcal{K} = \mathcal{K}_G \ominus (\mathcal{D}_+ \oplus \mathcal{D}_-)$ does have this property. We take a function $\varphi(t)$ of compact support in $\mathbf{R}_+$ and define $C = C_\varphi$ by

$$C_\varphi = \int_0^\infty Z(t)\varphi(t)\, dt. \tag{8.16}$$

We claim that C_φ is of trace class. For proof we refer to [6], where two demonstrations are given. In the second proof, contained in Appendix 2 to §9, it is shown that C_φ is an integral operator whose kernel is smooth except for a jump dicontinuity along the line $y = a$ (caused by the action of $P_\pm$) and that the kernel decays faster than any power of y as $y \to \infty$. The argument showing this decay is based, appropriately enough, on the Poisson summation formula.

We now determine the eigenvectors of C_φ. Aside from the null vectors of C_φ, which do not contribute to the trace, all of these are eigenvectors of $Z(t)$ restricted to $\mathcal{K}$. It is shown in the appendix to Chapter 9 of [6] that these eigenvectors are obtained, by P_+ projection, from the eigenvectors of $U(t)$ *which are orthogonal to* $\mathcal{D}_\ominus$.

We treat first the proper eigenvectors of $U(t)$. We saw in §7 that L has infinitely many eigenfunctions p corresponding to negative eigenvalues $-\nu^2$:

$$Lp = -\nu^2 p.$$

Integrating this relation with respect to x, we get for the zero Fourier coefficient $p^{(0)}$ the ordinary differential equation:

$$y^2 \frac{d^2}{dy^2} p^{(0)} + \frac{1}{4} p^{(0)} = -\nu^2 p(0),$$

all of whose solutions are of the form

$$p^{(0)} = a y^{1/2 - i\nu} + b y^{1/2 + i\nu}.$$

Since $-\nu^2$ is in the point spectrum, p belongs to $L_2(F)$ and this requires both a and b to be zero, i.e. $p^{(0)} = 0$.

We saw in §7 that if p is an eigenfunction of L, then

$$g_\pm = \{p, \pm i\nu p\}$$

are eigenvectors of $U(t)$, with eigenvalues $\exp(\pm i\nu t)$. The condition $p^{(0)} = 0$ implies that $g_\pm$ are orthogonal to both $\mathcal{D}_-$ and $\mathcal{D}_+$, i.e. that

$$P_- g_\pm = P_+ g_\pm = g_\pm.$$

It follows that $g_\pm$ are eigenvectors of $Z(t)$ as well, with eigenvalues $\exp(\pm i\nu t)$; hence $g_\pm$ are also eigenvectors of C_φ, with eigenvalues

$$\hat{\varphi}(\pm \nu_j). \tag{8.17}$$

We have noted in §5 that L has a finite number of positive eigenvales λ^2; denote the corresponding eigenfunction by q. Then $\{q, \lambda q\}$ is an eigenvector of $U(t)$, with eigenvalue $\exp(\lambda t)$. As remarked in (5.6), for $\lambda > 0$, $\{q, \lambda q\}$ is orthogonal to $\mathcal{D}_-$; from this one can deduce, using the relation (3.7), that $P_+\{q, \lambda q\}$ is an eigenvector of $Z(t)$ with eigenvalue $\exp(\lambda t)$, and therefore of C_φ with eigenvalue

$$\hat{\varphi}(-i\lambda_j). \tag{8.18}$$

178

The vector $\{q, -\lambda q\}$ is also an eigenvector of $U(t)$, in this case orthogonal to $\mathcal{D}_+$, but not in general orthogonal to $\mathcal{D}_-$. However if it happens also to be orthogonal to $\mathcal{D}_-$, then as above $\{q, -\lambda q\}$ is an eigenvector of $Z(t)$ with eigenvalue $\exp(-\lambda t)$, and of C_φ with eigenvalue

$$\hat{\varphi}(i\lambda_j). \tag{8.18}'$$

We turn now to the generalized eigenvectors of $U(t)$ which are orthogonal to $\mathcal{D}_-$. The corresponding eigenvalues appeared in the proof of Theorem 4.2 as poles of the meromorphic continuation of the scattering function $s(z)$ into the upper half plane. According to this theorem, these poles can occur only at points of $-i\sigma(B)$; here B denotes the infinitesimal generator of $Z(t)$ restricted to $\mathcal{K}$. Taking the residue of the Eisenstein series about such a point, it is clear from (4.7) that the resulting generalized eigenvector of A will be orthogonal to $\mathcal{D}_-$ and hence, after projection by P_+, will be a proper eigenvector of B. The corresponding eigenvalue of B is a complex number μ_j, $\mathrm{Re}\,\mu_j < 0$, that of $Z(t)$ is $\exp(\mu_j t)$ and that of C_φ is

$$\hat{\varphi}(-i\mu_j). \tag{8.19}$$

Finally we denote by η_0 the multiplicity of 0 as an eigenvalue of B; then $\hat{\varphi}(0)$ is the corresponding eigenvalue of C_φ and it has multiplicity η_0. Thus the sum of the eigenvalues of C_φ is obtained by adding (8.17) to (8.18), (8.18)', (8.19) and to the above. Altogether we get

$$\sum \hat{\varphi}(\pm \nu_j) + \sum \hat{\varphi}(-i\lambda_j) + \sum{}' \hat{\varphi}(i\lambda_j) + \sum \hat{\varphi}(-i\mu_j) + \eta_0 \hat{\varphi}(0). \tag{8.20}$$

Next we construct the kernel of C_φ as integral operator. This task is roughly similar to constructing the kernel of K_φ, mildly complicated by the fact that (i) C_φ acts on vectors with two components, and (ii) the underlying Hilbert space is the direct sum of $L_2(F)$ and a Dirichlet space over F. When the dust settles, we end up with the following formula:

$$(8.20) = \hat{\varphi}(0)/2 + (8.12)_{\mathrm{id}} + (8.12)_{\mathrm{hyp}} + (8.12)_{\mathrm{ell}} + (8.12)_{\mathrm{par}}.$$

It is instructive to compare this with our previous trace formula. We extend φ used in the definition of C_φ in (8.16) as an even function of t and use this extended φ in the definition of K_φ in (8.7). Subtracting the trace formula for K_φ from that for C_φ we obtain an identity of the following form

$$\sum \hat{\varphi}(-i\mu_j) - \sum{}'' \hat{\varphi}(i\lambda_j) + (\eta_0 - 2\eta)\hat{\varphi}(0)$$

$$= \frac{i}{4\pi} \int [\hat{\varphi}(\sigma) + \hat{\varphi}(-\sigma)] \frac{d}{d\sigma} \log s(\sigma)\, d\sigma. \tag{8.21}$$

Since the poles of $s(z)$ in the upper half plane coincide with the set $-i\mu_j$ and the so-called relevant $i\lambda_j$'s, this formula can be verified by deforming the integral on the right from the real axis off to infinity. This can be justified on the basis of appropriate estimates for $s(z)$ away from the poles; for the modular group such estimates are available. In order to verify (8.21) in general we use a "mini-trace formula" for the operator

$$C_\varphi'' = \int Z''(t)\varphi(t)\, dt$$

on $\mathcal{K}''$, where

$$\mathcal{K}'' = \mathcal{K}'_c \ominus (\mathcal{D}''_- \oplus \mathcal{D}''_+)$$

and

$$Z''(t) = P''_+ U(t) P''_-;$$

here $\mathcal{D}''_\pm$ denote that subspace of $\mathcal{D}_\pm$ which is orthogonal to the subspace $\mathcal{P}$ defined in (5.4) and $P''_\pm$ are orthogonal projections onto $\mathcal{K}'_c \ominus \mathcal{D}''_\pm$. As a by-product of this analysis we find that $\eta_0 = 2\eta$. For details see the appendix to §9 of [6].

Appendix: A new proof of Theorem 2.1. Theorem 2.1 is a restatement of Theorem 6.6 in [6]. Since the initial steps of the two proofs are the same we shall refer the reader to [6, pp. 126–130] for the proofs of some of the initial lemmas.

THEOREM 2.1. *For every λ in the resolvent set of A, $(\lambda I - A)^{-1}$ maps the unit ball in $\mathcal{K}$ into a compact subset of $\mathcal{K}_G$.*

REMARK. Since it suffices to prove the theorem for a single value of λ, we shall for convenience take λ to be real > 4.

Theorem 2.1 is a consequence of a simpler result:

THEOREM A.1. *Let φ be any C_0^∞ function on $\mathbf{R}_+$. Define the operator $M = M_\varphi$ by*

$$M_\varphi = \int_0^\infty \varphi(t) U(t) \, dt. \tag{A.1}$$

Then M_φ maps the unit ball in $\mathcal{K}$ into a compact subset of $\mathcal{K}_G$.

The proof of this theorem is based on the following two propositions:

LEMMA A.2. (a) *The set*

$$u = Mf, \qquad G(f) < 1, \tag{A.2}$$

is precompact with respect to the norm

$$G_Y(u) = \iint\limits_{F(Y)} \left\{ |u_{1,x}|^2 + |u_{1,y}|^2 + |u_1|^2 + |u_2|^2 \right\} dx \, dy \tag{A.3}$$

for any Y; here $F(Y)$ denotes the domain

$$F(Y) = F \cap \{ y < Y \}.$$

(b) *Given any $\varepsilon > 0$, there is a $Y > a$ such that*

$$G^Y(u) = \iint\limits_{y > Y} \left\{ |u_{1,x}|^2 + y \left| \partial_y \frac{u_1}{\sqrt{y}} \right|^2 + \frac{|u_2|^2}{y^2} \right\} dx \, dy < \varepsilon G(f) \tag{A.4}$$

for all u of the form (A.2) with f in $\mathcal{K}$.

Part (a) is a simple consequence of Rellich's compactness criterion.

PROOF OF PART (b). We note that the imposition of a finite number of linear continuous constraints on the domain of M does not affect the precompactness argument. Now $f_1^{(0)}(a) = 0$ is just such a constraint since by Lemma 4.2

in [6]

$$|f_1^{(0)}(a)|^2 \leqslant \int_{-1/2}^{1/2} |f_1(x, a)|^2 \, dx \leqslant \text{const } G(f).$$

Since for f in $\mathcal{K}$, $f_2^{(0)}(y) = 0$ for all $y > a$, it follows by (2.17) that we can assume without loss of generality that $f^{(0)}(y) = 0$ for all $y > a$.

Fixing φ, we choose T so that supp $\varphi \subset (0, T)$. Since the speed of propagation for the solution to the non-Euclidean wave equation is $\leqslant 1$, a signal which originates at a point below $y = 2be^T$ (or above $y = be^T$) will not get above $y = 2be^{2T}$ (or below $y = b$) during time $t \in (0, T)$. According to (A.1) and (A.2), u is a superposition of values of $U(t)f$, $0 < t < T$. It follows that values of u at points where $y > 2be^{2T}$ (or $y < b$) do not depend on values of f below $y = 2be^T$ (or above $y = be^T$). This suggests that we split f as

$$f = g + h$$

so that

$$g = 0 \quad \text{for } y < be^T,$$
$$h = 0 \quad \text{for } y > 2be^T$$

and so that

$$G(g) \leqslant \text{const } G(f).$$

We can do this in such a way that the constant is independent of our choice of $b > a$, which will be fixed later on.

Since part (b) of the proof is concerned only with values of u for large y, h plays no role in what follows. We therefore need only consider g and to keep our notation simple we shall hereafter denote g by f. We may therefore assume that f satisfies

$$f = 0 \quad \text{for all } y < be^T. \tag{A.5}$$

In this case $u = 0$ for all $y < b$. Moreover the various Fourier coefficients of u, that is the

$$u^{(k)}(y) = \int_{-1/2}^{1/2} e^{-ikx} u(x, y) \, dx \tag{A.6}$$

are obtained directly from M acting only on the corresponding Fourier coefficients of f; that is, the Fourier coefficients of $U(t)f$ remain uncoupled as long as the signal stays above $y = a$ as it does for $0 < t < T$. In particular $u^{(0)} = 0$.

We note that G and E are identical for data vanishing for $y < a$. Since $f^{(0)} = 0$ for all y, we have

$$G(f) = \int_a^\infty \int_{-1/2}^{1/2} \left\{ |\partial_x f_1|^2 + y \left| \partial_y \frac{f_1}{\sqrt{y}} \right|^2 + \frac{|f_2|^2}{y^2} \right\} dx \, dy$$

$$= \sum_{k \neq 0} \int_a^\infty \left\{ (2\pi k)^2 |f_1^{(k)}|^2 + y \left| \partial_y \frac{f_1^{(k)}}{\sqrt{y}} \right|^2 + \frac{|f_2^{(k)}|^2}{y^2} \right\} dy. \tag{A.7}$$

We now set

$$p = (\lambda I - A)u = \int (\lambda \varphi + \partial_t \varphi) U(t) f \, dt. \tag{A.8}$$

Clearly

$$G(p) \leqslant \text{const } G(f). \tag{A.9}$$

Further writing (A.8) in component form, we get

$$\lambda u_1 - u_2 = p_1, \qquad \lambda u_2 - L u_1 = p_2, \tag{A.8}'$$

so that

$$\lambda^2 u_1 - y^2 \Delta u_1 - u_1/4 = \lambda p_1 + p_2 \equiv q. \tag{A.8}''$$

Applying (A.7) to p_1 we get

$$\sum_{k \neq 0} \int_b^\infty |p_1^{(k)}|^2 \, dy \leqslant G(p) \tag{A.10}$$

and combining this with (A.9) we see that

$$\sum_{k \neq 0} \int_b^\infty \frac{|q^{(k)}|^2}{y^2} \, dy \leqslant \text{const } G(p) \leqslant \text{const } G(f). \tag{A.11}$$

Finally it is clear from (A.8)″ that the kth Fourier coefficient of u_1 satisfies the equation

$$-\frac{d^2}{dy^2} u_1^{(k)} + \left(4\pi^2 k^2 + \frac{\kappa}{y^2}\right) u_1^{(k)} = \frac{q^{(k)}}{y^2}, \tag{A.12}$$

where $\kappa = \lambda^2 - 1/4$.

LEMMA A.3. *Let $v \in C^1(\mathbf{R})$ and $\int_a^\infty |v|^2 \, dy < \infty$, then there exists a sequence $y_j \to \infty$ for which $\operatorname{Re} v'\bar{v}|_{y_j} \to 0$.*

PROOF. Obviously

$$\lim_{y \to \infty} \int_j^{j+1} |v|^2 \, dy = 0.$$

By the law of the mean there is y_j' in $(j, j+1)$ for which $v(y_j') \to 0$. Moreover

$$|v(y_{j+2}')|^2 - |v(y_j')|^2 = \int_{y_j'}^{y_{j+2}'} \frac{d}{dy} |v|^2 \, dy = 2 \int_{y_j'}^{y_{j+2}'} \operatorname{Re}(v'\bar{v}) \, dy.$$

Since this expression converges to zero and since $y_{j+2}' - y_j' > 1$, the law of the mean again furnishes us with a y_j in (y_j', y_{j+2}') satisfying the statement of the lemma.

We return now to the proof of part (b) of Lemma A.2. Multiply the relation (A.12) by $\overline{u_1^{(k)}}$ and integrate by parts. Take the real part of the resulting relation and use Lemma A.3 at the upper limit of integration and the fact $u_1^{(k)}(b) = 0$ at the lower limit. This gives us

$$\int_b^\infty \left\{ |u_1^{(k)'}|^2 + \left(4\pi^2 k^2 + \frac{\kappa}{y^2}\right)|u_1^{(k)}|^2 \right\} dy = 2 \operatorname{Re} \int_b^\infty \frac{q^{(k)}}{y^2} \, \overline{u_1^{(k)}} \, dy.$$

182

Applying the Schwarz inequality to the right member gives us

$$\int_b^\infty \left\{ |u_1^{(k)'}|^2 + 4\pi^2 k^2 |u_1^{(k)}|^2 \right\} dy < \frac{1}{b^2} \int_b^\infty \frac{|q^{(k)}|^2}{y^2} dy + \int_b^\infty |u_1^{(k)}|^2 dy;$$

and transposing the last term on the right we get

$$\int_b^\infty \left\{ |u_1^{(k)'}|^2 + 2\pi^2 k^2 |u_1^{(k)}|^2 \right\} dy < \frac{1}{b^2} \int_b^\infty \frac{|q^{(k)}|^2}{y^2} dy. \tag{A.13}$$

From the first relation in (A.8)' we get

$$\int_b^\infty \frac{|u_2^{(k)}|^2}{y^2} dy < 2\lambda^2 \int_b^\infty \frac{|u_1^{(k)}|^2}{y^2} dy + 2\int_b^\infty \frac{|p_1^{(k)}|^2}{y^2} dy$$

$$< \frac{2}{b^2} \left[\lambda^2 \int_b^\infty |u_1^{(k)}|^2 dy + \int_b^\infty |p_1^{(k)}|^2 dy \right]. \tag{A.14}$$

It now follows from

$$\int_b^\infty y \left| \partial_y \frac{u_1^{(k)}}{\sqrt{y}} \right|^2 dy < 2\int_b^\infty \frac{|u_1^{(k)}|^2}{y^2} dy + 2\int_b^\infty |u_1^{(k)'}|^2 dy$$

$$< 2\int_b^\infty \left(|u_1^{(k)}|^2 + |u_1^{(k)'}|^2 \right) dy$$

together with (A.7) as applied to u_1, (A.10), (A.11), (A.13) and (A.14) that

$$G(u) < \frac{\text{const}}{b^2} G(f),$$

where the constant is independent of our choice of b. Hence if we choose b so that $\text{const}/b^2 < \varepsilon$ and $Y > 2be^{2T}$, then we get the desired inequality for part (b) of Theorem A.1.

References

1. L. D. Faddeev, *Expansion in eigenfunctions of the Laplace operator in the fundamental domain of a discrete group on the Lobačevskiĭ plane*, Trudy Moscow. Mat. Obšč. **17** (1967), 323–350; see also English transl., Trans. Moscow Math. Soc. **17** (1967), 357–386.

2. L. D. Faddeev and B.S. Pavlov, *Scattering theory and automorphic functions*, Proc. Steklov Inst. Math. **27** (1972), 161–193.

3. I. C. Gohberg and M. G. Krein, *Introduction to the theory of linear non-selfadjoint operators*, Transl. Math. Monographs, vol. 18, Amer. Math. Soc., Providence, R. I., 1969.

4. T. Kubota, *Elementary theory of Eisenstein series*, Wiley, New York, 1973.

5. P. D. Lax and R. S. Phillips, *Scattering theory*, Academic Press, New York, 1967.

6. _____, *Scattering theory for automorphic functions*, Ann. of Math. Studies, no. 87, Princeton Univ. Press, Princeton, N. J., 1976.

7. _____, *The scattering of sound waves by an obstacle*, Comm. Pure Appl. Math. **30** (1977), 195–233.

8. _____, *Translation representations for the solution of the non-Euclidean wave equation*, Comm. Pure Appl. Math. **32** (1979), 617–667.

9. H. P. McKean, *Selberg's trace formula as applied to a compact Riemann surface*, Comm. Pure Appl. Math. **25** (1972), 225–246.

10. A. Selberg, *Harmonic analysis and discontinuous groups in weakly symmetric Riemannian spaces with applications to Dirichlet series*, J. Indian Math. Soc. **20** (1956), 47–87.

11. M. A. Semenov-Tian-Shansky, *Harmonic analysis on Riemannian symmetric spaces of negative curvature and scattering theory*, Math. USSR Izvestija, vol. 10 (1976), 535–563.

COURANT INSTITUTE OF MATHEMATICAL SCIENCES, NEW YORK UNIVERSITY, NEW YORK, NEW YORK 10012

DEPARTMENT OF MATHEMATICS, STANFORD UNIVERSITY, STANFORD, CALIFORNIA 94305

Translation Representations for the Solution of the Non-Euclidean Wave Equation. II

PETER D. LAX

Courant Institute

AND

RALPH S. PHILLIPS

Stanford University

Introduction

This is a continuation of our paper [3]. The present paper begins with Section 5; references to Sections 1 through 4 pertain to the original paper. In Section 5 we derive Friedlander's formula for the outgoing translation representation of solutions of the non-Euclidean wave equation and prove the eventual equipartition of energy for such solutions. In Section 6 we study the Radon transform of Kf, where K is an invariant integral operator, and show that K can be expressed as a superposition of the operators $U(t)$ with the Selberg transform of K as weight factor. In Section 7 we correct a small lapse in the proof of Theorem 3.8.

In a subsequent paper we shall extend many of the results in these two papers to automorphic solutions of the wave equation.

5. Friedlander's Formula

Let $f = (f_1, f_2)$ be initial data of finite energy for the non-Euclidean wave equation in hyperbolic three-space Π and let $k = T_+ f$ be its outgoing translation representer defined by formula (3.27). Denote by j the origin chosen in Π and let (t, γ) be non-Euclidean polar coordinates around j, i.e., $w(t, \gamma)$ is that point on the geodesic issuing from j in the direction γ whose distance from j is t. The direction γ can be thought of as a point on the infinity set B bounding Π.

THEOREM 5.1. *Let* $u(w, t)$ *denote the solution of the non-Euclidean wave equation* (3.22) *with initial data* f. *Then*

$$(5.1) \qquad \lim_{t \to \infty} k(s, \gamma; t) = k(s, \gamma),$$

where

$$(5.2) \qquad k(s, \gamma; t) = \sqrt{2} \sinh(t + s) u_t(w(t + s, \gamma), t) \quad for \quad s > -t$$

and equals 0 *otherwise; here the limit is in the* $L_2(\mathbb{R}, B)$ *sense.*

Communications on Pure and Applied Mathematics, Vol. XXXIV, 347–358 (1981)

Proof: We shall use formula $(3.41)_+$ expressing u in terms of its outgoing translation representer k defined by equation (3.27):

$$(5.3) \qquad u(w,t) = |c| \int_B \exp\langle w, \beta \rangle k(\langle w, \beta \rangle - t, \beta)\, d\beta.$$

Here it is convenient to use the unit ball model of Π with $j = 0$; in this model

$$(5.4) \qquad \langle w, \beta \rangle = \log \frac{1 - |w|^2}{1 - 2w \cdot \beta + |w|^2},$$

where $|w|$ denotes the Euclidean absolute value and $w \cdot \beta$ the Euclidean scalar product (see (1.16)). In terms of Euclidean polar coordinates this becomes

$$(5.4)' \qquad \langle w, \beta \rangle = \log \frac{1 - r^2}{1 - 2r\cos\theta + r^2},$$

where $r = |w|$ and θ is the angle between w and β.

The relation between the Euclidean distance r from j to w and the non-Euclidean distance R is given by

$$(5.5) \qquad e^R = \frac{1 + r}{1 - r}.$$

For $w = w(t, \gamma)$ in $(5.4)'$, we have $R = t$ so that

$$(5.5)' \qquad e^t = \frac{1 + r}{1 - r},$$

and θ is the angle between β and γ. As β varies over B, $\langle w, \beta \rangle$ varies between

$$\log \frac{1 - r}{1 + r} \quad \text{and} \quad \log \frac{1 + r}{1 - r},$$

i.e., in view of $(5.5)'$, between $-t$ and t.

We introduce in B spherical polar coordinates around γ as pole; then

$$(5.6) \qquad d\beta = \sin\theta\, d\theta\, d\varphi = -d\rho\, d\varphi,$$

where

$$(5.7) \qquad \rho = \cos\theta \quad \text{and} \quad \varphi \text{ is the angle of rotation around } \gamma.$$

Furthermore we introduce the new variable

$$(5.8) \qquad \tau = \langle w, \beta \rangle = \log \frac{1 - r^2}{1 - 2r\rho + r^2}.$$

Clearly,

$$\frac{d\tau}{d\rho} = \frac{2r}{1 - 2r\rho + r^2},$$

so that

(5.9) $$e^{\langle w, \beta \rangle} d\beta = -\frac{1 - r^2}{1 - 2r\rho + r^2} \frac{d\rho}{d\tau} d\tau \, d\varphi = -\frac{1 - r^2}{2r} d\tau \, d\varphi.$$

Differentiating (5.3) with respect to t, we obtain:

(5.10) $$u_t(w, t) = -|c| \int e^{\langle w, \beta \rangle} k'(\langle w, \beta \rangle - t, \beta) \, d\beta$$

which by (5.9) is equal to

$$|c| \int \int \frac{1 - r^2}{2r} k'(\tau - t, \beta) \, d\tau \, d\varphi.$$

We shall show first, when k is C_0^1, that the limit (5.1) holds pointwise uniformly if s is restricted to a compact set. To this end we may replace, in (5.1), t by $t - s$ and $\sinh(t + s)$ by $\frac{1}{2} e^t$, and using (5.5)' and (5.10) write

(5.11)
$$e^t u_t(w(t, \gamma), t - s) = |c| \int_{-\pi}^{\pi} \int_{-t}^{t} \frac{1 + r}{1 - r} \frac{1 - r^2}{2r} k'(\tau - t + s, \beta) \, d\tau \, d\varphi$$

$$= |c| \int_{-\pi}^{\pi} \int_{-t}^{t} \frac{(1 + r)^2}{2r} k'(\tau - t + s, \beta) \, d\tau \, d\varphi.$$

Suppose that $k(s, \beta)$ is supported on $-l \leqq s \leqq l$; then the τ-integration in (5.11) is restricted to $(t - l - s, t)$. It follows easily from (5.5)' and (5.8) that if τ lies in this interval, then ρ tends to 1 as t tends to ∞. Since ρ is the cosine of the angle between β and γ it follows that, for $k \in C_0'$, β in (5.11) is restricted to a neighborhood of γ which shrinks to γ as $t \to \infty$. Hence for such k we can replace β by γ in (5.10) at the cost of an error that approaches 0 as $t \to \infty$. Also, as $t \to \infty$, r tends to 1; thus we can replace $(1 + r)^2 / 2r$ by 2, again at the cost of a small error. Consequently, (5.11) can be written as

$$e^t u_t(w(t, \gamma), t - s) = 4\pi |c| \int_{t-l-s}^{t} k'(\tau - t + s, \gamma) \, d\tau + \epsilon(t)$$

$$= 4\pi |c| k(s, \gamma) + \epsilon(t).$$

Since[1] $c = -(2\sqrt{2}\,\pi)^{-1}$, this shows that the relation (5.1) holds when k is C_0^1, uniformly for s restricted to a bounded set.

To complete the proof of Theorem 5.1 we need

LEMMA 5.2. *Denote by* $k(s,\gamma,t)$ *the function defined in (5.2) and denote the* $L_2(\mathbb{R}, B)$ *norm by* $|\cdot|$. *Then*

(i) $$|k(t)|^2 \leqq 2|k|^2,$$

(ii) $$|k(t)| \leqq (1 + \epsilon(t))|k|,$$

where the function $\epsilon(t)$ *tends to* 0 *as* $t \to \infty$; $\epsilon(t)$ *depends on* k.

Before giving the proof of Lemma 5.2 we show how it can be used to complete the proof of Theorem 5.1. So far we have seen that

(5.1)′ $$\lim_{t \to \infty} k(s, \gamma; t) = k(s, \gamma)$$

uniformly on compact subsets of $\mathbb{R} \times B$. According to part (ii) of Lemma 5.2,

$$\int\int |k(s, \gamma; t)|^2 \, ds \, d\gamma \leqq (1 + \epsilon(t))^2 \int\int |k(s, \gamma)|^2 \, ds \, d\gamma,$$

where $\epsilon(t) \to 0$ as $t \to \infty$. This combined with (5.1)′ implies that (5.1) holds in the L_2 sense for initial data f in C_0^1.

To prove (5.1) in the L_2 sense for all k in L_2 we merely observe that C_0^1 is a dense subset of L_2 and that by part (i) of Lemma 5.2 the L_2 norm of $k(t)$ is uniformly bounded by twice that of k. The result then follows by the principle of dense convergence.

We turn now to the proof of Lemma 5.2. Using the definition of $k(t)$ we have

(5.12)
$$|k(t)|^2 = \int\int_{-t}^{\infty} 2\sinh^2(t + s)|u_t(w(t + s, \gamma), t)|^2 \, ds \, d\gamma$$

$$= 2\int\int_0^{\infty} \sinh^2 s \, |u_t(w(s, \gamma), t)|^2 \, ds \, d\gamma.$$

---

[1] According to (3.27), $c = -(\sqrt{2}\,\pi)^{-1}$; this is because $m(B)$ was chosen to be π. If a factor 4 is introduced in (1.17) so that the measure of B if 4π, then the factor in (3.1), (3.7) and (3.10) becomes $(2\pi)^{-3/2}$, the factor in (3.12) becomes $-(2\pi)^{1/2}$, the factor in (3.13) and (3.14) becomes $(8\pi^2)^{-1}$ and c becomes $-(2\sqrt{2}\,\pi)^{-1}$.

Now the formula for the volume element in non-Euclidean polar coordinates is

$$\sinh^2 s \, ds \, d\gamma.$$

Hence (5.12) can be rewritten as

(5.12)′ $$|k(t)|^2 = 2|u_t(t)|^2,$$

i.e., as twice the integral of u_t squared over Π.

According to the definition of energy $|u|_E$ in (3.23),

$$|u|_E^2 = |u|_L^2 + |u_t|^2.$$

In view of the non-negativity of energy, see (3.24)′, we conclude that $|u_t|^2 \leqq |u|_E^2$. Finally, according to Theorem 3.4, formula (3.29), the translation representation is an isometry, i.e.,

(5.13) $$|u|_E^2 = |k|^2;$$

this completes the proof of part (i).

To prove part (ii) we need the following result on the equipartition of energy: Let $u(t)$ be a solution of the non-Euclidean wave equation which has finite energy. Then the total energy, defined by (3.23),

$$|u|_E^2 = |u|_L^2 + |u_t|^2$$

is an invariant of the motion; however the individual constituents $|u(t)|_L^2$ and $|u_t(t)|^2$ are not. Nevertheless the following asymptotic relation holds:

THEOREM 5.3. *As $t \to \infty$,*

(5.14) $$|u(t)|_L^2 / |u_t(t)|^2 \to 1.$$

Note that (5.14) implies that $2|u_t(t)|^2 \to |u|_E^2$ and this combined with (5.12)′ and (5.13) proves part (ii) of Lemma 5.2. The Euclidean version of Theorem 5.3 is known, see Corollary 2.3, Chapter IV of [1]; the key point of the proof is the orthogonality of $\mathcal{D}_-$ and $\mathcal{D}_+$. Since this no longer holds in the non-Euclidean case, an additional argument is needed here.

It suffices to prove Theorem 5.3 for a set of solutions dense in the sense of the energy norm. We choose for this dense subset the set of all data f whose outgoing translation representer $k = k(s, \beta)$ vanishes for $s < -b$, b some number.

We decompose such a k into

(5.15) $$k = k_N + k^{(N)},$$

where k_N contains all spherical harmonic components of order at most N and $k^{(N)}$ contains all of the rest. N is chosen so large that

$$(5.16) \qquad |k^{(N)}| < \epsilon.$$

We further decompose $k_N(s - t)$ into incoming and outgoing parts (see Theorem 3.9) as follows:

$$(5.17) \qquad k_N(s - t) = k_N^{(1)}(s, t) + k_N^{(2)}(s, t),$$

where $k_N^{(2)}$ is of the form

$$(5.18) \qquad k_N^{(2)}(s, \beta; t) = \begin{cases} \displaystyle\sum_{0 < m \leqq n \leqq N} a_{m,n}(t) e^{-ms} Y_n(\beta), & s \geqq 0, \\ 0, & s < 0, \end{cases}$$

the coefficients $a_{m,n}$ being chosen so that $k_N^{(2)}$ is orthogonal to $k_N^{(1)}$:

$$(5.19) \qquad \left(k_N^{(1)}(t), k_N^{(2)}(t) \right) = 0.$$

This can be accomplished by applying the inverse of the Gram matrix of the functions $e^{-ms} Y_n(\beta)$, $0 < m \leqq n \leqq N$ over $\mathbb{R}_+$ to the array of numbers

$$(5.20) \qquad \int \int_0^\infty k(s - t, \beta) e^{-ms} Y_n(\beta) \, ds \, d\beta.$$

Changing variables to $p = s - t$, and using the fact that $k(p) = 0$ for $p < -b$, we can rewrite (5.20) for $t > b$ as

$$(5.20)' \qquad e^{-mt} \int \int_{-b}^\infty k(p) e^{-mp} Y_n(\beta) \, dp \, d\beta.$$

Obviously the quantities (5.20)' tend to 0 as $t \to \infty$; consequently $a_{m,n}(t) \to 0$ as $t \to \infty$. It follows then from (5.18) that

$$(5.21) \qquad \lim_{t \to \infty} |k_N^{(2)}(t)| = 0.$$

We recall now the definitions of outgoing and incoming solutions of the non-Euclidean wave equations as ones vanishing in the respective forward and backward light cones: $\mathrm{dist}(w, j) < t$ and $\mathrm{dist}(w, j) < -t$, respectively. Clearly, if $u(t)$ is outgoing then $u(-t)$ is incoming. The subspaces $\mathcal{D}_+$ and $\mathcal{D}_-$ are defined as the Cauchy data of outgoing and incoming data, respectively.

Theorem 3.9 characterizes the outgoing translation representers of $\mathcal{D}_+$ as consisting of those $k(s, \beta)$ which vanish on $s < 0$; and the outgoing representers of $\mathcal{D}_-$ are characterized as those k which are of the form (5.18) for $s > 0$.

Since $k(s, \beta)$ is assumed to vanish for $s < -b$, $k(s - t_0, \beta)$, which is the representer of $(u(t_0), u_t(t_0))$, vanishes on $s \leqq 0$ when $t_0 \geqq b$. The same is true then of $k_N(s - t_0, \beta)$ and of $k_N^{(1)}(s, \beta; t_0)$.

Denote by $u_N^{(1)}(w, t; t_0)$ the solution whose initial data is represented by $k_N^{(1)}(s, \beta; t_0)$. Then it follows from Theorem 3.9 that

(i) $u_N^{(1)}(t_0)$ is outgoing,

(ii) the initial data of $u_N^{(1)}(t_0)$ is orthogonal to $\mathcal{D}_-$.

As noted earlier, it follows from (i) that $u_N^{(1)}(w, -t; t_0)$ is incoming; therefore by (ii) its initial data and those of $u_N^{(1)}(w, t; t_0)$ are orthogonal in the energy norm:

$$\{ u_N^{(1)}(w, 0; t_0), -u_{N,t}^{(1)}(w, 0; t_0) \} \perp \{ u_N^{(1)}(w, 0; t_0), u_{N,t}^{(1)}(w, 0, t_0) \}.$$

Thus the energy scalar product of these two is zero; i.e.,

$$(5.22) \qquad |u_N^{(1)}(0; t_0)|_L^2 - |u_{N,t}^{(1)}(0; t_0)|^2 = 0.$$

In other words, the energy contained in $u_N^{(1)}(t_0)$ is equipartitioned at $t = 0$.

By construction of $k^{(N)}$, $k_N^{(1)}$ and $k_N^{(2)}$,

$$k(s - t_0) = k_N^{(1)}(s, t) + k_N^{(2)}(s, t_0) + k^N(s - t_0).$$

Hence the same relation holds for the data represented by these functions, and for the solutions corresponding to these data:

$$(5.23) \qquad u(t + t_0) = u_N^{(1)}(t; t_0) + u_N^{(2)}(t; t_0) + u^{(N)}(t + t_0).$$

It follows from (5.16) that

$$(5.24)_1 \qquad\qquad\qquad |u^N(t_0)|_E < \epsilon.$$

Relation (5.21) implies that, for t_0 large enough,

$$(5.24)_2 \qquad\qquad\qquad |u_N^{(2)}(t_0)|_E < \epsilon.$$

Finally, by (5.22), energy is equipartitioned in $u_N^{(1)}(t; t_0)$ at $t = 0$; Therefore, we conclude from (5.23) that

$$\|u(t_0)\|_L - |u_t(t_0)\| < 2\epsilon.$$

This proves Theorem 5.3.

6. Invariant Integral Operators

Invariant integral operators were introduced by Selberg in [4]. These are operators which commute with motions. An integral operator K is of the form

(6.1) $$g(w) = Kf = \int_\Pi K(w, w') f(w') \, dw',$$

where dw' is the non-Euclidean volume element. Commutation requires that, for every motion τ,

$$K(f^\tau) = g^\tau,$$

where

$$f^\tau(w) = f(\tau^{-1}w).$$

It is easy to show that K commutes with motions if and only if its kernel satisfies the relation

$$K(\tau w, \tau w') = K(w, w').$$

Such a kernel is called a two-point invariant function by Selberg and is simply a function of the non-Euclidean distance $d(w, w')$. Recall that

$$\cosh d(w, w') = 1 + \frac{|w - w'|^2}{2zz'},$$

where $|\cdot|$ denotes Euclidean distance. We can therefore write

(6.2) $$K(w, w') = k\left(\frac{|w - w'|^2}{2zz'} \right)$$

so that (6.1) can be expressed as

(6.3) $$g(w) = \int_\Pi k\left(\frac{|w - w'|^2}{2zz'} \right) f(w') \, dw'.$$

In view of the above, the right side of (6.1) is like a convolution. In the Euclidean case the Radon transform of the convolution of two functions is the convolution of their Radon transforms with respect to the scalars; see [2] for an application of this observation. We now prove that in the non-Euclidean case the Radon transform of (6.1) is the convolution of the so-called *Selberg transform* of K with the Radon transform of f.

The simplest way to compute the general form of the Radon transform of Kf is to evaluate it at $\beta = \infty$; in this case the horospheres are the planes:

(6.4)
$$z = e^s.$$

Hence using (6.3) we get

$$\tilde{g}(s, \infty) = \int_{\mathbb{R}^2} g(w) \frac{dx \, dy}{z^2}$$

$$= \int \frac{dx \, dy}{z^2} \int k\left(\frac{(x - x')^2 + (y - y')^2 + (z - z')^2}{2zz'} \right) f(w') \, dw'$$

$$= \int_\Pi \left\{ \int_0^{2\pi} \int_0^\infty \frac{r \, dr \, d\theta}{z^2} k\left(\frac{r^2 + (z - z')^2}{2zz'} \right) \right\} f(w') \, dw',$$

where $r^2 = (x - x')^2 + (y - y')^2$. Introducing

$$v = \frac{r^2 + (z - z')^2}{2zz'}$$

as new variable, we get

(6.5)
$$\hat{g}(s, \infty) = \int_\Pi \left\{ 2\pi \int_{(z - z')^2/2zz'}^\infty k(v) \, dv \, \frac{z'}{z} \right\} f(w') \frac{dx' \, dy' \, dz'}{(z')^3}.$$

Performing the x', y'-integrations, making use of (6.4), and setting

(6.4)'
$$z' = e^t,$$

this becomes

(6.6)
$$\hat{g}(s, \infty) = 2\pi \int_{-\infty}^\infty \left\{ \int_{(z - z')^2/2zz'}^\infty k(v) \, dv \right\} \hat{f}(t, \infty) e^{t - s} \, dt.$$

Now

$$\frac{(z - z')^2}{2zz'} = \frac{1}{2}\left(\frac{z}{z'} + \frac{z'}{z} \right) - 1$$

$$= \cosh(s - t) - 1.$$

If we define the Selberg transform of K to be

$$(6.7) \qquad Q(s) = 2\pi \int_{\cosh s - 1}^{\infty} k(v) \, dv,$$

then we can write (6.6) as a convolution:

$$e^s \hat{g}(s, \infty) = \int_{-\infty}^{\infty} Q(s - t) e^t \hat{f}(t, \infty) \, dt.$$

By invariance this holds for any β:

$$(6.8) \qquad e^s \hat{g}(s, \beta) = \int_{-\infty}^{\infty} Q(s - t) e^t \hat{f}(t, \beta) \, dt.$$

Finally differentiating with respect to s and integrating the resulting expression on the right by parts, we obtain

$$(6.9) \qquad \frac{d^j}{ds^j} e^s \hat{g} = \int Q(s - t) \frac{d^j}{dt^j} (e^t \hat{f}) \, dt.$$

We define the action of the operator K on data $F = \{f_1, f_2\}$ by letting it act independently on each of the component functions f_1 and f_2. Denote by R the operator which assigns to F its, say, outgoing translation representation. From the definition (3.27) of R and formula (6.9) we deduce the simple relation

$$(6.10) \qquad \begin{aligned} RKF &= \int Q(s - t) RF \, dt \\ &= \int Q(t)(RF)(s - t) \, dt. \end{aligned}$$

Since R is a translation representation for the unitary group U (see (3.28)), we deduce from (6.10) that

$$(6.11) \qquad RKF = \int Q(t) R(U(t) F) \, dt,$$

which can be stated as:

THEOREM 6.1. *For the operator K we have the relation*

$$(6.11)' \qquad K = \int Q(t) U(t) \, dt,$$

where the integral is to be taken in the strong sense.

Thus every invariant integral operator can be expressed as a superposition of the operators $U(t)$ with the Selberg transform as weight factor.

7. Correction to the Proof of Theorem 3.8

The correct version of the argument goes as follows: Set (3.47) into (3.46); then, since $v(w) \equiv 0$, we get instead of (3.49)

$$(7.1) \qquad 0 = \partial_w^N v(w)|_{w=j} = \int k(s, \beta) Q_N(\partial_s, \beta) \phi(-s) \, ds \, d\beta$$

and instead of (3.51)

$$(7.2) \qquad 0 = \partial_w^N v(\rho w)|_{w=j} = \int k(s, \rho\gamma) Q_N(\partial_s, \gamma) \phi(-s) \, ds \, d\gamma.$$

Expand k into spherical harmonics:

$$(7.3) \qquad k(s, \beta) = \sum_{n,i} k_i^n(s) Y_n^i(\beta).$$

For any rotation ρ about j we have

$$(7.4) \qquad Y_n^i(\rho\gamma) = \sum_m e_{im}^n(\rho) Y_n^m(\gamma).$$

Combining (7.3) and (7.4) gives us the spherical harmonic expansion of $k(s, \rho\gamma)$. We set this expansion into (7.2), multiply by $e_{jh}^n(\rho)$ and integrate over all rotations ρ. In view of the orthogonality of the functions $e_{im}^n(\rho)$, we get

$$(7.5) \qquad 0 = \int k_j^n(s) Y_n^h(\gamma) Q_N(\partial_s, \gamma) \phi(-s) \, ds \, d\gamma.$$

Carrying out the γ integration, we obtain

$$(7.6) \qquad 0 = \int k_j^n(s) P_n(\partial_s) \phi(-s) \, ds,$$

where

$$(7.7) \qquad P_n(\partial_s) = \int Y_n^h(\gamma) Q_N(\partial_s, \gamma) \, d\gamma.$$

Because of (3.48) the leading coefficient of $P_N(\partial_s)$ is

$$\int \gamma^N Y_n^h(\gamma) \, d\gamma;$$

if we choose N with $|N| = n$ so that this integral is not 0, then $P_N(\partial_s) \neq 0$. We then proceed in the way it was done at the bottom of page 652 of [3].

Acknowledgement.   The work of both authors was supported in part by the National Science Foundation, that of the first author under Grant No. NSF MCS-76-07039 and that of the second under Grant No. NSF MCS-77-04908.

Bibliography

[1] Lax, P. D., and Phillips, R. S., *Scattering Theory*, Academic Press, New York, 1967.
[2] Lax, P. D., and Phillips, R. S., *The Paley–Wiener theorem for the Radon transform*, Comm. Pure Appl. Math. 23, 1970, pp. 409–424.
[3] Lax, P. D., and Phillips, R. S., *Translation representations for the solutions of the non-Euclidean wave equation*, Comm. Pure Appl. Math. 32, 1979, pp. 617–667.
[4] Selberg, A., *Harmonic analysis and discontinuous groups in weakly symmetric Riemannian spaces with applications to Dirichlet series*, Jr. Indian Math. Soc. 20, 1956, pp. 47–87.

Received October, 1980.

Reprinted from JOURNAL OF FUNCTIONAL ANALYSIS
All Rights Reserved by Academic Press, New York and London

Vol. 46, No. 3, May 1982
Printed in Belgium

The Asymptotic Distribution of Lattice Points in Euclidean and Non-Euclidean Spaces*

PETER D. LAX

Courant Institute of Mathematics, New York University,
New York, New York 10012

AND

RALPH S. PHILLIPS

Department of Mathematics, Stanford University,
Stanford, California 94305

Communicated by the Editors

Received October 22, 1981

DEDICATED TO ATLE SELBERG WITH ADMIRATION AND TREPIDATION

The asymptotic distribution of orbits for discrete subgroups of motions in Euclidean and non-Euclidean spaces are found; our principal tool is the wave equation. The results are new for the crystallographic groups in Euclidean space and for those groups in non-Euclidean spaces which have fundamental domains of infinite volume. In the latter case we show that the only point spectrum of the Laplace–Beltrami operator lies in the interval $(-((m-1)/2)^2, 0]$; furthermore we show that when the subgroup is nonelementary and the fundamental domain has a cusp, then there is at least one eigenvalue in this interval.

1. INTRODUCTION

Early on Gauss observed that the number of integer lattice points in a circle of radius s is equal to the area of this circle πs^2 to within an error of no more than the circumference, that is, $O(s)$. Since then lattice problems of this sort have been extensively studied.

The lattice points are the orbit of the origin under the action of the group of integer translations. This group can be replaced by any discrete subgroup of motions, for example, the group generated by any two linearly

* The work of the first author was supported in part by Department of Energy under Contract DE-AC02-76 ERO 3077 and the second by the National Science Foundation under Grant MCS-80-01943.

280

0022-1236/82/060280-71$02.00/0

independent vectors or the crystallographic groups; and the plane can be replaced by a Euclidean space of any dimension. The number of these orbital points inside a ball of large radius can then be computed asymptotically.

In the non-Euclidean analogue, we count the number of orbit points in the (non-Euclidean) ball of radius s about a point w which are generated by a discrete subgroup Γ of the motions of hyperbolic space:

$$N(s; w, w_0) = \#[\tau \in \Gamma; \operatorname{dist}(w, \tau w_0) \leqslant s]. \tag{1.1}$$

If Γ has a fundamental domain F of finite volume $|F|$, the main term in $N(s)$ is again equal to the volume of the ball of radius s divided by $|F|$:

$$\frac{\omega_m}{2^{m-1}(m-1)|F|} e^{(m-1)s}.$$

However, since the surface of the ball is also $O(e^{(m-1)s})$, a surface error approximation is useless. Moreover when $|F|$ is infinite, the above way of even estimating N is useless. However, it turns out that both $N(s)$ and the error term can be expressed in terms of the eigenvalues and the eigenfunctions of the Laplace–Beltrami operator, even when $|F| = \infty$.

The study of the counting number in the non-Euclidean setting was initiated by Huber [5, 6], who treated Fuchsian subgroups with compact fundamental domains in 2-dimensions. Somewhat later Patterson [12] treated all discrete Fuchsian subgroups in 2-dimensions with fundamental domains of finite area. Selberg (see [1]) solved the same problem in real hyperbolic spaces of arbitrary dimensions, again for fundamental domains of finite volume. Selberg's error estimates are significantly better than those of Huber and Patterson.

In the present work we extend the above non-Euclidean results to all discrete subgroups with the *finite geometric property*, that is, to all subgroups for which the polygonal representation of the fundamental domain has only a finite number of sides; the volume may be finite or infinite. For Γ a discrete subgroup with the finite geometric property, the Laplace–Beltrami operator has at most a finite number of eigenvalues $\mu_1 \geqslant \mu_2 \geqslant \cdots \geqslant \mu_N$ lying above the continuous spectrum: that is, $\mu_i > -((m-1)/2)^2$. Denote the corresponding normalized eigenfunctions by $\{\varphi_j; j = 1, ..., N\}$; $\varphi_1(w) > 0$ and is constant if and only if $|F| = \operatorname{vol}(F) < \infty$.

Our main result requires the existence of at least one μ_i; it can be stated as follows:

THEOREM 1. *Set*

$$\lambda_j = \sqrt{\mu_j + \left(\frac{m-1}{2}\right)^2} \quad and \quad \lambda = \lambda_1. \tag{1.2}$$

Define $\Sigma = \Sigma(s; w, w_0)$ *by*

$$\Sigma(s) = \frac{\omega_m}{2\sqrt{\pi}} \frac{m-2}{2}! \sum \frac{(\lambda_j - 1)!}{(\lambda_j + (m-1)/2)!} \varphi_j(w)\, \varphi_j(w_0)\, e^{((m-1)/2 + \lambda_j)s}, \qquad (1.3)$$

summed over the $\lambda_j > \lambda(m-1)/(m+1)$. *Then as* $s \to \infty$

$$|N(s; w, w_0) - \Sigma(s)|$$

$$\leqslant O\left(s^{3/(m+1)} \exp\left[(m-1)\left(\frac{1}{2} + \frac{\lambda}{m+1}\right)s\right]\right) \qquad \text{for} \quad m > 2,$$

$$\leqslant O\left(s^{5/6} \exp\left[\left(\frac{1}{2} + \frac{\lambda}{3}\right)s\right]\right) \qquad \text{for} \quad m = 2.$$

(1.4)

In the case where $|F| < \infty$ our error estimate differs from that of Selberg by a power of s. Selberg's proof uses a particular class of integral operators that commute with non-Euclidean motions; we use a different class, derived from the wave equation. This enables us to base our error estimate on the principle of conservation of energy for the wave equation.

It is clear from definition (1.1) that $N(s; w, w_0) = N(s, \gamma w, \gamma w_0)$ for any isometry γ for which $\gamma^{-1} \Gamma \gamma = \Gamma$. In the non-Euclidean problem the motions do not commute. It is not surprising therefore to see, when $|F| = \infty$ and hence when φ_1 is not constant, that the leading term in $N(s; w, w_0)$ depends on w and w_0.

In order to familiarize the reader with our method in a simpler setting, we prove the Euclidean analogue of Theorem 1 in Section 2:

THEOREM 2. *Let* Γ *be one of the crystallographic groups and set*

$$N(s; x, x_0) = \#[\tau \in \Gamma; |x - \tau x_0| \leqslant s]. \qquad (1.5)$$

The number of orbital points in a ball of radius s *in* $\mathbb{R}^m$ *about a point* x *is*

$$N(s; x, x_0) = V_m(s) + O(s^{m(m-1)/(m+1)}) \qquad \text{for} \quad m > 2, \qquad (1.6)$$

where

$$V_m(s) = \frac{\omega_m}{m\,|F|}\, s^m,$$

$\omega_m = $ *surface of the sphere of radius* 1 *in* $\mathbb{R}^m$, $|F| = $ *volume of the fundamental domain* F *of the group* Γ. *When* $m = 2$,

$$N(s; x, x_0) = \frac{\pi s^2}{|F|} + O(s^{2/3}(\log s)^{1/2}). \qquad (1.6)'$$

As far as we know, Theorem 2 is new for the crystallographic groups, except for groups of translations. In the latter case (1.6), without the logarithmic factor when $m = 2$, was proved by Landau [7, 8] in 1915. For integer (or "near" integer) lattices in dimensions $\geqslant 4$, considerably better estimates have been obtained by number-theoretic arguments; see the book [15] by Walfisz.

It might be of interest to know if better error estimates than the ones derived in this paper hold for the distribution of orbits of crystallographic groups whose translation subgroup is integral. The same question can be raised about arithmetic subgroups of non-Euclidean motions.

The organization of this paper is as follows: Section 2 contains the proof of Theorem 2. Sections 3 and 4 deal with automorphic solutions of the non-Euclidean wave operator. The main result here is that if Γ has the finite geometric property, then the energy form E is positive definite on a subspace of finite codimension. This is equivalent to the statement that the spectrum of the Laplace–Beltrami operator in the interval $[-((m-1)/2)^2, 0]$ consists of a finite number of eigenvalues. We also show in Therem 4.8 that if the fundamental domain has infinite volume, there are no point eigenvalues $\leqslant -((m-1)/2)^2$.

In Section 5 we prove Theorem 1, which requires that the Laplace–Beltrami operator have eigenvalues in the interval $(-((m-1)/2)^2, 0]$. If this is not the case then, as shown in Theorem 5.7,

$$N(s; w, w_0) = \mathrm{o}(se^{((m-1)/2)s}). \tag{1.7}$$

Section 6 contains examples of groups for which the energy form is positive definite, nonnegative and indefinite. We prove in Theorem 6.4 that if the fundamental domain contains a cusp of maximal rank and if Γ is nonelementary, then E is indefinite and the Laplace–Beltrami operator has eigenvalues in the interval $(-((m-1)/2)^2, 0]$. This extends to higher dimensions a result of Patterson [13].[1] Section 7 presents the derivation of explicit formulas for the solution to the non-Euclidean wave equation that are used in Section 5.

In a first reading of this paper the reader may find it helpful to limit himself to the three dimensional case.

---

[1] We thank Dennis Sullivan for acquainting us with this result of Patterson. A proof for $m \geqslant 2$ due to Sullivan is soon to appear [14].

2. The Density of Orbits for a Crystallographic Group in Euclidean Space

The main tool in our approach to the lattice point problem is the wave equation, which in Euclidean space of dimension m is

$$u_{tt} = \Delta u = \sum_1^m \partial_{x_i}^2 u. \tag{2.1}$$

This equation has solutions of the form (see [2, pp. 223, 226])

$$u(x, t) = c_m \left(\frac{1}{t} \partial_t\right)^{(m-3)/2} \frac{1}{t} \int_{|y|=t} f(x+y)\, d\sigma \qquad \text{for } m \text{ odd}, \tag{2.2}_0$$

$$u(x, t) = c_m \left(\frac{1}{t} \partial_t\right)^{(m-2)/2} \int_{|y|<t} \frac{f(x+y)}{\sqrt{t^2 - y^2}}\, dy \qquad \text{for } m \text{ even}, \tag{2.2}_e$$

where

$$c_m = \frac{\sqrt{\pi}}{2^{(m-1)/2}((m-2)/2)!} \qquad \text{for } m \text{ odd},$$

$$= \frac{1}{2^{(m-2)/2}((m-2)/2)!} \qquad \text{for } m \text{ even}. \tag{2.2}'$$

The initial values of u as given by (2.2) are

$$u(x, 0) = 0, \qquad u_t(x, 0) = \omega_m f(x); \tag{2.3}$$

here $\omega_m = 2\pi^{m/2}/((m-2)/2)!$ is the area of the unit sphere in $\mathbb{R}^m$.

If f is a function of x automorphic with respect to the group Γ, then so is $u(x, t)$ for each value of t. We shall write the automorphic function f in the form

$$f(x) = \sum h^\tau(x), \qquad h^\tau(x) = h(\tau^{-1}x). \tag{2.4}$$

We choose h of the form

$$h(x) = \frac{1}{\alpha^m} h_1 \left(\frac{x - x_0}{\alpha}\right), \tag{2.5}$$

where h_1 is smooth, nonnegative, supported in the ball $\{|x| < 1\}$ and normalized as

$$\int h_1(x)\, dx = 1. \tag{2.6}$$

It follows from this that

$$\int_{|\tau z - x| < T} h(z)\, dz = 1 \quad \text{if} \quad |x - \tau x_0| \leqslant T - \alpha$$

$$= 0 \quad \text{if} \quad |x - \tau x_0| \geqslant T + \alpha \tag{2.7}$$

and that the value of the integral lies between 0 and 1 otherwise.

We shall study the expression

$$I(T, \alpha) = c'_m \int_0^T (T^2 - t^2)^{(m-3)/2}\, tu(x, t)\, dt, \tag{2.8}$$

where

$$c'_m = \frac{2((m-2)/2)!}{\sqrt{\pi}\,((m-3)/2)!} \quad \text{for all } m. \tag{2.8}'$$

LEMMA 2.1. *The following identities hold:*

$$I_o = \int_0^T (T^2 - t^2)^{(m-3)/2}\, t \left(\frac{1}{t} \partial_t\right)^{(m-3)/2} \frac{1}{t} \int_{|y| = t} f(y)\, d\sigma\, dt$$

$$= 2^{(m-3)/2} \frac{m-3}{2}! \int_{|y| < T} f(y)\, dy \quad \text{for } m \text{ odd}, \tag{2.9}_o$$

$$I_e = \int_0^T (T^2 - t^2)^{(m-3)/2}\, t \left(\frac{1}{t} \partial_t\right)^{(m-2)/2} \int_{|y| < t} \frac{f(y)}{\sqrt{t^2 - y^2}}\, dy\, dt$$

$$= 2^{(m-4)/2} \frac{m-3}{2}!\, \sqrt{\pi} \int_{|y| < T} f(y)\, dy \quad \text{for } m \text{ even}. \tag{2.9}_e$$

Proof. Making use of the substitution $s = t^2$, $S = T^2$, I_o can be rewritten as

$$I_o = 2^{(m-5)/2} \int_0^S (S - s)^{(m-3)/2} \partial_s^{(m-3)/2} s^{-1/2} \int_{|y| = \sqrt{s}} f(y)\, d\sigma\, ds.$$

Integrating by parts $(m - 3)/2$ times, this becomes

$$I_o = 2^{(m-5)/2} \frac{m-3}{2}! \int_0^S s^{-1/2} \int_{|y| = \sqrt{s}} f(y)\, d\sigma\, ds$$

$$= 2^{(m-3)/2} \frac{m-3}{2}! \int_{|y| < T} f(y)\, dy.$$

We treat I_e similarly; setting $s = t^2$, $S = T^2$ and $r = y^2$,

$$I_e = 2^{(m-4)/2} \int_0^S (S-s)^{(m-3)/2} \, \partial_s^{(m-2)/2} \int_{y^2 < s} \frac{f(y)}{\sqrt{s-y^2}} \, dy \, ds$$

$$= 2^{(m-4)/2} \frac{((m-3)/2)!}{\sqrt{\pi}} \int_0^S \frac{1}{\sqrt{S-s}} \int_{y^2 < s} \frac{f(y)}{\sqrt{s-y^2}} \, dy \, ds.$$

Interchanging the order of integration, we get

$$I_e = 2^{(m-4)/2} \frac{((m-3)/2)!}{\sqrt{\pi}} \int_{y^2 < S} f(y) \left(\int_{y^2}^S \frac{ds}{\sqrt{(S-s)(s-y^2)}} \right) dy.$$

The inner integral is easily transformed into the Beta function $B(\frac{1}{2}, \frac{1}{2}) = \pi$ and hence

$$I_e = 2^{(m-4)/2} \frac{m-3}{2}! \sqrt{\pi} \int_{|y| < T} f(y) \, dy.$$

This concludes the proof of Lemma 2.1.

Setting (2.2) into (2.8) and making use of (2.4) and (2.9), we get

$$I(T) = \int_{|y-x| < T} f(y) \, dy = \sum \int_{|y-x| < T} h^\tau(y) \, dy$$

$$= \sum \int_{|\tau z - x| < T} h(z) \, dz. \qquad (2.10)$$

It follows from (2.7) and the definition (1.5) of $N(s; x, x_0)$ that

$$N(T-\alpha) \leqslant I(T) \leqslant N(T+\alpha), \qquad (2.11)$$

which is equivalent with

$$I(T-\alpha) \leqslant N(T) \leqslant I(T+\alpha). \qquad (2.11)'$$

In order to prove Theorem 2 we need an independent asymptotic evaluation of $I(T)$. The main term is obtained by splitting off the mean value of u. Define

$$m(t) = \frac{1}{|F|} \int_F u(x, t) \, dx; \qquad (2.12)$$

here $|F|$ is the volume of the fundamental domain F. Making use of Eq. (2.1), we see that for automorphic solutions of the wave equation

$$\frac{d^2}{dt^2} m(t) = \frac{1}{|F|} \int_F u_{tt} \, dx = \frac{1}{|F|} \int_F \Delta u \, dx = 0;$$

Thus m is a linear function:

$$m(t) = at + b. \tag{2.13}$$

From (2.13) we see that

$$a = \frac{1}{|F|} \int_F u_t(x, 0) \, dx, \qquad b = \frac{1}{|F|} \int_F u(x, 0) \, dx. \tag{2.13}'$$

For α small, the sum (2.4) has only a single nonzero term in F; so from (2.3)

$$u_t(x, 0) = \omega_m h(x), \qquad u(x, 0) = 0. \tag{2.14}$$

Setting this into (2.13)′ and using (2.6), we find that

$$a = \frac{\omega_m}{|F|} \int h(x) \, dx = \frac{\omega_m}{|F|}, \qquad b = 0,$$

and hence

$$m(t) = \frac{\omega_m}{|F|} t. \tag{2.15}$$

Next we decompose u as

$$u(x, t) = \frac{\omega_m}{|F|} t + v(x, t). \tag{2.16}$$

Clearly v satisfies the wave equation

$$v_{tt} = \Delta v, \tag{2.17}$$

and the mean value of v is zero for all t:

$$\int_F v(x, t) \, dx = 0. \tag{2.17}'$$

Note also that

$$v(x, 0) = u(x, 0) = 0. \tag{2.17}''$$

We set (2.16) into expression (2.8) for I and after a straightforward calculation we obtain

$$I(T, \alpha) = \frac{\omega_m}{m \, |F|} T^m + V(x, T), \tag{2.18}$$

where

$$V = V(x, T) = c'_m \int_0^T (T^2 - t^2)^{(m-3)/2} \, tv(x, t) \, dt. \tag{2.19}$$

We shall estimate V with the aid of the following version of Sobolev's inequality:[2]

$$|V(x)| \leqslant c \, \|V\|_{(m-1)/2}^{1/2} \, \|V\|_{(m+1)/2}^{1/2} \qquad \text{a.e.,} \qquad (2.20)$$

where $\|V\|_s$ denotes the H_s norm of V over F. For automorphic functions V whose mean value is zero

$$\|V\|_s = \|\varDelta^{s/2} V\|, \qquad (2.21)$$

where $\| \; \|$ denotes the L_2 norm over F.

LEMMA 2.2. *For $m > 2$,*

$$\|V\|_{(m+1)/2} \leqslant O(T^{(m-1)/2} \alpha^{-m/2}), \qquad (2.22)$$

$$\|V\|_{(m-1)/2} \leqslant O(T^{(m-1)/2} \alpha^{(2-m)/2}). \qquad (2.22)'$$

Combining this with (2.20) and making use of the fact that $V(x)$ is smooth, we obtain

$$|V(x)| \leqslant O(T^{(m-1)/2} \alpha^{(1-m)/2}). \qquad (2.23)$$

Setting this into (2.18) we get

$$I(T \pm \alpha, \alpha) = \frac{\omega_m}{m \, |F|} \, T^m + O(T^{m-1}\alpha) + O(T^{(m-1)/2} \alpha^{(1-m)/2}).$$

Combining this with (2.11)' gives

$$\left| N(T) - \frac{\omega_m}{m \, |F|} \, T^m \right| \leqslant O(T^{m-1}\alpha) + O(T^{(m-1)/2} \alpha^{(1-m)/2}). \qquad (2.24)$$

We now set

$$\alpha = T^{(1-m)/(1+m)}.$$

The resulting inequality is (1.6). This proves Theorem 2 for $m > 2$.

The proof of Lemma 2.2 eventually splits up into four cases depending on the residue of m mod 4. We set

$$m = 4q + r, \qquad r = -1, 0, 1, 2. \qquad (2.25)$$

We begin deriving some identities. Applying $\varDelta$ to (2.19), using Eq. (2.17) and integrating by parts gives

---

[2] In the original version of this paper we used a slightly different form of Sobolev's inequality, which led to less precise results (see the case $m = 2$ discussed at the end of this section). We thank Haim Brezis, Louis Nirenberg and Jaak Peetre for suggesting the use of (2.20) in its stead.

$$\Delta V = \int_0^T (T^2 - t^2)^{(m-3)/2} \, t \, \Delta v \, dt = \int_0^T (T^2 - t^2)^{(m-3)/2} \, tv_{tt} \, dt$$

$$= (T^2 - t^2)^{(m-3)/2} \, tv_t \big|_0^T - \int_0^T [\partial_t (T^2 - t^2)^{(m-3)/2} \, t] v_t \, dt \qquad \text{for} \quad m > 2$$

$$= - \int_0^T [\partial_t (T^2 - t^2)^{(m-3)/2} \, t] v_t \, dt \qquad \text{for} \quad m > 3$$

$$= -[\partial_t (T^2 - t^2)^{(m-3)/2} \, t] v \big|_0^T + \int_0^T [\partial_t^2 (T^2 - t^2)^{(m-3)/2} \, t] v \, dt \qquad \text{for} \quad m > 4.$$

$$\tag{2.26}$$

Recall that $v(0) = 0$. Hence

$$\Delta V = \int_0^T [\partial_t^2 (T^2 - t^2)^{(m-3)/2} \, t] v \, dt \qquad \text{for} \quad m > 5. \tag{2.26$'$}$$

Applying Δ to (2.26)$'$, using (2.17) and integrating by parts

$$\Delta^2 V = [\partial_t^2 (T^2 - t^2)^{(m-3)/2} \, t] v_t \big|_0^T$$

$$- \int_0^T [\partial_t^3 (T^2 - t^2)^{(m-3)/2} \, t] v_t \, dt \qquad \text{for} \quad m > 6. \tag{2.26$''$}$$

Notice that $(T^2 - t^2)^{(m-3)/2} t$ is odd in t so that any even number of differentiations will vanish at $t = 0$. Consequently

$$\Delta^2 V = - \int_0^T [\partial_t^3 (T^2 - t^2)^{(m-3)/2} \, t] v_t \, dt \qquad \text{for} \quad m > 7. \tag{2.26$'''$}$$

After q applications of Δ the following expressions hold:

$$\Delta^q V = [\partial_t^{2q-2} (T^2 - t^2)^{(m-3)/2} \, t] v_t \big|^T$$

$$- \int_0^T [\partial_t^{2q-1} (T^2 - t^2)^{(m-3)/2} \, t] v_t \, dt \qquad \text{for} \quad r = -1,$$

$$= - \int_0^T [\partial_t^{2q-1} (T^2 - t^2)^{(m-3)/2} \, t] v_t \, dt \qquad \text{for} \quad r = 0,$$

$$= -[\partial_t^{2q-1} (T^2 - t^2)^{(m-3)/2} \, t] v \big|^T \qquad \tag{2.27}$$

$$+ \int_0^T [\partial_t^{2q} (T^2 - t^2)^{(m-3)/2} \, t] v \, dt \qquad \text{for} \quad r = 1,$$

$$= \int_0^T [\partial_t^{2q} (T^2 - t^2)^{(m-3)/2} \, t] v \, dt \qquad \text{for} \quad r = 2.$$

Next we derive some inequalities for the function $v(x, t)$.

LEMMA 2.3. *The ineqality*

$$\|\Delta^p v_t\| \leqslant O(\alpha^{-2p-m/2}) \tag{2.28}$$

holds for all t and

$$
\begin{aligned}
-1 &\leqslant p \leqslant 1 & \text{when} \quad m > 4, \\
-\tfrac{1}{2} &\leqslant p \leqslant 1 & \text{when} \quad m = 3 \text{ and } 4.
\end{aligned}
\tag{2.29}
$$

When $m = 2$, (2.28) holds for $0 \leqslant p \leqslant 1$; whereas for $-\tfrac{1}{2} \leqslant p < 0$

$$\|\Delta^p v_t\| \leqslant O(\alpha^{-2p-1} |\log \alpha|^{-p}). \tag{2.28}'$$

The inequality

$$\|\Delta^p v\| \leqslant O(\alpha^{-2p+1-m/2}) \tag{2.30}$$

holds for all t and

$$
\begin{aligned}
-\tfrac{1}{2} &\leqslant p \leqslant 1 & \text{when} \quad m > 4, \\
0 &\leqslant p \leqslant 1 & \text{when} \quad m = 3 \text{ and } 4, \\
\tfrac{1}{2} &\leqslant p \leqslant 1 & \text{when} \quad m = 2.
\end{aligned}
\tag{2.31}
$$

Proof. In the proof of this lemma we shall use the conservation of energy principle: For any automorphic solution w of the wave equation, the energy in F is conserved; that is,

$$E(w) = \int_F (w_x^2 + w_t^2) \, dx \tag{2.32}$$

is independent of t.

Since v is an automorphic solution of the wave equation with mean value zero, so is $\Delta^p v$. From the definition of energy and its independence of t, we deduce that

$$\|\Delta^p v_t\|^2 \leqslant E(\Delta^p v) = E(\Delta^p v(0)) = \|\Delta^p v_t(0)\|^2; \tag{2.33}$$

in the last step we have used the fact that $v(0) = u(0) = 0$. Inequality (2.33) shows that it suffices to prove (2.28) at $t = 0$.

At $t = 0$ we know v_t explicitly. Combining (2.5), (2.14) and (2.16), we see that in the fundamental domain F

$$u_t(x, 0) = \alpha^{-m} h_1(x/\alpha), \tag{2.34}$$

$$v_t(x, 0) = \alpha^{-m} h_1(x/\alpha) - \text{const.} \tag{2.34}'$$

Consequently in F

$$|u_t(x, 0)| \leqslant \text{const. } \alpha^{-m} \qquad \text{for} \quad |x| < \alpha,$$
$$= 0 \qquad\qquad \text{elsewhere in } F. \tag{2.35}$$

Since v_t has mean value 0 and differs from u_t only by a constant, we deduce that

$$\|v_t(0)\| \leqslant \|u_t(0)\| = O(\alpha^{-m/2}). \tag{2.36}$$

This gives inequality (2.28) for $p = 0$; (2.28) for $p = 1$ follows immediately from (2.34)' by differentiation. To prove (2.28) for intermediate values between 0 and 1 we appeal to the following convexity principle:

Let A be any nonnegative operator in Hilbert space and f any vector in its domain. Then for $0 \leqslant p \leqslant 1$

$$\|A^p f\| \leqslant \|f\|^{1-p} \|Af\|^p. \tag{2.37}$$

This is easily proved using Hölder's inequality on the spectral representation for A.

Next consider negative values of p; we remark that the operator $\varDelta^{-1}$ is defined on functions whose mean value is zero. Nevertheless it is convenient to introduce the invertible operator M:

$$M = 1 - \varDelta. \tag{2.38}$$

Since 0 is an isolated point of the spectrum of $\varDelta$ with constant as eigenfunction, and since v_t is obtained from u_t by removing the constant component, it follows that

$$\|\varDelta^{-p} v_t\| \leqslant c \|M^{-p} v_t\| \leqslant c \|M^{-p} u_t\|. \tag{2.39}$$

Now M^{-1} is an integral operator whose kernel has a mild singularity along the diagonal; in fact the kernel is bounded by

$$\frac{\text{const.}}{|x - y|^{m-2}} \text{ for } m > 2 \qquad \text{and} \qquad \text{const.} |\log|x - y|| \text{ for } m = 2. \tag{2.40}$$

M^{-2} is obtained by composing M^{-1} with itself; an easy calculation shows that the kernel for M^{-2} is bounded by

$$\frac{\text{const.}}{|x - y|^{m-4}} \qquad \text{for} \quad m > 4. \tag{2.40'}$$

Using expression (2.34) for $u_t(x, 0)$, we deduce for $|x| < \alpha$ that

$$|M^{-2}u_t(0)|\,(x) \leqslant \frac{\text{const.}}{\alpha^m} \int_{|y|<\alpha} \frac{dy}{|x-y|^{m-4}}$$

$$\leqslant O(\alpha^{4-m}) \qquad \text{for} \quad m > 4,$$

$$|M^{-1}u_t(0)|\,(x) \leqslant \frac{\text{const.}}{\alpha^m} \int_{|y|<\alpha} \frac{dy}{|x-y|^{m-2}} \tag{2.41}$$

$$\leqslant O(\alpha^{2-m}) \qquad \text{for} \quad m > 2,$$

$$|M^{-1}u_t(0)|\,(x) \leqslant \frac{\text{const.}}{\alpha^2} \int_{|y|<\alpha} |\log|x-y||\,dy$$

$$\leqslant O(|\log \alpha|) \qquad \text{for} \qquad m = 2.$$

According to (2.6)

$$\int_F |u_t(x,0)|\,dx = \omega_m.$$

Consequently

$$\|M^{-1}u_t(0)\|^2 = (M^{-2}u_t(0), u_t(0))$$

$$\leqslant \text{const.} \max_{|x|<\alpha} |M^{-2}u_t(0)|\,(x) \int_F |u_t(x,0)\,dx$$

$$\leqslant O(\alpha^{4-m}) \qquad \text{for} \quad m > 4, \tag{2.42}$$

$$\|M^{-1/2}u_t(0)\|^2 = (M^{-1}u_t(0), u_t(0)) \leqslant O(\alpha^{2-m}) \qquad \text{for} \quad m = 3 \text{ and } 4,$$

$$\|M^{-1/2}u_t(0)\|^2 = (M^{-1}u_t(0), u_t(0)) \leqslant O(|\log \alpha|) \qquad \text{for} \quad m = 2.$$

Finally, applying the convexity principle with $A = M^{-1}$ for $m > 4$ and $A = M^{-1/2}$ for $m = 2, 3$ and 4, we obtain inequalities (2.28) and (2.28)'.

The proof of (2.30) is now an easy matter. Integration by parts, together with (2.32), gives

$$\|\Delta^{1/2}w\|^2 = (\Delta^{1/2}w, \Delta^{1/2}w) = -(\Delta w, w) = \int_F w_x^2\,dx \leqslant E(w).$$

Apply this to $w = \Delta^{p-1/2}v$ and we get

$$\|\Delta^p v\|^2 = \|\Delta^{1/2}\Delta^{p-1/2}v\|^2 \leqslant E(\Delta^{p-1/2}v).$$

Inequality (2.33) then gives

$$\|\Delta^p v\| \leqslant \|\Delta^{p-1/2}v_t(0)\|.$$

Thus (2.30) is a consequence of (2.28). This completes the proof of Lemma 2.3.

We turn now to the proof of Lemma 2.2. We treat the four cases in (2.25) separately.

Case (i): $m = 4q - 1$. To prove relation (2.22), we use (2.21) with $s = (m + 1)/2$:

$$\|V\|_{(m+1)/2} = \|\Delta^{(m+1)/4}V\| = \|\Delta^q V\|. \tag{2.43}$$

Next we use formula (2.27) for $\Delta^q V$, with $r = -1$:

$$\Delta^q V = [\partial_t^{2q-2}(T^2 - t^2)^{2q-2}\, t]\, v_t\,|^T - \int_0^T [\partial_t^{2q-1}(T^2 - t^2)^{2q-2}\, t]\, v_t\, dt. \tag{2.44}$$

Clearly

$$|\partial_t^{2q-2}(T^2 - t^2)^{2q-2}\, t| \leqslant \text{const. } T^{2q-1} = \text{const. } T^{(m-1)/2}$$

and

$$|\partial_t^{2q-1}(T^2 - t^2)^{2q-2}\, t| \leqslant \text{const. } T^{(m-3)/2}.$$

So we obtain from (2.44) the estimate

$$\|\Delta^q V\| \leqslant O(T^{(m-1)/2} \|v_t\|),$$

which by (2.28), $p = 0$, is $O(T^{(m-1)/2}\alpha^{-m/2})$. Combined with (2.43) this yields inequality (2.22) of Lemma 2.2.

As for (2.22)′, we see by (2.21) that

$$\|V\|_{(m-1)/2} = \|\Delta^{(m-1)/4}V\| = \|\Delta^{q-1/2}V\|. \tag{2.43′}$$

We now apply $\Delta^{-1/2}$ to (2.44) and, arguing as before, we get

$$\|\Delta^{q-1/2}V\| \leqslant O(T^{(m-1)/2} \|\Delta^{-1/2}v_t\|)$$

which by (2.28), $p = -1/2$, is $O(T^{(m-1)/2}\alpha^{(2-m)/2})$. Combined with (2.43)′, this proves (2.22)′.

Case (ii): $m = 4q + 1$. By (2.21)

$$\|V\|_{(m+1)/2} = \|\Delta^{(m+1)/4}V\| = \|\Delta^{q+1/2}V\|. \tag{2.45}$$

We now use formula (2.27), $r = 1$:

$$\Delta \quad ' = -[\partial_t^{2q-1}(T^2 - t^2)^{2q-1}t]v\,|^T + \int_0^T [\partial_t^{2q}(T^2 - t^2)^{2q-1}\, t]v\, dt. \tag{2.46}$$

Applying $\Delta^{1/2}$ to (2.46) and arguing as before, we get

$$\|\Delta^{q+1/2}V\| \leqslant O(T^{(m-1)/2}\|\Delta^{1/2}v\|),$$

which by (2.30), $p=\tfrac{1}{2}$, is $O(T^{(m-1)/2}\alpha^{-m/2})$. Combined with (2.45), this proves (2.22).

To prove (2.22)', we again appeal to (2.21):

$$\|V\|_{(m-1)/2} = \|\Delta^{(m-1)/2}V\| = \|\Delta^q V\|. \qquad (2.45)'$$

Arguing as before we see from (2.46) that

$$\|\Delta^q V\| \leqslant O(T^{(m-1)/2}\|v\|),$$

and using (2.30), $p=0$, this is $\leqslant O(T^{(m-1)/2}\alpha^{(2-m)/2})$. Combining this with (2.45)' we obtain (2.22)'.

Case (iii): $m = 4q$. In this case

$$\|V\|_{(m+1)/2} = \|\Delta^{(m+1)/4}V\| = \|\Delta^{q+1/4}V\|. \qquad (2.47)$$

We apply $\Delta^{1/4}$ to formula (2.27), $r = 0$:

$$\Delta^{q+1/4}V = -\int_0^T [\partial_t^{2q-1}(T^2 - t^2)^{2q-3/2}\, t]\, w_t\, dt$$

$$= \int_0^S + \int_S^T \equiv W_1 + W_2, \qquad (2.48)$$

where $w = \Delta^{1/4}v$; here S is some parameter $0 < S < T$ to be specified later. We shall estimate W_1 and W_2 separately.

In the case of W_1, we integrate by parts, obtaining

$$W_1 = -[\partial_t^{2q-1}(T^2 - t^2)^{2q-3/2}\, t]w\, |^S - \int_0^S [\partial_t^{2q}(T^2 - t^2)^{2q-3/2}\, t]w\, dt.$$

Now $[\partial_t^{2q-1}(T^2 - t^2)^{2q-3/2}\, t]$ can be expanded as a sum of terms of the kind

$$[\partial_t^k(T-t)^{2q-3/2}] \cdot [\partial_t^{2q-1-k}(T+t)^{2q-3/2}\, t], \qquad 0 \leqslant k \leqslant 2q-1.$$

The largest terms comes from $k = 2q - 1$ for which a suitable bound is const. $T^{2q-1/2}(T-t)^{-1/2}$. Likewise

$$|\partial_t^{2q}(T^2 - t^2)^{2q-3/2}\, t| \leqslant \text{const. } T^{2q-1/2}(T-t)^{-3/2}.$$

Using these bounds, an easy calculation shows that

$$\|W_1\| \leqslant O(T^{(m-1)/2}(T-S)^{-1/2})\|\Delta^{1/4}v\|.$$

Applying (2.30) with $p = \frac{1}{4}$ and setting

$$T - S = \alpha, \tag{2.49}$$

we obtain

$$\|W_1\| \leqslant O(T^{(m-1)/2}\alpha^{-m/2}).$$

We estimate W_2 directly, using (2.28) with $p = \frac{1}{4}$:

$$\|W_2\| \leqslant O\left((T^{(m-1/2)})\|\Delta^{1/4}v_t\|\int_S^T (T-t)^{-1/2}\, dt\right)$$

$$= O(T^{(m-1)/2}\alpha^{-(1+m)/2})(T-S)^{1/2}),$$

which by (2.47) is $O(T^{(m-1)/2}\alpha^{-m/2})$. This proves (2.22).

To prove (2.22)′, we write

$$\|V\|_{(m-1)/2} = \|\Delta^{(m-1)/4}V\| = \|\Delta^{q-1/4}V\|.$$

We apply $\Delta^{-1/4}$ to (2.27), $r = 0$, obtaining an expression for $\Delta^{q-1/4}V$ similar to (2.48) with $w = \Delta^{-1/4}v$ on the right. To this expression for $\Delta^{(q-1)/4}V$ we apply the same argument as above. This time we use (2.30) and (2.28) with $p = -\frac{1}{4}$. The resulting estimates for $\|W_1\|$ and $\|W_2\|$ yield (2.22)′.

Case (iv): $m = 4q + 2$. By (2.21)

$$\|V\|_{(m+1)/2} = \|\Delta^{(m+1)/4}V\| = \|\Delta^{q+3/4}V\|$$

and

$$\|V\|_{(m-1)/2} = \|\Delta^{(m-1)/4}V\| = \|\Delta^{q+1/4}V\|.$$

We therefore apply Δ^s, $s = \frac{3}{4}$ and $\frac{1}{4}$, to (2.27), $r = 2$:

$$\Delta^{q+s}V = \int_0^T [\partial_t^{2q}(T^2 - t^2)^{2q-1/2}\, t]\,\Delta^s v\, dt$$

$$= \int_0^S + \int_S^T \equiv W_1 + W_2. \tag{2.50}$$

We now rewrite W_1. Using the fact that v satisfies the wave equation and performing an integration by parts, we get

$$W_1 = \int_0^S [\partial_t^{2q}(T^2 - t^2)^{2q-1/2} t] \, \Delta^{s-1} \, \Delta v \, dt$$

$$= \int_0^S [\partial_t^{2q}(T^2 - t^2)^{2q-1/2} t] \, \Delta^{s-1} \, v_{tt} \, dt$$

$$= [\partial_t^{2q}(T^2 - t^2)^{2q-1/2} t] \, \Delta^{s-1} \, v_t |^S$$

$$- \int_0^S [\partial_t^{2q+1}(T^2 - t^2)^{2q-1/2} t] \, \Delta^{s-1} \, v_t \, dt. \qquad (2.51)$$

Estimating as before, using (2.28) with $p = -\frac{1}{4}$ when $s = \frac{3}{4}$ and $p = -\frac{3}{4}$ when $s = \frac{1}{4}$, and choosing S according to (2.49), we obtain for $m \geqslant 6$

$$\|W_1\| \leqslant O(T^{(m-1)/2}\alpha^{-m/2}) \qquad \text{when} \quad s = \tfrac{3}{4},$$

$$\|W_1\| \leqslant O(T^{(m-1)/2}\alpha^{1-m/2}) \qquad \text{when} \quad s = \tfrac{1}{4}.$$

Making use of (2.30) with $p = s$, $s = \frac{3}{4}$ and $\frac{1}{4}$, we obtain directly identical estimates for $\|W_2\|$. Combining all of the above inequalities we can verify (2.22) and (2.22)' for all values of $m \equiv 2 \pmod 4$ which are > 2.

This completes the proof of Lemma 2.2 and thereby proves Theorem 2, except in the case of $m = 2$. To treat $m = 2$ we use a different form of Sobolev's inequality:

$$|V(x)| \leqslant c(\varepsilon) \|V\|_{m/2 + 2\epsilon}, \qquad \varepsilon > 0, \qquad (2.52)$$

where the constant c satisfies

$$c(\varepsilon) = O(\varepsilon^{-1/2}). \qquad (2.52)'$$

By (2.21)

$$\|V\|_{1 + 2\epsilon} = \|\Delta^{1/2 + \epsilon} V\|. \qquad (2.53)$$

This leads us again to (2.50) with $q = 0$ and $s = \frac{1}{2} + \varepsilon$. We proceed as above, using (2.28)' with $p = -\frac{1}{2} + \varepsilon$ to estimate $\|W_1\|$ and (2.30) with $p = \frac{1}{2} + \varepsilon$ to estimate $\|W_2\|$. This gives

$$\|W_1\| \leqslant O(T^{1/2}(T - S)^{-1/2} \, \alpha^{-2\epsilon} \, |\log \alpha|^{1/2 - \epsilon}),$$

$$\|W_2\| \leqslant O(T^{1/2}(T - S)^{1/2} \, \alpha^{-1 - 2\epsilon}).$$

Combining these estimates with (2.52) and (2.53), we get

$$|V(x)| \leqslant \text{const.} \, \frac{T^{1/2}}{\alpha^{1/2}} \frac{\alpha^{-2\epsilon}}{\varepsilon^{1/2}} \left(\frac{\alpha^{1/2} \, |\log \alpha|^{1/2}}{(T - S)^{1/2}} + \frac{(T - S)^{1/2}}{\alpha^{1/2}} \right). \qquad (2.54)$$

In this case the optional choice for S is

$$T - S = \alpha |\log \alpha|^{1/2}.$$

With $\varepsilon = |\log \alpha|^{-1}$, (2.54) becomes

$$|V(x)| \leqslant \text{const.} \frac{T^{1/2}}{\alpha^{1/2}} |\log \alpha|^{3/4}.$$

Setting this into (2.18) we get

$$I(T \pm \alpha, \alpha) = \frac{\pi T^2}{|F|} + O(T\alpha) + O(T^{1/2}\alpha^{-1/2} |\log \alpha|^{3/4}).$$

Combining this with (2.11)′ gives

$$\left| N(T) - \frac{\pi T^2}{|F|} \right| \leqslant O(T\alpha) + O(T^{1/2}\alpha^{-1/2} |\log \alpha|^{3/4}).$$

Finally, choosing $\alpha = T^{-1/3}(\log T)^{1/2}$ yields

$$\left| N(T) - \frac{\pi T^2}{|F|} \right| \leqslant O(T^{2/3}(\log T)^{1/2}). \tag{2.55}$$

Relation (2.55) is inequality (1.6)′ and with this we conclude the proof of Theorem 2 in its entirety.

3. The Energy Form for Automorphic Data

In the rest of this paper we study the distribution of lattice points in a non-Euclidean setting, using methods analogous to those of the previous section. For purposes of exposition we shall limit our considerations in this and the next section to three space dimensions; however, the results hold and the methods of proof work for real hyperbolic spaces of any dimension.

The model $\mathbb{H}_3$ we shall use consists of points $\{x_1, x_2, y\}$, $y > 0$, with the Riemannian metric

$$ds^2 = \frac{dx_1^2 + dx_2^2 + dy^2}{y^2}. \tag{3.1}$$

The isometries for $\mathbb{H}_3$ can be described by means of the 2×2 complex-valued matrices

$$\begin{pmatrix} a & b \\ c & d \end{pmatrix}, \qquad \text{where} \qquad ad - bc = 1 \tag{3.2}$$

in the following fashion: To each point (x_1, x_2, y) of $\mathbb{H}_3$ we assign a quaternion w with three components

$$w = x_1 + x_2 i + yj,$$

where $i^2 = j^2 = -1$ and $ij = -ji$. Every isometry of $\mathbb{H}_3$ can then be written as

$$\tau w = (aw + b)(cw + d)^{-1}. \tag{3.3}$$

The invariant volume element is

$$dw = \frac{dx_1\, dx_2\, dy}{y^3}; \tag{3.4}$$

the invariant Dirichlet integral can be written as

$$\int y^2 \, |\partial u|^2 \, dw; \tag{3.5}$$

and the invariant Laplace–Beltrami operator is

$$L_0 u = y^3 \, \partial \frac{1}{y} \, \partial u. \tag{3.6}$$

Automorphic functions are defined by means of a discrete subgroup Γ of the isometries by the property

$$u(\gamma w) = u(w) \tag{3.7}$$

for all γ in Γ. Such a function is uniquely determined by its values in a *fundamental domain*: $F = \mathbb{H}_3/\Gamma$. We shall limit our considerations to subgroups for which the fundamental domains satisfy the *finite geometric property*. This means that F may be chosen as a polyhedron with a finite number of sides. On the other hand F may have either *finite* or *infinite volume*.

The fundamental domain F can be regarded as a Riemannian manifold; if Γ contains elliptic motions, F is a manifold with a finite number of singular geodesics; these are the fixed points of the elliptic motions in Γ. This manifold can be described by a finite number of conformal charts. We divide the charts for such a manifold into three categories:

(1) Interior charts mapped onto the ball: $\operatorname{dist}(w, (0, 0, 1)) < 1$ or, if they correspond to points on the fixed geodesic of an elliptic motion, onto a sector of the ball: $\tan^{-1} x_2/x_1 < 2\pi/n$, n an integer $\geqslant 2$.

(2) Regular charts at infinity mapped onto the half-ball: $\{x_1^2 + x_2^2 + y^2 < 1, \, y > 0\}$ or, if they correspond to a fixed point at infinity of an elliptic motion, onto a sector of the half-ball: $\tan^{-1} x_2/x_1 < 2\pi/n$, n an integer $\geqslant 2$.

(3)   Charts of neighborhoods of points β at ∞ which are fixed points. The shape of the fundamental domain near β depends on the structure of the subgroup Γ_β leaving β fixed. If Γ_β is generated by a single parabolic motion, the chart is mapped onto the portion of the slab: $\{-1/2 < x_2 < 1/2\} \cdot$ (or the half-slab: $\{-1/2 < x_2 < 1/2, x_1 > 0\}$) exterior to the sphere $\{x_1^2 + x_2^2 + y^2 = 1\}$ in the upper half-plane. For a doubly parabolic point the chart is mapped onto the portion of a prism with vertical generators lying above $y = a$. Such a prism may have as cross section a parallelogram, a triangle or a hexagon.

Since all charts are conformal, where two charts overlap the mapping from one to the other can be realized by a fractional linear transformation of the kind described in (3.3).

We shall treat the lattice point problem in non-Euclidean space by means of the wave equation

$$u_{tt} = Lu = L_0 u + u. \tag{3.8}$$

With the zero order term u included on the right, the solutions satisfy Huygens' Principle. We denote the initial conditions:

$$u(w, 0) = f_1(w), \qquad u_t(w, 0) = f_2(w)$$

by the data pair $f = \{f_1, f_2\}$. One can either suppose that the automorphic wave acts on automorphic data defined on all of $\mathbb{H}_3$, or on data defined on the fundamental domain F with suitably periodic boundary conditions or simply on data defined on F considered as a manifold. We shall take the latter point of view in what follows.

We need a suitable Hilbert space framework for our analysis. Our starting point is the energy form for (3.8), that is,

$$E(f, g) = -(Lf_1, g_1) + (f_2, g_2). \tag{3.9}$$

Here (u, v) denotes the $L_2(F)$ inner product. It is easy to see that E is constant in time for solutions of the wave equation. For f, g in $C_0^\infty(F)$ an integration by parts brings (3.9) into a more symmetric form:

$$E(f, g) = \int_F (y^2 \, \partial f_1 \, \bar{\partial} g_1 - f_1 \bar{g}_1 + f_2 \bar{g}_2) \, dw. \tag{3.10}$$

As such E is obviously locally indefinite and it will be shown by examples in Section 6 to be globally indefinite for some F.

We now construct a related locally definite form

$$G = E + K \tag{3.11}$$

216

which differs from E by a form K which will be shown to be compact with respect to G. As we shall prove, all such G forms are equivalent. Our Hilbert space $\mathcal{H}$ is then obtained as the completion with respect to G of $C_0^\infty(F)$ data. The following calculations involve only $C_0^\infty(F)$ data.

We first choose a finite $C^\infty(\bar{F})$ partition of unity for $\bar{F}$ subordinate to charts of F; by $\bar{F}$ we mean F plus the points at infinity with the Euclidean topology of the charts. We denote these functions by $\{\omega_j\}$, $\{\varphi_j\}$ and $\{\psi_j\}$ according to whether they are supported on interior, regular infinite and parabolic charts, respectively. We require that the support of each of the φ's and ψ's overlap with that of at least one of the ω's. We also require that the ω's, φ's and ψ's be nonnegative, that their supports have connected interiors and that the ψ's be *identically one near the parabolic fixed points*.

Clearly, the indefiniteness of E occurs only in the first component; the contribution of the second component is positive definite. Therefore in what follows up to Theorem 3.4, we shall restrict our attention to data whose second component is zero. This being the case, we can omit the subscript 1 and let f stand for the first component.

We now set

$$E_j^\omega(f) = \int \omega_j(y^2 |\partial f|^2 - |f|^2) \, dw \qquad (3.12)_\omega$$

with similar expressions for E_j^φ and E_j^ψ. We also set

$$K_j^\omega(f) = 2 \int \omega_j |f|^2 \, dw \qquad (3.13)$$

and we define

$$G_j^\omega(f) = E_j^\omega(f) + K_j^\omega(f) = \int \omega_j(y^2 |\partial f|^2 + |f|^2) \, dw. \qquad (3.14)$$

Clearly G_j^ω is $\geqslant 0$.

Next we use the local chart coordinates to bring E^φ and E^ψ into a more convenient, but no longer invariant, form. We begin by writing

$$\int \varphi y \left| \partial_y \left(\frac{f}{y} \right) \right|^2 dy = \int \varphi y \left| \frac{\partial_y f}{y} - \frac{f}{y^2} \right|^2 dy$$

$$= \int \varphi \left(\frac{|\partial_y f|^2}{y} - \frac{\partial_y |f|^2}{y^2} + \frac{|f|^2}{y^3} \right) dy.$$

An integration by parts gives

$$\int \varphi \frac{\partial_y |f|^2}{y^2} \, dy = -\int \left(\varphi_y \frac{|f|^2}{y^2} - 2\varphi \frac{|f|^2}{y^3} \right) dy.$$

Combining the above and inserting the resulting expression into E^φ we get

$$E_j^\varphi(f) = \int \varphi_j y^4 \left| \partial \left(\frac{f}{y} \right) \right|^2 dw - K_j^\varphi(f), \qquad (3.15)_\varphi$$

where

$$K_j^\varphi(f) = \int (\partial_y \varphi_j) y \, |f|^2 \, dw; \qquad (3.16)_\varphi$$

in each of these formulas the integration is over the corresponding chart. We also define

$$G_j^\varphi(f) = E_j^\varphi(f) + K_j^\varphi(f) = \int \varphi_j y^4 \left| \partial \left(\frac{f}{y} \right) \right|^2 dw. \qquad (3.17)_\varphi$$

Finally we define

$$K = \sum K_j^\omega + \sum K_j^\varphi + \sum K_j^\psi \qquad (3.18)$$

and

$$G = E + K = \sum G_j^\omega + \sum G_j^\varphi + \sum G_j^\psi. \qquad (3.19)$$

It is evident from (3.14) and $(3.17)_{\varphi,\psi}$ that G is a locally positive form. We need an estimate on the local L_2 norm of the data:

LEMMA 3.1. *For any compact set S in F there is a positive constant c_S such that*

$$\int_S |f|^2 \, dw \leqslant c_S \, G(f). \qquad (3.20)$$

Proof. The following is an easy variant of the Poincaré inequality:

Let D be a connected domain in $\mathbb{R}^3$, whose closure is compact, and D_0 a nonempty subdomain of D. Then there exist constants a and b depending only on D and D_0 such that for every smooth function g defined on D

$$\int_D |g|^2 \, dx \leqslant a \int_D |\partial g|^2 \, dx + b \int_{D_0} |g|^2 \, dx. \qquad (3.21)$$

Using this fact, we now show that for δ small enough there is a constant c_δ such that or every ball B_δ radius δ and center a point of S,

$$\int_{B_\delta} |f|^2 \, dw \leqslant c_\delta \, G(f). \qquad (3.22)$$

Since S is compact, it can be covered by a finite number of balls B_δ of this kind; so (3.20) is a consequence of (3.22).

As a preliminary step we enlarge S to ensure that, while still compact, it has the following additional properties:

(i) S is connected,

(ii) S contains an interior point in the support of one of the functions ω_j.

We then choose δ so small that in each ball B_δ at least one of the functions ω_j, φ_j or ψ_j is $\geqslant \kappa > 0$. This is possible since

$$\sum \omega_j + \sum \varphi_j + \sum \psi_j = 1$$

is a finite partition of unity and since each of these functions is uniformly continuous on the compact set S. By property (ii), for δ small enough, there is at least one ball B_δ^0 on which $\omega_j \geqslant \kappa$. It follows then from (3.14) that for this ball (3.22) holds with $c_\delta = 1/\kappa$.

Next let B_δ be a ball in which, say, $\varphi_j \geqslant \kappa$. It follows then from $(3.17)_\varphi$ that

$$\int_{B_\delta} |\partial g|^2 \, dx \leqslant \frac{1}{\kappa Y} G(f),\tag{3.23}$$

where

$$g = \frac{f}{y}\tag{3.24}$$

and $Y = \min y$ in B_δ. Suppose B_δ intersects another ball B_δ' for which (3.22) holds. Applying (3.21) with $D = B_\delta$, $D_0 = B_\delta \cap B_\delta'$ we conclude, using (3.23), that (3.22) holds for B_δ as well, possibly with a larger constant c_δ. Since S is connected, each ball B_δ can be linked by a finite chain of balls to B_δ^0. This proves (3.22) and thereby Lemma 3.1.

LEMMA 3.2.   *Let S be any compact subset of F; then*

$$\int_S |f|^2 \, dw\tag{3.25}$$

is compact with respect to $G(f)$.

Proof.   As before, it suffices to prove this for $S = B_\delta$. This follows from inequalities (3.20) and (3.23), by means of the Rellich compactness criterion.

The following is the basic result of this section:

THEOREM 3.3.   *Let K and G be the quadratic forms defined in* (3.18), (3.19). *Then*

(i)   *K is compact with respect to G,*

(ii)   *E is bounded with respect to G.*

Proof.   Since $G = E + K$, (ii) follows from (i). To prove (i) we show that each term in the sum K is compact. We note that all terms K_j^ω, and each K_j^ψ associated with a multiple parabolic point is of the form (3.25) and therefore compact by Lemma 3.2.

To show the compactness of the remaining K_j^ψ and of the K_j^φ we break each up into two parts, depending on a parameter a: In the first part the y integration is extended over $y > a$, in the second over $y < a$. By Lemma 3.2, the first part is compact; to complete the proof of Theorem 3.3 we shall show that the second part, K_a, satisfies

$$K_a(f) \leqslant \varepsilon_a G(f), \tag{3.26}$$

where ε_a tends to zero as a tends to 0.

It follows from $(3.16)_{\varphi,\psi}$ that K_a is bounded by a sum of quadratic forms

$$\iiint_0^a \frac{|f|^2}{y^2} \, dy \, dx_1 \, dx_2, \tag{3.27}$$

the integrals being confined to the support of $\partial\varphi$ and $\partial\psi$.

Let b denote any positive number $>a$;

$$\left| \frac{f(y)}{y} \right|^2 = \left| \int_y^b \partial_y \left(\frac{f}{y} \right) dy - \frac{f(b)}{b} \right|^2$$

$$\leqslant 2 \left| \log \frac{b}{y} \right| \int_y^b y \left| \partial_y \left(\frac{f}{y} \right) \right|^2 dy + 2 \left| \frac{f(b)}{b} \right|^2. \tag{3.28}$$

Let S_a denote a tube of the form

$$|x_1 - x_1^0|^2 + |x_2 - x_2^0|^2 < \delta, \qquad 0 < y < a.$$

Integrating (3.28) over S_a gives

$$\int_{S_a} \left| \frac{f(w)}{y} \right|^2 dx_1 \, dx_2 \, dy \leqslant 2 \int_0^a \log \frac{b}{y} dy \int_{S_b} y \left| \partial_y \left(\frac{f}{y} \right) \right|^2 dx_1 \, dx_2 \, dy$$

$$+ 2a \int_{D_0} \left| \frac{f(x_1, x_2, b)}{b} \right|^2 dx_1 \, dx_2 \, dy,$$

where D_0 is the base of the tube S_a. Integrating this with respect to b from b to $2b$ and divide by b; we get

$$\int_{S_a} \left| \frac{f(w)}{y} \right|^2 dx_1 \, dx_2 \, dy \leqslant \varepsilon \int_{S_{2b}} y \left| \partial_y \left(\frac{f}{y} \right) \right|^2 dx_1 \, dx_2 \, dy$$

$$+ \frac{2a}{b^3} \int_{S_{2b} \setminus S_b} |f|^2 \, dx_1 \, dx_2 \, dy, \tag{3.29}$$

where

$$\varepsilon = 2 \int_0^a \log \frac{2b}{y} \, dy. \tag{3.30}$$

220

Suppose that $\varphi \geqslant c > 0$ on S_{2b}; then using $(3.17)_\varphi$ we see that the first quadratic form on the right in (3.29) is bounded by

$$\frac{\varepsilon}{c} G^\varphi(f). \tag{3.31}_1$$

By (3.20), the second term on the right in (3.29) is bounded by

$$a \text{ const. } G(f). \tag{3.31}_2$$

The sum of these two is $\leqslant \varepsilon_a G(f)$, where ε_a tends to 0 as a tends to 0.

Since we are dealing with a finite partition of unity there is a $c > 0$ such that in a neighborhood of any nonparabolic point at infinity at least one φ or ψ is greater than c. It may of course happen that the calculation (3.29) has to be made in one coordinate system in which the φ or ψ is greater than c and then transformed into another overlapping coordinate system. The change of coordinates is accomplished by means of a fractional linear mapping

$$w' = \tau(w) = (aw + b)(cw + d)^{-1},$$

where $ad - bc = 1$ and $w = -d/c$ is not in the overlap. Thus the Jacobian of the map is well behaved on S_a and since $y' = y |cw + d|^{-2}$ we see that for w in S_a

$$\alpha y < y' < \frac{1}{\alpha} y,$$

for some $\alpha > 0$. Thus for any tube S'_a contained in $\tau(S_a)$ we obtain from (3.29)

$$\int_{S'_a} \left| \frac{f}{y'} \right|^2 dx'_1 dx'_2 dy' \leqslant C \int_{S_a} \left| \frac{f}{y} \right|^2 dx_1 dx_2 dy \leqslant \varepsilon'_a G(f).$$

Since the infinite boundary of F, excluding a neighborhood of the parabolic points where the ψ's are identically 1, can be covered by a finite set of such tubes, it follows that K restricted to $y < a$ in the charts tends to zero as $a \to 0$, uniformly for all data f of G-norm $\leqslant 1$. This completes the proof of Theorem 3.3.

Remark. It is clear from the proof of Theorem 3.3 that the assertion also holds if we replace K by $\bar{K}$, defined by

$$\bar{K} = \sum \bar{K}_j^\omega + \sum \bar{K}_j^\varphi + \sum \bar{K}_j^\psi, \tag{3.18}'$$

where $\bar{K}_j^\omega = K_j^\omega$ but

$$\bar{K}_j^\circ(f) = \int |\partial_y \varphi_j| \, y \, |f|^2 \, dw. \tag{3.16$)'_\omega$}$$

The equivalence of two G forms obtained by different partitions of unity follows from the following general result:

THEOREM 3.4. *Suppose that G_0 and G_1 are positive forms on a Hilbert space $\mathscr{H}_1$ complete with respect to G_1. If G_0 is continuous with respect to G_1 and $K_1 = G_1 - G_0$ is compact with respect to G_1, then G_0 and G_1 are equivalent forms on $\mathscr{H}_1$.*

Proof. It follows from the hypothesis of the theorem that there is a compact selfadjoint operator T associated with K_1 defined as

$$K_1(f, g) = G_1(Tf, g). \tag{3.32}$$

We claim that all of the eigenvalues of T are less than 1. For if

$$Tf = \lambda f,$$

then by (3.32)

$$G_0(f) = (1 - \lambda) \, G_1(f).$$

Since $G_0(f) > 0$ by assumption, it follows that $\lambda < 1$. Since T is compact its eigenvalues can accumulate only at 0. Denote by λ the largest of these. As shown above $\lambda < 1$ and since

$$K_1(f) \leqslant \lambda G_1(f)$$

we see that $G_0(f) = G_1(f) - K_1(f) \geqslant (1 - \lambda) \, G_1(f)$. Since G_0 is assumed to be continuous with respect to G_1. This proves the equivalence of G_0 and G_1.

We will say that a positive form G' is *admissible* relative to the energy form E if E is continuous with respect to G' and if $K' = G' - E$ is compact with respect to G'. Let $\mathscr{H}'$ denote the completion of $C_0^\infty(F)$ data in the G'-norm. It follows from Theorem 3.3 that G defined as above is an admissible form. We now show that two different partitions of unity result in equivalent forms.

COROLLARY 3.5. *If G and G' are both admissible relative to E, then the corresponding Hilbert spaces $\mathscr{H}$ and $\mathscr{H}'$ are the same and G and G' are equivalent.*

Proof. Define a third form G_1 as

$$G_1 = \tfrac{1}{2}(G + G')$$

and let $\mathscr{H}_1$ denote the completion of $C_0^\infty(F)$ data with respect to G_1. Obviously $\mathscr{H}_1$ is contained in $\mathscr{H} \cap \mathscr{H}'$ and both G and G' are continuous with respect to G_1. Also since K and K' are compact with respect to G and G', respectively, they are *a fortiori* compact with respect to G_1. Finally we note that

$$G_1 = E + (K + K')/2 = G + (K' - K)/2$$

and hence that $K_1 = G_1 - G$ is compact with respect to G_1. It then follows by Theorem 3.4 that $\mathscr{H}_1 = \mathscr{H}$ and that G and G_1 are equivalent on $\mathscr{H}_1$. Likewise $\mathscr{H}_1 = \mathscr{H}'$ and G and G' are equivalent on H_1. We conclude that $\mathscr{H} = \mathscr{H}'$ and that G and G' are equivalent.

THEOREM 3.6. *There is a space of finite codimension on which E is positive.*

Proof. Since K is compact with respect to G, there is a subspace of $\mathscr{H}$ of finite codimension on which

$$|K(f)| \leqslant \tfrac{1}{2}G(f).$$

On this subspace

$$E(f) = G(f) - K(f) \geqslant \tfrac{1}{2}G(f). \tag{3.33}$$

THEOREM 3.7. *If E is positive on a closed subspace $\mathscr{H}''$ of $\mathscr{H}$, then E and G are equivalent on H''.*

Proof. This is a direct consequence of Theorem 3.4.

In Appendix 6 we give examples of subgroups for which E is positive, nonnegative and indefinite.

Remark. The following result is occasionally useful: Let θ be a smooth function, uniformly bounded with uniformly bounded first derivatives, which is constant in a neighborhood of each parabolic point. Then multiplication by θ maps $\mathscr{H}_G$ into itself. This can be proved by using the same techniques as those employed in the proof of Theorem 3.3.

4. AUTOMORPHIC SOLUTIONS OF THE NON-EUCLIDEAN WAVE EQUATION

The non-Euclidean wave equation

$$u_{tt} = Lu \tag{4.1}$$

was defined in Eq. (3.8). Recall that

$$L = L_0 + I, \tag{4.2}$$

where L_0 denotes the Laplace–Beltrami operator. We are interested in the automorphic solutions of (4.1). To construct such solutions we solve the initial value problem

$$u(w, 0) = f_1(w), \qquad u_t(w, 0) = f_2(w) \tag{4.3}$$

for w in all of non-Euclidean 3-space $\mathbb{H}_3$ and for initial data $f = \{f_1, f_2\}$ that are automorphic with respect to some given discrete subgroup Γ. Since L is invariant under Γ and since solutions of (4.1) are uniquely determined by their initial data (4.3), it follows that for τ in Γ

$$u^\tau(w, t) = u(w, t);$$

here $u^\tau(w, t) = u(\tau^{-1}w, t)$.

Now the initial value problem (4.1), (4.3) with $C_0^\infty(\mathbb{H}_3)$ initial data can be solved by standard methods from the theory of hyperbolic equations; alternately one can write an explicit formula expressing the solution $u(w, t)$ as an integral of f_1 and f_2 (see Section 7 for such formulas). The solutions thus constructed are C^∞ functions of w and t which, because of finite speed of propagation, are of compact support in w for each t.

To construct a solution to the initial value problem with $C^\infty(\mathbb{H}_3)$ automorphic data, one makes use of a locally finite C^∞ partition of unity: $\sum \varphi_i = 1$. Solving for $u_i(w, t)$ with $C_0^\infty(\mathbb{H}_3)$ data $f_i = \varphi_i f$, the automorphic solution is then

$$u(w, t) = \sum u_i(w, t);$$

because of finite speed of propagation this sum has only a finite number of nonzero terms on any compact subset of $\mathbb{H}_3 \times \mathbb{R}$. It is clear that if we start with automorphic data in $C_0^\infty(F)$, then the solution will be C^∞ with compact support in F at each fixed t. As already observed in Section 3, energy is conserved, that is,

$$E(u(t)) = -(Lu, u) + (u_t, u_t) \tag{4.4}$$

is independent of t; here $u(t)$ on the left stands for the data $\{u, u_t\}$ at time t and $(\ ,\ )$ denotes the L_2 scalar product on F.

THEOREM 4.1. *Let $u(w, t)$ be any C^∞ automorphic solution of (4.1), of compact support in F for each t. There is a constant c, independent of u, such that for all solutions of this kind*

$$G(u(t)) \leqslant e^{c|t|} G(u(0)). \tag{4.5}$$

Proof. By definition (3.19), $G = E + K$. Therefore

$$\frac{d}{dt} G(u(t)) = \frac{d}{dt} E(u(t)) + \frac{d}{dt} K(u(t))$$

$$= \frac{d}{dt} K(u(t)) \tag{4.6}$$

since, as remarked above, $E(u(t))$ is independent of t. We shall make use of the following description of K.

PROPOSITION 4.2. *The quadratic form K, defined in Section 3, can be written as*

$$K(u) = \int_F k(w) |u(w)|^2 \, dw, \tag{4.7}$$

where

(i) $k(w)$ *is bounded,*

(ii) $k(w)$ *tends to 0 as w tends to ∞.*

Proof. Recall that $k = K^\omega + K^\varphi + K^\psi$, where K^ω, K^φ and K^ψ are defined as in (3.13) and (3.16)$_{\varphi, \psi}$. Each term is of the form (4.7) and since $\psi \equiv 1$ near the parabolic points at ∞, we conclude that each of these satisfies (i) and (ii). Therefore so does their sum.

We note that $k(w)$ need not be positive. Because of this we introduce still another form

$$\hat{K}(u) = \int_F |k(w)| |u(w)|^2 \, dw. \tag{4.7}$$

According to the remark following the proof of Theorem 3.3

$$\hat{K}(u) \leqslant \bar{K}(u) \leqslant \text{const. } G(u). \tag{4.8}$$

From (4.7) and the Schwarz inequality

$$\frac{d}{dt} K(u(t)) = 2 \, \text{Re} \int_F k u \, \bar{u}_t \, dw$$

$$\leqslant \int_F (k^2 |u|^2 + |u_t|^2) \, dw$$

$$\leqslant |k|_{\max} \hat{K}(u(t)) + \int_F |u_t|^2 \, dw, \tag{4.9}$$

where $|k|_{max}$ denotes the maximum of $|k(w)|$ on F. Recall that $G = G^\omega + G^\varphi + G^\psi$; by definitions (3.14) and (3.17)$_{\varphi, \psi}$ of G^ω, G^φ and G^ψ, we conclude that

$$\int_F |u_t|^2 \, dw \leqslant G(u(t)).$$

Combining this with (4.6), (4.8) and (4.9) we see that

$$\frac{d}{dt} G(u(t)) \leqslant \text{const. } G(u(t)),$$

from which (4.5) follows by Gronwall's inequality.

The Hilbert space $\mathscr{H}$ was defined as the completion in the G-norm of all $C_0^\infty(F)$ data. It follows from (4.5) that the operator $U(t)$, mapping $C_0^\infty(F)$ initial data into $C_0^\infty(F)$ data at time t, is bounded in the G-norm. Therefore we can, by continuity, extend the operator $U(t)$ to all data in the Hilbert space $\mathscr{H}$; we denote the so extended operator also by $U(t)$. For the extended operator, $E(U(t)f)$ is again indepedent of t.

The exponential increase indicated by inequality (4.5) is absent when E is positive definite. For in that case Theorem 3.7 implies that E and G are equivalent forms and as noted above $E(u(t))$ is constant in t. On the other hand when E takes on negative values, the G-form will grow at an exponential rate due, as we shall show, to the existence of solutions of the form

$$u(w, t) = e^{\pm \lambda t} \varphi(w), \qquad \lambda > 0. \tag{4.10}$$

Setting (4.10) into (4.11) shows that φ satisfies

$$L\varphi = \lambda^2 \varphi \tag{4.11}$$

and that the initial data for the solution (4.10) are

$$\{\varphi, \pm \lambda \varphi\}.$$

The second component of data in $\mathscr{H}$ belongs to $L_2(F)$; since $\lambda > 0$, it follows that φ above belongs to $L_2(F)$. Thus exponentially increasing (or decreasing) solutions of the kind (4.10) correspond to the positive point eigenvalues of L in $L_2(F)$.

We are thus led to study L as a selfadjoint operator on $L_2(F)$; for definiteness we will denote this operator by L'. For this purpose it is convenient to introduce the Sobolev space defined by means of the norm

$$\|u\|_1 = \left[\int_F (y^2 |\partial u|^2 + \kappa^2 |u|^2) \, dw \right]^{1/2} \tag{4.12}$$

for fixed $\kappa > 0$. Let $\mathscr{H}'$ denote the completion of $C_0^\infty(F)$ with respect to this norm. $\mathscr{H}'$ is obviously independent of $\kappa > 0$. Note that $\mathscr{H}'$ is contained in the set of first components of $\mathscr{H}$ and that

$$\|u\|_1^2 = E'(u) + (\kappa^2 + 1)\|u\|^2, \tag{4.12}'$$

where $E'(u)$ abbreviates the energy of $\{u, 0\}$.

In terms of $\mathscr{H}'$ we set

$$D(L') = [u \in \mathscr{H}'; Lu \in L_2(F) \text{ in the weak sense}]. \tag{4.13}$$

THEOREM 4.3. *L' is a selfadjoint operator on $L_2(F)$.*

Proof. Since $D(L')$ contains $C_0^\infty(F)$, it is clear that $D(L')$ is dense in $L_2(F)$. Further for u in $D(L')$ and θ in $C_0^\infty(F)$ we can write

$$(L'u, \theta) = (u, L\theta) = -\int (y^2 \, \partial u \, \overline{\partial \theta} - u\bar\theta) \, dw,$$

as in (3.10). Let v be any function in $\mathscr{H}'$; taking a sequence of θ_n's in $C_0^\infty(F)$ converging to v in $\mathscr{H}'$, we deduce from the above identity that

$$-(L'u, v) + (\kappa^2 + 1)(u, v) = (u, v)_1. \tag{4.14}$$

It follows from this that L' is closed; for if u_n in $D(L')$ converges to u and $L'u_n$ converges to g, both in the $L_2(F)$ norm, then we see from (4.14) that u_n converges weakly in the $\mathscr{H}'$ sense. This proves that u belongs to $\mathscr{H}'$ and obviously $L'u = g$ in the weak sense. Moreover if v also belongs to $D(L')$, then it follows from the symmetry of the right side of (4.14) that $(L'u, v) = (u, L'v)$, that is, that L' is symmetric.

It therefore suffices to show that the range of $\lambda - L'$ is dense in $L_2(F)$ for some $\lambda > 1$. Clearly

$$|(g, v)| \leqslant \|g\| \|v\| \leqslant \frac{1}{\kappa} \|g\| \|v\|_1.$$

According to the Riesz representation theorem there is a u in $\mathscr{H}'$ satisfying the relation

$$(u, v)_1 = (g, v)$$

for all v in $\mathscr{H}'$. In particular for v in $C_0^\infty(F)$ we can integrate by parts and write this as

$$(u, (\kappa^2 + 1)v - L'v) = (g, v),$$

from which it follows that $L'u$ exists in the weak sense. Thus u belongs to $D(L')$ and

$$(\kappa^2 + 1)u - L'u = g.$$

This completes the proof of Theorem 4.3.

Remark. If $\{\theta_n\} \subset C_0^\infty(F)$ converges to u in $\mathcal{H}'$, then it is clear from (3.10) and (4.7) that the data $\{\theta_n, 0\}$ converge to $\{u, 0\}$ in the G norm. Consequently $\{u, 0\}$ belongs to the Hilbert space of data $\mathcal{H}$. Moreover for u in $D(L')$, it is clear from (4.12)' and (4.14) that

$$E(\{u, 0\}) = -(L'u, u). \tag{4.15}$$

THEOREM 4.4.   *The positive spectrum of L' is finite dimensional; it has at least one positive eigenvalue iff E takes on negative values in $\mathcal{H}$.*

Proof. It is clear from (4.15) that if L' has a positive eigenvalue, then E takes on negative values in $\mathcal{H}$. Conversely, suppose that E takes on negative values in $\mathcal{H}$. According to Theorem 3.6, E is positive definite in a subspace of $\mathcal{H}$ of finite codimension. If the positive spectrum of L' were infinite dimensional then, since $D(L') \subset \mathcal{H}'$ and since $\mathcal{H}'$ is contained in the first component space of $\mathcal{H}$, it would follow that E would be negative on an infinite dimensional subspace of $\mathcal{H}$, contrary to the assertion of Theorem 3.6. Hence the positive spectrum of L' is finite dimensional and thus pure point spectrum.

Denote the positive eigenvalues of L', in decreasing order by λ_j^2, $j = 1, ..., N$, and the corresponding eigenfunctions by φ_j. Then we conclude from Theorem 4.4 that

COROLLARY 4.5.   *If f belongs to $\mathcal{H}'$ and is L_2-orthogonal to $\{\varphi_j, j = 1, ..., N\}$, then $E'(f) \geqslant 0$.*

Remark. It also follows from Theorem 3.6 that the null space of L' is finite dimensional.

LEMMA 4.6. $\lambda_1 = 1$ *if and only if* $|F| < \infty$.

Proof. It is clear from (4.14) and (4.12) that for u in $D(L')$

$$((u - L'u), u) = \int_F y^2 |\partial u|^2 \, dw \geqslant 0. \tag{4.16}$$

Hence in any case

$$(\varphi_1, \varphi_1) \geqslant (L'\varphi_1, \varphi_1) = \lambda_1(\varphi_1, \varphi_1)$$

so that $\lambda_1 \leqslant 1$. Now for $|F| < \infty$, $\varphi = \text{const.}$ is clearly an eigenfunction with eigenvalue 1. Conversely suppose that $\lambda_1 = 1$; then it follows from (4.16) that

$$\int_F y^2 |\partial \varphi_1|^2 \, dw = 0$$

and hence that $\varphi_1 = \text{const.}$ In order to be in $L_2(F)$ and nontrivial we see that $|F| < \infty$. This concludes the proof of the lemma.

We now restrict attention to initial data in $\mathscr{H}$ whose first component lies in $L_2(F)$ and hence in $\mathscr{H}'$. These form a subspace $\mathscr{H}_1$ of $\mathscr{H}$ which is not closed in the G-norm. We denote by $\mathscr{H}_2$ that subspace of $\mathscr{H}_1$ whose first and second components are both $L_2(F)$-orthogonal to all of the eigenfunctions $\{\psi_j, j = 1, ..., N\}$ of L' with positive eigenvalues.

THEOREM 4.7.   (i)   *The operators $U(t)$ map $\mathscr{H}_1$ onto itself*;

  (ii) $U(t)$ *maps $\mathscr{H}_2$ onto itself*;

  (iii)   *Let $u(w, t)$ be an automorphic solution of the wave equation* (4.1) *whose initial data belong to $\mathscr{H}_2$. Then*

$$\|u(t)\| \leqslant \|u(0)\| + t E^{1/2}(u). \tag{4.17}$$

Proof.   (i)   It is clear from (3.14) and (3.17)$_{\psi, \varphi}$ that

$$\|u_t\|^2 \leqslant G(u). \tag{4.18}$$

We now differentiate, using the Schwarz inequality and (4.18):

$$2 \|u\| \frac{d}{dt} \|u\| = \frac{d}{dt} \|u\|^2 = 2 \operatorname{Re}(u, u_t)$$

$$\leqslant 2 \|u\| \|u_t\| \leqslant 2 \|u\| \, G^{1/2}(u),$$

from which we obtain

$$\frac{d}{dt} \|u(t)\| \leqslant G^{1/2}(u(t)).$$

Integrating with respect to t and using (4.5) gives

$$\|u(t)\| \leqslant \|u(0)\| + \int_0^t G^{1/2}(u(s)) \, ds$$

$$\leqslant \|u(0)\| + \frac{1}{c} e^{c|t|} G^{1/2}(u, (0)). \tag{4.19}$$

This proves part (i).

(ii) Suppose next that $u(w, t)$ is a $C_0^\infty(F)$ solution of (4.1); its initial data belong to $\mathscr{H}_1 = \mathscr{H}' \times L_2(F)$. Projecting the initial data into $\mathscr{H}_2$ results in the solution

$$v(w, t) = u(w, t) - \sum a_j \varphi_j e^{\lambda_j t} - \sum b_j \varphi_j e^{-\lambda_j t}, \qquad (4.20)$$

where

$$\begin{aligned} a_j &= \tfrac{1}{2}[(u(0), \varphi_j) + (u_t(0), \varphi_j)/\lambda_j], \\ b_j &= \tfrac{1}{2}[(u(0), \varphi_j) - (u_t(0), \varphi_j)/\lambda_j]. \end{aligned} \qquad (4.20)'$$

To show that $v(t)$ remains in $\mathscr{H}_2$ we consider the function

$$c(t) = (v(t), \varphi),$$

where φ is any one of the eigenfunctions of L' with positive eigenvalue. Differentiating and using (4.1), the symmetry of L' and (4.11), we get

$$\begin{aligned} c_{tt} &= (v_{tt}, \varphi) = (L'v, \varphi) = (v, L'\varphi) \\ &= \lambda^2(v, \varphi) = \lambda^2 c. \end{aligned}$$

Since $\{v(0), v_t(0)\}$ belong to $\mathscr{H}_2$, it follows that $c(0) = 0 = c_t(0)$. Since $c(t)$ satisfies a second order ordinary differential equation we conclude that $c(t) = 0$ (and $c_t(t) = 0$) for all t; in other words, $v(t)$ belongs to $\mathscr{H}_2$ for all t.

To obtain assertion (ii) for arbitrary initial data in $\mathscr{H}_2$, we approximate these data in the $\mathscr{H}' \times L_2(F)$ metric by $C_0^\infty(F)$ data. Projecting the $C_0^\infty(F)$ data into $\mathscr{H}_2$ as above, we see that the projected solutions remain in $\mathscr{H}_2$ and converge to the solution of the given initial data.

(iii) The proof in this case is essentially the same as that of part (i), except that we exploit the fact that according to Corollary 4.5 energy is nonnegative for data in $\mathscr{H}_2$. Consequently (4.18) can be replaced by

$$\|u_t\|^2 \leqslant E(u).$$

We use part (ii) which assures us that $\{u(t), u_t(t)\}$ remains in $\mathscr{H}_2$ for all t; unlike $G(u(t))$, which grows exponentially, $E(u(t))$ remains constant. The resulting inequality, replacing (4.19), is (4.17). This complete the proof of Theorem 4.7.

When F is of finite volume, 1 is an eigenvalue with eigenfunction $\equiv 1$ and multiplicity one. L' may have further positive eigenvalues all <1 but finite in number. In addition L' can have negative eigenvalues interspersed with its continuous spectrum (see [9]). We now show that when F is of infinite volume, L' has no nonpositive eigenvalues. A similar result was obtained by Patterson [11] in the two dimensional case when Γ is a nonelementary subgroup of $SL(2, R)$ of the second kind. A Euclidean analogue of this result involving only a partial neighborhood of infinity has recently been proved by Littman [10].

THEOREM 4.8. *If u is a solution of*

$$Lu = -\mu^2 u, \qquad \mu \geqslant 0,$$

in a connected domain in $\mathbb{H}_3$ *containing a regular neighborhood of a point at* ∞ *and if u is square integrable in this neighborhood, then* $u \equiv 0$. *In particular* L' *has no nonpositive eigenvalues.*

Proof. One can map the half-ball $\{x_1^2 + x_2^2 + y^2 < 1, y > 0\}$ of a regular neighborhood at ∞ into the half-space $\{x_2 > 0, y > 0\}$ by an isometry. In this half-space we introduce cylindrical coordinates:

$$x = x_1, \qquad \rho = \sqrt{x_2^2 + y^2}, \qquad \theta = \tan^{-1} y/x_2. \tag{4.21}$$

It is convenient to make a change of scale for θ:

$$s = \int_\theta^{\pi/2} \frac{d\theta}{\sin \theta}; \tag{4.21'}$$

here s is arclength along circular arcs centered on the x-axis. In terms of these coordinates the half-space is $\{x \in \mathbb{R}; \rho, s > 0\}$ and the relation $L'u = -\mu^2 u$ becomes

$$\rho^2 \operatorname{sech}^2 s \left(\partial_x^2 u + \partial_\rho^2 u + \partial_s \frac{1}{\rho^2 \operatorname{sech}^2 s} \partial_s u \right) = -\mu^2 u. \tag{4.22}$$

Finally we make a change in the dependent variable:

$$u = (\operatorname{sech} s)v$$

in terms of which (4.22) becomes

$$\partial_s^2 v + \rho^2 \operatorname{sech}^2 s \, (\partial_x^2 v + \partial_\rho^2 v) = -\mu^2 v. \tag{4.22'}$$

The invariant L_2 integral transforms as

$$\int u^2 \, dw = \int v^2 \frac{dx \, d\rho}{\rho^2} \, ds. \tag{4.23}$$

We now take the non-Euclidean Fourier transform in the $\{x, \rho\}$ plane (see Helgason [4]):

$$\tilde{v}(\lambda, b, s) = \frac{1}{2\pi} \int e^{(1/2 - i\lambda)\langle\{x,\rho\}, b\rangle} v(x, \rho, s) \frac{dx \, d\rho}{\rho^2}; \tag{4.24}$$

here b corresponds to a point on the x-axis with x coordinate b and

$$\langle\{x,\rho\},b\rangle = \log\frac{1+b^2}{(x-b)^2+\rho^2}\rho.$$

There is a Plancherel theorem relating v and $\tilde{v}$:

$$\int|v(x,\rho,s)|^2\frac{dx\,d\rho}{\rho^2} = \int|\tilde{v}(\lambda,b,s)|^2\,dm,\qquad(4.25)$$

where $dm = 2\lambda\tanh\pi\lambda(db/(1+b^2))\,d\lambda$ and the range of integration of the right member of (4.25) is $\{0<\lambda,\ b\in\mathbb{R}\}$. It follows from (4.23) and (4.25) that $\tilde{v}(\lambda,b,s)$ belongs to $L_2(\mathbb{R}_+)$ in s for almost all λ,b.

The action of the Laplace–Beltrami operator $\rho^2(\partial_x^2+\partial_\rho^2)$ on the Fourier transform is multiplication by $-(\lambda^2+1/4)$. It follows from (4.22)$'$ that $\tilde{v}$ satisfies (at least in the weak sense) the ordinary differential equation

$$\frac{d^2\tilde{v}}{ds^2}+[\mu^2-(\tfrac{1}{4}+\lambda^2)\operatorname{sech}^2 s]\tilde{v}=0,\qquad\text{a.e. in }\lambda,b.\qquad(4.26)$$

Since $\operatorname{sech} s$ decays exponentially at infinity, a perturbation argument shows when $\mu>0$ that $\tilde{v}(s)$ behaves like

$$ae^{i\mu s}+be^{-i\mu s}$$

for large s. Consequently $\int|\tilde{v}(s)|^2\,ds$ is either 0 or infinite. We proved above that this integral is finite and hence 0 for almost all λ,b. It follows that $v(x,\rho,s)$ and hence $u(w)$ vanishes almost everywhere in this neighborhood of ∞.

When $\mu=0$ we argue somewhat differently. In this case if $\int|\tilde{v}(s)|^2\,ds<\infty$ one can easily show that (4.26) is equivalent to the integral equation

$$\tilde{v}(s)=\int_s^\infty(\tfrac{1}{4}+\lambda^2)(s'-s)\operatorname{sech}^2 s'\tilde{v}(s')\,ds'.\qquad(4.27)$$

A change of variables:

$$r=e^{-s},\qquad\tilde{V}(r)=r^{-1/2}\tilde{v}(\log 1/r)$$

transforms (4.27) into a regular Volterra integral operator equation on $L_2(0,1)$:

$$\tilde{V}(r)=(1+4\lambda^2)\int_0^r\sqrt{\frac{r'}{r}}\log\frac{r}{r'}\frac{r'}{(1+r'^2)^2}\,\tilde{V}(r')\,dr',\qquad(4.27)'$$

form which it follows that the only solution is the trivial one. So again v and hence u vanishes in this neighborhood of infinity.

Finally making use of unique continuation for the elliptic operator L, it follows that if u vanishes in a neighborhood of ∞ and if u satisfies $Lu = -\mu^2 u$, then u vanishes throughout its domain of definition. This completes the proof of Theorem 4.8. It should be noted that Theorem 4.8 is valid in any number of dimensions.

5. THE DENSITY OF ORBITS FOR A DISCRETE GROUP IN NON-EUCLIDEAN SPACE

Let Γ be a discrete group of motions of the m-dimensional hyperbolic space with the *finite geometric property*. In this case, as we have shown in Section 4, the modified Laplace–Beltrami operator L has only a finite number N of positive eigenvalues and L is nonpositive on the L_2-orthogonal complement of these eigenfunctions φ_j:

$$(v, Lv) \leqslant 0 \qquad \text{if} \quad (v, \varphi_j) = 0 \quad \text{for} \quad j = 1, 2, ..., N. \tag{5.1}$$

In the proof of Theorem 5.1 we *assume* that the set of positive eigenvalues is nonempty; we denote them by $\lambda_1^2, ..., \lambda_N^2$, arranged in decreasing order. We choose each λ_j to be positive and set $\lambda = \max \lambda_j$.

Note that when the fundamental domain F has finite volume $|F|$, then $\varphi_1 \equiv |F|^{-1/2}$ is in $L_2(F)$ and is an eigenfunction with eigenvalue $((m-1)/2)^2$. According to Lemma 4.6, $\lambda^2 = ((m-1)/2)^2$ is the largest eigenvalue when $|F| < \infty$ and when $|F| = \infty$, $\lambda^2 < ((m-1)/2)^2$.

Since Γ is discrete, the orbits $\{\tau w\}$ of any point w of $\mathbb{H}_m$ tend to infinity. The rate at which they tend to ∞ is measured by the *counting number* N defined as:

$$N(T; w, w_0) = \#[\tau; \operatorname{dist}(w_0, \tau w) \leqslant T]. \tag{5.2}$$

It turns out that the asymptotic behavior of N for large T is governed by the positive eigenvalues of L' and the corresponding eigenfunctions, normalized as usual by

$$\int_F |\varphi_j|^2 \, dw = 1. \tag{5.3}$$

THEOREM 5.1. *Define* $\Sigma = \Sigma(T; w, w_0)$ *by*

$$\Sigma(T) = \frac{\omega_m}{2\sqrt{\pi}} \frac{m-2}{2}! \sum \frac{(\lambda_j - 1)!}{(\lambda_j + (m-1)/2)!} \varphi_j(w) \, \varphi_j(w_0) \, e^{((m-1)/2 + \lambda_j)T}; \tag{5.4}$$

233

summed over the $\lambda_j > [(m-1)/(m+1)]\lambda$. Then as $T \to \infty$

$$|N(T) - \Sigma(T)| \leqslant O(T^{3/(m+1)}) \exp\left[(m-1)\left(\frac{1}{2} + \frac{\lambda}{m+1}\right)T\right] \qquad \text{for } m > 2,$$

(5.5)

$$|N(T) - \Sigma(T)| \leqslant O(T^{5/6}) \exp\left[\left(\frac{1}{2} + \frac{\lambda}{3}\right)T\right] \qquad \text{for } m = 2.$$

Our proof follows the Euclidean analogue, with a few changes where necessary. In this case we employ the non-Euclidean wave equation

$$u_{tt} = Lu. \tag{5.6}$$

This equation has, in analogy with (2.2), solutions of the form (see Section 7)

$$u(w, t) = c_m \left(\frac{1}{\sinh t}\partial_t\right)^{(m-3)/2} \frac{1}{\sinh t} \int_{d(w, w') = t} f(w')\, d\sigma$$

for m odd,

(5.7)$_o$

$$u(w, t) = c_m \left(\frac{1}{\sinh t}\partial_t\right)^{(m-2)/2} \int_{d(w, w') < t} \frac{f(w')}{\sqrt{\cosh t - \cosh d}}\, dw',$$

for m even,

(5.7)$_e$

where

$$c_m = \frac{\sqrt{\pi}}{2^{(m-1)/2}((m-2)/2)!} \qquad \text{for } m \text{ odd}$$

(5.7)′

$$= \frac{1}{2^{(m-1)/2}((m-2)/2)!} \qquad \text{for } m \text{ even.}$$

The initial values of u are

$$u(w, 0) = 0, \qquad u_t(w, 0) = \omega_m f(w), \qquad \text{where} \quad \omega_m = 2\pi^{m/2} \left/ \left|\frac{m-2}{2}\right.\right.! \tag{5.8}$$

If f is an automorphic function of w, so is $u(w, t)$. In this case we write f in the form

$$f(w) = \sum h^\tau(w), \tag{5.9}$$

where as usual

$$h^\tau(w) = h(\tau^{-1}w). \tag{5.10}$$

234

We shall study the integral

$$I(T) = c'_m \int_0^T (\cosh T - \cosh t)^{(m-3)/2} \sinh t\, u(w, t)\, dt, \qquad (5.11)$$

where

$$c'_m = \frac{2^{(m-1)/2}((m-2)/2)!}{\sqrt{\pi}((m-3)/2)!} \qquad \text{for all } m. \qquad (5.11)'$$

LEMMA 5.2.   *The following identities hold*:

$$I_0 = \int_0^T (\cosh T - \cosh t)^{(m-3)/2} \sinh t \left(\frac{1}{\sinh t}\, \partial_t\right)^{(m-3)/2}$$

$$\times \frac{1}{\sinh t} \int_{d(w, w_0)=t} f(w)\, d\sigma\, dt$$

$$= \frac{m-3}{2}! \int_{d(w, w_0)<T} f(w)\, dw \qquad \text{for } m \text{ odd}, \qquad (5.12)_0$$

$$I_e = \int_0^T (\cosh T - \cosh t)^{(m-3)/2} \sinh t \left(\frac{1}{\sinh t}\, \partial_t\right)^{(m-2)/2}$$

$$\times \int_{d(w, w_0)<t} \frac{f(w)}{\sqrt{\cosh t - \cosh d}}\, dw\, dt$$

$$= \frac{m-3}{2}! \sqrt{\pi} \int_{d(w, w_0)<T} f(w)\, dw \qquad \text{for } m \text{ even}. \qquad (5.12)_e$$

Proof.   We introduce the following change of variables:

$$r = \cosh t - 1, \qquad R = \cosh T - 1, \qquad \rho = \cosh d - 1. \qquad (5.13)$$

The expression for I_0 becomes

$$I_0 = \int_0^R (R-r)^{(m-3)/2} \partial_r^{(m-3)/2} \left((r^2 + 2r)^{-1/2} \int_{\rho=r} f(w)\, d\sigma\right) dr.$$

Integrating by parts $(m-3)/2$ times yields

$$I_0 = \frac{m-3}{2}! \int_0^R (r^2 + 2r)^{-1/2} \int_{\rho=r} f(w)\, d\sigma\, dr = \frac{m-3}{2}! \int_{d(w, w_0)<T} f(w)\, dw.$$

Likewise

$$I_e = \int_0^R (R-r)^{(m-3)/2} \, \partial_r^{(m-2)/2} \left(\int_{\rho<r} \frac{f(w)}{\sqrt{r-\rho}} \, dw \right) dr$$

$$= \frac{((m-3)/2)!}{\sqrt{\pi}} \int_0^R \frac{1}{\sqrt{R-r}} \int_{\rho<r} \frac{f(w)}{\sqrt{r-\rho}} \, dw \, dr.$$

Interchanging the order of integration, we get

$$I_e = \frac{((m-3)/2)!}{\sqrt{\pi}} \int_{\rho<R} f(w) \left(\int_\sigma^R \frac{dr}{\sqrt{(R-r)(r-\rho)}} \right) dw$$

$$= ((m-3)/2)! \, \sqrt{\pi} \int_{d(w,w_0)<T} f(w) \, dw.$$

Setting (5.7) into (5.11) and making use of (5.9) and (5.12) gives

$$I(T) = \int_{d(w,w')<T} f(w') \, dw' = \sum \int_{d(w,\ w')<T} h^\tau(w') \, dw'.$$

Using (5.10) and introducing $z = \tau^{-1} w'$ as new variable of integration, we get

$$I(T) = \sum \int_{d(\tau z, w)<T} h(z) \, dz. \tag{5.14}$$

We now choose the function h so that it satisfies the analogues of (2.5) and (2.6):

(a) $\quad \int h(w) \, dw = 1,$ $\hspace{3cm}$ (5.14)′

(b) $\quad h(w) \geqslant 0,$

(c) $\quad h(w) = 0 \quad$ if $\quad d(w, w_0) > \alpha.$

It follows from these requirements that

$$\int_{d(\tau z, w)<T} h(z) \, dz = 1 \quad \text{if} \quad d(\tau w_0, w) < T - \alpha$$

$$= 0 \quad \text{if} \quad d(\tau w_0, w) > T + \alpha \tag{5.15}$$

and that the value of the integral lies between 0 and 1 otherwise. It follows from this, (5.2) and (5.14) that

$$N(T-\alpha) \leqslant I(T) \leqslant N(T+\alpha) \tag{5.16}$$

or, equivalently,

$$I(T - a) \leqslant N(T) \leqslant I(T + a). \tag{5.17}$$

In terms of local coordinates z, where $z = 0$ corresponds to w_0, we shall set

$$h(z) = c(a) \, h_1 \left(\frac{z}{a} \right), \tag{5.18}$$

where

$$\lim_{a \to 0} a^m c(a) = 1. \tag{5.18}'$$

Here h_1 is a C^∞ function satisfying (5.14)(b) and (c), normalized so as to satisfy (5.14)'(a) in the limit as $a \to 0$.

Following the Euclidean case we now try to obtain an asymptotic evaluation of $I(T)$ by splitting off those components of u that correspond to the positive eigenvalues of L. That is, we write u as

$$u(w, t) = \sum_1^N (a_j e^{\lambda_j t} + b_j e^{-\lambda_j t}) \, \varphi_j(w) + v(w, t), \tag{5.19}$$

where the coefficients a_j, b_j are chosen so that the initial data for v are orthogonal to the φ_j, $j = 1, \ldots, N$. Since by (5.8), $u(w, 0) = 0$ the formulas are

$$a_j = \frac{1}{2\lambda_j} \, (u_t(0), \varphi_j), \qquad b_j = -a_j, \tag{5.20}$$

$u_t(0)$ being given by (5.8) and (5.9). If w_0 is not an elliptic point, then the sum (5.9) contains only one nonzero term in F. Using (5.8) and the normalization (5.14)', we deduce that

$$a_j = \frac{\omega_m}{2\lambda_j} \, \varphi_j(w_0) + O(a). \tag{5.21}$$

A moment's reflection (or rather rotation) shows that (5.21) is valid at elliptic points as well. Substituting (5.20) and (5.21) into (5.19) gives

$$u(w, t) = \frac{\omega_m}{2} \sum_i \frac{e^{\lambda_j t}}{\lambda_j} \, \varphi_j(w) \, \varphi_j(w_0) + O(a e^{\lambda_1 t}) + v(w, t). \tag{5.22}$$

Note that $v(w, t)$ is a solution of the wave equation and according to Theorem 4.7 both v and v_t remain orthogonal to the φ_j for all t.

In order to estimate $I(T)$ we set (5.22) into expression (5.11). This requires that we evaluate terms of the type

$$\int_0^T (\cosh T - \cosh t)^{(m-3)/2} \sinh t\, e^{\lambda t}\, dt;$$

making the change of variables (5.13), this becomes

$$2^\lambda \int_0^R (R-r)^{(m-3)/2}\, r^\lambda \left(1 + O\left(\frac{1}{r}\right)\right) dr$$

$$= R^{(m-1)/2+\lambda} 2^\lambda \int_0^1 (1-\theta)^{(m-3)/2}\, \theta^\lambda + O(R^{(m-3)/2+\lambda})$$

$$= \frac{((m-3)/2)!\, \lambda!}{2^{(m-1)/2}((m-1)/2+\lambda)!}\, e^{((m-1)/2+\lambda)T} + O(e^{((m-3)/2+\lambda)T}).$$

Taking the value of c'_m (see (5.11)') into account, we have

$$I(T, \alpha) = \Sigma(T) + O(\alpha e^{((m-1)/2+\lambda)T}) + O(e^{((m-3)/2+\lambda)T}) + V, \quad (5.23)$$

where $\Sigma(T)$ is defined by (5.4) and

$$V = V(T, \alpha) = \int_0^T (\cosh T - \cosh t)^{(m-3)/2} \sinh t\, v(w, t)\, dt. \quad (5.24)$$

In order to estimate $V(w)$ we use Sobolev's inequality:

$$|V(w)| \leqslant c(\varepsilon) \|V\|_{m/2+\epsilon}^{\text{loc}}, \qquad \varepsilon > 0, \quad (5.25)$$

where

$$c(\varepsilon) = O\left(\frac{1}{\varepsilon^{1/2}}\right). \quad (5.25)'$$

A convenient way of defining the local Sobolev norm is as follows: Let M denote some positive definite second order elliptic operator defined in a neighborhood of w, and χ a C_0^∞ cutoff function, that is $\chi \equiv 1$ in some smaller neighborhood of w. Then we set

$$M_0 = \chi M \chi \quad (5.26)$$

and define

$$\|V\|_s^{\text{loc}} = \|M_0^{s/2} V\|. \quad (5.26)'$$

Next we define the operator M as

$$M = k^2 - L, \qquad k > \frac{m-1}{2}. \tag{5.27}$$

LEMMA 5.3. *For* $0 \leqslant s \leqslant m$,

$$\|M_0^s V\| \leqslant \text{const.} \|M^s V\|. \tag{5.28}$$

The quantity $\|M^{s/2}V\|$ can be thought of as a *global* Sobolev norm. Combining (5.28) with (5.25) and (5.26) we get

$$|V(w)| \leqslant c(\varepsilon) \|M^{m/4+\epsilon} V\|. \tag{5.29}$$

For the proof of Lemma 5.3 we use the following theorem of Heinz (see Satz 3 on p. 426 of [3]), which is itself a generalization of a theorem of Loewner:

THEOREM (Heinz). *Let* A_0 *and* A *be two selfadjoint operators, both* $\geqslant 0$, *such that* $D(A) \subset D(A_0)$ *and*

$$\|A_0 f\| \leqslant \|Af\| \tag{5.30}$$

for all f *in* $D(A)$. *Then for all* p, $0 \leqslant p \leqslant 1$,

$$\|A_0^p f\| \leqslant \|A^p f\|. \tag{5.31}$$

In order to prove the lemma, it suffices to show that $A = M^m$ and $A_0 = M_0^m$ satisfy the hypothesis of the theorem. Now M^m is a $2m$th order elliptic operator, positive definite when $k > (m-1)/2$: By elliptic theory

$$\|f\|_{2m}^{\text{loc}} \leqslant \text{const.} \|M^m f\|. \tag{5.32}$$

On the other hand, since $M_0^m = (\chi M \chi)^m$ is a $2m$th order differential operator whose coefficients have compact support, it follows that

$$\|M_0^m f\| \leqslant \text{const.} \|f\|_{2m}^{\text{loc}}. \tag{5.33}$$

Inequality (5.30) follows from (5.32), (5.33), after rescaling M_0 if necessary.

By elliptic theory every function in the domain of M^m has locally L_2 derivatives of order $\leqslant 2m$. Since the coefficients of M_0^m have compact support, every such function is in $D(M_0^m)$. This shows that Heinz's theorem is applicable and completes the proof of Lemma 5.3.

LEMMA 5.4. *For* $m > 2$

$$\|M^{m/4+\epsilon} V\| \leqslant O(Te^{(m-1)T/2}/\alpha^{(m-1)/2+2\epsilon}). \tag{5.34}$$

Putting this estimate into (5.25) and setting $\varepsilon = |\log \alpha|^{-1}$, we obtain

$$|V(w)| \leqslant O\left(Te^{(m-1)T/2}\frac{|\log \alpha|^{1/2}}{\alpha^{(m-1)/2}}\right).$$

Substituting this into (5.23), we get

$$I(T \pm \alpha, \alpha) = \Sigma(T) + O(\alpha e^{((m-1)/2+\lambda)T})$$

$$+ O(e^{((m-3)/2+\lambda)T}) + O\left(Te^{(m-1)T/2}\frac{|\log \alpha|^{1/2}}{\alpha^{(m-1)/2}}\right).$$

It now follows from (5.17) that

$$|N(T) - \Sigma(T)| \leqslant O(\alpha e^{((m-1)/2+\lambda)T})$$

$$+ O(e^{((m-3)/2+\lambda)T}) + O\left(Te^{(m-1)T/2}\frac{|\log \alpha|^{1/2}}{\alpha^{(m-1)/2}}\right). \qquad (5.35)$$

An optimal choice for α is

$$\alpha = T^{3/(m+1)}\exp[-2\lambda T/(m+1)].$$

In this case the first and last terms on the right in (5.35) are

$$O(T^{3/(m+1)}\exp[(m-1)(\tfrac{1}{2} + \lambda/(m+1))T], \qquad (5.36)$$

whereas the middle term is of lower order since $\lambda \leqslant (m-1)/2$. The error estimate (5.36) is the one stated in Theorem 5.1 when $m > 2$.

The proof of Lemma 5.4 will occupy much of the remainder of this section. As in the Euclidean case, we begin by writing

$$m = 4q + r, \qquad r = -1, 0, 1 \text{ or } 2. \qquad (5.37)$$

Noting from (5.27) that

$$M^q = \sum_{j=0}^{q} c_j^q L^j, \qquad (5.38)$$

we see that

$$\|M^{m/4+\epsilon}V\| \leqslant \sum_{j=0}^{q} \|M^{r/4+\epsilon}L^j V\|, \qquad r = -1, 0, 1, 2. \qquad (5.39)$$

In order to estimate the right hand side of (5.39) we proceed as in the Euclidean case, applying L to (5.24) q times, using the fact that v satisfies the wave equation and integrating by parts. In analogy with (2.27), we now obtain the relations

$$L^q V = [\partial_t^{2q-2}(\cosh T - \cosh t)^{(m-3)/2} \sinh t] \, v_t|^T$$

$$- \int_0^T [\partial_t^{2q-1}(\cosh T - \cosh t)^{(m-3)/2} \sinh t] \, v_t \, dt \quad \text{for} \quad r = -1,$$

$$= - \int_0^T [\partial_t^{2q-1}(\cosh T - \cosh t)^{(m-3)/2} \sinh t] \, \dot{v}_t \, dt \quad \text{for} \quad r = 0,$$

$$\tag{5.40}$$

$$= - [\partial_t^{2q-1}(\cosh T - \cosh t)^{(m-3)/2} \sinh t] \, v|^T$$

$$+ \int_0^T \partial_t^{2q}[(\cosh T - \cosh t)^{(m-3)/2} \sinh t] \, v \, dt \quad \text{for} \quad r = 1,$$

$$= \int_0^T [\partial_t^{2q}(\cosh T - \cosh t)^{(m-3)/2} \sinh t] \, v \, dt \quad \text{for} \quad r = 2.$$

Applying $M^{r/4+\epsilon}$ to both sides of (5.40) and employing the notation

$$M^{r/4+\epsilon} v = z, \tag{5.41}$$

the resulting relations for $M^{r/4+\epsilon} L^q V$ are the same as (5.40) with v and v_t on the right side replaced by z and z_t, respectively. In order to estimate $\|M^{r/4+\epsilon} L^q V\|$ we need estimates for $\|z\|$ and $\|z_t\|$ for all t.

LEMMA 5.5. *The following inequalities hold for all t:*

$$\|M^p v_t\| \leqslant O(\alpha^{-2p-m/2}) \tag{5.42}$$

when

$$-\tfrac{1}{2} \leqslant p \leqslant 1 \quad \text{for} \quad m \geqslant 3,$$
$$0 \leqslant p \leqslant 1 \quad \text{for} \quad m = 2;$$

and in the case $m = 2$

$$\|M^p v_t\| \leqslant O(\alpha^{-2p-1} |\log \alpha|^{-p}) \quad \text{when} \quad -\tfrac{1}{2} \leqslant p \leqslant 0. \tag{5.42}'$$

Likewise

$$\|M^p v\| \leqslant (1 + |t|) O(\alpha^{-2p+1-m/2}) \tag{5.43}$$

when

$$0 \leqslant p \leqslant 1 \quad for \quad m \geqslant 3,$$
$$\tfrac{1}{2} \leqslant p \leqslant 1 \quad for \quad m = 2.$$

Furthermore

$$\|M^p v\| \leqslant (1 + |t|) O(\alpha^{-2p-m/2}) \tag{5.43}'$$

when $-\tfrac{1}{2} \leqslant p \leqslant 0$ for $m \geqslant 3$; and in the case $m = 2$

$$\|M^p v\| \leqslant (1 + |t|) O(\alpha^{-2p-1} |\log \alpha|^{-p}) \quad when \quad -\tfrac{1}{2} \leqslant p \leqslant 0. \tag{5.43}''$$

Proof. Recall that v satisfies the wave equation and together with v_t is L_2-orthogonal to the φ_j, $j = 1,...,N$. Since M^p commutes with the wave operator, $z = M^p v$ will also have these properties. We now make use of the fact that the energy of solutions is preserved in time, energy being defined as

$$E(z) = -(Lz, z) + \|z_t\|^2. \tag{5.44}$$

Recall also that $E(z)$ is nonnegative when z and z_t are orthogonal to the φ_j (see Corollary 4.5). Further since $u(w, 0) = 0$ by (5.8), so are $v(w, 0)$ and $z(w, 0)$. It follows from this, (5.44) and the invariance of E that

$$\|M^p v_t(t)\|^2 \leqslant E(M^p v(t)) = E(M^p v(0)) = \|M^p v_t(0)\|^2. \tag{5.45}$$

Thus it suffices to prove (5.42) only at $t = 0$.

A bound for $z(t)$ is obtained from Theorem 4.7(iii), which asserts that

$$\|z(t)\| \leqslant (1 + |t|) E^{1/2}(z(t)). \tag{5.46}$$

This relation together with (5.45) shows that

$$\|M^p v(t)\| \leqslant (1 + |t|) \|M^p v_t(0)\|. \tag{5.46}'$$

Thus (5.43)' and (5.43)'' are immediate consequences of (5.42) and (5.42)'. Moreover using (5.27), (5.44), (5.45) and (5.46) we can write

$$\|z(t)\|^2 = (MM^{-1/2}z(t), M^{-1/2}z(t))$$
$$= k^2 \|M^{-1/2}z(t)\|^2 - (LM^{-1/2}z(t), M^{-1/2}z(t))$$
$$\leqslant k^2(1 + |t|)^2 E(M^{-1/2}z(t)) + E(M^{-1/2}z(t))$$
$$\leqslant \text{const.}(1 + |t|)^2 \|M^{-1/2}z_t(0)\|^2.$$

Replacing z by $M^p v$, we see that (5.43) also is a consequence of (5.42).

It remains to prove (5.42). We note first of all that since $M^p v$ and $M^p v_t$ are L_2-orthogonal to the φ_j and differ from $M^p u$ and $M^p u_t$ only by linear combinations of the φ_j, we have

$$\|M^p v\| \leqslant \|M^p u\| \quad and \quad \|M^p v_t\| \leqslant \|M^p u_t\|. \tag{5.47}$$

Now

$$|u_t(w, 0)| \leqslant \text{const. } \alpha^{-m} \quad \text{for} \quad d(w, w_0) < \alpha,$$
$$= 0 \qquad \text{elsewhere in } F. \tag{5.48}$$

Making use of (5.47) and (5.48), we see that

$$\|v_t(0)\| \leqslant \|u_t(0)\| = O(\alpha^{-m/2}). \tag{5.49}$$

Using (5.47) and the fact that M is a second order differential operator, we conclude that

$$\|Mv_t(0)\| \leqslant \|Mu_t(0)\| \leqslant O(\alpha^{-2-m/2}). \tag{5.50}$$

Inequalities (5.49) and (5.50) establish (5.42) for $p = 0$ and $p = 1$; as in Lemma 2.3, we can now use the convexity principle to verify (5.42) for the in-between values of p.

To treat negative values of p, we estimate $\|M^{-1/2}v_t\|$. According to (5.47)

$$\|M^{-1/2}v_t\|^2 \leqslant \|M^{-1/2}u_t\|^2 = (M^{-1}u_t, u_t). \tag{5.51}$$

We now require an estimate on the operator M^{-1}:

LEMMA 5.6. *For* $k > (m-1)/2$, M^{-1} *is an integral operator whose kernel* G *has a singularity bounded by*

$$|G(w, w_0)| \leqslant \begin{cases} \dfrac{\text{const.}}{d(w, w_0)^{m-2}} & \text{for} \quad m > 2, \\[3mm] \text{const.} \, |\log d(w, w_0)| & \text{for} \quad m = 2. \end{cases} \tag{5.52}$$

Proof. Considering M as an operator on the whole space, i.e., $\Gamma = id$, we show in Lemma 7.2 that the kernel G_0 is bounded by

$$|G_0(w, w_0)| \leqslant \begin{cases} \dfrac{\text{const. } \exp[-((m-1)/2 + k)\, d(w, w_0)]}{d(w, w_0)^{m-2}} & \text{for } m > 2, \\[4mm] \text{const.} \, |\log d(w, w_0)| \exp\left[-\left(\dfrac{m-1}{2} + k\right) d(w, w_0)\right] \end{cases}$$
$$\text{for } m = 2. \tag{5.53}$$

The kernel of M^{-1} over F is now obtained by summing over the group:

$$G(w, w_0) = \sum_{\Gamma} G_0(\tau^{-1}w, w_0). \tag{5.54}$$

To prove the convergence of this sum we need a crude estimate on the distribution of orbital points:

LEMMA 5.7.

$$N(T; w, w_0) \leqslant O(e^{((m-1)/2+\lambda)T}). \tag{5.55}$$

Proof. We denote by $\bar{N}(T, w, w_0)$ the average of N over the ball of, say, unit radius centered at w:

$$\bar{N}(T; w, w_0) = \frac{1}{\text{vol}} \int_{d(w',w)<1} N(t; w', w_0) \, dw'. \tag{5.56}$$

Clearly if $d(w', w) \leqslant 1$ then

$$N(T-1; w, w_0) \leqslant N(T; w', w_0)$$

and so

$$N(T-1; w, w_0) \leqslant \bar{N}(T; w, w_0). \tag{5.57}$$

Taking $\alpha = 1$ in (5.17) and integrating over the unit ball about w, we get

$$\bar{N}(T; w, w_0) \leqslant \bar{I}(T+1; w, w_0). \tag{5.58}$$

Using (5.23) we deduce that

$$\bar{I}(T; w, w_0) \leqslant O(e^{((m-1)/2+\lambda)T}) + \bar{V}. \tag{5.59}$$

By the Schwarz inequality

$$|\bar{V}| \leqslant \text{const.} \|V\|$$

and from (5.24)

$$\|V\| \leqslant \int_0^T (\cosh T - \cosh t)^{(m-3)/2} \sinh t \, \|v(t)\| \, dt$$

$$\leqslant \text{const.} \int_0^T e^{(m-1)t/2} \|v(t)\| \, dt. \tag{5.60}$$

Now (5.46) holds with z replaced by v; inserting this estimate into (5.60) gives

$$|\bar{V}| \leqslant \text{const.} \|V\| \leqslant O(Te^{(m-1)T/2}).$$

Combining this with (5.59), (5.58) and (5.57) gives (5.55), as asserted in Lemma 5.7.

We can now prove the convergence of (5.54). Using the bound on G_0 given by (5.55) we have

$$\sum_{\tau \neq id} |G_0(\tau^{-1}w, w_0)| \leqslant \text{const.} \sum_{\tau \neq id} \exp\left[-\left(\frac{m-1}{2} + k\right) d(\tau^{-1}w, w_0)\right]$$

$$= \text{const.} \int_0^\infty e^{-((m-1)/2 + k)s} \, dN(s) \qquad (5.61)$$

$$= \text{const.} \left(\frac{m-1}{2} + k\right) \int_0^\infty e^{-((m-1)/2 + k)s} N(s) \, ds.$$

It follows from (5.55) that this integral converges for $k > (m-1)/2 \geqslant \lambda$. If we include the identity term in the sum we get the bound (5.52), as required by Lemma 5.6.

The proof of (5.42) for $p = -\frac{1}{2}$ now follows from relation (5.51) and the bound (5.52) exactly as in (2.41). The in-between values $-\frac{1}{2} \leqslant p \leqslant 0$ can again be obtained from the convexity principle. This concludes the proof of Lemma 5.5.

We are now in a position to complete the proof of Lemma 5.4. The four cases described in (5.37) will be treated in turn.

Case (i): $m = 4q - 1$. According to (5.39) we require estimates on $\|M^{-1/4 + \epsilon} L^j V\|$ for $0 \leqslant j \leqslant q$. By (5.40), with j in place of q on the left side, we have

$$M^{-1/4 + \epsilon} L^j V = [\partial_t^{2j-2}(\cosh T - \cosh t)^{2q-2} \sinh t] \, z_t |^T$$

$$- \int_0^T [\partial_t^{2j-1}(\cosh T - \cosh t)^{2q-2} \sinh t] \, z_t \, dt,$$

$$j = 1, ..., q,$$

where $z = M^{-1/4 + \epsilon} v$. Now

$$|\partial_t^k(\cosh T - \cosh t)^{2q-2} \sinh t|$$

$$\leqslant \text{const.} \sum_{l=0}^{2q-2-k} e^{(2q-2-l)T} e^{(l+1)t} \qquad (5.62)$$

and it follows that for $1 \leqslant j \leqslant q$

$$\|M^{-1/4 + \epsilon} L^j V\| \leqslant \text{const.} \, e^{((m-1)/2)T} \sup \|z_t(t)\|.$$

Using (5.42) with $p = -\frac{1}{4} + \epsilon$, we get

$$\|M^{-1/4 + \epsilon} L^j V\| \leqslant \text{const.} \, T e^{((m-1)T/2}/\alpha^{(m-1)/2 + 2\epsilon}. \qquad (5.63)$$

We also need an estimate for

$$M^{-1/4 + \epsilon} V = \int_0^T (\cosh T - \cosh t)^{(m-3)/2} (\sinh t) z \, dt.$$

Here we use estimate (5.43)' with $p = -\frac{1}{4} + \varepsilon$ to obtain

$$\|M^{-1/4+\varepsilon}\| \leqslant \int_0^T (\cosh T - \cosh t)^{(m-3)/2} (\sinh t) t \, dt$$

$$\times O(\alpha^{(1-m)/2 - 2\varepsilon}). \tag{5.64}$$

An integration by parts gives

$$\int_0^T (\cosh T - \cosh t)^{(m-3)/2} (\sinh t) t \, dt$$

$$= \frac{2}{m-1} \int_0^T (\cosh T - \cosh t)^{(m-1)/2} \, dt \tag{5.65}$$

and this is obviously $O(T \exp((m-1)T/2))$. Inserting inequalities (5.63), (5.64) and (5.65) into (5.39), we get the desired estimate (5.34).

Case (ii): $m = 4q + 1$. We now need estimates on $\|M^{-1/4+\varepsilon} L^j V\|$ for $0 \leqslant j \leqslant q$. According to (5.40)

$$M^{1/4+\varepsilon} L^j V = -[\partial_t^{2j-1}(\cosh T - \cosh t)^{2q-1} \sinh t] z \,|^T$$

$$+ \int_0^T [\partial_t^{2j}(\cosh T - \cosh t)^{2q-1} \sinh t] z \, dt, \qquad j = 1,\ldots, q,$$

where $z = M^{1/4+\varepsilon} v$. Consequently

$$\|M^{1/4+\varepsilon} L^j V\| \leqslant |\partial_t^{2j-1}(\cosh T - \cosh t)^{2q-1} \sinh t|^T \|z(T)\|$$

$$+ \int_0^T |\partial_t^{2j}(\cosh T - \cosh t)^{2q-1} \sinh t| \, \|z(t)\| \, dt. \tag{5.66}$$

Using the estimates (5.43) with $p = \frac{1}{4} + \varepsilon$ and (5.62), we again get for $1 \leqslant j \leqslant q$ the bound

$$\|M^{1/4+\varepsilon} L^j V\| \leqslant \text{const. } T e^{(m-1)T/2} / \alpha^{(m-1)/2 + 2\varepsilon}.$$

The case $j = 0$ is handled as before (see (5.65)) as is the deduction of (5.34) from these estimates.

Case (iii): $m = 4q$. According to (5.40) we must now consider

$$W \equiv M^\varepsilon L^j V$$

$$= -\int_0^T [\partial_t^{2j-1}(\cosh T - \cosh t)^{2q-3/2} \sinh t] z_t \, dt, \qquad j = 1,\ldots, q, \tag{5.67}$$

where $z = M^\varepsilon v$. As in the Euclidean analogue, we split W into two terms:

$$W = -\int_0^T [\partial_t^{2j-1}(\cosh T - \cosh t)^{2q-3/2} \sinh t]\, z_t\, dt$$

$$= \int_0^S + \int_S^T \equiv W_1 + W_2. \tag{5.68}$$

Integrating W_1 by parts, we get

$$W_1 = -[\partial_t^{2j-1}(\cosh T - \cosh t)^{2q-3/2} \sinh t]z \,|^S$$

$$+ \int_0^S [\partial_t^{2j}(\cosh T - \cosh t)^{2q-3/2} \sinh t]z\, dt.$$

We estimate W_1 for $j = q$; the same estimate holds also for $1 \leqslant j < q$. The first term on the right is bounded by

$$\text{const.} \frac{e^{2qT}}{(\cosh T - \cosh S)^{1/2}} \|z(S)\|.$$

The integrand in the second term is bounded by

$$\text{const.}\, e^{2qT} \frac{\sinh t}{(\cosh T - \cosh t)^{3/2}} \|z(t)\|.$$

Making use of estimate (5.43) with $p = \varepsilon$, we see that

$$\|W_1\| \leqslant \text{const.} \frac{Te^{mT/2}}{(\cosh T - \cosh S)^{1/2}} \frac{1}{\alpha^{(m-2)/2+2\varepsilon}}. \tag{5.69}$$

We treat W_2 as in the Euclidean case:

$$\|W_2\| \leqslant \int_S^T |\partial_t^{2j-1}(\cosh T - \cosh t)^{2q-3/2} \sinh t|\, dt \sup \|z_t\|.$$

The integrands are bounded for $1 \leqslant j \leqslant q$ by

$$\text{const.}\, e^{(2q-1)T} \frac{\sinh t}{(\cosh T - \cosh t)^{1/2}},$$

from which it follows by (5.42) with $p = \varepsilon$ that

$$\|W_2\| \leqslant \text{const.}\, e^{(m-2)T/2}(\cosh T - \cosh S)^{1/2} \frac{1}{\alpha^{m/2+2\varepsilon}}. \tag{5.70}$$

The case $j = 0$ is handled as before. Using (5.65) together with (5.42), we get

$$\| M^\epsilon V \| \leqslant \text{const. } Te^{(m-1)T/2} \frac{1}{\alpha^{(m-2)/2+2\epsilon}} . \tag{5.71}$$

Substituting (5.69), (5.70) and (5.71) into (5.39) gives

$$\| M^{m/4+\epsilon} V \| \tag{5.72}$$

$$\leqslant \text{const. } Te^{(m-1)T/2} \frac{1}{\alpha^{(m-1)/2+2\epsilon}}$$

$$\times \left[\frac{e^{T/2}\alpha^{1/2}}{(\cosh T - \cosh S)^{1/2}} + \alpha^{1/2} + e^{-T/2} \frac{(\cosh T - \cosh S)^{1/2}}{\alpha^{1/2}} \right].$$

Note that the middle term in the bracket is less than the first term if we choose S so that

$$e^{-T}(\cosh T - \cosh S) = \alpha.$$

The resulting inequality is (5.34).

Case (iv): $m = 4q + 2$. From (5.40) we get for all $j \leqslant q$

$$W = M^{1/2+\epsilon}L^j V = \int_0^T [\partial_t^{2j}(\cosh T - \cosh t)^{2q-1/2} \sinh t] z \, dt$$

$$= \int_0^S + \int_S^T \equiv W_1 + W_2,$$

where $z = M^{1/2+\epsilon}v$. We begin by applying L to W_1, using $Lz = z_{tt}$ and integrating the resulting expression by parts:

$$LW_1 = \int_0^S [\partial_t^{2j}(\cosh T - \cosh t)^{2q-1/2} \sinh t] z_{tt} \, dt$$

$$= [\partial_t^{2j}(\cosh T - \cosh t)^{2q-1/2} \sinh t] z_t |^S$$

$$- \int_0^S [\partial_t^{2j+1}(\cosh T - \cosh t)^{2q-1/2} \sinh t] z_t \, dt. \tag{5.73}$$

The worst case $(j = q)$ bounds are

$$|\partial_t^{2q}(\cosh T - \cosh t)^{2q-1/2} \sinh t|$$

$$\leqslant \text{const. } e^{(2q+1)T}/(\cosh T - \cosh t)^{1/2},$$

$$|\partial_t^{2q+1}(\cosh T - \cosh t)^{2q-1/2} \sinh t|$$

$$\leqslant \text{const. } e^{(2q+1)T} \frac{\sinh t}{(\cosh T - \cosh t)^{3/2}} .$$

Applying M^{-1} to (5.73) and making use of these bounds and (5.42) with $p = -\frac{1}{2} + \varepsilon$, we get

$$\|M^{-1}LW_1\| \leqslant \text{const. } e^{mT/2}(\cosh T - \cosh S)^{-1/2} \sup \|M^{-1}z_t\|,$$

$$= \text{const. } e^{mT/2}(\cosh T - \cosh S)^{-1/2} \frac{1}{\alpha^{(m-2)/2+2\varepsilon}} . \quad (5.74)$$

Next we apply M^{-1} to W_1; estimating as before we get

$$\|M^{-1}W_1\| \leqslant \int_0^S |\partial_t^{2j}(\cosh T - \cosh t)^{2q-1/2} \sinh t| \|M^{-1}z\| \, dt.$$

Using the bounds

$$|\partial_t^{2j}[(\cosh T - \cosh t)^{2q-1/2} \sinh t]| \leqslant \text{const. } e^{2qT} \frac{\sinh t}{(\cosh T - \cosh t)^{1/2}},$$

and (5.43)′ with $p = -\frac{1}{2} + \varepsilon$, we get for $m \geqslant 6$

$$\|M^{-1}W_1\| \leqslant Te^{(m-2)T/2}(\cosh T - \cosh S)^{1/2} \frac{1}{\alpha^{(m-2)/2+2\varepsilon}} . \quad (5.75)$$

Combining (5.74) and (5.75), we see that

$$\|W_1\| = \|MM^{-1}W_1\| \leqslant k^2 \|M^{-1}W_1\| + \|M^{-1}LW_1\| \quad (5.76)$$

$$\leqslant \text{const. } e^{(m-1)T/2}(Te^{-T/2}(\cosh T - \cosh S)^{1/2}$$

$$+ e^{T/2}(\cosh T - \cosh S)^{-1/2}) \frac{1}{\alpha^{(m-2)/2+2\varepsilon}} \quad \text{for} \quad m \geqslant 6;$$

if $m = 2$ the last factor is to be replaced by $|\log \alpha|^{1/2} \alpha^{-2\varepsilon}$.

Similarly

$$\|W_2\| \leqslant \int_S^T e^{2qT} \frac{\sinh t}{(\cosh T - \cosh t)^{1/2}} \|z(t)\| \, dt. \quad (5.77)$$

Using (5.43) with $p = \frac{1}{2} + \varepsilon$, we see that

$$\|W_2\| \leqslant \text{const. } Te^{(m-2)T/2}(\cosh T - \cosh S)^{1/2} \frac{1}{\alpha^{m/2+2\varepsilon}} . \quad (5.78)$$

Substituting (5.76) and (5.78) into (5.39) gives for $m \geqslant 6$

$$\|M^{m/4+\epsilon}V\| \leqslant \text{const. } Te^{((m-1)/2)T} \frac{1}{\alpha^{(m-1)/2+2\epsilon}} \quad (5.79)$$

$$\times \left[\frac{e^{(1/2)T}\alpha^{1/2}}{(\cosh T - \cosh S)^{1/2}} + e^{-(1/2)T} \frac{(\cosh T - \cosh S)^{1/2}}{\alpha^{1/2}} \right].$$

This is the same as (5.72) and the rest of the argument proceeds as in part (iii). This completes the proof of Lemma 5.4.

If $m = 2$, we have, instead of (5.79),

$$\|M^{1/2+\epsilon}V\| \leqslant \text{const. } e^{(1/2)T} \frac{1}{\alpha^{1/2+2\epsilon}} \tag{5.80}$$

$$\times \left[\frac{e^{(1/2)T}\alpha^{1/2}|\log \alpha|^{1/2}}{(\cosh T - \cosh S)^{1/2}} + Te^{-(1/2)T} \frac{(\cosh T - \cosh S)^{1/2}}{\alpha^{1/2}} \right].$$

In this case the optimal choice for S is

$$Te^{-T}(\cosh T - \cosh S) = \alpha|\log \alpha|^{1/2}.$$

Inserting this into (5.80) and applying the Sobolev inequality (5.29) with $\epsilon = |\log \alpha|^{-1}$, we get

$$|V(w)| \leqslant \text{const. } T^{1/2}e^{(1/2)T} \frac{|\log \alpha|^{3/4}}{\alpha^{1/2}}.$$

Setting this into (5.23) we get

$$I(T, \alpha) = \Sigma(T) + O(\alpha e^{(1/2+\lambda)T}) + O\left(T^{1/2}e^{(1/2)T} \frac{|\log \alpha|^{3/4}}{\alpha^{1/2}}\right).$$

We now set

$$\alpha = T^{5/6}e^{-(2\lambda/3)T}$$

and obtain for our error estimate

$$\left|N(T) - \Sigma(T)\right| \leqslant O(T^{5/6}e^{(1/2+\lambda/3)T}).$$

This completes the proof of Theorem 5.1.

By making use of the Sobolev inequality (2.20) instead of (5.25) and following more closely the analysis of Section 2, we can improve somewhat the error estimate in Theorem 5.1. Unfortunately this does not enable us to eliminate the power of T in (5.5); all that this accomplishes is to reduce it from $3/(m+1)$ to $2/(m+1)$.

Theorem 5.1 holds only if L' has positive eigenvalues. According to Theorem 4.4 this is equivalent with the energy form being indefinite on $\mathscr{H}$. If this condition is not met, we have

THEOREM 5.8. *Suppose that $E \geqslant 0$; then*

$$N(T; w, w_0) = O(Te^{(m-1)T/2}). \tag{5.81}$$

The proof of this theorem is based on a series of lemmas. We denote by $\mathfrak{Z}$ the null set of E:

$$\mathfrak{Z} = [f \in \mathscr{H}; E(f) = 0]. \tag{5.82}$$

Since $E \geqslant 0$, $\mathfrak{Z}$ is a linear subspace of $\mathcal{H}$. We define $\dot{\mathcal{H}}$ as the quotient space:

$$\dot{\mathcal{H}} = \mathcal{H} \quad (\text{mod } \mathfrak{Z}). \qquad (5.83)$$

Denote the infinitesimal generator of U by A. It is clear from our construction of the solution group in Section 4 that $C_0^\infty(F)$ is contained in $D(A)$ and that for $\{u_1(t), u_2(t)\} = U(t)f$, $f \in C_0^\infty(F)$, we have

$$A U(t)f = \partial_t \{u_1(t), u_2(t)\} = \{u_2(t).\ L u_1(t)\}.$$

The relation $\partial_t u_1 = u_2$ extends upon completion to all data in $\mathcal{H}$, at least in the weak sense. In particular for f in $D(A)$, we see that

$$[Af]_1 = \partial_t u_1(0) = u_2(0) = f_2. \qquad (5.84)$$

Further it follows from the invariance of $E(U(t)f)$ that A is skewsymmetric with respect to E. Hence for f in $D(A)$ and g in $C_0^\infty(F)$ we have

$$E(Af, g) = -E(f, Ag).$$

Setting $g = \{0, \varphi\} \in C_0^\infty(F)$ this becomes

$$([Af]_2, \varphi) = -(f_1, L\varphi). \qquad (5.84)'$$

The relations (5.84) and $(5.84)'$ show that A is a restriction of

$$A_w = \begin{pmatrix} 0 & 1 \\ L & 0 \end{pmatrix}, \qquad (5.85)$$

where L is to be taken in the weak sense. Actually $A = A_w$, but we do not need this fact. The null space of A, that is, $N(A)$, consists of all f in the domain of A such that $Lf_1 = 0$ and $f_2 = 0$.

LEMMA 5.9. $\mathfrak{Z} = N(A)$.

Proof. $\mathfrak{Z}$ is finite dimensional by Theorem 3.6 and, by conservation of energy, $U(t)$ maps $\mathfrak{Z}$ into itself. It follows that $\mathfrak{Z}$ is contained in $D(A)$. Further for z in $\mathcal{H}$ and f in $C_0^\infty(F)$, we have by (5.39)

$$E(z, f) = -(z_1, Lf_1) + (z_2, f_2). \qquad (5.86)$$

For z in $\mathfrak{Z}$, $E(z, f) = 0$ and since this holds for all f in $C_0^\infty(F)$ we conclude that $Lz_1 = 0$ and $z_2 = 0$; that is, z belongs to $N(A)$. Conversely if z lies in $N(A)$, then (5.86) implies that $E(z, f) = 0$ for a dense subset of $\mathcal{H}$ and hence that z belongs to $\mathfrak{Z}$.

LEMMA 5.10. (i) $\mathscr{H}$ *is a Hilbert space under the energy norm*;

(ii) $U(t)$ *defines a unitary group on* $\mathscr{H}$;

(iii) $U(t)$ *has no point spectrum over* $\mathscr{H}$.

Proof. Since $\mathfrak{Z}$ is finite dimensional, it is a closed subspace of $\mathscr{H}$. Clearly if f_1 and f_2 lie in the same coset $\hat{f}$ of $\mathscr{H}$, then $E(f_1) = E(f_2)$; we define $E(\hat{f})$ as this common value. Denoting the G-orthogonal complement of $\mathfrak{Z}$ by $\mathscr{H}''$, we can represent each $\hat{f}$ in $\mathscr{H}$ by the element of H'' belonging to $\hat{f}$, and the mapping

$$\tau: \hat{f} \in \hat{\mathscr{H}} \to f \in \mathscr{H}'' \cap \hat{f}$$

is unitary with respect to the E-norm. Clearly $\mathscr{H}''$ is a closed subspace of $\mathscr{H}$ and E is positive on $\mathscr{H}''$. It therefore follows by Theorem 3.7 that E and G are equivalent norms on $\mathscr{H}''$. Consequently $\mathscr{H}''$ is complete in the E-norm and, since τ is unitary, so is $\hat{\mathscr{H}}$. This proves part (i) of the lemma.

Since $\mathfrak{Z}$ is invariant under the action of U, it follows that $U(t)$ is well defined on $\hat{\mathscr{H}}$ and obviously unitary. This proves part (ii).

We have assumed that $E \geqslant 0$ on $\mathscr{H}$. It therefore follows from Theorem 4.4 that L' has no positive eigenvalues. Consequently F must have infinite volume since otherwise the constant function would be an eigenfunction of L' with positive eigenvalue $((m-1)/2)^2$. By the finite geometric property, F has only a finite number of sides and hence the boundary of F must contain an open neighborhood of B. It therefore follows by Theorem 4.8 that L' has no point spectrum.

Suppose that $U(t)$ over $\hat{\mathscr{H}}$ had an eigenvector: $U(t)\hat{f} = e^{i\sigma t}\hat{f}$. Then for any f in $\hat{f}$, this can be rewritten as

$$U(t)f = e^{i\sigma t}f + z(t) \tag{5.87}$$

for some $z(t)$ in $\mathfrak{Z}$. The linear space of the coset $\hat{f}$ is obviously invariant under the action of U and since this subspace is finite dimensional it belongs to $D(A)$. We can therefore differentiate (5.87) to obtain

$$Af = i\sigma f + z$$

for some z in $\mathfrak{Z}$.

Since U is unitary on $\hat{\mathscr{H}}$, σ will be real. Suppose first that $\sigma \neq 0$ and set $g = f + z/i\sigma$. Then $Ag = i\sigma g$ and by (5.85)

$$g_2 = i\sigma g_1 \quad \text{and} \quad Lg_1 = i\sigma g_2.$$

Thus g_1 lies in $L_2(F)$, hence in $D(L')$ by (4.13), and $L'g_1 = -\sigma^2 g_1$. Since L' has no point spectrum, we conclude that $g = 0$ and therefore that f belongs to $\mathfrak{Z}$; so that $\hat{f} = 0$. If $\sigma = 0$, then $Af = z$ and hence $A^2 f = 0$. This means that

$Lf_1 = 0 = Lf_2$. Since f_2 is square integrable, $L'f_2 = 0$ and hence by Theorem 4.8 $f_2 = 0$. Consequently f belongs to $N(A) = 3$ and $\hat{f} = 0$. This proves that U has no point spectrum over $\mathcal{H}$ and completes the proof of Lemma 5.10.

LEMMA 5.11. *If $E \geqslant 0$ then for any solution u in F of the wave equation whose initial data belong to $\mathcal{H}$,*

$$\lim_{t \to \infty} \frac{1}{t} \int_0^t \| u_t(s) \|_{B_1} \, ds = 0, \tag{5.88}$$

where $\| \cdot \|_{B_1}$ is the L_2-norm over the unit ball B_1 centered at w.

Proof. Assume first that the initial data for u are smooth with compact support. In this case the data $\{ L'u, L'u_t \}$ will belong to $\mathcal{H}$ for all t. By the conservation of energy and the nonnegativity of E, it follows that both $\| u_t(s) \|$ and $\| L'u_t \|$ are uniformly bounded in t. Making use of elliptic estimates and the Rellich compactness criterion, we see that the functions $\{ u_t(s); -\infty < s < \infty \}$ form a compact set in $L_2(B_1)$. It follows that for any $\varepsilon > 0$ there is a finite dimensional subspace $\mathcal{R}$ of $L_2(B_1)$ such that for all s

$$\| u_t(s) \|_{B_1}^2 \leqslant \| Ru_t(s) \|_{B_1}^2 + \varepsilon^2; \tag{5.89}$$

here R denotes the orthogonal projection of $L_2(B_1)$ onto $\mathcal{R}$. Take any orthogonal basis in $\mathcal{R}$: $g_1, ..., g_r$; then (5.89) can be rewritten as

$$\| u_t(s) \|_{B_1}^2 \leqslant \sum_1^r |(u_t(s), g_j)_{B_1}|^2 + \varepsilon^2. \tag{5.89$'$}$$

Setting $g_j = 0$ outside of B_1 we can rewrite (5.89)$'$ as

$$\| u_t(s) \|_{B_1}^2 \leqslant \sum_1^r |(u_t(s), g_j)|^2 + \varepsilon^2. \tag{5.89$''$}$$

Next let $P(\lambda)$ be the spectral resolution for the unitary group U on $\mathcal{H}$:

$$U(t) = \int e^{i\lambda t} \, dP(\lambda).$$

Then for any f and g in $\mathcal{H}$

$$E(U(s)f, g) = \int e^{i\lambda t} \, dE(P(\lambda)f, g). \tag{5.90}$$

According to Lemma 5.10(iii), U has no point spectrum over $\mathscr{H}$ so it follows that

$$dm = dE(P(\lambda)f, g)$$

is a measure without any point mass. According to a classical result of Wiener the Fourier transform of such a measure tends to zero in the mean:

$$\lim_{t \to \infty} \frac{1}{t} \int_0^t |\tilde{m}(s)|^2 \, ds = 0. \tag{5.91}$$

We see by (5.90) that in our case $\tilde{m}(s) = E(U(s)f, g)$ and we deduce from (5.91) that

$$\lim_{t \to \infty} \frac{1}{t} \int_0^t |E(U(s)f, g)|^2 \, ds = 0. \tag{5.92}$$

Taking f to be the initial data of u and g to be the data $\{0, g_j\}$, it follows from (5.92) that

$$\lim_{t \to \infty} \frac{1}{t} \int_0^t |(u_t(s), g_j)|^2 \, ds = 0. \tag{5.92}'$$

Applying this to (5.89)″ we conclude that

$$\overline{\lim_{t \to \infty}} \frac{1}{t} \int_0^t \|u_t(s)\|_{B_1}^2 \, ds < \varepsilon^2. \tag{5.92}''$$

Using the inequality

$$\|u_t(s)\|_{B_1} \leqslant \varepsilon + \frac{\|u_t(s)\|_{B_1}^2}{4\varepsilon},$$

we deduce from (5.92)″ that

$$\overline{\lim_{t \to \infty}} \frac{1}{t} \int_0^t \|u_t(s)\|_{B_1} \, ds \leqslant \varepsilon + \frac{\varepsilon}{4}.$$

Since ε is arbitrary, this proves (5.88) for u with $C_0^\infty(F)$ initial data.

To prove (5.88) for u with arbitrary initial data f in $\mathscr{H}$, we approximate f by $C_0^\infty(F)$ data: $f^{(n)} \to f$ in the G-norm. It is clear that

$$E(U(s)f - U(s)f^{(n)}) = E(f - f^{(n)}) \to 0;$$

and setting $U(s) f^{(n)} = \{u^{(n)}(s), u_t^{(n)}(s)\}$ we see in particular that for n large enough

$$\|u_t(s) - u_t^{(n)}(s)\| < \varepsilon \qquad (5.93)$$

for all s. Since (5.88) holds for $u^{(n)}$, it follows from (5.93) that it also holds for u. This complete the proof of Lemma 5.11.

COROLLARY 5.12.

$$\lim_{t \to \infty} \frac{1}{t} \|u(t)\|_{B_1} = 0. \qquad (5.94)$$

Proof. We use the formula

$$u(t) = u(0) + \int_0^t u_t(s)\, ds$$

to estimate u. This gives

$$\|u(t)\|_{B_1} \leqslant \|u(0)\|_{B_1} + \int_0^t \|u_t(s)\|_{B_1}\, ds.$$

Using (5.88) to estimate the right side of this inequality, we deduce (5.94).

We are now ready to prove Theorem 5.8; we proceed along the lines of the proof of Lemma 5.7. We denote by u the solution (5.7) of the wave equation, with f as in (5.9), h as in (5.14)' and $\alpha = 1$. We denote by $I(T)$ the quantity defined in (5.11):

$$I(w, T) = c_m' \int_0^T (\cosh T - \cosh t)^{(m-3)/2} \sinh t\, u(w, t)\, dt.$$

We denote by $\bar{u}$ and $\bar{I}$ the mean values of u and I over the ball B_1. Combining (5.57) and (5.58) we get

$$N(T - 2) \leqslant \overline{I(T)} = c_m' \int_0^T (\cosh T - \cosh t)^{(m-3)/2} \sinh t\, \bar{u}(t)\, dt. \qquad (5.95)$$

By the Schwarz inequality we have

$$|\bar{u}(t)| \leqslant (\operatorname{vol} B_1)^{1/2} \|u(t)\|_{B_1}.$$

Using this to estimate the right side of (5.95) gives

$$N(T - 2) \leqslant \operatorname{const.} \int_0^T (\cosh T - \cosh t)^{(m-3)/2} \sinh t\, \|u(t)\|_{B_1}\, dt. \qquad (5.96)$$

The integrand on the right is bounded by

$$e^{(m-3)T/2} e^t \|u(t)\|_{B_1} \qquad \text{for } m \geqslant 3. \tag{5.97}$$

According to (4.17), $\|u(t)\|_{B_1}$ grows at most linearly in t and we see from (5.94) that given $\varepsilon > 0$, there is a T_ε such that

$$\|u(t)\|_{B_1} \leqslant \varepsilon t \qquad \text{for } t \geqslant T_\varepsilon. \tag{5.98}$$

Hence splitting the integral in (5.92) into two parts:

$$\int_0^T = \int_0^{T_\varepsilon} + \int_{T_\varepsilon}^T,$$

and making use of the above estimates, an easy calculation yields

$$N(T-2) \leqslant \text{const.} (e^{(m-1)T/2} + \varepsilon T e^{(m-1)T/2})$$

for $T > T_\varepsilon$. This implies inequality (5.81) of Theorem 5.8.

6. EXAMPLES

We now show by example that E can be positive, nonnegative or indefinite when vol. F is infinite. An interesting source of examples is obtained from discrete Fuchsian subgroups, that is, isometries of the form (3.3) with real coefficients.

LEMMA 6.1. *Suppose Γ is a Fuchsian subgroup with 2-dimensional fundamental domain $F_2 = \mathbb{H}_2/\Gamma$. Then F can be obtained by rotating F_2 about the x_1-axis in the upper half-space $y > 0$.*

Proof. Define the projection map π as

$$\pi: \{x_1, x_2, y\} \in \mathbb{H}_3 \to \{x_1, \sqrt{x_2^2 + y^2}\} \in \mathbb{H}_2. \tag{6.1}$$

We claim that for any τ with real coefficients

$$\pi\tau = \tau\pi. \tag{6.2}$$

An easy calculation gives

$$\begin{aligned}
\tau w &= (aw + b)(cw + d)^{-1} \\
&= \frac{[(ax_1 + b)(cx_1 + d) + ac(x_2^2 + y^2)] + x_2 i + yj}{\delta},
\end{aligned} \tag{6.3}$$

where

$$\delta = (cx_1 + d)^2 + (cx_2)^2 + (cy)^2.$$

Hence

$$\pi\tau w = \left\{ \frac{(ax_1 + b)(cx_1 + d) + ac(x_2^2 + y^2)}{\delta}, \frac{\sqrt{x_2^2 + y^2}}{\delta} \right\}$$

$$= \tau\pi w.$$

The lemma asserts that $F = \pi^{-1}F_2$. It is clear that $\bigcup (\pi^{-1}\gamma\bar{F}_2) = \mathbb{H}_3$. For take any $w \in \mathbb{H}_3$, then $\pi w \in \mathbb{H}_2$ and there exists a γ in Γ such that $\gamma\pi w$ lies in $\bar{F}_2$. Consequently $\pi\gamma w$ belongs to $\bar{F}_2$ and γw to $\bar{F}$. Next suppose that w and γw belong to F for some γ in Γ, $\gamma \neq id$. Then πw and $\pi\gamma w = \gamma(\pi w)$ belong to F_2 which is impossible. This shows that F is indeed a fundamental domain for Γ.

In view of this result, the natural coordinate system to use in considering such fundamental domains is the cylindrical coordinates defined in (4.21). We also introduce a change of the dependent variable

$$f = \operatorname{sech} s \, g \tag{6.4}$$

in terms of which

$$E(f) = \int_F [\rho^2 \operatorname{sech}^2 s \, (|\partial_x g_1|^2 + |\partial_\rho g_1|^2) + |\partial_s g_1|^2 + |g_2|^2] \frac{dx \, d\rho \, ds}{\rho^2}. \tag{6.5}$$

The E form is clearly nonnegative. In view of Theorem 3.4 and Lemma 3.2, a suitable G form for these examples is

$$G(f) = E(f) + K(f), \tag{6.6}$$

where

$$K(f) = \int_S |g_1|^2 \frac{dx \, d\rho \, ds}{\rho^2}, \tag{6.6}'$$

S being a compact subset of F.

LEMMA 6.2. *If Area $F_2 < \infty$, then $f = \operatorname{sech} s\{1, 0\}$ is a null vector for E.*

Proof. Since $E(f) = 0$ it is only necessary to show that f belongs to $\mathscr{H}$, that is, that f can be approximated in the G-norm by $C_0^\infty(F)$ data. To this

end we choose a sequence of data $f_n = \text{sech}\, s\{\varphi_n, 0\}$ in $C_0^\infty(F)$, $\varphi_n = \varphi_1^n(x, \rho)$
$\varphi_2^n(s)$; here $\varphi_1^n(x, \rho) = 1$ except in cusps and $\varphi_2^n(s) = \xi(s/n)$,

$$\xi(s) = 1 \qquad |s| < 1$$
$$= 0 \qquad |s| > 2.$$

Typically a cusp in F_2 is of the form $\{|x| < 1/2, \rho > a\}$ and on such a cusp
we set $\varphi_1^n(x, \rho) = \xi(\rho/n)$. It is now easy to show that

$$G(f - f_n) \to 0 \qquad \text{and} \qquad E(f_n) \to 0. \tag{6.7}$$

LEMMA 6.3. *If Area $F_2 = \infty$, then E is positive on $\mathscr{H}$.*

Proof. If Area $F_2 = \infty$, then F_2 has to contain a neighborhood of infinity
in $\mathbb{H}_2$, and hence a rectangle Q of the form $\{0 \leqslant x \leqslant a, 0 \leqslant \rho \leqslant b\}$. For
$v \in C_0^\infty(\mathbb{R})$, a simple integration by parts shows that

$$\int_0^b \rho \left| \partial_\rho \left(\frac{v}{\rho^{1/2}} \right) \right|^2 d\rho = \int_0^b \left(|\partial_\rho v|^2 - \frac{1}{4\rho^2} |v|^2 \right) d\rho - \frac{|v(b)|^2}{2b}.$$

Hence

$$\int_0^b |\partial_\rho v|^2 \, d\rho \geqslant \frac{1}{4} \int_0^b \frac{|v|^2}{\rho^2} \, d\rho$$

and, multiplying through by $\text{sech}^2 s$, we see by (6.5) that

$$E(f) \geqslant \frac{1}{4} \iiint |f_1|^2 \frac{dx \, d\rho \, ds}{\rho^2}, \tag{6.8}$$

the range of integration being $\pi^{-1}Q$.

If $E(f) = 0$, then we see from (6.5) that $g_2 = 0$ and $\partial g_1 = 0$; so that
$f = \text{sech}\, s \{1, 0\}$. If such an f is in $\mathscr{H}$, then it can be approximated in the G-
norm by a sequence f_n of $C_0^\infty(F)$ data: $f_n = \text{sech}\, s\{\varphi_n, 0\}$ satisfying (6.7).
According to Lemma 3.1, φ_n converges to 1 locally in L_2. In this case
$E(f_n) \to 0$ while the integral on the right side in (6.8) tends to infinity. In
view of (6.8) this is impossible.

Next we present an example for which $\text{vol}\, F = \infty$ and E is strictly
indefinite. We take for the subgroup Γ the group generated by

$$\begin{pmatrix} 1 & 2\lambda \\ 0 & 1 \end{pmatrix}, \qquad \begin{pmatrix} 1 & 2\lambda i \\ 0 & 1 \end{pmatrix}, \qquad \begin{pmatrix} 0 & -1 \\ 1 & 0 \end{pmatrix},$$

for $\lambda \geqslant 1$. It can be shown that a fundamental domain F for Γ is determined by the inequalities

$$|x_1| < \lambda, \qquad |x_2| < \lambda, \qquad x_1^2 + x_2^2 + y^2 > 1, \qquad y > 0. \qquad (6.9)$$

Since each of the generating elements of Γ take the y coordinate of a boundary point into itself, any continuous function of y alone will be automorphic. A simple integration by parts described in Section 3 brings the energy form (3.10) into

$$E(f) = \int_F \left[\frac{|\partial_{x_1} f_1|^2 + |\partial_{x_2} f_1|^2}{y} + y \left| \partial_y \left(\frac{f_1}{y} \right) \right|^2 + \frac{|f_2|^2}{y^3} \right] dx_1 \, dx_2 \, dy$$

$$- \int_S \frac{|f_1|^2}{y^2} \, dx_1 \, dx_2, \qquad (6.10)$$

where the boundary integral is over the unit hemisphere S:

$$\{x_1^2 + x_2^2 + y^2 = 1, y > 0\}.$$

To prove that E is indefinite it is enough to exhibit a sequence of data $\{f_n\}$, automorphic and piecewise $C_0^1(F)$, for which

$$E(f_n) < 0 \quad \text{for } n \text{ large enough.} \qquad (6.11)$$

To this end choose $\varphi(s) \in C_0^\omega(\mathbb{R})$ so that

$$\varphi(s) = 1 \qquad \text{for} \quad |s| < 1$$

$$= 0 \qquad \text{for} \quad |s| > 2.$$

We then set

$$f_n = \{ y\varphi_n(y), 0 \}, \qquad (6.12)$$

when

$$\varphi_n(y) = \varphi \left(\frac{\log y}{n} \right).$$

We see from (6.10) that

$$E(f_n) = \int_F y \, |\varphi_n'(y)|^2 \, dx_1 \, dx_2 \, dy - \int_S |\varphi_n(y)|^2 \, dx_1 \, dx_2.$$

An easy estimate shows that

$$\int_F y \, |\varphi_n'(y)|^2 \, dx_1 \, dx_2 \, dy \leqslant \frac{c}{n^2} \left(\int_{e^{-2n}}^{e^{-n}} + \int_{e^n}^{e^{2n}} \right) \frac{dy}{y} = \frac{2c}{n}.$$

It follows that

$$\lim E(f_n) = -\int_S dx_1\, dx_2 = -\pi,$$

as desired. Note that the energy form (3.9) is not applicable because the function y is not in $D(L)$; this is so because y has discontinuous first derivatives along S when considered as a function on the manifold F.

We show now that the last example is typical for all nonelementary groups whose fundamental domains have a cusp. We call Γ nonelementary if not all of its elements have a common fixed point.

THEOREM 6.4. *Suppose Γ is nonelementary and that its fundamental domain F has a cusp of maximal rank. Then the energy form over F is indefinite.*

A cusp of maximal rank of F at ∞ has the following structure in the $\{x, y\}$ parameterization of $\mathbb{H}_m$:

$$F_\infty \times (a, \infty), \tag{6.13}$$

where F_∞ is the fundamental domain of a crystallographic group Γ_∞ in $\mathbb{R}^{m-1}$ of finite area. Γ_∞ is a subgroup of Γ. According to Theorem 1.1, the counting number $N_\infty(s)$ for Γ_∞ in $\mathbb{R}^{m-1}$ is

$$N_\infty(s) = \frac{\omega_{m-1}}{(m-1)|F_\infty|} s^{m-1} + O(s^\alpha) \tag{6.14}$$

when α is any number $> (m-1)(m-2)/m$.

Γ is nonelementary if it contains a τ for which ∞ is not a fixed point:

$$\tau(\infty) \neq \infty. \tag{6.15}$$

In order to prove Theorem 6.4, we shall show that the contribution of elements of the kind $\{\gamma\tau\mu;\ \gamma, \mu \in \Gamma_\infty\}$ to the counting number $N(T)$ for Γ is enough to make

$$N(T) \geqslant cTe^{(m-1)T/2} \tag{6.16}$$

for some $c > 0$. If the energy form were $\geqslant 0$ over F, then according to Theorem 5.8, $N(T)$ would be $o(T\exp((m-1)T/2)$. This, being contrary to (6.16), proves the theorem.

Denote by Γ_e those elliptic maps in Γ which keep ∞ and $\tau(\infty)$ fixed; and by Γ_e' those elliptic maps in Γ which keep ∞ and $\tau^{-1}(\infty)$ fixed. Since Γ is discrete, both Γ_e and Γ_e' are *finite* subgroups of Γ.

260

LEMMA 6.5. *Let γ, γ', μ, μ' be elements of Γ_∞. Two elements of the form $\gamma\tau\mu$ and $\gamma'\tau\mu'$ are equal only if $\gamma \in \gamma'\Gamma_e$ and $\mu \in \Gamma'_\epsilon\mu'$.*

Proof. We rewrite

$$\gamma\tau\mu = \gamma'\tau\mu'$$

as

$$\gamma_1\tau = \tau\mu_1, \tag{6.17}$$

where

$$\gamma_1 = {\gamma'}^{-1}\gamma \quad \text{and} \quad \mu_1 = \mu'\mu^{-1}. \tag{6.18}$$

Clearly γ_1 and μ_1 belong to Γ_∞.

Next let Σ be any horosphere through ∞; then $\tau\Sigma$ and $\tau^{-1}\Sigma$ are horospheres through $\tau(\infty)$ and $\tau^{-1}(\infty)$, respectively. We now show that

(i) γ_1 map $\tau\Sigma$ into itself

(ii) μ_1^{-1} maps $\tau^{-1}\Sigma$ into itself.

In fact if w is any point in Σ, then $\mu_1 w$ also lies in Σ and $\tau\mu_1 w$ lies in $\tau\Sigma$. Using (6.17), we deduce that $\gamma_1\tau w = \tau\mu_1 w$ lies in $\tau\Sigma$. This proves part (i); part (ii) can be proved in the same way.

This shows that γ_1 maps horospheres through ∞, and those through $\tau(\infty)$ into themselves. Such a map is elliptic; thus γ_1 belongs to Γ_e; using (6.18) we deduce that γ belongs to $\gamma'\Gamma_e$. Similarly we deduce that μ_1^{-1} belongs to Γ'_e and therefore that μ belongs to $\Gamma'_e\mu'$. This completes the proof of the lemma.

It follows from this lemma that as γ runs through all the elements of Γ_∞ and μ through $\Gamma'_e\backslash\Gamma_\infty$, the elements $\gamma\tau\mu$ are distinct elements of Γ. We may suppose that $j = \{0, 1\}$ is not a fixed point of Γ. Set $\mu j = \{x_\mu, 1\}$ and from each set of $\Gamma'_e\backslash\Gamma_\infty$ choose the μ which minimizes $|x_\mu|$. Then since the cardinality of Γ'_e is finite, say, M, there will be at least $N_\infty(s)/M$ orbital points x_μ, with μ chosen as above, lying in the Euclidean ball of radius s about the origin; here N_∞ is given by (6.14). Each of these is mapped by τ into $\tau\mu j$ which is of the form

$$\left\{ \frac{l(x_\mu)}{q(x_\mu)}, \frac{1}{q(x_\mu)} \right\}: \tag{6.19}$$

here l is linear and q quadratic in x_μ with

$$q(x_\mu) \geqslant c(|x_\mu|^2 + 1) \tag{6.20}$$

for some $c > 0$. Next applying the elements of Γ_∞ to $\tau\mu j$ we get

$$\gamma\tau\mu j = \left\{ \frac{l(x_\mu)}{q(x_\mu)} + x_\gamma, \frac{1}{q(x_\mu)} \right\}. \tag{6.21}$$

According to (6.14)

$$N_\infty(s) = [\gamma \in \Gamma_\infty, |x_\gamma| \leqslant s] \simeq \text{const. } s^{m-1}. \tag{6.22}$$

Now the non-Euclidean distance of a point $\{x, y\}$ from j is given by

$$\log \frac{|x|^2}{y} \tag{6.23}$$

plus a term that tends to zero with y. Hence the distance of the points (6.21) from j is less than

$$\log(\text{const.} + \text{const.} |x_\gamma|^2 \, q(x_\mu)) \simeq \log \text{const.} |x_\gamma|^2 |x_\mu|^2.$$

It follows that the number $N(T)$ of points of Γ_j whose non-Euclidean distance from j is $\leqslant T$ is not less than the number of γ in Γ_∞ and μ in Γ_e'/Γ_∞ for which

$$|x_\gamma|^2 |x_\mu|^2 \leqslant \text{const. } e^T. \tag{6.24}$$

Fixing μ so that

$$|x_\mu|^2 \leqslant \text{const. } e^T, \tag{6.25}$$

the number of $\gamma \in \Gamma_\infty$ for which (6.24) holds is by (6.14) greater than

$$\text{const. } \left(\frac{e^T}{|x_\mu|^2} \right)^{(m-1)/2} \tag{6.26}$$

We sum this over the range (6.25) taking one μ from each coset of $\Gamma_e'\backslash\Gamma_\infty$; this gives

$$N(T) \geqslant \text{const. } e^{(m-1)T/2} \sum \frac{1}{|x_\mu|^{m-1}}.$$

Taking the cardinality of Γ_e' into account, we get

$$N(T) \geqslant \frac{\text{const.}}{M} e^{(m-1)T/2} \int \frac{1}{s^{m-1}} \, dN_\infty(s)$$

$$\simeq \frac{\text{const.}}{M} (m-1) e^{(m-1)T/2} \int \frac{N_\infty(s)}{s^m} \, ds,$$

262

where the range of integration is $[1, \text{const. } e^{T/2}]$. Using (6.14) this gives

$$N(T) \geqslant \text{const. } T\, e^{((m-1)/2)T}.$$

This is inequality (6.16) from which Theorem 6.4 follows.

We saw in Theorem 4.4 that the energy form for F is indefinite iff L' has positive eigenvalues. An immediate consequence of this and Theorem 6.4 is

COROLLARY 6.6. *If Γ is nonelementary and its fundamental domain contains a cusp of maximal rank, then L' has positive eigenvalues.*

Corollary 6.6 generalizes to dimensions greater than 2 a result due to Patterson [13]. Another proof is given by Sullivan in [14].

7. SOLUTIONS OF THE NON-EUCLIDEAN WAVE EQUATIONS

In this section we shall construct a solution for the non-Euclidean wave equation in the m-dimensional Poincaré half space $\mathbb{H}_m$ consisting of points (x, y), $x \in \mathbb{R}^{m-1}$, $y > 0$, with the metric

$$ds^2 = \frac{dx^2 + dy^2}{y^2}. \tag{7.1}$$

The wave equation we shall consider is

$$u_{tt} = L_m u = y^2 \Delta_x u + y^m \partial_y \frac{1}{y^{m-2}} \partial_y u + \left(\frac{m-1}{2}\right)^2 u. \tag{7.2}$$

In polar coordinates this becomes

$$u_{tt} = u_{ss} + (m-1)\frac{\cosh s}{\sinh s} u_s + \Lambda u + \left(\frac{m-1}{2}\right)^2 u, \tag{7.3}$$

where Λ is a second order differential operator in the sphere of radius s.

LEMMA 7.1. *For m odd*

$$g_m(s, t) = \left(\frac{1}{\sinh s} \partial_s\right)^{(m-3)/2} \cdot \frac{h(s-t)}{\sinh s} \tag{7.4}$$

is a spherically symmetric solution of (7.3).

Proof. For the case $m = 3$, a straightforward verification suffices. We prove the general case by induction, assuming that it holds for m. The commutant

$$\left[L_m, \frac{1}{\sinh s} \partial_s \right] = -2 \frac{\cosh s}{\sinh^2 s} \partial_s^2 + \left[2 \frac{\cosh^2 s}{\sinh^3 s} - \frac{m}{\sinh s} \right] \partial s. \quad (7.5)$$

It is also readily seen that

$$L_{m+2} g_{m+2} = L_m \left(\frac{1}{\sinh s} \partial_s g_m \right) + 2 \frac{\cosh s}{\sinh s} \partial_s \left(\frac{1}{\sinh s} \partial_s g_m \right)$$

$$+ \frac{m}{\sinh s} \partial_s g_m$$

$$= \left[L_m, \frac{1}{\sinh s} \partial_s \right] g_m + \frac{1}{\sinh s} \partial_s (L_m g_m)$$

$$+ 2 \frac{\cosh s}{\sinh^2 s} \partial_s^2 g_m - 2 \frac{\cosh^2 s}{\sinh^3 s} \partial_s g_m + \frac{m}{\sinh s} \partial_s g_m. \quad (7.6)$$

Combining (7.5) and (7.6) we get

$$L_{m+2} g_{m+2} = \frac{1}{\sinh s} \partial_s (L_m g_m) = \frac{1}{\sinh s} \partial_s \partial_t^2 g_m$$

$$= \partial_t^2 g_{m+2},$$

as desired.

Taking s to be the non-Euclidean distance between two points w and w',

$$\cosh s = 1 + \frac{|w - w'|^2}{2yy'}, \quad (7.7)$$

where $|w - w'|$ is the Euclidean distance between w and w', we obtain a solution, spherically symmetric about w'. Clearly the function obtained by a superposition of spherical waves, that is,

$$u(w, t) = c_m \left(\frac{1}{\sinh s} \partial_s \right)^{(m-3)/2} \int f(w') \frac{h(s-t)}{\sinh s} dw' \quad (7.8)$$

is a solution of (7.2). Taking h to be the delta function, we obtain the solution

$$u(w, t) = c_m \left(\frac{1}{\sinh t} \partial_t \right)^{(m-3)/2} \frac{1}{\sinh t} \int_{d(w, w') = t} f(w') \, d\sigma; \quad (7.9)$$

here $d\sigma$ denotes the non-Euclidean surface element. With

$$c_m = \frac{\sqrt{\pi}}{2^{(m-1)/2}((m-2)/2)!} \tag{7.9}'$$

this solution satisfies the initial conditions

$$u(w, 0) = 0, \qquad u_t(w, 0) = \omega_m f(w). \tag{7.10}$$

We obtain a solution for the even dimensional problem by the method of descent. This method is somewhat complicated in the non-Euclidean case by the fact that a solution u of $u_{tt} = L_{m+1} u$ which does not depend on x_m is itself not a solution of the m-dimensional wave equation but instead

$$v = \frac{u}{y^{1/2}} \tag{7.11}$$

is a suitable solution. With this substitution, formula (7.9) becomes

$$v(w, t) = c_{m+1} \frac{1}{y^{1/2}} \left(\frac{1}{\sinh t} \partial_t\right)^{(m-2)/2} \frac{1}{\sinh t} \int_{d(w,w')=t} (y')^{1/2} f(w') \, d\sigma.$$

Taking f to be independent of x_m, the integral on the right can be written as an integral over the m-dimensional ball $d(w, w') < t$:

$$v(w, t) = c_{m+1} \sqrt{2} \left(\frac{1}{\sinh t} \partial_t\right)^{(m-2)/2} \int_{d(w,w')<t} \frac{f(w')}{\sqrt{\cosh t - \cosh d}} \, dw'. \tag{7.12}$$

Replacing the constant $\sqrt{2} c_{m+1}$ by

$$c_m = \sqrt{2} c_{m+1} \frac{\omega_m}{\omega_{m+1}} = \frac{1}{2^{(m-1)/2}((m-2)/2)!} \tag{7.13}$$

and v by u, the resulting function satisfies the initial conditions (7.10).

Finally we obtain bounds on the Green's function G_0 for $M = k^2 - L$ as an operator on the entire hyperbolic space $\mathbb{H}_m$. Since $G_0(w, w_0)$ is spherically symmetric about w_0, it satisfies the ordinary differential equation

$$\frac{d^2}{ds^2} G_0 + (m-1) \frac{\cosh s}{\sinh s} \frac{d}{ds} G_0 + \left[\left(\frac{m-1}{2}\right)^2 - k^2\right] G_0 = 0, \tag{7.14}$$

where $s = d(w, w_0)$. For m odd set $h(s) = e^{-ks}$ in formula (7.4); Lemma 7.1 shows that

$$G_0 = \text{const.} \left(\frac{1}{\sinh s} \partial_s\right)^{(m-3)/2} \frac{e^{-ks}}{\sinh s} \tag{7.15}$$

satisfies (7.14). For m even Green's function can be defined in terms of Legendre functions (see Woo [16, pp. 102–103]). However, to obtain the estimates that we require, a cruder analysis suffices.

LEMMA 7.2. *The whole space Green's function G_0 for $M = k^2 - L$, $k > (m-1)/2$, is bounded by*

$$|G_0(w, w_0)| \leqslant \begin{cases} \dfrac{\text{const.} \exp[-((m-1)/2 + k)\, d(w, w_0)]}{d(w, w_0)^{m-2}} & \text{for } m > 2, \\[2ex] \text{const.} \left| \log d(w, w_0) \right| \exp[-(\tfrac{1}{2} + k)\, d(w, w_0)] & \text{for } m = 2. \end{cases}$$

$$(7.16)$$

Proof. Near $s = 0$ the differential equation (7.14) is of the regular singular types with indicial equation

$$\gamma(\gamma - 1) + (m - 1)\gamma = 0.$$

Since the roots of this equation are $\gamma = 0$ and $\gamma = 2 - m$, the corresponding solutions to (7.14) behave like

$$s^\gamma \sum_{n=0}^\infty a_n s^n, \qquad a_0 \neq 0.$$

However, when $m = 2$, $\gamma = 0$ is a double root and there is in this case a singular solution of the type

$$\log s \sum_{n=0}^\infty a_n s^n, \qquad a_0 \neq 0.$$

This establishes the local behavior of G_0.

For large s Eq. (7.14) approaches

$$\frac{d^2}{ds^2} G_0 + (m-1) \frac{d}{ds} G_0 + \left[\left(\frac{m-1}{2} \right)^2 - k^2 \right] G_0 = 0, \qquad (7.14)'$$

which has solutions of the type

$$\exp\left[-\left(\frac{m-1}{2} \pm k \right) s \right].$$

The solution which behaves like $\exp[-((m-1)/2 + k)s]$ at infinity can not be regular at the origin since such a solution would be an L_2 null function of M, contrary to the fact that M is positive definite when $k > (m-1)/2$.

Hence the solution which behaves like $\exp[-((m-1)/2+k)s]$ at infinity must be singular at the origin and this singularity is proper for a Green's function.

REFERENCES

1. P. J. COHEN AND P. SARNAK, "Lectures on the trace formula," to appear.
2. G. B. FOLLAND, "Introduction to Partial Differential Equations," Mathematical Notes, Princeton Univ. Press, Princeton, N. J., 1976.
3. E. HEINZ, Beiträge zur Störungstheorie der Spektralzerlegung, *Math. Ann.* **123** (1951), 415–438.
4. S. HELGASON, "Lie Groups and Symmetric Spaces" (C. M. DeWitt and J. A. Wheeler, Eds.) pp. 1–71, Battelle Rencontres, 1967 Lectures in Math. and Physics, Benjamin, New York, 1968.
5. H. HUBER, Über eine neue Klasse automorpher Functionen und eine Gitterpunktproblem in der hyperbolischen Ebene, *Comment. Math. Helv.*, **30** (1956), 20–62.
6. H. HUBER, Zur analytischen Theorie hyperbolischen Raumformen und Bewegungs-gruppen I, *Math. Ann.* **138** (1959), 1–26; II, *Math. Ann.* **142** (1961), 385–398; **143** (1961), 463–464.
7. E. LANDAU, Über die Gitterpunkte in einem mehrdimensionalen Ellipsoid, *Berliner Akademieberichte* (1915), 458–476.
8. E. LANDAU, Über eine Aufgabe aus der Theorie der quadratischen Formen; *Wiener Akademieberichte* **124** 1915, 445–468.
9. P. D. LAX AND R. S. PHILLIPS, "Scattering Theory for Automorphic Functions," Ann. of Math. Studies, No. 87, Princeton Univ. Press, Princeton, N. J., 1976.
10. W. LITTMAN, Décroissance à l'infini des solutions à l'extérieur d'un cône d'equations aud dérivées partielles à coefficients constants, *C. R. Acad. Sci. Paris Sér. A-B* **287** (1978), no. 1, A15–A17.
11. S. J. PATTERSON, The Laplacian operator on a Riemann surface, *Compositio Math.* **31** (1975), 83–107.
12. S. J. PATTERSON, A lattice point problem in hyperbolic space, *Mathematika* **22** (1975), 81–88.
13. S. J. PATTERSON, The limit set of a Fuchsian group, *Acta Math.* **136** (1976), 241–273.
14. D. SULLIVAN, Entropy, Hausdorff measures old and new, and limit sets of geometrically finite Kleinian groups, *Acta Math.*, in press.
15. A. WALFISZ, "Gitterpunkte in mehrdimensionalen Kugeln," Monog. Mat., Polska Akad. Nauk., Warszawa, 1957.
16. A. C. WOO, "Scattering Theory on Real Hyperbolic Spaces and Their Compact Perturbations," dissertation, Stanford University, 1980.

Printed by the St. Catherine Press Ltd., Tempelhof 41, Bruges, Belgium

A Local Paley-Wiener Theorem for the Radon Transform of L_2 Functions in a Non-Euclidean Setting

PETER D. LAX

Courant Institute

AND

RALPH S. PHILLIPS

Stanford University

1. Introduction

Paley-Wiener theorems for the Radon transforms in a non-Euclidean setting deal with functions whose Radon transforms vanish on the set of horospheres contained in some neighborhood $\mathfrak{N}$ of infinity, $\mathfrak{B}$, or of a point β_0 in $\mathfrak{B}$, and conclude from this that the functions themselves vanish on $\mathfrak{N}$. Results of this sort differ in the choice of $\mathfrak{N}$ and the class of functions considered.

The theorem proved in the present paper is for L_2 functions in a spherical cap neighborhood of a point in $\mathfrak{B}$. This result complements our previous result [5] which requires that the function vanish near a point β_0 of $\mathfrak{B}$ but treats a wider class of neighborhoods; namely $\mathfrak{N}$ consists of the set of all horospheres of radius less than b through all points β in a neighborhood of β_0.

The Euclidean analogues of these theorems are global in nature and similar to a result due to S. Helgason [3] which treats exponentially rapidly decreasing functions whose Radon transforms vanish on all horospheres outside a given ball. In both spirit and method the present paper has much in common with our in-between result in [4], Theorem 3.14 (see also A. Woo [6]) which holds for L_2 functions in the exterior of a fixed horosphere.

In order to state our result more precisely we first have to recall some properties of the Fourier and Radon transform in a non-Euclidean space. We shall work in the n-dimensional real hyperbolic space $\mathbf{H}_n$ in the model $z = \{x, y\}$, where $x \in \mathbb{R}^{n-1}$ and $y > 0$. The Riemannian metric for this space is

$$(1.1) \qquad ds^2 = \frac{dx^2 + dy^2}{y^2}$$

and we shall denote the non-Euclidean volume and surface elements by dV and dS.

The isometries for $\mathbf{H}_n$ are generated by

(i)   translations in the x-directions,

Communications on Pure and Applied Mathematics, Vol. XXXV, 531–554 (1982)
CCC 0010-3640/82/040531-24$03.40

(ii) rotations about the y-axis,

(iii) homothetic transformations,

(iv) inversions through the unit sphere about the origin followed by a reflection in an x-direction.

The n-dimensional non-Euclidean Fourier transform is defined by

$$(1.2) \qquad \tilde{f}(\mu, \beta) = \int \exp\left\{ \left(\frac{n-1}{2} - i\mu \right) \langle z, \beta \rangle \right\} f(z)\, dV,$$

where $\langle z, \beta \rangle$ denotes the non-Euclidean distance from $z_0 = \{0, 1\}$ to the horosphere through z and tangent at $\beta \in \mathcal{B}$. In particular, if $\beta = \infty$ and $z = \{x', y\}$, then

$$(1.3) \qquad \langle z, \infty \rangle = \log y.$$

We note that for an isometry of $\mathbf{H}_n$ for which $\gamma z_0 = z_0$, we have

$$(1.4) \qquad \langle z, \gamma^{-1}\beta \rangle = \langle \gamma z, \beta \rangle.$$

Defining

$$(1.5) \qquad f^\gamma(z) = f(\gamma^{-1}z),$$

it is easy to see that if $\gamma z_0 = z_0$

$$(1.6) \qquad \tilde{f}^\gamma(\mu, \beta) = \tilde{f}(\mu, \gamma^{-1}\beta).$$

The Plancherel formula for the non-Euclidean Fourier transform, due to Gelfand and Graev [1], is

$$(1.7) \qquad \int_{\mathbf{H}_n} |f(z)|^2\, dV = c_n' \int_{\mathcal{B}} \int_{\mathbf{R}_+} |\tilde{f}(\mu, \beta)|^2 \frac{d\mu}{|c_n(\mu)|^2} \frac{d\beta}{(1 + |\beta|^2)^{n-1}},$$

where c_n' is a constant depending on n and the Harish-Chandra c function is defined by

$$(1.8) \qquad c_n(\mu) = B\left(\tfrac{1}{2}(n-1), i\mu\right) / B\left(\tfrac{1}{2}(n-1), \tfrac{1}{2}(n-1)\right),$$

B being the Beta function.

The horosphere $\xi(r, \beta)$ tangent to the x-plane at $\{\beta, 0\}$ and with radius r is described by the relation

$$(1.9) \qquad |x - \beta|^2 + |y - r|^2 = r^2.$$

269

The Radon transform of f, obtained by integrating over horospheres, is defined as

$$(1.10) \qquad \hat{f}(r, \beta) = \int_{\xi(r,\beta)} f(z)\, dS.$$

The integral (1.10) is well defined for C^∞ functions of compact support in $\mathbf{H}_n$ but need not exist for L_2 functions. To extend the Radon transform to this wider class of functions we use Cauchy sequences of C^∞ functions with compact support together with a Plancherel relation due to Helgason [2]:

$$(1.11) \qquad \int |f(z)|^2\, dV = c_n \int_{\mathbf{R}^{n-1}} \int_{\mathbf{R}_+} |J_n \hat{f}(r, \beta)|^2 \frac{dr}{r}\, d\beta;$$

here c_n is a constant depending on n, $d\beta = d\beta_1 \cdots d\beta_{n-1}$, and J_n is a differential operator when n is odd and a composite differential-convolution operator when n is even (see [6] pp. 48–49):

$$(1.12) \qquad J_n = r\partial_r \Xi r^{(1-n)/2} \quad \text{for} \quad n \quad \text{odd},$$

where

$$(1.12)' \qquad \Xi = (1 + r\partial_r) \cdots (\tfrac{1}{2}(n-3) + r\partial_r);$$

and

$$(1.13) \qquad J_n = r\partial_r \Xi N r^{(1-n)/2} \quad \text{for} \quad n \quad \text{even},$$

where now

$$\Xi = (1 + r\partial_r) \cdots (\tfrac{1}{2}(n-2) + r\partial_r),$$

$$(1.13)' \qquad Nh(r) = \int_0^1 (1 - \sigma)^{-1/2} \sigma^{(n-1)/2} h(\sigma r)\, \frac{d\sigma}{\sigma}.$$

A precise statement of our result can now be given.

THEOREM 1.1. *Let $\mathcal{C}$ be a spherical cap in $\mathbf{H}_n$, i.e., a set of the form*

$$(1.14) \qquad |x - x_0|^2 + |y - y_0|^2 \leq c^2, \qquad\qquad y > 0, |y_0| < c.$$

If f is square integrable in $\mathcal{C}$ and if $J_n \hat{f}$ vanishes on almost all horospheres contained in $\mathcal{C}$, then $f = 0$ a.e. in $\mathcal{C}$.

The proof of this theorem is somewhat involved. The reader may find it helpful to first look at a simpler version of the proof to be found in Theorem 3.14 of [4] or to limit himself to the two-dimensional case of the proof of Theorem 1.1.

In a forthcoming paper we shall show how to use Theorem 1.1 to prove the following: Let F be a discrete subgroup of isometries on $\mathbf{H}_n$ whose fundamental domain $\mathcal{F}$ has a finite number of sides and infinite volume; then the Laplace-Beltrami operator on $\mathcal{F}$ has no singular spectrum.

2. Geometrical Properties of a Spherical Cap

We begin with the observation that a spherical cap can be mapped by an isometry of $\mathbf{H}_n$ onto a wedge of the form

$$(2.1) \qquad\qquad W: \tan^{-1} y/x_{n-1} < \theta_0.$$

Such a wedge turns out to be a convenient setting for our analysis.

It is clear that W will be left invariant by the subgroup F of isometries generated by

(i)′   translations in the $(x_1, \cdots, x_{n-2})$-directions,
(ii)′   rotations in the $(x_1, \cdots, x_{n-2})$-space,
(iii)′   homothetic transformations,
(iv)′   inversions through the unit sphere about the origin followed by a reflection in an $(x_1, \cdots, x_{n-2})$-direction.

In analogy with the $n = 3$ case in which the isometries of $\mathbf{H}_3$ correspond to $SL(2, \mathbb{C})/\pm id$ and $F = SL(2, \mathbb{R})/\pm id$, we call F the Fuchsian subgroup of isometries. Each of the isometries of F leaves invariant every half-plane of the form $\tan^{-1} y/x_{n-1} = \theta > 0$. (In fact such a half-plane itself constitutes a real hyperbolic space of dimension $n - 1$ under the metric induced by the imbedding in $\mathbf{H}_n$, and with respect to this metric every γ in F is an isometry mapping each half-space onto itself.)

Let f be a function defined on $\mathbf{H}_n$ expressed in terms of cylindrical coordinates:

$$f(x', \rho, \theta);$$

here x', ρ and θ are given by

$$(2.2) \quad x' = (x_1, \cdots, x_{n-2}), \qquad \rho = (x_{n-1}^2 + y^2)^{1/2}, \qquad \tan \theta = y/x_{n-1}.$$

It follows from the above discussion that for γ in F

$$(2.3) \qquad\qquad f^\gamma(x', \rho, \theta) = f(\gamma^{-1}\{x', \rho\}, \theta).$$

Notice that if $\gamma \in F$ leaves $z_0 = \{0, 1\}$ fixed, then it also leaves $\{0', 1, \theta\}$ fixed for all θ.

We define the *partial* Fourier transform $\tilde{f}$ of f as the $(n-1)$-dimensional non-Euclidean Fourier transform of $f(x', \rho, \theta)$ with respect to the variables $w = \{x', \rho\}$, regarding θ as a parameter. Thus replacing n by $n-1$ in (1.2) we get

$$(2.4) \qquad \tilde{f}(\mu, \chi, \theta) = \int \int \exp\left\{ \left(\frac{n-2}{2} - i\mu \right) \langle w, \chi \rangle \right\} f(x', \rho, \theta) \frac{dx' \, d\rho}{\rho^{n-1}},$$

where now $\langle w, \chi \rangle$ denotes the non-Euclidean distance from $w_0 = \{0', 1\}$ in $\mathbf{H}_{n-1}$ to the $(n-1)$-dimensional horosphere through w and tangent at χ in $\mathcal{B}_{n-1}$; in this case, $\mathcal{B}_{n-1}$ corresponds to the $(x_1, \cdots, x_{n-2})$-space.

Making use of (2.3) we see that the analogue of (1.6) for $\gamma \in F$ with $\gamma z_0 = z_0$ is simply

$$(2.4)' \qquad\qquad\qquad \tilde{f}^\gamma(\mu, \chi, \theta) = \tilde{f}(\mu, \gamma^{-1}\chi, \theta).$$

Using the $(n-1)$-dimensional analogue of the Plancherel relation (1.7) we derive the following relation for the partial Fourier transform:

$$\int |f|^2 \, dV = \int \int \int \left| \frac{f(x', \rho, \theta)}{\sin^{n/2}\theta} \right|^2 \frac{dx' \, d\rho}{\rho^{n-1}} \, d\theta$$

$$(2.5)$$

$$= c'_{n-1} \int \int \int \left| \frac{\tilde{f}(\mu, \chi, \theta)}{\sin^{n/2}\theta} \right|^2 \frac{d\mu}{|c_{n-1}(\mu)|^2} \frac{d\chi}{(1+|\chi|^2)^{n-2}} \, d\theta.$$

This shows that $\tilde{f}(\theta)/\sin^{n/2}\theta$, regarded as a vector-valued function whose values lie in an appropriately weighted $L_2(\mathbb{R}^{n-2} \times \mathbb{R}_+)$ space, is an L_2 function of θ.

The horosphere $\xi(r, \beta_0)$ of radius r and point of tangency $\beta_0 = \{0', 1\}$ is tangent to the half-plane $\theta = \theta_r \equiv 2\tan^{-1}r$ (see Figure 1). The images of $\xi(r, \beta_0)$ under the Fuchsian subgroup F fill out the set of all horospheres of the form

$$(2.6) \qquad\qquad \xi(\lambda r, \{\beta', \lambda\}), \qquad\qquad \beta' \in \mathbb{R}^{n-2}, \lambda > 0,$$

which are tangent to the half-plane $\theta = \theta_r$.

The surface element on these horospheres will be invariant under the action of F and we introduce surface coordinates to take advantage of this fact. One of these is of course θ, defined as before by

$$(2.7) \qquad\qquad\qquad \tan\theta = y/x_{n-1}.$$

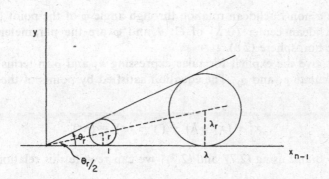

Figure 1

On the intersection of the horosphere $\xi(\lambda r, \{\beta', \lambda\})$ with the half-plane $\theta = \text{const.}$, which is an $(n-2)$-sphere, we introduce non-Euclidean spherical coordinates within the half-plane. We first explain how this is done for the case $n = 3$ (see Figure 2).

In each θ half-plane, defined now by

$$(2.7)' \qquad \tan \theta = y/x_2,$$

we introduce as coordinates $\{x_1, \rho\}$, where

$$(2.7)'' \qquad \rho = \left(x_2^2 + y^2\right)^{1/2}.$$

We set $\beta' = 0$ in (2.6), i.e., we consider the special horospheres

$$(2.8) \qquad \xi(\lambda r, \{0, \lambda\}).$$

These horospheres intersect each θ half-plane in a circle C. The intersection of C with the ρ-axis consists of two points ρ_1 and ρ_2, $\rho_1 > \rho_2$. Every point of C can be

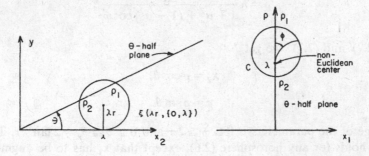

Figure 2

273

obtained by a non-Euclidean rotation through angle φ of the point $\{0, \rho_1\}$ about the non-Euclidean center $\{0, \lambda\}$ of C; θ and φ are the parameters we use to describe the horosphere (2.8).

We now give the explicit formulas expressing x_1 and ρ in terms of θ and φ. First we calculate ρ_1 and ρ_2. The equation satisfied by points of the horosphere, see (1.9), is

$$(2.8)' \qquad\qquad x_1^2 + (x_2 - \lambda)^2 + (y - \lambda r)^2 = (\lambda r)^2.$$

Setting $x_1 = 0$ and using (2.7)′ and (2.7)″ we can rewrite this relation as

$$(2.8)'' \qquad\qquad \rho_j^2 - 2\lambda\tau\rho_j + \lambda^2 = 0, \qquad\qquad j = 1, 2,$$

where

$$(2.9) \qquad\qquad \tau = \cos\theta + r\sin\theta.$$

The solution of (2.8)″ is given by

$$(2.10) \qquad\qquad \rho_1 = \lambda\alpha, \qquad \rho_2 = \lambda\alpha^{-1},$$

where

$$(2.10)' \qquad\qquad \alpha = \tau + (\tau^2 - 1)^{1/2}$$

As a check we note that the non-Euclidean center of C, namely $\{0, \lambda\}$, is the geometric mean of ρ_1 and ρ_2. A simple calculation shows that the non-Euclidean rotation of $\{0, \rho_1\}$ through the angle φ around $\{0, \lambda\}$ is given by

$$(2.11) \qquad\qquad x_1 = \lambda \frac{(\alpha^2 - 1)\sin\varphi}{1 + \alpha^2 + (1 - \alpha^2)\cos\varphi},$$

$$\rho = \lambda \frac{2\alpha}{1 + \alpha^2 + (1 - \alpha^2)\cos\varphi}.$$

Using (2.7)′ and (2.7)″ we get

$$x_2 = \rho\cos\theta,$$

$$(2.11)' \qquad\qquad y = \rho\sin\theta.$$

The range of the parameters is $0 \leq \varphi \leq 2\pi$ and $0 \leq \theta \leq \theta_r = 2\tan^{-1}r$. The same formula holds for any horosphere (2.6), except that x_1 has to be augmented by β_1.

In the general n-dimensional case for $n > 3$, the sphere of intersection is obtained by rotating the circles (2.11) about the ρ-axis. This yields

$$
(2.12) \qquad v = \begin{cases}
x_1 = p \cos \psi_1, \\
x_2 = p \sin \psi_1 \cos \psi_2, \\
\cdots \cdots \cdots, \\
x_{n-2} = p \sin \psi_1 \sin \psi_2 \cdots \sin \psi_{n-3}, \\
x_{n-1} = \rho \cos \theta, \\
\quad y = \rho \sin \theta;
\end{cases}
$$

here p and ρ are the same functions of φ and α as x_1 and ρ in (2.11), and $0 \le \varphi \le \pi$. In the case $n = 2$, the sphere of intersection reduces to the points $\{0, \rho_1\}$ and $\{0, \rho_2\}$.

It is convenient in computing the surface element to note that v in (2.12) decomposes as

$$
(2.12)' \qquad v = \begin{pmatrix} pv^{\mathrm{I}} \\ \rho v^{\mathrm{II}} \end{pmatrix},
$$

where v^{I} is a function of ψ, v^{II} of θ. The surface element for the horosphere $\xi(\lambda r, \{0', \lambda\})$ is given by

$$
(2.13) \quad dS = \left[\mathrm{Gramian}\,(\partial_\theta v, \partial_\varphi v, \partial_{\psi_1} v, \cdots, \partial_{\psi_{n-3}} v) \right]^{1/2} \frac{d\theta \, d\varphi \, d\psi_1 \cdots d\psi_{n-3}}{(\rho \sin \theta)^{n-1}}.
$$

It is readily verified that both $\partial_\theta v$ and $\partial_\varphi v$ are orthogonal to the $\partial_\psi v$. Hence the Gramian splits into the product of two determinants:

$$
(2.14) \qquad \mathrm{Gramian}\,(\partial_\theta v, \partial_\varphi v) \times \mathrm{Gramian}\,(\partial_{\psi_1} v, \cdots, \partial_{\psi_{n-3}} v)
$$

It is also readily verified that $\partial_\theta v$ and $\partial_\varphi v$ are orthogonal. Hence a straightforward calculation gives

$$
(2.15) \qquad \mathrm{Gramian}\,(\partial_\theta v, \partial_\varphi v) = r^2 \left(\frac{2\lambda\alpha}{1 + \alpha^2 + (1 - \alpha^2)\cos\varphi} \right)^4.
$$

For $n > 3$, the second Gramian in (2.14) is clearly the square of the surface element on an $(n - 3)$-dimensional sphere of radius p in a Euclidean space, that is, $p^{n-3} d\omega_{n-3}$ where

$$
(2.16) \qquad d\omega_{n-3} = \left| \sin^{n-4}\psi_1 \sin^{n-5}\psi_2 \cdots \sin\psi_{n-4} \right| d\psi_1 \cdots d\psi_{n-3}.
$$

Substituting (2.11), (2.15) and (2.16) into (2.13), we obtain, for $n > 3$,

$$(2.17) \qquad dS = \frac{r}{\sin^{n-1}\theta} (\sin\varphi)^{n-3} \left(\frac{\alpha^2 - 1}{2\alpha} \right)^{n-3} d\theta \, d\varphi \, d\omega_{n-3}.$$

For $n = 3$ the second Gramian in (2.14) does not appear and

$$(2.17)' \qquad dS = \frac{r}{\sin^2\theta} d\theta \, d\varphi.$$

Finally in the case $n = 2$, a simple calculation gives

$$(2.18) \qquad dS = \frac{r}{\sin\theta} \frac{2\alpha}{\alpha^2 - 1} d\theta.$$

These surface elements are obviously invariant under translations in the β' direction and hence hold for all of the horospheres in the set (2.6).

3. Proof of Theorem 1.1

We are now ready to start the proof of Theorem 1.1. As explained in the previous section we can replace the spherical cap in the statement of the theorem by the wedge $W = \{\theta < \theta_0\}$. The hypothesis of the theorem is now that $J\hat{f}(r, \beta) = 0$ for almost all horospheres contained in W. We shall transform this condition into a homogeneous Volterra integral equation with respect to the parameter θ of the partial non-Euclidean Fourier transform of $f(x', \rho, \theta)$ and conclude from this that the partial Fourier transform of f, and hence f itself, vanishes in W.

Step 1. A partial Fourier transform of $\hat{f}$. Let f be a C_0^∞ function defined on H_n and $\hat{f} = \hat{f}(r, \beta)$ its Radon transform defined in (1.10). We denote by R_λ the operation of stretching the r variable by the factor λ:

$$(3.1) \qquad (R_\lambda h)(r) = h(\lambda r).$$

We decompose $\beta \in \mathcal{B}$ as in Section 2,

$$\beta = \{\beta', \lambda\}, \qquad\qquad \lambda > 0,$$

and denote by R the operator

$$(3.1)' \qquad R\hat{f} = \hat{f}(\lambda r, \{\beta', \lambda\}).$$

Finally we denote by I the $(n-1)$-dimensional non-Euclidean Fourier transform with respect to $\beta = \{\beta', \lambda\}$; thus

$$(3.2) \qquad (IR\hat{f})(\mu, \chi, r) = \int \exp\left\{\left(\frac{n-2}{2} - i\mu\right)\langle\beta, \chi\rangle\hat{f}(\lambda r, \beta)\,dV(\beta),\right.$$

where

$$(3.2)' \qquad\qquad\qquad dV(\beta) = \frac{d\beta'\,d\lambda}{\lambda^{n-1}}\,.$$

We recall from Section 1 that, for any isometry γ, the function f^γ is defined by

$$(3.3) \qquad\qquad\qquad f^\gamma(z) = f(\gamma^{-1}z).$$

LEMMA 3.1. *For any γ in the Fuchsian subgroup F that keeps $z_0 = \{0, 1\}$ fixed,*

$$(3.4) \qquad\qquad (IR\hat{f})(\mu, \gamma^{-1}\chi, r) = (IR\hat{f^\gamma})(\mu, \chi, r).$$

Proof: Any isometry of $\mathbf{H}_n$ maps $\mathcal{B}$ into itself. However if γ belongs to F, then it also maps

$$(3.5) \qquad\qquad \mathcal{B}_+ \equiv \left[\beta = \{\beta', \lambda\}, \lambda > 0\right]$$

into itself and we can treat $\mathcal{B}_+$ as an $(n-1)$-dimensional hyperbolic space with its own infinity set

$$(3.5)' \qquad\qquad \mathcal{B}_{n-1} \equiv \left[\chi = \{\beta', 0\}, \beta' \in \mathbb{R}^{n-2}\right].$$

Moreover if $\gamma \in F$ keeps z_0 fixed, then it also keeps $\beta_0 = \{0', 1\}$ fixed and the $(n-1)$-dimensional analogue of (1.4) is

$$(3.6) \qquad\qquad\qquad \langle\beta, \gamma^{-1}\chi\rangle = \langle\gamma\beta, \chi\rangle,$$

which holds for every $\beta \in \mathcal{B}_+$ and χ in $\mathcal{B}_{n-1}$.

Using this and the definition (3.2) of I we have

$$(3.7) \quad (IR\hat{f})(\mu, \gamma^{-1}\chi, r) = \int \exp\left\{\left(\frac{n-2}{2} - i\mu\right)\langle\gamma\beta, \chi\rangle\right\}\hat{f}(\lambda r, \beta)\,dV(\beta),$$

where again λ is the $(n-1)$-st component of β. We introduce $\eta = \gamma\beta$ as new variable of integration and rewrite the right side of (3.7) as

$$(3.7)' \qquad \int \exp\left\{\left(\frac{n-2}{2} - i\mu\right)\langle\eta,\chi\rangle\right\} \hat{f}(\lambda r, \gamma^{-1}\eta)\, dV(\eta),$$

where λ is the $(n-1)$-st component of $\beta = \gamma^{-1}\eta$.

Next we compute the Radon transform of f^γ; by definition (1.10),

$$(3.8) \qquad \hat{f}^\gamma(s,\eta) = \int_{\xi(s,\eta)} f(\gamma^{-1}z)\, dS.$$

Introducing $w = \gamma^{-1}z$ as new variable of integration, we get

$$(3.8)' \qquad \hat{f}^\gamma(s,\eta) = \int_{\gamma^{-1}\xi(s,\eta)} f(w)\, dS.$$

The image of $\xi(s,\eta)$ under γ^{-1} is the form

$$\gamma^{-1}\xi(s,\eta) = \xi(t, \gamma^{-1}\eta);$$

and since γ is Fuchsian, the Euclidean radii s and t satisfy the relation

$$(3.9) \qquad \frac{s}{\eta_{m-1}} = \frac{t}{(\gamma^{-1}\eta)_{n-1}}.$$

We now choose

$$t = \lambda r,$$

where, as before,

$$\lambda = (\gamma^{-1}\eta)_{m-1}.$$

Combining the last two relations with (3.9), we get

$$s = \eta_{n-1} r.$$

Inserting this into (3.8) and (3.8)' we obtain

$$(3.10) \qquad \hat{f}^\gamma(\eta_{n-1}r, \eta) = \int_{\xi(t,\gamma^{-1}\eta)} f(w)\, dS = \hat{f}\left((\gamma^{-1}\eta)_{n-1}r, \gamma^{-1}\eta\right),$$

which inserted into (3.7)' yields (3.4) of the lemma.

We now show how to use Lemma 3.1 to express $IR\hat{f}$ in terms of the partial Fourier transform $\tilde{f}$ of f. We choose γ in F so that

$$(3.11) \qquad \gamma z_0 = z_0 \quad \text{and} \quad \gamma\chi = \infty.$$

The $(n-1)$-dimensional analogue of (1.3) is

$$(3.11)' \qquad\qquad \langle \{ \beta', \lambda \}, \infty \rangle = \log \lambda.$$

Using the second relation in (3.11) gives

$$(3.12) \qquad\qquad (IR\hat{f})(\mu, \chi, r) = (IR\hat{f})(\mu, \gamma^{-1}\infty, r).$$

We rewrite the right side, making use of (3.4), (3.2) and (3.11)':

$$(3.12)' \qquad (IR\hat{f})(\mu, \chi, r) = \int \lambda^{(n-2)/2 - i\mu} \hat{f}^{\gamma}(\lambda r, \beta) \, dV(\beta).$$

If one uses (3.2)' and the definition (1.10) of the Radon transform, the right side of (3.12)' becomes

$$(3.12)'' \qquad \int\int_{\xi(\lambda r, \beta)} \lambda^{(n-2)/2 - i\mu} f^{\gamma}(\overset{x'}{\chi'}, \rho, \theta) \, dS \, \frac{d\beta' d\lambda}{\lambda^{n-1}} .$$

We now make use of the parametrization of the horosphere given in (2.12):

$$(3.13) \qquad\qquad \begin{aligned} x' &= \beta' + p\omega, \\ x_{n+1} &= \rho \cos\theta, \\ y &= \rho \sin\theta, \end{aligned}$$

where ω is an $(n-2)$-dimensional unit vector, p and ρ are functions of φ and θ given in (2.11) and dS is defined by (2.17) and (2.18).

We interchange the dS and $d\beta'$ integrations; we denote the result of the $d\beta'$ integration by

$$(3.14) \qquad\qquad \int_{\mathbb{R}^{n-2}} f^{\gamma}(\beta' + p\omega, \rho, \theta) \, d\beta' \equiv \hat{f}^{\gamma}(\rho, \theta).$$

Inserting this into (3.12)'' and using (3.12), we get

$$(3.15) \qquad (IR\hat{f})(\mu, \chi, r) = \int \lambda^{(n-2)/2 - i\mu} \hat{f}^{\gamma}(\rho, \theta) \, dS \, \frac{d\lambda}{\lambda^{n-1}} .$$

According to (2.11),

$$\rho = \lambda \frac{2\alpha}{1 + \alpha^2 + (1 - \alpha^2) \cos\varphi} .$$

We interchange the dS and $d\lambda$ integration on the right in (3.15); we may do this because the domain over which θ, φ and ψ vary is independent of λ. Then we introduce ρ in place of λ as a new variable of integration. We obtain

(3.15)′
$$(IR\hat{f})(\mu,\chi,r) = \int\int \rho^{(n-2)/2-i\mu}\left(\frac{2\alpha}{1+\alpha^2+(1-\alpha^2)\cos\varphi}\right)^{(n-2)/2+i\mu}$$

$$\times \overset{\blacktriangle}{f^\gamma}(\rho,\theta)\frac{d\rho}{\rho^{n-1}}\,dS.$$

Next we carry out the ρ integration. Using the definition (3.14) of $\overset{\blacktriangle}{f^\gamma}$, with x' in place of β', and (3.11)′ we get

$$\int \rho^{(n-2)/2-i\mu}\overset{\blacktriangle}{f^\gamma}(\rho,\theta)\frac{d\rho}{\rho^{n-1}}$$

(3.16)
$$= \int\int \rho^{(n-2)/2-i\mu}f^\gamma(x',\rho,\theta)\frac{dx'\,d\rho}{\rho^{n-1}}$$

$$= \int\int \exp\left\{\left(\frac{n-2}{2}-i\mu\right)\langle\{x',\rho\},\infty\rangle\right\}f^\gamma(x',\rho,\theta)\frac{dx'\,d\rho}{\rho^{n-1}}.$$

Now we use the definition (2.4) of the partial Fourier transform to recognize the right side of (3.16) as

(3.17)
$$\tilde{f}^\gamma(\mu,\infty,\theta).$$

By (2.4)′, (3.17) equals $\tilde{f}(\mu,\gamma^{-1}\infty,\theta)$ which by (3.11) is equal to $\tilde{f}(\mu,\chi,\theta)$. Setting this into the right side of (3.15)′ one obtains

(3.18)
$$(IR\hat{f})(\mu,\chi,r) = \int\left(\frac{2\alpha}{1+\alpha^2+(1-\alpha^2)\cos\varphi}\right)^{(n-2)/2+i\mu}\tilde{f}(\mu,\chi,\theta)\,dS.$$

We now use (2.17):

$$dS = \frac{r}{\sin^{n-1}\theta}(\sin\varphi)^{n-3}\left(\frac{\alpha^2-1}{2\alpha}\right)^{n-3}d\theta\,d\varphi\,d\omega_{n-3}.$$

Carrying out the $d\varphi$ and $d\omega_{n-3}$ integrations in (3.18) gives

(3.19)
$$(IR\hat{f})(\mu,\chi,r) = \int_0^{\theta_r}G(\mu,r,\theta)\frac{\tilde{f}(\mu,\chi,\theta)}{\sin^{n/2}\theta}\,d\theta,$$

where

$$G(\mu, r, \theta) = \frac{r\omega_{n-3}}{(\sin\theta)^{(n-2)/2}} \left(\frac{\alpha^2 - 1}{2\alpha}\right)^{n-3}$$

(3.20)

$$\times \int_0^\pi \left(\frac{2\alpha}{1 + \alpha^2 + (1 - \alpha^2)\cos\varphi}\right)^{(n-2)/2 + i\mu} (\sin\varphi)^{n-3}\,d\varphi.$$

Relation (3.19) is the sought after expression of $IR\hat{f}$ in terms of $\tilde{f}$ when $n \geqq 3$.
When $n = 2$, the relation (3.2) reduces to

$$(3.21) \qquad (IR\hat{f})(\mu, r) = \int_0^\infty \lambda^{-i\mu}\hat{f}(\lambda r, \lambda)\frac{d\lambda}{\lambda}.$$

Substituting for $\hat{f}$, this can be expressed as

$$(3.21)' \qquad (IR\hat{f})(\mu, r) = \int \lambda^{-i\mu} \int_{\xi(\lambda r, \lambda)} f(\rho, \theta)\,dS\,\frac{d\lambda}{\lambda},$$

where ρ is given in the two branches of the horocycle $\xi(\lambda r, \lambda)$ by (2.10):

$$\rho_1 = \lambda\alpha \quad \text{and} \quad \rho_2 = \lambda\alpha^{-1},$$

and dS is given by (2.18). Inserting this into (3.21)' we obtain

$$(3.22) \quad (IR\hat{f})(\mu, r) = \int_0^\infty \lambda^{-i\mu} \int_0^{\theta_r} \left[f(\lambda\alpha, \theta) + f(\lambda\alpha^{-1}, \theta)\right]\frac{r}{\sin\theta}\frac{2\alpha}{\alpha^2 - 1}\,d\theta\,\frac{d\lambda}{\lambda}.$$

The partial Fourier transform of f is now

$$(3.23) \qquad \tilde{f}(\mu, \theta) = \int_0^\infty \rho^{-i\mu}f(\rho, \theta)\frac{d\rho}{\rho}.$$

An obvious change of variable brings (3.22) into the form

$$(3.24) \qquad (IR\hat{f})(\mu, r) = \int_0^{\theta_r} G(\mu, r, \theta)\frac{\tilde{f}(\mu, \theta)}{\sin\theta}\,d\theta,$$

where now

$$(3.25) \qquad G(\mu, r, \theta) = 2r\frac{2\alpha}{\alpha^2 - 1}\cos(\mu\log\alpha).$$

Step 2. A composite Bessel inequality. The next lemma concerns the operators J_n defined by (1.12) and (1.13) and R_λ defined in (3.1). The operators J_n act as functions of the single variable r and have the following form:

(3.26)
$$J_n = r\partial_r \Xi_n r^{(1-n)/2} \quad \text{for} \quad n \quad \text{odd},$$

$$J_n = r\partial_r \Xi_n N_n r^{(1-n)/2} \quad \text{for} \quad n \quad \text{even}.$$

Here Ξ_n is a polynomial in $(r\partial_r)$ and N_n, defined by (1.13)′, is a multiplicative-convolution operator. R_λ is the operation of stretching the variable r by the factor λ.

LEMMA 3.2. *The following relation holds*:

(3.27)
$$R_\lambda J_n = \lambda^{(1-n)/2} J_n R_\lambda.$$

Proof: The operators R_λ form a multiplicative semigroup whose generator is $r\partial_r$. Consequently, $r\partial_r$ commutes with R_λ and therefore so does $r\partial_r \Xi_n$. Since N_n is a superposition of the operators R_λ, it also commutes with R_λ. Clearly the operation of multiplication by r^α satisfies

$$R_\lambda r^\alpha = \lambda^\alpha r^\alpha R_\lambda.$$

Combining these facts with the definition (3.26) of J_n we deduce (3.27).

We return now to the Plancherel relation (1.11):

(3.28)
$$\int |f(z)|^2 dV = c_n \int_{\mathbb{R}^{n-1}} \int_0^\infty |J_n \hat{f}(r, \beta)|^2 \frac{dr}{r} d\beta.$$

We write, as before, $\beta = \{\beta', \lambda\}$ and replace on the right the variable r by λr. Since dr/r is invariant under such a stretching, we get, using the notation (3.1)′ and restricting the β integration to $\lambda > 0$,

(3.29)
$$\int |f(z)|^2 dV \geq c_n \int_{\mathbb{R}^{n-1}} \int_0^\infty |RJ_n \hat{f}|^2 \frac{dr}{r} d\beta.$$

Using (3.27) we can rewrite this as

(3.29)′
$$\int |f(z)|^2 dV \geq c_n \int_0^\infty \int_0^\infty \int_{\mathbb{R}^{n-2}} |J_n R\hat{f}|^2 \frac{d\beta' d\lambda}{\lambda^{n-1}} \frac{dr}{r}.$$

Next we apply the operator I, i.e., the $(n-1)$-dimensional non-Euclidean Fourier transform with respect to $\{\beta', \lambda\}$ and we use the Plancherel formula (1.7) on $IJ_n R \hat{f}$. We get

$$(3.29)'' \qquad \int |f(z)|^2 dV \geqq c'_{n-1} c_n \int_0^\infty \int_{B_{n-1}} \int_0^\infty |IJ_n R\hat{f}|^2 \frac{d\mu}{|c_{n-1}(\mu)|^2} \frac{d\chi}{(1+|\chi|^2)^{n-2}} \frac{dr}{r}.$$

Finally, we observe that since the operator J_n acts on the variable r and I acts on the variables $\{\beta', \lambda\}$ and since the domains of these variables are fixed, the operators I and J_n commute. Consequently, we arrive at the following composite Bessel inequality:

$$(3.30) \qquad \int |f(z)|^2 dV \geqq c'_{n-1} c_n \int_0^\infty \int |r^{-1/2} J_n IR\hat{f}|^2 dM(\mu, \chi) dr,$$

where

$$(3.30)' \qquad dM(\mu, \chi) = \frac{d\mu\, d\chi}{|c_{n-1}(\mu)|^2 (1+|\chi|^2)^{n-2}}.$$

Step 3. The Volterra equation. The composite Bessel inequality (3.30) relates the L_2 norm of f over $\mathbf{H}_n$ with a weighted L_2 norm of $r^{-1/2} J_n IR\hat{f}$. We shall now express the latter in terms of $\tilde{f}$.

Formula (3.19) expresses $IR\hat{f}$ in terms of an integral operator acting on $\tilde{f}$. The desired relation is obtained by letting $r^{-1/2} J_n$ act on (3.19). We begin by bringing the kernel G of this integral operator, defined by (3.20), into a more tractable form.

According to (2.10)′,

$$(3.31) \qquad \alpha = \tau + (\tau^2 - 1)^{1/2} \quad \text{and} \quad \alpha^{-1} = \tau - (\tau^2 - 1)^{1/2},$$

where

$$(3.32) \qquad \tau = \cos\theta + r\sin\theta.$$

Combining the two relations (3.31), we see that

$$\tfrac{1}{2}(\alpha + \alpha^{-1}) = \tau,$$

$$(3.33)$$

$$\tfrac{1}{2}(\alpha - \alpha^{-1}) = \frac{\alpha^2 - 1}{2\alpha} = (\tau^2 - 1)^{1/2} = ((\tau-1)(\tau+1))^{1/2}.$$

Moreover, by (3.32),

$$\tau - 1 = \cos\theta + r\sin\theta - (\cos\theta + \tan\tfrac{1}{2}\theta\sin\theta)$$

(3.34)

$$= \sin\theta(r - \tan\tfrac{1}{2}\theta).$$

Clearly, $\tau - 1 \geqq 0$ for $0 \leqq \theta \leqq \theta_r = 2\tan^{-1}r$. It follows from this and (3.31) that $\alpha \geqq 1$ for $0 \leqq \theta \leqq \theta_r$ and that $\alpha = 1$ in this range only for $\theta = 0$ or $\theta = \theta_r$.

Making use of (3.33) and (3.34) we now express G, defined in (3.20) and (3.25), as

$$(3.35) \qquad r^{(1-n)/2}G = \frac{\omega_{n-3}}{\sin^{1/2}\theta}\left(1 - \frac{\tan\tfrac{1}{2}\theta}{r}\right)^{(n-3)/2} K(\alpha),$$

where

$$K(\alpha) = \left(1 + \tfrac{1}{2}(\alpha + \alpha^{-1})\right)^{(n-3)/2}$$

(3.36)

$$\times \int_0^\pi \left(\frac{2\alpha}{1 + \alpha^2 + (1 - \alpha^2)\cos\varphi}\right)^{(n-2)/2+i\mu} (\sin\varphi)^{n-3}d\varphi \quad \text{for} \quad n \geqq 3,$$

$$K(\alpha) = \left(1 + \tfrac{1}{2}(\alpha + \alpha^{-1})\right)^{-1/2}\cos(\mu\log\alpha) \quad \text{for} \quad n = 2;$$

to make the formulas come out correctly we have set $\omega_0 = 2 = \omega_{-1}$. Notice that, for $\alpha \geqq 1$, $K(\alpha)$ is a smooth function of α and that

$$\frac{d}{d\alpha}\left(1 + \tfrac{1}{2}(\alpha + \alpha^{-1})\right)\big|_{\alpha=1} = 0, \qquad \frac{d}{d\alpha}\cos(\mu\log\alpha)\big|_{\alpha=1} = 0,$$

$$\frac{d}{d\alpha}\int_0^\pi \left(\frac{2\alpha}{1 + \alpha^2 + (1 - \alpha^2)\cos\varphi}\right)^{(n-2)/2+i\mu} (\sin\varphi)^{n-3}d\varphi\,\big|_{\alpha=1}$$

$$= \tfrac{1}{2}((n - 2) + i\mu)\int_0^\pi \cos\varphi(\sin\varphi)^{n-3}d\varphi = 0.$$

It follows that $K'(\alpha) = 0$ when $\alpha = 1$.

LEMMA 3.3. *Let $K_1(\alpha)$ be a smooth function, j any integer greater than 0. Define K_2 by*

$$(3.37) \qquad (j + r\partial_r)\left(1 - \frac{\tan\tfrac{1}{2}\theta}{r}\right)^j K_1(\alpha) = \left(1 - \frac{\tan\tfrac{1}{2}\theta}{r}\right)^{j-1} K_2(\alpha).$$

The function $K_2(\alpha)$ is smooth for $\alpha \geqq 1$, and if $K_1'(1) = 0$, then $K_2'(1) = 0$.

Proof: Carrying out the indicated differentiations, we find that

$$(3.38) \qquad K_2(\alpha) = jK_1(\alpha) + (r - \tan \tfrac{1}{2}\theta)K_1'(\alpha)\frac{\partial \alpha}{\partial r}.$$

Now by (3.31) and (3.32),

$$(3.39) \qquad \frac{\partial \alpha}{\partial r} = \frac{\partial \alpha}{\partial \tau}\frac{\partial \tau}{\partial r} = \left(1 + \frac{\tau}{(\tau^2 - 1)^{1/2}}\right)\sin\theta = \frac{\alpha}{(\tau^2 - 1)^{1/2}}\sin\theta.$$

Using (3.33) and (3.34), this can be expressed as

$$\frac{\partial \alpha}{\partial r} = \frac{\alpha^2 - 1}{r - \tan \tfrac{1}{2}\theta}\frac{1}{2 + \alpha + \alpha^{-1}}.$$

Inserting this into (3.38) we get

$$K_2(\alpha) = jK_1(\alpha) + \frac{\alpha^2 - 1}{2 + \alpha + \alpha^{-1}}K_1'(\alpha),$$

from which the assertion of the lemma is easily verified.

We introduce the abbreviation

$$(3.40) \qquad h = r^{-1/2}JIR\hat{f}.$$

In evaluating h we first treat the case $n \geqq 3$ and odd. Inserting (3.19) and (3.35) into (3.40), we get

$$(3.41) \quad h = \text{const. } r^{1/2}\partial_r \Xi \int_0^{\theta_r}\left(1 - \frac{\tan \tfrac{1}{2}\theta}{r}\right)^{(n-3)/2} K(\alpha)\frac{\tilde{f}(\mu, \chi, \theta)}{\sin^{n/2}\theta}\frac{d\theta}{\sin^{1/2}\theta},$$

where by (1.12)'

$$\Xi = (1 + r\partial_r)\cdots(\tfrac{1}{2}(n - 3) + r\partial_r).$$

Since $1 - r^{-1}\tan \tfrac{1}{2}\theta = 0$ when $\theta = \theta_r = 2\tan^{-1}r$ and since Ξ is of order $\tfrac{1}{2}(n - 3)$, it is clear that the action of Ξ commutes with the integration. Applying Lemma 3.3 inductively and using the fact that $K'(1) = 0$ we see that

$$(3.42) \qquad \Xi\left(1 - \frac{\tan \tfrac{1}{2}\theta}{r}\right)^{(n-3)/2} K(\alpha) = K_n(\alpha),$$

where K_n is smooth for $\alpha \geqq 1$ and $K_n'(\alpha) = 0$ at $\alpha = 1$. Consequently,

$$(3.43) \qquad h = \text{const. } r^{1/2} \partial_r \int_0^{\theta_r} K_n(\alpha) \frac{\tilde{f}(\mu, \chi, \theta)}{\sin^{n/2}\theta} \frac{d\theta}{\sin^{1/2}\theta} \, .$$

The final differentiation introduces a non-zero contribution from the limit of integration, namely

$$(3.44) \qquad \frac{2}{1 + r^2} \left(\frac{r}{\sin\theta_r} \right)^{1/2} K_n(1) \frac{\tilde{f}(\mu, \chi, \theta_r)}{\sin^{n/2}\theta_r} \, ;$$

the first factor comes from

$$\theta_r = 2 \tan^{-1} r \quad \text{and} \quad \frac{d\theta_r}{dr} = \frac{2}{1 + r^2} \, .$$

It is easy to see from (3.36) and (3.42) that

$$(3.45) \quad K_n(1) = (\tfrac{1}{2}(n-3))! K(1) = (\tfrac{1}{2}(n-3))! 2^{(n-3)/2} \int_0^\pi (\sin\varphi)^{n-3} d\varphi \neq 0.$$

Using the relation $r = \tan \tfrac{1}{2}\theta_r$, we get for the first factor in (3.44)

$$(3.46) \qquad \frac{2}{1 + r^2} \left(\frac{r}{\sin^{1/2}\theta_r} \right)^{1/2} = \sqrt{2} \cos \tfrac{1}{2}\theta_r \, ,$$

which is bounded away from zero in the wedge $W = \{ \theta < \theta_r < \theta_0 < \pi \}$.

The kernel in the integrand in (3.43) resulting from the last differentiation is

$$(3.47) \qquad K_0(\mu, r, \theta) = \left(\frac{r}{\sin\theta} \right)^{1/2} K_n'(\alpha) \frac{\partial\alpha}{\partial r} \, .$$

According to (3.33) and (3.39),

$$\frac{\partial\alpha}{\partial r} = \frac{2\alpha^2}{\alpha^2 - 1} \sin\theta$$

and hence

$$(3.48) \qquad K_0(\mu, r, \theta) = (r\sin\theta)^{1/2} \left[\frac{2\alpha^2}{\alpha^2 - 1} K_n'(\alpha) \right].$$

Since $K_n'(\alpha) = 0$ when $\alpha = 1$, it follows that the expression in the bracket is smooth in $\alpha \geqq 1$. This proves that the kernel K_0 is a smooth function of θ for $0 \leqq \theta \leqq \theta_r < \theta_0$ uniformly in $r \leqq \tan \tfrac{1}{2}\theta_0$.

We have now shown that

$$(3.49) \qquad h = k(r) \frac{\tilde{f}(\mu, \chi, \theta_r)}{\sin^{n/2}\theta_r} + \int_0^{\theta_r} K_0(\mu, r, \theta) \frac{\tilde{f}(\mu, \chi, \theta)}{\sin^{n/2}\theta} \, d\theta,$$

where $k(r)$ is smooth and bounded away from zero and K_0 is well behaved.

The above analysis was done under the assumption that f was a function of class $C_0^\infty(\mathbf{H}_n)$. Consider now a sequence $\{f_m\}$ of C_0^∞ functions converging in the $L_2(\mathbf{H}_n)$ topology to a limit f. It follows from the Plancherel relation (2.5) that then the partial Fourier transforms $\{\tilde{f}_m(\mu, \chi, \theta)/\sin^{n/2}\theta\}$ form a Cauchy sequence in a weighted L_2 space, where the weight is independent of θ. It follows then that a subsequence of $\{\tilde{f}_m(\mu, \chi, \theta)/\sin^{n/2}\theta\}$ converges for almost all $\{\mu, \chi\}$, in the $L_2(d\theta)$ sense, to $\tilde{f}(\mu, \chi, \theta)/\sin^{n/2}\theta$.

According to the composite Bessel inequality (3.30) the sequence

$$\{h_m\} = \{r^{-1/2} JIR\hat{f}_m\}$$

also forms a Cauchy sequence in a weighted L_2 space, where the weight $(3.30)'$ does not depend on r. It follows that a further subsequence of $\{h_m\}$ converges for almost all $\{\mu, \chi\}$, in the $L_2(dr)$ sense, to $h = r^{-1/2} JIR\hat{f}$. The limit satisfies for almost all $\{\mu, \chi\}$ the Volterra equation (3.49).

According to the hypothesis of Theorem 1.1, $(J\hat{f})(r, \beta) = 0$ for all horospheres in the wedge $W = \{\theta < \theta_0\}$, that is

$$(J\hat{f})(r, \beta) = 0 \quad \text{for} \quad r < \beta_{n-1} \tan \tfrac{1}{2}\theta_0.$$

We deduce from this that

$$RJ\hat{f} = (J\hat{f})(\lambda r, \{\beta', \lambda\}) = 0 \quad \text{for} \quad r < \tan \tfrac{1}{2}\theta_0, \qquad \lambda > 0;$$

from this we can conclude that

$$h = r^{-1/2} JIR\hat{f} = 0 \quad \text{for} \quad r < \tan \tfrac{1}{2}\theta_0.$$

Thus the Volterra integral equation which $\tilde{f}/\sin^{n/2}\theta$ satisfies is homogeneous for $\theta_r < \theta_0$. Since $\tilde{f}/\sin^{n/2}\theta$ is $L_2(d\theta)$ for almost all $\{\mu, \chi\}$ and since the kernel K_0 is a smooth function and $k(r)$ is smooth and bounded away from zero, it follows from the classical uniqueness theorem for homogeneous Volterra equations that $\tilde{f}(\mu, \chi, \theta) = 0$ a.e. for $\theta < \theta_0$. By Fourier inversion we conclude that $f(x', \rho, \theta) = 0$ a.e. for $\theta < \theta_0$, i.e., that f vanishes a.e. in the wedge. This proves Theorem 1.1 for n odd.

When n is even,

(3.50)
$$h = r^{-1/2}JIR\hat{f} = \text{const. } r^{1/2}\partial_r \Xi N \int_0^{\theta_r}\left(1 - \frac{\tan\frac{1}{2}\theta}{r}\right)^{(n-3)/2}$$

$$\times K(\alpha)\frac{\tilde{f}(\mu,\chi,\theta)}{\sin^{n/2}\theta}\frac{d\theta}{\sin^{1/2}\theta},$$

where now, by (1.13)′, N is a multiplicative convolution operator and

$$\Xi = (1 + r\partial_r)\cdots\left(\frac{1}{2}(n-2) + r\partial_r\right);$$

$K(\alpha)$, described in (3.36), is a smooth function of α for $\alpha \geqq 1$ with $K'(1) = 0$.

The action of N on the integral in (3.50) can be written as

(3.51)
$$\int_0^1 (1-\sigma)^{-1/2}\sigma^{(n-1)/2}\int_0^{2\tan^{-1}\sigma r}\left(1 - \frac{\tan\frac{1}{2}\theta}{\sigma r}\right)^{(n-3)/2}$$

$$\times K(\alpha(\sigma))\frac{\tilde{f}(\mu,\chi,\theta)}{\sin^{1/2}\theta}\frac{d\theta}{\sin^{1/2}\theta}\frac{d\sigma}{\sigma},$$

where

(3.52) $\qquad \alpha(\sigma) = \tau(\sigma) + \left(\tau(\sigma)^2 - 1\right)^{1/2}$ and $\tau(\sigma) = \cos\theta + \sigma r\sin\theta.$

On interchanging the order of integration this becomes

(3.53)
$$\int_0^{\theta_r}\left[\frac{1}{\sin^{1/2}\theta}\int_{r^{-1}\tan\theta/2}^1 (1-\sigma)^{-1/2}\sigma^{(n-1)/2}\left(1 - \frac{\tan\frac{1}{2}\theta}{\sigma r}\right)^{(n-3)/2}K(\alpha(\sigma))\frac{d\sigma}{\sigma}\right]$$

$$\times \frac{\tilde{f}(\mu,\chi,\theta)}{\sin^{n/2}\theta}d\theta.$$

The bracket expression $\mathcal{B}_1$ can be brought into a more convenient form by making the substitution

(3.54) $\qquad\qquad\qquad \sigma = v + (1 - v)r^{-1}\tan\frac{1}{2}\theta.$

In this case

$$(3.55) \qquad \mathscr{B}_1 = \frac{1}{\sin^{1/2}\theta} \left(1 - \frac{\tan\frac{1}{2}\theta}{r}\right)^{(n-2)/2} L(\alpha),$$

where

$$(3.56) \qquad L(\alpha) = \int_0^1 \frac{\nu^{(n-3)/2}}{(1-\nu)^{1/2}} K(\alpha(\sigma))\, d\nu.$$

Implicit in the above is the claim that the right side of (3.56) is a function of α alone. To see that this is so, it suffices to show that $\alpha(\sigma)$ can be expressed as a function of α and ν; since $\alpha(\sigma)$ is a function of $\tau(\sigma)$, it suffices to show that $\tau(\sigma)$ can be so expressed:

$$\tau(\sigma) = \cos\theta + \sigma r \sin\theta$$

$$(3.57) \qquad = \nu(\cos\theta + r\sin\theta) + (1-\nu)(\cos\theta + \tan\tfrac{1}{2}\theta\sin\theta)$$

$$= \nu\tau + 1 - \nu = \nu\tfrac{1}{2}(\alpha + \alpha^{-1}) + 1 - \nu;$$

in the last step we have made use of (3.33).

It follows easily from the expression for K in (3.36) together with (3.56) and (3.57) that $L(\alpha)$ is a smooth function of α for $\alpha \geq 1$. We now show that $L'(\alpha) = 0$ at $\alpha = 1$. To this end we compute $\partial\alpha(\sigma)/\partial\alpha$. Combining (3.52) and (3.57), we get

$$\tfrac{1}{2}\big(\alpha(\sigma) + \alpha(\sigma)^{-1}\big) = \tau(\sigma) = \nu\tfrac{1}{2}(\alpha + \alpha^{-1}) + 1 - \nu,$$

from which we obtain by implicit differentiation

$$(3.58) \qquad \frac{\partial\alpha(\sigma)}{\partial\alpha} = \nu \frac{\alpha(\sigma)^2}{\alpha(\sigma)^2 - 1} \cdot \frac{\alpha^2 - 1}{\alpha^2}.$$

We differentiate (3.56) with respect to α:

$$(3.59) \qquad L'(\alpha) = \frac{\alpha^2 - 1}{\alpha^2} \int_0^1 \frac{\nu^{(n-1)/2}}{(1-\nu)^{1/2}} K'(\alpha(\sigma)) \frac{\alpha(\sigma)^2}{\alpha^2(\sigma) - 1}\, d\nu.$$

Recalling that $K'(\alpha)$ vanishes when $\alpha = 1$, we see that the integrand in (3.59) is a smooth function of ν and hence that $L'(1) = 0$.

We now proceed exactly as in the case n odd, with $K(\alpha)$ now replaced by $L(\alpha)$. In place of (3.43) we have

$$(3.60) \qquad h = \text{const. } r^{1/2} \partial_r \int_0^{\theta_r} L_n(\alpha) \frac{\tilde{f}(\mu, \chi, \theta)}{\sin^{n/2}\theta} \frac{d\theta}{\sin^{1/2}\theta},$$

where

$$(3.61) \qquad L_n(\alpha) = \Xi \left(1 - \frac{\tan\frac{1}{2}\theta}{r} \right)^{(n-2)/2} L(\alpha).$$

Again the final differentiation introduces a non-zero contribution from the limit of integration:

$$(3.62) \qquad \frac{2}{1 + r^2} \left(\frac{r}{\sin\theta_r} \right)^{1/2} L_n(1) \frac{\tilde{f}(\mu, \chi, \theta_r)}{\sin^{n/2}\theta_r}.$$

Now, $\alpha = 1$ when $\theta = \theta_r$ and from (3.57) we see that $\tau(\sigma) = 1$ and hence $\alpha(\sigma) = 1$ when $\alpha = 1$. Hence by (3.56) and (3.61) we have

$$(3.63) \quad L_n(1) = \left(\tfrac{1}{2}(n - 2) \right)! L(1) = \left(\tfrac{1}{2}(n - 2) \right)! \int_0^1 \frac{\nu^{(n-3)/2}}{(1 - \nu)^{1/2}} K(1)\, d\nu.$$

In view of (3.36),

$$K(1) = \begin{cases} 2^{(n-3)/2} \int_0^\pi (\sin\varphi)^{n-3} d\varphi & \text{for } n \geq 3, \\ 2^{-1/2} & \text{for } n = 2; \end{cases}$$

thus again the coefficient of $\tilde{f}(\mu, \chi, \theta_r)/\sin^{n/2}\theta_r$ is smooth and bounded away from zero in the wedge $W = \{\theta < \theta_0 < \pi\}$.

The kernel in the integrand resulting from the last differentiation is

$$(3.64) \qquad L_0(\mu, s, \theta) = \left(\frac{r}{\sin\theta} \right)^{1/2} L_n'(\alpha) \frac{\partial\alpha}{\partial r}.$$

This is of the same form as (3.47) and in fact the rest of the argument for n even follows exactly as in the odd-dimensional case.

Acknowledgment. The work of the first author was supported in part by the Department of Energy under Contract No. DE-AC02-76ER003077 and that of the second author by the National Science Foundation under Grant No. MCS-80-01943.

Bibliography

[1] Gelfand, I. M., and Graev, M. I., *Geometry of homogeneous spaces, representations of groups in homogeneous spaces and related questions of integral geometry.* I, Trudy Moskov. Mat. Obsc. 8, 1959, pp. 321–390; English transl., Amer. Math. Soc. Transl., Ser. 2 37, 1964, pp. 351–429.

[2] Helgason, S., *Duality and Radon transform for symmetric spaces*, Amer. J. Math. 85, 1963, pp. 667–692.

[3] Helgason, S., *The surjectivity of invariant differential operators on symmetric spaces.* I, Ann. Math. 98, 1973, pp. 451–479.

[4] Lax, P. D., and Phillips, R. S., *Translation representations for the solution of the non-Euclidean wave equation*, Comm. Pure Appl. Math. 32, 1979, pp. 617–667.

[5] Lax, P. D., and Phillips, R. S., *A local Paley-Wiener theorem for the Radon transform in real hyperbolic spaces*, Math. Analysis and Applic., Part B, Adv. in Math., Suppl. Studies 7B, 1981, pp. 483–487.

[6] Woo, A. C., *Scattering theory on real hyperbolic spaces and their compact perturbations*, Stanford Univ., Thesis, 1980.

Received November, 1981.

Translation Representations for Automorphic Solutions of the Wave Equation in Non-Euclidean Spaces. I

PETER D. LAX

Courant Institute

AND

RALPH S. PHILLIPS

Stanford University

Abstract

This paper deals with the spectral theory of the Laplace–Beltrami operator Δ acting on automorphic functions in n-dimensional hyperbolic space H^n. We study discrete subgroups Γ which have a fundamental polyhedron F with a finite number of sides and infinite volume. Concerning these we have shown previously that the spectrum of Δ contains at most a finite number of point eigenvalues in $[-(\frac{1}{2}(n-1))^2, 0]$, and none less than $(\frac{1}{2}(n-1))^2$. Here we prove that the spectrum of Δ is absolutely continuous and of infinite multiplicity in $(-\infty, -(\frac{1}{2}(n-1))^2)$. Our approach uses the non-Euclidean wave equation introduced by Faddeev and Pavlov,

$$u_n - Lu = 0, \qquad L = \Delta + (\tfrac{1}{2}(n-1))^2.$$

Energy E_F is defined as $(u_t, u_t) - (u, Lu)$, where the bracket is the L_2 scalar product over a fundamental polyhedron with respect to the invariant volume of the hyperbolic metric. Energy is conserved under the group of operator $U(t)$ relating initial data to data at time t. We construct two isometric representations of the space of automorphic data by $L_2(\mathbf{R}, N)$ which transmute the action of $U(t)$ into translation. These representations are given explicitly in terms of integrals of the data over horospheres. In Part II we shall show the completeness of these representations.

Introduction

This paper deals with the spectral theory of the Laplace–Beltrami operator Δ acting on automorphic functions in n-dimensional hyperbolic space H^n. The principal tool is the wave equation.

A few remarks on the history of this subject are in order. The modern theory started with the 1949 paper of Maass [10], although the influence of Hecke and Siegel on Maass is discernable. In this approach Δ is treated as an operator on a Hilbert space, semibounded and selfadjoint with respect to the invariant L_2 scalar product over a fundamental domain $F = \mathsf{H}^n/\Gamma$, where Γ is a discrete subgroup of motions of H^n. The domain of Δ, constructed by the Friedrichs extension, is a subspace of automorphic functions on H^n. This idea has been successfully exploited by many mathematicians, including Roelcke [14] and [15], Patterson [11] and Elstrodt [1]; the research of all three is especially pertinent to the present paper.

Communications on Pure and Applied Mathematics, Vol. XXXVII, 303–328 (1984)

© 1984 John Wiley & Sons, Inc.   CCC 0010-3640/84/030303-26$04.00

The fundamental paper of Selberg [16] is also connected with these ideas. In this paper Selberg works with automorphic functions on symmetric spaces where the role of the Laplace–Beltrami operator is replaced by an algebra of differential operators that commute with the motions. Information on the eigenvalues of such operators is one of the byproducts of his trace formula.

In the present paper, as in most of the work in this field, only geometrically finite groups are considered, that is groups whose fundamental polyhedrons are bounded by a finite number of sides. Among these the following cases can be distinguished:

(i) F compact,

(ii) vol (F) finite,

(iii) vol (F) infinite.

In case (i) the standard theory of elliptic partial differential equations can be used to prove that the resolvent of Δ is a compact operator, from which it follows that Δ has a pure point spectrum accumulating at $-\infty$. In case (ii), part of the spectrum of Δ is continuous. Selberg has shown that the continuous part of the spectrum extends from $-\infty$ to $-(\frac{1}{2}(n-1))^2$ and is absolutely continuous, of uniform multiplicity N; here N is the number of cusps of maximal rank. His proof proceeds by explicit construction of the spectral representation of the continuous part of the spectrum, using the analytic continuation of the Eisenstein series. Another version of the proof has been given by Faddeev in [2] and two other proofs by the authors in [4] and [5]. In addition to the continuous spectrum there is also, in this case, a discrete spectrum which always includes $\lambda = 0$, with eigenfunction $\varphi \equiv 1$. The only general result about the extent of the point spectrum is Theorem 8.6 on page 205 of [4].

We turn now to case (iii), the subject of this paper. In [8] the authors have shown the following:

(a) The spectrum of Δ in $(-(\frac{1}{2}(n-1))^2, 0]$ consists of a finite number of point eigenvalues.

(b) Δ has no point eigenvalues $\mu \leq -(\frac{1}{2}(n-1))^2$.

In this paper we prove:

(c) The spectrum of Δ in $(-\infty, -(\frac{1}{2}(n-1))^2]$ is absolutely continuous and of infinite uniform multiplicity.

We note that if instead of being geometrically finite, Γ is merely finitely generated, then (a) need not hold. Charles Epstein has recently shown that for such a group, acting on $\mathbb{H}^3$, the interval $(-1, 0)$ can have infinite spectral content (New York University Thesis, 1983).

Earlier studies of the continuous spectrum of Δ proceeded by constructing explicitly a spectral representation of Δ; the generalized eigenfunctions of Δ entering this spectral representation are Eisenstein series, constructed by analytic continuation.

Our approach is entirely different; it is applicable to representing operators whose continuous spectrum has uniform multiplicity on the whole line. Let A

be an *anti-selfadjoint* operator whose spectrum is of *uniform multiplicity* on the whole imaginary axis. Then the spectral representation for A can be thought of as representing the underlying Hilbert space H by $L^2(R, N)$, N some auxiliary Hilbert space whose dimension equals the multiplicity of the spectrum of A.

In a spectral representation the action of the operator A is transmuted into multiplication by $i\sigma$, $\sigma \in \mathbb{R}$. The Fourier transform of a spectral representation is another representation, that transmutes the action of A into $-d/ds$. The action of $U(t) = \exp\{At\}$ is then transmuted into translation.

In this paper we construct directly a translation representation for a unitary group whose generator has spectrum of uniform multiplicity, and which is simply related to Δ. From this translation representation one can readily derive a spectral representation for Δ; we shall carry this out in Part II.

Our approach uses the non-Euclidean wave equation, to which we were introduced by Faddeev and Pavlov [3]:

$$u_{tt} - Lu = 0, \qquad L = \Delta + (\tfrac{1}{2}(n-1))^2.$$

The initial value problem with $u(0)$ and $u_t(0)$ as prescribed initial data is a local problem and hence is properly posed for the non-Euclidean wave equation. If the initial data are automorphic, so is the solution at time t.

We denote by $U(t)$ the operator mapping $\{u(0), u_t(0)\}$ into $\{u(t), u_t(t)\}$; these operators form a one-parameter group. The energy E_F of given data $f^\Gamma = \{f_1, f_2\}$ is defined classically as

$$E_F(f^\Gamma) = \|f_2\|^2 - (f_1, Lf_1);$$

here $(\ ,\ )$ denotes the L_2 inner product over F. Energy is independent of t, i.e., it is conserved under the action of $U(t)$.

Now if Δ has a nonempty point spectrum, then we see from (a) that $L = \Delta + (\tfrac{1}{2}(n-1))^2$ has positive eigenvalues; it follows that in this case the energy form takes on negative values. Denote by $\mathcal{H}_c$ the space of data $f^\Gamma = \{f_1, f_2\}$ for which f_1 and f_2 are both orthogonal to the eigenfunctions of L. Clearly, $\mathcal{H}_c$ is invariant under $U(t)$ and $E_F(f^\Gamma) \geq 0$ over $\mathcal{H}_c$; in fact, $E_F(f^\Gamma) > 0$ unless f^Γ is of the form $\{n, 0\}$, where $Ln = 0$ and f^Γ has finite energy. The space $\mathcal{N}$ of such functions n is finite-dimensional.

In [6] we have constructed two translation representations R_- and R_+, called incoming and outgoing, for $U(t)$ acting on the space $\mathcal{H}$ of all data f with finite energy in the whole space $\mathbb{H}^3$. R_- and R_+ map $\mathcal{H}$ onto $L_2(\mathbb{R} \times B)$ unitarily, B being the points at infinity of $\mathbb{H}^3$, that is,

$$\|R_\pm f\|^2 = E(f),$$

and they transmute the action of U into translation:

$$R_\pm U(t)f = T(t)R_\pm f;$$

here $T(t)$ denotes translation.

In [5] we constructed translation representations for $U(t)$ acting on automorphic data f^Γ for a fundamental domain F of *finite volume*. In this case, R_- and R_+ map f^Γ onto $L_2(\mathbb{R}\oplus\cdots\oplus\mathbb{R})$, the number of components being equal to the number of cusps. For f^Γ orthogonal to the eigenfunctions of Δ, the representation is again unitary:

$$\|R_\pm f^\Gamma\|^2 = E(f^\Gamma),$$

and $R_\pm$ transmutes the action of $U(t)$ into translation.

In this paper we define translation representations R_-^F and R_+^F when F has infinite volume; our definition is the direct sum of the two kinds of representations defined above. In Part I we prove that these are *partial isometric translation representations*, decomposing $\mathcal{H}_c$ as

$$\mathcal{H}_c = \mathcal{H}_+ \oplus \mathcal{H}_+^\perp \quad \text{and} \quad \mathcal{H}_c = \mathcal{H}_- \oplus \mathcal{H}_-^\perp,$$

so that R_-^F is a unitary translation representation of $U(t)$ on $\mathcal{H}_-$, and is zero on $\mathcal{H}_-^\perp$. Similarly, R_+^F is a unitary translation representation of $U(t)$ on $\mathcal{H}_+$, and is zero on $\mathcal{H}_+^\perp$. In Part II we shall prove that these representations are not merely partial but, in fact, complete. This implies:

If a wave $u(t)$ is orthogonal to the eigenfunctions of Δ, then its energy propagates to infinity in the sense that the energy content of any compact subset of F tends to zero as $|t|$ tends to ∞.

The remarkable thing about our translation representations is that they are expressed very explicitly and very geometrically in terms of integrals of the data over horospheres in F.

Fourier transformation changes a translation representation into a spectral representation. In [4] and [5] we have shown that in the case $\mathrm{vol}\,(F) < \infty$, this spectral representation is expressible in terms of Eisenstein series.

When R_-^F and R_+^F are complete, then $R_+^F (R_-^F)^{-1}$ defines a natural scattering operator. As we have shown in [6], this scattering operator is nontrivial, i.e., not equal to identity, already in the case where $F = \mathbb{H}^3$.

It is known that both the geodesic flow and the horocyclic flow are useful tools for studying discrete groups and their fundamental polyhedra. Since signals for the wave equation propagate both along geodesics (rays) and horospheres (plane waves), it is not astonishing to find that the wave equation is a natural tool for the study of discrete groups.

The organization of this paper is as follows: Section 1 reviews previous results that are needed in the sequel. Section 2 contains the detailed definition of the translation representations and establishes their basic properties.

We would like to thank T. Jorgensen for encouraging one of us in 1975 to consider fundamental domains of infinite volume.

1. Review

The representations $R_\pm^F$ will be built out of the incoming and outgoing representation of solutions of the non-Euclidean wave equation in all of H^n, as described in [6] for $n = 3$. We review those parts of this theory that are needed here, limiting ourselves for ease of exposition to the case $n = 3$.

We use the Poincaré model of hyperbolic 3-space, i.e., the half-space

$$[w = (x, y); x = (x_1, x_2), y > 0]$$

with the line element

$$(1.1) \qquad ds^2 = \frac{dx^2 + dy^2}{y^2}.$$

As before, B is the set of points at infinity; in this model, B is the union of the plane $y = 0$ and ∞. L is the augmented Laplace–Beltrami operator:

$$(1.2) \qquad L = y^2(\partial_x^2 + \partial_y^2) - y\partial y + 1$$

and

$$(1.3) \qquad u_{tt} - Lu = 0$$

is the non-Euclidean wave equation. It follows from (1.3) that the energy defined by

$$(1.4) \qquad E(u) = (u_t, u_t) - (u, Lu)$$

is conserved for solutions of the wave equation; here $(\ ,\ )$ denotes the L_2 inner product for the metric (1.1). It was shown in [6] (see also [18]) that the energy form is positive for u of compact support. The initial value problem

$$(1.5) \qquad u(w, 0) = f_1(w), \qquad u_t(w, 0) = f_2(w)$$

has a unique solution for all smooth initial data $f = \{f_1, f_2\}$.

We denote by $U_0(t)$ the operator relating smooth initial data of compact support to data at time t of solutions of (1.3). We denote by $\mathcal{H}$ the Hilbert space obtained by completing such data in the energy norm, which can be written in more symmetrical form as

$$(1.4)' \qquad E(f) = \int [y^2(|\partial_x f_1|^2 + |\partial_y f_1|^2) - |f_1|^2 + |f_2|^2]\, dw.$$

As shown in [6], E is actually positive definite and the first component of data in $\mathcal{H}$ is locally L_2. The operators $U_0(t)$ form a strongly continuous one-parameter group of unitary operators on $\mathcal{H}$.

The horosphere $\xi(s, \beta)$ through the point β of B at a distance s from the origin $j = (0, 0, 1)$ is the non-Euclidean analogue of a plane. It has the shape of the Euclidean sphere

(1.6) $$|x-\beta|^2+y^2-y(1+|\beta|^2)\,e^{-s}=0;$$

here $|\cdot|$ denotes Euclidean distance.

The Radon transform $\hat{u}$ of a function u on $\mathbb{H}^3$ is defined as its integral over horospheres:

(1.7) $$\hat{u}(s,\beta)=\int_{\xi(s,\beta)} u(w)\,dS,$$

dS being the non-Euclidean surface element over $\xi(s,\beta)$.

The outgoing and incoming translation representations R_+ and R_- of data f in $\mathcal{H}$ are defined as follows:

(1.8) $$R_\pm f=\frac{1}{2\sqrt{2}\,\pi}[\partial_s^2 e^s \hat{f}_1 \mp \partial_s e^s \hat{f}_2].$$

THEOREM 1.1. R_+ is a translation representation of $U_0(t)$ over $\mathcal{H}$, i.e.,
   (i) R_+ is a unitary map of $\mathcal{H}$ onto $L_2(\mathbb{R}\times B)$:

(1.9)$_+$ $$E(f)=\|R_+f\|^2;$$

here the norm on the right is

(1.9)$'_+$ $$\|R_+f\|^2=\int_B\int_{-\infty}^{\infty}|R_+f(s,\beta)|^2\,ds\,dm(\beta),$$

(1.10) $$dm(\beta)=\frac{4d\beta}{(1+|\beta^2|)^2}.$$

   (ii) R_+ transmutes the action of $U(t)$ into translation:

(1.11)$_+$ $$R_+U_0(t)f=T(t)R_+f,$$

where $T(t)$ is right translation by t:
$$T(t)k(s,\beta)=k(s-t,\beta).$$

   Similarly, R_- is a left-translation representation, i.e.,

(1.9)$_-$ $$\|R_-f\|^2=E(f)$$

and

(1.11)$_-$ $$R_-U_0(t)f=T(-t)R_-f.$$

The following relation of $R_\pm$ to non-Euclidean motions is relevant:
Let τ be a motion, and f^τ, as usual,

$$f^\tau(w)=f(\tau^{-1}w).$$

Denote $R_+ f$ by k; then

$$(1.12) \qquad\qquad R_+ f^\tau = k^\tau,$$

where k^τ is

$$(1.12)' \qquad\qquad k^\tau(s, \beta) = e^{\langle \tau j, \beta \rangle} k(s - \langle \tau j, \beta \rangle, \tau^{-1}\beta).$$

Here j is the origin, and $\langle w, \beta \rangle = s$ is defined by the stipulation that $\xi(s, \beta)$ pass through w.

The inverses of R_+ and R_- can be written down explicitly:

$$J_\pm k = \{f_1, f_2\},$$

$$f_1(w) = \frac{1}{2\sqrt{2}\,\pi} \int e^{\langle w, \beta \rangle} k(\langle w, \beta \rangle, \beta)\, dm(\beta),$$

$$(1.13)$$

$$f_2(w) = \mp \frac{1}{2\sqrt{2}\,\pi} \int e^{\langle w, \beta \rangle} k'(\langle w, \beta \rangle, \beta)\, dm(\beta),$$

where $k'(s, \beta) = \partial_s k(s, \beta)$.

We turn now to the automorphic case. Let Γ denote a discontinuous subgroup of the group of motions of H^3. We assume that H^3/Γ can be represented by a fundamental polyhedren F with a *finite number of sides* and *infinite volume*. The last condition implies that some sides of F are in B. We denote the union of these sides by B_F. In addition, F may possess a finite number of cusps of maximal rank, with associated parabolic fixed points π_j, $j = 1, \cdots, N$.

We shall study automorphic data f^Γ, satisfying for every τ in Γ the condition

$$(1.14) \qquad\qquad f^\Gamma(\tau w) = f^\Gamma(w).$$

The domain F can be regarded as a Riemannian manifold (possibly with a finite number of singular internal edges), and f^Γ as data defined on the manifold F.

We shall study automorphic solutions of the wave equation (1.3). Since the wave operator is invariant under motions and since solutions are uniquely determined by their initial data, it follows that if f is automorphic, and $u(w, t)$ is the solution of (1.3) with initial data f, then $u(w, t)$ is automorphic for all t. We use the symbol $U(t)$ for the operator relating automorphic initial data to data at t. We denote by A the infinitesimal generator of this group of operators.

For automorphic solutions of (1.3) the energy contained in one fundamental domain is conserved. We denote this by $E_F(u)$:

$$(1.15) \qquad\qquad E_F(u) = \int_F [u_t^2 + y^2(u_x^2 + u_y^2) - u^2]\, dw.$$

The quadratic form E_F is not positive definite; however, as shown in [8], E_F can be rendered positive definite by adding to it a positive but relatively compact form $K(f^\Gamma)$:

(1.16) $$K(f^\Gamma) = \int_F k(w)|f_1^\Gamma(w)|^2 \, dw.$$

It was shown in Section 3 of [8] that $\dot{K}$ can be so chosen that

(1.17) $$G(f^\Gamma) = E_F(f^\Gamma) + K(f^\Gamma)$$

is locally as well as globally positive definite, and that K is compact with respect to G. The completion of C_0^∞ data on the manifold F with respect to the G norm is denoted by $\mathscr{H}_G$.

We have shown in [8]:

THEOREM 1.2. (a) *The space $\mathscr{H}_G$ is independent of the specific choice of G.*
(b) $U(t)$ *is a strongly continuous group of operators on* $\mathscr{H}_G$.
(c) $G(f^\Gamma) \geqq c \int_S |f_1^\Gamma|^2(w) \, dw, c > 0$, *for any compact S in F.*
(d) *The spectrum of L is nonpositive except for a finite number of positive eigenvalues.*

We denote these eigenvalues of L by

$$\lambda_1^2, \cdots, \lambda_M^2, \qquad\qquad \lambda_j > 0$$

and the corresponding eigenfunctions by φ_j:

(1.18) $$L\varphi_j = \lambda_j^2 \varphi_j;$$

the φ_j are in $L_2(F)$.

The functions

(1.18)′ $$\exp\{\pm\lambda_j t\}\varphi_j$$

are automorphic solutions of the wave equation; we denote their initial data by

(1.18)″ $$p_j^+ = \{\varphi_j, \lambda_j\varphi_j\}, \qquad p_j^- = \{\varphi_j, -\lambda_j\varphi_j\}.$$

It follows from (1.18)′ that

(1.19) $$U(t)p_j^+ = \exp\{\lambda_j t\}p_j^+, \qquad U(t)p_j^- = \exp\{-\lambda_j t\}p_j^-.$$

The data $p_j^\pm$ have finite energy, and satisfy

(1.20) $$E_F(p_j^+, p_k^+) = 0, \qquad E_F(p_j^-, p_k^-) = 0,$$

(1.20)′ $$E_F(p_j^+, p_k^-) = -2\lambda_j\lambda_k\delta_{jk};$$

here $E_F(u, v)$ is the bilinear form associated with the quadratic form (1.15). As (1.20), (1.20)′ show, the Gram matrix of the $p^\pm$ with respect to the energy form is nondegenerate.

In the rest of this paper, orthogonality shall mean orthogonality with respect to the E_F form, unless otherwise specified.

LEMMA 1.3.   *Suppose that f^Γ in $\mathcal{H}_G$ is orthogonal to all $p_j^\pm$:*

$$(1.21) \qquad\qquad E_F(f^\Gamma, p_j^\pm) = 0, \qquad\qquad j = 1, \cdots, M.$$

Then

$$(1.21)' \qquad\qquad E_F(f^\Gamma) \geqq 0.$$

Proof: Suppose f_1^Γ is in $L_2(F)$; then it follows from $(1.18)''$ that orthogonality to $p_j^\pm$ implies that both f_1^Γ and f_2^Γ have zero components with respect to the φ_j. Since the spectrum of L as a selfadjoint operator on $L_2(F)$ is at most 0 except for the eigenvalues λ_j^2, $(1.21)'$ follows from the expression (1.4) for E_F. To prove the lemma for any f^Γ in $\mathcal{H}_G$ we use approximation by data in $L_2(F)$.

We denote by $\mathcal{H}_c$ the space of all f^Γ in $\mathcal{H}_G$ orthogonal to all $p_j^\pm$. Since the span of the $p_j^\pm$ is invariant under $U(t)$ so is $\mathcal{H}_c$. We denote by $\mathcal{N}$ the set of all f^Γ in $\mathcal{H}_c$ which have zero energy.

LEMMA 1.4.   (a) $\mathcal{N}$ *is a finite-dimensional subspace of $\mathcal{H}_c$, consisting of data n^Γ of the form*

$$(1.22) \qquad\qquad n^\Gamma = \{h^\Gamma, 0\}, \qquad Lh^\Gamma = 0.$$

(b) *For the continuous group we have*

$$(1.23) \qquad\qquad U(t)n^\Gamma = n^\Gamma \quad for\ all \quad t.$$

(c) E_F *and G are equivalent on the G-orthogonal complement of $\mathcal{N}$ in $\mathcal{H}_c$; i.e., there is a constant c such that*

$$(1.24) \qquad\qquad G(f^\Gamma) \leqq cE_F(f^\Gamma)$$

for all f^Γ in $\mathcal{H}_c$ that are G-orthogonal to $\mathcal{N}$.

(d) $\mathcal{H}_c/\mathcal{N}$ *is complete in the E_F norm.*

Proof: (a) Since E is non-negative on $\mathcal{H}_c$ and since $\mathcal{N}$ is the null set of E_F, we see from the Schwarz inequality that $\mathcal{N}$ is orthogonal to all of $\mathcal{H}_c$ and forms a linear space. It follows from (1.17) that for n^Γ in $\mathcal{N}$

$$G(n^\Gamma) = K(n^\Gamma);$$

since K is compact with respect to the G form, $\dim(\mathcal{N})$ must be finite.

Since $\mathcal{N}$ is a subset of $\mathcal{H}_c$, it is E_F-orthogonal to $p_j^\pm$; therefore $\mathcal{N}$ is E_F-orthogonal to all of $\mathcal{H}_G$. In particular, using form (1.4) of E_F, we conclude that, for any f_1 and f_2 in $C_0^\infty(F)$ and any n^Γ in $\mathcal{N}$,

$$(n_1, Lf_1) = 0, \qquad (n_2, f_2) = 0;$$

relations (1.22) follow from this.

(b) It follows from conservation of energy that $U(t)$ maps $\mathcal{N}$ into $\mathcal{N}$. It follows from (1.22) that

$$(d/dt)U(t)n^\Gamma = 0;$$

this implies (1.23).

(c) Suppose (1.24) were false; then there would be a sequence $\{f_i^\Gamma\}$ in $\mathscr{H}_c$, each f_i^Γ being G-orthogonal to $\mathscr{N}$, and satisfying

$$(1.25) \qquad\qquad\qquad G(f_i^\Gamma) = 1,$$

$$(1.25)' \qquad\qquad\qquad E_F(f_i^\Gamma) \to 0.$$

Since $E_F \geqq 0$ on $\mathscr{H}_c$, it follows from (1.25)' that

$$(1.26) \qquad\qquad\qquad E_F(f_i^\Gamma - f_j^\Gamma) \to 0.$$

Using (1.17),

$$(1.27) \qquad\qquad G(f_i^\Gamma - f_j^\Gamma) = E_F(f_i^\Gamma - f_j^\Gamma) + K(f_i^\Gamma - f_j^\Gamma).$$

By (1.26), the first term on the right of (1.27) tends to zero. By (1.25) and the compactness of K, the second term on the right of (1.27) tends to zero for a subsequence. It follows therefore that $G(f_i^\Gamma - f_j^\Gamma)$ tends to zero. Since $\mathscr{H}_G$ is complete,

$$G\text{-}\lim f_i^\Gamma = f^\Gamma$$

exists; from the first relation in (1.25),

$$G(f^\Gamma) = 1.$$

Since each f_i^Γ is G-orthogonal to $\mathscr{N}$ so is f^Γ.

Since the G norm dominates the E_F norm, it follows from (1.25)' that

$$E_F(f^\Gamma) = \lim E(f_i^\Gamma) = 0,$$

i.e., that f^Γ belongs to $\mathscr{N}$. So f^Γ is G-orthogonal to itself, a contradiction. Thus (1.24) must be true.

Part (d) of Lemma 1.4 is a corollary of part (c). This completes the proof of Lemma 1.4.

Remark 1. We have shown in Theorem 4.8 of [8] that h^Γ appearing in (1.22) is not in L_2.

Remark 2. If $E_F \geqq 0$, then it is not hard to show that $h^\Gamma(w) > 0$, and that $\mathscr{N}$ is at most one-dimensional.

Whenever it causes no confusion, we shall refer to $\mathscr{H}_c/\mathscr{N}$ simply as $\mathscr{H}_c$, and call it a Hilbert space in the E_F norm.

It follows from (1.20), (1.20)' that E_F is nondegenerate on the space $\mathscr{P}$ spanned by the $p_j^\pm$. We denote by Q the orthogonal projection that removes $\mathscr{P}$ components; Q maps $\mathscr{H}_G$ onto $\mathscr{H}_c$.

We denote the infinitesimal generator of U in $\mathcal{H}_G$ by A:

(1.28)
$$A\{f_1, f_2\} = \{f_2, Lf_1\},$$

and the domain of A by $\mathcal{D}(A)$.

2. Basic Properties of the Translation Representations

In this section we define the translation representations for automorphic data and derive their basic properties. We start with some technical preliminaries.

A subset B_F' of B_F is called *compact* if it is compact and contained in the interior of B_F. Note that a compact B_F' contains no parabolic points of Γ. We call a subset of $\mathbb{R}$ *compact* if it is closed and bounded away from $-\infty$. In particular, all intervals $y \geqq a$ are compact.

LEMMA 2.1. *If (s, β) is restricted to a compact subset of $\mathbb{R} \times B_F$, then the horospheres $\xi(s, \beta)$ are contained in a finite number of replicas of F.*

The reader can easily convince himself of the validity of the lemma by drawing a picture of $\xi(s, \beta)$.

We define the *outgoing* (and *incoming*) translation representation $R_{\pm}^F$ of f^Γ in $\mathcal{H}_G$ as a direct sum:

(2.1)+
$$R_+^F f^\Gamma = \sum_0^N \oplus \{R_+^j f^\Gamma\},$$

where

(2.1)'
$$R_+^0 f^\Gamma = R_+ f^\Gamma|_{B_F}$$

is the restriction to β in B_F of the translation representation defined by (1.8) for f^Γ regarded as a function in $\mathbf{H}^3$. The rest of the terms $R_+^j f^\Gamma$, $j = 1, \cdots, N$, are associated with the cusps of maximal rank and are defined as follows:

Map the parabolic point π_j of a cusp of maximal rank to ∞. Then the parabolic subgroup Γ_j leaving ∞ fixed is one of the crystallographic groups of $\mathbb{R}^2$. Let F_j be a fundamental polygon of Γ_j in $\mathbb{R}^2$; the transformed cusp is a half-cylinder of the form $F_j \times (a, \infty)$. We set

(2.2)+
$$R_+^j f^\Gamma(s) = \frac{1}{(2|F_j|)^{1/2}} [\partial_s e^{-s} \overline{f_1}(e^s) - e^{-s} \overline{f_2}(e^s)],$$

where

(2.3)
$$\overline{f_k}(y) = \int_{F_j} f_k(w) \, dx, \qquad\qquad k = 1, 2.$$

We first show that this definition makes sense.

LEMMA 2.2. *For any f^Γ in $\mathcal{H}_G$, $R_+^F f^\Gamma$ is L_2 on any compact subset of* $\mathbb{R} \times (B_F \cup \{\pi_j\})$.

Proof: According to Lemma 2.1, $R_+ f^\Gamma$ on $B_F' \times (a, \infty)$ depends only on values of f^Γ in $\bigcup_\sigma \tau F$, σ a finite subset of Γ. Since f^Γ has finite energy on F, it has finite energy on $\bigcup_\sigma \tau F$; it follows then from Theorem 1.1 that the first component of $R_+^F f^\Gamma$ belongs to $L^2(B_F' \times (a, \infty))$.

Next we look at the energy of f^Γ in a cusp. As shown in Section 3 of [8],

$$\int_a^\infty \int_{F_j} \left[y \left| \partial_y \left(\frac{f_1}{y} \right) \right|^2 + \frac{|f_2|^2}{y^3} \right] dx\, dy \leq G_F(f^\Gamma)$$

which implies that

$$(2.4) \qquad \int_a^\infty \left[y \left| \partial_y \left(\frac{\overline{f_1}}{y} \right) \right|^2 + \frac{|f_2|^2}{y^3} \right] dy \leq G_F(f^\Gamma).$$

Introducing $s = \log y$ as new variable, we see that (2.4) implies that $R_+^j f^\Gamma$ is in $L^2(a, \infty)$. This completes the proof of Lemma 2.2.

The incoming representation R_-^F is analogously

$$(2.1)_- \qquad\qquad R_-^F f^\Gamma = \sum_{j=0}^N \oplus \{R_-^j f^\Gamma\};$$

the operators R_-^j are defined by a change in sign in $(2.2)_+$.

The main result of this section is

THEOREM 2.3. R_+^F *as given by* $(2.1)_+$ *is a partial isometric translation representation of $U(t)$ on $\mathcal{H}_c$, i.e., there is an orthogonal decomposition*

$$(2.5)_+ \qquad\qquad \mathcal{H}_c = \mathcal{H}_+ \oplus \mathcal{H}_+^\perp$$

such that R_+^F is a unitary translation representation of $U(t)$ on $\mathcal{H}_+$ and is zero on $\mathcal{H}_+^\perp$. That is:

(i) *For all h^Γ in $\mathcal{H}_+$,*

$$(2.6) \qquad\qquad E_F(h^\Gamma) = \|R_+^F h^\Gamma\|^2,$$

where the norm on the right is the L_2 norm:

$$(2.6)' \qquad \|R_+^F h^\Gamma\|^2 = \int_{-\infty}^\infty \left[\int_{B_F} |R_+ h^\Gamma|^2 \, dm(\beta) + \sum_1^N |R_+^j h^\Gamma|^2 \right] ds.$$

(ii) *For all f^Γ orthogonal to $\mathcal{H}_+$,*

$$(2.7) \qquad\qquad R_+^F f^\Gamma = 0.$$

(iii) *For all f^Γ in $\mathcal{H}_G$,*

$$(2.8) \qquad\qquad R_+^F U(t) f^\Gamma = T(t) R_+^F f^\Gamma.$$

Similarly, R_-^F is a partial left translation representation, i.e., $\mathcal{H}_c$ can be decomposed as

$$(2.5)_- \qquad\qquad \mathcal{H}_c = \mathcal{H}_- \oplus \mathcal{H}_-^\perp;$$

R_-^F is a unitary left translation representation of $\mathcal{H}_-$, and is zero on $\mathcal{H}_-^\perp$.

First we prove part (iii), i.e., that R_+^F is a translation representation of $\mathcal{H}_G$:

$$(2.8)' \qquad\qquad (R_+^F U(t) f^\Gamma)(s, \beta) = (R_+^F f^\Gamma)(s - t, \beta).$$

Let B_F' be a compact subset of B_F. We construct a cutoff function φ in $\mathbf{H}^3$ with the following properties:

$$(2.9) \qquad \varphi(w) = \begin{cases} 1 & \text{at all } w \text{ whose distance from } \xi(s, \beta), \\ & s > a, \beta \text{ in } B_F' \text{ is at most } t. \\[6pt] 0 & \text{at all } w \text{ whose distance from } \xi(s, \beta), \\ & s > a, \beta \text{ in } B_F' \text{ is greater than } 2t. \end{cases}$$

We choose φ to be C^∞ in $\mathbf{H}^3$.

It follows from Lemma 2.1 that the support of φ is contained in a finite number of replicas of F. We define

$$(2.10) \qquad\qquad f = \varphi f^\Gamma.$$

Since f^Γ has finite energy in F, it follows that f has finite energy in $\mathbf{H}^3$. By part (ii) of Theorem 1.1, equation $(1.11)_+$,

$$(2.11) \qquad\qquad R_+ U(t) f = T(t) R_+ f.$$

By construction of φ, $f = f^\Gamma$ at all w on $\xi(s, \beta)$, $s > a$, $\beta \in B_F'$; it follows that the zero-th component of $R_+^F f^\Gamma$ equals $R_+ f$ on $(a, \infty) \times B_F'$. Since for solutions of the wave equation (1.3) signals propagate with speed ≤ 1, it follows from the construction of φ that $U(t) f = U(t) f^\Gamma$ at all w on $\xi(s, \beta)$, $s > a$, $\beta \in B_F'$. This shows that the first component of $R_+^F U(t) f^\Gamma$ equals $R_+ U(t) f$. Combining these results with (2.11), we deduce that the first component of $R_+^F f^\Gamma$ satisfies (2.8).

To show the same about the components associated with the j-th cusp we first map π_j into ∞, and then integrate the wave equation (1.3) with respect to x over the cross section F_j of the cusp. We obtain

$$(2.12) \qquad\qquad \bar{u}_{tt} - y^2 \bar{u}_{yy} + y \bar{u}_y - \bar{u} = 0.$$

We introduce new dependent and independent variables:

$$(2.13) \qquad\qquad y = e^s, \qquad \bar{u} = yv,$$

and obtain

(2.14) $$v_{tt} - v_{ss} = 0.$$

This can be written as

(2.14)' $$(\partial_t + \partial_s)(v_t - v_s) = 0,$$

which proves that $v_t - v_s$ is a function of $s - t$. Definition (2.3) of $R_+^j f^\Gamma$, when compared with (2.13), shows that

(2.15) $$R_+^j f^\Gamma = -(v_t - v_s)\frac{1}{(2|F_j|)^{1/2}}.$$

This proves (2.8) for the cusp components of $R_+^F f^\Gamma$ and completes part (iii) of Theorem 2.3.

An interesting application of (2.8) is when we take $f^\Gamma = p_j^-$, defined in (1.18)''. According to (1.19), p_j^- defined in (1.18)'' satisfies

$$U(t)p_j^- = \exp\{-\lambda_j t\}p_j^-.$$

Setting $f^\Gamma = p_j^-$ in (2.8) and $t = s$ we get

$$\exp\{-\lambda_j s\}R_+^F p_j^-(s, \beta) = \theta(\beta),$$

where $\theta(\beta)$ denotes the value of $R_+^F p_j^-$ at $s = 0$. Thus

(2.16) $$R_+^F p_j^-(s, \beta) = \exp\{\lambda_j s\}\theta(\beta).$$

According to Lemma 2.2, R_+^F maps $\mathcal{H}_G$ into $L_2((a, \infty) \times B_F \cup \{\pi_j\})$; but since $\lambda_j > 0$, the right side of (2.16) is in $L_2(a, \infty)$ only if $\theta \equiv 0$. This, and an analogous argument applied to $\mathcal{N}$, proves

LEMMA 2.4.

(2.17) (a) $R_+^F p_j^- = 0.$

(2.18) (b) $R_+^F n^\Gamma = 0$ for all n^Γ in $\mathcal{N}.$

An analogous calculation shows that

(2.19) (c) $R_+^\Gamma p_j^+ = \exp\{-\lambda_j s\}\theta_j(\beta).$

Parts (i) and (ii) of Theorem 2.3 will be proved through a series of lemmas. We start by constructing a right inverse for R_+^F.

Let B_F' be any compact subset of B, $l(s, \beta)$ any C_0^∞ function on $(a, \infty) \times B_F'$. We denote by g_0 the data:

(2.20) $$g_0 = J_+ l_0,$$

where the operator J_+ is defined in (1.13). It follows from Lemma 2.1 that g is supported on a finite number of replicas of F. We now define

$$(2.21) \qquad\qquad J_+^F l_0 = g_0^\Gamma,$$

where

$$(2.21)' \qquad\qquad g_0^\Gamma = \sum_\Gamma g_0^\tau, \qquad g_0^\tau(w) = g_0(\tau^{-1} w).$$

Clearly, g_0^Γ is automorphic. Note that if w is restricted to F, then the sum $(2.21)'$ can be restricted to a finite subset of Γ.

We show now that g_0^Γ as defined by $(2.21)'$ belongs to $\mathcal{H}_G$. By a partition of unity we can reduce the discussion to the case where the β support of l_0 is in a small open subset of B contained in the interior of B_F.

We choose a so large that in the sum $(2.21)'$ the only term that is non-zero in F is g_0, and that the support of g_0 is contained in a single chart of F. We recall now the construction of G, defined by equations (1.16), (1.17). As shown in [8], Section 3, the function k appearing in (1.16) can be made zero on any compact subset of a chart, such as on the support of g_0. With such a choice, $G(g_0^\Gamma) = E_F(g_0^\Gamma)$. In general, g_0^Γ is a sum of terms, each supported in a different chart of F, so each requires a different choice of G. But since all G forms are equivalent, $G(g_0^\Gamma) < \infty$ follows.

To treat arbitrary a, we anticipate the fact, see (2.27), that J_+^F transmutes translation into $U(t)$. Since $U(t)$ maps $\mathcal{H}_G$ into $\mathcal{H}_G$, we conclude that g_0^Γ belongs to $\mathcal{H}_G$.

We turn now to $j \neq 0$; let $l(s) = l_j(s)$ be any C_0^∞ function. We define $\{g_1, g_2\}$ by

$$(2.22) \qquad\qquad g_1(w) = \frac{1}{(2|F_j|)^{1/2}} y m(s), \qquad g_2 = \frac{-1}{(2|F_j|)^{1/2}} y \partial_s m(s),$$

where

$$(2.22)' \qquad\qquad s = \log y, \qquad m(s) = \int_{-\infty}^s l(r)\, dr.$$

Then we set

$$(2.23) \qquad\qquad g_{(j)}(w) = g(\gamma_j^{-1} w),$$

where γ_j is the mapping that carries π_j to ∞.

Next we define

$$(2.24) \qquad\qquad J_{+j}^F l_j = g_{(j)}^\Gamma,$$

where

$$(2.24)' \qquad\qquad g_{(j)}^\Gamma = \sum_{\Gamma_j \backslash \Gamma} g_{(j)}^\tau;$$

here Γ_j is the parabolic subgroup keeping π_j fixed, and the summation is over all right cosets of Γ_j. Note that since $g_{(j)}$ is automorphic with respect to Γ_j, $g_{(j)}^\Gamma$ is automorphic with respect to Γ.

We have assumed that the function l_j has compact support; from this it is easy to deduce that the sum $(2.24)'$ has only a finite number of non-zero terms. In fact, if $l_j(s) = 0$ for $y < a$, a large enough, then the only non-zero term in the sum $(2.24)'$ is $g_{(j)}$.

The same argument used in the case $j = 0$ shows that $g_{(j)}^\Gamma$ as defined by $(2.24)'$ belongs to $\mathcal{H}_G$.

We define now the operator

$$(2.25) \qquad J_+^F = \sum_0^N J_{+j}^F.$$

J_+^F takes C_0^∞ functions l with compact support into $\mathcal{H}_G$. We shall write $l = \{l_0, l_1, \cdots, l_N\}$,

$$(2.25)' \qquad J_+^F l = g^\Gamma = \sum_0^N g_{(j)}^\Gamma.$$

LEMMA 2.5. (i) J_+^F *is a right inverse for* R_+^Γ:

$$(2.26) \qquad R_+^F J_+^F l = l.$$

(ii) J_+^Γ *transmutes translation into the group* U:

$$(2.27) \qquad J_+^F T(t) l = U(t) J_+^F l.$$

(iii) J_+^F *is an isometry*:

$$(2.28) \qquad E_F(J_+^F l) = \|l\|^2,$$

where

$$\|l\|^2 = \iint |l_0|^2 \, dm(\beta) \, ds + \sum_1^N \int_{-\infty}^\infty |l_j|^2 \, ds.$$

Proof: We shall verify (2.26) component by component. We show first that

$$(2.29) \qquad R_+ J_{+0}^F l_0 = l_0.$$

We use definition (2.21), $(2.21)'$ of J_{+0}^F; as remarked there, if w is restricted to a compact subset of H, all but a finite number of terms in the sum $(2.21)'$ are zero. Using the linearity of R_+ we have

$$(2.30) \qquad R_+ g_0^\Gamma = \sum_\tau R_+ g_0^\tau.$$

Using (1.12) and the fact that R_+ is the inverse of J_+ we deduce that

$$R_+ g_0^\tau = R_+ J_+ l_0^\tau = l_0^\tau;$$

combining this with (2.30) we get

$$(2.30)' \qquad\qquad R_+ g_0^\Gamma = \sum_\tau l_0^\tau.$$

Since the support of l_0 is contained in B_F, it follows from (1.12) that the support of l_0^τ is contained in $\tau^{-1} B_F$; therefore, if we restrict (2.30)' to β in $B_{\tilde{F}}$, the only non-zero term on the right is l_0. This proves (2.29).

Next we show that

$$(2.31) \qquad\qquad R_+ g_{(j)}^\Gamma = 0, \qquad\qquad j = 1, \cdots, N.$$

To see this we write the equation of the horosphere (1.6) as

$$(2.32) \qquad\qquad \begin{aligned} |x - \beta|^2 + |y - r|^2 &= r^2, \\ r &= \tfrac{1}{2}(1 + |\beta|^2) \, e^{-s}. \end{aligned}$$

We parametrize the sphere as

$$x = \beta + (2ry - y^2)^{1/2}\{\cos \varphi, \sin \varphi\};$$

the Euclidean surface element on the sphere is $dy \, r \, d\varphi$, so the non-Euclidean surface element is

$$\frac{r \, dy \, d\varphi}{y^2}.$$

We take $\pi_j = \infty$; using the definition (2.22), (2.22)' of g_1, g_2 and (1.7), we have

$$\hat{g}_1(s, \beta) = cr \int_0^{2r} m \, (\log y) \, \frac{dy}{y},$$

$$\hat{g}_2(s, \beta) = -cr \int_0^{2r} m'(\log y) \, \frac{dy}{y}.$$

Introducing as new variable of integration

$$\log y = t,$$

we get

$$\hat{g}_1(s, \beta) = cr \int_0^{\log 2r} m(t) \, dt,$$

$$\hat{g}_2(s, \beta) = -cr \int_0^{\log 2r} {}' m'(t) \, dt.$$

We use now definition (1.8) of $R_+ g$, and relate r to s by (2.32); the result is

$$(2.33) \qquad\qquad R_+ g_{(j)} = 0.$$

308

Using (2.24)' we see that

$$(2.34) \qquad R_+ g_{(j)}^{\Gamma} = \sum R_+ g_{(j)}^{\tau}.$$

Now by (1.12),

$$R_+ g^{\tau} = (R_+ g)^{\tau}.$$

Using (2.33) we see that $R_+ g_{(j)}^{\tau} = 0$ for all τ, and so, by (2.34), $R_+ g_{(j)}^{\Gamma} = 0$.

Next we show that

$$(2.35) \qquad R_+^k g_{(j)}^{\Gamma} = \delta_{jk} l_j.$$

We base the proof on

LEMMA 2.6.   *For t large enough positive,*

$$(2.36) \qquad U(t) g_{(j)}^{\Gamma} = U(t) g_{(j)} \quad in \quad F$$

for $j = 0, \cdots, N$.

Proof: The case $j = 0$ was treated in the discussion following (2.20). For $j > 0$, it follows from (2.22) that

$$(2.37) \qquad u(w, t) = cym(\log y - t)$$

has initial data $\{g_1, g_2\}$ defined in (2.22). Since the support of m is bounded from below, (2.37) implies that $u(w, t)$ is supported in $y > \text{const.} \, e^t$. From this and (2.23) it follows that $U(t) g_{(j)}$ is supported in a small neighborhood of the parabolic fixed point π_j; this implies that $g_{(j)}^{\tau}$ is supported in a small neighborhood of $\tau \pi_j$. Thus if $\tau \notin \Gamma_j$, then for t large enough none of these neighborhoods intersect F. Now apply $U(t)$ to (2.24)':

$$U(t) g_{(j)}^{\Gamma} = \sum_{\Gamma_j \backslash \Gamma} U(t) g_{(j)}^{\tau}.$$

We have shown that for t large enough only one term in the above sum is not equal to 0 in F as asserted in (2.36).

Suppose $k \neq j$; to establish (2.35) we assume that $\pi_k = \infty$, so that π_j is located on $y = 0$. As observed above, for t large, $U(t) g_{(j)}$ is supported in a small neighborhood of π_j. Using definition (2.2), (2.3) of R_+^k we conclude that, for $k \neq j$,

$$R_+^k U(t) g_{(j)} = 0 \quad \text{for} \quad s > 0, t \gg 1.$$

Since, according to part (i) of Theorem 2.3, R_+^k is a translation representation, we conclude that

$$R_+^k U(t) g_{(j)} = R_+^k g_{(j)}(s - t) = 0 \quad \text{for} \quad s > 0, t \gg 1.$$

Letting $t \to \infty$ we deduce that $R_+^k g_{(j)} = 0$ for $k \neq j$. The same argument shows that $R_+^k g_{(j)}^{\tau} = 0$ for all τ; using (2.24)' we obtain (2.35) in case $k \neq j$.

When $k = j$ we argue as before except that now

$$U(t)g_{(j)} = \frac{1}{(2|F_j|)^{1/2}}\{ym(\log y - t), -ym'(\log y - t)\}.$$

Setting this into (2.2), (2.3) gives

$$R_+^j U(t)g_{(j)} = \partial_s m_j(s - t) \quad \text{for} \quad s > 0, t \gg 1.$$

Using the fact that R_+^j is a translation representation as well as relation (2.22)', we deduce that for all s

$$(R_+^j g_{(j)})(s) = l_j(s).$$

The same argument that was used in the case $k \neq j$ shows that $R_+^j g_{(j)}^\tau \equiv 0$ when $\tau \notin \Gamma_j$; (2.35) follows then from (2.24)'.

We turn now to part (ii) of Lemma 2.5. We shall use the fact that, for functions on $\mathbb{H}^3$, J_+ transmutes translation into the action of $U(t)$, i.e.,

$$J_+ T(t)l_0 = U(t)J_+ l_0.$$

Similarly,

$$J_+ T(t)l_0^\tau = U(t)J_+ l_0^\tau.$$

Summing over τ and using definition (2.21), (2.21)' we get

$$J_{+0}^F T(t)l_0 = U(t)J_{+0}^F l_0.$$

An entirely similar argument yields the analogous relation for J_{+j}^F; summing over $j = 0, \cdots, N$ we obtain (2.27).

We turn now to part (iii) of Lemma 2.5. It is deduced from

LEMMA 2.7.

(2.38) $E_F(g_j^\Gamma) = \|l_j\|^2,$ $j = 0, \cdots, N,$

and

(2.38)' $E_F(g_j^\Gamma, g_k^\Gamma) = 0 \quad \text{for} \quad j \neq k.$

Clearly, (2.28) follows from (2.25)' combined with (2.38) and (2.38)'. To prove (2.38) we choose t so large that (2.36) holds for $j = 0, \cdots, N$; this is possible by Lemma 2.6. Since J_+ is the inverse of R_+, the unitarity of R_+, see part (i) of Theorem 1.1, and (2.20), implies

$$E(g_0) = \|l_0\|^2.$$

Replacing l_0 by $T(t)l_0$ we see in view of the translation property that

(2.39) $E(U(t)g_0) = \|T(t)l_0\|^2.$

Using relation (2.36) for $j = 0$ we obtain

$$E_F(U(t)g_0^{\Gamma}) = E(U(t)g_0)$$

for t large enough. Combining this with (2.39) yields

$$E_F(U(t)g_0^{\Gamma}) = \|T(t)l_0\|^2.$$

But since E_F is conserved under $U(t)$, and the L_2 norm is conserved under $T(t)$, relation (2.38) for $j = 0$ follows.

To prove (2.38) for $j > 0$ we show first that

$$E_{F_j}(g_{(j)}) = \|l_j\|^2, \qquad\qquad j > 0,$$

where F_j is the fundamental domain of the parabolic subgroup Γ_j. This is easily shown by calculation, carried out when $\pi_j = \infty$; just set formulas (2.22) into (1.15), with $F = F_j = \mathbb{R} \times F_{\infty}$ and carry out the indicated integration, using $s = \log y$ as variable of integration. Then we use relation (2.24)$'$ and finish off the proof as in the case $j = 0$.

To prove (2.38)$'$ we observe that for t large enough, (2.36) holds; furthermore, $U(t)g_{(j)}$ is supported in a small neighborhood of $\pi_j, j = 1, \cdots, N$, while $U(t)g_0$ is supported in a sufficiently small neighborhood of B'_F. Thus, for t large enough and $j \neq k$, $U(t)g_j$ and $U(t)g_k$ have their supports on disjoint sets of F, and therefore are orthogonal. Relation (2.38)$'$ follows from this, since $U(t)$ is an isometry.

This completes the proof of Lemma 2.7.

DEFINITION. The range of the operator J_+^F, acting on smooth functions of compact support, is denoted by $\mathcal{D}'_+$.

We can express Lemma 2.7 as follows:

R_+^{Γ} is a translation representation of $U(t)$ over $\mathcal{D}'_+$.

Our next result goes beyond this.

LEMMA 2.8. For every f^{Γ} in $\mathcal{H}_G$ and every g^{Γ} in $\mathcal{D}'_+$,

$$(2.40) \qquad\qquad E_F(f^{\Gamma}, g^{\Gamma}) = (R_+^F f^{\Gamma}, R_+^F g^{\Gamma}).$$

Proof: We assume at first that the support of f^{Γ} in F is compact. We claim that there is a C^{∞} non-negative function φ defined in $\mathbb{H}^3$, of compact support, such that

$$(2.41) \qquad \sum_{\Gamma} \varphi^{\tau} = 1 \quad \text{on the support of} \quad f^{\Gamma}.$$

The construction of such a φ is a standard partition of unity.

We define now f in H^3 by

$$(2.42) \qquad\qquad f = \varphi f^\Gamma.$$

Since φ has compact support in H^3 so does f; and on account of (2.41),

$$(2.43) \qquad\qquad \sum_\Gamma f^\tau = f^\Gamma.$$

We note that for w in F, the sum in (2.43) is finite.

By definition of $\mathcal{D}'_+$, every g^Γ in $\mathcal{D}'_+$ can be written as in (2.25)′:

$$(2.44) \qquad\qquad g^\Gamma = \sum_0^N g^\Gamma_{(j)},$$

where g^Γ_0 is given by (2.21)′ and $g^\Gamma_{(j)}, j = 1, \cdots, N$, by (2.24)′. We shall verify (2.40) for each $g^\Gamma_{(j)}$ separately, starting with $j = 0$.

For (s, β) restricted to a compact set, the union of the horospheres $\xi(s, \beta)$ is contained in a finite number of replicas of F. Over such a set only a finite number of terms on the left in (2.43) are non-zero. Therefore we may apply R_+ to (2.43) and obtain

$$(2.45) \qquad\qquad R_+ f^\Gamma = \sum R_+ f^\tau$$

on any compact set of (s, β).

We use now (2.21)′ to represent g^Γ_0 and to write

$$(2.46) \qquad E_F(f^\Gamma, g^\Gamma_0) = \sum_\Gamma E_F(f^\Gamma, g^\tau_0) = \sum_\Gamma E_{\tau^{-1}F}(f^\Gamma, g_0) = E(f^\Gamma, g_0).$$

The support of g_0 also belongs to a finite number of replicas of F; in this set only a finite number of terms in (2.43) are non-zero. Thus

$$(2.47) \qquad\qquad E(f^\Gamma, g_0) = \sum E(f^\tau, g_0);$$

using the bilinear form of the Plancherel relation $(1.9)_+$ we can write

$$(2.48) \qquad\qquad E(f^\tau, g_0) = (R_+ f^\tau, R_+ g_0).$$

Since by Lemma 2.5, part (i), $R_+ g_0 = l_0$, and since l_0 is not equal to 0 only when $\beta \in B_F$, we may in the integral on the right in (2.48) restrict the β integration to B_F:

$$(2.48)' \qquad\qquad E(f^\tau, g_0) = (R_+ f^\tau, R_+ g_0)_{B_F}.$$

Setting (2.48)′ into the right side of (2.47) and combining this with (2.46) and (2.45) we get

$$(2.49) \qquad\qquad E_F(f^\Gamma, g^\Gamma_0) = (R_+ f^\Gamma, R_+ g_0)_{B_F}.$$

It follows from Lemma 2.5 part (i) that the last N components of $R^F_+ g_0$ are zero; we see from definition $(2.1)_+$ of R^F_+ that the right side of (2.49) is $(R^F_+ f^\Gamma, R^F_+ g_0)$. This completes the proof of (2.40) for $g^\Gamma = g^\Gamma_0$.

The proof for $j > 0$ is analogous. Instead of $(2.21)'$ we use $(2.24)'$, so in place of (2.46) we get

$$(2.50) \qquad \begin{aligned} E_F(f^\Gamma, g_{(j)}^\Gamma) &= \sum_{\Gamma_\infty \backslash \Gamma} E_F(f^\Gamma, g_{(j)}^\tau) \\ &= \sum_{\Gamma_\infty \backslash \Gamma} E_{\tau^{-1}F}(f^\Gamma, g_{(j)}) = E_{F_j}(f^\Gamma, g_{(j)}), \end{aligned}$$

where F_j is the fundamental domain of Γ_j. For $\pi_j = \infty$, F_j is a cylinder:

$$F_j = \mathbb{R} \times F'_\infty.$$

Integration over F_j is x integration followed by y integration. Since, by (2.22), $g_{(j)}$ is independent of x, the x integration effects only f^Γ. Using the notation (2.3) we can therefore rewrite (2.50) as

$$(2.50)' \qquad \begin{aligned} E_F(f^\Gamma, g_{(j)}^\Gamma) &= E_{F_j}(\overline{f^\Gamma}, g_{(j)}) \\ &= \int_0^\infty [\overline{f_2} g_2 + y^2 \overline{f_{1_y}} g_{1_y} - \overline{f_1} g_1] \frac{dy}{y^3}, \end{aligned}$$

where in the last step we have used definition (1.15) of energy. Introducing $s = \log y$ as new variable of integration and differentiation, and using the definition (2.22) of g_1 and g_2 and definition $(2.2)_+$ of R_+^j, we see that the right side of $(2.50)'$ equals

$$(2.51) \qquad \int_0^\infty (R_+^j f^\Gamma) l_j(s) \, ds = (R_+^j f^\Gamma, l_j).$$

Since according to Lemma 2.5, part (i), the j-th component of $R_+^F g_{(j)}$ is l_j and all others are zero, the right side of (2.51) is

$$(R_+^F f^\Gamma, R_+^F g_{(j)}).$$

Combining this with (2.50) we obtain (2.40) for $j = 1, \cdots, N$. This completes the proof of Lemma 2.8 for all f^Γ whose support in F is compact.

To remove this restriction on f^Γ we take any f^Γ in $\mathcal{H}_G$ and approximate it in the sense of the G norm by a sequence $\{f_n^\Gamma\}$, each member of which has compact support. Therefore by the foregoing,

$$(2.52) \qquad E_F(f_n^\Gamma, g^\Gamma) = (R_+^F f_n^\Gamma, R_+^F g^\Gamma)$$

holds for each n. Since the E_F form is dominated by the G form, and since the G form is positive, the E_F-scalar product satisfies

$$(2.53) \qquad |E_F(f^\Gamma, g^\Gamma)| \leq G^{1/2}(f^\Gamma) G^{1/2}(g^\Gamma).$$

It follows that the left side of (2.52) tends to $E_F(f^\Gamma, g^\Gamma)$. On the other hand, we saw in Lemma 2.1 that R_+^F is a bounded map of $\mathcal{H}_G$ into L_2 over any compact subset. Since by Lemma 2.5, $R_+^F g^\Gamma = l$, and by choice, l is supported on a compact subset, the right side of (2.52) tends to $(R_+^F f^\Gamma, R_+^F g^\Gamma)$. This completes the proof of Lemma 2.8.

We apply now relation (2.40) of Lemma 2.8 to $f = p_j^-$, defined by (1.18)''. Since according to relation (2.17) of Lemma 2.4, $R_+^F p_j^- = 0$, we deduce

LEMMA 2.9.

$$(2.54) \qquad E_F(p_j^-, g^\Gamma) = (R_+^F p_j^-, R_+^F g^\Gamma) = 0, \qquad\qquad j = 1, \cdots, M,$$

for every g^Γ in $\mathscr{D}_+'$.

We recall now the space $\mathscr{H}_c$ defined as the set of f in $\mathscr{H}_G$ that are orthogonal to p_j^+ and p_j^-, $j = 1, \cdots, M$, and the fact that Q is the orthogonal projection of $\mathscr{H}_G$ onto $\mathscr{H}_c$.

LEMMA 2.10.   For any g^Γ in $\mathscr{D}_+'$

$$(2.55) \qquad\qquad g^\Gamma = Qg^\Gamma + p^-,$$

where p^- is a linear combination of p_j^-.

$$(2.56) \qquad\qquad E_F(Qg^\Gamma) = E_F(g^\Gamma).$$

Proof: By definition of the projection Q,

$$(2.57) \qquad\qquad g^\Gamma = Qg^\Gamma + p^+ + p^-,$$

and Qg^Γ is orthogonal to $p_j^\pm$. Thus, for any p_j^-,

$$(2.58) \qquad E_F(g^\Gamma, p_j^-) = E_F(p^+, p_j^-) + E_F(p^-, p_j^-).$$

According to (2.54) of Lemma 2.9, the left side of (2.58) is zero, and according to relation (1.20), $E_F(p^-, p_j^-) = 0$; thus

$$E_F(p^+, p_j^-) = 0.$$

In view of relation (1.20)', $p^+ = 0$. Inserting this into (2.57) yields (2.55).

To prove (2.56), we use (2.55) and the quadratic nature of E_F. By (2.55),

$$E_F(Qg^\Gamma) = E_F(g^\Gamma - p^-) = E_F(g^\Gamma) - 2E_F(g^\Gamma, p^-) + E_F(p^-).$$

The second term on the right is zero by (2.54), the third term is zero by (1.20); this gives (2.56).

The following is a simple corollary of Lemma 2.8 and 2.10:

LEMMA 2.11.   For every f^Γ in $\mathscr{H}_c$ and every h^Γ in $Q\mathscr{D}_+'$,

$$(2.59) \qquad\qquad E_F(f^\Gamma, h^\Gamma) = (R_+^F f^\Gamma, R_+^F h^\Gamma).$$

Proof: Since f^Γ is in $\mathcal{H}_c$, it is orthogonal to p^-; it follows from this and (2.55) that, for every g^Γ in $\mathcal{D}'_+$,

$$(2.60) \qquad E_F(f^\Gamma, g^\Gamma) = E_F(f^\Gamma, Qg^\Gamma).$$

On the other hand, applying R_+^F to (2.55) and using (2.17) of Lemma 2.4 we get

$$(2.61) \qquad R_+^F g^\Gamma = R_+^F Qg^\Gamma.$$

Using (2.60) on the left of (2.40), (2.61) on the right, and setting $h^\Gamma = Qg^\Gamma$, gives (2.59).

We are now ready to prove Theorem 2.3. We take the space $\mathcal{H}_+$ to be as follows:

$\mathcal{H}_+$ *is the closure of* $Q\mathcal{D}'_+$ *in the energy norm.*

To prove part (i) of Theorem 2.3 we combine (2.26) and (2.28) of Lemma 2.5 to deduce, for $g^\Gamma = J_+^F l$, that

$$E_F(g^\Gamma) = \|R_+^F g^\Gamma\|^2.$$

Combining the left side of this with (2.56) of Lemma 2.10 and using (2.61) on the right side gives relation (2.6) for $h^\Gamma = Qg^\Gamma$.

To prove the remaining part (ii), of Theorem 2.3 let f^Γ be orthogonal to $\mathcal{H}_+$. We deduce from (2.59) that then $R_+^F f^\Gamma$ is orthogonal to every $R_+^F h^\Gamma$, h^Γ in $Q\mathcal{D}'_+$. By Lemma 2.7 and (2.61) the functions $R_+^F h^\Gamma$ include all L_2 functions on compact subsets of $\mathbb{R} \times (B_F \cup \{\pi_j\})$. Since by Lemma 2.2, $R_+^F f^\Gamma$ is L_2 on every such compact subset, it follows that $R_+^F f^\Gamma$ is zero on every compact subset, and hence identically 0. This proves (2.7) and completes the proof of Theorem 2.3.

Remark. Since $\mathcal{H}_+$ is defined as a completion in the E_F norm, its elements are defined only in $\mathcal{H}_c/\mathcal{N}$. We show now by a simple argument that elements of $\mathcal{H}_+$ can be realized as data in $\mathcal{H}_G$. For, let $\{h_i\}$ be a Cauchy sequence in the E_F norm of elements in $Q\mathcal{D}'_+$. Define h_i' by

$$h_i' = h_i + n_i,$$

where n_i lies in $\mathcal{N}$, so chosen that h_i' is G-orthogonal to $\mathcal{N}$. The new sequence h_i' is again a Cauchy sequence in the E_F norm; since $h_i' - h_j'$ is G-orthogonal to $\mathcal{N}$, it follows from part (c) of Lemma 1.4 that h_i' is a Cauchy sequence also in the G norm. Therefore it tends, in the sense of the G norm, to a limit that lies in $\mathcal{H}_G$.

Relation (2.59) holds for all f^Γ in $\mathcal{H}_G$ and all h^Γ in $\mathcal{H}_+$.

We have remarked after formulating Theorem 2.3 that R_+^F is not only a partial but a complete translation representation, in the sense that

$$(2.62) \qquad \mathcal{H}_+ = \mathcal{H}_c/\mathcal{N}.$$

We state now a consequence of (2.62).

THEOREM 2.12. $R_+^F f^\Gamma = 0$ for f^Γ in $\mathscr{H}_G$ holds only for f^Γ of the form

(2.63)
$$f^\Gamma = p^- + n.$$

Proof: Suppose $R_+^F f^\Gamma = 0$; denote by h^Γ the projection $Q f^\Gamma$ of f^Γ into $\mathscr{H}_c$:

(2.64)
$$f^\Gamma = h^\Gamma + p^- + p^+.$$

Apply R_+^F; using (2.17) and (2.19) we get

$$0 = R_+^F h^\Gamma + \sum a_j \exp\{-\lambda_j s\} \theta_j.$$

Since according to Theorem 2.3, $R_+^F h^\Gamma$ is L_2, it follows that

$$R_+^F p^+ = 0 \quad \text{and} \quad R_+^F h^\Gamma = 0.$$

As we shall prove in Part III, $R_+^F p^+ = 0$ implies that $p^+ = 0$; relation (2.62) shows that h^Γ belongs to $\mathcal{N}$. Setting this into (2.64) we obtain (2.63).

The development in this section goes through with slight modifications for real hyperbolic spaces of all dimensions. The background material in H^n can be found in [18]. In general the translation representation is written as

(2.65)
$$R_\pm f = \partial_s K_n e^{(n-1)s/2} \hat{f}_1 \mp K_n e^{(n-1)s/2} \hat{f}_2,$$

where K_n is a differential operator when n is odd and a composite differential-convolution operator when n is even. The inverses of R_+ and R_- can again be written down explicitly as

$$J_\pm k = (f_1, f_2),$$

(2.66)
$$f_1(w) = \int e^{(n-1)\langle w, \beta \rangle/2} J_n k(\langle w, \beta \rangle, \beta) \, dm(\beta),$$

$$f_2(w) = \mp \int e^{(n-1)\langle w, \beta \rangle/2} J_n \partial_s k(\langle w, \beta \rangle, \beta) \, dm(\beta).$$

Here again J_n is a differential operator when n is odd and a differential-convolution operator when n is even. The relevant fact is that J_n averages over the past, taking functions with support on $[a, \infty)$ into themselves. This was the essential property of $J_\pm$ which was exploited in our treatment of the translation representation for automorphic functions. Finally we remark that the definition $R_\pm^j$ and $J_{\pm j}^F$ for cusps of maximal rank is exactly the same as before.

Acknowledgments. The work of the first author was supported in part by the Department of Energy under Contract DE-AC02-76 ERO 3077 and that of the second by the National Science Foundation under Grant MCS-80-01943.

Bibliography

[1] Elstrodt, J., *Die Resolvente zum Eigenwertproblem der automorphen Formen in der hyperbolischen Ebene*, Teil I, Math. Ann. 203, 1973, pp, 295–330; Teil II, Math. Zeitschr. 132, 1973, pp. 99–134; Teil III, Math. Ann. 208, 1974, pp. 99–132.

[2] Faddeev, L., *Expansion in eigenfunctions of the Laplace operator in the fundamental domain of a discrete group on the Lobacevskii plane*, Trudy Moscov, Mat. Obsc. 17, 1967, pp. 323–350.

[3] Faddeev, L., and Pavlov, B. S., *Scattering theory and automorphic functions*, Seminar of Steklov Math. Inst. of Leningrad 27, 1972, pp. 161–193.

[4] Lax, P. D., and Phillips, R. S., *Scattering theory for automorphic functions*, Ann. of Math. Studies, No. 87, Princeton University Press, Princeton, NJ, 1976.

[5] Lax, P. D., and Phillips, R. S., *Scattering theory for automorphic functions*, Bulletin of the A.M.S. 2, 1980, pp. 261–295.

[6] Lax, P. D., and Phillips, R. S., *Translation representations for the solution of the non-Euclidean wave equation*, Comm. Pure Appl. Math. 32, 1979, pp. 617–667.

[7] Lax, P. D., and Phillips, R. S., *Translation representations for the solution of the non-Euclidean wave equation, II*, Comm. Pure Appl. Math. 34, 1981, pp. 347–358.

[8] Lax, P. D., and Phillips, R. S., *The asymptotic distribution of lattice points in Euclidean and non-Euclidean spaces*, Journal of Funct. Anal. 46, 1982, pp. 280–350.

[9] Lax, P. D., and Phillips, R. S., *A local Paley–Wiener theorem for the Radon transform of L_2 functions in a non-Euclidean setting*, Comm. Pure Appl. Math. 35, 1982, pp. 531–554.

[10] Maass, H., *Über eine neue Art von nichtanalytischen automorphen Funktionen und die Bestimmung Dirichletscher Reihen durch Functionalgleichungen*, Math. Ann. 121, 1949, pp. 141–183.

[11] Patterson, S. J., *The Laplace operator on a Riemann surface, I*, Compositio Mathematica 31, 1975, pp. 83–107; Part II, 32, 1976, pp. 71–112; Part III, 33, 1976, pp. 227–259.

[12] Patterson, S. J., *The limit set of a Fuchsian group*, Acta Math. 136, 1976, pp. 241–273.

[13] Phillips, R. S., *Scattering theory for the wave equation with a short range perturbation, II*, to appear.

[14] Roelcke, W., *Über die Wellengleichung bei Grenzkreisgruppen erster Art*, S. B. Heidelberger Akad. Wiss. Math.-Nat. Kl., 1956, pp. 159–267.

[15] Roelcke, W., *Das eigenwertproblem der automorphen Formen in der hyperbolischen ebene*, Teil I, Math. Ann. 1966, pp. 292–337; Teil II, Math. 168, 1967, pp. 261–324.

[16] Selberg, A., *Harmonic analysis and discontinuous groups in weakly symmetric Riemannian spaces with applications to Dirichlet series*, Journal Indian Math. Soc. 20, 1956, pp. 47–87.

[17] Sullivan, D., *Entropy, Hausdorff measures old and new, and limit sets of geometrically finite Kleinian groups*, Acta Math., in press.

[18] Woo, A. C., *Scattering theory on real hyperbolic spaces and their compact perturbations*, Dissertation, Stanford University, 1980.

Received December, 1982.

Translation Representations for Automorphic Solutions of the Wave Equation in Non-Euclidean Spaces. II

PETER D. LAX

Courant Institute

AND

RALPH S. PHILLIPS

Stanford University

Abstract

In this part we prove that the translation representation defined in Part I for the non-Euclidean wave equation is complete. We show how one can derive from this translation representation a spectral representation for the Laplace-Beltrami operator over geometrically finite fundamental polyhedra with infinite volume, without the use of Eisenstein series.

3. Introduction

The incoming and outgoing translation representations R_-^F and R_+^F of Γ-automorphic solutions with finite energy of the non-Euclidean wave equation were defined in Section 2 of Part I of [5]; the superscript F refers to the fundamental domain of the group Γ. There we have shown (see Theorem 2.3) that these representations are partial isometries onto $L_2(\mathbb{R}, N)$, where N denotes the square integrable functions on the boundary points at infinity of F. Thus there are closed subspaces $\mathcal{H}_-$ and $\mathcal{H}_+$ of data with finite energy which are respectively mapped by R_-^F and R_+^F isometrically onto $L_2(\mathbb{R}, N)$, while their orthogonal complements in $\mathcal{H}_c$ are mapped into zero; here $\mathcal{H}_c$ denotes the space of all data with finite energy which are orthogonal to the eigenfunctions of the Laplace-Beltrami operator over the fundamental domain. In Part II we show that these representations are complete isometries, i.e., that both $\mathcal{H}_-$ and $\mathcal{H}_+$ are equal to $\mathcal{H}_c$.

Our principal result is

THEOREM 3.1. *The translation representations R_-^F and R_+^F are complete, i.e., they are unitary maps of $\mathcal{H}_c$ (modulo the null space of the energy form) onto $L_2(R \times B_F) \times \prod_1^N L_2(\mathbb{R})$; here the infinite boundary of F consists of B_F plus N cusps of maximal rank.*

The proof of this theorem relies rather heavily on our previous paper [4] and Part I of this article [5]. The sections and formulas of the two parts are numbered consecutively.

Communications on Pure and Applied Mathematics, Vol. XXXVII, 779–813

We denote the operator relating Γ-automorphic initial data $f^\Gamma = \{f_1, f_2\}$ of solutions of the non-Euclidean wave equation to their data at time t by $U(t)$; we denote the infinitesimal generator of $U(t)$ by A, and $U(t)f^\Gamma$ by $f^\Gamma(t)$. Γ is assumed to have the finite geometric property and F is assumed to have infinite volume.

THEOREM 3.2. *Suppose f^Γ belongs to $\mathcal{H}_c$.*
  (i) *Let π_j be a cusp of maximal rank for F and suppose that*

$$(3.1) \qquad R_+^j f^\Gamma(s) \equiv 0,$$

where R_+^j is the j-th component of the outgoing translation representation defined by $(2.2)_+$, (2.3). Then the energy content of the cusp π_j tends to zero as t tends to infinity; more precisely, given $\varepsilon > 0$, we can choose a neighborhood N_j of the cusp so small that

$$(3.2) \qquad \int_{N_j} |f_2(t)|^2 \, dV < \varepsilon \quad \text{for} \quad t \geq 0.$$

  (ii) *Let D be a disk at infinity, i.e., a set of points β satisfying*

$$(3.3) \qquad |\beta - \beta_0| \leq \rho,$$

and assume that D contains no parabolic points of Γ. Suppose that

$$(3.4) \qquad R_+^0 f^\Gamma(s, \beta) = 0 \quad \text{for all} \quad s \quad \text{and all} \quad \beta \in D,$$

where R_+^0 is the zeroth component of the outgoing translation representation, defined by $(2.1)'$. Then the energy content of any spherical cap $\mathcal{C}$ through D tends to zero as t tends to infinity. A spherical cap $\mathcal{C}$ over D is a set of points (x, y), $y > 0$, satisfying, for some y_0,

$$(3.5) \qquad |x - \beta_0|^2 + |y - y_0|^2 \leq \rho^2 + y_0^2.$$

We claim that, for any $\varepsilon > 0$ and any y_0, there is T such that

$$(3.6) \qquad \int_{\mathcal{C}} |f_2(t)|^2 \, dV < \varepsilon \quad \text{for} \quad t > T.$$

  (iii) *Let ν_j be a cusp of F, not of maximal rank. Then, given $\varepsilon > 0$, there exists a neighborhood $V_j(\varepsilon)$ of ν_j in F such that for all t*

$$(3.7) \qquad \int_{V_j} |f_2(t)|^2 \, dV < \varepsilon.$$

In the proof of Theorem 3.2 we can assume without loss of generality that f^Γ belongs to $\mathcal{H}_c \cap \mathcal{D}(A^2)$. For we can approximate any f^Γ in $\mathcal{H}_c$ by

$$f_\varphi = \int \varphi(t) V(t) f^\Gamma \, dt,$$

where φ is in C^2, $\varphi \geqq 0$, $\int \varphi \, dt = 1$. Since R_+^F is a translation representation, $R_+^F f^\Gamma = 0$ implies that $R_+^F f_\varphi^\Gamma = 0$. Since f^Γ, f_φ^Γ lie in $\mathcal{H}_c$,

$$(3.8) \qquad \int |f_2^\Gamma - f_{\varphi 2}^\Gamma|^2 \, dV \leqq E_F(f^\Gamma - f_\varphi^\Gamma).$$

If the support of φ lies in a small enough interval around $t = 0$, the right side of (3.8) is less than ε. Therefore, if inequalities (3.2) and (3.6) hold for f_φ^Γ, they also hold for f^Γ.

We show now that in the proof of part (iii) we may take f^Γ of the form

$$(3.9) \qquad f^\Gamma = Ag^\Gamma, \qquad\qquad g^\Gamma \text{ in } \mathcal{D}(A^2).$$

For the range of A on $\mathcal{H}_c \cap \mathcal{D}(A^2)$ is dense in the energy norm of $\mathcal{H}_c$; therefore, any f^Γ in $\mathcal{H}_c$ can be approximated by elements of the form (3.9). Using (3.8) it follows that if (3.7) holds for all elements of form (3.9), it holds for all f^Γ in $\mathcal{H}_c$.

We now show that Theorem 3.1 follows from Theorem 3.2.

Suppose that R_+^F is not complete. Then since, by Theorem 2.3, R_+^F is a partial isometry, there is a f^Γ in $\mathcal{H}_c$ with $E_F(f^\Gamma) \neq 0$ whose outgoing representation is zero:

$$R_+^F f^\Gamma \equiv 0.$$

Our proof consists in showing that (3.8) implies that $E_F(f^\Gamma) = 0$.

The parabolic points π_j of B_F for cusps of maximal rank are isolated. We cover the parabolic points ν_j of B_F for cusps of less than full rank by small neighborhoods in B_F, chosen in accordance with part (iii) of Theorem 3.2. We can then cover the rest of B_F by a finite number of discs D_j that contain no parabolic points. It follows that neighborhoods N_j of π_j, V_j of ν_j and sufficiently large spherical caps $\mathcal{C}_j$ over D_j cover all of F. We may conclude from (3.2), (3.6) and (3.7) of Theorem 3.2 that

$$(3.10) \qquad \int_F |f_2(t)|^2 \, dV < \varepsilon \quad \text{for} \quad t > T.$$

For the rest of the argument we need only consider the zeroth component of $R_\pm^F$. These components are defined by

$$(3.11) \qquad R_\pm^0 f^\Gamma = \partial_s P \hat{f}_1 \mp P \hat{f}_2,$$

where P is in general an integral-differential operator (see [7]), which in the case of $n = 3$ is given by (2.1) and (1.8).

H   The condition $R_+^0 f^\Gamma = 0$ means then that

$$(3.12) \qquad \partial_s P \hat{f}_1 = P \hat{f}_2.$$

As we noted earlier, $U(t)f^\Gamma = \{f_1(t), f_2(t)\}$ belongs to the nullspace of R^0_+ if f^Γ does; therefore, it satisfies the same relation

$$(3.12)' \qquad\qquad \partial_s P \hat{f}_1(t) = P \hat{f}_2(t).$$

Let us denote $R^0_- f^\Gamma$ by k_-; it follows from (3.11) and (3.12) that

$$(3.13) \qquad\qquad k_- = 2 P \hat{f}_2.$$

Since R^0_- is an anti-translation representation, we deduce that

$$(3.13)' \qquad\qquad k_-(s + t) = 2 P \hat{f}_2(t).$$

We define now

$$(3.14) \qquad\qquad h^\Gamma(t) = \{0, f_2(t)\}.$$

Using the definition (3.11) of R^0_- and the relation (3.13) we get

$$(3.15) \qquad\qquad R^0_- h^\Gamma(t) = P f_2(t) = \tfrac{1}{2} k_-(s + t).$$

We use the fact that R^0_- is one component of the representation R^F_-, and that R^F_-, being a partial representation, is a contraction on $\mathcal{H}_c$; we conclude from this that

$$(3.16) \qquad\qquad \| R^0_- h^\Gamma(t) \|^2 \leq \| R^F_- h^\Gamma(t) \|^2 \leq E_F(h^\Gamma(t)).$$

Apply now definition (1.15) of energy to $h^\Gamma(t)$, defined by (3.14):

$$E_F(h^\Gamma(t)) = \int_F |f_2(t)|^2 \, dV.$$

Use this to express the right side of (3.16), and (3.15) to express the left side. The resulting inequality is

$$(3.17) \qquad\qquad \tfrac{1}{4} \int\int |k_-(s + t)|^2 \, ds \, dm(\beta) \leq \int_F |f_2(t)|^2 \, dV.$$

According to (3.10), the right side tends to zero as $t \to \infty$. Since the left side is independent of t, it follows that $\| k_- \| = 0$. Thus we deduce from (3.13) that

$$(3.18) \qquad\qquad P \hat{f}_2 = \tfrac{1}{2} k_- = 0.$$

Until now we have parametrized horospheres by s, their non-euclidean distance from the origin; see equation (1.6). We switch now to r, their Euclidean radius; see equation (1.9) of [4]. The relation of the two is

$$(3.19) \qquad\qquad r = \tfrac{1}{2}(1 + |\beta|^2) e^{-s}.$$

321

Under this change of coordinates, the operator P becomes

(3.20) $$P = c_n(1 + |\beta|)^2)^{(n-1)/2}J,$$

where J is defined by (1.12) and (1.13) of [4]. In terms of the coordinate r, equation (3.18) becomes

(3.21) $$J\hat{f}_2 = 0.$$

We appeal now to Theorem 1.1 of [4], which asserts:

A function f defined and square integrable in a spherical cap $\mathscr{C}$ which satisfies $J\hat{f} = 0$ for all horospheres in $\mathscr{C}$ is zero in $\mathscr{C}$.

It follows then from (3.21) that $f_2 = 0$ in every spherical cap $\mathscr{C}$ in which f_2 is square integrable. Every point of F belongs to such a spherical cap; therefore, $f_2(t) = 0$ in F.

We denote the first component of $U(t)f^\Gamma$ by $u(t)$. By definition, $f_2(t) = u_t(t)$. It therefore follows that $U(t)f^\Gamma$ is independent of t. This implies that $Af^\Gamma = 0$. According to Lemma 5.9 of [3], this implies that $E(f^\Gamma) = 0$; i.e., modulo the null space of E, $f^\Gamma = 0$. This completes the proof of Theorem 3.1 for all groups Γ which have geometrically finite fundamental polyhedron.

In Section 4 we show how to construct a spectral representation for L out of a translation representation for U.

The proof of Theorem 3.2 occupies the rest of the paper. Part (i) is proved in Section 5, part (ii) in Section 6 and part (iii) in Section 7.

We acknowledge with pleasure the help in Section 7 of Hans Samelson; he acquainted us with the relevant fact about Euclidean subgroups.

4. Spectral Representation for L

In this section we show how to construct a spectral representation for L out of a translation representation for U. We start with the observation that the Fourier transform of a translation representation for $U(t)$ yields a spectral representation for A, the generator of U. I.e., for f^Γ in $\mathscr{H}_c$ we define

(4.1) $$\tilde{f}(\sigma) = \frac{1}{\sqrt{2\pi}} \int e^{i\sigma s} R_+^F f^\Gamma(s) \, ds.$$

A consequence of Theorem 3.1. and of the unitary character of the Fourier transform is

THEOREM 4.1. *The mapping $f^\Gamma \to \tilde{f}$ defined by (4.1) is a unitary mapping of $\mathscr{H}_c$ onto $L_2(\mathbb{R} \times \beta_F) \times L_2(\mathbb{R})^N$.*

We decompose $\mathcal{H}_c$ into two orthogonal subspaces $\mathcal{H}_1$ and $\mathcal{H}_2$; $\mathcal{H}_1$ consists of all data whose 2-*nd* component is zero; $\mathcal{H}_2$ visa versa. Clearly, $\mathcal{H}_1$ and $\mathcal{H}_2$ are orthogonal. If $\{f_1, f_2\}$ belongs to $\mathcal{H}_c$, so do $\{f_1, 0\}$ and $\{0, f_2\}$. This shows that

$$(4.2) \qquad \mathcal{H}_c = \mathcal{H}_1 \oplus \mathcal{H}_2.$$

The representation (4.1) carries $\mathcal{H}_1$ and $\mathcal{H}_2$ into the orthogonal subspaces $\tilde{\mathcal{H}}_1$ and $\tilde{\mathcal{H}}_2$. It follows from Theorem 4.1 and (4.2) that

$$(4.3) \qquad \tilde{\mathcal{H}}_1 \oplus \tilde{\mathcal{H}}_2 = L_2(\mathbb{R} \times \beta_F) \times L_2^N(\mathbb{R}).$$

We define the operator J to be multiplication by $\mathrm{sgn}\, m(\sigma)$, i.e.,

$$(4.4) \qquad J\tilde{f}(\sigma) = \begin{cases} \tilde{f}(\sigma) & \text{for } \sigma > 0, \\ -\tilde{f}(\sigma) & \text{for } \sigma < 0. \end{cases}$$

LEMMA 4.2. *J maps $\tilde{\mathcal{H}}_1$ and $\tilde{\mathcal{H}}_2$ into each other.*

Before presenting the proof of Lemma 4.2, we show how to use it to prove

LEMMA 4.3. (a) *For every $\tilde{f}$ in $\tilde{\mathcal{H}}_2$',*

$$(4.5) \qquad \int_{-\infty}^0 |\tilde{f}(\sigma)|^2\, d\sigma = \int_\infty^0 |\tilde{f}(\sigma)|^2\, d\sigma.$$

(b) *Denote by $\tilde{\mathcal{H}}_2^+$ the restriction of functions $\tilde{f}$ in $\tilde{\mathcal{H}}_2$ to the positive axis $\mathbb{R}_+$. We claim that*

$$(4.6) \qquad \tilde{\mathcal{H}}_2^+ = L_2(\mathbb{R}_+ \times B_F) \times L_2(\mathbb{R}_+)^N.$$

Proof: According to Lemma 4.2, if f lies in $\tilde{\mathcal{H}}_2$, then Jf lies in $\tilde{\mathcal{H}}_1$. Since these are orthogonal subspaces, we obtain using (4.4)

$$(\tilde{f}, J\tilde{f}) = \int_0^\infty |\tilde{f}|^2\, d\sigma - \int_{-\infty}^0 |\tilde{f}|^2\, d\sigma = 0;$$

this proves (4.5) of part (a).

It follows form (4.5) that restriction to $\mathbb{R}_+$ of $\tilde{\mathcal{H}}_2$ is $1/\sqrt{2}$ times an isometry. This shows that $\tilde{\mathcal{H}}_2^+$ is a closed suspace. To prove (4.6) it suffices therefore to show that any function g on $\mathbb{R}_+$ which lies in $L_2(\mathbb{R}_+ \times \beta_F) \times L_2^N(\mathbb{R}_+)$ and is orthogonal to $\tilde{\mathcal{H}}_2^+$ is zero. To see this, we extend g to be zero on $\mathbb{R}_-$; g thus extended is then orthogonal to $\tilde{\mathcal{H}}_2$, and so by (4.3), belongs to $\tilde{\mathcal{H}}_1$. According to Lemma 4.2, if g belongs to $\tilde{\mathcal{H}}_1$, then Jg belongs to $\tilde{\mathcal{H}}_2$; but since g is zero on $\mathbb{R}_-$, it follows from (4.4) that $Jg = g$. Thus g belongs to both $\tilde{\mathcal{H}}_1$ and $\tilde{\mathcal{H}}_2$; since these are orthogonal subspaces, $g \equiv 0$. This proves part (b).

We turn now to Lemma 4.2. We note that since R_+^F is a translation representation, the action of the infinitesimal generator A is represented by $-d/ds$. It follows that in the representation $f^\Gamma \to \tilde{f}$ defined by (4.1), the action of A is represented by multiplication by $i\sigma$. Consequently, the action of A^2 is represented by multiplication by $-\sigma^2$.

We recall that $A = d/dt$ acting on data can be written as the matrix operator

$$(4.7) \qquad\qquad A = \begin{pmatrix} 0 & I \\ L & 0 \end{pmatrix}, \qquad A^2 = \begin{pmatrix} L & 0 \\ 0 & L \end{pmatrix};$$

thus A^2 acts like L on each component, and therefore like L on the subspace $\mathcal{H}_2$.

It follows from Lemma 4.2 that A maps $\mathcal{H}_1 \cap D(A)$ into $\mathcal{H}_2$ and $\mathcal{H}_2 \cap D(A)$ into $\mathcal{H}_1$. Similarly, A^2 maps $\mathcal{H}_i \cap D(A^2)$ into $\mathcal{H}_i$. Thus

$$(4.8) \qquad\qquad A(k^2 - A^2)^{-1}$$

maps $\mathcal{H}_1$ and $\mathcal{H}_2$ into each other. Using the previously given description of how A and A^2 act in the representation (4.1), we deduce that multiplication by

$$(4.9) \qquad\qquad \sigma(k^2 + \sigma^2)^{-1}$$

maps $\tilde{\mathcal{H}}_1$ and $\tilde{\mathcal{H}}_2$ into each other. It follows from this that multiplication by any odd plynomial in the functions (4.9) maps $\tilde{\mathcal{H}}_1$ and $\tilde{\mathcal{H}}_2$ into each other, and so does any uniform limit of such polynomials on $\mathbb{R}$.

The functions (4.9) are odd, separate points of $\mathbb{R}$ and vanish at ∞. From the Stone-Weierstrass theorem we know that any odd continuous function that vanishes at ∞ is a uniform limit of odd polynomials; in particular,

$$j_n(\sigma) = \begin{cases} \operatorname{sgn} \sigma & \text{for} \quad 1/n < |\sigma| < n, \\ 0 & \text{for} \quad |\sigma| > n+1, \\ \text{linear in between.} \end{cases}$$

So J_n, multiplication by j_n, maps $\mathcal{H}_1$ and $\mathcal{H}_2$ into each other. Now J_n tends in the strong operator topology to J defined in (4.4); therefore J has the same property, as asserted in Lemma 4.2.

The desired spectral representation is the mapping of $\{0, f\}$ in $\mathcal{H}_2$ onto $\tilde{\mathcal{H}}_2^+$, multiplied by $\sqrt{2}$. On $\mathcal{H}_2$ the energy norm is the L_2 norm of f, and A^2 acts as multiplication by $-\sigma^2$ in $\tilde{\mathcal{H}}_2^+$. The fact that this representation is unitary follows from Lemma 4.3.

Formula (4.1), applied to elements $f^\Gamma = \{0, f\}$ in $\mathcal{H}_2$, defines a kind of Fourier transform for $L_2(F)$ functions f which are orthogonal to the finite number of eigenfunctions of the Laplace–Beltrami operator.

This result extends that of S. J. Patterson, who in [6] proved the completeness of the spectral representation of the Laplace–Beltrami operator for finitely generated discrete Fuchsian subgroups with no parabolic elements, acting on H^2. Patteron's approach depends on the construction of appropriate Eisenstein series and their analytic continuation; our construction is purely geometrical.

In [8], Mandou-valos shows how to handle some groups acting on H^3.

5. The Energy Content of Cusps of Maximal Rank

We map the parabolic point π_j of the cusp to ∞. The subgroup Γ_j leaving ∞ fixed is then one of the crystallographic groups of $\mathbb{R}^{n-1}$, with compact fundamental polyhedron F_j. The cusp is a half-cylinder of the form

$$F_j \times (\alpha, \infty).$$

The corresponding component of the translation representation is given by formulas $(2.2)_+$, (2.3):

$$(5.1) \qquad R_+^j f^\Gamma(s) = \frac{1}{\sqrt{2}} [\partial_s{}^{(1-n)s/2} \overline{f}_1(e^s) - e^{(1-n)s/2} \overline{f}_2(e^s)],$$

where

$$(5.1)' \qquad\qquad \overline{f}_k = |F_j|^{-1/2} \int_{F_j} f_k(w) \, dx, \qquad\qquad k = 1, 2.$$

Since F_j is a compact fundamental polygon for Γ_j in $\mathbb{R}^{n-1}$, the Euclidean Laplace operator Δ, acting on functions in F_j that are automorphic with respect to Γ_j, has a standard discrete spectrum. Furthermore, $-\Delta$ is non-negative, and zero is an eigenvalue of multiplicity one; the corresponding eigenfunction is constant. Denote the eigenvalues and eigenfunctions of $-\Delta$ by

$$(5.2) \qquad\qquad 0 = \omega_0 < \omega_1 \leqq \omega_2 \leqq,$$

and

$$\psi_0 \equiv \text{const}, \ \psi_1, \psi_2, \cdots.$$

Again let u be a solution of the wave equation with initial data f^Γ. Expand $u(x, y, t)$ into a Fourier series with respect to these eigenfunctions:

$$(5.3) \qquad\qquad u(x, y, t) = \sum_0^\infty u^{(k)}(y, t) \psi_k(x).$$

Since $\psi_0 = |F_j|^{-1/2}$, it follows from $(5.1)'$ that

$$(5.3)' \qquad\qquad u^{(0)}(y, 0) = \overline{f}_1, \qquad u_t^{(0)}(y, 0) = \overline{f}_2.$$

Differentiate (5.3) with respect to t; by Parseval's relation,

$$\int_{F_j} |u_t(x, y, t)|^2 \, dx = \sum_{k=0}^\infty |u_t^{(k)}(y, t)|^2.$$

Multiplying by y^{-n} and integrating from a to ∞ gives

$$(5.4) \qquad \int_a^\infty \int_{F_j} |u_t(x, y, t)|^2 \frac{dx \, dy}{y^n} = \sum_{k=0}^\infty \int_a^\infty |u_t^{(k)}(y, t)|^2 \frac{dy}{y^n}.$$

To prove inequality (3.2) we shall estimate the right side of (5.4). We do this in two steps, first estimating the terms $k > 0$.

LEMMA 5.1. *Suppose f^Γ and Af^Γ belong to $\mathcal{H}_c$. Then the solution u with initial data f^Γ satisfies*

$$(5.5) \qquad \sum_1^\infty \int_a^\infty |u_t^{(k)}(y, t)|^2 \frac{dy}{y^n} \leq \frac{\text{const.}}{a^2}$$

for all t.

Proof: Energy is defined by equation (1.15):

$$(5.6) \qquad E_F(u) = D_F(u) - (\tfrac{1}{2}(n-1))^2 H_F(n) + H_F(u_t),$$

where D_F denotes the invariant Dirichlet integral and H_F the invariant square integral over F:

$$(5.7) \qquad D_F(u) = \int_F y^2(u_x^2 + u_y^2)\, dV,$$

$$(5.7)' \qquad H_F(u) = \int_F u^2\, dV.$$

The initial data f^Γ of u were assumed to belong to $\mathcal{H}_c$, i.e., to be orthogonal to the eigenfunctions of the Laplace-Beltrami operator over F. It follows from Lemma 1.3 that energy is non-negative over F; in particular,

$$E_F(u) \geq H_F(u_t).$$

As noted in Section 3, we may assume that f^Γ belongs to the domain of A. Since $Af^\Gamma = \{u_t, u_{tt}\}$, we see from (5.6) that

$$E_f(u_t) \geq D_F(u_t) - (\tfrac{1}{2}(n-1))^2 H_F(u_t).$$

Setting

$$(5.8) \qquad E = E_F(Af^\Gamma) + (\tfrac{1}{2}(n-1))^2 E_F(f^\Gamma)$$

and combining the last two inequalities we get

$$(5.8)' \qquad E \geq D_F(u_t).$$

By Green's formula,

$$\int_{F_j} |u_{tx}|^2\, dx = - \int_{F_j} u_t \overline{\Delta u_t}\, dx.$$

Differentiating (5.3) with respect to t and applying Δ, we have from Parseval's relation

$$\int_{F_j} |u_{tx}|^2\, dx = \sum \omega_k |u_t^{(k)}(y, t)|^2.$$

We multiply this by y^{2-n} and integrate with respect to y from a to ∞. By (5.7) the left side is less than or equal to $D_F(u_t)$; therefore using (5.8)' we deduce the inequality

$$(5.8)'' \qquad \sum \omega_k \int_a^\infty |u_t^{(k)}(y, t)|^2 \frac{dy}{y^{n-2}} \leq E,$$

where the constant E is determined by (5.8).

326

Since by (5.2), for $k > 0$, $\omega_k \geqq \omega_1 > 0$, and since $y \geqq a$ in the range of integration, we deduce from (5.8)″ that (5.5) holds with constant E/ω_1. To estimate the contribution of the component $k = 0$ to (5.4) we need

LEMMA 5.2. *Suppose that f^Γ in $\mathcal{H}_c$ satisfies*

$$(5.9) \qquad R_+^j f^\Gamma \equiv 0.$$

Denote by $u(t)$ the solution with initial data f^Γ, and by $\bar{u} = \bar{u}(y, t)$ the zeroth mode of u. Define

$$(5.10) \qquad h(y, t) = y^{(1-n)/2} \bar{u}_t(y, t).$$

Then

$$(5.11) \qquad \text{(a)} \quad h(y, t) = h(y\, e^t),$$

$$(5.12) \qquad \text{(b)} \quad \int_a^\infty h^2(y)\, \frac{dy}{y} < \infty.$$

Proof: The non-Euclidean wave equation satisfied by u is

$$(5.13) \qquad u_{tt} - y^2(u_{yy} + \Delta u) - (n-2)yu_y + (\tfrac{1}{2}(n-1))^2 u = 0.$$

Integrating this equation with respect to x over F_j yields the following equation for $\bar{u}$:

$$(5.13)' \qquad \bar{u}_{tt} - y^2 \bar{u}_{yy} - (n-2)y\bar{u}_y + (\tfrac{1}{2}(n-1))^2 \bar{u} = 0.$$

We introduce a new dependent variable v by

$$(5.14) \qquad \bar{u} = y^{(n-1)/2} v,$$

and a new independent variable by

$$(5.14)' \qquad y = e^s,$$

a change of variables originally employed by Faddeev and Pavlov [1]. Then (5.13)′ becomes

$$(5.15) \qquad v_{tt} - v_{ss} = 0.$$

Setting v in place of $\bar{u}$ in formula (5.1) yields

$$(5.16) \qquad R_+^j f^\Gamma = \frac{1}{\sqrt{2}} (v_s - v_t).$$

Assumption (5.9) can therefore be expressed as:

$$(5.17) \qquad v_s = v_t.$$

We write now (5.15) as

$$(\partial_t - \partial_s)(v_t + v_s) = 0$$

from which it follows that

(5.18) $$v_t + v_s = k(s + t),$$

k some function of a single variable. Using (5.17) we conclude that

$$v_t = v_s = \tfrac{1}{2}k(s + t).$$

Using the relation (5.14) of v to $\bar{u}$, the relation (5.14)' of y to s, we conclude that v_t is of the form (5.10), and that h is of the form (5.11). This proves part (a) of Lemma 5.2.

To prove part (b) we set $t = 0$ in (5.10), (5.11) and write

(5.19) $$\int_a^\infty |\bar{u}_t(y, 0)|^2 \frac{dy}{y^n} = \int_a^\infty h(y)^2 \frac{dy}{y}.$$

Since f^Γ has finite energy, it follows that $u_t(0)$ is square integrable over F; so by (5.4), the left side of (5.19) is finite. But then so is the right side, as asserted in (5.12). This completes the proof of Lemma 5.2.

Analogously to (5.19), we have, for any t,

(5.19)' $$\int_a^\infty |\bar{u}_t(y, t)|^2 \frac{dy}{y^n} = \int_a^\infty h(y\,e^t)^2 \frac{dy}{y} = \int_{ae^t}^\infty h(y)^2 \frac{dy}{y}.$$

Clearly, we can choose a so large that (5.19)' is less than ε for all $t \geq 0$. We choose next a in inequality (5.5) so large that the right side is less than ε. Combining this with the above estimate for (5.19)', and recalling $\bar{u} \equiv u^{(0)}$, we deduce that, for all $t > 0$,

$$\sum_{k=0}^\infty \int_a^\infty |u_t^{(k)}(y, t)|^2 \frac{dy}{y^n} < 2\varepsilon.$$

In view of (5.4) this implies inequality (3.2), with $N_j = F_j \times (a, \infty)$. This completes the proof of part (i) of Theorem 3.2.

6. The Energy Content of Spherical Caps

Our proof of part (ii) of Theorem 3.2 is based squarely on the notions and results of [4].

We start by mapping the disk D of the theorem onto the half-space $\beta_{n-1} \geq 0$; the spherical cap $\mathscr{C}$ is mapped thereby onto a wedge $W : \theta \leq \theta_0$, where

$$\theta = \arctan \frac{y}{x_{n-1}}, \qquad \rho = (x_{n-1}^2 + y^2)^{1/2}$$

are cylindrical coordinates. The rest of the x-variables we denote by $x' = (x_1, \cdots, x_{n-2})$. In terms of these coordinates $y = \rho \sin \theta$; so over the wedge W

the invariant L_2 norm (5.7)' and Dirichlet integral (5.7) take the form

$$(6.1) \qquad H_W(f) = \int_W |f|^2 \frac{dx' \, d\rho \, d\theta}{\rho^{n-1} \sin^n \theta},$$

$$(6.1)' \qquad D_W(f) = \int_W \left(f_\rho^2 + \frac{1}{\rho^2} f_\theta^2 + f_{x'}^2 \right) \frac{dx' \, d\rho \, d\theta}{\rho^{n-3} \sin^{n-2} \theta}.$$

We take now the *partial Fourier transform of f*, i.e., the non-Euclidean Fourier transform with respect to the variables x' and ρ (see equation (2.4) of [4]):

$$(6.2) \qquad \tilde{f}(\mu, \chi, \theta) = \iint \exp\{(\tfrac{1}{2}(n-2) - i\mu)\langle w', \chi \rangle\} f(x', \rho, \theta) \frac{dx' \, d\rho}{\rho^{n-1}},$$

here $\langle w', \chi \rangle$ denotes the non-Euclidean distance of $w_0 = (0, 1)$ in H^{n-1} to the horosphere through $w' = (x', \rho)$ and the point χ at infinity. The variable θ is merely a parameter.

According to the Plancherel relation (see (1.7) of [4]),

$$(6.3) \qquad \int |f|^2 \frac{d\rho \, dx'}{\rho^{n-1}} = \int |\tilde{f}|^2 \, dM(\mu, \chi),$$

and similarly (see (3.6) of [2])

$$(6.3)' \qquad \int (|f_\rho|^2 + |f_{x'}|^2) \frac{d\rho \, dx'}{\rho^{n-3}} = \int \mu^2 |\tilde{f}|^2 \, dM(\mu, \chi).$$

Dividing (6.3) by $\sin^n \theta$ and integrating with respect to θ we get, using (6.1),

$$(6.4) \qquad H_W(f) = \int_0^{\theta_0} \int |\tilde{f}|^2 \frac{dM \, d\theta}{\sin^n \theta}.$$

Similarly we get from (6.3)' and (6.1)'

$$(6.4)' \qquad D_W(f) \geq \int_0^{\theta_0} \int \mu^2 |\tilde{f}|^2 \frac{dM \, d\theta}{\sin^{n-2} \theta}.$$

The form of the Plancherel measure does not matter here; see however (2.5) of [4].

In these coordinates the spherical cap is the wedge $\theta \leq \theta_0$; θ_0 relates to the parameter y_0 in (3.5). To prove part (ii) of Theorem 3.2 we have to estimate the square integral of u_t over the wedge. Using (6.4) that integral can be written as

$$(6.5) \qquad \int_0^{\theta_0} \int |\tilde{u}_t(t)|^2 \frac{dM(\mu, \chi) \, d\theta}{\sin^n (\theta)}.$$

We divide the range of integration in (9.5) into two parts:

$$(6.6) \qquad (\mu\eta)^{-1} < \theta \quad \text{or} \quad (\mu\eta)^{-1} \geqq \theta,$$

η being an arbitrary positive parameter which later will be chosen suitably small.

LEMMA 6.1.   *Suppose f^Γ and Af^Γ belong to $\mathcal{H}_c$; then the solution u with initial data f^Γ satisfies, for all t and all η,*

$$(6.7) \qquad \iint\limits_{(\mu\eta)^{-1} < \theta < \theta_0} |\tilde{u}_t(t)|^2 \frac{dM\, d\theta}{\sin^n \theta} < c\eta^2.$$

Proof: We saw in the proof of Lemma 5.1 that if f^Γ and Af^Γ have finite energy, then

$$(6.8) \qquad D_F(u_t(t)) \leqq E \quad \text{for all} \quad t.$$

Obviously, $D_W(u_t) \leqq D_F(u_t)$. By calculus, $\sin\theta \geqq c\theta$ for all θ in $[0, \theta_0]$. Inserting this into the left side of (6.7) we deduce that

$$(6.8)' \qquad \iint\limits_{(\mu\eta)^{-1} < \theta < \theta_0} |\tilde{u}_t|^2 \frac{dM\, d\theta}{\sin^n \theta} \leqq \iint \left(\frac{\mu\eta}{c}\right)^2 |\tilde{u}_t|^2 \frac{dM\, d\theta}{\sin^{n-2} \theta}.$$

According to (6.4)$'$ the right side is at most $\eta^2 D_W(u_t)/c^2$; combining (6.8)$'$ with (6.8) gives (6.7). This completes the proof of Lemma 6.1.

To estimate the integral (6.5) over the second part of the range (6.6) we shall use the Volterra integral equation derived in [4] for any $\tilde{f}$ that is the partial Fourier transform of a function f square integrable in the wedge. That equation, (3.49) in [4], is

$$(6.9) \qquad k(r)\frac{\tilde{f}(\mu, \chi, \theta_r)}{\sin^{n/2} \theta_r} + \int_0^{\theta_r} K(\mu, r, \varphi)\frac{\tilde{f}(\mu, \chi, \varphi)}{\sin^{n/2} \varphi}\, d\varphi = r^{-1/2} l(\mu, x, r).$$

The variable r in this equation is the Euclidean radius of horospheres introduced in equation (3.18) of Section 3.

The function θ_r, introduced in [4], is

$$(6.10) \qquad \theta_r = 2 \arctan r.$$

The function $k(r)$ is defined in (3.46) of [4]:

$$(6.11) \qquad k(r) = \text{const.} \cos \tfrac{1}{2}\theta_r.$$

For $\theta_r \leqq \theta_0 < \pi$, $k(r)$ is bounded away from 0 and ∞.

The kernel K is given by (3.48) in [4]; and is denoted there by K_0.

The inhomogeneous term h in (3.49) of [4] is denoted in (6.9) by $r^{-1/2}l$. Since h is defined in [4] by equation (3.40) as $r^{-1/2}JIR\hat{f}$, it follows that

$$(6.12) \qquad l(\mu, \chi, r = JIR\hat{f}.$$

Here $\hat{f}$ is the Radon transform of f regarded as function of r; the operators R and I are defined in Section 3 of [4], and J is defined by (1.12), (1.13) of [4]. We have already noted in Section 3 that J is related to the operator P by (3.20).

We state now those properties of l and of K that are needed to complete the proof of Theorem 3.2:

LEMMA 6.2. *Let f^{Γ} be data belonging to $\mathcal{H}_c$ that satisfy*

$$(6.13) \qquad R_+^0 f^{\Gamma} = 0 \quad \text{for} \quad \beta \quad \text{in} \quad D \equiv \{\beta_{n-1} \geqq 0\}.$$

We denote as before

$$(6.14) \qquad U(t)f^{\Gamma} = \{f_1(t), f_2(t)\}$$

and set

$$f = f_2(t);$$

define l by (6.12). We claim that

$$(6.15) \qquad \text{(a)} \quad l \text{ is of the form } l(\mu, \chi, r e^{-t}),$$

$$(6.16) \qquad \text{(b)} \quad \int_0^{\infty} \int l^2(\mu, \chi, r) \, dM(\mu, \chi) \frac{dr}{r} < \infty.$$

LEMMA 6.3.

$$(6.17) \qquad |K(\mu, r, \theta)| \leqq c\mu(\mu\delta)^{(n-1)/2},$$

for

$$(6.18) \qquad 0 \leqq \theta \leqq \delta = 2 \arctan r \leqq \theta_0.$$

Before presenting the proof of these lemmata, we show that they imply the estimate for that part of the integral (6.5) that is not covereed by Lemma 6.1.

LEMMA 6.4. *Let f^{Γ} denote data in $\mathcal{H}_c$ that satisfy (6.13). Denote by u the solution with initial data f^{Γ}. Then, for any given η and $\theta_0 < \pi$,*

$$(6.19) \qquad \lim_{t \to \infty} \iint_{\theta \leqq \delta(\mu)} |\tilde{u}_t(t)|^2 \frac{dM(\mu, \chi) \, d\theta}{\sin^n \theta} = 0,$$

where

$$(6.19)' \qquad \delta(\mu) = \min \, (\mu\eta)^{-1}, \, \theta_0).$$

Proof of Lemma 6.4:   We rewrite the Volterra integral equation (6.9) regarding r as function of θ rather than the other way around:

$$(6.20) \quad k(r)\frac{\tilde{f}(\mu, \chi, \theta)}{\sin^{n/2} \theta} + \int_0^{\theta} K(\mu, r(\theta), \varphi)\frac{\tilde{f}(\mu, \chi, \varphi)}{\sin^{n/2} \varphi} \, d\varphi = (r(\theta))^{-1/2}l(\mu, \chi, r(\theta)),$$

where $r(\theta)$ is obtained by inverting (6.10):

(6.21) $$r(\theta) = \tan \tfrac{1}{2}\theta.$$

We now invert the integral equation (6.20). We shall make use of the following well-known result:

THEOREM 6.5. *Let*

(6.22) $$V = I + K$$

be a Volterra operator on the half-line, K an integral operator whose kernal $K(\theta, \varphi)$ satisfies the Volterra property:

$$K(\theta, \varphi) = 0 \quad for \quad \theta < \varphi.$$

We assume that K is uniformly bounded and denote

(6.23) $$\max_{0 \leqq \varphi \leqq \theta} K(\theta, \varphi) = K_{max}.$$

Then V is invertible on $(0, \delta)$ for any δ, and its $L^2(0, \delta)$ norm is bounded by

(6.24) $$\|(1 + K)^{-1}\| \leqq \exp\{\delta K_{max}\}.$$

We remind the reader that the proof rests on the Neumann series

$$(I + K)^{-1} = \sum_0^\infty (-K)^m,$$

and the estimate

$$\|K^m\| \leqq \frac{K_{max}^m \delta^m}{m!}.$$

Applying (6.24) to the operator (6.20), we get

(6.24)′ $$\int_0^\delta |\tilde{f}(\mu, \chi, \theta)|^2 \, d\theta \leqq c \exp\{\delta K_{max}\} \int_0^\delta |l(\mu, \chi, r)|^2 \frac{d\theta}{r}.$$

According to (6.17),

$$K_{max} \leqq c\mu(\mu\delta)^{(n-1)/2};$$

by (6.21), $d\theta \leqq c \, dr$. Inserting these estimates on the right in (6.24)′ gives

(6.25) $$\int_0^\delta |\tilde{f}(\mu, \chi, \theta)|^2 \, d\theta \leqq c \exp\{c(\mu\delta)^{(n+1)/2}\} \int_0^{r_0} |l(\mu, \chi, r)|^2 \frac{dr}{r},$$

where $r_0 = \tan \tfrac{1}{2}\theta_0$, the largest possible value of r.

We take δ in (6.25) to be $\delta(\mu)$ as defined in (6.19)', and we integrate with respect to $dM(\mu, \chi)$:

$$(6.25)' \quad \iint_{\theta \leq \delta(\mu)} |\tilde{f}(\mu, \chi, \theta)|^2 \, d\theta \, dM \leq c \exp\{c\eta^{-(n+1)/2}\} \int_0^{r_0} \int |l(\mu, \chi, r)|^2 \frac{dr}{r} \, dM.$$

Now we let f be $f_2(t)$. According to Lemma 6.2 the time dependence of l is of form (6.15); therefore we can write the right side of (6.25)' as

$$c \exp\{c\eta^{-(n+1)/2}\} \int_0^{r_0 \exp\{-t\}} \int |l(\mu, \chi, r)|^2 \frac{dr}{r} \, dM.$$

By (6.16), for fixed η and r_0, this tends to zero as $t \to \infty$, but then by (6.25), so does (6.19), as asserted in Lemma 6.4.

We can complete the proof of part (ii) of Theorem 3.2 by combining Lemmas 6.1 and 6.4. Given any $\varepsilon > 0$, we can choose η so small that the right side of (6.7) is less than ε. With this choice of η we take t so large that the integral, whose limit according to (6.19) is zero, is less than ε. Adding these inequalities gives us inequality (3.6) and completes the proof of Theorem 3.2(ii).

We turn now to the proof of Lemma 6.2. We start with relation (3.13):

$$\tfrac{1}{2}k_-(s+t) = P\hat{f}_2(t).$$

We introduce the variable r in place of s by equation (3.19). As already remarked in (3.20) of Section 3, in terms of the variable r the action of P is expressed by the operator J. So we obtain

$$\tfrac{1}{2}k_-(t - \log r + c(\beta)) = c_n(1 + |\beta|^2)^{(n-1)/2} J\hat{f}_2(t).$$

This shows that $J\hat{f}_2(t)$ is a function of $r e^{-t}$ for each β.

In (6.12) we have defined l as

$$l = JIR\hat{f}_2(t);$$

as observed in Step 2 of Section 3 of [4], the operators J and I commute, and the operators J and R transmute according to equation (3.27) of [4]. Making use of these facts we obtain

$$(6.26) \qquad\qquad l = I\lambda^{(n-1)/2} RJ\hat{f}_2(t).$$

We have already shown that $J\hat{f}_2(t)$ is a function of $r e^{-t}$. None of the operators R, $\lambda^{(n-1)/2}$ or I alter this property. This proves part (a) of Lemma 6.2.

The proof of part (b) is based on inequality (3.17), with t set equal to 0 and k_- replaced according to (3.13) by $2P\hat{f}_2$:

$$\int |P\hat{f}_2|^2 \, ds \, dm(\beta) \leq \int_F |f_2|^2 \, dV \leq E_F(f^\Gamma).$$

We switch to r as variable of integration in place of s. From (3.19),

(6.27)
$$ds = \frac{dr}{r};$$

using (3.20) we get

$$c_n^2 \int |J\hat{f}_2|^2 (1 + |\beta|^2)^{n-1} \, dm(\beta) \frac{dr}{r} \leqq E_F(f^\Gamma).$$

It is easy to show (see (1.10) for the case $n = 3$) that

(6.27)′
$$c_n^2 (1 + |\beta|^2)^{n-1} \, dm(\beta) = d\beta = d\beta_1 \cdots d\beta_{n-1};$$

using this on the left above gives

$$\int |J\hat{f}_2|^2 \, d\beta_1 \cdots d\beta_{n-1} \frac{dr}{r} \leqq E_F(f^\Gamma).$$

We proceed now exactly as in Step 2 of Section 3 of [4], which there led to inequality (3.30) and here leads to inequality (6.16). This completes the proof of Lemma 6.2.

We turn now to proving Lemma 6.3; we shall closely follow Step 3 in Section 3 of [4]. We take first the case when n is odd.

The kernel K is defined by formula (3.48) as

(6.28)
$$K(\mu, r, \theta) = (r \sin \theta)^{1/2} \frac{2\alpha^2}{\alpha^2 - 1} K'_{(n-3)/2}(\mu, \alpha).$$

Here α is a function of r and θ defined by (3.31) and (3.32) of [4] as

(6.29)
$$\tau = \cos \theta + r \sin \theta, \qquad \alpha = \tau + (\tau^2 - 1)^{1/2},$$

(6.29)′
$$\tan \tfrac{1}{2}\theta \leqq r \leqq \tan \tfrac{1}{2}\theta_0.$$

The symbol ′ denotes differentiation with respect to α; the function $K_{(n-3)/2}$ is defined recursively on page 548 of [4] by

(6.30)
$$(\tfrac{1}{2}(n-3) - j) K_j + \frac{\alpha^2 - 1}{2 + \alpha + \alpha^{-1}} K'_j = K_{j+1},$$

$j = 0, 1, \cdots, \tfrac{1}{2}(n-3)$, starting with K_0 defined by (3.36) of [4]:

(6.30)′
$$K_0(\mu, \alpha) = \left(1 + \frac{\alpha + \alpha^{-1}}{2}\right)^{(n-3)/2}$$
$$\times \int_0^\pi \left(\frac{2\alpha}{1 + \alpha^2 + (1 - \alpha^2) \cos \varphi}\right)^{(n-2)/2 + i\mu} (\sin \varphi)^{n-3} \, d\varphi.$$

LEMMA 6.6.   *Denote by δ the quantity defined in* (6.18):

$$\delta = 2 \arctan r.$$

We claim that in the range (6.29)', *and* $j = 1, \cdots, \frac{1}{2}(n-3)$,

(6.31) $$|K_j(\mu, \alpha)| \leqq c(\mu\delta)^j,$$

(6.32) $$K_j'(\mu, 1) = 0.$$

Proof: We do an induction on j. Certainly (6.31) holds for $j = 0$, since K_0 as defined by (6.30)' is a continuous function of α, and α is bounded in the range (6.29)'.

It follows from (6.29) and (6.18) that

(6.33) $$\alpha - 1 \leqq c(\tau - 1)^{1/2} \leqq c(\theta^2 + r\theta)^{1/2} \leqq c\delta.$$

Formula (6.30)' shows that the dependence of K_0 on μ is imaginary exponential; differentiation with respect to α brings in a factor μ. In the recursion relation (6.30) differentiation with respect to α is combined with multiplication by $\alpha - 1$. In view of the estimate (6.33) we see that each factor μ introduced is accompanied by a factor δ. This proves (6.31).

As already noted in [4], an explicit calculation shows that (6.32) holds for $j = 0$, and follows for all j by induction in (6.30).

We turn now to (6.28). By the mean value theorem and (6.32),

$$\frac{1}{\alpha - 1} K'_{(n-3)/2}(\mu, \alpha) = K''_{(n-3)/2}(\mu, \bar{\alpha})$$

for some $\bar{\alpha}$ between α and 1. Since each differentiation brings in a factor of μ, we see from (6.31) that

$$\frac{1}{\alpha - 1} K'_{(n-3)/2}(\mu, \alpha) \leqq c(\mu\delta)^{(n-3)/2} \mu^2.$$

The first factor $(r \sin \theta)^{1/2}$ on the right in (6.28) is bounded by δ; therefore we conclude from (6.28) that

$$K(\mu, r, \theta) \leqq c\mu(\mu\delta)^{(n-1)/2},$$

as asserted in (6.17). This completes the proof for n odd. The even case goes analogously (see pages 551–553 of [4]).

7. The Energy Content of Cusps of Less than Maximal Rank

In this section we prove part (iii) of Theorem 3.2. We start, as in Section 5, by mapping the cusp into ∞. The subgroup Γ_∞ leaving ∞ fixed is a crystallographic group of motions of $\mathbb{R}^{n-1}$, the assumption that the cusp is of less than maximal rank means that Γ_∞ has no compact fundamental polyhedron in $\mathbb{R}^{n-1}$. According to the theory of crystallographic groups, in this case Γ_∞ has a fundamental polyhedron F_∞ of the form

$$F_\infty = F_c \times \mathbb{R}^{m-1},$$

where F_c is a compact set in some l-dimensional subspace $\mathbb{R}^l$ of $\mathbb{R}^{n-1}$, and $\mathbb{R}^{m-1}$ is the product of $m-1$ lines or half-lines. In all cases, the cusp is contained in

$$(7.1) \qquad F_c \times \mathbb{R}^{m-1} \times \mathbb{R}_+.$$

Conversely, if we remove the set $|x|^2 + y^2 < b$, b large enough, the remaining points of the set (7.1) are contained in a finite number of replicas of F.

We introduce x-coordinates so that F_c is contained in $(x_1, \cdots, x_l) = x'$-space, and $\mathbb{R}^{m-1}$ is the $(x_{l+1}, \cdots, x_{n-1}) = x''$-space.

The neighborhoods V of ∞ that appear in (3.7) will be taken to be of the form

$$V(b) = F_c \times \{|x''|^2 + y^2 > b^2\}.$$

As noted in Section 3, in the proof of part (iii) of Theorem 3.2, we may assume that the initial data f^Γ of the solution u is of the form $f^\Gamma = Ag^\Gamma$, where $g^\Gamma \in \mathcal{H}_c$ is in the domain of A^2. The theorem asserts that for any $\epsilon > 0$ we can choose b so large that, for all t,

$$(7.2) \qquad \int_{V(b)} |u_t(t)|^2 \, dV < \varepsilon.$$

Our proof of (7.2) contains no new ideas essentially different from the ones used in Sections 5 and 6, but they are mixed together more elaborately. We begin by breaking up $V(b)$ into two parts:

$$(7.3) \qquad V(b) = V^a(b) \cup V_a(b),$$

where

$$(7.3)^a \qquad V^a(b) = V(b) \cap \{y > a\}$$

and

$$(7.3)_a \qquad V_a(b) = V(b) \cap \{y \leq a\}.$$

We shall estimate separately the integral (7.2) over $V^a(b)$ and over $V_a(b)$.

PROPOSITION 7.1. *For u as above and any $\varepsilon > 0$ we can choose a and b so that, for all t,*

$$(7.4) \qquad \int_{V^a(b)} |u_t(t)|^2 \, dV \leq \varepsilon.$$

To prove this we introduce, as we did in Section 5, the eigenfunctions ψ_k and the eigenvalues ω_k of the negative of the Euclidean-Laplace operator over F_c, automorphic with respect to Γ_∞. Note that

(7.5) $$0 = \omega_0 < \omega_1 \leq \omega_2 \cdots$$

and that

(7.5)' $$\psi_0 \equiv \text{cost.}$$

Moreover, $u(x, y, t)$ can be expanded in a Fourier series:

(7.6) $$u(x', x'', y, t) = \sum_0^\infty u^{(k)}(x'', y, t)\psi_k(x').$$

Differentiating both sides with respect to t and then using the Parseval relation gives

$$\int_{F_c} |u_t(x', x'', y, t)|^2 \, dx' = \sum_0^\infty |u_t^{(k)}(x'', y, t)|^2.$$

Integrating this with respect to $dx'' dy/y^n$ over $\mathbb{R}^{m-1} \times (a, \infty)$ gives

(7.7) $$\int_a^\infty \int_{\mathbb{R}^{m-1} \times F_c} |u_t(x, y, t)|^2 \, dx \frac{dy}{y^n} = \sum_{k=0}^\infty \int_a^\infty \int_{\mathbb{R}^{m-1}} |u_t^{(k)}(x'', y, t)|^2 \, dx'' \frac{dy}{y^n}.$$

It follows from definitions (7.2) and (7.3) that, for $b > a$,

$$V^a(b) \subset F_c \times \mathbb{R}^{m-1} \times (a, \infty).$$

Hence, the left side of (7.7) is $\geq$ the left side of (7.4). Therefore to prove (7.4) it suffices to show that the right side of (7.7) is $\leq$ the right side of (7.4).

Just as in Section 5, the component $k = 0$ has to be treated differently from the others. For $k \geq 1$ we claim

LEMMA 7.2.   *For u as above,*

(7.8) $$\sum_1^\infty \int_a^\infty \int_{\mathbb{R}^{m-1}} |u_t^{(k)}(x'', y, t)|^2 \frac{dx'' \, dy}{y^n} \leq \frac{c}{a^2}$$

for all a and all t.

Proof: Proceeding as in the proof of Lemma 5.1 and using the fact that the initial data of u belong to the domain of the generator, we can derive the analogue of inequality (5.8)'':

$$\sum \omega_k \int_a^\infty \int |u_t^{(k)}(x'', y, t)|^2 \, dx'' \frac{dy}{y^{n-2}} \leq E.$$

Since, for $k \geq 1$, $\omega_k \geq \omega_1 > 0$, and since $y > a$ in the range of integration, (7.8) follows with

$$c = E/\omega_1.$$

We turn now to the zeroth component; since $\psi_0 \equiv$ const.,

$$(7.9) \qquad u^{(0)}(x'', y, t) = \frac{1}{|F_c|^{1/2}} \int_{F_c} u(x, y, t)\, dx'.$$

LEMMA 7.3. *For u as above and any $\varepsilon > 0$, we can choose a so that, for all t,*

$$(7.10) \qquad \int_a^\infty \int_{\mathbb{R}^{m-1}} |u_t^{(0)}(x'', y, t)|^2\, dx'' \frac{dy}{y^n} < \varepsilon.$$

Proof: The function u is a solution of the non-Euclidean wave equation:

$$u_{tt} - y^2(\Delta + \partial_y^2)u + (n-2)yu_y - (\tfrac{1}{2}(n-1))^2 u = 0,$$

where Δ is the Euclidean-Laplace operator with respect to the x variables; so

$$\Delta = \Delta' + \Delta''.$$

Integrating the wave equation over F_c we obtain

$$(7.11) \qquad u_{tt}^{(0)} - y^2(\Delta'' + \partial y^2)u^{(0)} + (n-2)yu_y^{(0)} - (\tfrac{1}{2}(n-1))^2 u^{(0)} = 0.$$

We define now v by

$$(7.12) \qquad u^{(0)} = y^{(n-m)/2} v.$$

Setting this into (7.11) shows that v satisfies the m-dimensional wave equation

$$(7.11)' \qquad v_{tt} - y^2(\Delta'' + \partial_y^2)v + (m-2)yv_y - (\tfrac{1}{2}(m-1))^2 v = 0.$$

Let S be any set in $\mathbb{H}^m$; it follows from the definition (7.12) of v that

$$(7.13) \qquad \int_S |v_t|^2 \frac{dx''\, dy}{y^m} = \int_S |u_t^{(0)}|^2 \frac{dx''\, dy}{y^n}.$$

From the definition (7.9) of $u^{(0)}$ as the zero component of u, we deduce that the right side of (7.13) is less than or equal to

$$\int_{S \times F_c} |u_t|^2 \frac{dx\, dy}{y^n}.$$

In the case the set $S \times F_c$ is contained in A replicas of the fundamental polyhedron F, it follows from (7.12) that

$$(7.13)' \qquad \int_S |v_t|^2 \frac{dx''\, dy}{y^m} \leqq AE_F(u).$$

In what follows we shall take S to be a wedge, specifically sets of the form

$$(7.14) \qquad a \leqq x_{n-1} + y \quad \text{or} \quad x_{n-1} - y \leqq -a.$$

We shall denote either of these wedges by W, or by $W_+(a)$ and $W_-(a)$, respectively, when we wish to emphasize their dependence on the parameter.

It is obvious that the union of the two wedges $W_+(a)$ and $W_-(a)$ includes the set

$$\mathbb{R}^{m-1} \times (a, \infty).$$

Therefore inequality (7.10) of Lemma 7.3 would follow from (7.13) if we could show that, for all t,

$$(7.15) \qquad \int_W |v_t(t)|^2 \frac{dx'' \, dy}{y^m} < \varepsilon$$

for both of the wedges (7.14), whenever a is chosen large enough.

Section 6 was devoted to estimating integrals of the form (7.15) over wedges; we shall mimic the technique used there. First we need

LEMMA 7.4.   *Choose a_0 so large·that*

$$(7.16) \qquad W(a_0) \times F_c \subset F \cup F',$$

where F' denotes the reflection of F through $x_{n-1} = 0$. Then, for $a \geqq a_0$,

$$(7.17) \qquad H_W(v_t) \leqq 2 H_F(u_t)$$

and

$$(7.17)' \qquad D_W(v_t) \leqq 4 D_F(u_t) + \text{const.} \, H_F(u_t).$$

Proof: The inequality (7.17) follows from (7.13)' upon setting $S = W(a_0)$ and $A = 2$. To prove (7.17)', we start with the observation that since $u^{(0)}$ is the mean value of u (see (7.9)), we have

$$(7.18) \qquad \begin{aligned} \int_W (|u^{(0)}_{tx''}|^2 + |u^{(0)}_{ty}|^2) \frac{dx'' \, dy}{y^{n-2}} &\leqq \int_{W \times F_c} (|u_{tx''}|^2 + |u_{ty}|^2) \frac{dx \, dy}{y^{n-2}} \\ &\leqq D_{W \times F_c}(u_t) \leqq 2 D_F(u_t); \end{aligned}$$

here in the last inequality we have used (7.16). Using the definition (7.12) of v we can rewrite the left side of (7.18) as

$$(7.19) \qquad \int_W \left(|v_{tx''}|^2 + \left| v_{ty} + \tfrac{1}{2}(n-m) \frac{v_t}{y} \right|^2 \right) \frac{dx'' \, dy}{y^{m-2}}.$$

The second term can be estimated by a difference of squares:

$$(7.20) \qquad \left| v_{ty} + \tfrac{1}{2}(n-m) \frac{vt}{y} \right|^2 \geqq \tfrac{1}{2} v_{ty}^2 - (\tfrac{1}{2}(n-m))^2 \frac{v_t^2}{y^2}.$$

Combining (7.19) and (7.20) together with (7.17) gives (7.17)'.

COROLLARY 7.5. *Suppose both the square integral and the Dirichlet integral of u are finite over F. Then, for $a \geqq a_0$,*

$$(7.21) \qquad H_W(v) \leqq 2H_F(u)$$

and

$$(7.21)' \qquad D_W(v) \leqq 4D_F(u) + \text{const. } H_F(u).$$

In what follows, $a \geqq a_0$. In working with the wedge W_+ (or W_-) it is convenient to extend $v(x'', y, t)$ over the exterior of the wedge so that the extended function is of finite energy:

For (x'', y) in $W_+(a_0)$ we take v as defined in (7.12). In the complement of $W_+(a_0)$ we define v by horizontal reflection; i.e., for $x_{n-1} < a_0 - y$ we set

$$(7.22) \qquad v(x''', x_{n-1}, y) = v(x''', 2a_0 - x_{n-1} - 2y, y),$$

where $x''' = (x_{l+1}, \cdots, x_{n-2})$. The Jacobian of this transformation is 1.

The reason for requiring f^Γ to be of the form $f^\Gamma = Ag^\Gamma$ was to insure the square integrability of $u(t)$ in F. In fact,

$$U(t)f^\Gamma = U(t)Ag^\Gamma = AU(t)g^\Gamma$$

shows that $u(t)$ is the second component of $U(t)g^\Gamma$. As a consequence

$$(7.23) \qquad H_F(u(t)) \leqq E_F(U(t)g^\Gamma) = E_F(g^\Gamma)$$

and

$$(7.24) \qquad \begin{aligned} D_F(u(t)) &\leqq E_F(U(t)f^\Gamma) + (\tfrac{1}{2}(n-1))^{1/2} H_F(u(t)) \\ &\leqq E_F(f^\Gamma) + (\tfrac{1}{2}(n-1))^{1/2} E(g^\Gamma). \end{aligned}$$

It follows dierectly from (7.22) that the square integral and the Dirichlet integral of the extended v and v_t over the complement of $W_+(a_0)$ are bounded by a constant multiple of the corresponding integrals over $W_+(a_0)$. This together with (7.23), (7.24), Lemma 7.4 and Corollary 7.5 implies that

$$(7.25) \qquad H(v), H(v_t), D(v), D(v_t) \leqq C$$

for all t, C some constant. The unsubscripted H and D mean that the integration is extended over all of $\mathbb{H}^m$.

For (x'', y) in $W_+(a_0)$ and for all t, v satisfies the m-dimensional non-Euclidean wave equation:

$$v_{tt} - Lv = 0.$$

Consider now all horospheres which are contained in $W_+(a_0)$; these are characterized by an inequality of the form

$$(7.26) \qquad s > a_0 + s_0(\beta), \qquad \beta_{n-1} > a_0,$$

s_0 some function. Take the Radon transform of the wave equation for horospheres satisfying (7.26):

$$\hat{v}_{tt} - \widehat{Lv} = 0.$$

This can be turned into the one-dimensional wave equation by applying the operator P and using the identity

$$P\widehat{Lv} = \partial_s^2 P\hat{v}.$$

We get

(7.26)′ $$(P\hat{v})_{tt} - (P\hat{v})_{ss} = 0,$$

valid when (7.26) holds. We can factor this wave equation in two different ways:

(7.27)$_+$ $$(\partial_t + \partial_s)(P\hat{v}_t - \partial_s P\hat{v}) = 0$$

and

(7.27)$_-$ $$(\partial_t - \partial_s)(P\hat{v}_t + \partial_s P\hat{v}) = 0.$$

We define now the translation representations of $\{v, v_t\}$ (see equation (1.8) for $n = 3$):

(7.28)$_\pm$ $$k_\pm(s, \beta, t) = \partial_s P\hat{v} \mp P\hat{v}_t.$$

It follows from (7.27)$_\pm$ and (7.28)$_\pm$ that there are two functions h_+ and h_- such that

(7.29)$_+$ $$k_+(s, \beta, t) = h_+(s - t, \beta),$$

and

(7.29)$_-$ $$k_-(s, \beta, t) = h_-(s + t, \beta)$$

for all t, and for all s satisfying (7.26).

According to the translation representation theorem (see Theorem 1.1) which holds for all initial data on $\mathbf{H}^m$,

$$\int_B \int_\mathbf{R} k_\pm^2(s, \beta, t)\, ds\, dm(\beta) = E(\{v, v_t\}).$$

Since the energy $E(\{v, v_t\}) \leqq D(v) + H(v_t)$, it follows from (7.25) that, for all t,

$$\iint k_\pm^2(s, \beta, t)\, ds\, dm(\beta) \leqq C.$$

Using (7.29)$_+$ we deduce

$$\int_{a_0 < \beta_{n-1}} \int_{a_0 + s_0(\beta)}^\infty h_+^2(s - t, \beta)\, ds\, dm(\beta) \leqq C.$$

Introducing $s - t$ as new variable of integration, we conclude, letting $t \to \infty$, that

$$(7.30)_+ \qquad \int_{a_0 < \beta_{n-1}} \int_{\mathbf{R}} h_+^2(s, \beta) \, ds \, dm(\beta) \leqq C.$$

Similarly, letting $t \to -\infty$ we see that

$$(7.30)_- \qquad \int_{a_0 < \beta_{n-1}} \int_{\mathbf{R}} h_-^2(s, \beta) \, ds \, dm(\beta) \leqq C.$$

It follows from $(7.30)_\pm$ that, given any $\zeta > 0$, we can choose a so large that

$$(7.31) \qquad \int_{a < \beta_{n-1}} \int_{\mathbf{R}} h_\pm^2(s, \beta) \, ds \, dm(\beta) < \zeta.$$

Having fixed a, we introduce polar coordinates, as in Section 6:

$$(7.32) \qquad \tan \theta = y/(x_{n-1} - a); \qquad \rho^2 = (x_{n-1} - a)^2 + y^2.$$

We denote by $\tilde{f}$ the partial Fourier transform of f (see (6.2)) with n replaced by m. The Plancheral relations (6.4) and (6.4)' are satisfied, for $f = v_t$:

$$(7.33) \qquad \int_W |v_t|^2 \frac{dx'' \, dy}{y^m} = \int\!\!\!\int_0^{3\pi/4} |\tilde{v}_t(t)|^2 \frac{dM \, d\theta}{\sin^m \theta},$$

and

$$(7.34) \qquad D_W(v_t) \geqq \int\!\!\!\int |\mu \tilde{v}_t(t)|^2 \frac{dM \, d\theta}{\sin^{m-2} \theta}.$$

The following analogue of Lemma 6.1 holds.

LEMMA 7.6. *Let η be an arbitrary positive number; then*

$$(7.35) \qquad \int\!\!\!\int_{(\mu\eta)^{-1} < \theta < 3\pi/4} \tilde{v}_t^2(t) \frac{dM \, d\theta}{\sin^m \theta} < c\eta^2$$

for all t; here c is a constant.

The proof is the same as that of Lemma 6.1; the analogue of (6.8) is obtained by combining (7.34) with (7.25).

To estimate the integral of $\tilde{v}_t^2$ over the remaining portion of W we proceed as in Section 6: we use the Volterra integral equation satisfied by $\tilde{v}_t$ (see (6.20)):

$$(7.36) \qquad k(r) \frac{\tilde{v}_t(\theta)}{\sin^{m/2} \theta} + \int_0^\theta K(r, \varphi) \frac{\tilde{v}_t(\varphi)}{\sin^{m/2} \varphi} \, d\varphi = r^{-1/2} l(r).$$

Here r is a function of θ, given by (6.21). The kernel K, $\tilde{v}_t$ and l depend on μ; in addition $\tilde{v}_t$ and l depend on t. The function l is expressed in terms of the Radon transform of v_t (see (6.12)):

$$l = JIR\hat{v}_t,$$

where $\hat{v}_t$ is regarded as function of r, the radius of the horosphere, in place of s.

LEMMA 7.7. *For a so chosen that* (7.31) *holds*,

$$(7.37) \qquad \int\limits_0^\infty\!\!\int l^2(\mu, \chi, r)\, dM \frac{dr}{r} < O(\zeta).$$

Before giving the proof we note that, when combined with Lemma 6.3 and Theorem 6.5, Lemma 7.7 leads to the following estimate for all t:

$$(7.38) \qquad \iint\limits_{\theta \leq \delta(\mu)} |\tilde{v}_t(t)|^2 \frac{dM\, d\theta}{\sin^m \theta} \leq c\zeta \exp\{c\eta^{-(m+1)/2}\},$$

$$\delta(\mu) = \min((\mu\eta)^{-1}, \tfrac{3}{4}\pi).$$

The proof of (7.38) is the same as that of Lemma 6.4; inequality (7.38) is the same as (6.25)′ combined with estimate (7.37).

To prove Lemma 7.7 we use the expression (6.26) for l:

$$(7.39) \qquad l = I\lambda^{(m-1)/2} RJ\hat{v}_t.$$

By (3.20),

$$(7.40) \qquad c_m(1+|\beta|^2)^{(m-1)/2} J\hat{v}_t = P\hat{v}_t.$$

Using $(7.28)_\pm$ we can write

$$(7.41) \qquad P\hat{v}_t = \tfrac{1}{2}k_-(s, \beta, t) - \tfrac{1}{2}k_+(s, \beta, t).$$

From the definition of the operator R (see (3.1)′ of [4]) it follows that, for

$$r(\theta) = \tan\tfrac{1}{2}\theta \leq \tan\tfrac{3}{8}\pi,$$

the horospheres that figure in l all lie in the wedge W. Therefore we may use $(7.29)_\pm$ on the right in (7.41):

$$(7.42) \qquad P\hat{v}_t = \tfrac{1}{2}h_-(s+t, \beta) - \tfrac{1}{2}h_+(s-t, \beta) \equiv k(s, \beta, t).$$

Using (7.40) and (7.42), and setting these into (7.39) we get the following representation for l:

$$(7.43) \qquad l = cI\lambda^{(m-1)/2}g,$$

343

where

$$g(\lambda r, \beta_\lambda, t) = (1 + |\beta_\lambda|^2)^{-(m-1)/2} k(s_\lambda, \beta_\lambda, t),$$

$$\beta_\lambda = \{\beta_{l+1}, \cdots, \beta_{n-2}, \lambda + a_0\} \quad \text{and} \quad s_\lambda = \log((1 + |\beta_\lambda|^2)/2\lambda r),$$

since R transforms s into s_λ and β into β_λ. As we have seen in (6.27) and (6.27)′,

$$dm(\beta) = c \frac{d\beta}{(1 + |\beta|^2)^{m-1}}$$

and

$$ds = -\frac{dr}{r}.$$

Changing the order of integration and making the substitutions $r \to \rho/\lambda$ and $\lambda \to \beta_{n-1}$ (this essentially inverts R) we deduce from these relations and (7.31) that

$$\int_0^{r(\theta_0)} \int_0^\infty \int |\lambda^{(m-1)/2} g|^2 \, d\beta_{l+1} \, d\beta_{l+2} \cdots d\beta_{n-2} \frac{d\lambda}{\lambda^{m-1}} \frac{dr}{r}$$

(7.44)
$$= \int_{\beta_{n-1} > a_0} \int |g(\rho, \beta, t)|^2 \frac{d\rho}{\rho} d\beta$$

$$= \int_{\beta_{n-1} > a_0} \int |h_-(s + t, \beta) - h_+(s - t, \beta)|^2 \, ds \, dm(\beta)$$

$$\leq \text{const. } \zeta.$$

The range of the inner integrals in the second and third expressions in (7.44) are $(0, (\beta_{n-1} - a_0)r(\theta_0))$ and $s > a_0 + s_0(\beta)$, respectively. Using the Plancherel relation (6.3), we deduce from (7.43) and (7.44) that (7.37) holds. This completes the proof of Lemma 7.7.

Given any $\varepsilon > 0$, we choose now η so small that $c\eta^2$ appearing on the right in (7.35) is less than ε. Then we choose ζ in (7.31) so small that the right side of (7.38) is less than ε; adding (7.35) and (7.38) we conclude that

$$\int\int_0^{3\pi/4} |\tilde{v}_t(t)|^2 \frac{dM \, d\theta}{\sin^m \theta} < 2\varepsilon.$$

This together with (7.33) implies that (7.15) holds. Inserting this into (7.13) with $S = W_\pm(a)$, we conclude that, for all t,

$$\int_{W_- \cup W_+} |u_t^{(0)}(t)|^2 \frac{dx'' \, dy}{y^n} < 4\varepsilon.$$

Since $W_- \cup W_+$ contains $\mathbb{R}^{m-1} \times (a, \infty)$, inequality (7.10) of Lemma 7.3 follows.

We now choose a so large that the right side of (7.8) is less than ε and (7.10) holds; adding (7.8) and (7.10) we obtain (7.4). This proves Proposition 7.1.

We turn now to estimating the square integral of u_t over $V_a(b)$.

PROPOSITION 7.8.   *For u as above, and any ε, $a > 0$, we can choose b so large that, for all t,*

$$\tag{7.45} \int_{V_a(b)} |u_t(t)|^2 \, dV < \varepsilon.$$

Proof: Let φ denote a smooth cutoff such that

$$\tag{7.46} \varphi(p) = \begin{cases} 1 & \text{for } p > \tfrac{1}{2}b, \\ 0 & \text{for } p < \tfrac{1}{2}b - 1. \end{cases}$$

For $b > \sqrt{\tfrac{4}{3}}a$, the function $\varphi(|x''|)$ equals 1 in $V_a(b)$. Therefore,

$$\tag{7.47} \int_{V_a(b)} |u_t(t)|^2 \, dV \leq H_S(\varphi u_t),$$

where S denotes the slab

$$\tag{7.47\('\)} S = F_c \times \mathbb{R}^{m-1} \times (0, a).$$

Since φ is smooth, $H_S(\varphi u_t)$ and $D_S(\varphi u_t)$ are bounded in terms of $H_F(u_t)$ and $D_F(u_t)$:

$$H_S(\varphi u_t) + D_S(\varphi u_t) \leq c(H_F(u_t) + D_F(u_t)),$$

c independent of b.

We assume now — an assumption that will be removed later — that F_c is a parallelogram. Then the eigenfunctions of Δ' over F_c are exponentials $\exp\{i\alpha \cdot x'\}$; the corresponding eigenvalue is $-|\alpha|^2 = -\omega$.

We define the vector ζ_α by

$$\tag{7.48} \zeta_\alpha = \{\alpha, \xi\}, \qquad\qquad\qquad \xi \in \mathbb{R}^{m-1},$$

and the Fourier transform of any function $v(x)$ by

$$\tag{7.49} \tilde{v}(\zeta_\alpha) = \int_{F_c \times \mathbb{R}^{m-1}} v(x) \exp\{-ix \cdot \zeta_\alpha\} \, dx.$$

We take in particular $v = \varphi u_t$; the Plancherel formula gives

$$\tag{7.50} H_S(\varphi u_t) = \sum_\alpha \int_0^a \!\!\! \int |\widetilde{\varphi u_t}(\zeta_\alpha, y)|^2 \, d\xi \frac{dy}{y^n},$$

$$\tag{7.50\('\)} D_S(\varphi u_t) \geq \sum_\alpha \int_0^a \!\!\! \int |\zeta_\alpha|^2 |\widetilde{\varphi u_t}(\zeta_\alpha, y)|^2 \, d\xi \frac{dy}{y^{n-2}}.$$

We wish to estimate $H_S(\varphi u_t)$; to this end we break the domain of integration on the right of (7.50) into two parts.

345

LEMMA 7.9. *For u as above, and any η, b and t,*

$$(7.51) \qquad \sum_\alpha \iint\limits_{(|\zeta_\alpha|\eta)^{-1}<y<a} |\widehat{\varphi u_t}|^2 \, d\xi \frac{dy}{y^n} < c\eta^2.$$

The proof is the same as that of Lemma 6.1; the role of (6.4)′ is taken by the analogous inequality (7.50)′.

We turn now to the remaining part of the integral.

LEMMA 7.10. *For u as above, given any t, η and ε, we can choose b′ so that, for some $b < b'$,*

$$(7.52) \qquad \sum_\alpha \iint\limits_{y \leq \delta_\alpha(\xi)} |\widehat{\varphi u_t}|^2 \, d\xi \frac{dy}{y^n} < \varepsilon,$$

where

$$(7.53) \qquad \delta_\alpha(\xi) = \min\left((|\zeta_\alpha|\eta)^{-1}, a\right).$$

This is an analogue of Lemma 6.4, and has an analogous proof, with some new twists, incorporated in two lemmata:

LEMMA 7.11. *Let $f(x, y)$ be a square integrable function in S, automorphic with respect to Γ_∞. Denote by $\tilde{f}$ the Fourier transform of f:*

$$(7.54) \qquad \tilde{f}(\zeta_\alpha, y) = \int_{\mathbf{R}^{m-1}} \int_{F_c} f(x, y) e^{ix \cdot \zeta_\alpha} \, dx' \, dx'';$$

then $\tilde{f}$ satisfy a Volterra integral equation of the form

$$(7.55) \qquad \frac{\tilde{f}(\zeta, y)}{y^{n/2}} + \int_0^y K(|\zeta|, y, z)\frac{\tilde{f}(\zeta, z)}{z^{n/2}} \, dz = cy^{-1/2}l(\zeta, \tfrac{1}{2}y).$$

Here the kernal K is continuous and satisfies the inequality

$$(7.56) \qquad |K(|\zeta|, y, z)| \leq c|\zeta|(|\zeta|y)^{(n-1)/2}$$

and l has the form

$$(7.57) \qquad l = JF\hat{f},$$

where $\hat{f}(r, \beta)$ is the Radon transform of f, parametrized by the Euclidean radius r (see (3.19)), F is the Fourier transform with respect to β, and J is the operator in (3.20), with the identification

$$y = 2r.$$

Proof: The integral equation (7.55) was derived in [2] on pages 658–659 (equation (3.63)) in the case when $n = 3$ and Γ_∞ trivial. As Alex Woo (see Section 5 of [7]) has shown, an adaptation of this derivation also works in the general case.

The estimate (7.56) is entirely analogous to inequality (6.17); the proof given after Lemma 6.6 carries over to the present case.

We combine now Theorem 6.5 and inequality (7.56) to obtain the following estimate for the inverse of $I + K$ in the $L_2(0, \delta)$ norm:

$$\|(I + K)^{-1}\| \leq \exp \{c(|\zeta|\delta)^{(n+1)/2}\}.$$

In particular it follows that

$$(7.58) \qquad \int_0^\delta |\tilde{f}(\zeta, y)|^2 \frac{dy}{y^n} \leq \exp \{c(|\zeta|\delta)^{(n+1)/2}\} \int_0^{\delta/2} |l(\zeta, r)|^2 \frac{dr}{r},$$

where l is defined by (7.57).

LEMMA 7.12. *Given any $\gamma > 0$ and t, we can choose b' so that, for some $b < b'$,*

$$(7.59) \qquad \|JF\widehat{\varphi u}_t(t)\|^2 < \gamma,$$

where the norm is the L_2 norm:

$$(7.59)' \qquad \|l\|^2 = \sum_\alpha \int_0^{a/2} \int_{\mathbf{R}^{m-1}} |l(\zeta_\alpha, r)|^2 \, d\xi \frac{dr}{r},$$

ζ_α the vector defined in (7.48).

Proof: Since J commutes with F, and F is isometric, the left side of (7.59) equals

$$(7.60) \qquad \int_0^{a/2} \int_{F_c \times \mathbf{R}^{m-1}} |J\widehat{\varphi u}_t(t)|^2 \, d\beta \frac{dr}{r}.$$

To estimate (7.60) we break up the range of the β'' integration into three parts:

$$B_1(b): |\beta''| < \tfrac{1}{2}b - \tfrac{1}{2}a - 1,$$
$$(7.61) \qquad B_2(b): \tfrac{1}{2}b - \tfrac{1}{2}a - 1 \leq |\beta''| < \tfrac{1}{2}b + \tfrac{1}{2}a,$$
$$B_3(b): \tfrac{1}{2}b + \tfrac{1}{2}a \leq |\beta''|.$$

We recall that the variable r is the Euclidean radius of the horosphere over which the integration is performed in forming the Radon transform. It follows that, for any v, $\hat{v}(r, \beta)$ does not depend on values of $v(x, y)$ outside the set

$$(7.62) \qquad |x - \beta| \leq r.$$

For β'' in B_1 and $r \leq \tfrac{1}{2}a$, all points of the set (7.62) satisfy

$$|x''| < \tfrac{1}{2}b - 1.$$

Since by construction (see (7.46)), $\varphi(|x''|)$ is zero in this set, we conclude that

$(7.63)_1$ $\qquad\qquad\qquad \widehat{\varphi u_t} = 0$ when $\beta'' \in B_1$ and $r \leqq \frac{1}{2} a$.

To deal with B_2 we introduce two additional cutoff functions:

(7.64) $\qquad\qquad\qquad \psi(|x''|) = \begin{cases} 1 & \text{for } |x''| \leqq \frac{1}{2} b + a, \\ 0 & \text{for } |x''| > \frac{1}{2} b + a + 1, \end{cases}$

and

(7.65) $\qquad\qquad\qquad \chi(|x'|) = \begin{cases} 1 & \text{for } |x'| < d + \frac{1}{2} a, \\ 0 & \text{for } |x'| > d + \frac{1}{2} a + 1; \end{cases}$

here d denotes the Euclidean diametr of F_c. Arguing as above we conclude that

$(7.63)_2$ $\qquad\qquad \widehat{\varphi u_t} = \widehat{\chi \psi \varphi u_t}$ when $\beta'' \in B_2, \beta' \in F_c, r < \frac{1}{2} a$.

and

$(7.63)_3$ $\qquad\qquad\qquad \widehat{\varphi u_t} = \hat{u}_t$ when $\beta'' \in B_3, r < \frac{1}{2} a$.

We also have

$(7.63)_4$ $\qquad\qquad\qquad \hat{u} = \widehat{\chi \varphi u}$ when $\beta'' \in B_3, \beta' \in F_c, r < \frac{1}{2} a$.

We can now write the integral (7.60) as the sum of two integrals:

$(7.66)_2$ $\qquad\qquad\qquad \displaystyle\int_0^{a/2} \int_{F_c \times B_2} |J \widehat{\chi \psi \varphi u_t}|^2 \, d\beta \, \frac{dr}{r}$

and

$(7.66)_3$ $\qquad\qquad\qquad \displaystyle\int_0^{a/2} \int_{F_c \times B_3} |J \hat{u}_t|^2 \, d\beta \, \frac{dr}{r}.$

We estimate these in turn; $(7.66)_2$ is less than

(7.67) $\qquad\qquad\qquad \displaystyle\int_0^{a/2} \int_{\mathbf{R}^{n-1}} |J \widehat{\chi \psi \varphi u_t}|^2 \, d\beta \, \frac{dr}{r}.$

By construction of the cutoff functions, the support of $\psi \varphi$ is confined to

(7.68) $\qquad\qquad\qquad \frac{1}{2} b - 1 \leqq |x''| \leqq \frac{1}{2} b + a + 1;$

the support of χ lies in

$(7.68)'$ $\qquad\qquad\qquad |x'| \leqq d + \frac{1}{2} a + 1.$

Denote the intersection of (7.68), (7.68)' by $K(b)$. For $b \geqq b_0$, large enough, $K(b)$ lies in a finite number A of replicas of F; therefore, since u has finite energy in F, $\chi \psi \varphi u_t$ is square integrable in $\mathbb{H}^n$.

348

It follows from the Plancheral relation for the Radon transform (see (1.11) of [4]) that (7.67) is equal to

$$(7.69) \qquad \int_{H^n} |\chi \psi \varphi u_t|^2 \, dV;$$

clearly this is less than

$$(7.69)' \qquad \int_{K(b)} |u_t|^2 \, dV.$$

Define b_j for j an integer by

$$b_j = b + j(2a + 4).$$

It follows from (7.68) that the sets $K(b_j)$ are disjoint; since their union is contained in A replicas of F, it follows that

$$\sum_1^N \int_{K(b_j)} |u_t(t)|^2 \, dV \leq A \int_F |u_t(t)|^2 \, dV \leq AE_F(u) = C.$$

Thus, for any t, there is an integer j between 1 and N such that

$$(7.70) \qquad \int_{K(b_j)} |u_t(t)|^2 \, dV \leq \frac{C}{N}.$$

Since $(7.69)'$ is an upper bound for $(7.66)_2$, the integral $(7.66)_2$ is less than C/N for any t and some $b = b_j$.

We turn now to estimating $(7.66)_3$; we follow the steps of the proof of Lemma 7.7.

For b large enough, say $b > b_0$, $h = \{\chi \varphi u, \chi \varphi u_t\}$ has finite energy H^n:

$$E(h) \leq \text{const. } E_F(u).$$

The translation representations of h on H^n are (see [2] and [7])

$$(7.71) \qquad R_{\pm}^0 h(t) = \partial_s P \widetilde{\chi \varphi u} \mp P \widetilde{\chi \varphi u_t} \overset{\text{def}}{=} k_{\pm}(s, \beta, t).$$

Since the translation representation in H^n is isometric,

$$(7.72) \qquad \iint |k_{\pm}(s, \beta, t)|^2 \, dm(\beta) \, ds = E(h(t)).$$

In analogy with (7.25),

$$(7.72)' \qquad E(h) \leq \text{const. } \{D(h) + H(h) + H(h_t)\} \leq C.$$

It follows from $(7.63)_4$ that, for $\beta' \in F_c$, $\quad \beta'' \in B_3(b)$, $\quad r < \tfrac{1}{2}a$,

$$(7.73) \qquad \widehat{\chi \varphi u} = \hat{u}, \qquad \widehat{\chi \varphi u_t} = \hat{u}_t.$$

The function u satifies the non-Euclidean wave equation; thus (see (7.26)′) $P\hat{u}$ satisfies the one-dimensional wave equation. From this is follows (see (7.29)) that

$$(7.74) \qquad \begin{aligned} \partial_s P\hat{u} - P\hat{u}_t &= h_+(s-t, \beta), \\ \partial_s P\hat{u} + P\hat{u}_t &= h_-(s+t, \beta). \end{aligned}$$

Since the variables s and r are related by (3.19), the condition $r < \frac{1}{2}a$ is equivalent to

$$(7.75) \qquad s > \log \tfrac{1}{2}(1 + |\beta|^2) - \log \tfrac{1}{2}a.$$

Since (7.73) is true when $\beta' \in F_c$, $\beta'' \in B_3(b)$ and (7.75) holds, we see, comparing (7.74) and (7.71), that

$$(7.76) \qquad k_\pm(s, \beta, t) = h_\pm(s \mp t, \beta)$$

for this range of s and β. Setting this into (7.72), (7.72)′ we conclude that

$$(7.77) \qquad \int_{F_c \times B_3(b)} \int_{s > \log((1+|\beta|^2)/a)} |h_\pm(s \mp t, \beta)|^2 \, ds \, dm(\beta) \leqq C.$$

The right side is independent of t; letting $t \to \pm\infty$ we conclude that

$$(7.78) \qquad \int_{F_c \times B_3(b)} \int_{-\infty}^{\infty} |h_\pm(s, \beta)|^2 \, ds \, dm(\beta) \leqq C.$$

It follows from (7.78) that for any $\tau > 0$ we can find $b_1' > b_0$ such that, for all $b > b_1'$,

$$(7.78)' \qquad \int_{F_c \times B_3(b)} \int_{-\infty}^{\infty} |h_\pm(s, \beta)|^2 \, ds \, dm(\beta) < \tau.$$

In view of (7.74) we can express $P\hat{u}_t$ as

$$P\hat{u}_t = \tfrac{1}{2}h_-(s+t, \beta) - \tfrac{1}{2}h_+(s-t, \beta)$$

in the region (7.75). From this and (7.78)′ we conclude that, for $b > b_1'$ and all t,

$$(7.79) \qquad \int_{s > \log((1+|\beta|^2)/a)} \int_{F_c \times B_3(b)} |P\hat{u}_t^2 \, dm(\beta) \, ds \leqq \tau.$$

According to (3.20),

$$P = c_n(1 + |\beta|^2)^{(n-1)/2} J,$$

and, according to (6.27) and (6.27)′,

$$ds = \frac{dr}{r}, \qquad c_n^2(\tfrac{1}{2}(1 + |\beta|^2))^{(n-1)/2} \, dm(\beta) = d\beta.$$

Setting these expressions into (7.79) we conclude that, for $b > b_1'$ and all t,

$$\int_0^{a/2} \int_{F_c \times \{|\beta''| > (a+b)/2\}} |J\hat{u}_t|^2 \, d\beta \frac{dr}{r} < \tau.$$

Consequently, the integral $(7.66)_3$ is also less than τ.

We now choose $\tau = \frac{1}{2}\gamma$ and N so that, in (7.70), $C/N < \frac{1}{2}\gamma$. We further set

$$b_j = b_1' + j(2a + 4), \qquad\qquad j = 1, \cdots, N.$$

Then given any value of t, as we saw above, there is a value b_j for which $(7.66)_2$ is less than $C/N < \frac{1}{2}\gamma$. Since $(7.66)_3$ is less than $\frac{1}{2}\gamma$, the sum of these integrals, (7.60), is less than γ. Since the left side of (7.59) equals (7.60), the proof of Lemma 7.12 is complete with $b' = b_1' + N(2a + 4)$.

We now apply the inequality (7.58) to $f = \varphi u_t$. Since $y = 2r$, we obtain

$$(7.80) \qquad \int_0^\delta |\widehat{\varphi u_t}(\zeta, y, t)|^2 \frac{dy}{y^n} \leq \exp\{c(|\zeta|\delta)^{(n+1)/2}\} \int_0^{\delta/2} |l(\zeta, r)|^2 \frac{dr}{r},$$

where

$$(7.81) \qquad\qquad\qquad l = JF\widehat{\varphi u_t}(t).$$

We choose now $\delta = \delta_\alpha(\xi)$ as defined in (7.53). For this choice,

$$|\zeta_\alpha|\delta \leq \min(\eta^{-1}, a|\zeta_\alpha|)$$

and (7.80), with $\zeta = \zeta_\alpha$, becomes

$$\int_0^{\delta_\alpha(\xi)} |\widehat{\varphi u_t}(\zeta_\alpha, y, t)|^2 \frac{dy}{y^n} \leq e^{c/\eta} \int_0^{a/2} |l(\zeta_\alpha, r)|^2 \frac{dr}{r}.$$

We integrate with respect to ξ and sum with respect to α:

$$(7.82) \qquad \begin{aligned} &\sum_\alpha \int_{\mathbf{R}^{m-1}} \int_0^{\delta_\alpha(\xi)} |\widehat{\varphi u_t}(\zeta_\alpha, y, t)|^2 \frac{dy}{y^2} \, d\xi \\ &\qquad \leq e^{c/\eta} \sum_\alpha \int_{\mathbf{R}^{m-1}} \int_0^{a/2} |l(\zeta_\alpha, r)|^2 \frac{dr}{r} \, d\xi. \end{aligned}$$

According to (7.59)$'$, the right side is $e^{c/\eta}\|l\|^2$; since l is given by (7.81), it follows from (7.59) that $\|l\|^2 < \gamma$. Given η and any $\varepsilon > 0$ we can choose γ so small that $e^{c/\eta}\gamma < \varepsilon$. Inequality (7.52) is then a consequence of (7.82). This completes the proof of Lemma 7.10.

We combine now Lemmas 7.9 and 7.10. First we choose η so small that $c\eta^2$, that is the right side of (7.51), is less than ε. Then we choose b' so large that (7.52) holds; we add (7.51) and (7.52) and obtain

$$\sum_\alpha \int_0^a \int_{\mathbf{R}^{m-1}} |\widehat{\varphi u_t}(\zeta_\alpha, y, t)|^2 \, d\xi \frac{dy}{y^n} < 2\varepsilon.$$

Using the identity (7.50) we conclude that

$$H_S(\varphi u_t) < 2\varepsilon.$$

Combining this with (7.47) gives the desired inequality (7.45) with $b = b'$. Since (7.45) is free of φ, which depended on our choice of b_j, it holds for all t. The proof of Proposition 7.8 is now complete provided F_c is a parallelogram.

If F_c is not a parallelogram, that is if Γ_∞ is not a free Abelian subgroup of Euclidean motions, then there exists a free abelian subgroup Γ'_∞ of finite index in Γ_∞ with a parallelogram F'_c as fundamental domain. Any eigenfunction over F_c can be obtained from one over F'_c by averaging over the cosets of $\Gamma_\infty / \Gamma'_\infty$. In particular, all of the ψ_k can be obtained in this fashion. Since Γ_∞ consists of Euclidean motions, ψ_k is a sum of exponentials $\exp \{i\alpha \cdot x'\}$, each with the same absolute value $|\alpha(k)|$. Each contributes to the kernel K, but since $|\zeta_a|^2 = \xi^2 + |\alpha(k)|^2$, the resulting K is precisely the same as before. The rest of the argument therefore goes through unchanged.

Finally we add inequality (7.4) of Proposition 7.1 and inequality (7.45) of Proposition 7.8. In view of (7.3) what we obtain is the inequality (7.2). This completes the proof of part (iii) of Theorem 3.2.

Acknowledgment. The work of the first author was supported in part by the Department of Energy under Contract DE-AC02-76 ERO 3077 and that of the second by the National Science Foundation under Grant MCS-83-04317.

Bibliography

[1] Faddeev, L., and Pavlov, B. S., *Scattering theory and automorphic functions*, Seminar of Steklov Math. Inst. of Leningrad, Vol. 27, 1972, pp. 161–193. (In Russian).

[2] Lax, P. D., and Phillips, R. S., *Translation representations for the solution of the non-Euclidean wave equation*, I, Comm. Pure and Appl. Math., Vol. 32, 1979, pp. 617–667 and II, Vol. 34, 1981, pp. 347–358.

[3] Lax, P. D., and Phillips, R. S., *The asymptotic distribution of lattice points in Euclidean and non-Euclidean spaces*, Jr. of Functional Analysis, Vol. 46, 1982, pp. 280–350.

[4] Lax, P. D., and Phillips, R. S., *A local Paley-Wiener theorem for the Radon transform of L_2 functions in a non-Euclidean setting*, Comm. Pure and Appl. Math., Vol. 35, 1982, pp. 531–554.

[5] Lax, P. D., and Phillips, R. S., *Translation representations for automorphic solutions of the wave equation in non-Euclidean spaces*, I, Comm. Pure and Appl. Math., Vol. 37, 1984, pp. 303–328.

[6] Patterson, S. J., *The Laplace operator on a Riemann surface*, I, Compositio Mathematica, Vol. 31, 1975, pp. 83–107; II, Vol. 32, 1976, pp. 71–112; III, Vol. 33, 1976, pp. 227–259.

[7] Woo, A. C., *Scattering theory on real hyperbolic spaces and their compact perturbations*, Dissertation, Stanford University, 1980.

[8] Mandouvalos, N. H., *The theory of Eisenstein series and spectral theory for Kleinian groups*, Cambridge Univ. Dissertation, 1983.

Received March, 1984.

Translation Representations for Automorphic Solutions of the Wave Equation in Non-Euclidean Spaces. III

PETER D. LAX

Courant Institute

AND

RALPH S. PHILLIPS

Stanford University

Abstract

In Part III we show that the translation representations defined in Part I, and shown to be complete in Part II, are incoming and outgoing in the sense of propagation of signals along rays. We give a new proof of the absolute continuity of the spectrum below $-\frac{1}{4}n^2$, and point out its implications for local energy decay. We give a new proof of completeness in dimensions 2 and 3 when energy is positive. Finally, we define and prove completeness of the translation representations when the metric is perturbed on a compact set.

Introduction

This is Part III of a series [5] on the translation representations for solutions of the wave equation in non-Euclidean spaces which are automorphic with respect to a discrete subgroup Γ of motions. It is assumed throughout that the fundamental domain F of Γ has only a finite number of sides and is of infinite volume. In Part I the incoming and outgoing translation representations R^F_- and R^F_+ were defined and shown to be isometric mappings on closed subspaces $\mathcal{H}_-$ and $\mathcal{H}_+$ of the space $\mathcal{H}_c$ of all data with finite energy orthogonal to the eigenfunctions of the Laplace-Beltrami operator over F. The completeness of these representations was established in Part II; this amounted to showing that both $\mathcal{H}_-$ and $\mathcal{H}_+$ were equal to $\mathcal{H}_c$.

In Part III we give some further insights into the translation representations. In Section 8 we show that R^F_- is incoming and R^F_+ is outgoing in the sense of propagation of signals along rays. Section 9 contains a proof, based only on [4], that $\mathcal{H}_- + \mathcal{H}_+$ is dense in $\mathcal{H}_c$, from which follows the absolute continuity of the spectrum of the Laplace-Beltrami operator below $-(\frac{1}{2}(n-1))^2$ and the local energy decay. In Section

Communications on Pure and Applied Mathematics, Vol. XXXVIII, 179–207 (1985)
CCC 0010-3640/85/020179-29$04.00

10 we give an independent proof of the completeness of the translation representations when the energy form E_F is positive definite. Finally in Section 11 we prove the completeness of the translation representation of automorphic solutions of a perturbed wave equation when the support of the perturbation is compact in F. This is the only section in Part III which depends in an essential way on Part II.

For simplicity in exposition we restrict ourselves to the three-dimensional hyperbolic space unless otherwise stated. The results are valid for arbitrary dimensions.

8. Incoming and Outgoing Solutions

In this section we shall derive a property of the translation representations constructed in Section 2 of Part I that justifies the names "incoming" and "outgoing." We start with the observation that translation representations are by no means unique, just as spectral representations are not. E.g., if

$$(8.1) \qquad\qquad f \leftrightarrow l(s,\beta)$$

is a translation representation, so is

$$(8.2) \qquad\qquad f \leftrightarrow l(s - \varphi(\beta),\beta),$$

φ an arbitrary measurable function. The representations constructed in Section 2 depend on the particular choice of a fundamental domain; it is easy to see that two different choices of F lead to two different representations that are related to each other as (8.1) is to (8.2).

For purposes of this section we shall make a particular, natural choice of F, obtained by the so-called polygonal method.

F is defined to consist of all points w whose distance to j is less than its distance to any image of j under Γ:

$$(8.3) \qquad\qquad d(w,j) < d(w,\tau j), \qquad\qquad\qquad \tau \text{ in } \Gamma, \quad \tau \neq \text{identity.}$$

Such a fundamental domain is necessarily convex. We call this fundamental domain *centered at j*. We assume that j is not a fixed point of Γ.

We define: A solution $u(w,t)$ of the non-Euclidean wave equation (1.3) defined in $\mathbb{H}^3$ is called *outgoing from j* if $u(w,t)$ vanishes in the ball $b_t(j)$ centered at j and of radius t, $t > 0$.

Incoming solutions are defined analogously, with $-t$ in place of t.

In this section we shall characterize outgoing solutions in terms of their outgoing translation representations. Incoming solutions can be similarly characterized in terms of their incoming translation representations.

We shall call the initial data of outgoing solutions *outgoing data*.

THEOREM 8.1. *Let Γ be a discrete subgroup, F its fundamental domain centered at j. Let f^Γ be data in $\mathcal{H}_G$ which is outgoing from j. Then the outgoing translation representation of f^Γ is supported on $\mathbb{R}_+$, i.e.,*

(8.4) $R_+^F f^\Gamma(s) = 0 \quad for \quad s < 0.$

The proof relies on *Friedlander's formula* for the translation representation of f^Γ in terms of u^Γ. We shall deduce the automorphic version of the formula needed here from the analogous formula for data defined in all $\mathbb{H}^3$, derived in Section 5 of [2].

Given data f of finite energy in $\mathbb{H}^3$, denote by $u(w,t)$ the solution of (1.3) with initial data f. Define $w(r,\beta)$ as that point on the geodesic issuing from j toward β whose distance from j is r. Define the function $k(s,\beta;t)$ by

$$(8.5) \qquad k(s,\beta;t) = \begin{cases} \sqrt{2}\,\sinh\,(s + t)u_t(w(s + t,\beta),t) & \text{for } s > -t, \\ 0 & \text{for } s < -t. \end{cases}$$

Then, as shown in [7], Theorem 5.1,

$$(8.6) \qquad \lim_{t \to \infty} k(s,\beta;t) = R_+ f,$$

the limit being in the sense of $L_2(\mathbb{R} \times B)$.

LEMMA 8.2. *Let f^Γ denote data in $\mathbb{H}_G$, denote by $u^\Gamma(w,t)$ the solution of the non-Euclidean wave equation with initial data f^Γ. Define*

$$(8.7) \qquad k^\Gamma(s,\beta;t) = \begin{cases} \sqrt{2}\,\sinh(s + t)\,u_t^\Gamma(w(s + t,\beta),t) & \text{for } s > -t, \\ 0 & \text{for } s < -t. \end{cases}$$

Then

$$(8.8) \qquad \lim_{t \to \infty} k^\Gamma(s,\beta;t) = R_+^0 f^\Gamma,$$

in the sense of $L_2((a,\infty) \times B_F')$ for any compact B_F'.

Proof: We argue similarly to the proof of part (iii) of Theorem 2.3. We denote by $\Sigma(a,B_0)$ the union of horospheres:

$$\Sigma(a,B_0) = \cup \{\xi(s,\beta); s \geqq a, \ \beta \in B_0\}.$$

Given a and B_F', we choose $b < a$ and B_F'' compact and containing B_F' in its interior. We choose a cutoff function φ as a $C^\infty(\mathbb{H}^3)$ function, satisfying

$$(8.9) \qquad \varphi(w) = \begin{cases} 1 & \text{on } \ \Sigma(a,B_F'), \\ 0 & \text{for } \ w \notin \Sigma(b,B_F''). \end{cases}$$

According to Lemma 2.1,

(8.9)' $f = \varphi f^\Gamma$

is zero except on a finite number of replicas of F. Thus f has finite energy on $\mathbb{H}^3$.

It follows from (8.9), (8.9)' that

(8.10) $f^\Gamma(w) = f(w)$ on $\Sigma(a, B_F')$.

This implies that

(8.10)' $R_+^0 f^\Gamma(s, \beta) = R_+ f(s, \beta)$ for $s > a$, $\beta \in B_F'$.

We claim now that for $t \geqq 0$

(8.11) $u^\Gamma(w, t) = u(w, t)$ for w in $\Sigma(a + t, B_F')$.

To see this we recall that for solutions of the wave equation (1.3), signals propagate with speed 1. Thus the values of $u^\Gamma(w, t)$ and $u(w, t)$ depend on values of f^Γ and f, respectively, only at those points whose distance from w is t. For w in $\Sigma(a + t, B_F')$, these points are contained in $\Sigma(a, B_F')$. According to (8.10), $f^\Gamma = f$ on this set, therefore (8.11) follows.

Recalling that $w(r, \beta)$ has distance r from j along the geodesic issuing from j toward β, we conclude that $w(r, \beta)$ lies in $\Sigma(a + t, B_F')$ for $r > a + t$; it follows then from (8.11) that, for $s > a$ and $\beta \in B_F'$,

$$u^\Gamma(w(s + t, \beta), t) = u(w(s + t, \beta), t).$$

Setting this into (8.5) and (8.7) we conclude that

$$k^\Gamma(s, \beta; t) = k(s, \beta; t) \quad \text{for} \quad s > a, \, \beta \in B_F'.$$

We deduce then from (8.6) that

$$\lim_{t \to \infty} k^\Gamma(s, \beta; t) = R_+ f(s, \beta), \qquad\qquad \beta \in B_F'.$$

In view of (8.10)', (8.8) follows.

We turn to the proof of Theorem 8.1. Since u^Γ is outgoing from j,

$$u^\Gamma(w(r, \beta), t) = 0 \quad \text{for} \quad r < t.$$

Inserting this into (8.7) we see that

$$k^\Gamma(s, \beta; t) = 0 \quad \text{for} \quad s < 0.$$

Using this in (8.8) we conclude that

$$R_+^0 f^\Gamma(s, \beta) = 0 \quad \text{for} \quad s < 0,$$

i.e., that the first component of $R_+^F f^\Gamma$ is zero on $\mathbb{R}_-$.

We show now that all the other components vanish on $\mathbb{R}_-$. We shall use the translation property (2.8),

$$(8.12) \qquad R_+^F U(t) f^\Gamma(s,\beta) = R_+^F f^\Gamma(s - t,\beta).$$

Assume that $\beta = \pi_i = \infty$, something that can be accomplished by a rotation keeping j fixed; in this case $y = e^s$. According to formulas $(2.2)_+$, (2.3) applied to $U(t)f^\Gamma$,

$$(8.13) \qquad R_+^i U(t) f^\Gamma(s) = \frac{1}{\sqrt{2|F_i'|}}\Big[\partial_s \, e^{-s} \, \overline{f_1^\Gamma}(e^s,t) - e^{-s} \overline{f_2^\Gamma}(e^s,t)\Big],$$

where

$$(8.13)' \qquad \overline{f_k^\Gamma}(y,t) = \int_{F_i} f_k^\Gamma(w,t) \, dx, \qquad\qquad k = 1, 2,$$

with

$$(8.14) \qquad w = (x,y), \quad \{f_1,f_2\} = \{u^\Gamma(w,t), u_t^\Gamma(w,t)\}.$$

Using the triangle inequality we get

$$d(w,j) \leq \log y + \frac{|x|}{y}.$$

In particular, if x is restricted to F_i',

$$(8.15) \qquad d(w,j) \leq \log y + \frac{c}{y}.$$

Now let n denote any negative number; set $s = t + n$, and thus $y = e^{t+n}$. Inserting this into (8.15) we obtain

$$d(w,j) \leq t + n + c \, e^{-(t+n)}.$$

Clearly, for t large enough,

$$d(w,j) < t.$$

We have assumed that u^Γ is outgoing; that means that

$$u^\Gamma(w,t) = 0$$

for all such w; setting this into (8.13)', (8.14) we conclude that

$$\overline{f_k^\Gamma}(y,t) = 0, \qquad\qquad k = 1, 2,$$

which if inserted into (8.13) yields

$$R_+^i \, U(t) f^\Gamma(s) = 0.$$

We use now (8.12) with $\beta = \pi_i$ and the fact that $s = t + n$ to conclude that

$$R^i_+ f^\Gamma(n,\beta) = R^i_+ f^\Gamma(s - t,\beta) = R^i_+ U(t) f^\Gamma(s,\beta),$$

from which it follows that $R^i_+ f^\Gamma(n,\beta) = 0$ for all $n < 0$. This completes the proof of Theorem 8.1.

We give now a partial converse of Theorem 8.1.

THEOREM 8.3. *Suppose, for automorphic data* f^Γ *belonging to the space* $\mathcal{H}_G$, *that the outgoing translation representation of* f^Γ *satisfies condition* (8.4), *i.e., is supported on* $\mathbb{R}_+$. *Then there exist* $n^\Gamma \in \mathcal{N}$ *and* p^- *such that the solution* u^Γ *with initial data* $f^\Gamma + n^\Gamma + p^-$ *is outgoing from* j. *Here* p^- *is a linear combination of the eigendata* (1.18)″.

Proof: We show first that if $l = \{l_0, l_1, \cdots, l_N\}$ lies in $L_2(\mathbb{R}_+ \times B'_F) \times L_2(\mathbb{R}_+)^N$, where B'_F is a compact subset of B_F, then $J^F_+ l = g^\Gamma$, defined by (2.25), (2.25)′, is outgoing from j. According to (2.25)′,

$$g^\Gamma = \sum_0^N g^\Gamma_{(i)};$$

we shall show that each $g^\Gamma_{(i)}$ separately is outgoing from j.

We start with $i = 0$. By (2.21), (2.21)′, and (2.20),

(8.16) $$g_0^\Gamma = \sum_\Gamma g_0^\tau, \qquad g_0 = J_+ l_0,$$

where the operator J_+ is defined by (1.13). Since J_+ transmutes translation into U_0, the function $u_0(w,t)$ with initial data g_0 is, by (1.13),

(8.17) $$u_0(w,t) = C \int_{B_F} e^{\langle w,\beta \rangle} l_0(\langle w,\beta \rangle - t,\beta) \, dm(\beta).$$

By assumption, $l_0(s,\beta) = 0$ for $s < 0$; therefore it follows from (8.17) that $u_0(w,t) = 0$ for all w for which

(8.18) $$\langle w,\beta \rangle < t \quad \text{for all} \quad \beta.$$

Clearly, w satisfies (8.18) if the distance of w from j is less than t. Thus, denoting by $b_r(w)$ the ball of radius r centered at w,

(8.19) $$u_0(w,t) = 0 \quad \text{when} \quad w \in b_{t-\varepsilon}(j).$$

Next we show: For all τ in Γ,

(8.20) $$u_0(w,t) = 0 \quad \text{when} \quad w \in b_{t-\varepsilon}(\tau^{-1}j).$$

Again we use formula (8.17), and we note that the β-integration in (8.17) is over B_F. Therefore, (8.20) follows from

(8.21) $\langle w, \beta \rangle < t$ for all β in B_F when $w \in b_{t-\varepsilon}(\tau^{-1}j)$.

Recalling that the horosphere $\xi(s, \beta)$ is the set of w where $\langle w, \beta \rangle = s$, we can restate (8.21) as follows:

The horospheres

(8.21)′ $\xi(s, \beta)$, β in B_F, $s \geqq t$,   are disjoint from $b_{t-\varepsilon}(\tau^{-1}j)$.

The horosphere $\xi(s, \beta)$ is the limit of spheres through $w(s, \beta)$ whose center z approaches β along the geodesic from j to β. It follows from (8.18) that when the center z is close enough to β, these spheres are disjoint from $b_{t-\varepsilon}(j)$. Since F is convex, the geodesic joining j to $\beta \in B_F$ lies in F; in particular, the centers z of the spheres belong to F. Since F is centered at j, it follows that

$$d(z, j) \leqq d(z, \tau^{-1}, j).$$

It follows from this that since the sphere around z does not intersect $b_{t-\varepsilon}(j)$, it does not intersect $b_{t-\varepsilon}(\tau^{-1}j)$ either. Letting z tend to β we conclude that (8.21)′ holds. This proves (8.20).

From the first relation in (8.16) we have

(8.22) $u_0^\Gamma(w, t) = \Sigma u_0^\tau(w, t) = \Sigma u_0(\tau^{-1}w, t)$.

Suppose w is in $b_{t-\varepsilon}(j)$; then $\tau^{-1}w$ is in $b_{t-\varepsilon}(\tau^{-1}j)$. We apply now (8.20), with $\tau^{-1}w$ in place of w, and conclude that $u_0(\tau^{-1}w, t) = 0$. It follows then from (8.22) that $u_0^\Gamma(w, t) = 0$ for w in $b_{t-\varepsilon}(j)$, i.e., that u_0^Γ is outgoing from j.

We show now that the $u_{(i)}^\Gamma$, with initial data $g_{(i)}^\Gamma$, $i = 1, \cdots, N$, are outgoing. For map the ith cusp π_i to ∞ by a rotation around j. $g_{(i)}$ is defined by (2.22), therefore

(8.23) $u_{(i)}(w, t) = cy\, m(\log y - t)$.

By assumption, $l(s) = 0$ for $s < 0$; it follows from (2.22)′ that $m(s) = 0$ for $s < 0$. From this and (8.23) we draw the conclusion:

(a) $u_{(i)}(w, t) = 0$ for w in $b_t(j)$.

As in the case $i = 0$,

(b) $u_{(i)}(w, t) = 0$ for w in $b_t(\tau^{-1}j)$.

Therefore,

(c) $u_{(i)}(\tau^{-1}w, t) = 0$ in $b_t(j)$.

According to (2.24)′,

$$g_{(i)}^\Gamma = \sum_{\Gamma_\infty \backslash \Gamma} g_{(i)}^\tau$$

and so

$$u_{(i)}^\Gamma = \sum u_{(i)}^\tau.$$

From (a) and (c) above we conclude that $u_{(i)}^\Gamma(w,t) = 0$ for w in $b_t(j)$.

Next we consider a Cauchy sequence $l^{(i)}$, with each $l_0^{(i)}$ supported on a compact subset of $\mathbb{R}_+ \times B_F$, such that

(8.24) $l^{(i)} \to l$ in $L_2(\mathbb{R}_+ \times B_F).$

Set

$$g^{\Gamma(i)} = J_+^F l^{(i)}.$$

Relation (2.40) of Lemma 2.8 implies that $\lim E_F(k^\Gamma, g^{\Gamma(i)})$ exists for every k^Γ in $\mathcal{H}_G$. From (2.28) of Lemma 2.5 and (8.24) we conclude that the $g^{\Gamma(i)}$ form a Cauchy sequence in the E_F form.

LEMMA 8.4.   *The $g^{\Gamma(i)}$ form a Cauchy sequence in the G norm and the limit g^Γ can be decomposed as*

(8.25) $g^\Gamma = h^\Gamma + p^- + n,$

where h^Γ belong to $\mathcal{H}_c$.

Remark.   The limit g^Γ is clearly outgoing from j.

Proof: To establish convergence we project each $g^{\Gamma(i)}$ into $\mathcal{H}_c$. According to (2.55),

(8.26) $g^{\Gamma(i)} = h^{\Gamma(i)} + p^{-(i)},$ $h^{\Gamma(i)} \in \mathcal{H}_c.$

Since $E_F(p_j^+, g^{\Gamma(i)}) = E_F(p_j^+, p^{-(i)})$ and since $\lim E_F(p_j^+, g^{\Gamma(i)})$ exists, we see from (1.20)' that the $p^{-(i)}$ converge in every norm, in particular in the G norm. It follows then from Lemma 2.10 that $h^{\Gamma(i)}$ are a Cauchy sequence in the E_F norm.

Decompose each $h^{\Gamma(i)}$ as

(8.27) $h^{\Gamma(i)} = h_{(i)}^\Gamma + n_{(i)}^\Gamma,$

where $n_{(i)}^\Gamma \in \mathcal{N}$ and $h_{(i)}^\Gamma$ is G-orthogonal to $\mathcal{N}$. According to Lemma 1.4, the $h_{(i)}^\Gamma$ form a Cauchy sequence in the G norm; thus

$$G \text{ - } \lim h_{(i)}^\Gamma = h^\Gamma.$$

We insert (8.27) into (8.26):

(8.28) $g^{\Gamma(i)} = h_{(i)}^\Gamma + n_{(i)}^\Gamma + p^{-(i)},$

and apply $U = U(1)$ to this relation:

(8.28)' $U g^{\Gamma(i)} = U h_{(i)}^\Gamma + U n_{(i)}^\Gamma + U p^{-(i)}.$

We claim that the first components of all terms in (8.28)' converge in the $L_2(b)$ norm, where b is the unit ball around j. For $h_{(i)}^\Gamma$ and $p^{-(i)}$ converge in the G norm; therefore, according to part (b) of Theorem 1.2, so do $Uh_{(i)}^\Gamma$ and $Up^{-(i)}$. But then, according to part (c) of Theorem 1.2, $Uh_{(i)}^\Gamma$ and $Up^{-(i)}$ converge in the $L_2(b)$ norm.

We have shown above that each $g^{\Gamma(i)}$ is outgoing from j; therefore, $Ug^{\Gamma(i)}$ is zero in b. We conclude then from (8.28)' that $Un_{(i)}^\Gamma$ also converges in the $L_2(b)$ norm. By (1.23), $n_{(i)}^\Gamma = Un_{(i)}^\Gamma$ so that $n_{(i)}^\Gamma$ also converges in the $L_2(b)$ norm. Since $\mathcal{N}$ is finite-dimensional, it follows that $n_{(i)}^\Gamma$ converges in any other norm, in particular in the G norm. using (8.28)' once more we conclude that $Ug^{\Gamma(i)}$ is G-convergent, and so, by Lemma 1.2, $g^{\Gamma(i)}$ is G-convergent.

We can now complete the proof of Theorem 8.3. Suppose f^Γ belongs to $\mathcal{H}_G$ and $R_+^F f^\Gamma = 0$ for $s < 0$. We denote the projection of f^Γ in $\mathcal{H}_c$ by h_0^Γ:

$$(8.29) \qquad f^\Gamma = h_0^\Gamma + p_0^- + p_0^+.$$

Apply R_+^F to (8.29); using (2.17) we obtain

$$(8.30) \qquad R_+^F f^\Gamma = R_+^F h_0^\Gamma + R_+^F p_0^+.$$

According to Theorem 2.3, $R_+^F h_0^\Gamma$ lies in $L_2(\mathbb{R} \times B_F) \times L_2(\mathbb{R})^N$; according to (2.19), $R_+^F p_0^+$ is a linear combination of exponentials $\exp\{-\lambda_j s\} \theta_j$. Since $R_+^F f^\Gamma$ vanishes for $s < 0$, it follows that $R_+^F p_0^+ \equiv 0$. As remarked after Theorem 9.1 on page 188, this implies that

$$(8.31) \qquad p_0^+ = 0.$$

It now follows from (8.30) that

$$(8.32) \qquad R_+^F f^\Gamma = R_+^F h_0^\Gamma \equiv l$$

is square integrable with support in $\mathbb{R}_+$. As we have shown in Lemma 8.4, there exists a g^Γ in $\mathcal{H}_G$ which is outgoing from j such that

$$(8.33) \qquad g^\Gamma = h^\Gamma + p^- + n, \qquad h^\Gamma \text{ in } \mathcal{H}_c,$$

with

$$(8.34) \qquad R_+^F g^\Gamma = R_+^F h^\Gamma = l.$$

Combining (8.32) and (8.34) we see that $R_+^F(h^\Gamma - h_0^\Gamma) = 0$. It now follows from the completeness of R_+^F on $\mathcal{H}_c$, which was established in Part II, that $h^\Gamma = h_0^\Gamma$ modulo $\mathcal{N}$. On combining this with (8.29), (8.31) and (8.33), we obtain the assertion of Theorem 8.3.

9. Absolute Continuity of the Spectrum and Energy Decay

In Part II we proved that $\mathcal{H}_+ = \mathcal{H}_- = \mathcal{H}_c$. In this section we give a simpler independent proof of a lesser result, namely that $\mathcal{H}_+ + \mathcal{H}_-$ is dense in $\mathcal{H}_c$. This suffices to prove that (i) the spectrum of L is absolutely continuous except for the finite set of eigenfunctions $\{\varphi_j\}$ introduced in (1.18) and (ii) a local decay theorem holds for solutions of the wave equation in $\mathcal{H}_c$.

The principle tool here is a local Paley-Wiener theorem for the Radon transform in a non-Euclidean setting proved in [4] for functions f defined on a spherical cap C at infinity:

$$(9.1) \qquad |x - x_0|^2 + |y - y_0|^2 < c^2, \qquad\qquad |y_0| < c.$$

If f is square integrable in C and if

$$(9.2) \qquad \partial_s e^s \hat{f}(s,\beta) = 0$$

for all horospheres $\xi(s,\beta)$ contained in C, then $f \equiv 0$ in C.

THEOREM 9.1. $\overline{\mathcal{H}_- + \mathcal{H}_+} = \mathcal{H}_c$ *(modulo $\mathcal{N}$).*

Proof: Suppose the statement of the theorem were not true. Then since, by Lemma 1.4, $\mathcal{H}_c/\mathcal{N}$ is complete in the E_F norm, there would exist a f^Γ in $\mathcal{H}_c\backslash\mathcal{N}$ which is E_F-orthogonal to both $\mathcal{H}_-$ and $\mathcal{H}_+$. It follows by part (ii) of Theorem 2.3 that, for such an f^Γ,

$$(9.3) \qquad R_+^F f^\Gamma = 0 = R_-^F f^\Gamma.$$

Using the zeroth component (2.1)' of $R_\pm^F$, we see that (9.2) implies

$$(9.4) \qquad R_\pm^0 f^\Gamma = c(\partial_s^2 e^s \hat{f}_1^\Gamma \mp \partial_s e^s \hat{f}_2^\Gamma) = 0.$$

If follows that

$$(9.5) \qquad \partial_s\, e^s \hat{f}_2^\Gamma(s,\beta) = 0$$

for all s,β. Next choose a circle S in the interior of B_F and let C be any spherical cap intersecting B in S. Then C will lie in a finite number of replicas of F so that f_2 is square integrable in C. Combining this with (9.5), we conclude by the above described Paley-Wiener theorem that $f_2 = 0$ in C. Since any point in F is contained in a spherical cap of this kind, we see that $f_2^\Gamma = 0$ everywhere in F.

Since the $R_\pm^F$ are translation representations, it follows from (9.3) that

$$R_+^F U(t)f^\Gamma = 0 = R_-^F U(t)f^\Gamma$$

for all t. Then according to the foregoing, the second component of $U(t)f^\Gamma$ is zero for all t. Since that second component of $U(t)f^\Gamma$ is $u_t(t)$, we see that $U(t)f^\Gamma$ is independent of t; but then f^Γ belongs to the domain of the infinitesimal generator A of U and $Af = 0$. It was proved in Lemma 5.9 of [3], that $\mathcal{N}$ is the null space of A. Consequently, f^Γ belongs to $\mathcal{N}$ which is contrary to our choice of f^Γ. This concludes the proof of Theorem 9.1.

Remark. The proof presented above applies just as well to any f^Γ in $\mathcal{H}_G$ and in particular to the eigendata p_j^+ of $U(t)$ defined in (1.18)″. According to (2.17) of Lemma 2.4, with $+$ and $-$ interchanged, $R_-^F p_j^+ = 0$; while according to (2.19), $R_+^F p_j^+ = \exp\{-\lambda_j s\}\, \theta_j(\beta)$. Since it was shown above that $R_-^F p_j^+$ and $R_+^F p_j^+$ cannot both be zero, it follows that θ_j in (2.19) is different from zero. An analogous assertion holds for p_j^-.

It will clarify the discussion in the rest of this section if we introduce the notation

$$(9.6) \qquad \overset{\wedge}{\mathcal{H}}_c = \mathcal{H}_c/\mathcal{N},$$

$\hat{f}{}^\Gamma$ for the coset containing $f^\Gamma \in \mathcal{H}_c$, $\hat{U}(t)$ for the operator induced by $U(t)$ on these cosets, and $\hat{A}$ for the corresponding generator.

THEOREM 9.2. *The spectrum of $\hat{A}$ is absolutely continuous on $\overset{\wedge}{\mathcal{H}}_c$ with uniform multiplicity on the entire real line.*

Proof: We remind the reader that the Fourier transform of a translation representation results in a spectral representation which is obviously absolutely continuous and of uniform multiplicity on $\mathbb{R}$. Since R_+^F is a (complete) translation representation on $\overset{\wedge}{\mathcal{H}}_+$, it follows that the spectrum of $\hat{A}$ restricted to $\overset{\wedge}{\mathcal{H}}_+$ is absolutely continuous with uniform multiplicity on the entire real line. The same is true of $\hat{A}$ restricted to $\overset{\wedge}{\mathcal{H}}_-$. An element with singular spectrum would therefore be orthogonal to both $\overset{\wedge}{\mathcal{H}}_+$ and $\overset{\wedge}{\mathcal{H}}_-$. Since, by Theorem 9.1,

$$(9.7) \qquad \overset{\wedge}{\mathcal{H}}_c = \overline{\overset{\wedge}{\mathcal{H}}_+ + \overset{\wedge}{\mathcal{H}}_-},$$

we conclude that $\hat{A}$ has an absolutely continuous spectrum on $\overset{\wedge}{\mathcal{H}}_c$ with uniform infinite multiplicity on the entire real line.

COROLLARY 9.3. *Aside from the point spectrum $\lambda_j^2 > 0$, $j = 1, \cdots, N$, the operator L restricted to $L_2(F)$ has an absolutely continuous spectrum with uniform multiplicity on $(-\infty, 0]$.*

Proof: The set of vectors of the form $\{0, f_2\}$ is an invariant subspace of A^2; the action of A^2 on such data is precisely that of L restricted to $L_2(F)$. The only square integrable eigenfunctions are the φ_j, described in (1.18); $L_2(F) \ominus \{\varphi_j\}$ is just the second component space $\mathcal{H}_c$. Since by Lemma 1.4 the second component of data in $\mathcal{N}$ vanishes, this is also the second component space of $\overset{\wedge}{\mathcal{H}}_c$. The assertion of the corollary is therefore a simple consequence of Theorem 9.2.

Next we prove a local decay theorem for the second component of $U(t)f^\Gamma$ as $t \to \infty$ when f^Γ lies in $\mathcal{H}_c$.

THEOREM 9.4. *Suppose f^Γ belongs to $\mathcal{H}_c$. Then for any compact subset S of F,*

$$(9.8) \qquad \lim_{t \to \infty} \|[U(t)f^\Gamma]_2\|^S = 0;$$

here $\|\cdot\|^s$ denotes the $L_2(S)$ norm.

We shall need

LEMMA 9.5. *For all f^Γ in $\mathcal{H}_c \cap \mathcal{D}(A)$,*

$$(9.9) \qquad \sum_{|\alpha| \leqq 1} \|\partial^\alpha f_2^\Gamma\|^s \leqq c(E_F^{1/2}(f^\Gamma) + E_F^{1/2}(Af^\Gamma).$$

Proof: Since $E_F \geqq 0$ on $\mathcal{H}_c$, it follows that

$$(9.10) \qquad \|f_2^\Gamma\| \leqq E_F^{1/2}(f^\Gamma).$$

According to (1.15) and (1.28),

$$(9.11) \qquad E_F(Af^\Gamma) = \int_F [(Lf_1^\Gamma)^2 + y^2(\partial f_2^\Gamma)^2 - (f_2^\Gamma)^2] \, dw.$$

The relation (9.9) is a direct consequence of (9.10) and (9.11).

We are now ready to prove Theorem 9.4. Since the L_2 norm of the second component of data in $\mathcal{H}_c$ is dominated by the E_F form and since U is an isometry relative to E_F, it suffices to establish (9.8) for a subset of data dense in $\mathcal{H}_c$ relative to the E_F form; $\mathcal{H}_c \cap \mathcal{D}(A)$ will do as this dense subset.

The spectal resolution of $\overset{\blacktriangle}{A}$ gives the following representation for $\hat{U}(t)$:

$$\overset{\blacktriangle}{U}(t) = \int e^{i\lambda t} \, d\overset{\blacktriangle}{P}(\lambda).$$

In particular, for any $\overset{\blacktriangle}{f}{}^\Gamma$, $\overset{\blacktriangle}{g}{}^\Gamma$ in $\overset{\blacktriangle}{\mathcal{H}}_c$,

$$(9.12) \qquad E_F(\overset{\blacktriangle}{U}(t)\overset{\blacktriangle}{f}{}^\Gamma, \overset{\blacktriangle}{g}{}^\Gamma) = \int e^{i\lambda t} \, d(\overset{\blacktriangle}{P}(\lambda)\overset{\blacktriangle}{f}{}^\Gamma, \overset{\blacktriangle}{g}{}^\Gamma).$$

Since $\overset{\blacktriangle}{A}$ has absolutely continuous spectrum over $\overset{\blacktriangle}{\mathcal{H}}_c$, the measure on the right in (9.12) is absolutely continuous. Therefore, by the Riemann-Lebesgue lemma,

$$(9.13) \qquad \lim_{t \to \infty} E_F(\overset{\blacktriangle}{U}(t)\overset{\blacktriangle}{f}{}^\Gamma, \overset{\blacktriangle}{g}{}^\Gamma) = 0$$

for all $\overset{\blacktriangle}{g}{}^\Gamma$ in $\overset{\blacktriangle}{\mathcal{H}}_c$. Since $\mathcal{N}$ is the null set of E_F, we can replace the cosets by any of their elements and get

$$(9.13)' \qquad \lim_{t \to \infty} E_F(U(t)f^\Gamma, g^\Gamma) = 0$$

for all g^Γ in $\mathcal{H}_c$. Since f^Γ lies in $\mathcal{H}_c$, so does $U(t)f^\Gamma$. Thus $U(t)f^\Gamma$ is E_F-orthogonal to the $p_j^\pm$ of (1.18)''. We may therefore add arbitrary linear combinations of $p_j^\pm$ to g^Γ in (9.13)'. Since

$$\mathcal{H}_G = \mathcal{H}_c \oplus \{p_j^\pm\},$$

we see that (9.13)' holds for all g^Γ in $\mathcal{H}_G$. In particular, it holds for all g^Γ of the form $\{0, g_2^\Gamma\}$ so that

(9.13)''
$$\lim_{t \to \infty} ([U(t)f^\Gamma]_2, g_2^\Gamma) = 0.$$

Writing $f^\Gamma(t) = U(t)f^\Gamma$ and again using the isometric property of U with respect to E_F, we get

(9.14)
$$E_F(A^j f^\Gamma(t)) = E_F(A^j f^\Gamma) \leqq C'$$

for $j = 0, 1$. Combining this with (9.9) yields

(9.15)
$$\sum_{|\alpha| \leqq 1} \|\partial^\alpha f^\Gamma(t)_2\|^S \leqq C''$$

for all t. We conclude from this and the Rellich compactness criterion that the $f^\Gamma(t)_2$ are compact in the $L_2(S)$ norm. According to (9.13)'', the $f^\Gamma(t)_2$ converge weakly to zero as $t \to \infty$ in L_2; this combined with compactness in the $L_2(S)$ norm shows that they converge strongly to zero in the $L_2(S)$ norm. This completes the proof of Theorem 9.4.

We note that data in the null space $\mathcal{N}$ remain invariant under the action of $U(t)$ so that we can not expect local decay for the first component of $U(t)f^\Gamma$ for arbitrary f^Γ in $\mathcal{H}_c$. However if we project out the null space, we can prove local energy decay.

We denote by N the G-orthogonal projection of $\mathcal{H}_c$ onto $\mathcal{N}$; $P = 1 - N$ is then projection onto the G-orthogonal complement of $\mathcal{N}$ in $\mathcal{H}_c$. By part (c) of Lemma 1.4, the E_F and G forms are equivalent in $P\mathcal{H}_c$.

THEOREM 9.6. $PU(t)f^\Gamma$ tends to zero as $t \to \infty$ in the local L_2 norm.

We shall use the following simple but useful result.

LEMMA 9.7.   Set

(9.16)
$$W(t) = PU(t)P.$$

Then $W(t)$ is a strongly continuous group of operators on $P\mathcal{H}_c$ which is unitary with respect to the E_F form.

Proof: Since $Pf^\Gamma = f^\Gamma - n$ for some n in $\mathcal{N}$, we see that $U(t)Pf^\Gamma = U(t)f^\Gamma - n$. Applying P to this, one obtains

(9.17)
$$PU(t)Pf^\Gamma = PU(t)f^\Gamma.$$

Hence

$$W(t)W(s) = PU(t)P^2U(s)P = PU(t)U(s)P = W(t + s);$$

that is, $W(t)$ is a group of operators. Moreover, since P only projects out null vectors of E_F, we have for f^Γ in $P\mathcal{H}_c$:

$$(9.18) \qquad E_F(W(t)f^\Gamma) = E_F(PU(t)f^\Gamma) = E_F(U(t)f^\Gamma) = E_F(f^\Gamma).$$

It follows that W is unitary on $P\mathcal{H}_c$.

We return to the proof of Theorem 9.6. Since $W(t)$ is unitary in the E_F norm, it is bounded in any equivalent G norm. Hence it is enough to prove decay for a dense subset of $P\mathcal{H}_c$.

LEMMA 9.8. *Denote by $R(A)$ the range of A on $\mathcal{H}_c$. We claim that $PR(A)$ is dense in $P\mathcal{H}_c$.*

Proof: Suppose this is not the case; then there is a non-zero g^Γ in $P\mathcal{H}_c$ that is E_F-orthogonal to $PR(A)$:

$$(9.19) \qquad E_F(g^\Gamma, PAf^\Gamma) = 0$$

for all f^Γ in $\mathcal{D}(A) \cap \mathcal{H}_c$. Since P removes only a null vector,

$$(9.19)' \qquad E_F(g^\Gamma, Af^\Gamma) = E_F(g^\Gamma, PAf^\Gamma) = 0.$$

Since the subspace $\mathcal{P}$ generated by the eigendata $\{p_j^\pm\}$ is E_F-orthogonal to $\mathcal{H}_c$, $(9.19)'$ holds for all f^Γ in $\mathcal{D}(A)$. Writing $(9.19)'$ in component form we see that in particular

$$-(g_1^\Gamma, Lf_2^\Gamma) + (g_2^\Gamma, Lf_1^\Gamma) = 0$$

for all f^Γ in $C_0^\infty(F)$. This shows that

$$Lg_1^\Gamma = 0 = Lg_2^\Gamma,$$

in the sense of distributions. By part (c) of Theorem 1.2, L has no square-integrable null function, so that $g_2 = 0$. By part (a) of Lemma 1.4, it follows that g^Γ belongs to $\mathcal{N}$. Since g^Γ is in $P\mathcal{H}_c$, $g^\Gamma = 0$; this completes the proof of Lemma 9.8.

Since the second components of data in $\mathcal{N}$ vanish, it is clear that

$$(9.20) \qquad U(t)f^\Gamma|_2 = PU(t)f^\Gamma|_2.$$

Thus it follows from Theorem 9.4 that

$$(9.21) \qquad PU(t)f^\Gamma|_2 \to 0$$

in the $L_2(S)$ norm. Likewise,

$$(9.22) \qquad AU(t)f^\Gamma|_1 = U(t)f^\Gamma|_2 \to 0$$

in the $L_2(S)$ norm. We choose now

$$(9.23) \qquad G = K + E_F,$$

with

$$(9.23)' \qquad\qquad K(f^\Gamma) = \int_S |f_1^\Gamma|^2 \, dw,$$

S a compact subset of F. Since $K > 0$ on $\mathcal{N}$, it follows by Theorem 3.4 of [3] that (9.23) defines an equivalent G form on $\mathcal{H}_c$ (but not necessarily on all of $\mathcal{H}_G$).

We recall now that Ng^Γ is the G-orthogonal projection of g^Γ into $\mathcal{N}$; thus for any g^Γ in $\mathcal{H}_c$, $n = Ng^\Gamma$,

$$G(g^\Gamma, n) = G(n, n).$$

Using the definition (9.23) of G, this can be rewritten as

$$K(g^\Gamma, n) + E_F(g^\Gamma, n) = K(n) + E_F(n).$$

Since n is a null vector for E_F, the E_F terms above are zero; consequently,

$$K(g^\Gamma, n) = K(n).$$

Since K is non-negative, the Schwarz inequality holds and we get

$$(9.24) \qquad\qquad K(g^\Gamma) \geqq K(n) = K(Ng^\Gamma).$$

We choose now

$$(9.25) \qquad\qquad g^\Gamma = g^\Gamma(t) = AU(t)f^\Gamma,$$

f^Γ any element of $\mathcal{H}_c \cap \mathcal{D}(A)$. By (9.22),

$$(9.26) \qquad\qquad \lim_{t \to \infty} K(g^\Gamma(t)) = 0.$$

Using this in (9.24) we deduce that

$$(9.26)' \qquad\qquad \lim_{t \to \infty} K(Ng^\Gamma(t)) = 0.$$

Since by definition $Pg^\Gamma = g^\Gamma - Ng^\Gamma$, (9.26) and (9.26)' imply that

$$\lim_{t \to \infty} K(Pg^\Gamma(t)) = 0.$$

By definition (9.25) of $g^\Gamma(t)$, this means

$$(9.27) \qquad\qquad \lim_{t \to \infty} K(PAU(t)f^\Gamma) = 0.$$

Using the commutativity of A and U, and relation (9.17), we get

$$PU(t)PAf^\Gamma = PU(t)Af^\Gamma = PAU(t)f^\Gamma.$$

So, by (9.27),

(9.27)' $$\lim_{t\to\infty} K(PU(t)PAf^\Gamma) = 0$$

for all f^Γ in $\mathcal{H}_c \cap \mathcal{D}(A)$. Now according to Lemma 9.8, every element of $P\mathcal{H}_c$ can be approximated in the energy norm by elements of the form PAf^Γ, f^Γ in $\mathcal{H}_c \cap \mathcal{D}(A)$. Since E_F and G are equivalent on $P\mathcal{H}_c$, K is continuous in the energy norm; therefore it follows from (9.27)' that

$$\lim_{t\to\infty} K(PU(t)f^\Gamma) = 0$$

for every f^Γ in $P\mathcal{H}_c$. This proves Theorem 9.6.

Even more is true:

THEOREM 9.9. *The first component of $PU(t)f^\Gamma$ tends to zero as $t \to \infty$ in the local H_1 norm, i.e., its first derivatives also tend to zero in the local L_2 norm.*

Proof: Take f^Γ in $\mathcal{H}_c \cap \mathcal{D}(A)$; applying Theorem 9.4 to Af we deduce that

$$U(t)Af|_2 \to 0$$

in the local L_2 norm. Since

$$U(t)Af^\Gamma|_2 = AU(t)f^\Gamma|_2 = L[U(t)f^\Gamma]_1,$$

it follows that

$$L[U(t)f^\Gamma]_1 \to 0$$

in the local L_2 norm. Since $\mathcal{N}$ consists of data whose first component is annihilated by L, it follows that

$$L[PU(t)f^\Gamma]_1 \to 0$$

in the local L_2 norm. Combining this with Theorem 9.6 we deduce by elliptic estimates that the first derivative of $PU(t)f^\Gamma|_1$ tends to zero in the local L_2 norm. The set of data for which this holds, that is $\mathcal{H}_c \cap \mathcal{D}(A)$, is clearly dense in $\mathcal{H}_c$.

It was shown in [3] (relations (3.14) and (3.19)) that there exists a G form for which

(9.28) $$\int_S |\partial f_1^\Gamma|^2 \, dw \leq \text{const. } G(f^\Gamma)$$

for all f^Γ in $\mathcal{H}_G$. By Lemmas 9.7 and 1.4, $W(t)$ is uniformly bounded in the G norm, so that, applying (9.28) to $W(t)f^\Gamma$ and using (9.17), we have

(9.29) $$\int_S |\partial PU(t)f^\Gamma|_1|^2 \, dw \leq \text{const. } G(W(t)f^\Gamma) \leq \text{const. } G(f^\Gamma).$$

Since the left-hand side tends to zero on $\mathcal{H}_c \cap \mathcal{D}(A)$, a dense subset of $\mathcal{H}_c$, it follows from (9.29) that this remains true for all f^Γ in $\mathcal{H}_c$. This completes the proof of Theorem 9.9.

Denote by $N(T)$ the number of points of an orbit generated by a discrete group Γ that are contained inside a ball of radius T, centered at w_0. In Theorem 5.8 of [3], see inequality (5.81) on page 333, we show that if the energy E_F is non-negative for functions automorphic with respect to Γ, then

$$(9.30) \qquad N(T) = o(Te^T).$$

(The O on the right in (5.81) of [3] is a misprint for o). We shall show

THEOREM 9.10. *Suppose that the energy E_F is positive definite for functions automorphic with respect to Γ; then*

$$(9.31) \qquad N(T) = o(e^T).$$

Proof: If the energy is positive definite, the null space $\mathcal{N}$ is empty, and the projection $P = I - N$ is the identity. In this case, Theorem 9.6 asserts that, for any ball B_1,

$$(9.32) \qquad \lim_{t \to a} \|U(t)f^\Gamma\|_{B_1} = 0,$$

where $\| \ \|_{B_1}$, denotes the L_2 norm over B_1. We argue now as on pp. 338–339 of [3], except that in place of (5.94) there we use (9.32) above. This proves inequality (9.31) of Theorem 9.10.

For groups in n-dimensional hyperbolic space the right side of (9.31) has to be replaced by

$$(9.33) \qquad o(\exp\{\tfrac{1}{2}(n-1)T\}).$$

10. Completeness when $E_F > 0$

In this section we prove the completeness of the translation representations for dimensions 2 and 3 when $E_F > 0$. As explained in Section 1 of Part I, if $E_F > 0$, then L can have no eigenfunctions. As a consequence, $\mathcal{H}_G = \mathcal{H}_c$, $\mathcal{N}$ is trivial, and the E_F and G forms are equivalent on $\mathcal{H}_G$. The Hilbert space $\mathcal{H}_G$ can therefore be obtained as the completion of $C_0^\infty(F)$ data in the E_F norm and we denote it by $\mathcal{H}_E$. Part (c) of Theorem 1.2 can now be restated as

$$(10.1) \qquad \int_S |f_1^\Gamma|^2 \, dw \leq c \, E_F(f^\Gamma),$$

where S is any compact subset of F and c is a constant depending on S.

THEOREM 10.1. *If $E_F > 0$ and $n = 2$ or 3, then $R_\pm^0$ defines an isometry on $\mathcal{H}_E$ to $L_2(\mathbb{R} \times B_F)$ and hence the translation representations are complete.*

Thus the zeroth component of $R_\pm^F$ by itself defines an isometry on $\mathcal{H}_E$; this can happen only if the other components (i.e., the $R_\pm^j$, $j > 0$) vanish. As a consequence we conclude that

COROLLARY 10.2. *If $E_F > 0$ and $n = 2$ or 3, then F contains no cusp of maximal rank.*

Even more is true:

If $E_F \geqq 0$, n arbitrary, then F contains no cusp of maximal order, unless Γ is elementary, i.e., consists of parabolic motions with a common fixed point.

For $n = 2$ this is due to Patterson [7], for $n > 2$ to Sullivan [8]; in [3], Theorem 6.4, we gave a new proof of this result.

We break up the proof of Theorem 10.1 into four lemmas; only the first depends on the dimension of the space. To simplify the exposition we begin with the case $n = 3$; the modifications needed for $n = 2$ are presented at the end of this section.

We recall that the map $J_{+,0}^F$, described in (2.21) of Part I on C_0^∞ functions $l(s,\beta)$ with support in compact subsets of $\mathbb{R} \times B_F$, is defined as

$$(10.2) \qquad\qquad J_{+,0}^F l = g^\Gamma;$$

here

$$(10.3) \qquad\qquad g^\Gamma = \sum_\Gamma g^\tau \quad \text{and} \quad g = J_+ l.$$

Now

$$(10.4) \qquad\qquad g^\tau = J_+ l^\tau,$$

where, as in (1.12)′,

$$(10.5) \qquad\qquad l^\tau(s,\beta) = e^{\langle \tau j, \beta \rangle} l(s - \langle \tau j, \beta \rangle, \tau^{-1}\beta).$$

The support of l^τ is clearly contained in $\mathbb{R} \times (\tau B_F)$ so that the supports for different τ's in Γ are disjoint. Summing over Γ we obtain

$$(10.6) \qquad\qquad l^\Gamma = \sum_\Gamma l^\tau,$$

in terms of which (10.3) can be rewritten as

$$(10.7) \qquad\qquad g^\Gamma = J_+ l^\Gamma.$$

We shall discuss the interpretation of this below.

According to Theorem 2.3,

$$(10.8) \qquad\qquad E_F(g^\Gamma) = \iint_{B_F} |l^\Gamma(s,\beta)|^2 \, dm(\beta) \, ds.$$

By taking Cauchy sequences we can extend the above class of l's to all square integrable functions on $\mathbb{R} \times B_F$. The first step in our proof of the theorem is

LEMMA 10.3.   *Suppose $l^\Gamma|_{B_F}$ belongs to $L_2(\mathbb{R} \times B_F)$; then, for any finite interval $I = (-a,a) \subset \mathbb{R}$, $l^\Gamma|_I \in L_1(I \times B)$; in fact there is a constant c depending only on a such that*

$$(10.9) \qquad \int_I \int_B |l^\Gamma| \, dm \, ds \leqq c \left[\int_\mathbb{R} \int_{B_F} |l^\Gamma|^2 \, dm \, ds \right]^{1/2}.$$

Proof: We may assume that $l^\Gamma \geqq 0$ since any l^Γ can be decomposed into its positive and negative parts. We start with an l in C^∞ having compact support in $\mathbb{R} \times B_F$. According to (1.13), for $n = 3$ the first component of $J_+ l$ can be written as

$$(10.10) \qquad g_1(w) = |c| \int_B e^{\langle w, \beta \rangle} l(\langle w, \beta \rangle, \beta) \, dm(\beta).$$

It follows from Lemma 2.1 that except for a finite set of τ's in Γ, $g_1^\tau(w)$ vanishes on F. Since $l^\tau \geqq 0$, we see from (10.10) that $g_1^\tau(w)$ vanishes on F if and only if $l^\tau(\langle w, \beta \rangle, \beta)$ vanishes on F. Summing (10.10) over Γ we deduce that, for w in F,

$$(10.10)' \qquad g_1^\Gamma(w) = |c| \int_B e^{\langle w, \beta \rangle} l^\Gamma(\langle w, \beta \rangle, \beta) \, dm(\beta)$$

is well defined on F.

The completion of the proof of Lemma 10.3 is based on

LEMMA 10.4.   *Let $u(w,t)$ be a solution of the wave equation in dimension 2 or 3 with initial data f^Γ in $\mathcal{H}_E$ and suppose φ is $C_0^2(\mathbb{R})$. Define*

$$(10.11) \qquad v_\varphi(w) = \int u(w,t) \varphi(t) \, dt.$$

For any compact subset S of F there is a constant such that

$$\text{(i)} \qquad \int_S |v_\varphi|^2 \, dw \leqq \text{const.} \; E_F(f^\Gamma),$$

$$\text{(ii)} \qquad |v_\varphi(w)|^2 \leqq \text{const} \; E_F(f^\Gamma),$$

for all f^Γ in $\mathcal{H}_E$.

Proof: Part (i) follows directly from (10.1), (10.11) and the conservation of energy. To see part (ii) write

$$Lv_\varphi = \int (Lu)\varphi(t) \, dt = \int u_{tt} \varphi(t) \, dt$$

$$= \int u \varphi''(t) \, dt = v_{\varphi''}.$$

It follows from this and part (i) that

$$(10.12) \qquad \int_S |Lv_\varphi|^2 \, dw = \int_S |v_{\varphi'}|^2 \, dw \leqq \text{const. } E_F(f^\Gamma).$$

For any compact S' contained in the interior of S, we obtain from elliptic estimates

$$(10.13) \qquad \int_{S'} |\partial^2 v|^2 \, dw \leqq c(S') \int_S [|Lv|^2 + |v|^2] \, dw.$$

By the Sobolev inequality, for $w \in S'$,

$$(10.14) \qquad |v(w)|^2 \leqq \text{const. } \int_S [|\partial^2 v|^2 + |v|^2] \, dw.$$

The assertion (ii) follows from (10.12), (10.13) and (10.14).

We now return to the proof of Lemma 10.3. Since J_+ transmutes translation into the action of U, $U(t)g^\Gamma$ is represented by $l^\Gamma(s - t)$. Thus if $u(w,t)$ is a solution of the wave equation with initial data $g^\Gamma = J_+ l^\Gamma$, where l is C^∞ and of compact support in $\mathbb{R} \times B_F$, it follows from (10.10) that

$$(10.15) \qquad u(w,t) = \int e^{\langle w,\beta \rangle} l^\Gamma(t - \langle w,\beta \rangle, \beta) \, dm(\beta).$$

Set $w = j$ in (10.15), multiply by $\varphi(t)$, and integrate with respect to t. We get, since $\langle j,\beta \rangle = 0$,

$$(10.15)' \qquad v_\varphi(j) = \iint \varphi(t) l^\Gamma(-t,\beta) \, dm(\beta) \, dt.$$

Choosing $\varphi \geqq 0$ in C_0^2 with $\varphi(t) = 1$ for $|t| < a$, we see from (10.15)' that

$$(10.15)'' \qquad v_\varphi(j) \geqq \int_{-a}^a \int l^\Gamma(t,\beta) \, dm(\beta) \, dt.$$

Combining this with (10.8) and part (ii) of Lemma 10.4 we see that

$$\iint_I \int_B l^\Gamma(t,\beta) \, dm(\beta) \, dt \leqq \text{const. } \int_\mathbb{R} \int_{B_F} |l^\Gamma(s,\beta)|^2 \, dm(\beta) \, ds.$$

This proves the assertion (10.9) for C^∞ compactly supported l. For any l in $L_2(\mathbb{R} \times B_F)$ the result holds by completion.

LEMMA 10.5.   *Suppose $k(s,\beta)$ is a bounded function, supported in $I \times B$, $I = (-a,a)$. We define, for any finite subset π of Γ,*

$$(10.16) \qquad k^\pi = \sum_{\tau \in \pi} k^\tau.$$

We claim that

(10.17)
$$k^\pi|_{B_F} \to k^\Gamma \quad in \quad L_2(\mathbb{R} \times B_F)$$

and

(10.17)′
$$k^\pi|_l \to k^\Gamma \quad in \quad L_1(I \times B),$$

where $\pi \to \Gamma$ in the sense of generalized limit with respect to the ordering of π by inclusion.

Proof: Again it suffices to treat only $k \geqq 0$. For any L_2 function l supported on $\mathbb{R} \times B_F$,

(10.18)
$$(k^\pi, l)_{B_F} = (k^\pi, l)_B = \sum_\pi (k^\tau, l)_B$$
$$= \sum_\pi (k, l^{\tau^{-1}})_B = (k, l^{\pi'})_B,$$

where

$$l^{\pi'} = \sum_\pi l^{\tau^{-1}}$$

and $(\ ,\ )_B$ is the $L_2(\mathbb{R} \times B)$ scalar product.

We choose now

$$l = \begin{cases} k^\pi & in \ \mathbb{R} \times B_F, \\ 0 & elsewhere, \end{cases}$$

and set this into (10.18). Since k is supported on $I \times B$ and bounded there, we can estimate the right-hand side as follows:

(10.19)
$$\int_\mathbb{R}\int_{B_F} |k^\pi|^2 \, dm \, ds \leqq (\max_{I \times B} k) \int_I\int_B l^{\pi'} \, dm \, ds.$$

Since k was taken to be $\geqq 0$, so is k^π and hence so is l. It follows then that $l^{\pi'} \leqq l^\Gamma$, and therefore

(10.20)
$$\int_I\int_B l^{\pi'} \, dm \, ds \leqq \int_I\int_B l^\Gamma \, dm \, ds.$$

Since $l^\Gamma|_{B_F} = l|_{B_F} = k^\pi$ obviously belongs to $L_2(\mathbb{R} \times B_F)$, Lemma 10.3 is applicable. Combining (10.9) with (10.19) and (10.20), we get

(10.21)
$$\left(\int_\mathbb{R}\int_{B_F} |k^\pi|^2 \, dm \, ds \right)^{1/2} \leqq c \max_{I \times B} k.$$

Since k is non-negative, k^π is an increasing function of π; denote its limit by k^Γ. Applying Fatou's lemma to the sequence $|k^\pi|^2$, we conclude from (10.21) that

$$\left(\int_\mathbb{R} \int_{B_F} |k^\Gamma|^2 \, dm \, ds \right)^{1/2} \leq c \max_{I \times B} k.$$

We now apply the monotone convergence theorem to $(k^\Gamma - k^\pi)^2$ to deduce that (10.16) holds.

Since $k^\Gamma|_{B_F}$ belongs to $L_2(\mathbb{R} \times B_F)$, it follows from Lemma 10.3 that k^Γ belongs to $L_1(I \times B)$. Since $k^\pi \leq k^\Gamma$ and $k^\pi \to k^\Gamma$ monotonically, we see by Fatou's lemma that $k^\pi \to k^\Gamma$ in $L_1(I \times B)$. This completes the proof of Lemma 10.5.

LEMMA 10.6.   *Suppose f^Γ is smooth with compact support in F and set $R^0_+ f^\Gamma = k^\Gamma$. Then*

$$(10.22) \qquad\qquad E_F(f^\Gamma, f^\Gamma) = (k^\Gamma, k^\Gamma)_{B_F}.$$

Proof: Choose $\varphi \in C^\infty_o(\mathbb{H}^n)$, $n = 2$ or 3, so that $\{\varphi^\tau\}$ defines a partition of unity on the support of f^Γ. Then supp φ is disjoint from all but a finite set of the τF's. Set

$$f = \varphi f^\Gamma \quad \text{and} \quad k = R_+ f.$$

Since f is smooth with compact support in $\mathbb{H}^n$, k will be smooth with support in $I \times B$ for some interval I. Obviously,

$$f^\Gamma = \Sigma_\tau f^\tau \quad \text{and} \quad R^0_+ f^\Gamma = k^\Gamma|_{B_F} = \Sigma_\tau k^\tau|_{B_F}.$$

Each of the sums in the following relation has only a finite number of non-zero terms. Using the bilinear version of (1.9), we get

$$(10.23) \qquad \begin{aligned} E_F(f^\Gamma, f^\Gamma) &= \Sigma_\tau E_F(f^\Gamma, f^\tau) = \Sigma_\tau E_{\tau F}(f^\Gamma, f) \\ &= E(f^\Gamma, f) = \Sigma_\tau E(f^\tau, f) = \Sigma_\tau (k^\tau, k)_B = (k^\pi, k)_B \end{aligned}$$

for π sufficiently large; here k^π is defined as in (10.15). Since k is bounded with support in a set of the type $I \times B$, it follows from (10.17)' of Lemma 10.5 that $(k^\pi, k)_B$ tends to $(k^\Gamma, k)_B$ as $\pi \to \Gamma$. We conclude therefore from (10.23) that

$$(10.24) \qquad\qquad E_F(f^\Gamma, f^\Gamma) = (k^\Gamma, k)_B.$$

We rewrite the right side as

$$(10.25) \qquad (k^\Gamma, k)_B = \Sigma_\tau (k^\Gamma, k)_{B_{\tau F}} = \Sigma_\tau (k^\Gamma, k^\tau)_{B_F} = \lim_\pi (k^\Gamma, k^\pi)_{B_F}.$$

Using the relation (10.17) and the fact that $k^\Gamma|_{B_F}$ is in $L_2(\mathbb{R} \times B_F)$ we see that

$$\lim_\pi (k^\Gamma, k^\pi)_{B_F} = (k^\Gamma, k^\Gamma)_{B_F}.$$

Combining this with (10.24) and (10.25), we get the desired relation (10.22).

Since $C_0^\infty(F)$ data are dense in $\mathcal{H}_E$ and since the map $R_+^0 : \mathcal{H}_E$ to $L_2(\mathbb{R} \times B_F')$ is continuous for compact subsets B_F' of B_F, the isometry (10.20) holds for all f^Γ in $\mathcal{H}_E$. This proves Theorem 10.1 for $n = 3$.

To complete the proof of the theorem, we merely have to verify Lemma 10.3 for $n = 2$. As we noted at the end of Section 2, in this case the first component of

$$(10.26) \qquad J_\pm k = \{f_1, f_2\}$$

is given by

$$(10.26)' \qquad f_1(w) = \int e^{\langle w, \beta \rangle / 2} J_2 k(\langle w, \beta \rangle, \beta) \, dm_2(\beta),$$

where

$$dm_2(\beta) = \frac{2 d\beta}{1 + \beta^2}.$$

J_2 is a convolution operator:

$$(10.27) \qquad J_2 k = \mathcal{J} * k$$

whose precise form has been computed by A. Woo in Lemma 5.5 of [9]:

$$(10.27)' \qquad \mathcal{J}(s) = \begin{cases} e^{-s/2}(1 - e^{-s})^{-1/2} & \text{for } s \geq 0, \\ 0 & \text{for } s < 0. \end{cases}$$

The positivity of $\mathcal{J}$ is essential for the proof.

We argue now just as before. Take an $l^\Gamma \geq 0$ for which $l^\Gamma|_{B_F}$ belongs to C_0 with support in a compact subset of $\mathbb{R} \times B_F$ and set $g^\Gamma = J_+ l^\Gamma$. Again we replace g^Γ by

$$\int \varphi(t) U(t) g^\Gamma \, dt,$$

the first component of which is of the form v_φ in (10.11). We use (10.26)' and (10.27) to obtain

$$(10.28) \qquad \begin{aligned} v_\varphi(j) &= |c| \iint_B \varphi(t) J_2 l^\Gamma(-t, \beta) \, dm_2(\beta) \, dt \\ &= |c| \iint_B \varphi(t) \left[\int_{-\infty}^{-t} \mathcal{J}(-t - \sigma) l^\Gamma(\sigma, \beta) \, d\sigma \right] dm_2(\beta) \, dt. \end{aligned}$$

We choose $\varphi \geq 0$ in C_0^2, equal to 1 on the interval $(-2a, 2a)$. Then

$$(10.28)' \qquad v_\varphi(j) \geq |c| \int_{-2a}^{2a} \int_B \int_{-\infty}^s \mathcal{J}(s - \sigma) l^\Gamma(\sigma, \beta) \, d\sigma \, dm_2(\beta) \, ds.$$

Interchanging the order of the s and σ integrations and truncating the σ range of integration, we get

$$(10.28)'' \qquad v_\varphi(j) \geqq |c| \int_B \int_{-a}^a \left[\int_\sigma^{2a} \mathcal{J}(s - \sigma)\, ds \right] l(\sigma,\beta)\, d\sigma\, dm_2(\beta).$$

The bracket term on the right is clearly bounded from below for σ in $I = (-a,a)$ and it follows that $(10.15)''$ holds when $n = 2$. Combining this with (10.8) and part (ii) of Lemma 10.4 proves (10.9) and hence Lemma 10.3.

11. Perturbations with Compact Support

In this section we show how to modify the development in Parts I and II of the translation representations so as to treat automorphic solutions of a perturbed wave equation when the support of the perturbation is compact. As before we work with a discrete subgroup Γ of motions having the finite geometric property and a fundamental domain F of infinite volume. The analogous treatment of the finite volume case can be found in [6].

It is convenient to choose F to be j-centered. In place of the usual hyperbolic metric, we now take

$$(11.1) \qquad\qquad ds^2 = g_{ij}\, dx_i\, dx_j;$$

here (g_{ij}) is positive definite, Γ-automorphic, of class $C^{(1)}$ and on F is equal to δ_{ij}/x_n^2 outside the ball B_a of radius a about j. Note that we have replaced y by x_n.

We shall work with the associated wave equation

$$(11.2) \qquad\qquad u_{tt} = Lu,$$

where

$$(11.3) \qquad\qquad L = g^{-1}\partial_{x_i} g^{ij} g \partial_{x_j} + (\tfrac{1}{2}(n - 1))^2 - q;$$

here g denotes the square root of $\det(g_{ij})$ and q is L_∞, Γ-automorphic and on F vanishes outside of B_a. The L_2 and Dirichlet integrals are now

$$(11.4) \qquad H_F(u) = \int_F |u|^2 g\, dx \quad \text{and} \quad D_F(u) = \int_F g^{ij}\partial_{x_i}u\, \overline{\partial_{x_j}u}\, g\, dx,$$

the energy form for data $f^\Gamma = (f_1, f_2)$ is

$$(11.5) \qquad E_F(f) = -H_F(f_1, Lf_1) + H_F(f_2)$$

$$= D_F(f_1) - (\tfrac{1}{2}(n - 1))^2 H_F(f_1) + \int_F q|f_1|^2 g\, dx + H_F(f_2).$$

Since E_F can be indefinite, we introduce the form

$$(11.6) \qquad\qquad K(f) = \int_F k(x)\, |f_1|^2 g\, dx,$$

constructed as in Section 3 of [3] so that

$$G = E_F + K$$

is locally definite and K is compact with respect to G. In adapting this construction to the present situation, we note that the perturbation effects only the interior patches where the negative values of q can be easily offset by changing the value of k over B_a.

As before we set $\mathcal{H}_G$ equal to the completion with respect to G of C^∞-automorphic data with compact support in F. The solution operators $U(t)$ define a strongly continuous group of operators on $\mathcal{H}_G$ with infinitesimal generator

$$A = \begin{pmatrix} 0 & I \\ L & 0 \end{pmatrix}.$$

LEMMA 11.1. *There are no nontrivial square-integrable automorphic solutions of $Lu = -\lambda^2 u$ when $\lambda \geqq 0$.*

Proof: It follows from Theorem 4.8 of [3] that u vanishes identically in a neighborhood of B_F. Since q is bounded, the unique continuation property of solutions of $Lu = -\lambda^2 u$ forces u to be identically zero.

Making use of these preliminary remarks, Section 1 of Part I can be taken over verbatim for the perturbed problem. We again define $p_j^\pm$ by means of the square-integrable eigenfunctions of L; in view of Lemma 11.1 these eigenfunctions have positive eigenvalues. We denote the span of the $(p_j^\pm)$ by $\mathcal{P}$ and its E_F-orthogonal complement by $\mathcal{H}_c$. $E_F \geqq 0$ on $\mathcal{H}_c$ and the null space $\mathcal{N}$ of E_F in $\mathcal{H}_c$ is finite-dimensional. Further the quotient space $\mathcal{H}_c/\mathcal{N}$ is complete in the E_F norm.

The development in Section 2 of Part I requires some modifications. To begin with, we define the translation representations $R_\pm^F$ as before, but only for $s > a$, that is for horospheres which do not intersect B_a. The argument used in the proof of part (iii) of Theorem 2.3 shows that

$$(11.7) \qquad \begin{aligned} R_+^F U(t) f^\Gamma &= T(t) R_+^F f^\Gamma \quad \text{for} \quad s > \max(a, a + t), \\ R_-^F U(t) f^\Gamma &= T(-t) R_-^F f^\Gamma \quad \text{for} \quad s > \max(a, a - t). \end{aligned}$$

To obtain (11.7) for the entire range of s we merely *define*

$$(11.8) \qquad \begin{aligned} R_+^F f^\Gamma &= T(-t) R_+^F U(t) f^\Gamma \quad \text{for} \quad s > a - t, \\ R_-^F f^\Gamma &= T(t) R_-^F U(t) f^\Gamma \quad \text{for} \quad s > a + t. \end{aligned}$$

Using the relations (11.7) it is easy to see that $R_+^F f^\Gamma$ and $R_-^F f^\Gamma$ are well defined by (11.8) and satisfy (11.7) for all s.

The rest of the proof of Theorem 2.3 now goes through essentially as before. In fact, if we set

(11.9) $\mathcal{D}_\pm = [f^\Gamma = J^F_\pm l, l(s,\beta) \text{ in } C^\infty_o \text{ with support in } (a,\infty) \times B'_F]$,

then it follows from the proof of Theorem 8.3 that the support of $J^F_\pm l$ is disjoint from B_a. Consequently all of the arguments in the proof of Theorem 2.3 go through for f^Γ in $\mathcal{D}_\pm$. Using the invariance of E_F and $\|R^F_\pm f^\Gamma\|$ under the action of $U(t)$, the proofs readily extend to $\mathcal{D}'_\pm$ and hence to $\mathcal{H}_\pm$.

Theorem 3.2 holds in the present setting since it deals only with neighborhoods of infinity, which can be chosen to be disjoint from B_a. Thus if f^Γ belongs to $\mathcal{H}_c \cap \mathcal{D}(A)$ and $R^F_+ f^\Gamma = 0$, then setting $f(t) = U(t)f^\Gamma$, there exists for any given $\varepsilon > 0$ and any point at the infinite set of F a neighborhood N such that

(11.10) $\varlimsup_{t\to\infty} H_N(f_2(t)) < \varepsilon$.

To prove completeness for $R^F_\pm$, that is to prove the analogue of Theorem 3.1, we also need local energy decay.

THEOREM 11.2.   *Suppose f^Γ belongs to $\mathcal{H}_c \cap \mathcal{D}(A^3)$ and set $f^{(j)}(t) = U(t)A^j f^\Gamma$. Then for any compact subset S of F, there exists a sequence $t_k \to \infty$ such that*

(11.11) $\lim_{k\to\infty} H_S(f^{(j)}_2(t_k)) = 0 \quad for \quad j = 0, 1, 2$.

We prove this by means of the next two lemmas.

LEMMA 11.3.   *A has no point spectrum in $\mathcal{H}_c/\mathcal{N}$.*

Proof: Since U is unitary on $\mathcal{H}_c/\mathcal{N}$, A is skew-selfadjoint with purely imaginary spectrum. Suppose first that $A\varphi = i\sigma\varphi$ in $\mathcal{H}_c/\mathcal{N}$, $\sigma \neq 0$. Then componentwise

$$L\varphi_1 = i\sigma\varphi_2 \quad and \quad \varphi_2 = i\sigma\varphi_1 + \eta, \quad where \quad (\eta,0) \in \mathcal{N}.$$

Setting $\varphi'_1 = \varphi_1 - \eta/i\sigma$, this becomes

$$L\varphi'_1 = i\sigma\varphi_2 \quad and \quad \varphi_2 = i\sigma\varphi'_1.$$

Thus φ'_1 lies in $L_2(F)$ and $L\varphi'_1 = -\sigma^2\varphi'_1$. According to Lemma 11.1, $\varphi'_1 \equiv 0$ so that φ is actually a null vector in $\mathcal{N}$. Next suppose that $A\varphi = 0 \mod \mathcal{N}$. Then $A^2\varphi = 0$. This requires that

$$L\varphi_1 = 0 \quad and \quad L\varphi_2 = 0.$$

Again since φ_2 is square integrable on F, Lemma 11.1 implies that $\varphi_2 \equiv 0$, so that again φ belongs to $\mathcal{N}$.

LEMMA 11.4.   *If f^Γ lies in $\mathcal{H}_c \cap \mathcal{D}(A^3)$, then there exists a sequence $t_k \to \infty$ such that, for all g^Γ in $\mathcal{H}_G$,*

(11.12) $\lim_{k\to\infty} E_F(f^{(j)}(t_k), g^\Gamma) = 0 \quad for \quad j = 0, 1, 2$.

Proof: Since the spectrum of A on $\mathcal{H}_c/\mathcal{N}$ is continuous by Lemma 11.3, a standard application of Wiener's theorem on the Fourier-Stieltjes transform of continuous measures gives the weak convergence of a sequence on $\mathcal{H}_c/\mathcal{N}$ (see pages 145–147 of [1]). The extension to all of $\mathcal{H}_c$ is trivial and since $\mathcal{P}$ is E_F-orthogonal to $\mathcal{H}_c$, the assertion (11.12) holds for all g^{Γ} in $\mathcal{H}_G$.

A simple consequence of (11.12), obtained by setting $g^{\Gamma} = (0, g_2)$, is that the $f_2^{(j)}(t_k)$ converge weakly to zero in $L_2(F)$ for $j = 0, 1, 2$. To complete the proof of Theorem 11.2, we have merely to note that the inequality (9.9) holds for the $f_2^{(j)}(t)$, $j = 0, 1, 2$, and hence these functions are compact on any $L_2(S)$. This together with the weak convergence yields the assertion of the theorem.

We are now in a position to prove

THEOREM 11.5. *If f^{Γ} belongs to $\mathcal{H}_c \cap \mathcal{D}(A^3)$ and $R_+^F f^{\Gamma} = 0$, then there exists a sequence $t_k \to \infty$ such that*

$$(11.13) \qquad \lim_{k \to \infty} H_F(f_2^{(j)}(t_k)) = 0 \quad for \quad j = 0, 1, 2,$$

where $f^j(t) = U(t)Af^{\Gamma}$.

Proof: We note that

$$R_+^F A^j f = (-\partial_s)^j R_+^F f^{\Gamma} = 0.$$

Hence according to Theorem 3.2, given $\varepsilon > 0$ there exists a neighborhood N_ε of infinity for F, say the complement of the compact set S_ε, on which

$$\overline{\lim_{t \to \infty}} H_{N_\varepsilon}(f_2^{(j)}(t)) < \varepsilon \quad for \quad j = 0, 1, 2.$$

Choosing t_k as in Theorem 11.2 and setting $S = S_\varepsilon$ in (11.11), it follows that

$$\overline{\lim_{k \to \infty}} H_F(f_2^{(j)}(t_k)) < \varepsilon \quad for \quad j = 0, 1, 2.$$

Since ε is arbitrary, the assertion of Theorem 11.5 follows.

We now prove the analogue of Theorem 3.1.

THEOREM 11.6. *The translation representations $R_\pm^F$ are complete, i.e., they are unitary maps of $\mathcal{H}_c/\mathcal{N}$ onto $L_2(\mathbb{R}, L_2(B_F)) \times L_2(\mathbb{R})^N$; here the infinite boundary of F consists of B_F plus the N cusps of maximal rank.*

Proof: Suppose that R_+^F is not complete. Then it follows by Theorem 2.3 that there is a g^{Γ} in $\mathcal{H}_c$ with $E_F(g^{\Gamma}) \neq 0$ such that

$$(11.14) \qquad R_+^F(g^{\Gamma}) = 0.$$

Since R_+^F is a translation representation, $g^{\Gamma}(t) = U(t)g^{\Gamma}$ also belongs to the null space of R_+^F, and so does

$$(11.15) \qquad g_\varphi^\Gamma = \int \varphi(t) U(t) g^\Gamma \, dt$$

for any φ in $C_0^\infty(\mathbb{R})$. Approximating the δ-function with such φ's, the corresponding g_φ^Γ will converge to g^Γ in the G norm. Hence we can choose φ so that $E_F(g_\varphi^\Gamma) \neq 0$. Moreover for φ in $C_0^3(\mathbb{R})$, g_φ^Γ will belong to $\mathcal{H}_c \cap \mathcal{D}(A^3)$.

Setting $f^{(j)}(t) = U(t) A^j g_\varphi^\Gamma$, we see that Theorem 11.5 applies. Hence there exists a sequence $t_k \to \infty$ such that (11.13) holds. Setting $v(t) = f_2^{(0)}(t)$, this can be rewritten as

$$\lim H_F(v(t_k)) = 0,$$

$$(11.16) \qquad \lim H_F(v_t(t_k)) = 0,$$

$$\lim H_F(Lv(t_k)) = 0.$$

Now

$$E_F(f^{(1)}) = -H_F(Lv, v) + H_F(v_t).$$

Therefore it follows from (11.16) that

$$\lim_{k \to \infty} E_F(f^{(1)}(t_k)) = 0.$$

Using the invariance of E_F, we get

$$E_F(A g_\varphi^\Gamma) = E_F(f^{(1)}(0)) = \lim_{k \to \infty} E_F(f^{(1)}(t_k)) = 0.$$

Consequently, $A g_\varphi^\Gamma$ belongs to $\mathcal{N}$ and since, by Lemma 11.3, A has no point spectrum in $\mathcal{H}_c / \mathcal{N}$, it follows that g_φ^Γ belongs to $\mathcal{N}$, contrary to our choice of g_φ^Γ. This concludes the proof of Theorem 11.6.

Finally we remark that Section 4 of Part II also carries over to furnish us with a characterization of the spectrum of the Laplace-Beltrami operator.

Acknowledgment. The work of the first author was supported in part by the Department of Energy under Contract DE-AC02-76 ERO 3077 and that of the second by the National Science Foundation under Grant MCS-83-04317.

Bibliography

[1] Lax, P. D., and Phillips, R. S., *Scattering Theory*, Academic Press, New York, 1967.
[2] Lax, P. D., and Phillips, R. S., *Translation representations for the solution of the non-Euclidean wave equation*, II, Comm. Pure and Appl. Math., 34, 1981, pp. 347–358.
[3] Lax, P. D., and Phillips, R. S., *The asymptotic distribution of lattice points in Euclidean and non-Euclidean spaces*, Jr. of Functional Anal., 46, 1982, pp. 280–350.
[4] Lax, P. D., and Phillips, R. S., *A local Paley-Wiener theorem for the Radon transform of L_2 functions in a non-Euclidean setting*, Comm. Pure and Appl. Math., 35, 1982, pp. 531–554.

[5] Lax, P. D., and Phillips, R. S., *Translation representations for automorphic solutions of the wave equation in non-Euclidean spaces*, I and II, Comm. Pure and Appl. Math, 37, 1984, pp. 303–328, 779–813.

[6] Lax, P. D., and Phillips, R. S., *Translation representations for automorphic solutions of the wave equation in non-Euclidean spaces; the case of finite volume*, Trans. Amer. Math. Soc., in press.

[7] Patterson, S. J., *The limit set of a Fuchsian group*, Acta Math., 136, 1976, pp. 241–273.

[8] Sullivan, D., *Entropy, Hausdorff measures old and new, and limit sets of geometrically finite Kleinian groups*, Acta Math., in press.

[9] Woo, A. C., *Scattering theory on real hyperbolic spaces and their compact perturbations*, Dissertation, Stanford University, 1980.

Received April, 1984.

Translation Representation
for Automorphic Solutions of the
Wave Equation in Non-Euclidean Spaces, IV

PETER D. LAX
Courant Institute

AND

RALPH S. PHILLIPS
Stanford University

Abstract

In Part I of this series of papers we have defined the incoming and outgoing translation representations for automorphic solutions of the hyperbolic wave equations; in Part II we have proved the completeness of these representations when the fundamental polyhedron F has a finite number of sides with a finite or infinite volume, but is not compact. In Part IV we present a proof of completeness which is simpler than our original proof contained in Section 7 of Part II for the case when F has cusps of less than maximal rank; and we supply a proof for the case, not covered in Section 7, when the parabolic subgroup associated with such cusps contains twists.

12. Introduction

In Part I of our treatment of the translation representations for automorphic solutions of the wave equation (see [4]) we obtained the basic properties of the translation representors $R_\pm$ and in Part II we proved completeness, assuming that the stability subgroups for cusps of intermediate rank were without twist. In Part IV we drop this assumption and treat cusps of intermediate rank having stability subgroups with twist. To do this it suffices to reprove part (iii) of Theorem 3.2, taking twist into account.

THEOREM 3.2 (iii). *Suppose f belongs to $\mathcal{H}_c$ and let ν_j be a cusp of the fundamental domain F not of maximal rank. Then, given $\varepsilon > 0$, there exists a neighborhood $V_j(\varepsilon)$ of ν_j in F such that for all t*

$$(12.1) \qquad \int_{V_z(\varepsilon)} |f_2(t)|^2 \, dV < \varepsilon \,.$$

As before, the setting for this problem is the manifold H^n/Γ, where H^n is the hyperbolic n-space and Γ is a discrete subgroup of motions on H^n. The stability subgroup Γ_∞ for a cusp at ∞ maps each horosphere ($y = y_0$) into itself, and within such a horosphere it acts like a discrete group of Euclidean

Communications on Pure and Applied Mathematics, Vol. XLV, 179–201 (1992)

isometries. Moreover any discrete group of Euclidean isometries contains a maximal normal free abelian subgroup Γ_a of finite index.

In the case of a cusp of intermediate rank at ∞ the fundamental domain will be of the form $F_\infty \times \mathbf{R}_+$ near the cusp; here $F_\infty \subset \mathbf{R}^{n-1}$ is a fundamental domain of Γ_∞ and is not compact. $\mathbf{R}^{n-1}$ has an orthogonal decomposition $\mathbf{R}^q \times \mathbf{R}^{n-q-1}$, in terms of which F_∞ may be chosen to take the form $F_c \times \mathbf{R}^{n-q-1}$, where F_c is a compact polyhedron. In the coordinates (x, y, z), where $x \in \mathbf{R}^q$, $z \in \mathbf{R}^{n-q-1}$, $y > 0$, a motion γ in Γ_∞ is of the form

$$(12.2) \qquad \gamma : (x, z, y) \to (Ax + b, Rz, y) ;$$

here A and R are rotations. It is the R that creates the twist. Finally, for the maximal free abelian subgroup Γ_a the motions are of the form

$$(12.3) \qquad \gamma : (x, z, y) \to (x + b, Rz, y) .$$

In this case F_c is a parallelepiped and the R's form a commuting set of rotations.

Now any solution of the wave equation on $\mathbf{H}^n / \Gamma_\infty$ defines a solution of the same equation of the larger domain $\mathbf{H}^n / \Gamma_a$ and it is clear that if the inequality (12.1) holds in the larger domain it necessarily holds on the smaller one. Hence it suffices to prove Theorem 2.3 (iii) for $F = \mathbf{H}^n / \Gamma_a$. Our proof is in two steps. The first is a simplified version of the proof without twist in Section 7 of Part II. It is based on the Plancherel relation for the non-Euclidean Fourier transform on the slab of dimension $n - 1$, which, by induction, is a consequence of the main argument for the case $n - 1$ dimensions. The second step takes care of the twist.

It is possible to avoid the induction argument by a direct proof of the Plancherel relation for the Fourier transform on the slab. In Section 16 we show how the proof goes in the case of a cusp of maximal rank in three dimensions.

We shall continue to use the notation introduced in Parts I and II, except that in order to simplify the Plancherel relation, we have included the normalizing factor P in the definition of the (augmented) Radon transform. The reader will be well advised to familiarize himself with Parts I and II of [4].

13. The Non-Euclidean Fourier Transform on the Slab

The by now classical Fourier transform $\tilde{g}$ for square integrable functions g on $\mathbf{H}^n$ is connected to the Radon transform $\hat{g}$ of g as follows:

$$
\begin{aligned}
(13.1) \qquad \tilde{g}(\mu, \beta) &= \int_{\mathbf{H}^n} \exp\left\{((n-1)/2 + i\mu)\langle w, \beta\rangle\right\} g(w)\, dV \\
&= \frac{1}{c(\mu)} \int_{\mathbf{R}} \exp\left\{i\mu_s\right\} \hat{g}(s, \beta)\, ds ;
\end{aligned}
$$

here $\mu > 0$, β is a point at infinity (i.e., $(y = 0)$), $\langle w, \beta \rangle$ is the signed hyperbolic distance of the horosphere through β and w from some arbitrary point, say $(0,1)$, $c(\mu)$ is the Harish-Chandra c-function, and $\hat{g}(s, \beta)$ is the (augmented) Radon transform:

$$(13.2) \qquad \hat{g}(s, \beta) = P \int_{\langle w, \beta \rangle = s} g(w) \, dS ,$$

where P is an integro-differential operator. For $n = 3$, $P = (\sqrt{2}\pi)^{-1} \partial_s e^s$.
The corresponding Plancherel formulae are

$$(13.3) \qquad H(g) = \int_{\mathbf{H}^n} |g(w)|^2 \, dV = \int_B \int_{\mathbf{R}_+} |\tilde{g}(\mu, \beta)|^2 \, dm_1(\mu) \, dm(\beta)$$

$$= \int_B \int_{\mathbf{R}} |\hat{g}(s, \beta)|^2 \, ds \, dm(\beta) ,$$

where B is the set of points at infinity and m and m_1 are suitable measures.

In Section 4 of Part II we constructed a spectral representation for the shifted Laplacian on square integrable automorphic functions g; essentially it is just the Fourier transform but now defined in terms of the Radon transform (13.2), as in (13.1). A Plancherel formula analogous to (13.3) holds when the fundamental domain has no cusps of maximal rank and when the Laplace operator acting on automorphic functions has no point spectrum:

$$(13.3)' \qquad H_F(g) = \int_F |g(w)|^2 \, dV = \int_{B_F} \int_{\mathbf{R}_+} |\tilde{g}(\mu, \beta)|^2 \, dm_1(\mu) \, dm(\beta)$$

$$= \int_{B_F} \int_{\mathbf{R}} |\hat{g}(s, \beta)|^2 \, ds \, dm(\beta) ,$$

where now B_F is the set of boundary points of F lying on B.

When F contains cusps of maximal rank, the Fourier transform must be augmented by additional components which contribute to the Plancherel formula. For instance in the case of a single cusp of maximal rank located at $w = \infty$ and of the form $F_c \times (a, \infty)$ with F_c compact, the added component is

$$(13.4) \qquad \tilde{g}^1(\mu) = \int_{\mathbf{R}} \exp\{i\mu_s\} \, \hat{g}^1(s) \, ds ,$$

where

$$(13.5) \qquad \hat{g}^1(s) = \frac{1}{\sqrt{2}} \exp\{(1 - n)s/2\} \, \tilde{g}(e^s)$$

and

$$(13.5)' \qquad \bar{g}(y) = \frac{1}{\sqrt{|F_c|}} \int_{F_c} g(x, y) \, dx \, ;$$

here $|F_c|$ denotes the $(n-1)$-dimensional volume of F_c. The Plancherel formulae are now

$$(13.6)$$
$$H_F(g) = \int_F |g(w)|^2 \, dV = \int_{B_F} \int_{\mathbf{R}} |\hat{g}(s, \beta)|^2 \, ds \, dm(\beta) + \int_{\mathbf{R}} |\hat{g}^1(s)|^2 \, ds$$

$$= \int_{B_F} \int_{\mathbf{R_+}} |\tilde{g}(\mu, \beta)|^2 \, dm_1(\mu) \, dm(\beta) + \int_{\mathbf{R_+}} |\tilde{g}^1(\mu)|^2 \, dm_1(\mu) \, .$$

As a byproduct of the completeness we also obtain a relation for

$$(13.7) \qquad D_F(g) = \int_F y^2 |\partial g|^2 \, dV \, .$$

If there are no cusps of maximal rank then

$$(13.8) \qquad D_F(g) - \left(\frac{n-1}{2}\right)^2 H_F(g) = \int_{B_F} \int_{\mathbf{R}} |\partial_s \hat{g}(s, \beta)|^2 \, ds \, dm\beta)$$

$$= \int_{B_F} \int_{\mathbf{R_+}} \mu^2 |\tilde{g}(\mu, s)|^2 \, dm_1(\mu) \, dm(\beta) \, ,$$

and when there is a cusp of maximal rank

$$D_F(g) - \left(\frac{n-1}{2}\right)^2 H_F(g)$$

$$(13.9)$$
$$= \int_{B_F} \int_{\mathbf{R}} |\partial_s \hat{g}(s, \beta)|^2 \, ds \, dm(\beta) + \int_{\mathbf{R}} |\partial_s \hat{g}^1(s)|^2 \, ds$$

$$= \int_{B_F} \int_{\mathbf{R_+}} \mu^2 |\tilde{g}(\mu, \beta)|^2 \, dm_1(\mu) \, dm(\beta) + \int_{\mathbf{R_+}} \mu^2 |\tilde{g}^1(\mu)| \, dm_1(\mu) \, ,$$

Our proof of Theorem 3.2 (iii) requires the use of a Plancherel relation for the non-Euclidean Fourier transform on the slab: $S = F_c \times \mathbf{R}^{n-q-1} \times \mathbf{R^+}$, where now F_c is a rectangular parallelepiped without twist. To this end we consider functions automorphic with respect to the discrete group generated by the q motions

$$(13.10) \qquad \gamma_j: (x, z, y) \to (x + b_j, z, y) \, , \qquad j = 1, \dots, q \, ,$$

where the b_j's are linearly independent vectors in $\mathbf{R}^q$. Such functions are periodic in x. For g square integrable over S and $n > q+1$ the non-Euclidian Fourier transform is defined as in (13.1) via the Radon transform. When

$n = q + 1$ then ∞ is a cusp of maximal rank and S is a prism with cross section F_c. In this case the Fourier transform has the added component $\tilde{g}^1$.

In order that the Plancherel formulae hold we need to know that the translation representations are complete for the slab and further that all square integrable automorphic functions fill out the second component data in $\mathcal{H}_c$. As we showed in Theorem 3.7 of [3], the second assertion follows if the energy form E_S is non-negative. Now for data $f = (f_1, f_2)$ on the slab

$$
E_S(f) = D_S(f_1) - \left(\frac{n-1}{2}\right)^2 H_S(f_1) + H_S(f_2)
$$

(13.11)

$$
= \int_S \left(y^{n+1} \left| \partial \frac{f_1}{y^{(m-1)/2}} \right|^2 + \|f_2\|^2 \right) dV ,
$$

and it is clear from this expression that $E_S \geqq 0$.

It follows from Sections 5 and 6 of Part II that the translation representations are complete when $n = q + 1$ and hence that the formulae (13.6) and (13.9) are valid in this case. We shall prove by induction on n for fixed q that when $n > q + 1$ the translation representations are complete so that (13.3)′ and (13.8) also hold.

To prove completeness for the translation representations for slabs with successively larger n, beginning with $n = q + 2$, we need only establish Theorem 3.2 (iii) in these cases. This will be done in Section 14. The proof for dimension n uses only the partial Fourier transform on slabs of dimension $n - 1$ and this sets up the induction since the Plancherel relation has been established for $n = q + 1$. The added component $\tilde{g}^1$ occurs only in the first step of the induction, i.e., $n = q + 2$, and requires a somewhat more elaborate argument than the subsequent steps.

14. Energy Content for Cusps of Intermediate Rank Without Twist

In this section we give a new proof of Theorem 3.2 (iii) without twist, which replaces the somewhat cumbersome proof in Section 7 of Part II. The idea of the proof is essentially the same as that used in Section 7 for the zeroth component of the data. Here we shall make use of the partial Fourier transform on x-periodic data in H^{n-1}, whereas in Section 7 we used the partial Fourier transform on data defined on all of H^{n-1}.

For cusps of maximal rank, that is $n = q + 1$, the Plancherel relation for the slab follows from the completeness of the translation representation, which was established in Sections 3 through 6 of Part II; an independent proof of this is given in Section 16. We use the Plancherel formula for this case to treat the first step of the induction, i.e., when $n = q + 2$. Starting with this case, the induction argument proceeds on the dimension n for fixed q. It is only in the first step of the induction, i.e., when $n = q + 2$, that the partial

Fourier transform on H^{n-1} contains the mean value term $\check{g}^1$. In the first part of the argument we prove the necessary inequality for $\check{g}$, assuming the Plancherel relation for $n-1$. After this we prove the analogous inequality for $\check{g}^1$, which is required only in the first step of the induction. Once we have these estimates Theorem 3.2 (iii) follows for dimension n and from this the Parseval relation for the n-dimensional slab follows by Sections 3 through 4 of Part II.

For each $j \leqq n - q - 1$ we define the wedges

(14.1)
$$W_+(j,a) = (y \geqq a - z_j \, , \; x \text{ in } F_c) \, ,$$
$$W_-(j,a) = (y \geqq a + z_j \, , \; x \text{ in } F_c) \, .$$

For each choice of $a > 0$ the union of the $2(n - q - 1)$ wedges will cover a neighborhood of infinity for the cusp. For a given f in $\mathcal{H}_c$ and solution $u(t)$ of the automorphic wave equation with initial data f and $\varepsilon > 0$, we shall show that for a sufficiently large

(14.2)
$$\int_{W_\pm(j,a)} |u_t(t)|^2 \, dV < \varepsilon$$

for all j and t. It is, of course, sufficient to show how to do this for a particular wedge, say $W(a) = W_+(1,a)$.

First we choose a_0 so large that

(14.3)
$$W(a) \subset F \quad \text{for all} \quad a \geqq a_0 \, .$$

In this cases

(14.4)
$$H_W(u) \leq H_F(u) \, , \qquad H_w(u_t) \leq H_F(u_t) \qquad \text{and}$$
$$D_W(u) \leqq D_F(u) \, , \qquad D_W(u_t) \leqq D_F(u_t) \, .$$

As explained in Section 3 of Part II, we may assume that $f = Ag$ with g in $D(A^2) \cap \mathcal{H}_c$. In this case all four of the expressions on the right in (14.4) are bounded for all t.

For (x, z, y) in $W(a_0)$ we set $v = u$ and in the complement we define v by horizontal reflection; i.e., for $z_1 < a_0 - y$ we set

(14.5)
$$v(w', z_1, y) = u(w', 2a_0 - z_1 - 2y, y) \, ,$$

where $w' = (x, z_2, \ldots, z_{n-q-1})$. The Jacobian of this transformation is 1. It follows from the above that over the entire slab S

(14.6)
$$H_S(v) \, , \; H_s(v_t) \, , \; D_S(v) \, , \; D_S(v_t) \leqq C$$

for all t and some constant C.

For (x, z, y) in $W(a_0)$ and for all t, v satisfies the non-Euclidian wave equation

$$(14.7) \qquad v_{tt} = Lv = \Delta v + \left(\frac{n-1}{2}\right)^2 v .$$

All of the horospheres contained in $W(a_0)$ are characterized by an inequality of the form

$$(14.8) \qquad s > a_0 + s_0(\beta) , \qquad \beta_{q+1} > a_0 ,$$

for some function s_0. Next take the Radon transform of (14.7) for those horospheres satisfying (14.8):

$$(14.9) \qquad \hat{v}_{tt} = \widehat{Lv} .$$

This can be rewritten as

$$(14.10) \qquad \hat{v}_{tt} - \hat{v}_{ss} = (\partial_t - \partial_s)(\partial_t + \partial_s)\hat{v} = 0 .$$

The translation representations of (v, v_t) on the slab is given by

$$(14.11) \qquad k_\pm(s, \beta, t) = \partial_s \hat{v} \mp \hat{v}_t .$$

It follows from this and (14.10) that there exist two functions h_+ and h_- for which

$$(14.12)_+ \qquad k_+(s, \beta, t) = h_+(s - t, \beta)$$

and

$$(14.12)_- \qquad k_-(s, \beta, t) = h_-(s + t, \beta)$$

for s satisfying (14.8) and for all t.

According to Theorem 2.3 of Part I, the L^2 norm of the translation representation is less than or equal to the energy norm. Hence by (14.6)

$$(14.13) \qquad \int_{B_F} \int_{\mathbf{R}} |k_t(s, \beta, t)|^2 \, ds \, dm(\beta) \leqq E_S((v, v_t)) \leqq C .$$

We conclude from this and (14.12) that

$$(14.14) \qquad \int_{B_a} \int_{\mathbf{R}} |h_+(s - t, \beta)|^2 \, ds \, dm(\beta) \leqq C ,$$

where $B_a = B_F \cap (\beta_{q+1} > a)$. Finally introducing $s - t$ as a new variable of integration and letting $t \to \infty$, we see that

$$(14.15)_+ \qquad \int_{B_{a_0}} \int_{\mathbf{R}} |h_+(s, \beta)|^2 \, ds \, dm(\beta) \leqq C .$$

A similar argument (letting $t \to -\infty$) gives

$$(14.15)_- \qquad \int_{B_{a_0}} \int_{\mathbf{R}} |h_-(s, \beta)|^2 \, ds \, dm(\beta) \leqq C .$$

Finally it follows from this that given any $\varepsilon > 0$ we can choose a so large that

$$(14.16) \qquad \int_{B_a} \int_{\mathbf{R}} |h(s, \beta)|^2 \, ds \, dm(\beta) < \varepsilon .$$

This inequality plays a basic role in what follows.

Next we introduce polar coordinates:

$$(14.17) \qquad \tan \theta = y/(z_1 - a) , \qquad \rho^2 = (z_1 - a)^2 + y^2 ,$$

and we denote by $\tilde{g}'$ the partial Fourier transform of g on the $(n - 1)$-dimensional slab $S' = (\theta = \text{const.}) = F_c \times \mathbf{R}^{n-q-2} \times \mathbf{R}_+$; that is

$$(14.18)$$
$$\tilde{g}'(\mu, \chi; \theta) = \int_{\mathbf{H}^{n-1}} \exp \{((n - 2)/2 + i\mu)\langle (w, \beta), \chi \rangle\} g(w', \rho; \theta) \frac{dw' \, d\rho}{\rho^{n-1}}$$
$$= \int_{\mathbf{R}} \exp \{i\mu s\} \, \hat{g}(s, \, ; \theta) \, ds ,$$

where now χ lies in the set $F_c \times \{\rho = 0\}$ and $\hat{g}'$ is the partial Radon transform:

$$(14.19) \qquad \hat{g}'(s, \chi; \theta) = P' \int_{\langle w', \chi \rangle = s} g(w', \, ; \theta) \, dS'$$

in the $\theta = \text{const.}$ plane. The analogue of $(13.3)'$ holds and implies

$$(14.20) \qquad H_W(g) = \int_0^{3\pi/4} \int |\tilde{g}'|^2 \frac{dM \, d\theta}{\sin^n \theta} ,$$

and

$$(14.21) \qquad D_W(g) \geqq \int_0^{3\pi/4} \int |\mu g'(g)|^2 \frac{dM \, d\theta}{\sin^{n-2} \theta} ,$$

Here $M(\mu, \chi) = m'(\chi), m'_1(\mu)$. Combining (14.6) and (14.21) for $g = v_t$ we obtain the analogue of Lemma 7.6 of Part II:

LEMMA 14.1. *Let η be an arbitrary positive number; then*

$$(14.22) \qquad \int_{(\mu\eta)^{-1}<\theta<3\pi/4} \int |\tilde{v}'_t(\mu,\chi,\theta,t)|^2 \frac{dM\,d\theta}{\sin^n \theta} < \dot{C}\eta^2$$

for all t.

To continue with the proof as described in the middle portion of Section 7 we need to show that the Volterra integral equation satisfied by v_t holds in the periodic setting. For a suitable function g this equation is

$$(14.23) \qquad k(r)\frac{\tilde{g}'(\mu,\chi;\theta)}{\sin^{n/2}\theta} + \int_0^\theta K(\mu,r,\phi)\frac{\tilde{g}'(\mu,\chi;\phi)}{\sin^{n/2}\phi}\,d\phi = r^{-1/2}\,l(\mu,\chi,r)\;;$$

here r is a function of θ:

$$(14.24) \qquad\qquad\qquad r(\theta) = \tan\theta/2\;,$$

and

$$(14.25) \qquad \begin{aligned} &l(\mu,\chi,r) \\ &= \int_{\mathbf{H}^{n-1}} \exp\left\{\left(\frac{n-1}{2}+i\mu\right)\langle\beta_\lambda,\chi\rangle\right\}\lambda^{(n-1)/2}\overset{\blacktriangle}{g}(\lambda r,\beta_\lambda)\frac{d\beta'd\lambda}{\lambda^{n-1}}\;, \end{aligned}$$

where $\beta_\lambda = (\beta,\lambda+a)$, $\lambda+a$ being the z_1 coordinate of β_λ, and $\overset{\blacktriangle}{g}$ is the Radon transform in $\mathbf{H}^n$ with the s-parameter replaced by the radius r of the horosphere. The two Radon transforms are related by

$$(14.26) \qquad\qquad \hat{g}(s,\beta) = c_n(1+|\beta|^2)^{(n-1)/2}\overset{\blacktriangle}{g}(r,\beta)\;,$$

where

$$s = \log\left((1+|\beta|^2)/2r\right)\;.$$

It was proved in [2] that equation (14.23) is valid for all functions g square integrable on all of $\mathbf{H}^n$.

LEMMA 14.2. *The Volterra integral equation is valid for x-periodic functions square integrable on the slab S.*

Proof: It is not hard to verify that the derivation of the integral equation in [2] remains valid for x-periodic C^∞ functions with compact (z,y) support. This is so because integration over horospheres is well defined for such functions. Since x-periodic functions which are square integrable in the wedge are the L^2 limits of such functions, it is enough to bound $\tilde{g}'$ and l by the L^2 norm of g in the wedge. The Parseval relation (14.20) furnishes us with such a bound for $\tilde{g}$.

According to (14.25), for each r the function l is the partial Fourier transform of an x-periodic function of β. It therefore follows from (13.4) as applied to the $(n-1)$-dimensional slab S' that

$$(14.27) \quad \begin{aligned} \int_{B'} \int_{\mathbf{R}_+} |l(\mu, \chi, r)|^2 \, dm'_1(\mu) \, dm'(\chi) \\ = \int \int_{S'} |\lambda^{(n-1)/2} \, \overset{\blacktriangle}{g}(\lambda r, \beta_\lambda)|^2 \frac{d\beta' \, d\lambda}{\lambda^{n-1}} , \end{aligned}$$

where as before $\beta_\lambda = (\beta', \lambda + a)$. Dividing by r, integrating this expression with respect to r from 0 to $r_1 = r(3\pi/4)$ (see (14.24)) and making the substitution $r \to \rho/\lambda$ after interchanging the λ, r order of integration, we get

$$(14.28) \quad \begin{aligned} \int_0^{r_1} \int_{B'} \int_{\mathbf{R}_+} |l(\mu, \, , r)/r^{1/2}|^2 \, dm'_1(\mu) \, dm'(\chi) \, dr \\ = \int_0^{r_1} \int \int_{S'} |\overset{\blacktriangle}{g}(\lambda r, \beta_\lambda)|^2 \, d\beta' \, d\lambda \frac{dr}{r} \\ \leqq \int_{B_a} \int_0^\infty |\overset{\blacktriangle}{g}(\rho, \beta)|^2 \frac{d\rho}{\rho} \, d\beta \\ = \int_{B_a} \int_{\mathbf{R}} |\hat{g}(s, \beta)|^2 \, ds \, dm(\beta) \leqq H_S(g) ; \end{aligned}$$

here the last two relations follow from (14.26) and (14.24). This completes the proof of Lemma 14.2.

Next we apply Lemma 14.2 to v_t. We see from (14.11) and (14.12) that for all β in B_a and s satisfying (14.8) and all t

$$(14.29) \quad \begin{aligned} \hat{v}_t(s, \beta, t) = k_-(s, \beta, t) - k_+(s, \beta, t) \\ = h_-(s + t, \beta) - h_+(s - t, \beta) . \end{aligned}$$

By (14.16) we get for all t and our choice of a

$$(14.30) \quad \int_{B_a} \int_{\mathbf{R}} |\hat{v}_t(s, \beta, t)|^2 \, ds \, dm(\beta) \leqq 4\varepsilon .$$

Set $g = v_t$ in (14.28); then this combined with (14.30) furnishes us with a suitable bound for the right-hand side of the integral equation (14.23).

The inverse of the Volterra integral equation is easily estimated (see Lemma 6.3 and Theorem 6.5 of Part II) and gives us the following estimate which holds for all t

$$(14.31) \quad |\tilde{v}'_t(\mu, \chi, \theta, t)|^2 \frac{dM \, d\theta}{\sin^n \theta} \leqq c \varepsilon \exp\left\{ c' \eta^{(n+1)/2} \right\} ,$$

where $\delta(\mu) = \min(\mu\eta)^{-1}, 3\pi/4)$. Finally (14.31) combined with Lemma 14.1 gives the desired estimate

$$(14.32) \qquad \int_{W(a)} |v_t(t)|^2 \, dV = O(\varepsilon) .$$

The Parseval relation $(13.3)'$ for slabs of dimension $n - 1$, which we used in (14.27) holds only when $n - 1 > q + 1$. To treat the case $n - 1 = q + 1$ we must use (13.6) and this requires a suitable bound on the mean value component of the Fourier transform. Once we have done this we can complete the induction by proving the Parseval relation for the n-dimensional slab. This, however, will follow from the completeness of the translation representations in dimension n (Theorem 3.1 of Part II), which, in turn, follows from Theorem 3.2.

The mean value is

$$(14.33) \qquad \bar{u}(z, y) = \frac{1}{\sqrt{|F_c|}} \int_{F_c} u(x, z, y) \, dx .$$

Integrating the wave equation with respect to the x variable over F_c, we obtain

$$(14.34) \qquad \bar{u}_{tt} = y^2(\partial_z^2 + \partial_y^2)\bar{u} - (n - 2)y\,\bar{u}_y + \left(\frac{n-1}{2}\right)^2 \bar{u} ,$$

We now set

$$(14.35) \qquad v = y^{(2-n)/2}\,\bar{u} .$$

Substituting this into (14.34) we see that v satisfies the two-dimensional wave equation:

$$(14.36) \qquad v_{tt} = y^2 \left(\partial_z^2 + \partial_y^2\right) v + \frac{1}{4}v .$$

For any subset W of H^2 it follows from (14.35) that

$$(14.37) \qquad \int_W |v_t|^2 \frac{d\tau\,dy}{y^2} = \int_W |\bar{u}_t|^2 \frac{dz\,dy}{y^n} .$$

From the definition of $\bar{u}$ as the zero Fourier component over F_c we deduce that the right side of (14.37) is less than or equal to

$$\int_{W \times F_c} |u_t|^2 \, dV \leqq E_S(u) .$$

In what follows we shall take W to be a wedge of the form

$$(14.38) \qquad y \geqq a - z \quad \text{or} \quad y \geqq a + z .$$

It is obvious that the union of these two wedges includes a neighborhood of infinity for H^2. Given $\varepsilon > 0$, we need to prove that for all t

$$(14.39) \qquad \int_{W(a)} |v_t|^2 \frac{d\tau\,dy}{y^2} < \varepsilon$$

for both wedges (14.38) when a is chosen sufficiently large. This is essentially the same problem that we treated in the first part of this section. The only difference is that in this case the partial Fourier transform over the leaves of the wedge are to be taken over H^1, whereas before it was over a prism (a degenerate slab) in H^{q+1}. The Parseval relation for the Fourier transform over H^1 is well established and we may therefore assert on the basis of the previous development that (14.39) holds. This completes the first step in the induction for the Fourier transform on the slab and hence completes the proof of Theorem 3.2 (iii).

15. Energy Content for Cusps of Intermediate Rank with Twist

As explained in the introduction, it suffices to treat a cusp of intermediate rank for which the stability group is generated by the motions:

$$(15.1) \qquad \gamma_j \colon (x, z, y) \to (x + b_j,\, R_j z,\, y), \qquad\qquad j = 1, \ldots, q,$$

where the b_j's are linear independent vectors in R^q and the R_j's are commuting rotations in R^{n-q-1}. The previous theory will not apply directly in this situation (unless some power of all the R_j's is the identity) because that proof used the Plancherel relation for functions periodic in the slab, and because the derivation of the Volterra integral equation made use of this periodicity. Nevertheless we shall, by expansion into eigenfunctions, untwist and reduce the general case to the periodic one.

We denote the parallelepiped generated by the b_j's in R^q by F_c, and denote by S the slab $F_c \times R^{n-q-1} \times R_+$. S contains a fundamental polyhedron F for the discrete group under discussion. We denote by S_R the subset of those points (x, z, y) of S for which $|z|^2 + y^2 > R^2$; for R sufficiently large, S_R is contained in F.

Next we introduce cylindrical coordinates:

$$(15.2) \qquad \begin{aligned} &r^2 = |z|^2 + y^2,\qquad \sin\theta = y/r,\\ &\phi = \text{coordinates on the unit } m = n - q - 2 \text{ sphere } S^m. \end{aligned}$$

In terms of these coordinates the metric is

$$(15.3) \qquad ds^2 = \left(dr^2 + dx^2 + r^2\,d\theta^2 + r^2\cos^2\theta\,d\phi^2\right) \Big/ \left(r^2\sin^2\theta\right)$$

with volume element

$$(15.4) \qquad dV = \frac{\cos^m \theta}{r^{q+1} \sin^n \theta} \, dr \, dx \, d\theta \, d\phi \,.$$

The H and D forms on the slab minus the half-ball are
(15.5)

$$H_R(g) = \int_{S_R} |g|^2 \, dV \leq H_F(g) \,,$$

$$D_R(g) = \int_{S_R} r^2 \sin^2 \theta \left[|d_r g|^2 + |\partial_x g|^2 + \frac{1}{r^2} |\partial_\theta g|^2 + \frac{1}{r^2 \cos^2 \theta} |\partial_\phi g|^2 \right] dV$$

$$\leq D_F(g) \,.$$

The Laplacian is

$$(15.6) \qquad \begin{aligned} \Delta &= r^{q+1} \sin^2 \theta \, \partial_r \frac{1}{r^{q-1}} \, \partial_p + r^2 \sin^2 \theta \, \Delta_x \\ &+ \frac{\sin^2 \theta}{\cos^m \theta} \, \partial_\theta \frac{\cos^m \theta}{\sin^{n-2} \theta} \, \partial_\theta + \frac{\sin^2 \theta}{\cos^2 \theta} \, \Delta_\phi \,, \end{aligned}$$

where Δ_x and Δ_ϕ are the usual Euclidean Laplacians on $\mathbf{R}^q$ and S^m, respectively.

Under the identifications by (γ_i) given in (15.1), the twisted Cartesian product $F_c \times S^m$ is a manifold M. The operators Δ_x and Δ_ϕ acting on functions on M are selfadjoint, and they commute. Furthermore, the operator Δ_ϕ commutes with all rotations, and Δ_x commutes with all rotations that commute with the R_j, in particular with the R_j themselves. It follows that the Δ_x and Δ_ϕ have a common set of eigenfunctions on M of the form

$$(15.7) \qquad w(x) \, Y(\phi) \,,$$

where

$$(15.8) \qquad \Delta_\phi Y = -\nu Y \,, \qquad \nu \geq 0 \,,$$

and

$$(15.9) \qquad Y(R_j \phi) = \exp\{i\tau_j\} \, Y(\phi) \,, \qquad \tau_j \text{ real.}$$

Furthermore

$$(15.10) \qquad \Delta_x w = -\kappa w \,, \qquad \kappa \geq 0 \,,$$

and w is automorphic under (15.1); in view of (15.9) this requires that

$$(15.11) \qquad w(x + b_j) = \exp\{-i\tau_j\} \, w(x) \,.$$

We see from (15.8) that Y is a spherical harmonic. It follows from (15.10) and (15.11) that w is of the form

$$(15.12) \qquad\qquad w(x) = c \exp\{ix \cdot \xi\},$$

where ξ satisfies

$$(15.13) \qquad\qquad b_j \cdot \xi = -\tau_j + 2\pi b',$$

for all j and b' is any element of the lattice dual to the lattice generated by the b_j's. The eigenvalue κ is given by

$$(15.14) \qquad\qquad \kappa = |\xi|^2.$$

The constant c is chosen so that the L^2 norm of w over F_c is equal to one.

Since Δ_x is selfadjoint under the identification (15.11), it follows that the eigenfunctions (15.7) are orthogonal on the twisted product M. Furthermore, since M is compact, the eigenfunctions are complete on M so that any function on M can be expanded into a series of these eigenfunctions. In particular any solution $u = u(x, \phi, r, \theta, t)$ of the automorphic wave equation

$$(15.15) \qquad u_{tt} = Lu, \qquad L = \Delta + \left(\frac{n-1}{2}\right)^2,$$

can be so expanded for $r > R$:

$$(15.16) \qquad\qquad u = \sum v^{\kappa\nu} w_\kappa Y_\nu = \sum u_{\kappa\nu}.$$

The coefficients $v^{\kappa\nu}(r, \theta, t)$ satisfy the wave equations:

$$(15.17) \qquad
\begin{aligned}
v_{tt}^{n\nu} = {}& r^{q+1} \sin^2\theta\, \partial_r \frac{1}{r^{q-1}} \partial_r v^{\kappa\nu} + \frac{\sin^n\theta}{\cos^m\theta} \partial_\theta \frac{\cos^m\theta}{\sin^{n-2}\theta} \partial_\theta v^{\kappa\nu} \\
& - \kappa r^2 \sin^2\theta\, v^{\kappa\nu} - \nu \frac{\sin^2\theta}{\cos^2\theta} v^{\kappa\nu} + \left(\frac{n-1}{2}\right)^2 v^{\kappa\nu},
\end{aligned}$$

again for $r > R$. Moreover each component in (15.16) satisfies the original wave equation $u_{tt} = Lu$ for $r > R$.

Since the eigenfunctions are orthogonal, it follows that for every t,

$$(15.18) \qquad
\begin{aligned}
H_R(u) &= \sum H_R(u_{\kappa\nu}), \\
H_R(u_t) &= \sum H_R(\partial_t u_{\kappa\nu}), \\
D_R(u) &= \sum D_R(u_{\kappa\nu}), \\
D_R(u_t) &= \sum D_R(\partial_t u_{\kappa\nu}).
\end{aligned}$$

As remarked in Section 3 of Part II, we may take the intial data $f = (u, u_t)$ of the solution under investigation to be of the form

$$f = Ag, \quad g \text{ in } \mathcal{H}_c \cap D\left(A^2\right) .$$

This implies that the L^2 and the Dirichlet integrals of u and u_t are uniformly bounded for all t:

$$(15.19) \qquad H_R(u), H_R(u_t), D_R(u), D_R(u_t) \leqq C .$$

The reduction of the twisted to the untwisted case is accomplished as follows: For each $\kappa \neq 0$, we define W_κ to be a function of the form

$$(15.20) \qquad V_\kappa(x) = c_\kappa \exp\{ix \cdot \alpha_\kappa\} ,$$

where α_κ satisfies

$$(15.20') \qquad\qquad |\alpha_\kappa|^2 = \kappa ,$$

and we define F_κ to be a parallelepiped in $\mathbf{R}^q$ over which w_κ is *periodic*. The constant c_κ is chosen so that w_κ is normalized over F_κ:

$$(15.21) \qquad\qquad \int_{F_\kappa} |W_\kappa|^2 \, dx = 1 .$$

The x-Dirichlet integral is then

$$(15.22) \qquad\qquad \int_{F_\kappa} |\partial_x W_\kappa|^2 \, ds = |\alpha_\kappa|^2 = \kappa .$$

If $\kappa = 0$, then $w_\kappa = \text{const.}$ and we set $W_\kappa = w_\kappa$ and $F_\kappa = F_c$.

It follows that w_κ and W_κ have the same L^2 and Dirichlet integrals over F_c and F_κ, respectively:

$$(15.23) \quad \int_{F_c} |w_\kappa|^2 \, dx = \int_{F_\kappa} |W_\kappa|^2 \, dx \text{ and } \int_{F_c} |\partial_x w_\kappa|^2 \, dx = \int_{F_\kappa} |\partial_x W_\kappa|^2 \, dx ,$$

Furthermore w_κ and W_κ both satisfy

$$(15.24) \qquad\qquad \Delta_x w_\kappa = -\kappa w_\kappa, \qquad \Delta_x W_\kappa = -\kappa W_\kappa$$

over F_c and F_κ, respectively.

We now define

$$(15.25) \qquad\qquad U_{\kappa\nu} = v^{\kappa\nu} W_\kappa Y_\nu ,$$

where $v^{\kappa\nu}$ is the coefficient appearing in the expansion (15.16). Since $v^{\kappa\nu}$ satisfies equation (15.17), it follows that $U_{\kappa\nu}$ satisfies the hyperbolic wave equation

$$U_{tt} = LU .$$

Furthermore $U_{\kappa\nu}$ is periodic in x, and for all t

$$(15.26) \qquad H_R'(\partial_t U_{\kappa\nu}) = H_R(\partial_t u_{\kappa\nu}) \quad \text{and} \quad D_R'(U_{\kappa\nu}) = D_R(u_{\kappa\nu}) ,$$

where H_R' denotes the L^2 norm over S_R^κ, which is the part of the slab $F_\kappa \times \mathbf{R}^{n-q-1} \times \mathbf{R}_+$ outside of the cylinder $|z|^2 + y^2 = R^2$; D_R' is defined similarly.

We proceed now as in Section 14, and look at the $2(n-q-1)$ wedges W:

$$z_i > a - y \quad \text{or} \quad z_i < y - a , \qquad i = 1, \dots, (n-q-1) .$$

For $a > 2R$, each wedge W belongs to S_R^κ. Conversely, the union of all $2(n-q-1)$ wedges contains S_a^κ.

We denote by $h_\pm^{\kappa\nu}(s \mp t, \beta)$ the translation representation of $(U_{\kappa\nu}(t), \partial_t U_{\kappa\nu}(t))$. As in (14.13) we deduce that

$$\int_{B_a^\kappa} \int_{S > a + s(\beta)} \left| h_\pm^{\kappa\nu}(s \mp t, \beta) \right|^2 ds\, dm(\beta) \leq D_R'(U_{\kappa\nu}(t)) + H_R'(\partial_t U_{\kappa\nu}(t)) .$$

Summing over κ, ν, using (15.18), (15.19), and (15.26), we deduce, upon letting t tend to ∞, that

$$(15.27) \qquad \sum \int_{B_a^\kappa} \int_{s > a + s(\beta)} \left| h_\pm^{\kappa\nu}(s, \beta) \right|^2 ds\, dm(\beta) \leq C ,$$

where C is some constant. Taking a large enough we can make C as small as we wish.

Using the Volterra integral equation (14.23) with $g = \partial_t U_{\kappa\nu}$, we obtain the analogue of the inequality (14.31):

$$\int_{\theta < \delta(\mu)} \int \left| \partial_t \tilde{U}_{\kappa\nu}'(t) \right|^2 \frac{dM\, d\theta}{\sin^n \theta}$$

$$c \exp\left\{ c' \eta^{(n+1)/2} \right\} \int_{B_a^\kappa} \int_{\mathbf{R}_+} \left[|h_+^{\kappa\nu}(s, \beta)|^2 + |h_-^{\kappa\nu}(s, \beta)|^2 \right] dx\, dm(\beta) ,$$

where $\delta(\mu) = \min(\eta\mu)^{-1}, 3\pi/4)$ and η is an arbitrary constant. Summing over κ, ν we get, using (15.27), that for all t

$$(15.28) \qquad \sum \int \int_{\theta < \delta(\mu)} \left| \partial_t \tilde{U}_{\kappa\nu}'(t) \right|^2 \frac{dM\, d\theta}{\sin^n \theta} \leq c \exp\left\{ c' \eta^{(n+1)/2} \right\} C ,$$

where C is as small as desired, providing that a is large enough.

The integral over the complementary range $\theta > \delta(\mu)$ is estimated as before by Lemma 14.1:

$$(15.29) \qquad \int\int_{(\mu\eta)^{-1} < \theta < 3\pi/4} \left| \partial_t \tilde{U}'_{\kappa\nu}(t) \right|^2 \frac{dM \, d\theta}{\sin^n \theta} \leqq \eta^2 D'_R(\partial_t U_{\kappa\nu}) \, .$$

Sum over κ, ν; then using (15.18), (15.19), and (15.26) we deduce that for all t

$$(15.30) \qquad \sum \int\int_{(\mu\eta)^{-1} < \theta < 3\pi/4} \left| \partial_t \tilde{U}'_{\kappa\nu}(t) \right|^2 \frac{dM \, d\theta}{\sin^n \theta} \leqq c\eta^2 \, .$$

We now choose η so small that the right side of (15.30) is less than ε, and then choose a so large that the right side of (15.28) is also less than ε. Adding these two relations we get

$$(15.31) \qquad \int\int_{\theta < 3\pi/4} \left| \partial_t \tilde{U}'_{\kappa\nu}(t) \right|^2 \frac{dM \, d\theta}{\sin^n \theta} < 2\varepsilon$$

for all t. Since $U_{\kappa\nu}$ is periodic in x, the Plancherel theorem for the partial Fourier transform holds and we get from (15.31) that

$$\sum H'_W(\partial_t U_{\kappa\nu}(t)) < 2\varepsilon$$

for all t. Using (15.26) we deduce that

$$\sum H_W(\partial_t u_{\kappa\nu}(t)) < 2\varepsilon$$

for all t. Summing over all $2(n - q - 1)$ wedges we get

$$\sum H_a(\partial_t u_{\kappa\nu}(t)) < 4(n - q - 1)\varepsilon \, .$$

So, by (15.18), it follows that

$$H_a(u_t) < 4(n - q - 1)\varepsilon \, .$$

This concludes the proof of Theorem 3.2 (iii) for the cusp of intermediate rank with twist.

16. The Plancherel Relation for the Radon Transform on a Slab

In Section 14 the Plancherel formula for the Fourier transform in $\mathbf{H}^n$ for x-periodic functions was deduced by an induction argument as part of the proof of completeness of the translation representation for arbitrary discrete subgroups without twist. Here we show how to deduce the Plancherel relation for the Radon transform directly from the Plancherel formula for the Radon transform in all of hyperbolic space. To simplify the exposition we treat only the case of a parabolic group of full rank acting on $\mathbf{H}^3$. The proof is based on an earlier result (Theorem 3.14 in [1]), which we shall now exploit more fully.

Let f denote an L^2 function defined in all of $\mathbf{H}^3$, and denote by $\hat{f}(x, \beta)$ its (augmented) Radon transform, defined in (13.2) as

$$(16.1) \qquad \hat{f}(s, \beta) = P \int_{\langle w, \beta \rangle = s} f(w) \, dS \,, \qquad P = \frac{1}{\sqrt{2\pi}} \, d_s \, e^s \,.$$

In what follows it is convenient to replace the s-parameter of the horosphere by its Euclidean radius r, which is related to s by $s = \log(1 + |\beta|^2)/2r)$. In terms of r the operator P has the form

$$(16.2) \qquad P = r \frac{1 + |\beta|^2}{2^{3/2}\pi} \, \partial_r \frac{1}{r} \,,$$

and the resulting (augmented) Radon transform $\overset{\blacktriangle}{f}$ is related to $\hat{f}$ by

$$(16.3) \qquad \overset{\blacktriangle}{f}(r, \beta) = \frac{1}{1 + |\beta|^2} \, \hat{f}(s, \beta) \,.$$

Recall that

$$dm(\beta) = \frac{d\beta}{(1 + |\beta|^2)} \quad \text{and} \quad ds = \frac{dr}{r} \,;$$

and hence by (13.3)

$$(16.4) \qquad \int_{\mathbf{R}^2} \int_{\mathbf{R}_+} \left| \overset{\blacktriangle}{f}(r, \beta) \right|^2 \frac{dr}{r} \, d\beta = \int_{\mathbf{R}^2} \int_{\mathbf{R}} \left| \hat{f}(s, \beta) \right|^2 \, ds \, dm(\beta)$$

$$= \int \int_{\mathbf{H}^3} |f(x, y)|^2 \frac{dx \, dy}{y^2} \,.$$

In terms of spherical coordinates:

$$(16.5) \qquad \begin{aligned} x = (x_1, x_2) &= \beta + r \sin\theta \, e(\phi) \,, \quad e(\phi) = (\cos\phi, \sin\phi) \,, \\ y &= r(1 - \cos\theta) \,, \end{aligned}$$

the formulae (16.1)–(16.3) combine to give

$$
\begin{aligned}
&\overset{\blacktriangle}{f}(r, \beta) \\
(16.6) \qquad &= \frac{r}{2^{3/2}\pi} \, \partial_r \frac{1}{r} \int_0^\pi \int_0^{2\pi} f(\beta + r\sin\theta \, e(\phi), r(1 - \cos\theta)) \\
&\qquad \frac{\sin\theta}{(1 - \cos\theta)^2} \, d\phi \, d\theta \,.
\end{aligned}
$$

Next we take the Euclidean Fourier transform $\mathcal{F}$ of $\hat{f}(r,\beta)$ with respect to β alone, and denoting the Euclidean Fourier transform of $f(x,y)$ with respect to x by $F(\xi,y)$, we get

$$r^{-1/2}\left(\mathcal{F}\hat{f}(r)\right)(\xi) = \frac{1}{2\pi r^{1/2}}\int \exp\{-i\xi\cdot\beta\}\,\hat{f}(r,\beta)\,d\beta$$

(16.7)
$$= \frac{r^{1/2}}{2^{3/2}\pi}\,\partial_r\frac{1}{r}\int\int \exp\{r\sin\theta\cdot e(\phi)\}$$

$$\times F\left(\xi, r(1-\cos\theta)\right)\frac{\sin\theta}{(1-\cos\theta)^2}\,d\phi\,d\theta\ .$$

The integration with respect to ϕ can be carried out explicitly. Recalling that

$$J_0(u) = \frac{1}{2\pi}\int_0^{2\pi}\exp\{iu\cos\mu\}\,d\mu\ ,$$

and using the abbreviation $|\xi| = \rho$, the right side of (16.7) becomes

(16.8) $$\left(\frac{r}{2}\right)^{1/2}\partial_r\frac{1}{r}\int_0^\pi J_0(r\rho\sin\theta)\,F\left(\xi, r(1-\cos\theta)\right)\frac{\sin\theta}{(1-\cos\theta)^2}\,d\theta\ .$$

Finally introducing $z = r(1-\cos\theta)$ as new variable of integration, (16.8) becomes

$$\left(\frac{r}{2}\right)^{1/2}\partial_r\int_0^{2\eta} J_0\left(\rho(2zr-z^2)^{1/2}\right) F(\xi, z)\frac{dz}{z^2}\ ,$$

and carrying out the differentiation we get
(16.9)
$$\frac{F(\xi, 2r)}{(2r)^{3/2}} + \frac{\rho}{2^{1/2}}\int_0^{2r} J_1\left(\rho\left(2zr-z^2\right)^{1/2}\right)\left(\frac{zr}{2zr-z^2}\right)^{1/2}\frac{F(\xi,z)}{z^{3/2}}\,dz\ ;$$

here J_1 denotes the Bessel function of order 1 $(J_0' = J_1$ and $J_1(0) = 0)$.

We can bring the above into a more tractable form by setting

(16.10) $$G(\xi, y) = F(\xi\cdot y)/y^{3/2}\ ,$$

and denoting by V_ρ the following Volterra integral operator:
(16.11)

$$(V_\rho(G)(t) = G(t) + \frac{1}{2}\rho\int_0^t J_1\left(\rho\left(zt-z^2\right)^{1/2}\right)\left(\frac{zt}{zt-z^2}\right)^{1/2} G(z)\,dz\ .$$

Then (16.9) can be expressed as $(V_\rho G(\xi))(2r)$ and since this is the right side of (16.7), we have

(16.12) $$r^{-1/2}\left(\mathcal{F}\hat{f}(r)\right)(\xi) = (V_\rho G(\xi))(2r)\ .$$

The two-dimensional Euclidean Fourier transform $\mathcal{F}$ preserves the L^2 norm. Hence it follows from (16.4) and (16.10) that

(16.13)

$$\int \int \left| \frac{\mathcal{F}\hat{f}}{r^{1/2}} \right|^2 d\xi \, dr = \int \int \left| \hat{\hat{f}}(r,\beta) \right|^2 \frac{dr}{r} \, d\beta = \int \int |f(x,y)|^2 \frac{dx\,dy}{y^3}$$

$$= \int \int \left| \frac{F(\xi,y)}{y^{3/2}} \right|^2 d\xi \, dy = \int \int |G(\xi,y)|^2 \, d\xi \, dy \, .$$

Referring back to (16.12), we see that the Volterra operator is an isometry:

(16.14)

$$\int \int |(V_\rho G(\xi))\,(2r)|^2 \, d\xi \, dr = \int \int |G(\xi,y)|^2 \, d\xi \, dy \, .$$

Introducing $2r = t$ as new variable of integration on the left, we can rewrite (16.14) as

(16.15)

$$\frac{1}{2} \int \|V_\rho G(\xi)\|^2 \, d\xi = \int \|G(\xi)\|^2 \, d\xi \, ,$$

where $\| \cdot \|$ denotes the L^2 norm with respect to the second variable:

(16.16) $\|G(\xi)\|^2 = \displaystyle\int_{\mathbf{R}_+} |G(\xi,y)|^2 \, dy$ and $\|V_\rho G(\xi)\|^2 = \displaystyle\int_{\mathbf{R}_+} |V_\rho G(\xi)|^2 \, dt \, .$

As the following theorem shows, the equality (16.15) actually holds for each ξ. This turns out to be the key to the present development.

THEOREM 16.1. *The Volterra operator* $2^{-1/2} V_\rho$, *defined by* (16.11) *is an isometry on* $L^2(R_+)$ *for* $\rho \neq 0$:

(16.17)

$$\frac{1}{\sqrt{2}} \|V_\rho h\| = \|h\| \, .$$

Proof: We denote by D_ρ the unitary dilation operator on $L^2(\mathbf{R}_+)$:

(16.18)

$$(D_\rho h)(y) = \rho^{1/2} h(\rho y) \, , \qquad \rho \neq 0 \, .$$

A straightforward calculation shows the operators V_ρ are similar via dilation, that is

(16.19)

$$V_\rho D_\rho = D_\rho V_1 \, , \qquad \rho \neq 0 \, .$$

Setting $G = D_\rho H$ and using (16.19) together with the dilation property of D_ρ allows us to rewrite (16.15) as

$$(16.20) \qquad \frac{1}{2} \int \|V_1 H(\xi)\|^2 \, d\xi = \int \|H(\xi)\|^2 \, d\xi \; .$$

It is easy to see that the functions G defined by (16.10) fill out $L^2(\mathbf{R}^2 \times \mathbf{R}_+)$ and so do the functions H, related to G via D_ρ. We can therefore choose H to be of the form:

$$
\begin{aligned}
H(\xi, y) &= h(y) \quad \text{for} \quad |\xi| \leqq 1 \\
&= 0 \qquad \text{for} \quad |\xi| > 1 \; ,
\end{aligned}
$$

where h is an arbitrary function in $L^2(\mathbf{R}_+)$. Setting this choice of H into (16.20) yields

$$\frac{1}{\sqrt{2}} \|V_\rho h\| = \frac{1}{\sqrt{2}} \|D_\rho V_1 h\| = \frac{1}{\sqrt{2}} \|V_1 h\| = \|h\| \; ,$$

and completes the proof of the theorem.

We turn now to the proof of the Plancherel formula for the Radon transform in a slab of the form

$$S = I \times I \times \mathbf{R}_+ \; , \qquad I = (-\pi, \pi) \; .$$

S is the fundamental polyhedron for the parabolic group Γ_∞, generated by translations of 2π in the x_1 and x_2 directions.

Let g be a function in $\mathbf{H}^3$, automorphic with respect to Γ_∞ and square integrable on S. The (augmented) Radon transform for the slab of full rank has two components. The first component is associated with the points β at infinity in $B_S = I \times I$ and is simply the Radon transform $\hat{g}$, defined as in (16.1). As before, we replace s by r and work with $\overset{\blacktriangle}{\hat{g}}(r, \beta)$, which is related to $\hat{g}(s, \beta)$ by (16.3). These transforms clearly satisfy the analogue of the first equality in (16.4):

$$(16.21) \qquad \int_{B_s} \int_{\mathbf{R}_+} \left| \overset{\blacktriangle}{\hat{f}}(r, \beta) \right|^2 \frac{dr}{r} \, d\beta = \int_{B_s} \int_{\mathbf{R}} |\hat{g}(s, \beta)|^2 \, ds \, dm(\beta) \; .$$

Now $\overset{\blacktriangle}{\hat{g}}(r, \beta)$ is a periodic function of β in B_S and we can expand it into Fourier series whose n-th term is

$$(16.22) \qquad a_n(r) = \frac{1}{2\pi} \int_{B_s} \exp\{-in \cdot \beta\} \overset{\blacktriangle}{\hat{g}}(r, \beta) \, d\beta \; .$$

402

As before, we can express $\overset{\blacktriangle}{g}$ in terms of g by formula (16.6). Substituting this into (16.22) we obtain the analogue of (16.12):

$$(16.23) \qquad r^{-1/2} a_n(r) = (V_\rho b_n)(2r) \; ;$$

here $\rho = |n|$ and, in analogy with (16.10),

$$(16.24) \qquad b_n(y) = \frac{1}{2\pi y^{3/2}} \int_{Bs} \exp(-in \cdot x) \, g(x, y) \, dx \; .$$

V_ρ is the integral operator defined by (16.11).

By (16.23) and Theorem 16.1

$$(16.25) \qquad \begin{aligned} \int_{\mathbf{R_+}} |a_n(r)|^2 \frac{dr}{r} &= \int_{\mathbf{R_+}} |(V_\rho b_n)(2r)|^2 \, dr \\ &= \frac{1}{2} \int_{\mathbf{R_+}} |(V_\rho b_n)(t)|^2 \, dt = \|b_n\|^2 \quad \text{for} \quad n \neq 0 \, . \end{aligned}$$

Since V_0 is the identity we also have

$$(16.25)' \qquad \int_{\mathbf{R_+}} |a_0(r)|^2 \frac{dr}{r} = \frac{1}{2} \|b_0\|^2 \; .$$

The additional component by which the Radon transform has to be augmented for slabs of full rank is given by (13.5):

$$(16.26) \qquad \hat{g}^1(s) = \frac{e^{-s}}{2^{3/2}\pi} \int_{Bs} g(x, e^s) \, dx = \frac{e^{s/2}}{\sqrt{2}} \, b_0(e^s) \; .$$

Hence

$$(16.27) \qquad \begin{aligned} \int_{\mathbf{R}} \left| \hat{g}^1(s) \right|^2 ds &= \frac{1}{2} \int_{\mathbf{R}} e^s |b_0(e^s)|^2 \, ds \\ &= \frac{1}{2} \int_{\mathbf{R_+}} |b_0(y)|^2 \, dy = \frac{1}{2} \|b_0\|^2 \; . \end{aligned}$$

By the Parseval relation for Fourier series

$$\int_{Bs} \int_{\mathbf{R_+}} \left| \overset{\blacktriangle}{g}(r, \beta) \right|^2 \frac{dr}{r} \, d\beta = \sum \int_{\mathbf{R_+}} |a_n(r)|^2 \frac{dr}{r}$$

Combining this with (16.21), (16.25), and (16.27) we get

$$(16.28) \qquad \int_{Bs} \int_{\mathbf{R}} |\hat{g}(s, \beta)|^2 \, ds \, dm(\beta) + \int_{\mathbf{R}} \left| \hat{g}^1(s) \right|^2 ds = \sum \|b_n\|^2 \; .$$

On the other hand, we deduce from (16.24) that

$$(16.29) \qquad H_S(g) = \int_S |g(x,y)|^2 \frac{dx\,dy}{y^3} = \sum \|b_n\|^2 \,.$$

From (16.28) and (16.29) we conclude that

$$(16.30) \qquad H_S(g) = \int_{B_S} \int_{\mathbf{R}} |\hat{g}(s,\beta)|^2 \, ds\, dm(\beta) + \int_{\mathbf{R}} \left|\hat{g}^1(s)\right|^2 \, ds \,,$$

as asserted in (13.6). This completes the proof of the Plancherel formula for slabs of maximal rank in three dimensions.

Acknowledgment. The work of the first author was suported in part by the Applied Mathematical Sciences Program of the Department of Energy under Contract DE-FG02-88ER25053 and that of the second author by the National Science Foundation under Grant No. DMS 8903076.

Bibliography

[1] Lax, P. D., and Phillips, R. S., *Translation representations for the solution of the non-Euclidian wave equation*, Comm. Pure Appl. Math. 32, 1979, pp. 617–667.
[2] Lax, P. D., and Phillips, R. S., *A local Paley-Wiener theorem for the radon transform of L^2 functions in a non-Euclidian setting*, Comm. Pure Appl. Math. 35, 1982, pp. 531–554.
[3] Lax, P. D., and Phillips, R. S., *The asymptotic distribution of lattice points in Euclidian and non-Euclidian spaces*, J. Func. Anal. 46, 1982, pp. 280–350.
[4] Lax, P. D., and Phillips, R. S., *Translation representations for automorphic solutions of the wave equation in non-Euclidian spaces, I and II*, Comm. Pure Appl. Math. 37, 1984, pp. 303–328 and pp. 779–813.

Received January 1990.

TRANSACTIONS OF THE
AMERICAN MATHEMATICAL SOCIETY
Volume 289, Number 2, June 1985

TRANSLATION REPRESENTATIONS FOR
AUTOMORPHIC SOLUTIONS OF
THE WAVE EQUATION IN NON-EUCLIDEAN SPACES;
THE CASE OF FINITE VOLUME

BY

PETER D. LAX AND RALPH S. PHILLIPS[1]

ABSTRACT. Let Γ be a discrete subgroup of automorphisms of H^n, with fundamental polyhedron of finite volume, finite number of sides, and N cusps. Denote by Δ_Γ the Laplace-Beltrami operator acting on functions automorphic with respect to Γ. We give a new short proof of the fact that Δ_Γ has absolutely continuous spectrum of uniform multiplicity N on $(-\infty, ((n-1)/2)^2)$, plus a standard discrete spectrum. We show that this property of the spectrum is unchanged under arbitrary perturbation of the metric on a compact set. Our method avoids Eisenstein series entirely and proceeds instead by constructing explicitly a translation representation for the associated wave equation.

Introduction. Using the methods developed in Parts I and II of [5], we obtain a short proof for the existence and completeness of incoming and outgoing translation representations for the wave equation acting on automorphic functions with fundamental polyhedron of finite volume and a finite number of sides. As a by-product we show that the associated Laplace-Beltrami operator has a standard discrete spectrum plus an absolutely continuous spectrum of uniform multiplicity N on $(-\infty, ((n-1)/2)^2)$, where n is the dimension of the hyperbolic space and N the number of cusps. We also treat automorphic solutions of a perturbed wave equation when the support of the perturbation is compact.

A spectral theory for the Laplace-Beltrami operator in this setting was first obtained by A. Selberg in his seminal paper [7] of 1956. Since then several different approaches to this problem have been developed among which we note that of Faddeev [1] and Faddeev and Pavlov [2]. The latter paper made the connection with the Lax-Phillips theory of scattering which we exploited in our 1976 monograph [3]. All of these treatments arrived at the spectral theory via the Eisenstein functions. In the present paper we attack the problem directly through the translation representation, which can be given by an explicit integral formula; this method was first described in [6]. A more complete discussion of this problem can be found in Venkov [8].

Received by the editors July 6, 1984.

1980 *Mathematics Subject Classification*. Primary 30F40, 35J15, 47A70; Secondary 10D15.

[1]The work of the first author was supported in part by the Department of Energy under contract DE-AC02-76 ERO 3077 and the second author by the National Science Foundation under Grant MCS-8304317 and the Danish National Science Research Council under Grant 11-3601.

1. Elementary properties of the translation representations. Let Γ be a discrete group of motions in $\mathbf{H}^n$, having a fundamental polyhedron F with a finite number of sides as well as a finite volume. Thus F can have only cusps of maximal rank and no free sides. We denote the L_2 and Dirichlet integrals by H and D:

$$(1.1) \qquad H_F(u) = \int_F |u|^2 \frac{dx\,dy}{y^n}, \qquad D_F(u) = \int_F \left(|u_x|^2 + |u_y|^2 \right) \frac{dx\,dy}{y^{n-2}}.$$

Here we use the Poincaré upper half-space to model the real hyperbolic space $\mathbf{H}^n$.

We define the operator L to be the Laplace-Beltrami operator plus $((n-1)/2)^2$;

$$(1.2) \qquad L = y^2 \left(\partial_y^2 + \Delta_x \right) - (n-2)y\,\partial_y + \left(\frac{n-1}{2} \right)^2.$$

Our main tool is the corresponding non-Euclidean wave equation

$$(1.2)' \qquad\qquad\qquad u_{tt} = Lu.$$

The energy associated with this equation is

$$(1.3) \qquad E_F(u) = -H_F(u, Lu) + H_F(u_t)$$

$$= D_F(u) - \left(\frac{n-1}{2} \right) H_F(u) + H_F(u_t).$$

Energy is conserved with the passage of time by solutions of the wave equation.

When $\mathrm{vol}(F) < \infty$, E_F always takes on negative values. However, this can be compensated for by adding a term of the form

$$(1.4) \qquad K(u) = c \int_S |u|^2 \, dw = cH_S(u),$$

where S is a compact subset of F.

LEMMA 1.1. *The form*

$$(1.5) \qquad\qquad\qquad G = E_F + K$$

is positive definite provided the constant c and the compact set S in (1.4) *are large enough.*

PROOF. We decompose F into cusp neighborhoods N_j and a compact set S:

$$(1.6) \qquad\qquad\qquad S = F \backslash \bigcup N_j.$$

N_j is defined as follows: Map the jth cusp into ∞; the resulting cusp neighborhood will be of the form

$$(1.6)_j \qquad\qquad\qquad N_j = F_j x (a, \infty)$$

for a sufficiently large, where F_j is the compact cross-section of the transformed cusp. Next we perform an integration by parts:

$$\int_a^\infty y \left| \partial_y \left(\frac{u}{y^{(n-1)/2}} \right) \right|^2 dy = \int_a^\infty \left(\frac{|u_y|^2}{y^{n-2}} + \left(\frac{n-1}{2} \right)^2 \frac{|u|^2}{y^n} - \frac{n-1}{2} \frac{\partial_y |u|^2}{y^{n-1}} \right) dy$$

$$= \int_a^\infty \left(\frac{|u_y|^2}{y^{n-2}} - \left(\frac{n-1}{2} \right)^2 \frac{|u|^2}{y^n} \right) dy + \frac{n-1}{2} \frac{|u(a)|^2}{a^{n-1}}.$$

Integrating over F_j we get

$$(1.7) \quad \int_{F_j} \int_a^\infty |u_y|^2 \frac{dy\,dx}{y^{n-2}} - \left(\frac{n-1}{2}\right)^2 H_{N_j}(u)$$

$$= \int_{F_j} \int_a^\infty y \left| \partial_y \left(\frac{u}{y^{(n-1)/2}} \right) \right|^2 dy\,dx - \left(\frac{n-1}{2}\right) \int_{F_j} \frac{|u(x,a)|^2}{a^{n-1}} dx.$$

To estimate the last term on the right we set

$$u(a) = \int_\eta^a u_y\,dy + u(\eta), \qquad a - 1 \leqslant \eta \leqslant a;$$

by the Schwarz inequality

$$|u(a)|^2 \leqslant 2 \int_{a-1}^a |u_y|^2 \frac{dy}{y^{n-2}} \frac{a^{n-1} - (a-1)^{n-1}}{n-1} + 2|u(\eta)|^2.$$

Integrating with respect to η from $(a-1)$ to a and dividing by a^{n-1}, we get

$$\frac{|u(a)|^2}{a^{n-1}} \leqslant \frac{2(a^{n-1} - (a-1)^{n-1})}{(n-1)a^{n-1}} \int_{a-1}^a |u_y|^2 \frac{dy}{y^{n-2}} + 2a \int_{a-1}^a |u|^2 \frac{dy}{y^n}.$$

Integrate with respect to x over F_j; for sufficiently large a
(1.8)

$$\left(\frac{n-1}{2}\right) \int_{F_j} \frac{|u(x,a)|^2}{a^{n-1}} dx \leqslant \frac{1}{2} \int_{a-1}^a \int_{F_j} |u_y|^2 \frac{dx\,dy}{y^{n-2}} + a(n-1) \int_{a-1}^a \int_{F_j} |u|^2 \frac{dx\,dy}{y^n}.$$

Combining (1.7) and (1.8) and summing over the cusps, we see that

$$D_F(u) - \left(\frac{n-1}{2}\right)^2 H_F(u) \geqslant \sum_{j=1}^N \int_{a_j}^\infty \int_{F_j} \left[y \left| \partial_y \left(\frac{u}{y^{(n-1)/2}} \right) \right|^2 + \frac{|u_x|^2}{y^{n-2}} \right] dx\,dy$$

$$+ \frac{1}{2} D_S(u) - c' H_S(u),$$

where $c' = ((n-1)/2)^2 + a(n-1)$. The assertion (1.5) follows if we set $c = c' + 1$.

We now complete the set of all $C_0^\infty(F)$ data with respect to the G norm and denote the resulting space by $\mathcal{H}_G$. It is easy to show, by an argument similar to that used in the proof of Lemma 1.1, that for f in $\mathcal{H}_G$ the local L_2 and Dirichlet norms of f_1 are majorized by $G(f)$. Hence it follows from Rellich's compactness theorem that K is compact with respect to G. We list two direct consequences of this fact.

LEMMA 1.2. (a) E_F is positive definite on a subspace of $\mathcal{H}_G$ of finite codimension.

(b) The positive spectrum of L consists of a finite number of points $\lambda_1^2, \ldots, \lambda_m^2$ with corresponding eigenfunctions $\varphi_1, \ldots, \varphi_m$;

$$(1.9) \quad L\varphi_j = \lambda_j^2 \varphi_j,$$

where φ_j is automorphic and square integrable, and $\lambda_j > 0$.

PROOF. Since K is compact with respect to G, there is a subspace of finite codimension in which $K(f) \leqslant \frac{1}{2} G(f)$. On this subspace

$$E(f) = G(f) - K(f) \geqslant \frac{1}{2} G(f).$$

407

This proves (a). Next suppose that L is positive on a subspace $\mathscr{J}$ of $L_2(F)$. For φ in $\mathscr{J}$, $H_F(L\varphi, \varphi)$ is finite and an integration by parts shows that $D_F(\varphi) < \infty$. Thus $f = \{\varphi, 0\}$ belongs to $\mathscr{H}_G$ and, by (1.3), $E_F(f) < 0$. Part (a) of the lemma requires $\mathscr{J}$ to be finite dimensional and it follows from this that the positive spectrum of L is pure point and finite.

The functions $e^{\pm\lambda_j t}\varphi_j$ are automorphic solutions of the wave equation. Their initial data

$$(1.10) \qquad\qquad p_j^\pm = \{\varphi_j, \pm\lambda_j\varphi_j\}$$

span a space which we denote by $\mathscr{P}$. A simple calculation shows that

$$(1.11) \qquad E_F(p_j^+, p_k^+) = 0, \quad E_F(p_j^-, p_k^-) = 0 \quad \text{for all } j, k;$$

$$(1.11)' \qquad E_F(p_j^+, p_k^-) = \begin{cases} 0 & \text{for } j \neq k, \\ -\lambda_j^2 H_F(\varphi_j) & \text{for } j = k. \end{cases}$$

It follows from these expressions that the energy form is nondegenerate in $\mathscr{P}$; so there is an E-orthogonal projection P of $\mathscr{H}_G$ onto $\mathscr{P}$. We denote the complement $I - P$ by Q and denote $Q\mathscr{H}_G$ by $\mathscr{H}_Q$.

LEMMA 1.3. E_F is nonnegative on $\mathscr{H}_Q$ and the null space $\mathscr{Z}$ of E_F on $\mathscr{H}_Q$ is finite dimensional. Modulo this null space $\mathscr{H}_Q$ is complete with respect to E_F.

PROOF. For $f = \{f_1, f_2\}$ in $\mathscr{H}_Q$, f_1 in the domain of L, both components are orthogonal to the φ_j. It follows in particular that $(Lf_1, f_1) \leqslant 0$ so that, by (1.3), $E_F(f) \geqslant 0$. Since $C_0^\infty(F)$ data are dense in $\mathscr{H}_G$, it is easy to see that any data in $\mathscr{H}_Q$ can be approximated by data of this kind. That $\mathscr{Z}$ is finite dimensional follows directly from Lemma 1.2(a).

Now $\mathscr{H}_Q$ is a closed subspace and so it is the G-orthogonal complement $\mathscr{H}_1$ of $\mathscr{Z}$ in $\mathscr{H}_Q$. Any f in $\mathscr{H}_Q$ can be decomposed into G-orthogonal parts: $f = z + g$, $z \in \mathscr{Z}$ and $g \in \mathscr{H}_1$. Since z belongs to the null space of E_F we have $E_F(f) = E_F(g)$. Thus the mapping $f \to g$ is an isometry in the E_F norm. To prove the completeness of $\mathscr{H}_Q/\mathscr{Z}$ in the E_F norm, it therefore suffices to prove that E_F and G are equivalent norms on $\mathscr{H}_1$. Obviously $E_F \leqslant G$ on H_1. Suppose next that the opposite inequality is not true; then there is a subsequence $\{f_n\} \subset \mathscr{H}_1$ such that

$$(1.12) \qquad\qquad G(f_n) \equiv 1 \quad \text{and} \quad E_F(f_n) \to 0.$$

Since K is compact with respect to G we can find a subsequence (which we renumber) such that $K(f_n - f_m) \to 0$. Since $E_F \geqslant 0$ in $\mathscr{H}_1$, we see from (1.12) that $E_F(f_n - f_m) \to 0$ and it follows that the $\{f_n\}$ form a Cauchy sequence in the G norm. Thus f_n converges to some f in $\mathscr{H}_1$ which by (1.12) is of G norm 1. At the same time we see by (1.12) that $E_F(f) = 0$; i.e. f belongs to both $\mathscr{H}_1$ and its G-orthogonal complement $\mathscr{Z}$. This contradiction proves the last assertion in Lemma 1.3.

In what follows, the E_F seminorm will be employed exclusively in $\mathscr{H}_Q$. Thus orthogonality, isometry, etc. are all to be taken with respect to E_F. Of course for isometry to be meaningful we must limit ourselves to cosets modulo $\mathscr{Z}$.

Denote by $U(t)$ the solution operator relating initial data of solutions of the wave equation to their data at time t:

(1.13) $$U(t)\{u(0), u_t(0)\} = \{u(t), u_t(t)\}.$$

Since energy is conserved, the operators $U(t)$ are *unitary* on $\mathcal{H}_Q/\mathcal{L}$. They form a one-parameter group whose generator A,

(1.14) $$A = \begin{pmatrix} 0 & I \\ L & 0 \end{pmatrix},$$

is *skew-selfadjoint*. It is obvious from (1.9) and (1.10) that the $p_j^{\pm}$ are eigenvectors of A:

(1.15) $$Ap_j^{\pm} = \pm\lambda_j p_j^{\pm}.$$

It follows that $\mathcal{P}$ is an invariant subspace of $U(t)$ and hence so is its orthogonal complement $\mathcal{H}_Q$. As a consequence Q commutes with $U(t)$:

(1.16) $$QU(t) = U(t)Q.$$

Next we recall the definition of the incoming and outgoing translation representations R_-^F and R_+^F. Each has N components, one for each cusp, defined in terms of integrals over horospheres. To describe the jth component $R_{\pm}^j$ it is again convenient to map the jth cusp into ∞, in which case the neighborhoods of the jth cusp are of the form $(1.6)_j$; F_j is a compact fundamental polyhedron of the subgroup Γ_j of Γ keeping ∞ fixed.

With the jth cusp at ∞ and $f = \{f_1, f_2\}$ the given automorphic data, we define

(1.17) $$R_{\pm}^j f(s) = \frac{1}{\sqrt{2}}\left[\partial_s e^{(1-n)s/2}\overline{f_1}(e^s) \mp e^{(1-n)s/2}\overline{f_2}(e^s)\right],$$

where the bar denotes the mean value:

(1.17)' $$\bar{f}(y) = |F_j|^{-1/2}\int_F f(x, y)\, dx.$$

It is easy to verify that $R_{\pm}^j$ is a linear transformation, mapping $\mathcal{H}_G$ continuously into $L_2(a, \infty)$ for each a in $\mathbf{R}$.

LEMMA 1.4. *$R_{\pm}^F$ transmutes the action of U into translation*:

(1.18)$_+$ $$R_+^F U(t)f = T(t)R_+^F f,$$

(1.18)$_-$ $$R_-^F U(t)f = T(-t)R_-^F f;$$

here $T(t)$ denotes translation to the right by t units.

REMARK. Because of properties $(1.18)_{\pm}^*$ we call R_+^F a *translation representation* and R_-^F an *anti-translation representation*.

PROOF. Let $u(x, y, t)$ be a solution of the wave equation $(1.2)'$ with automorphic initial data $f = \{f_1, f_2\}$: $u(0) = f_1$, $u_t(0) = f_2$. The mean value $\bar{u}(y, t)$ will then satisfy the equation

(1.19) $$\bar{u}_{tt} = y^2\bar{u}_{yy} - (n - 2)\bar{u}_y + \left(\frac{n-1}{2}\right)^2\bar{u} \quad \text{for all } y > 0.$$

The change of variables $s = \log y$ and $v = \bar{u}/y^{(n-1)/2}$ transform (1.19) into the classical wave equation

$$(1.20) \qquad\qquad v_{tt} = v_{ss}$$

and the initial data goes over into

$$(1.21) \qquad v(0) = e^{(1-n)s/2}\overline{f_1}(e^s), \qquad v_t(0) = e^{(1-n)s/2}\overline{f_2}(e^s).$$

Setting

$$(1.21)_+ \qquad\qquad k_+(s,t) = v_s - v_t,$$

it follows from (1.17) and (1.21) that

$$(1.22)_0 \qquad\qquad k_+(s,0) = \sqrt{2}\, R_+^j f.$$

Since the solution at time t is $U(t)f$, we have similarly

$$(1.22)_t \qquad\qquad k_+(s,t) = \sqrt{2}\, R_+^j U(t)f.$$

It follows from (1.20) and (1.21)$_+$ that $\partial_t k_+ + \partial_s k_+ = 0$ and hence that

$$(1.23) \qquad\qquad k_+(s,t) = k_+(s-t,0).$$

Combining (1.22) and (1.23) we obtain the assertion of the lemma for R_+^j and hence for R_+^F. A similar argument with

$$(1.21)_- \qquad\qquad k_-(s,t) = v_t + v_s$$

proves (1.18)$_-$.

The isometry of R_+^F (R_-^F) is easy to establish on what we call outgoing (incoming) data. These are defined as follows: Let l be any C_0^∞ function and set m equal to its indefinite integral:

$$(1.24) \qquad\qquad m(s) = \int_{-\infty}^{s} l(r)\, dr.$$

As we have seen in the proof of Lemma 1.4, the functions

$$(1.25) \qquad\qquad u_\pm(y,t) = y^{(n-1)/2} m(\log y \mp t)$$

are solutions of the wave equations (1.2)'. Being independent of x, they are automatically automorphic with respect to Γ_j.

Suppose a_j is so large that the subset of F (with the jth cusp at ∞), consisting of those points whose y coordinate is $> a_j$, belongs entirely to the jth cusp; i.e. is of the form

$$(1.26) \qquad\qquad N_j = F_j \times (a_j, \infty).$$

Then if the *support of l is contained in $s > \log a_j$*, u_+ is automorphic with respect to the entire group Γ for all $t \geq 0$, and u_- is likewise automorphic with respect to Γ for all $t \leq 0$.

We now define the incoming and outgoing subspaces $\mathscr{D}_-^j$ and $\mathscr{D}_+^j$ as the initial data of all such solutions u_- and u_+, respectively, i.e. $\mathscr{D}_-^j$ consists of data of the form

$$(1.27) \qquad d_- = \{ y^{(n-1)/2} m(\log y),\ y^{(n-1)/2} l(\log y) \}$$

and $\mathscr{D}_+^j$ consists of data of the form

$$(1.28) \qquad d_+ = \{ y^{(n-1)/2} m(\log y),\ -y^{(n-1)/2} l(\log y) \},$$

where m and l are related by (1.24) and l is supported in

(1.29) $$s > \log a_j.$$

It follows from (1.27) and (1.28) that d_+ and d_- in $\mathscr{D}_\pm^j$ are supported in N_j. We denote by $\mathscr{D}_\pm$ the sum of the $\mathscr{D}^j$:

(1.30) $$\mathscr{D}_+ = \mathscr{D}_+^1 + \cdots + \mathscr{D}_+^N, \qquad \mathscr{D}_- = \mathscr{D}_-^1 + \cdots + \mathscr{D}_-^N.$$

We denote by $\mathscr{D}_\pm'$ the union

(1.31) $$\mathscr{D}_\pm' = \bigcup U(t)\mathscr{D}_\pm.$$

LEMMA 1.5. $R_\pm^F$ *is an isometry on* $\mathscr{D}_\pm'$.

PROOF. Since $U(t)$ is an isometry and $R_\pm^F$ are translation representations, it suffices to show that, say, R_+^F is an isometry on $\mathscr{D}_+$. We assume without loss of generality that the a_j are so large that the cusp neighborhoods N_j, defined in (1.26), are disjoint. In this case the linear spaces on the right in (1.30) are orthogonal, and every f in $\mathscr{D}_+$ can be decomposed as the orthogonal sum

(1.32) $$f = \sum_1^N d^j, \qquad d^j \in \mathscr{D}_+^j.$$

Since the N_j are disjoint

(1.32)' $$E_F(f) = \sum E_F(d^j).$$

We shall show that for d^j of form (1.28)

(1.33) $$R_+^k d^j = \delta_{jk}\left(2|F_j|\right)^{1/2} l,$$

where δ_{jk} is the Kronecker symbol, and l is the function entering the definition (1.28) of d_+^j.

A straightforward calculation using (1.28) and (1.17) shows that (1.33) holds for $k = j$.

We turn now to the case $k \neq j$. Since d^j is supported in N_j, it vanishes in N_k and so from the definition (1.17) of R_+^k we conclude that for $k \neq j$,

(1.34) $$\left(R_+^k d^j\right)^+(s) = 0 \quad \text{for } s > \log a_k.$$

It follows from $(1.18)_+$ and (1.28) that for $t > 0$, $U(t)d^j$ is supported in N_j. Therefore (1.34) holds for $U(t)d^j$ in place of d^j; using $(1.18)_+$ this asserts

$$\left(R_+^k U(t)d^j\right)(s) = \left(R_+^k d^j\right)(s-t) = 0 \quad \text{for } s > \log a_k.$$

Since t is an arbitrary positive number, it follows that for $k \neq j$, $R_+^k d^j \equiv 0$, as asserted in (1.33).

A brief calculation, using the definition (1.3) of E_F and the definition $(1.27)_+$ of d^j, combined with (1.33) for the case $k = j$, shows that

$$E_F(d^j) = 2|F| \int |l(s)|^2 \, ds = \left\| R_+^j d^j \right\|^2.$$

Combining this with (1.32)' we deduce that R_+^F is an isometry; this completes the proof of Lemma 1.5.

We now define the space $\mathcal{H}_+$ to be the closure in the energy norm of $Q\mathcal{D}'_+$, and similarly $\mathcal{H}_-$ to be the closure of $Q\mathcal{D}'_-$. It is clear from the definitions (1.30) and (1.31) that $\mathcal{D}'_+$ is invariant under $U(t)$. Since $U(t)$ commutes with Q, by (1.16), it follows that $\mathcal{H}_+$ is also invariant under the action of $U(t)$, and so is $\mathcal{H}_-$.

LEMMA 1.6. R_+^F *is an isometry on* $\mathcal{H}_+$; *likewise* R_-^F *is an isometry on* $\mathcal{H}_-$.

PROOF. It suffices to show that R_+^F is an isometry on $Q\mathcal{D}'_+$. According to Lemma 1.5, R_+^F is an isometry on $\mathcal{D}'_+$; therefore Lemma 1.6 would follow from the observation that the projection Q does not alter the E_F norm of an element g of $\mathcal{D}'_+$ nor the value of $R_+^F g$. I.e., for every g in $\mathcal{D}'_+$,

$$(1.35) \qquad E_F(Qg) = E_F(g)$$

and

$$(1.36) \qquad R_+^F Qg = R_+^F g.$$

By definition of the projection Q,

$$(1.37) \qquad g = Qg + p_- + p_+,$$

where p_- is a linear combination of the $\{p_k^-\}$ and p_+ is a linear combination of the $\{p_k^+\}$ defined in (1.10). The key to (1.35) and (1.36) is the following result:

$$(1.37)' \qquad g = Qg + p_- \quad \text{for } g \text{ in } \mathcal{D}'_+.$$

To prove this we make use of the relations (1.11), (1.11)'; clearly, to show that there are no p_+ components in (1.37)', we have to show that for g in $\mathcal{D}'_+$

$$(1.38) \qquad E_F(g, p_k^-) = 0.$$

Every g in $\mathcal{D}'_+$ is of the form $g = U(t)f, f$ in $\mathcal{D}_+$. Using this in (1.38), as well as the invariance of E_F and the fact that $U(t)p_k^- = e^{-\lambda_k t}p_k^-$, we get

$$(1.38)' \quad E_F(g, p_k^-) = E_f(U(t)f, p_k^-) = E_F(f, U(-t)p_k^-) = e^{\lambda_k t}E_F(f, p_k^-).$$

Every f in $\mathcal{D}_+$ can, by definition, be written as a sum of elements d^j of $\mathcal{D}'_+$ (see (1.32)); therefore, in view of (1.38)', in order to prove (1.38) it suffices to show that

$$(1.39) \qquad E_F(d^j, p_k^-) = 0 \quad \text{for } d^j \text{ in } D_+^j.$$

By definition (1.28), d^j is independent of x in N_j, and zero outside N_j. Therefore

$$(1.39)' \qquad E_f(d^j, p_k^-) = |F_j|^{1/2} E_F(d^j, \overline{p_k^-}),$$

where the bar denotes the mean value defined in (1.17). We compute now $\overline{p_k^-}$.

By definition (1.10), and dropping the subscript k, $p^- = (\varphi, -\lambda\varphi)$. Thus

$$(1.40) \qquad \overline{p^-} = (\overline{\varphi}, -\lambda\overline{\varphi}).$$

As noted before, $e^{\lambda t}\varphi$ satisfies the wave equation; thus $e^{\lambda t}\overline{\varphi}$ satisfies the reduced equation (1.19):

$$\lambda^2\overline{\varphi} = y^2\overline{\varphi}_{yy} - (n-2)y\overline{\varphi}_y + \left(\frac{n-1}{2}\right)^2\overline{\varphi}.$$

This equation is satisfied by the powers $y^{(n-1)/2 \pm \lambda}$; all other solutions are linear combinations of them. Since φ is square integrable in N_j with respect to $dx\, dy/y^n$, $\bar{\varphi}$ is $L_2(a, \infty)$ with respect to dy/y^n; thus

$$(1.41) \qquad \bar{\varphi} = \text{const } y^{(n-1)/2 - \lambda}.$$

Setting this into (1.40) gives

$$(1.40)' \qquad \overline{p^-} = \text{const } \left(y^{(n-1)/2 - \lambda}, -\lambda y^{(n-1)/2 - \lambda} \right).$$

Substituting (1.40)' for $\overline{p_k^-}$ and (1.28) for d_+ into (1.39)', and using the definition (1.3) of E_F shows, after a brief calculation, that (1.39)' is zero. This proves (1.39), and thereby (1.38). From this (1.37)' follows.

We are now ready to prove (1.35) and (1.36). By (1.37)',

$$E_F(Qg) = E_F(g - p_-) = E_F(g) - 2E_F(g, p_-) + E_F(p_-).$$

The second term on the right is zero by (1.38), and the third team by (1.11); this proves (1.35).

We turn now to (1.36). By (1.37)',

$$(1.42) \qquad R_+^F Qg = R_+^F g - R_+^F p_-.$$

We claim that

$$(1.42)' \qquad R_+^F p_- = 0.$$

To see this we note that, by definition, p_- is a linear combination of the p_k^-. Setting formula (1.40)' for $\overline{p_-}$ into the definition $(1.17)_+$ of R_+^j, and using $y = e^s$ with $\partial_s = y \partial_y$, we obtain

$$R_+^j p^- = \text{const}\left\{ y\, \partial_y\, y^{(1-n)/2} y^{(n-1)/2 - \lambda} + y^{(1-n)/2} \lambda y^{(n-1)/2 - \lambda} \right\} = 0.$$

This proves (1.42)'; combining (1.42)' with (1.42) we obtain (1.36). This completes the proof of Lemma 1.6.

It follows from (1.33) that the range of R_+^j on $\mathcal{D}_+$ includes all $C_0^\infty(\log a_j, \infty)$ functions l. Since R_+^j is a translation representation, it follows that the range of R_+^j on $\mathcal{D}_+'$ includes all $C_0^\infty(\mathbf{R})$ functions. From this and (1.36) it follows that the range of R_+^F on $Q\mathcal{D}_+'$ is dense in $L^2(\mathbf{R})^N$. Combining this with Lemma 1.6 we obtain

THEOREM 1.7. R_+^F is a unitary translation representation of $\mathcal{H}_+$ onto $L^2(\mathbf{R})^N$ and R_-^F is a unitary anti-translation representation of $\mathcal{H}_-$ onto $L^2(\mathbf{R})^N$. For f in $\mathcal{H}_+$

$$(1.43)_+ \qquad R_+^F U(t)f = T(t) R_+^F f, \qquad E_F(f) = \left\| R_+^F f \right\|^2;$$

and for f in $\mathcal{H}_-$

$$(1.43)_- \qquad R_-^F U(t)f = T(-t) R_-^F f, \qquad E_F(f) = \left\| R_-^F f \right\|^2.$$

Since Fourier transform changes a translation representation into a spectral representation, an immediate corollary of Theorem 1.7 is

COROLLARY 1.8. $U(t)$ has an absolutely continuous spectrum on $\mathcal{H}_+$ and on $\mathcal{H}_-$. The infinitesimal generator A of $U(t)$ has an absolutely continuous spectrum of uniform multiplicity N on $\mathbf{R}$.

We shall also require the following results:

LEMMA 1.9. *If g lies in $\mathcal{H}_Q$, then $R_+^F g \equiv 0$ iff g is orthogonal to $\mathcal{H}_+$. Similarly $R_-^F g \equiv 0$ iff g is orthogonal to $\mathcal{H}_-$.*

PROOF. g, being in $\mathcal{H}_Q$, is orthogonal to $\mathcal{P}$; hence $E_F(g, Qd_+') = E_F(g, d_+')$ for any d_+' in $\mathcal{H}_G$. For d_+ of the form (1.28), an explicit calculation, using (1.17) and (1.3), gives

$$(1.44) \qquad\qquad E_F(g, d_+) = \int R_+^j g(s) \bar{l}(s) \, ds.$$

So for $d_+' = U(t)d_+$, using (1.43)$_+$ and (1.44), we get

$$E_F(g, d_+') = E_F(g, U(t)d_+) = E_F(U(-t)g, d_+) = \int \left(R_+^j (U(-t)g) \right) \bar{l}(s) \, ds$$

$$= \int T(-t)(R_+^j g) \bar{l}(s) \, ds = \int (R_+^j g) \bar{l}(s - t) \, ds.$$

Since the index j is arbitrary and the translates of l are dense, Lemma 1.9 follows.

A consequence of this is

LEMMA 1.10. *For any h in $\mathcal{H}_Q$*

$$(1.45) \qquad\qquad E_F(h) \geqslant \| R_-^F h \|^2.$$

PROOF. Decompose h into orthogonal parts

$$(1.46) \qquad\qquad h = f + g, \quad f \text{ in } \mathcal{H}_- \text{ and } g \perp \mathcal{H}_-.$$

By Lemma 1.9, $R_-^F g \equiv 0$ so that

$$(1.47) \qquad\qquad R_-^F h = R_-^F f.$$

Since f and g are orthogonal by construction,

$$E_F(h) = E_F(f) + E_F(g) \geqslant E_F(f).$$

In the last step we used the fact that $E_F \geqslant 0$ on $\mathcal{H}_Q$. On the other hand, by (1.43)$_-$,

$$E_F(f) = \| R_-^F f \|^2 = \| R_-^F h \|^2;$$

in the last step we have used (1.47). Combining these last two relations, we obtain (1.45).

We denote by T the time reversal operator mapping $f = \{f_1, f_2\}$ into $Tf = \{f_1, -f_2\}$. It is obvious from (1.28) that $T\mathcal{D}_+ = \mathcal{D}_-$, and from this it follows that

$$(1.48) \qquad\qquad T\mathcal{H}_+ = \mathcal{H}_-.$$

LEMMA 1.11. *$\mathcal{Z} =$ null space of A.*

PROOF. By definition

$$(1.49) \qquad\qquad \mathcal{Z} = \left[z \in \mathcal{H}_Q, E_F(z, f) = 0 \text{ for all } f \text{ in } \mathcal{H}_Q \right].$$

Clearly $E_F(z, g) = 0$ for all g orthogonal to $\mathcal{H}_Q$ and hence, combining this with (1.49), for all g in $\mathcal{H}_G$. In particular, it holds for all g of the form $\{0, g_2\}$ with arbitrary g_2 in $L_2(F)$, so that $z_2 = 0$ for $z = \{z_1, z_2\}$. It also holds for all

$g = \{g_1, 0\}$ with arbitrary g_1 in $C_0^\infty(F)$, and hence by (1.3), $H_F(z_1, Lg_1) = 0$ for all g_1 in $C_0^\infty(F)$. Thus $Lz_1 = 0$ in the weak sense and hence $Az = 0$. Conversely if $Af = 0, f = \{f_1, f_2\}$, then $f_2 = 0$ and

$$E_F(f, g) = -H_F(f_1, Lg_1) = 0$$

for all g in $C_0^\infty(F)$. Since $C_0^\infty(F)$ data are dense in $\mathcal{H}_G$, it follows that $E_F(f, g) = 0$ for all g in $\mathcal{H}_G$ and hence that f belongs to $\mathcal{Z}$.

2. Completeness. With these preliminaries out of the way, we are now ready to establish the completeness of the translation representations. More precisely, we shall show that $\mathcal{H}_+ = \mathcal{H}_-$ and that this subspace is the orthogonal complement of the point spectrum of A in $\mathcal{H}_Q$.

THEOREM 2.1.

(2.1) $$\mathcal{H}_+ = \mathcal{H}_- \pmod{\mathcal{Z}}.$$

The proof relies on four lemmas.

Define the space $\mathcal{H}_c$ as

(2.2) $$\mathcal{H}_c = \overline{\mathcal{H}_- + \mathcal{H}_-},$$

the bar denoting completion with respect to the energy norm. Clearly $\mathcal{H}_c$ is invariant under $U(t)$ along with $\mathcal{H}_-$ and $\mathcal{H}_+$; thus $\mathcal{H}_c$ is an invariant subspace of the generator A.

LEMMA 2.2. *The spectrum of A is absolutely continuous on $\mathcal{H}_c$.*

PROOF. Since $\mathcal{H}_c$ is invariant under A, if the point or singular spectrum of A were nonempty, then there would be a nonzero element f in $\mathcal{H}_c$ whose spectrum is purely point or singular. Such an f is orthogonal to all elements having an absolutely continuous spectrum. In particular, f will be orthogonal to $\mathcal{H}_-$ and $\mathcal{H}_+$ by Corollary 1.8 and hence by (2.2) to $\mathcal{H}_c$. Thus f is the zero element of $\mathcal{H}_c$ or, more precisely, f belongs to $\mathcal{Z}$.

LEMMA 2.3. *Suppose f belongs to $\mathcal{H}_c$; denote by $u(t)$ the solution to the wave equation with initial data f. Let S be any compact subset of F. Then*

(2.3) $$\lim_{t \to \infty} H_S(u_t(t)) = 0.$$

PROOF. Since $E_F \geqslant 0$ on $\mathcal{H}_c$ we can write

(2.4) $$H_S(u_t(t)) \leqslant H_F(u_t(t)) \leqslant E_F(u(t)) = E_F(f).$$

It therefore suffices to prove (2.3) for a subset of $\mathcal{H}_c$ dense relative to E_F. $\mathcal{H}_c \cap \mathcal{D}(A)$ will serve as this dense subset; here $\mathcal{D}(A)$ denotes the domain of A.

For $g = \{g_1, g_2\}$ in $\mathcal{H}_c \cap \mathcal{D}(A)$ it follows from (1.3) that

(2.5) $$H_F(g_2) \leqslant E_F(g)$$

and

(2.5)' $$D_F(g_2) - \left(\frac{n-1}{2}\right)^2 H_F(g_2) \leqslant E_F(Ag).$$

It follows directly from (2.5) and (2.5)′ that

$$(2.6) \qquad \sum_{|\alpha| \leqslant 1} H_S(\partial^\alpha g_2) \leqslant c[E_F(g) + E_F(Ag)].$$

This inequality holds in particular for $u_t(t)$:

$$(2.6)' \qquad \sum_{|\alpha| \leqslant 1} H_S(\partial^\alpha u_t(t)) \leqslant c[E_F(f) + E_F(Af)].$$

The spectral resolution of A gives the following representation for $U(t)$:

$$U(t) = \int e^{i\lambda t}\, dP(\lambda).$$

In particular for any f, g in $\mathcal{H}_c$ we have

$$(2.7) \qquad E_F(U(t)f, g) = \int e^{i\lambda t}\, dE_F(P(\lambda)f, g).$$

Since A has absolutely continuous spectrum over $\mathcal{H}_c$, the measure on the right in (2.7) is absolutely continuous. It therefore follows by the Riemann-Lebesgue lemma that

$$(2.7)' \qquad \lim_{t \to 0} E_F(U(t)f, g) = 0$$

for all g in $\mathcal{H}_c$. Since $U(t)f$ belongs to $\mathcal{H}_c$ we can replace g in (2.7)′ by any element in $\mathcal{H}_G$. In particular (2.7)′ holds for all g of the form $\{0, g_2\}$ with g_2 any element of $L_2(F)$ so that

$$(2.7)'' \qquad \lim_{t \to \infty} (u_t(t), g_2) = 0 \quad \text{for all } g_2 \in L_2(F).$$

We see from (2.6)′ and the Rellich compactness criterion that the $u_t(t)$ are compact in the $L^2(S)$ norm. According to (2.7)″, the $u_t(t)$ converge weakly to zero as $t \to \infty$ in $L^2(F)$; this combined with compactness in $L^2(S)$ shows that the $u_t(t)$ converge strongly to zero in the $L^2(S)$ norm. This completes the proof of Lemma 2.3.

LEMMA 2.4. *If f is in $\mathcal{H}_c$ and $R_+^F f = 0$, then*

$$(2.8) \qquad \lim_{t \to \infty} H_F(u_t(t)) = 0.$$

PROOF. We saw earlier that if $R_+^F f = 0$, then $R_+^F U(t)f = 0$ for all t; it follows then also that

$$(2.9) \qquad R_+^F f_\varphi = 0,$$

where

$$(2.10) \qquad f_\varphi = \int \varphi(t) U(t) f\, dt.$$

As φ tends to the δ function, f_φ tends to f in the E_F norm. Now for f in $\mathcal{H}_c$

$$H_F(u_t(t)) \leqslant E_F(u(t)f) = E_F(f).$$

Hence it follows that if (2.8) holds for all f_φ, it also holds for f. Since f belongs to the domain of A for φ in C_0^1, it follows that it suffices to prove Lemma 2.4 for f in the domain of A.

For f in $\mathscr{D}(A)$ both $U(t)f$ and $AU(t)f$ lie in $\mathscr{H}_G$ and are of finite energy. Since energy is conserved, it follows from (1.3) that both $H_F(u_t(t))$ and $D_F(u_t(t))$ are uniformly bounded:

$$(2.11) \qquad H_F(u_t(t)) + D_F(u_t(t)) \leqslant \text{const.}$$

Next let $\{\psi_k\}$ be the normalized eigenfunctions of $-\Delta_x$, the negative of the Euclidean Laplace operator, for Γ_j-automorphic functions over F_j. Denote the corresponding eigenvalues by $\{\omega_k\}$. Since F_j is a compact domain, Δ_x has a standard discrete spectrum. The lowest eigenvalue ω_0 is zero with multiplicity one, and the corresponding eigenfunction ψ_0 is constant:

$$(2.12) \qquad \psi_0 = |F_j|^{-1/2}, \qquad \psi_1,\ldots,$$

and

$$(2.12)' \qquad 0 = \omega_0 < \omega_1 \leqslant \cdots.$$

Setting $g = u_t(t)$ for fixed t, we expand $g(x, y)$ as a function of x into a Fourier series with respect to these eigenfunctions:

$$(2.13) \qquad g(x, y) = \sum_0^\infty g^{(k)}(y)\psi_k(x).$$

We have the two Parseval relations

$$(2.14) \qquad \int_{F_j} |g|^2 \, dx = \sum_0^\infty |g^{(k)}|^2$$

and

$$(2.15) \qquad \int_{F_j} |g_x|^2 \, dx = -\int_{F_j} g\,\overline{\Delta g}\, dx = \sum_1^\infty \omega_k |g^{(k)}|^2.$$

Integrating (2.14) with respect to dy/y^n from a to ∞ and (2.15) with respect to dy/y^{n-2} gives

$$(2.14)' \qquad H_{N_j}(g) = \int_{N_j} |g|^2 \frac{dx\,dy}{y^n} = \sum_0^\infty \int_a^\infty |g^{(k)}(y)|^2 \frac{dy}{y^n}$$

and

$$(2.15)' \qquad \int_{N_j} |g_x|^2 \frac{dx\,dy}{y^{n-2}} = \sum_1^\infty \omega_k \int_a^\infty |g^{(k)}(y)|^2 \frac{dy}{y^{n-2}},$$

where again

$$(2.16) \qquad N_j = F_j x(a, \infty).$$

The left side of (2.15)' is $\leqslant D_{N_j}(g)$. Thus

$$(2.17) \qquad D_{N_j}(g) \geqslant \sum_1^\infty \omega_k \int_a^\infty y^2 |g^{(k)}(y)|^2 \frac{dy}{y^n} \geqslant a^2 \omega_1 \sum_1^\infty \int_a^\infty |g^{(k)}|^2 \frac{dy}{y^n}.$$

Combining (2.17) with (2.14)' we obtain

$$(2.18) \qquad H_{N_j}(g) \leqslant \int_a^\infty |g^{(0)}(y)|^2 \frac{dy}{y^n} + \frac{1}{a^2\omega_1} D_{N_j}(g).$$

We obtain an estimate for $g^{(0)}$ by making use of the relation $g^{(0)} = \bar{u}_t(t)$. Using the definition (1.17) of R^j_+, we see that if $R^j_+ f \equiv 0$, then

$$\partial_s e^{(1-n)s/2} \overline{f_1} = e^{(1-n)s/2} \overline{f_2}.$$

Using the analogous definition of R^j_- we conclude from this that

$$(2.19) \qquad k_-(s) \equiv R^j_- f(s) = 2e^{(1-n)s/2} \overline{f_2}(s),$$

where k_- abbreviates $R^j_- f$. Since R^F_+ transmutes $U(t)$ into translation (Lemma 1.4), it follows from the hypothesis $R^j_+ f = 0$ that $R^j_+ U(t)f \equiv 0$. Consequently (2.19) holds with f replaced by $U(t)f$. Further, since R^F_- is an anti-translation representation, we get

$$(2.19)' \qquad k_-(s+t) = 2e^{(1-n)s/2} \overline{u_t(t)}.$$

Next we define

$$(2.20) \qquad h(t) = \{0, u_t(t)\}.$$

Using the definition (1.17) of R^j_-, we get $R^j_- h(t) = e^{(1-n)s/2} \overline{u_t(t)}$. Combining this with (2.19)' we have

$$(2.21) \qquad k_-(s+t) = 2R^j_- h(t)(s).$$

It follows by (2.19)' that for all t

$$(2.22) \qquad \int_a^\infty |\bar{u}_t(y,t)|^2 \frac{dy}{y^n} = \int_{\log a}^\infty |e^{(1-n)s/2} \bar{u}_t(e^s, t)|^2 ds$$

$$= \frac{1}{4} \int_{\log a}^\infty |k_-(s+t)|^2 ds.$$

It follows from Lemma 1.10 that k_- is square integrable over R; therefore it follows from (2.22) that for any a fixed,

$$(2.23) \qquad \lim_{t \to \infty} \int_a^\infty |\bar{u}_t(y,t)|^2 \frac{dy}{y^n} = 0.$$

Hence choosing a sufficiently large and combining (2.23) with (2.18) we get

$$(2.24) \qquad \lim_{t \to \infty} H_{N_j}(u_t(t)) \leqslant \text{const}/a^2.$$

We decompose F as $F = \cup N_j \cup S$, S compact, and write

$$(2.25) \qquad H_F(u_t) = \sum_1^N H_{N_j}(u_t) + H_S(u_t).$$

Applying (2.3) and (2.24) to the right member of (2.25), we conclude,

$$\lim_{t \to \infty} H_F(u_t(t)) \leqslant \text{const}/a^2.$$

Since a can be chosen arbitrarily large, (2.8) follows. This concludes the proof of Lemma 2.4.

LEMMA 2.5. *If f is in $\mathcal{H}_c$ and $R^F_+ f = 0$, then f belongs to $\mathcal{Z}$.*

PROOF. We may as well assume that f belongs to $\mathcal{H}_c \cap \mathcal{D}(A^2)$ since any f in $\mathcal{H}_c$ can be approximated by data of the form f_φ, as in (2.10), which belongs to $\mathcal{H}_c \cap \mathcal{D}(A^2)$. For such data we again have $R^F_+ f_\varphi = 0$. We shall prove that f_φ lies in $\mathcal{Z}$. As φ tends to the δ-function, f_φ tends to f; and since $\mathcal{Z}$ is closed, f itself will belong to $\mathcal{Z}$.

Since $R^F_+ A^j f = (-\partial_s)^j R^F_+ f = 0$ for $j = 0, 1, 2$, it follows from Lemma 2.4, applied to the second components of $A^j U(t) f$, that

$$(2.26) \quad \lim_{t \to \infty} H_F(u_t(t)) = 0, \quad \lim_{t \to \infty} H_F(Lu(t)) = 0, \quad \lim_{t \to \infty} H_F(Lu_t(t)) = 0.$$

For $v = u_t$, we have $v_t = u_{tt} = Lu$ and $Lv = Lu_t$. By (1.3), $E_F(v) = -H_F(v, Lv) + H_F(v_t)$. It follows therefore from (2.26) that $\lim_{t \to \infty} E_F(v(t)) = 0$. Since E_F is invariant in t, we get

$$E_F(Af) = E_F(v(0)) = \lim_{t \to \infty} E_F(v(t)) = 0.$$

Consequently $Af = 0 \bmod \mathcal{Z}$. Now f belongs to $\mathcal{H}_c$ and by Lemma 2.2, A is absolutely continuous on $\mathcal{H}_c$; it follows that if f is a null vector of A (mod $\mathcal{Z}$), then f lies in $\mathcal{Z}$. This proves Lemma 2.5.

Theorem 2.1 is an immediate consequence of Lemma 2.5. For if $\mathcal{H}_+ \neq \mathcal{H}_- \bmod \mathcal{Z}$, then $\mathcal{H}_+$, say, would be a proper subspace of $\mathcal{H}_c$ and there would be a nonzero f in $\mathcal{H}_c$ orthogonal to $\mathcal{H}_+$. According to Lemma 1.9, for such an f we would have $R^F_+ f = 0$. According to Lemma 2.5, f lies in $\mathcal{Z}$. This completes the proof of Theorem 2.1.

We go back now to inequality (2.18) and note that it has the following immediate corollary:

LEMMA 2.6. *Suppose that the automorphic functions g have finite square and Dirichlet integrals over F and that $\bar{g} = 0$ in the jth cusp. Then given any $\varepsilon > 0$, we can choose a neighborhood N_j of the type (2.16) with a so large that for all such functions*

$$(2.27) \quad H_{N_j}(g) = \int_{N_j} |g|^2 \frac{dx\,dy}{y^n} < \varepsilon D_F(g).$$

In what follows we denote $H_+ = H_-$ by H_c. The second main result of this section is

THEOREM 2.7. *Denote by $\mathcal{H}_p$ the orthogonal complement of $\mathcal{H}_c$ in $\mathcal{H}_Q$:*

$$\mathcal{H}_p = \mathcal{H}_Q \ominus \mathcal{H}_c.$$

$\mathcal{H}_p$ is an invariant subspace of $U(t)$ and A has a standard pure point spectrum over $\mathcal{H}_p$; i.e. each eigenvalue has finite multiplicity and the eigenvalues have no finite point of accumulation.

REMARK. As before, the assertion of Theorem 2.7 must be taken modulo $\mathcal{Z}$. According to Lemma 1.11, $\mathcal{Z} = $ null space of A; i.e. $\mathcal{Z}$ consists of all data of the form $\{v, 0\}$ in $\mathcal{H}_G$ with $Lv = 0$ where v need not be square integrable. Thus if $Af = \lambda f$ modulo $\mathcal{Z}$, then

$$f_2 = \lambda f_1 + v \quad \text{and} \quad Lf_1 = \lambda f_2.$$

If $\lambda \neq 0$, then setting $\chi = f_1 + v/\lambda$, we see that $f_2 = \lambda\chi$, $L\chi = \lambda^2\chi$ and $A\{\chi, \lambda\chi\}$ $= \lambda\{\chi, \lambda\chi\}$; that is, $\{\chi, \lambda\chi\}$ is a genuine eigenvector of A. However if $\lambda = 0$, f may be of the form $\{v, \chi\}$, where $\chi \in L_2(F)$ and $L\chi = 0$. In this case $A\{v, \chi\} =$ $\{\chi, 0\}$ need not vanish but does lie in $\mathscr{L}$.

PROOF. Both $\mathscr{H}_Q$ and $\mathscr{H}_c$ are invariant under $U(t)$ and since $U(t)$ is unitary on $\mathscr{H}_Q$ it follows that the orthogonal complement $\mathscr{H}_p$ is also invariant under $U(t)$. It also follows that $\mathscr{H}_p$ is invariant under $(k - A)^{-1}$. Similar reasoning shows that $\mathscr{H}_p$ is invariant under the time reversal operator T. Thus if $\{g_1, g_2\}$ belongs to $\mathscr{H}_p$, then so does $\{g_1, -g_2\}$, and from this it follows that $\{0, g_2\}$ also belongs to $\mathscr{H}_p$.

DEFINITION. We denote by $\mathscr{C}$ the space of functions g for which $\{0, g\}$ belongs to $\mathscr{H}_p$. $\mathscr{H}_p$ being an orthogonal complement, is closed; and it follows that $\mathscr{C}$ is a closed subspace of $L_2(F)$.

LEMMA 2.8. $(L - k^2)^{-1}$ *maps* $\mathscr{C}$ *into* $\mathscr{C}$ *and is a compact operator on* $\mathscr{C}$ *for* k *real* $\neq 0$.

PROOF. Since $(k - A)^{-1}$ maps $\mathscr{H}_p$ into itself, so does $(k - A)^{-1} + (k + A)^{-1} =$ $2k(k^2 - A^2)^{-1}$. Now

$$A = \begin{pmatrix} 0 & I \\ L & 0 \end{pmatrix} \quad \text{and} \quad A^2 = \begin{pmatrix} L & 0 \\ 0 & L \end{pmatrix}.$$

So $(k^2 - A^2)^{-1}$ acts on each component $(k^2 - L)^{-1}$. Applying this to elements of $\mathscr{H}_p$ of the form $\{0, g\}$, we conclude that $(k^2 - L)^{-1}$ maps $\mathscr{C}$ into $\mathscr{C}$, boundedly.

To establish compactness, we note that by definition every element f of $\mathscr{H}_p$ is orthogonal to $\mathscr{H}_+$. According to Lemma 1.9, this implies that $R_+^F f = 0$. Taking f to be of the form $\{0, g\}$, g in $\mathscr{C}$, we conclude, using the definition (1.17) of the components of R_+^F, that for g in $\mathscr{C}$

(2.28) $$\bar{g} = 0 \quad \text{in every cusp.}$$

Next let h be any function in $\mathscr{C}$ of L_2 norm $\leqslant 1$:

(2.29) $$H_F(h) \leqslant 1,$$

and let g be its image under $(k^2 - L)^{-1}$:

(2.30) $$g = (k^2 - L)^{-1}h,$$

so that

$$k^2 g - Lg = h.$$

Taking the scalar product with g we get, using (1.2),

$$\left(k^2 - \left(\frac{n-1}{2}\right)^2\right)H_F(g) + D_F(g) = H_F(h, g);$$

by the Schwarz inequality,

$$|H_F(h, g)| \leqslant \tfrac{1}{2}H_F(h) + \tfrac{1}{2}H_F(g).$$

Hence choosing k so large that $k^2 - ((n - 1)/2)^2 > \tfrac{3}{2}$, we obtain, using (2.29), that

(2.31) $$H_F(g) + D_F(g) \leqslant \tfrac{1}{2}.$$

Since $(k^2 - L)^{-1}$ maps $\mathscr{C}$ into $\mathscr{C}$, g belongs to $\mathscr{C}$ and hence, by (2.28), $\bar{g} = 0$ in every cusp. Therefore Lemma 2.6 is applicable and the inequality (2.27) holds in every cusp. The inequality (2.31) shows that, given any ε, the same neighborhoods N_j may be chosen for all g of the form (2.30), provided h belongs to $\mathscr{C}$ and $H_F(h) \leqslant 1$.

The complement of the cusps, $S = F \setminus \cup N_j$, is a compact set. Therefore by (2.31) and the Rellich compactness theorem, every sequence of g_n's of the form (2.30) has a subsequences which converges in the norm H_S. Taking a sequence $\varepsilon_n \to 0$ and applying the diagonal process, we can select a subsequence that converges in the H_F norm. This completes the proof of Lemma 2.8.

Since E_F and hence $-L$ is nonnegative on $\mathscr{C}$, it follows from Lemma 2.8 that L has a complete set of eigenfunctions $\{\chi_n\}$ in $\mathscr{C}$ with nonpositive eigenvalues $\{-\kappa_n^2\}$;

$$(2.32) \qquad L\chi_n = -\kappa_n^2 \chi_n.$$

Since χ_n is a square integrable eigenfunction of L it also has a finite Dirichlet integral and is orthogonal to the eigenfunction φ_j in (1.9). Therefore $\{\chi_n, 0\}$ belongs to $\mathscr{H}_Q$. We use these eigenpairs to construct eigendata for A. Set

$$(2.33) \qquad \begin{aligned} q_n^{\pm} &= \{\chi_n, \pm\kappa_n\chi_n\} \quad \text{for } \kappa_n \neq 0, \\ \theta_n &= \{0, \chi_n\}, \quad \nu_n = \{\chi_n, 0\} \quad \text{for } \kappa_n = 0. \end{aligned}$$

Then

$$(2.34) \qquad \begin{aligned} Aq_n^{\pm} &= \pm\kappa_n q_n^{\pm}, \qquad A\nu_n = 0, \\ A\theta_n &= \nu_n \quad \text{and} \quad A^2\theta_n = 0. \end{aligned}$$

We now define $\mathscr{H}_p'$ to be the closure of the subspace spanned by the data (2.33). It obviously belongs to the point spectrum of A and hence is orthogonal to $\mathscr{H}_c$, which by Corollary 1.8, has an absolutely continuous spectrum; so by (2.27), $\mathscr{H}_p' \subset \mathscr{H}_p$. It is clear from (2.34) that $\mathscr{H}_p'$ is an invariant subspace for $U(t)$ and therefore so is its orthogonal complement in $\mathscr{H}_p$:

$$(2.35) \qquad \mathscr{Z}' = \mathscr{H}_p \ominus \mathscr{H}_p'.$$

Since the $\{\chi_n\}$ form a complete set of eigenfunctions in C and the second component of any vector in $\mathscr{Z}'$ belongs to C, we see that the second component of any f in $\mathscr{Z}'$ vanishes.

Suppose now that f belongs to $\mathscr{Z}'$. We can approximate f in $\mathscr{Z}'$ by an element f_φ of the form (2.10) in the domain of A; we denote f_φ by $g = \{g_1, g_2\}$. Since $\mathscr{Z}'$ is an invariant subspace, both g and Ag lie in $\mathscr{Z}'$, and, as noted above, have second components equal to 0; i.e. $g_2 = 0$ and $Lg_1 = 0$. Then g lies in $\mathscr{Z}$ by Lemma 1.11. Since such data are dense in the G norm in $\mathscr{Z}'$, we may conclude that $\mathscr{Z}' \subset \mathscr{Z}$ and hence that $\mathscr{H}_p' = \mathscr{H}_p$, modulo $\mathscr{Z}$. To complete the proof of Theorem 2.7, we have only to note that the finite dimensionality of $\mathscr{Z}$ was established in Lemma 1.3.

3. A spectral representation for L. As a corollary to Theorem 2.7 we obtain

THEOREM 3.1. *The operator L acting on $L_2(F)$ has a standard discrete spectrum plus an absolutely continuous spectrum of uniform multiplicity N on $(-\infty, 0]$.*

PROOF. Since $\mathscr{P}$, $\mathscr{H}_p$ and $\mathscr{H}_c$ are all invariant under the time reversal operator T, it follows that the second component of data in $\mathscr{H}_G$, which fill out $L_2(F)$, can be decomposed into point eigenfunctions of L (that is the φ's of (1.9) and the χ's of (2.32)) and

$$(3.1)_2 \qquad\qquad \mathcal{F}_2 = \left[\, g \text{ in } \mathcal{H}_c \text{ of the form } \{0, g_2\}\,\right].$$

According to Lemma 2.2, the spectrum of L in $\mathcal{F}_2$, which is the same as that of A^2 in $\mathcal{F}_2$, is absolutely continuous. We need only to prove that it is of uniform multiplicity N on $(-\infty, 0]$.

To this end we decompose $\mathcal{H}_c$ into two orthogonal parts: $\mathcal{F}_2$ and

$$(3.1)_1 \qquad\qquad \mathcal{F}_1 = \left[\, g \text{ in } \mathcal{H}_c \text{ of the form } \{g_1, 0\}\,\right].$$

As remarked after Theorem 1.7, a spectral representation of A in $\mathcal{H}_c$ is obtained by Fourier transforming R_+^F:

$$(3.2) \qquad\qquad \tilde{f}(\sigma) = \frac{1}{\sqrt{2\pi}} \int_{-\infty}^{\infty} e^{i\sigma s} R_+^F f(s)\, ds.$$

Under this map the actions of A and A^2 go into multiplication by $i\sigma$ and $-\sigma^2$, respectively. Since A^2 acts as L on each component we obtain in this way a spectral representation for L on $\mathcal{F}_2$. We denote the Fourier image of $\mathcal{F}_i$ by $\tilde{\mathcal{F}}_i$, $i = 1, 2$. It follows from Theorem 2.1 and Corollary 1.8 via the Plancherel theorem that

$$(3.3) \qquad\qquad \tilde{\mathcal{F}}_1 \oplus \tilde{\mathcal{F}}_2 = L_2(\mathbf{R})^N.$$

It is clear from the expression (1.14) for A that $A\mathcal{F}_1 \subset \mathcal{F}_2$ and $A\mathcal{F}_2 \subset \mathcal{F}_1$, whenever this is meaningful. Likewise $(k^2 - A^2)^{-1}\mathcal{F}_i \subset \mathcal{F}_i$ for $i = 1, 2$.

As a consequence

$$(3.4) \qquad A(k^2 - A^2)^{-1}\mathcal{F}_1 \subset \mathcal{F}_2 \quad \text{and} \quad A(k^2 - A^2)^{-1}\mathcal{F}_2 \subset \mathcal{F}_1.$$

The spectral representator of $A(k^2 - A^2)^{-1}$ is multiplication by $i\sigma/(k^2 + \sigma^2)$; so it follows that

$$(3.4)' \qquad \frac{\sigma}{k^2 + \sigma^2}\tilde{\mathcal{F}}_1 \subset \tilde{\mathcal{F}}_2 \quad \text{and} \quad \frac{\sigma}{k^2 + \sigma^2}\tilde{\mathcal{F}}_2 \subset \tilde{\mathcal{F}}_1.$$

This result can be strengthened.

LEMMA 3.2. *Let J denote multiplication by* $\operatorname{sgnm}(\sigma)$:

$$(3.5) \qquad\qquad J\tilde{f}(\sigma) = \begin{cases} \tilde{f}(\sigma) & \text{for } \sigma > 0, \\ -\tilde{f}(\sigma) & \text{for } \sigma < 0. \end{cases}$$

J *is a unitary map taking* $\tilde{F}_1$ *onto* $\tilde{F}_2$ *and* $\tilde{F}_2$ *onto* $\tilde{F}_1$.

Before proving this lemma, we show how it can be used to complete the proof of Theorem 3.1. For $\tilde{f}$ in $\mathcal{F}_2$, $J\tilde{f}$ lies in $\mathcal{F}_1$ and is therefore orthogonal to $\tilde{f}$. The relation $(\tilde{f}, J\tilde{f}) = 0$ can be rewritten as

$$(3.6) \qquad\qquad \int_0^\infty |\tilde{f}|^2\, d\sigma - \int_{-\infty}^0 |\tilde{f}|^2\, d\sigma = 0.$$

Denoting the restriction map of $L_2(\mathbf{R})^N$ onto $L_2(\mathbf{R}_+)^N$ by τ,

$$\tau : \tilde{f} \to \sqrt{2}\, \tilde{f}(\sigma) \quad \text{for } \sigma > 0,$$

it follows from (3.6) that τ maps $\mathscr{F}_2$ isometrically into $L_2(\mathbf{R}_+)^N$. It remains only to show that this map is onto $L_2(\mathbf{R}_+)^N$. Since τ is an isometry, its range is closed and hence if it is not onto there exists a nonzero element $\tilde{h}$ in $L_2(\mathbf{R}_+)^N$ orthogonal to $\tau\tilde{F}_2$. Extend $\tilde{h}$ to be zero for $\sigma < 0$. Then for the extended $\tilde{h}$, $(\tilde{f}, \tilde{h}) = (\tau\tilde{f}, \tau\tilde{h}) = 0$ for all $\tilde{f}$ in $\mathscr{F}_2$. Thus $\tilde{h}$ lies in $\mathscr{F}_1$. However since $\tilde{h}(\sigma)$ vanishes on $\mathbf{R}_-$, $\tilde{h} = J\tilde{h}$ and by the lemma, $J\tilde{h}$ belongs to $\mathscr{F}_2$. Thus $\tilde{h}$ belongs to two orthogonal sets and is therefore zero. This contradiction proves that $\tilde{F}_2 = L_2(\mathbf{R}_+)^N$ and hence proves the theorem.

PROOF OF LEMMA 3.2. As k varies the set of functions $\sigma/(k^2 + \sigma^2)$ separates points on $\mathbf{R}$. It follows from the Stone-Weierstrass theorem that odd polynomials in these functions are dense in the sup norm in the set of all odd continuous functions which vanish at ∞: such polynomial multipliers also have the property (3.4)'. The strong and uniform operator limits of such polynomials continue to have this property. In particular it will be true for

$$J_n = \begin{cases} \operatorname{sgn} \sigma & \text{for } 1/n < |\sigma| < n, \\ 0 & \text{for } |\sigma| > n + 1/n, \\ \text{linear} & \text{in between.} \end{cases}$$

Finally $J_n \to J$ in the strong operator topology so it also holds for J. Since J is obviously unitary the rest of the assertion of the lemma is immediate.

In §6 of [6] we have shown how to express in terms of Eisenstein series the spectral representation obtained from the translation representation.

4. Perturbations with compact support. The above development is easily adapted to handle automorphic solutions of a perturbed wave equation when the support of the perturbation is compact in F. In place of the usual hyperbolic metric in $\mathbf{H}^n$, we take

$$(4.1) \qquad\qquad ds^2 = g_{ij}\,dx_i\,dx_j;$$

here (g_{ij}) is positive definite, Γ-automorphic, of class $C^{(1)}$ and on F is equal to δ_{ij}/x_n^2 outside of a compact subset S of the type (1.6). Note that we have replaced y by x_n.

In place of the operator (1.2) we now treat

$$(4.2) \qquad\qquad L = g^{-1}\partial_{x_i} g^{ij} g\,\partial_{x_j} + \left(\frac{n-1}{2}\right)^2 - q,$$

where g denotes the square root of $\det(g_{ij})$ and q is L_∞, Γ-automorphic and on F vanishes outside of S. With a little care one could allow q to be of class $L_p(F)$, where $p = n/2$ for $n > 4$, $p > 2$ for $n = 4$ and $p = 2$ for $n < 4$. However to simplify the exposition we take q in L_∞.

The L_2 and Dirichlet integrals are now of the form

$$(4.3) \qquad H_F(u) = \int_F |u|^2 g\,dx \quad \text{and} \quad D_F(u) = \int_F g^{ij}\,\partial_{x_i} u\,\overline{\partial_{x_j} u}\,g\,dx.$$

The perturbed wave equation is

$$(4.4) \qquad\qquad u_{tt} = Lu,$$

and the associated energy form is

$$(4.5) \qquad E_F(u) = -H_F(u, Lu) + H_F(u_t)$$

$$= D_F(u) + \int_F q|u|^2 g \, dx - \left(\frac{n-1}{2}\right)^2 H_F(u) + H_F(u_t).$$

Notice that in the cusp neighborhoods both L and E_F are equal to their unperturbed counterparts.

Again defining K as in (1.4), we prove

LEMMA 4.1. *The form*

$$(4.6) \qquad\qquad\qquad G = E_F + K$$

is positive definite provided the constant c and the compact set S are sufficiently large.

The proof of this assertion is essentially the same as that of Lemma 1.1; the same integration by parts in the cusp neighborhoods is used. Further since q is bounded, the extra term $\int q|u|^2 g \, dx$ is obviously majorized by K.

We again complete the set of $C_0^\infty(F)$ automorphic data with respect to the G-norm and denote the resulting Hilbert space by $\mathscr{H}_G$. We note that for data f in $\mathscr{H}_G$, the local L_2 and Dirichlet norms of f_1 are majorized by $G(f)$.

We then proceed more-or-less as before. The principal difference occurs in the definition of the incoming and outgoing translation representations. To begin with we define $R_\pm^j$ by (1.17) only for $s > a_j$. The argument used in the proof of Lemma 1.4 now shows that

$$(4.7) \qquad \begin{aligned} R_+^j U(t)f &= T(t)R_+^j f \quad \text{for } s > \max(a_j + t, a_j), \\ R_-^j U(t)f &= T(-t)R_-^j f \quad \text{for } s > \max(a_j - t, a_j). \end{aligned}$$

We obtain the assertion of Lemma 1.4 over the entire range of s essentially by fiat; that is we define

$$(4.8) \qquad \begin{aligned} R_+^j f &= T(-t)R_+^j(U(t)f) \quad \text{for } s > a_j - t, \\ R_-^j f &= T(t)R_-^j(U(t)f) \qquad \text{for } s > a_j + t. \end{aligned}$$

Using the relations (4.7), it is easy to see that $R_\pm^j f$ is well defined by (4.8) and satisfies (1.18) for all s.

The subspaces $\mathscr{D}_\pm^j$, $\mathscr{D}_\pm$ and $\mathscr{D}'_\pm$ are defined as before by the relations (1.27)–(1.31). The remaining material in §1 carries over verbatim since in all of these proofs the definition (1.17) of $R_\pm^j f$ is used only for $s > a_{j_F}$.

The proof of completeness for $R_\pm$ is basically the same as in §2. Since E_f now contains the term $\int q|u|^2 g \, dx$, the relation (2.5)′ must be suitably modified. Nevertheless (2.6) continues to hold. Again in the proof of Lemma 2.4, the displayed relations hold only for $s > a_j$. However since we are only interested in sufficiently large values of a in (2.22) and (2.23), this restriction is of no consequence. Otherwise all of the material in §§2 and 3 carries over.

We have learned from Peter Sarnak that Peter Perry has also shown that local perturbations of the metric can create no singular spectrum. Perry employs the "geometric" methods of Schrödinger scattering theory.

In [9], Colin de Verdiere has shown how to construct local, conformal perturbations of the metric in F so that the operator L in (4.2), with $q \equiv 0$, has no negative point spectrum. It follows from our Theorem 3.1 that for such L, $\mathscr{H}_p' = \{0\}$, so that $\mathscr{H}_c = \mathscr{H}_Q$. It then follows from Lemma 2.3 that if the initial data f of a solution u of the wave equation are orthogonal to the positive eigenfunctions $\varphi_j, j = 1, \ldots, m$, then $H_S(u_t(t))$ tends to zero as $t \to \infty$ for any compact set S.

REFERENCES

1. L. D. Faddeev, *Expansion in eigenfunctions of the Laplace operator in the fundamental domain of a discrete group on the Lobačevskii plane*, Trudy Moscow. Math. Obšč. **17** (1967), 323–350; English transl., Trans. Moscow Math. Soc. **17** (1967), 357–386.

2. L. D. Faddeev and B. S. Pavlov. *Scattering theory and automorphic functions*, Proc. Steklov Inst. Math. **27** (1972), 161–193.

3. P. D. Lax and R. S. Phillips, *Scattering theory for automorphic functions*, Ann. of Math. Studies, no. 87, Princeton Univ. Press, Princeton, N. J., 1976.

4. _____, *The asymptotic distribution of lattice points in Euclidean and non-Euclidean spaces*, J. Funct. Anal. **46** (1982), 280–350.

5. _____, *Translation representations for automorphic solutions of the wave equation in non-Euclidean spaces*. I, II, Comm. Pure Appl. Math. **37** (1984), 303–328, 780–813.

6. _____, *Scattering theory for automorphic functions*, Bull. Amer. Math. Soc. **2** (1980), 261–295.

7. A. Selberg, *Harmonic analysis and discontinuous groups in weakly symmetric Riemannian spaces with applications to Dirichlet series*, J. Indian Math. Soc. **20** (1956), 47–87.

8. A. B. Venkov, *Spectral theory of automorphic functions*, Proc. Steklov Inst. Math. **153** (1981); English transl., Trudy Mat. Inst. Steklov. (1982).

9. Y. Colin de Verdiere, *Pseudo-Laplaciens*. II, Ann. Inst. Fourier (Grenoble) **33** (1983), 87–113.

DEPARTMENT OF MATHEMATICS, NEW YORK UNIVERSITY, COURANT INSTITUTE OF MATHEMATICAL SCIENCES, NEW YORK, NEW YORK 10012

DEPARTMENT OF MATHEMATICS, STANFORD UNIVERSITY, STANFORD, CALIFORNIA 94305

COMMENTARY ON PART VI
88, 91, 95, 98, 100, 105, 109, 125

The papers [88], [91], [95], [98], [100], [105], and [109] are joint work with R. Phillips. They are concerned with scattering theory for hyperbolic manifolds. Faddeev and Pavlov [F-P] first observed that the Lax–Phillips time-dependent axiomatic approach to scattering theory in Euclidean space can be applied in the non-Euclidean setting of finite-area noncompact hyperbolic surfaces. This led Lax and Phillips to consider their theory in this context, and the outcome was a series of far-reaching and foundational papers on this subject. The papers included here are a good sample of their achievements. For the finite-volume case, the corresponding spectral theory of the Laplacian had been developed by Selberg [Se]. His main results were the analytic continuation of the Eisenstein series and the trace formula. In papers [88], [91], [109], Lax and Phillips develop their scattering theory for the continuous spectrum on such manifolds. Their method is based on their semigroup $Z(t)$ and its infinitesimal generator B, and it yields a short and conceptual proof of the analytic continuation of the Eisenstein series. Inasmuch as it identifies the poles of the Eisenstein series in terms of the eigenvalues of the compact nonselfadjoint operator B, it is well suited for studying the behavior of these poles under deformations. Their theory was used in [P-S] as a basis for such a study. The cutoff Laplacian (associated with a cusp) that was introduced in the monograph version of paper [91] has, at the hands of Colin de Verdiere [C] and Mueller [M], been a fundamental tool in the proof of the trace class conjecture for the spectrum of a general locally symmetric Riemannian manifold. Paper [91] also gives a direct truncation free derivation of the trace formula. Paper [109] was written after the ones concerned with infinite volume, and it gives a treatment (in the finite-volume cases) of the scattering and spectral theory without using Eisenstein series.

The papers [95], [100] and the series [103] are concerned with the Radon transform in non-Euclidean spaces and related Payley–Wiener theorems. These are developed for the purpose of carrying out the scattering theory for infinite-volume but geometrically finite hyperbolic manifolds. For these, little was known before these series of papers. For the case of surfaces, Patterson [P] had developed the spectral theory (of the Laplacian), but his methods were special to dimension 2. Lax and Phillips first establish the basic spectral properties (in general), that is, the finiteness of the point spectrum below the continuum and the absolute continuity of the rest of the spectrum. Their series of papers on translation representations for solutions of the wave equation for these manifolds yield a complete spectral and scattering theory. They stop short of obtaining a trace formula for this infinite-volume case. Other approaches to the spectral theory of such infinite-volume manifolds have been developed, notably by Mazzeo and Melrose [M-M] and Perry [Per]. Regarding the trace formula for these cases, one may view recent work on the Selberg zeta function for such quotients ([P-P]) as a substitute for the trace formula. It is worth noting that Epstein [E] has shown that the geometrically finite assumption made throughout these series of papers is necessary. If it is dropped, the spectrum can be of an entirely different nature.

The paper [98] investigates the analogue of the well known problem of Gauss of counting asymptotically the number of lattice points in a large circle, in Euclidean and hyperbolic

27

spaces. It was shown quite some time ago by Delsarte [D] that for the case of the hyperbolic plane and a cocompact lattice, the spectrum of the Laplacian on the quotient may be used to obtain the leading term in the asymptotics of counting the number of lattice points in a large hyperbolic disk. In paper [98] Lax and Phillips use the wave equation and their spectral analysis to obtain the asymptotics of the number of lattice points in a large ball for any geometrically finite group. The main term involves the small eigenvalues of the Laplacian on the manifold (if such exist). The error term that they obtain is of the same quality in all cases and is the sharpest result known in all cases. Before their work, Selberg in unpublished work had obtained a similar quality bound for the error term for the case of finite-volume quotients. These counting problems can be generalized to counting cosets for higher-rank and higher-dimensional lattices and have natural Diophantine applications to counting integer points on fixed homogeneous varieties [D-R-S].

References

[C]   Colin de Verdiere, Y. Psuedo Laplacians. *Ann. Inst. Fourier* **32**, 275–286 (1982).

[D]   Delsarte, J. Sur le gitter fuchsien. *C. R. Acad. Sci. Paris* **214** (1942) 147–179.

[D-R-S] Duke, W.; Rudnick, Z.; Sarnak, P. Density of integer points on affine homogeneous varieties. *D. M. J.* **71**, (1993) 143–179.

[E]   Epstein, L. The spectral theory of geometrically periodic hyperbolic 3-manifolds. *Mem. AMS*, **58**, (1985).

[F-P]   Faddeev, L.; Pavlov, B. Scattering theory and automorphic functions. *Proc. Steklol Inst. Math.* **27**, 161–193 (1972).

[M]   Mueller, W. The trace class-conjecture in the theory of automorphic forms. *Ann. Math.* 130 (1989) 473–529.

[M-M]   Melrose, R.; Mazzeo, R. Meromorphic extension d, the resolvent on complete spaces with asymptotically constant curvature. *J. F. A.* **75**, 260–310 (1987).

[P]   Patterson, S. The Laplace operator on a Riemann surface, I, II, III. *Composito Math.* **32**, 81–107 (1975), **32**, 71–112 (1976), **33**, 227–259 (1976).

[Per]   Perry, P. The Laplace operator on a hyperbolic manifold. *J. F. A.* 1987 (161–187).

[P-P]   Patterson, S.J.; Perry, P. The divisor of Selberg's zeta function for Kleinian Groups. *Duke Math. J.* **106** (2001) 321–390.

[P-S]   Phillips, R.; Sarnak, P. Perturbation theory for the Laplacian on automorphic functions. *JAMS* 5 (1992) 1–32.

[Se]   Selberg, A. Harmonic analysis and discontinuous groups in weakly symmetric Riemannian spaces with applications to Dirichlet series. *J. Indian Math. Soc.* **20**, 47–87 (1956).

P. Sarnak

28

PART VII

FUNCTIONAL ANALYSIS

COMMUNICATIONS ON PURE AND APPLIED MATHEMATICS, VOL. IX, 747–766 (1956)

A Stability Theorem for Solutions of Abstract Differential Equations, and Its Application to the Study of the Local Behavior of Solutions of Elliptic Equations[*]

P. D. LAX

Introduction

Let D denote a nonpositive self-adjoint operator over a Hilbert space H. With the aid of the functional calculus for self-adjoint operators we can solve the initial value problem for the differential equation

$$(1) \qquad \frac{du}{dt} = Du, \qquad u(0) = u_0,$$

whenever the initial function u_0 lies in D. The solution is contracting,

$$\|u(t)\| \leqq \|u(0)\|,$$

and can thus be extended by closure to all of H. The solution operator, relating $u(t)$ to u_0, can be written down in terms of the spectral resolution $E(\lambda)$ of D as the integral

$$\int_{-\infty}^{0} e^{\lambda t} \, dE_\lambda(t).$$

The asymptotic behavior of this integral is like $e^{-\lambda_0 t}$, in the sense of being $o(e^{(\lambda_0 - \varepsilon)t})$ for all positive but no negative ε. The quantity λ_0 denotes the top of the spectrum of u_0, i.e. the infimum of those values of λ for which $E_\lambda u_0 = u_0$.

It follows from this asymptotic description that *no solution, except the one that is identically zero, tends to zero faster than some exponential.* Our aim in this paper is to show that solutions of differential equations which do not differ much from (1) also possess this property.

We shall consider perturbations of equation (1) of the following kind:

$$(2) \qquad \frac{du}{dt} = (D + K(t))u.$$

In this paper we shall consider bounded perturbations. It turns out that if the norm of $K(t)$ is not larger than the size of the gaps in the spectrum of

---

[*]The work for this paper was supported by the Office of Naval Research, United States Navy, under contract N6ori-201 T. O. No. 1.

D, then solutions of equation (2), unless identically zero, do not tend to zero faster than some exponential. The precise formulation of this statement is Theorem I of Section 1; it is not restricted to the perturbation of self-adjoint operators.

It should be remarked that it is more realistic to think of (2) as a differential inequality

$$(2') \qquad \left\| \frac{du}{dt} - Du \right\| \leqq k(t) \, \|u\|$$

than as a differential equation.

In the terminology of Phillips, see [20], we prove the stability of the behavior at infinity of solutions of (1) under a certain class of perturbations.

I was led to Theorem I by analyzing a proof due to Hörmander (contained in a letter to Louis Nirenberg) of a theorem on complex-valued functions satisfying differential inequalities. A weaker stability theorem (Theorem I'), which nevertheless suffices for the applications in Section 2, can be obtained from a theorem of Carleman. Its statement and proof are given in an appendix to this paper.

In Section 2 we apply Theorem I to the study of the local behavior of solutions of second order equations which differ from the Laplacian by lower order terms only.

We show that a solution of such an equation cannot have a zero of infinite order unless it is identically zero. This follows from our Theorem I if we make the following interpretations:

    i) H is the Hilbert space of square integrable harmonic functions in the solid unit sphere.

    ii) D is the operator $-\Sigma \, x_j \partial / \partial x_j$.

    iii) $u(t)$ is that harmonic function in the unit sphere whose boundary values are $u_0(e^{-t}x)$, $\|x\| = 1$, where u_0 is the solution of the elliptic equation under consideration.

It is not too difficult to show that in this setup a differential inequality of the form (2') (with $k(t) = \text{const.} \, e^{-t}$) holds. A new inequality, embodied in Lemma III, has to be employed. This inequality is also of use in studying the geometrical relation of the nullspaces of elliptic operators (Theorem II).

In Section 3 we show that the operator D defined above is self-adjoint and that its spectrum lies on the nonpositive integers. This is of course a classical result, asserting the completeness of the spherical harmonics. Nevertheless it is worthwhile to present a proof wholly within the framework of the theory of one-parameter semigroups of self-adjoint operators.

The result on the finiteness of zeros of solutions of elliptic equations is not new. C. Müller [19] has given the first proof; the more general result has been given by E. Heinz [13] and Hartman and Wintner [11]. Recently, N. Aronszajn [1] succeeded in settling the more difficult case of second order equations whose leading term is not the Laplacian.

The problem for functions of two independent variables has been solved by Carleman [5]. His method is function-theoretical and is essentially restricted to the case of two variables.

In Section 4 we investigate the *unique continuation* property of solutions of elliptic equations which is closely related to the property of finiteness of zeros. We show there that solutions of an elliptic equation have the unique continuation property if and only if they have the *Runge property*, i.e., that solutions in a given domain can be approximated arbitrarily closely by solutions defined in a given larger domain which is not disconnected by the smaller domain. It follows in particular that the Runge property in the small implies the Runge property in the large.

That the Runge property follows from the unique continuation property has independently been observed by Malgrange [18]. He has also extended the Runge property to noncompact domains; this is important in several of his applications.

§1. THEOREM 1. *Let D be an operator in Hilbert space which satisfies the condition*

i) *There is an infinite sequence of lines parallel to the imaginary axis whose abcissae λ_n tend to $-\infty$ and on which the resolvent $(D-\lambda)^{-1}$ is uniformly bounded by some constant d^{-1}:*

$$\|(D-\lambda)^{-1}\| \leqq \frac{1}{d}, \qquad \mathscr{R}e\,\lambda = \lambda_n.$$

Let $u(t)$ be a function of t mapping $0 < t < \infty$ into H whose values lie in the domain of D. Assume that Du is a continuous function of t, and that u has a continuous strong derivative with respect to t. Suppose that $u(t)$ satisfies the differential inequality

$$(1.1) \qquad \left\| \frac{du}{dt} - Du \right\| \leqq k\|u\|, \qquad 0 < t < \infty,$$

k any constant less than d.

Conclusion: Unless identically zero, $u(t)$ tends to zero as t goes to infinity not faster than some exponential; i.e., there exist positive constants A and b such that

$$\int \ell^{bt}\|u(t)\|\,dt \doteq \infty$$

Remark. For D self-adjoint, condition i) is equivalent to this:

There are infinitely many intervals $(\lambda_n-d, \lambda_n+d)$, $\lambda_n \to \infty$, which are free of the spectrum of D.

The restriction in Theorem I on the size of the constant k cannot be lessened in general, as illustrated by the following example:

Let $\{u_n\}$ be a complete orthonormal set in H, and D the operator whose eigenelements are u_n, with eigenvalue $-n$,

$$Du_n = -nu_n.$$

Clearly, D is self-adjoint and its spectrum has infinitely many gaps of size 1. Define $u(t)$ as

$$u(t) = a_n e^{-(n+\frac{1}{2})t}\{u_n \cos \varepsilon(t-t_n) + u_{n+1} \sin \varepsilon(t-t_n)\}, \qquad t_n < t < t_{n+1};$$

the sequence of constants t_n and a_n are defined as

$$t_n = n\frac{\pi}{2\varepsilon},$$

$$a_n = \exp\left\{\frac{(n^2+n)\pi}{4\varepsilon}\right\}.$$

It is easy to see that $u(t)$ has the properties:

i) It belongs to the domain of D, and has a piecewise continuous first derivative.

ii) It satisfies the differential inequality

$$\|Du - u_t\| \leqq (\tfrac{1}{2} + \varepsilon)\|u\|.$$

iii) It tends to zero like $\exp\{-2\varepsilon t^2/\pi\}$.

The proof of Theorem I is based on the inequality (1.2) of

LEMMA I. *Let D be an operator in Hilbert space whose resolvent is bounded by $1/d$ on the vertical line with abcissa λ,*

$$(1.2) \qquad \|(D-\mu)^{-1}\| \leqq \frac{1}{d}, \qquad \mathscr{R}e\,\mu = \lambda.$$

Let $u(t)$ be a function from $-\infty < t < \infty$ into H whose values lie in the domain of D — Du is locally in L_2 — and which has a first derivative locally in L_2. Then the following inequality holds:

$$(1.3) \qquad d^2 \int_{-\infty}^{\infty} \|u(t)\|^2 e^{-2\lambda t}\,dt \leqq \int_{-\infty}^{\infty} \|Du - u_t\|^2 e^{-2\lambda t}\,dt,$$

provided that the integral on the left is finite.

We shall prove (1.3) for functions $u(t)$ with compact support. In the general case, approximate $u(t)$ by the sequence $u_n(t) = \varphi_n(t)u(t)$, $\varphi_n(t)$ being a scalar function equal to 1 for $|t| \leqq n$, zero for $n+1 \leqq |t|$, and smooth for $n < t < n+1$. As n tends to infinity, inequality (1.3) for u_n goes over into (1.3) for u.

Proof of Lemma I: Introduce $v(t) = e^{-\lambda t}u(t)$ and denote its Fourier-Plancherel transform by $\hat{v}(\tau)$. The function $(Du - u_t)e^{-\lambda t}$ is equal to

$(D-\lambda)v-v_t$, and its Fourier transform is $(D-\lambda-i\tau)\hat{v}$. According to $_e$(1.2),

(1.4) $$d^2||\hat{v}(\tau)||^2 \leq ||(D-\lambda-i\tau)\hat{v}||^2$$

for each value of τ. Since the Fourier-Plancherel transformation is an isometry in the L_2 norm[1], we obtain, integrating (1.4) with respect to τ over $(-\infty, \infty)$, inequality (1.3). □

Notice that we only require $(D-\mu)^{-1}$ to be bounded on the line $\mathcal{R}e\ \mu = \lambda$, but not that it be everywhere defined.

Suppose now that $u(t)$ satisfies inequality (1.1) of Theorem I, and dies down faster than any exponential; we wish to show that $u(t)$ is identically zero. As the first step, extend it for negative values of t, e.g. by defining $u(t)$ for t negative as $e^t u(0)$. Apply inequality (1.3) to $u(t)$ thus extended, with λ equal to one of the abscissae λ_n on which, according to assumption ii) of Theorem I, the resolvent of D is bounded by $1/d$:

(1.5) $$d^2\int_{-\infty}^{\infty} ||u(t)||^2 e^{-2\lambda_n t} dt \leq \int_{-\infty}^{\infty} ||Du-u_t||^2 e^{-2\lambda_n t} dt.$$

The left side is finite for any value of λ_n since we are dealing with a function $u(t)$ that dies down faster than any exponential.

Inequality (1.1) is satisfied for $t > 0$. Hence we have

(1.6) $$\int_{-\infty}^{\infty} ||Du-u_t||^2 e^{-2\lambda_n t} dt \leq \int_{-\infty}^{0} + k^2\int_{0}^{\infty} ||u||^2 e^{-2\lambda_n t} dt.$$

Since λ_n is negative, the integrals of both $||Du-u_t||^2 e^{-2\lambda_n t}$ and $||u(t)||\ e^{-2\lambda_n t}$ from $-\infty$ to 0 remain uniformly bounded. Thus we have from (1.6)

(1.6') $$\int_{-\infty}^{\infty} ||Du-u_t||^2 e^{-2\lambda_n t} dt \leq K + k^2\int_{-\infty}^{\infty} ||u||^2 e^{-2\lambda_n t} dt,$$

K some constant independent of λ_n. Combining this with (1.5) we get

$$d^2\int_{-\infty}^{\infty} ||u||^2 e^{-2\lambda_n t} dt \leq K + k^2\int_{-\infty}^{\infty} ||u||^2 e^{-2\lambda_n t} dt;$$

since k was assumed smaller than d, we conclude that

$$\int_{-\infty}^{\infty} ||u||^2 e^{-2\lambda_n t} dt$$

is uniformly bounded as λ_n tends to $-\infty$ which is possible only if $u(t)$ is identically zero for t positive. This completes the proof of Theorem I.

For time-independent perturbation of regular operators, it is easy to show that the equation $u_t = (D+K)u$ has a non-zero solution decaying faster than any exponential if and only if the eigenelements of $D+K$ are not complete. By specializing Theorem I to this case we obtain another proof of a result of Jack Schwartz, see [22], on the completeness of the spectra of perturbed operators.

---

[1]Plancherel's theorem for functions with values in Hilbert space can be reduced to the scalar case by the introduction of an orthonormal basis.

§2.   In this section we shall deduce from our stability theorem Müller's

THEOREM.   *Let L be a differential operator of the form $L = \Delta + N$, N a first order operator, and u_0 a solution of $Lu_0 = 0$ which is not identically zero. Conclusion: the zeros of u_0 are of finite order.*

In our proof we require that the coefficients of N be twice differentiable.

For the proof we introduce the following spaces, operators and functions

i) H is the Hilbert space of square integrable harmonic functions in the solid unit sphere normed by the L_2·norm; it is easy to show that H is complete.

ii) D is the operator $-\Sigma x_i \, \partial/\partial x_i$; the domain of D is the set of all harmonic functions h for which Dh is square integrable. Note that Dh is harmonic, as evidenced by the identity

$$(2.1) \qquad\qquad\qquad \Delta D = D\Delta - 2\Delta.$$

As we shall show in Section 3, D is self-adjoint over H, and its spectrum lies on the nonpositive integers.

iii) $C(r)$ is the operator mapping an element $u(x)$ of H into $u(rx)$, $r \leq 1$. Note that the operators $C(r)$ form a one-parameter semigroup.

iv) B is the transformation which maps any continuous function u defined in the unit sphere into the harmonic function v with the same boundary values

$$\Delta v = 0,$$
$$v = u \text{ on the boundary.}$$

v) $S(r) = BC(r)$.

vi) $u(t) = S(r)u_0$, $r = e^{-t}$. (We assume that the solution u_0 is defined in the unit sphere.)

We shall show presently that $u(t)$ as defined in vi) satisfies inequality (1.1) of Theorem I. Therefore, according to the conclusion of Theorem I, $u(t)$ tends to zero at most like an exponential:

$$(2.2) \qquad\qquad ||S(r)u_0|| \geq \text{const. } e^{-bt} = \text{const. } r^b.$$

As it turns out (see Lemma II), the operator B acting on the function Cu_0 is a near-isometry, so that (2.2) is equivalent to

$$(2.2') \qquad\qquad ||C(r)u_0|| \geq \text{const. } r^b.$$

This is just the assertion that u_0 has a zero of order b at the origin.

We shall verify that $u(t)$, as defined under vi), satisfies even the differential inequality (A.1) of Theorem I'. We form $du/dt = -r \, du/dr$:

$$(2.3) \qquad\qquad \frac{du}{dt} = \frac{dS}{dt} u_0 = B \frac{dC}{dt} u_0 = BDC u_0 ,$$

where we have made use of the relation

$$\frac{dC}{dt} = DC.$$

Therefore, we have from (2.3)

$$(2.4) \qquad \frac{du}{dt} - Du = \{BD - DB\} Cu_0.$$

We shall show that the commutator $BD - DB$ reduces the norm of $C(r)u_0$ by a factor $O(r)$. Since, as already remarked before, the norm of $C(r)u_0$ is comparable to the norm of $S(r)u_0$, this proves the desired inequality

$$\left\| \frac{du}{dt} - Du \right\| \leq \text{const. } r\|u\| = \text{const. } e^{-t}\|u\|.$$

To estimate $\{BD - DB\} Cu_0$, we write B as $I + R$; from the definition of B it follows that $Ru = d = Bu - u$ is characterized by the equations

$$(2.5) \qquad \begin{aligned} \Delta d &= -\Delta u, \\ d &= 0 \quad \text{on the boundary of the unit sphere.} \end{aligned}$$

Clearly, $K = BD - DB$ is equal to $RD - DR$. We shall show that both operators RD and DR are, separately, bounded by $O(r)$.

LEMMA II. *Let u_0 be a solution of $Lu_0 = 0$, and let u_r abbrieviate $C(r)u_0 = u_0(xr)$. Then*

$$(2.6) \qquad \|DRu_r\| \leq \text{const. } r\|u_r\|,$$

and

$$(2.6') \qquad \|RDu_r\| \leq \text{const. } r\|u_r\|.$$

The value of the constants depends only on bounds for the coefficients of the operator N in (2.1) and their first and second derivatives.

Proof: The proof of (2.6) is based on the fact that if u_0 is a solution of $Lu_0 = \Delta u_0 + Nu_0 = 0$, then $u_r = u(xr)$ is a solution of an equation that is even closer to the Laplacian, namely of

$$(2.7) \qquad L_r u_r = 0,$$

where L_r is equal to $\Delta + N_r$, the coefficients of N_r and their first derivatives being $O(r)$. This is easily verified; it turns out even that the bound on the first derivatives of N_r is $O(r^2)$.

Now denote Ru_r by d; according to (2.5), d satisfies the relations

$$\begin{aligned} \Delta d &= -\Delta u_r, \\ d &= 0 \quad \text{on the boundary.} \end{aligned}$$

According to (2.7), $-\Delta u_r$ is equal to $N_r u_r$ and hence

(2.8) $$\Delta d = N_r u_r .$$

Multiply (2.8) by d and integrate both sides by parts. There are no boundary terms since d is zero on the boundary:

(2.9) $$||d||_1^2 = -(u_r , N_r^* d).$$

The symbol $||d||_1^2$ here denotes, as customary, the integral of the sum of squares of the first derivatives of d. Applying the Schwarz inequality to the right side of (2.9) we get the inequality

$$||d||_1^2 \leqq \text{const. } ||u_r|| \, ||d||_1 ;$$

upon division by $||d||_1$ we have

(2.9′) $$||d||_1 \leqq \text{const. } ||u_r||;$$

the value of the constant depends on the size of the coefficients of N_r and on their derivatives, and is therefore $O(r)$.

The operator D is of first order, and therefore $||Dd||$ is less than const. $||d||_1$. This observation, combined with (2.9′) yields inequality (2.6) of Lemma II.

As is well known, the square integral of a function d vanishing on the boundary of the unit sphere is bounded by its Dirichlet integral:

$$||d|| \leqq \text{const. } ||d||_1 .$$

From this and inequality (2.9′) we conclude that the operator R, applied to any u_r, is, for r small, norm decreasing. This shows that $B = I + R$, when applied to the same class of elements, is an approximate isometry.

We turn now to the proof of the second assertion in Lemma II, inequality (2.6′). Denote RDu_r by d; it is defined as the solution of the boundary value problem

(2.10) $$\begin{aligned} \Delta d &= -\Delta Du_r , \\ d &= 0 \quad \text{on the boundary.} \end{aligned}$$

Now Δ and D satisfy the identity (2.1),

$$\Delta D = D\Delta - 2\Delta;$$

furthermore, the class of functions u_r considered is precisely the set of solutions of $L_r u_r = 0$. So Δu_r equals $-N_r u_r$, and the right side of equation (2.10) can be written as

$$(DN_r + 2N_r)u_r,$$

which we shall abbreviate as $P_r u_r$. P_r here is some second order operator whose coefficients are $O(r)$. Equation (2.10) thus has become

(2.10′) $$\Delta d = P_r u_r .$$

To estimate the norm of d we shall make use of

LEMMA III. *Let* Q, L, P *be three second order operators, Q and L elliptic, P arbitrary, and Q positive definite over a given domain. Let d and u be a pair of functions related by*

$$(2.11) \qquad\qquad Qd = Pu,$$

and assume that the function d vanishes on the boundary of the domain in question. Then the following inequality holds:

$$(2.12) \qquad\qquad ||d|| \leqq O\{\alpha||u|| + ||Lu||\},$$

where the constant α is the sum of the C^2 norms of the coefficients of P.

In our case

$$Q = \Delta,$$
$$L = L_r,$$
$$P = P_r;$$

the relation (2.11) is just equation (2.10′) between d and $u = u_r$. Inequality (2.12) in this case reads

$$||d|| \leqq O(\alpha)||u_r||$$

since $L_r u_r = 0$. The constant α is $O(r)$, so we have inequality (2.6′) of Lemma II.

It is in Lemma III that we need the coefficients of N to be twice differentiable.

Proof of Lemma III: Define the auxiliary function w by the relations

$$(2.13) \qquad \begin{aligned} Q^*w &= d, \\ w &= 0 \quad \text{on the boundary.} \end{aligned}$$

Multiply (2.11) by w and integrate by parts. On the left we obtain (d, d) $= ||d||^2$ without boundary terms since both d and w vanish on the boundary. On the right we get (P^*w, u) plus one boundary term. So we have the identity

$$(2.14) \qquad ||d||^2 = (P^*w, u) + \int p \frac{dw}{dn} u \, dS,$$

where p is a linear combination of the coefficients of P, and dw/dn denotes the normal derivative of w. Here we make use of the fact that since w is zero on the boundary, the conormal derivative with respect to the operator P is equal to $\cos \theta$ times the normal derivative, θ denoting the angle between the normal and conormal.

Next apply Green's formula to the operator L and the functions u and z, where z is as yet unspecified except for the requirement that it vanish on the boundary. We get

$$(2.15) \qquad (Lu, z) - (u, L^*z) + \int l \frac{dz}{dn} u \, dS = 0.$$

Here, the coefficient l is *non zero* since L is elliptic. Extend p and l to the interior of the domain and choose z to be pw/l; since w vanishes on the boundary, $dz/dn = (p/l)dw/dn$ there. Substituting this choice of z into (2.15) and subtracting from (2.14) we get

$$||d||^2 = \left(\left\{ P^* + L^* \frac{p}{l} \right\} w, u \right) - (w, Lu).$$

Applying the Schwarz inequality we have

(2.16) $||d||^2 \leqq \alpha ||w||_2 ||u|| + ||w|| \, ||Lu||,$

where, as customary, $||w||_2$ stands for the quantity

$$||w||_2 = \sqrt{\iint w^2 + \sum w_{x^i}^2 + w_{x^i x^k}^2}.$$

The size of the constant α depends on the magnitude of the coefficients of the operator $P^* + L^* p/l$, i.e. on the C^2 norm of the coefficients of P.

To complete the proof we observe that the function w is related to d by (2.13), and therefore, according to a theorem of Lichtenstein (see e.g. Friedrichs [9]), $||w||_2$ is bounded by $||d||$.

The present proof of Müller's theorem applies without alteration to solutions of second order systems whose leading term is the Laplacian and which are coupled by first order terms only. In this way one can study the local behavior of solutions of equations which differ from the polyharmonic operator by lower terms of a certain kind. Curiously, the lower order terms permitted in this treatment are subject to limitations which are similar to those required by Louis Nirenberg in his recent investigations on the unique determination of solutions by their Cauchy data.

Another application of Lemma III is to study the relation of nullspaces of second order operators:

THEOREM II. *Let L be a positive definite elliptic operator of second order and denote by $\mathscr{L}$ the space of solutions of $Lu = 0$ which are square integrable over some given smoothly bounded domain. Let M denote any other elliptic operator, and $\mathscr{M}$ its nullspace.*

Assertion: Given any ε one can choose δ so small that if the difference of the coefficients of L and M considered in the C^2 norm is less than δ, the angle of the two nullspaces $\mathscr{L}$ and $\mathscr{M}$ measured with respect to the L_2 scalar product over the domain in question is less than ε.

This result complements the one obtained by Paul Berg and the author [2] concerning the relation of the nullspaces of two operators which differ in lower order terms only.

§3.   In this section we shall prove that the operator D defined in the first section by the formula

$$Du = v = -\sum x^i u_{x^i}$$

and acting on all square integrable harmonic functions u for which v is square integrable, is self-adjoint. We shall prove this by characterizing D as the infinitesimal generator of the semigroup of contractions over the space of harmonic functions.

As first step, we show that D as defined is symmetric; i.e., if u and v are within its domain of definition,

$$(3.1) \qquad\qquad (u, Dv) = (Du, v).$$

We shall show first that

$$(3.1_R) \qquad\qquad (u, Dv)_R = (Du, v)_R$$

for every R less than one, the subscripts in the scalar product indicating that the domain of integration is to be extended over the solid sphere of radius R. From Green's formula

$$\iint u\Delta v - v\Delta u = \int u\,\frac{\partial v}{\partial n} - v\,\frac{\partial u}{\partial n}$$

we conclude namely that the integral of $uDv - vDu$ over the *surface* of the sphere of radius R is zero. Integrating this result with respect to R we get (3.1_R), and letting R tend to one we obtain (3.1).

The self-adjointness of D will be proved indirectly, on the basis of a theorem of Hille [14].

HILLE'S THEOREM. *Let $C(t)$ be a strongly continuous one parameter semigroup of bounded operators over a Banach space, i.e. satisfying*

i) $C(t_1+t_2) = C(t_1)C(t_2)$, $\quad 0 \leqq t_1, t_2 < \infty$, $\quad C(0) = I$;

ii) $C(t)u$ depends continuously on t in the strong topology for each u of the space;

iii) $\|C(t)\| \leqq K$ for all t.

Define the infinitesimal generator G by the formula

$$(3.2) \qquad\qquad Gu = \lim_{h \to 0} \frac{C(h)u - u}{h},$$

the domain of G being the set of all u for which the above limit exists.

Conclusion: The half-plane $\mathcal{R}e\,\lambda > 0$ belongs to the resolvent set of G.

The proof consists of writing down a formula for the resolvent:

$$(3.3) \qquad\qquad (G-\lambda)^{-1}u = \int_0^\infty C(t)u\,e^{-\lambda t}\,dt;$$

the integral is taken in the strong sense. The integrand is for each u, by assumption ii), a continuous and therefore integrable function of t over each finite t interval; since by iii) Cu is uniformly bounded for all t, the integral over the whole real axis converges for every λ whose real part is negative. This shows that the integral on the right of (3.3) converges. That it is indeed the inverse of $G-\lambda$ can be seen by substitution into the definition of the infinitesimal generator.

In the special case of interest to us when the underlying space is the space of harmonic functions square integrable over the unit sphere and $C(t)$ the contraction operator defined in iii) of Section 2, $r = e^{-t}$, the limit on the right of (3.2), lim $[C(h)u-u]/h$, exists for all u of the underlying space in the topology of uniform convergence in every closed subset of the unit sphere, and the value of the limit is equal to $-\sum x_j \partial u/\partial x_j$. Since this topology is weaker than the normed topology, it follows that if u belongs to the domain of the infinitesimal generator as defined by Hille, Gu must be equal to $-\sum x_j \partial u/\partial x_j$. This shows that G is a *contraction* of the operator D defined before; from this and the symmetry of D we can conclude:

a) G is symmetric.

b) Since the resolvent set of G contains points of the upper and lower half-plane, G is self-adjoint.

c) Since a self-adjoint operator has no proper symmetric extensions, G and D are the same.

It is easy to show that if the infinitesimal generator G is self-adjoint, the operators $C(t)$ of the semigroup are also self-adjoint, and that if λ belongs to the spectrum of G, $e^{\lambda t}$ belongs[2] to the spectrum of $C(t)$. Hence we can locate the spectrum of G by determining the spectrum of $C(t)$.

The operators $C(t)$, $0 < t$, are completely continuous, on account of the interior compactness of harmonic functions.[3] Hence their spectra are pure point spectra, and, since they commute, they have the same eigenelement. Now let u be an eigenelement of the contraction operators $C(r)$. This means that u is a homogeneous function; since harmonic functions are analytic, the degree of homogeneity must be an *integer* n. The eigenvalue of $C(r)$ is then $r^n = e^{-nt}$. This proves that the spectrum of $C(t)$ consists of numbers of the form e^{-nt}; therefore the spectrum of G lies on the negative integers.

§4. *Unique continuation*. In the first three sections we have investigated the circumstances under which functions u satisfying differential equations (or inequalities) of the form

(4.1) $u_t = (D+K)u$

decrease not faster than some exponential, unless identically zero. In this

---

[2]According to an important theorem of Phillips [21], such a relation between the spectrum of the infinitesimal generator and the semigroup operators is valid in general.

[3]See [16] for a systematic use of the interior compactness of the space of solutions of an elliptic equation.

section we shall look at a slightly simpler problem: *Can a solution of such an equation be zero for all t greater than some T without being identically zero?* This is the backward uniqueness problem for solutions of equation (4.1); it is difficult because the backward Cauchy problem is *incorrectly posed*, i.e. solutions at earlier times do not depend continuously on their values at later times.

In the concrete situation discussed in the previous sections backward uniqueness means that two solutions of an elliptic equation in the unit sphere which are equal in some sphere, no matter how small its radius, are equal in the whole unit sphere. It is this concrete problem we shall be dealing with in the present section.

The question of unique continuation for solutions of elliptic equations is, as is well known, equivalent to the question whether solutions are uniquely determined by their Cauchy data. This question has been answered in the affirmative for linear equations with analytic coefficients by Holmgren's theorem, [15], for elliptic equations with non-analytic coefficients in two independent variables by Carleman [5], for elliptic systems with multiple (complex) characteristics by Douglis [6]. Recently, Louis Nirenberg has shown that solutions of a large class of not necessarily elliptic equations in any number of variables can be uniquely determined from their Cauchy data on convex hypersurfaces. In this section we shall establish the equivalence of the uniqueness problem with the problem of approximating solutions, defined in small domains, by a sequence of solutions whose domain of existence is some prescribed larger domain.

DEFINITION. *Solutions of an equation $Lu = 0$ are said to have the Runge approximation property if, whenever D_1 and D_2 are two simply connected domains, D_1 a subset of D_2, any solution in D_1 can be approximated uniformly in compact subsets of D_1 by a sequence of solutions which can be extended as solutions to D_2.*

Simple connectivity can be replaced by the requirement that D_2 is not disconnected by D_1.

The prototype of approximation in the above sense is the approximation of analytic functions by polynomials. A similar approximation theorem holds for harmonic functions in two variables; a particularly strong result in this direction is due to Farrell [8]. Bergman [3] has extended such results to solutions of certain types of elliptic equations in two variables with analytic coefficients by establishing a domain-preserving correspondence between solutions of such equations and analytic functions. Similar results were obtained by Eichler [7] by a different method. Recently, Malgrange, in [17] and [18], has investigated this problem for a very large class of equations.

The aim of this section is to prove the

EQUIVALENCE THEOREM. *Let L be a second order elliptic operator in any number of independent variables. Solutions of $Lu = 0$ have the Runge approximation property if and only if solutions of $Lu = 0$ are uniquely determined throughout their domain of existence by their Cauchy data along any smooth hypersurface.*

The restriction to second order operators was made for the sake of convenience and is not relevant.

According to a result of Friedrichs [10], the L_2 norm of a solution u over a domain bounds the maximum norm of u over any compact subset of this domain. Therefore, in order to show that a solution over D_1 can be approximated uniformly over any compact subset by solutions in D_2, it is sufficient to show that it can be approximated in the L_2 sense over any subdomain whose closure lies in D_1.

Let D_0 be such a subdomain; denote by S_1 the restriction of solutions in D_1 to D_0, by S_2 the restriction of solutions in D_2 to D_0. S_2 is a subspace of S_1; our aim is to show that it is a dense subspace in the L_2 topology. According to a classical criterion, S_2 *is dense in* S_1 *if and only if every function* v_0 *in* L_2 *over* D_0 *orthogonal to* S_2 *is also orthogonal to* S_1.

We shall make repeated use of Green's formula

$$(4.1) \qquad \iint_D uL^*w - wLu = \int_C u\frac{\partial w}{\partial n} - w\frac{\partial u}{\partial n} + cuw;$$

here $\partial/\partial n$ denotes the conormal derivative on C associated with L, and c is some function of the coefficients of L.

We shall also make repeated use of the classical criterion governing the solvability of the first boundary value problem.

EXISTENCE THEOREM. *The boundary value problem*

$$(4.2) \qquad \begin{aligned} Lu &= f \ in \ D, \\ u &= \phi \ on \ C \end{aligned}$$

has a solution if and only if the data satisfy the compatibility condition

$$(4.3) \qquad \iint_D fw = \int_C \phi\frac{\partial w}{\partial n}$$

for all solutions w *of the homogeneous adjoint problem*

$$(4.4) \qquad \begin{aligned} L^*w &= 0 \ in \ D, \\ w &= 0 \ on \ C. \end{aligned}$$

Furthermore, the set of solutions of (4.4) *is finite dimensional.*

Of course the necessity of the compatibility condition (4.3) is just a consequence of Green's formula.

COROLLARY. *The orthogonal complement on C of the space of the boundary values of all solutions of Lu = 0 is the finite dimensional space spanned by ∂w/∂n, w a solution of the homogeneous problem* (4.4).

We shall now deduce the Runge property from the uniqueness theorem for the Cauchy problem. Let v_0 be any function in L_2 over D_0, orthogonal to S_2. We wish to show that then v_0 is orthogonal to S_1. For this purpose we construct the auxiliary function w_0 as a solution of the boundary value problem.

(4.5)
$$L^* w_0 = \begin{cases} v_0 \text{ in } D_0, \\ 0 \text{ in } D_2 - D_0, \end{cases}$$
$$w_0 = 0 \text{ on } C_2^-, \text{ the boundary of } D_2.$$

Clearly, the compatibility condition is satisfied, since v_0 is orthogonal to all functions in S_2, therefore, especially to those which are zero on C_2.

Apply Green's formula to w_0 and any function u in S_2; using the orthogonality of v_0 to S_2, we deduce that $\partial w_0/\partial n$ is orthogonal to the boundary values of functions in S_2. According to the above corollary to the existence theorem this means that $\partial w_0/\partial n$ is equal to the conormal derivative of a function w which satisfies the homogeneous equations (4.4). Since w_0 is determined only modulo such a function, we conclude that the boundary value problem (4.5) has a solution w_0 for which $\partial w_0/\partial n$ is zero on C_2. Then, according to the uniqueness theorem for the Cauchy problem, we conclude that w_0 is zero in the connected domain $D_2 - D_0$.

Applying Green's Theorem to w_0 and any function u of S_1 over a domain that contains D_0 but is contained in D_1 we conclude that v_0 is orthogonal to any element of S_1. This completes the proof of the Runge Theorem.

Next we shall deduce[4] the Uniqueness Theorem from the Runge Theorem. Let w_0 be a solution of $L^* w_0 = 0$ with zero Cauchy data on a piece of a surface C; we aim to show that w_0 is zero wherever defined. We shall show this first in case C is a *closed* surface, and w_0 defined in a *boundary strip* of the domain D bounded by C. From this special case we can, following a suggestion of Louis Nirenberg, deduce the general Uniqueness Theorem.

Since uniqueness is a local problem, we shall choose the underlying domain D so small that L^* is positive definite over D. Then the homogeneous boundary value problem (4.4) over any subdomain of D has only trivial solutions.

We start by extending w_0 to the whole interior of C, not necessarily as a

---

[4]This question was raised by Lipman Bers.

solution of $L^*w_0 = 0$. Green's formula applied to w_0 and any solution u in D shows that L^*w_0 is orthogonal to such a u; according to the Runge Theorem, L^*w_0 is then orthogonal to any solution of $Lu = 0$ in a domain which contains the support of $v_0 = L^*w_0$. Let $\breve{D}$ be such a domain, and $\tilde{C}$ the boundary of $\breve{D}$.

Define $\tilde{w}_0$ as the solution of the following boundary value problem:

(4.6)
$$L^*\tilde{w}_0 = 0 \text{ in } \breve{D},$$
$$\tilde{w}_0 = w_0 \text{ on } \tilde{C}.$$

Apply Green's theorem to $w_0 - \tilde{w}_0$ and any solution u of $Lu = 0$ over $\breve{D}$:

$$\iint_{\breve{D}} u L^* w_0 = \int_{\tilde{C}} u \frac{\partial(w - \tilde{w}_0)}{\partial n}.$$

As stated above, L^*w_0 is orthogonal to u, hence the left side is zero. Since the boundary values of u on $\tilde{C}$ are arbitrary, we conclude that $\partial(w_0 - \tilde{w}_0)/\partial n$ is zero on $\tilde{C}$.

Define now w_1 as

$$w_1 = \begin{cases} w_0 \text{ in } D - \breve{D}, \\ \tilde{w}_0 \text{ in } \breve{D}. \end{cases}$$

The function w_1 satisfies $L^*w_1 = 0$ in both domains and has continuous first derivatives across $\tilde{C}$; thus it is a weak solution of $L^*w_1 = 0$ and, therefore, according to the property of solutions of elliptic equations (see e.g. Friedrichs [10]), a genuine solution. Thus w_1 is an extension of w_0 to the whole interior of C as solution of $L^*w_1 = 0$. Since we assumed that the domain D is so small that L^* is positive definite over D, w_1, being a solution of $L^*w_1 = 0$ and vanishing on C, is zero. Hence w_0 itself vanishes in $D - \breve{D}$; since the only restriction on $\breve{D}$ was that its boundary $\tilde{C}$ should be contained in the boundary strip in which w_0 was originally defined, we can conclude that w_0 is zero in the whole boundary strip.

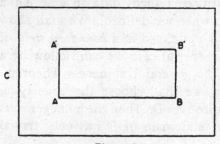

Figure 1

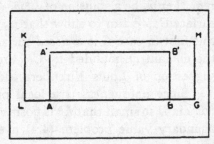

Figure 2

To complete the proof of the Uniqueness Theorem for the Cauchy problem, let F be any sufficiently small piece of a surface, denoted as the segment AB in Figure 1. Let w_0 be a solution of $L^*w_0 = 0$ on one side of F, the rectangle $ABA'B'$ in Figure 1 which has zero Cauchy data on F. Denote by C the boundary of the rectangle in Figure 1. Define w_0 as zero outside of the rectangle $ABA'B'$, and consider w_0 in the boundary strip consisting of those points inside C which lie outside of the octagon $AA'B'BGHKL$ of Figure 2. Since we have already shown on the basis of Runge's Theorem that solutions with zero Cauchy data on a closed surface are identically zero, the proof of the uniqueness theorem is complete.

Appendix

THEOREM I'. *Let H be a Hilbert space, D a self-adjoint operator whose spectrum lies on the nonpositive integers, $u(t)$ a strongly differentiable function of t in $0 < t < \infty$ taking its values in the domain of D and satisfying the differential inequality*

$$(A.1) \qquad \left\| \frac{du}{dt} - Du \right\| \leqq ce^{-t}\|u\|, \qquad\qquad 0 < t < \infty.$$

Conclusion: Unless identically zero, $u(t)$ tends to zero no faster than some exponential.

Proof: Let P_n denote projection into the eigenspace of the eigenvalue $\lambda = -n$ of D. The element $P_n u(t)$, and its norm $\|P_n u(t)\| = a_n(t)$ are differentiable functions of t, since $u(t)$ was assumed to be differentiable and P_n is a bounded operator. Denote $P_n u(t)/a_n(t)$ by $u_n(t)$, a normalized eigen-element of D:

$$Du_n = -nu_n.$$

u_n depends differentiably on t except possibly when $a_n(t) = 0$.

The Fourier expansion of $u(t)$ is

$$(A.2) \qquad\qquad u(t) = \sum a_n(t)u_n(t)$$

and that of du/dt is

$$\frac{du}{dt} = \sum \dot{a}_n u_n + a_n \dot{u}_n,$$

the dot denoting differentiation with respect to t; this expansion is valid when none of the coefficients $a_n(t)$ are zero.

Since the expansion for Du is

$$Du = - \sum na_n u_n,$$

we have

$$\frac{du}{dt} - Du = \sum (\dot{a}_n + na_n)u_n + a_n \dot{u}_n.$$

The element $\dot{u}_n$ lies in the n-th eigenspace and is orthogonal to u_n, since u_n

is normalized. We have, therefore,

$$(A.3) \qquad \left\| \frac{du}{dt} - Du \right\|^2 = \sum |\dot{a}_n + n a_n|^2 + \sum |a_n|^2 \|\dot{u}_n\|^2 \geqq \sum |\dot{a}_n + n a_n|^2.$$

From (A.2) we have

$$(A.4) \qquad \|u\|^2 = \sum |a_n|^2.$$

Substituting (A.3) and (A.4) into inequality (A.1) we get

$$(A.5) \qquad \sum |\dot{a}_n + n a_n|^2 \leqq c^2 e^{-2t} \sum |a_n|^2.$$

Our derivation of (A.5) is valid only when $a_n(t) \neq 0$ for all n. But if $a_n(t)$ is zero for $t = t_0$, it is easy to see that $|\dot{a}_n|$ is a *lower bound* for the norm of the n-th component of du/dt, and (A.5) follows as before.

Introduce $r = e^{-t}$ as a new independent variable and denote differentiation with respect to r by prime; then (A.5) becomes

$$(A.6) \qquad \sum \left| a_n' - \frac{n}{r} a_n \right|^2 \leqq c^2 \sum |a_n|^2.$$

Define the function w of the complex variable $z = \varrho e^{i\theta}$ as

$$(A.7) \qquad w(z) = \sum a_n(\varrho) e^{in\theta}.$$

Denote by $\partial/\partial \bar{z}$ the operator $\frac{1}{2}\{\partial/\partial x + i\partial/\partial y\}$; in polar coordinates it is $\frac{1}{2}\{\partial/\partial \varrho + i(1/\varrho)\partial/\partial \theta\}$. Form the $\bar{z}$ derivative of w

$$\frac{dw}{d\bar{z}} = \frac{1}{2} \sum \left\{ a_n' - \frac{n}{\varrho} a_n \right\} e^{-in\theta}.$$

Then our inequality (A.6) yields

$$(A.8) \qquad \left\| \frac{dw}{d\bar{z}} \right\|_\varrho \leqq c \|w\|_\varrho,$$

the symbol $\| \ \|_\varrho$ denoting the L_2 norm with respect to $d\theta$ on the boundary of the circle of radius ϱ. To analyze the implications of inequality (A.8) we recall that a function w whose $\bar{z}$ derivative vanishes in a domain is analytic. A function is called *approximately analytic* if its $\bar{z}$ derivative is bounded in terms of the function itself; i.e., if it satisfies an inequality of the form

$$(A.9) \qquad \left| \frac{\partial w}{\partial \bar{z}} \right| \leqq c|w|.$$

Carleman has shown that the zeros of an approximately analytic function are of finite order. Our inequality (A.8) is analogous to inequality (A.9) for approximately analytic functions, with the difference that the norm with respect to which $w_{\bar{z}}$ is bounded by w in (A.8) is the L_2 norm on concentric circles with center P, instead of the maximum norm. However, Carleman's theorem and method of its proof, based on Green's formula

$$w(u) = \int_{|s(s)|=R} \frac{w(z(s))}{z(s)-u} ds + \iint_{|s|\leq R} \frac{w_{\bar{z}}(z)}{z-u} dxdy,$$

can be easily carried over to this case.

Bibliography

[1] Aronszajn, N., *A unique continuation theorem for all partial differential equations and inequalities of second order*, Bull. Amer. Math. Soc., Vol. 62, Abstract 184, 1956, p. 154.

[2] Berg, P. W., and Lax, P. D., *Fourth order operators*, Rend. Sem. Mat. Fis. Milano, Vol. 11, 1951-1952, pp. 343-358.

[3] Bergman, S., *Partial Differential Equations, Advanced Topics*, Lecture notes, Brown University, Summer 1941.

[4] Bers, L., *Theory of Pseudo-analytic Functions*, Lecture notes, New York University, 1953.

[5] Carleman, M. F., *Sur les systèmes linéaires aux dérivées partielles du premier ordre à deux variables*, C. R. Acad. Sci. Paris, Vol. 197, 1933, pp. 471-474.

[6] Douglis, A., *Some existence theorems for hyperbolic systems of partial differential equations in two independent variables*, Comm. Pure Appl. Math., Vol. 5, No. 2, 1952, pp. 119-154.

[7] Eichler, M. M. E., *On the differential equation $u_{xx}+u_{yy}+N(x)u = 0$*, Trans. Amer. Math. Soc., Vol. 65, 1949, pp. 259-278.

[8] Farrell, O. J., *On approximations to analytic functions by polynomials*, Bull. Amer. Math. Soc., Vol. 40, 1934, pp. 908-914.

[9] Friedrichs, K. O., *A theorem of Lichtenstein*, Duke Math. J., Vol. 14, 1947, pp. 67-81.

[10] Friedrichs, K. O., *On the differentiability of the solutions of elliptic differential equations*, Comm. Pure Appl. Math., Vol. 6, No. 3, 1953, pp. 299-326.

[11] Hartman, P. and Wintner, A., *On the local behavior of solutions of non-parabolic partial differential equations (III)*, Amer. J. Math., Vol. 77, 1955, pp. 453-474.

[12] Hartman, P., *On the local behavior of solutions of $\Delta u = g(x, u, \nabla u)$*, Comm. Pure Appl. Math., Vol. 9, No. 3, 1956, pp. 435-445.

[13] Heinz, E., *Über die Eindeutigkeit beim Cauchyschen Anfangswertproblem einer elliptischen Differentialgleichung zweiter Ordnung*, Nachr. Akad. Wiss. Göttingen, Math.-Phys. Kl. IIa, Issue No. 1, 1955, pp. 1-12.

[14] Hille, E., *Functional Analysis and Semi-groups*, Proc. Amer. Math. Soc. Colloquium Publications, 1948.

[15] Holmgren, E., *Über Systeme von linearen partiellen Differentialgleichungen*. Ofversight af Kongl. Vetenskaps-Akademiens Forhandlinger, Vol. 58, 1901, pp. 91-103.

[16] Lax, P. D., *Spectral theory for functions on semi-groups, and the separation of variables*, University of Kansas, Dept. of Mathematics, Report 14, 1954, pp. 118-133.

[17] Malgrange, B., *Le théorème d'approximation pour les équations aux dérivées partielles à coefficients constants*, Séminaire Schwartz Fac. de Sci. de Paris, Vol. 2, 1954-1955, pp. 1-01-1-07.

[18] Malgrange, B., *Existence et approximation des solutions des équations aux dérivées partielles et des équations de convolution* (Thèse), Conference on Partial Differential Equations, Nancy, France, April 1956, to appear.

[19] Müller, C., *On the behavior of solutions of the equation $\Delta u = F(x, u)$ in the neighborhood of a point*, Comm. Pure Appl. Math., Vol. 7, No. 3, 1954, pp. 505-515.

[20] Phillips, R. S., *Perturbation theory for semi-groups of linear operators*, Trans. Amer. Math. Soc., Vol. 74, 1953, pp. 199-221.

[21] Phillips, R. S., *Spectral theory for semi-groups of linear operators*, Trans. Amer. Math. Soc. Vol. 71, 1951, pp. 393–415.

[22] Schwartz, J., *Perturbations of spectral operators and applications*, Pacific J. Math., Vol. 4, 1954, pp. 415–458.

Received May, 1956.

COMMUNICATIONS ON PURE AND APPLIED MATHEMATICS, VOL. X, 361–389 (1957)

A Phragmén-Lindelöf Theorem in Harmonic Analysis and Its Application to Some Questions in the Theory of Elliptic Equations *

P. D. LAX

Introduction

The classical Phragmén-Lindelöf principle can be illustrated by the following example: Let $u(x, y)$ denote a harmonic function in the half-strip $0 \leqq x \leqq 1, 0 \leqq y$, which vanishes on the semi-infinite lines forming part of the boundary, i.e., for which

$$u(0, y) = u(1, y) = 0 \qquad \text{for } y \geqq 0.$$

If such a function is bounded in the half-strip then it dies down exponentially.

The motivation for the present investigations was the attempt to extend this result to functions in half-cylinders $x \, \epsilon \, D_0$, $0 \leqq y$, which satisfy there an equation $Lu = 0$, L an elliptic operator of any order with coefficients independent of the variable y, u having zero Dirichlet data on those parts of the boundary contained in $y > 0$. In this paper we give an *intrinsic characterization* of function spaces whose elements have the Pragmén-Lindelöf property. This characterization has been abstracted from the concrete case of nullspaces of an elliptic operator and yields, when applied to it, the law of exponential decay.

Rather than dealing with scalar-valued functions defined over a half-cylinder, we shall regard the values of such functions over each cross-section as being elements of a linear space, and consider from now on vector-valued functions $u(y)$ defined over the half-line $0 \leqq y$, whose values lie in some Banach space.

The set of solutions of an elliptic equation of the kind considered before forms a linear space of vector-valued functions. This linear space is *translation invariant*, since the coefficients of the differential operator in question, as well as the boundary conditions, were independent of y. The elliptic character of the operator is related to the smoothness of solutions in the interior of their domain of definition. A quantitative version of this property

*This work was supported by the Office of Naval Research, United States Navy, under contract N6ori-201, T.O. No. 1.

is the *principle of interior compactness*. In this paper we abstract this property and show (Theorem 1.1) that elements of a translation invariant interior
compact subspace satisfy the Phragmén-Lindelöf principle; i.e., if for an
element u of such a space $\int_0^\infty \|u(y)\| e^{\alpha y} \, dy$ is finite, then already $\int_0^\infty \|u(y)\| e^{\beta y} \, dy$
is finite, where β is a constant *greater* than α, depending on α but not on the
individual function u. The interesting feature of this theorem is that it
relates the behavior of functions at infinity to their behavior over a finite
range.

A number of results follow from this theorem, such as the asymptotic
validity of a Fourier expansion in a series of exponentials contained in the
space, the unique continuation property and an abstract generalization of a
theorem of Weinstein. These results are equivalent to the discreteness of
the spectrum of the translation operators. On the other hand, answers to
further interesting questions depend on there being an adequate spectral
synthesis over the space.

Section 1 contains the spectral analysis of translation operators. In
Section 2 we show that the abstract theory developed in Section 1 is applicable to the nullspaces of elliptic operators. In addition, a concrete Phragmén-
Lindelöf principle is derived for solutions of *positive definite* elliptic equations.
An application of such a result to the theory of thin beams plus diverse
generalizations is contained in Peter Treuenfels' NYU dissertation [19];
this concrete problem has been suggested by K. O. Friedrichs.

Section 3 describes an important theorem of Beurling [1] on spectral
synthesis of translation invariant subspaces of scalar functions over the
positive reals. His result is extended to include spaces of functions whose
values lie in a finite-dimensional vector space. Finally we present examples
of translation invariant spaces of functions with values in infinite-dimensional vector spaces which do contain a sufficient number of exponentials,
and others which do not contain a single exponential.

This paper contains parts of two previous ones, [8] and [17]. Omitted
here is the proof of the extended principle of interior compactness (given in
Section 2 of [17]), and some observations on the determination of the
Fourier coefficients in the expansion of a solution u of an elliptic equation
in a half-cylinder into a series of exponentials (Section 3 of [17]), i.e. the
problem of relating the Fourier coefficients of u to the Dirichlet data of
u (or other data) given on the base of the half-cylinder.

These investigations grew out of a joint study with Paul Berg on the
biharmonic equation in a half-strip. I also had the benefit of several stimu-
lating discussions with Louis Nirenberg.

1. Spectral Analysis

1.1. *Definition of an interior compact space*: Let S be a linear space of functions $u(y)$ on the positive reals, $y > 0$, whose functional values belong to some Banach space B. Assume that S is translation invariant, i.e., that if $u(y)$ lies in S, so does $u(y+\eta)$ for η positive, and that elements of S belong to L_p, $1 \leq p < \infty$, over every subinterval of the positive real axis. Introduce the notation

$$\|u\|_a^b = \left\{ \int_a^b \|u(y)\|^p\, dy \right\}^{1/p}, \qquad 0 < a < b < \infty;$$

here $\|u(y)\|$ stands for the Banach norm of $u(y)$. We call such a space S *interior compact* if the unit sphere in the norm $\|u\|_a^b$ is precompact with respect to the norm $\|u\|_{a'}^{b'}$ whenever $a < a' < b' < b$.

A similar definition can be given in the maximum norm.

By *precompact* we mean that every bounded sequence in the norm $\|u\|_a^b$ contains a Cauchy sequence in the norm $\|u\|_{a'}^{b'}$.

The completion of an interior compact space S under L_p convergence in every finite subinterval of the real axis is still interior compact. Therefore we shall assume in what follows that S is already closed in this topology.

Functions defined in a half-cylinder in Euclidean space can be regarded as vector-valued functions over the positive reals. Our aim is to develop an abstract theory and apply it to such a situation. My motivation was the study of the space of solutions of linear elliptic equations over a half-cylinder, where the coefficients of the equation are independent of the vertical variable y, and the solutions have zero Dirichlet data on the vertical part of the boundary. Clearly this space is translation invariant; in Section 2 we shall show that it is *interior compact* in the sense of our definition if the Banach norm chosen is the square root of the Dirichlet integral over the base of the half-cylinder. In fact the notion of an interior compact space was abstracted from this example.

Other examples of interior compact spaces, this time of scalar-valued functions, are the spaces spanned by real exponentials $e^{-\mu_n y}$, where $\sum 1/\mu_n$ converges. These spaces were investigated by Laurent Schwartz in his monograph [12]; their interior compactness is easy to deduce from various properties investigated by Schwartz.

We wish to point out that if S is an interior compact space of vector-valued functions, the space of scalar-valued functions obtained from S by mapping each element $u(y)$ of S into $v(y) = [l, u(y)]$, where l is some given bounded linear functional over B, need *not* be interior compact.

The prototype of a space of solutions of an elliptic equation in a half-cylinder is the space of harmonic functions $u(x, y)$ in the half-strip

$0 \leqq x \leqq 1$, $0 \leqq y$, subject to the boundary conditions $u(0, y) = u(1, y) = 0$. The classical Phragmén-Lindelöf principle applied to this situation asserts that if such a harmonic function tends to infinity slower than $e^{\pi y}$, then it is in fact bounded in the half-strip, and even dies down like $e^{-\pi y}$. Our aim is to prove an abstract Phragmén-Lindelöf theorem, i.e. an analogue valid for any element in an interior compact space. A very general concrete Phragmén-Lindelöf theorem for solutions of elliptic equations then follows.

THEOREM 1.1. (*Abstract Phragmén-Lindelöf Principle*) *Let S be a translation invariant interior compact space; then there exists a positive number* α *such that for all elements of S, for which* $\int_0^\infty ||u(y)||^p \, dy$ *converges, the integral* $\int_0^\infty ||u(y)||^p \, e^{\alpha y} \, dy$ *also converges.*

THEOREM 1.1′. *The Phragmén-Lindelöf principle is still valid if the space is interior compact with respect to a single pair of intervals.*

This theorem makes an assertion only about those elements of S which are integrable over the whole interval. However, since multiplying every element of a translation invariant compact space by a fixed exponential e^{-sy} produces another interior compact space, a similar assertion can be made about those elements of S which grow at most exponentially at infinity.

A similar Phragmén-Lindelöf principle holds for spaces interior compact in the maximum norm; there the boundedness of $||u(y)||e^{-\alpha y}$ implies that of $||u(y)||$.

The proof of this theorem is somewhat involved. In subsection 1.2, we relate interior compactness to spectral properties of the translation operator; this is the basis of the proof of Theorem 1.1 given in subsection 1.3. Theorem 1.2 in subsection 1.2 is a consequence of Theorem 1.1 and establishes the equivalence between interior compactness of a space and complete continuity of the translation operator over that space.

From here on, we take $p = 1$; the proof for $p > 1$ is the same.

1.2. Denote by H the subspace of S consisting of those u for which $\int_0^\infty ||u(y)|| \, dy$ is finite. Denote the value of this integral by $||u||$, and use it as norm over H, which so normed is a Banach space; as remarked before, we may assume that H is complete. Denote by $T(\eta)$, $\eta \geqq 0$, the semigroup of translation operators

$$T(\eta)u(y) = u(y+\eta).$$

$T(\eta)$ is a bounded operator over H; in fact its norm does not exceed one.

We shall give now a reformulation of the Phragmén-Lindelöf theorem in terms of $T(\eta)$. It follows namely from the finiteness of $\int_0^\infty e^{\alpha y} ||u(y)|| \, dy$ that $e^{\alpha \eta} \int_\eta^\infty ||u(y)|| \, dy$ remains bounded for each u in H. In terms of $T(\eta)$ this means that

$$||T(\eta)u|| \leqq \text{const.} \; e^{-\alpha \eta}$$

for each u in H. From the principle of uniform boundedness, we now deduce that the norm of $T(\eta)$ is bounded by const. $e^{-\alpha\eta}$. Take η to be an integer n, and make use of the semigroup property of $T(\eta)$, i.e., that $T(n) = T^n(1)$:

$$||T(n)|| = ||T^n(1)|| \leq \text{const. } e^{-\alpha n}.$$

Take the n-th root of both sides and let n tend to infinity; we get the inequality

$$\lim_{n\to\infty} ||T^n(1)||^{1/n} \leq e^{-\alpha}.$$

The quantity on the left is, according to a well-known theorem of Gelfand (see e.g. [5]), just the spectral radius of $T(1)$. Thus we find that *the Phragmén-Lindelöf principle asserts: the spectral radius of $T(\eta)$ is, for $\eta > 0$, less than one*. The same reasoning shows that, *if the spectral radius of $T(\eta)$ is less than one, the Phragmén-Lindelöf principle holds*.

The proof of the Phragmén-Lindelöf principle given in [8] and reproduced here in subsection 1.3 is based on this equivalence. We show namely that the spectrum of $T(\eta)$ consists of a discrete set of points *inside* the unit circle, accumulating at most at the origin. This makes one suspect that the transformation $T(\eta)$ is completely continuous. Indeed, the following additional result is true:

THEOREM 1.2. *Let H be a translation invariant linear space of integrable vector-valued functions $u(y)$ over the positive y-axis, and define the norm of u in H as before by $||u|| = \int_0^\infty ||u(y)||dy$. Assertion: The translation operators $T(\eta)$, $\eta > 0$, are completely continuous if and only if H is an interior compact space.*

We shall deduce Theorem 1.2 from Theorem 1.1.

First we show that if H is interior compact, $T(\eta)$ is completely continuous. We rely on the following elementary lemma, already used in [8]:

LEMMA 1.1. *Let H be interior compact and $\{u\}$ a collection of elements of H satisfying the conditions*

i) *$\{u\}$ is bounded, i.e., $||u|| \leq$ const. for all u in the set,*

ii) *the elements of $\{u\}$ are uniformly small at infinity, i.e., to every $\varepsilon > 0$ there is an N such that*

$$||u||_N^\infty < \varepsilon$$

for all u of the set.

Conclusion: $T(\eta)$ maps such a set into a compact set[1].

Proof: We have to show that every sequence $\{u_n\}$ of our set contains a subsequence convergent in the norm $||u||_\eta^\infty$. According to the definition of an interior compact space and the assumed uniform boundedness, we can

---

[1] As before, $||u||_N^\infty$ stands for $\int_N^\infty ||u(y)||dy$. Under the weaker hypothesis of Theorem 1.1' we can still reach the same conclusion for η large enough.

select subsequences which converge in the norm $||u||_\eta^N$. If we let N tend to
infinity through a discrete sequence and perform a diagonal process, we obtain, on account of the uniform smallness of the functions at infinity, a convergent sequence[2].

We have previously deduced from the Phragmén-Lindelöf principle
that $||T(\eta)||$ tends to zero as η tends to infinity. This implies that the
elements of the unit sphere are uniformly small at infinity. According to
our lemma, $T(\eta)$ then maps the unit sphere of H into a compact set; i.e.,
$T(\eta)$ is completely continuous.

To show the converse, i.e., that the complete continuity of $T(\eta)$ implies
the interior compactness of the space H, we need the following

LEMMA 1.2. *If $T(\eta)$ is completely continuous for all positive η, $||T(\eta)||$
is less than one for all $\eta > 0$.*

Proof: Assume to the contrary that, for some η, $||T(\eta)|| = 1$. This
means that there is a sequence of functions u_n of unit norm such that
$||T(\eta)u_n||$ tends to one. $T(\eta)$ being completely continuous, a subsequence of
the u_n converges over the interval (η, ∞). But since $||u_n|| = 1$ and $||u||_\eta^\infty$
tends to one, the integral of $||u(y)||$ over $(0, \eta)$ tends to zero; i.e., the subsequence converges over the whole interval to an element u_0 of H which is
zero over the interval $(0, \eta)$. We shall show now that no such element u_0 can
exist[3]. For this purpose we consider the subspace H' of H consisting of those
functions $u(y)$ of H which vanish in the interval $(0, \eta/2)$. We claim that
H' is *finite-dimensional*. For, by assumption, $T(\eta/2)$ maps the unit sphere of
H into a compact set. *A fortiori* then it maps the unit sphere of H' into a
compact set. But over H', $T(\eta/2)$ is an isometry, hence the unit sphere of
H' is compact; from this it follows, according to a classical theorem, that
H' is finite-dimensional. Denote its dimension by n.

Consider now $n+1$ distinct real numbers d_i, $i = 1, \cdots, n+1$, lying
between 0 and $\eta/2$ and arranged according to magnitude, and the $n+1$
functions $u_i = u_0(y+d_i)$. Since u_0 is zero in $(0, \eta)$, u_i vanishes in $(0, \eta-d_i)$
and so belongs to H'. Since H' is of dimension n, there must be a nontrivial
linear relation between the functions u_i ; we write this in the form

---

[2]If, as in Theorem 1.1′, we make the weaker assumption that the set $\{||u||_a^b \leq 1\}$ is compact in the norm $||u||_{a'}^{b'}$ for a single pair (a, b), (a', b') of intervals, $a < a' < b' < b$, we can still
conclude that $T(\eta)$ maps the set $\{u\}$ satisfying the hypothesis of Lemma 1.1 into a compact
set for $\eta \geq a'$. For, it follows by translation that the elements of unit norm over $(A+a, A+b)$
are compact over the subinterval $(A+a', A+b')$ for $A > 0$, and by superposition of such
intervals that elements of unit norm over $(A+a, B+b)$ are compact over $(A+a', B+b')$,
$0 \leq A \leq B$. The rest of the proof goes as before.

[3]The question of whether such an element u_0 can exist was raised by Walter Rudin
(personal communication).

$$u_k = \sum_{i=1}^{k-1} a_i u_i .$$

Every function on the right vanishes over the interval $(0, \eta - d_{k-1})$. Therefore so does u_k; but this means that u_0 itself vanishes in the interval $(0, \eta + d_k - d_{k-1})$. Repeating this argument we can show recursively that u_0 vanishes over the whole infinite interval. This completes the proof of Lemma 1.2.

Denote the norm of $T(\eta)$ by d; according to Lemma 1.2, d is less than one. Therefore

$$||u||_0^\eta = \int_0^\eta |u(y)| dy = ||u|| - ||Tu|| \geqq (1-d)||u||.$$

This shows that *the quantity $||u||_0^\eta$ is a norm equivalent to the original one.* Consequently, since $T(\zeta)$ was assumed to be completely continuous, the unit sphere in the $||u||_0^\eta$ norm is mapped into a compact set in the sense of the original norm by $T(\zeta)$ for any positive ζ. The interior compact character of the underlying space is easily deduced from this fact.

1.3. In subsection 1.2, we have shown that the Phragmén-Lindelöf principle is equivalent to the following

PROPOSITION: *The spectrum of the translation operator $T(\eta)$ over an interior compact space H*

   (a) *is a pure point spectrum,*

   (b) *lies inside the unit circle,*

   (c) *accumulates at most at the origin.*

We note that, by F. Riesz' theorem, properties (a) and (c) would follow if we already knew that $T(\eta)$ is completely continuous. Since I have not been able to extend the existing proof of Riesz' theorem to our situation (in which the hypothesis is weaker), I shall first present a new proof of Riesz' theorem and then follow it closely in order to prove that our translation operator T has properties (a), (b) and (c).

The new proof relies on a well-known lemma (see [18]):

LEMMA 1.3. *Let K be any bounded transformation of a Banach space into itself and let λ_0 be a boundary point of its spectrum. Then λ_0 is an approximate point eigenvalue, i.e., there exists a sequence of elements u_i of unit norm such that*

$$||(K - \lambda_0)u_i|| \to 0.$$

Proof: By assumption, λ_0 is a limit point of a sequence of points $\{\lambda_i\}$ in the resolvent set. The norm of the resolvent at any point is not less than the reciprocal of the distance to any point of the spectrum:

$$||(K - \lambda_i)^{-1}|| \geqq |\lambda_0 - \lambda_i|^{-1};$$

hence there exists an element v_i such that

$$||v_i|| \leqq 2|\lambda_0 - \lambda_i|, \quad ||u_i|| = ||(K - \lambda_i)^{-1} v_i|| = 1.$$

But $\{u_i\}$ is just the sequence we need; for,

$$||(K - \lambda_0)u_i|| = ||v_i + (\lambda_i - \lambda_0)u_i|| \leqq 3|\lambda_0 - \lambda_i| \to 0.$$

Systematic use will be made of these two well-known results about the invariant subspaces N of a transformation K over a Banach space H:

LEMMA 1.4. *The resolvent set of K over H includes the intersection of the resolvent set of K over N with the resolvent set of K over H/N.*

LEMMA 1.5. *If N is finite-dimensional, the resolvent set of K over H/N includes the resolvent set of K over H.*

Suppose now that K is completely continuous; we wish to prove Riesz' theorem, i.e., that K has properties (a) and (c). Let λ_0 be a boundary point of its spectrum, $\lambda_0 \neq 0$. According to Lemma 1.3, λ_0 is an approximate point eigenvalue; since K is completely continuous, there exists a subsequence of the u_i such that Ku_i converges. Since $(K - \lambda_0)u_i$ tends to zero, and $\lambda_0 \neq 0$, the sequence $\{u_i\}$ also converges to an element u_0; u_0 is of unit norm and $(K - \lambda_0)u_0 = 0$. This shows that all boundary points of the spectrum of K are in the point spectrum.

Next we show that every boundary point $\lambda_0 \neq 0$ of the spectrum of K is isolated. From this it follows—via a simple set theoretic argument—that all points of the spectrum are boundary points, and that, therefore, they all belong to the point spectrum. We rely on a property of completely continuous operators: N_k, the nullspace of $(K - \lambda_0)^k$, is finite-dimensional and for some index k, $N_k = N_{k+1}$. Form now the quotient space H/N_k; K remains completely continuous over this quotient space. We claim that λ_0 does not lie in the spectrum of K over H/N_k. For, if it did, it would—according to Lemma 1.5—be a boundary point of the spectrum and thus—according to what we have already shown—in the point spectrum. But an eigenelement of K over H/N_k consists of elements of H *not* in N_k which are mapped into N_k by $K - \lambda_0$, and there are no such elements if $N_{k+1} = N_k$. Hence λ_0 lies in the resolvent set of K over H/N_k, and so does therefore every λ sufficiently close to λ_0. On the other hand, the spectrum of K over N_k consists entirely of the point λ_0; so according to Lemma 1.4 all points λ sufficiently close to λ_0, except λ_0 itself, belong to the resolvent set of K over H. This proves that all boundary points are isolated.

We turn now to the proof of Theorem 1.1. The norm of $T(\eta)$ over H does not exceed one; therefore the spectrum of $T(\eta)$ is contained inside and on the unit circle. Our proof of Theorem 1.1 is based on the following three properties of the spectrum of a translation over an interior compact space.

Proposition **A**: *Every boundary point of the spectrum inside the unit circle and $\neq 0$ belongs to the point spectrum.*

Proposition **B**: *Every boundary point of the spectrum inside the unit circle and $\neq 0$ is isolated.*

Proposition **C**: *Not all points of the unit circle are in the spectrum.*

In proving propositions **A**, **B**, **C**, in that order, we shall keep as close to the case of a completely continuous K, presented before, as possible.

SELECTION PRINCIPLE. *Let λ_0 be a complex number, $\lambda_0 \neq 0$ and $|\lambda_0| < 1$. Let $\{u_n\}$ be a uniformly bounded sequence of elements of an interior compact space and k an integer such that $(T - \lambda_0)^k u_n = v_n$ converges to some limit v. Then a subsequence of $\{u_n\}$ converges to a solution u of $(T - \lambda_0)^k u = v$.*

Proof: It is sufficient to prove this result for $k = 1$. Call $v_n = v + e_n$; then $\|e_n\| \to 0$. Using the identity

$$T^n = \lambda^n + \sum_{\nu=1}^{N} \lambda^{N-\nu} T^{\nu-1} (T - \lambda)$$

we get

$$u_n(y + N\eta) = \lambda_0^N u_n(y) + \sum_{\nu=1}^{N-1} \lambda_0^{N-\nu} v_n(y + (\nu-1)\eta)$$

$$= \lambda_0^N u_n(y) + \sum_{\nu=1}^{N-1} \lambda_0^{N-\nu} v(y + (\nu-1)\eta) + \sum_{\nu=1}^{N-1} \lambda_0^{N-\nu} e_n(y + (\nu-1)\eta);$$

from this one can read off that the u_n are uniformly small at infinity. According to Lemma 1.1 a subsequence of $\{T(\eta)u_n\}$ converges over (A, ∞); then the sequence $\{T(\eta)u_n\}$ converges over $(A - \eta, \infty)$. But, by assumption, $\{(T(\eta) - \lambda_0)u_n\}$ is a Cauchy sequence over $(0, \infty)$; consequently, since λ_0 is not zero, the sequence $\{u_n\}$ also converges over $(A - \eta, \infty)$. Repeating this argument $[A/\eta]$ times, we conclude that the subsequence $\{u_n\}$ itself converges over $(0, \infty)$.

Proposition **A** is an immediate consequence of the selection principle and Lemma 1.3. The proof of proposition **B** is much harder. The difficulty lies in proving that the *index* of $T(\eta)$ at λ_0 is finite. We shall need

LEMMA 1.6. *Let K be any bounded transformation, λ_0 a boundary point of its spectrum. Then the image space under $K - \lambda_0$ cannot be the whole space.*

Proof: If the image space were the whole space, $K - \lambda_0$ would map H/N one-to-one continuously onto H, where N is the nullspace of $K - \lambda_0$. By the closed graph theorem the inverse is then bounded. Translating from the language of quotient space, this means that $K - \lambda_0$ has over H a *bounded*, though not necessarily linear nor continuous, *right* inverse R:

$$(K - \lambda_0)R = I.$$

Then for all λ sufficiently close to λ_0, $K - \lambda$ will have a uniformly bounded right inverse*. But we have assumed that there are points λ of the resolvent set arbitrarily close to λ_0. For these, the right inverse of $K-\lambda$ is just the resolvent, and its bound being not less than $1/|\lambda-\lambda_0|$, tends to infinity.

Let $\lambda_0 \neq 0$ be a boundary point of the spectrum of $T(\eta)$ inside the unit circle. Let N_i be the nullspace of $(T-\lambda_0)^i$. The elements of N_1 die down exponentially and therefore the elements of unit norm in N_1 are uniformly small at infinity. By Lemma 1.3 the unit sphere in N_1 is mapped by $T(\eta)$ into a compact set. Since N_1 consists of eigenfunctions of $T(\eta)$, the image of the unit sphere in N_1 is the sphere of radius λ_0 in N_1. From its compactness it follows that N_1 must be finite-dimensional.

According to a classical result, the sequence $\{\dim (N_{i+1}/N_i)\}$ is non-increasing, so it follows that all the N_i are finite-dimensional too. Furthermore, from a certain point $i = k$ on, all the spaces N_{i+1}/N_i have the *same* dimension which can also be expressed in this way:

Denote $\cup_{i=1}^{\infty} N_i$ by N; *then* $T-\lambda_0$ *maps* N/N_{k+1} *one-to-one onto* N/N_k.

We claim that the inverse mapping is bounded; for, assume to the contrary that it is not, i.e., that there exists a sequence U_i of cosets in N/N_{k+1} such that $||U_i|| = 1$, while $||V_i|| \to 0$, V_i being $(T-\lambda_0)U_i$, a coset in N/N_k. In ordinary language this means that if u_i is any element in U_i, then

$$(T-\lambda_0)u_i = v_i + \eta_k, \qquad ||v_i|| \to 0,$$

η_k being some element in N_k. Applying $(T-\lambda_0)^k$ to both sides we get

$$(T-\lambda_0)^{k+1}u_i = (T-\lambda_0)^k v_i = w_i.$$

Since $(T-\lambda_0)^k$ is a bounded transformation, $||w_i||$ tends to zero with $||v_i||$ and so by the selection principle a subsequence of the u_i tends to an element of N_{k+1}. But then, for this subsequence, $||U_i||$ tends to zero, a contradiction.

The boundedness of the inverse permits one to extend the inverse of $T-\lambda_0$ by closure. Denote the closure of N by $\overline{N}$; the completions of N/N_k and N/N_{k+1} are, on account of the finite dimensionality of N_k and N_{k+1}, $\overline{N}/N_k$ and $\overline{N}/N_{k+1}$ and we have

LEMMA 1.7. $T-\lambda_0$ *maps* $\overline{N}/N_{k+1}$ *one-to-one onto* $\overline{N}/N_k$.

Next we show

LEMMA 1.8. λ_0 *is not in the spectrum of* T *over* $H/\overline{N}$.

Proof: Since λ_0 is a boundary point of the spectrum of T over H, there is a sequence λ_i tending to λ_0 and belonging to the resolvent set of T over H. There are two cases: either infinitely many of the λ_i belong to the resolvent set of T over $H/\overline{N}$, or not. In the first case, if λ_0 is a point of the spectrum of

---

Footnote added in proof: Given by the Neumann series

$$\sum_{n=0}^{\infty} (\lambda-\lambda_0)^n R^{n+1}.$$

T over $H/\overline{N}$, it is a boundary point and so by Lemma 1.3 there is a sequence of cosets U_i such that $\|U_i\| = 1$, $\|(T-\lambda_0)U_i\| \to 0$. Let u_i be in U_i, then

$$(T-\lambda_0)u_i = v_i + n,$$

$\|v_i\| \to 0$, $n \in \overline{N}$. Since, according to Lemma 1.7, the congruence

$$(T-\lambda_0)m \equiv n \pmod{N_k}$$

can be solved with m in $\overline{N}$, we have for $z_i = u_i - m$

$$(T-\lambda_0)z_i = v_i + n_k,$$

n_k being some element in N_k. Apply $(T-\lambda_0)^k$ to both sides; it follows as before from the selection principle that $\|U_i\|$ tends to zero. This is a contradiction.

In the second case, there is a sequence of λ_i tending to λ_0, such that these λ_i are in the resolvent set of T over H, but in the spectrum of T over $H/\overline{N}$. Such λ_i belong to the *point spectrum* of T over $H/\overline{N}$, since $T-\lambda_i$, having mapped H onto H, certainly maps $H/\overline{N}$ onto $H/\overline{N}$. Let U_i, $\|U_i\| = 1$, be an eigenelement of T over $H/\overline{N}$ with eigenvalue $\lambda_i : TU_i = \lambda_i U_i$. Since λ_i tends to λ_0 with increasing i, $\|(T-\lambda_0)U_i\|$ tends to zero, and we proceed as before.

Combining Lemmas 1.7 and 1.8 we get that $T-\lambda_0$ maps H/N_{k+1} one-to-one onto H/N_k, and therefore a fortiori H/N_k onto itself. By Lemma 1.6, λ_0 cannot be a boundary point of the spectrum of T over H/N_k. On the other hand, λ_0 was assumed to be a limit of points λ_i in the resolvent set of T over H. According to Lemma 1.5, these points λ_i belong to the resolvent set of T over H/N_k. Hence λ_0 must lie in the resolvent set of T over H/N_k. This proves that the *index* of λ_0 is k.

All λ close enough to λ_0 lie in the resolvent set of T over H/N_k; on the other hand, all λ but λ_0 lie in the resolvent set of T over N_k. Thus by Lemma 1.4 all λ close enough to λ_0 but not equal to λ_0 lie in the resolvent set of T over H. This proves proposition **B**: λ_0 is isolated.

We turn to the proof of proposition **C**. First we prove a lemma.

DEFINITION. Let S denote the completion of H under L_1-convergence over every finite subinterval.[4]

LEMMA 1.9. *If λ belongs to the spectrum of T over H, and $|\lambda| = 1$, then λ belongs to the point spectrum of T over S.*

Proof: Let λ be a point of the spectrum of T over H, $|\lambda| = 1$. Since the exterior of the unit circle is in the resolvent set, λ is an approximate eigenvalue in the sense of Lemma 1.3; i.e., there exists a sequence $\{u_i\}$ such that the quotient

$$\varepsilon_i = \frac{\|(T-\lambda)u_i\|}{\|u_i\|}$$

tends to zero.

---

[4]One of the consequences of the theory developed in this paper is that S is equal to H; but we cannot make use of this knowledge at this stage.

Let a be an arbitrary length. Write ε_i as the quotient of two sums:

$$\varepsilon_i = \frac{\sum\limits_{n=0}^{\infty} ||(T-\lambda)u_i||_{an}^{a(n+1)}}{\sum\limits_{n=0}^{\infty} ||u_i||_{an}^{a(n+1)}}.$$

Then there exists an integer n_i such that

(1.1) $$||(T-\lambda)u_i||_{an}^{a(n+1)} \leqq \varepsilon_i ||u_i||_{an}^{a(n+1)} \neq 0, \qquad\qquad n = n_i.$$

Define

$$v_i(y) = u_i(y+an_i)$$

and imagine the u_i so normalized that

$$||v_i||_0^a = 1.$$

It follows from (1.1) that

(1.2) $$||(T-\lambda)v_i||_0^a \leqq \varepsilon_i.$$

By interior compactness, there is a subsequence of the v_i which forms a Cauchy sequence over a subinterval (c, d).

LEMMA 1.10. *If $\{v_i\}$ is a sequence of functions which forms a Cauchy sequence over an interval (c, d), and is such that $\{(T-\lambda)v_i\}$, $\lambda \neq 0$, is a null-sequence over a larger interval $(0, a)$, then $\{v_i\}$ is a Cauchy sequence over the larger interval $(0, a)$.*

Proof: Since $\{v_i\}$ is a Cauchy sequence over (c, d), $\{Tv_i\}$ is a Cauchy sequence over $(c-\eta, d-\eta)$. But, by assumption, $\{(T-\lambda)v_i\}$ is a null-sequence over $(0, a)$; consequently, since λ is not zero, $\{v_i\}$ also is a Cauchy sequence over $(c-\eta, d-\eta)$. Repeating this argument a finite number of times, we conclude that $\{v_i\}$ forms a Cauchy sequence over $(0, d)$. In a similar manner we show that $\{v_i\}$ is a Cauchy sequence over (c, a), and hence over the whole interval $(0, a)$.

Denote the limit of v_i by v; v has unit norm in $(0, a)$ and satisfies

$$Tv = \lambda v \quad \text{in} \quad (0, a-\eta).$$

Next, we let the length a of our interval increase. For each a, we construct v_a, and normalize it to have unit norm over some fixed interval $(0, b)$:

$$||v_a||_0^b = 1.$$

These functions v_a are limits over the interval $(0, a)$ of elements of H; the principle of interior compactness applies to them. In particular, we can select a subsequence of them which converges over some subinterval (c, d) of $(0, b)$.

From the functional equation $Tv_a = \lambda v_a$ we can conclude that then the sequence $\{v_a\}$ converges over *every* finite subinterval on the positive reals, and that the limit v has unit norm over $(0, b)$ and satisfies the equation

$Tv = \lambda v$. Our function v is the limit of the v_a, and the v_a are limits over $(0, a)$ of the v_i; since the v_i belong to H, it follows that v belongs to S. This completes the proof of Lemma 1.9.

Now we show that $\lambda = -1$ (and, more generally, any λ of modulus $= 1$ except $\lambda = 1$) does not belong to the spectrum of $T(\eta)$ for small η. Assume to the contrary, that it does for a sequence of values $\eta = \eta_m$, $m = 1, 2, \cdots$, tending to zero. According to Lemma 1.9, there are corresponding eigenfunctions $v = v_m$ in S satisfying the functional equation

$$v_m(y+\eta_m) = -v_m(y),$$

all of unit norm over $(0, b)$. Such a sequence of more and more rapidly oscillating functions clearly violates interior compactness. Hence, $\lambda = -1$ is not in the spectrum of $T(\eta)$ for η small.

From this it follows that an open set around $\lambda = -1$ belongs to the resolvent set of $T(\eta)$; this is proposition **C** for η small. From propositions **B** and **C** we deduce via a simple set-theoretic argument that the spectrum of $T(\eta)$ inside the unit circle consists of isolated points, $\lambda = 0$ being the only possible point of accumulation inside the unit circle. Then, by the spectral mapping theorem, it follows that $T(\eta)$ has this character for all values of η.

We show next that the spectrum neither accumulates nor has any points on the unit circle. To each element $u \in S$ we define an element

$$u_s = e^{-sy} u, \qquad\qquad s > 0.$$

Denote the set of these u_s by S_s; S_s is a linear space which is again translation invariant[5] and interior compact. Moreover, if λ_0 is in the point spectrum of the translation operator $T(\eta)$ defined over S, then $\lambda_0 e^{-s\eta}$ is in the point spectrum of the operator $T(\eta)$ defined over S_s. Denote by H_s the subspace of S_s of those u_s which are integrable over $(0, \infty)$. If the spectrum of $T(\eta)$ over H accumulates on the unit circle, then the spectrum of $T(\eta)$ over H_s, for $s > 0$, $\eta > 0$, accumulates in the interior of the unit circle. But by proposition **A**, this point of accumulation is in the point spectrum, and hence, by proposition **B**, it is isolated, i.e. not a point of accumulation. This shows that the spectrum of T over H does not accumulate on the unit circle.

We want now to prove that the unit circle itself is free of the spectrum of T. First we show that the set of values of s for which the spectrum of T over H_s has a point on the unit circle is discrete. Assume to the contrary that there is an infinite and bounded set of values s_i such that the spectrum of T over H_{s_i} has a point λ_i on the unit circle. According to Lemma 1.9, there exists an eigenelement v_i,

---

[5]We use here the important fact that the exponential functions are *characters* of the semigroup formed by the positive reals.

$$(1.3) \qquad\qquad v_i(y+\eta) = \lambda_i v_i(y), \qquad\qquad |\lambda_i| = 1,$$

where v_i is in the completion of H_{s_i} in the L_1-norm over every finite sub-interval. It is easy to see that this completion of H_{s_i} is S_{s_i}.

In particular, v_i must be of the form

$$(1.4) \qquad\qquad v_i = e^{-s_i y} u_i, \qquad\qquad u_i \text{ in } S.$$

From the functional equation (1.3) satisfied by v_i it follows that $v_i e^{-\alpha y}$ is integrable over $(0, \infty)$ for any positive α, and therefore that, if s is greater than s_i,

$$w_i = e^{-sy} u_i = e^{(s_i - s)y} v_i$$

belongs to H_s. Clearly, w_i is an eigenelement of $T(\eta)$ over H_s with eigenvalue $\mu_i = \lambda_i e^{(s_i - s)\eta}$ and $|\mu_i| = e^{(s_i - s)\eta}$. Now if there were infinitely many values of s_i less than s, the spectrum of T over H_s would have a non-zero point of accumulation. This, as we saw before, is impossible.

Now let s be a positive number such that the spectrum of T over H_s has no point on the unit circle. Since we have already shown that the spectrum cannot accumulate on the unit circle, it follows that *the spectral radius of T is less than one*. This, according to subsection 1.2, implies that the Phragmén-Lindelöf principle holds for H_s. By Theorem 1.2, T is completely continuous over H_s.

Let ε be any small positive number, and denote by P the projection of H_s into the eigenspace of $T(1)$ over H_s corresponding to eigenvalues whose absolute value is not less than ε. Since T is completely continuous over H_s, the range of P is finite-dimensional.

Every element u_s in H_s can be decomposed as a sum

$$(1.5) \qquad\qquad u_s = v_s + w_s,$$

where $v_s = Pu_s$ lies in the range of P, and w_s is annihilated by P. From the definition of P it follows that the subspace $H_{s,\varepsilon}$ of elements w_s annihilated by P is mapped into itself by $T(1)$ and that the spectrum of $T(1)$ over $H_{s,\varepsilon}$ lies within a circle of radius ε. According to the Gelfand formula for the spectral radius, it follows that the norm over $H_{s,\varepsilon}$ of $T^n(1)$ is less than ε^n for n large. In particular, if w_s is any element of $H_{s,\varepsilon}$,

$$(1.6) \qquad\qquad ||T^n(1)w_s|| = ||w||_n^\infty \leqq \varepsilon^n$$

for n large.

By definition of the space H_s, every element w_s of H_s is of the form

$$w_s = e^{-sy} w, \qquad\qquad w \text{ in } S.$$

If w_s belongs to $H_{s,\varepsilon}$, it follows easily from inequality (1.6) that for ε less than e^{-s}, w is integrable and moreover dies down exponentially:

(1.6')
$$||w||_n^\infty \leqq r^n$$

for n large, where $r = \varepsilon e^s$.

We turn now to the range of P in H_s. *It is a translation invariant finite-dimensional subspace* of H_s. According to classical and well-known results going back to Cauchy, *a one-parameter semigroup of linear transformations over a finite-dimensional space is exponential*, i.e.,

$$T(\eta) = e^{A\eta},$$

where A is some linear transformation of the finite-dimensional space into itself.

According to the spectral theory of matrices, the eigenelements and generalized eigenelements of A span the underlying space. Since the semigroup operators are translations, it follows easily that the eigenelements of A must be exponential functions of y, the generalized eigenelements y^n times exponential functions. So we have shown:

The range of P is spanned by exponential polynomials, i.e. sums of functions of the form

(1.7)
$$v_s(y) = p(y)e^{\mu y}v(0),$$

where p is a rational polynomial, the exponents μ are of a finite set, and the degree of p does not exceed a certain number.

Take now an arbitrary element u of H. Construct $u_s = e^{-sy}u$ in H_s, and decompose u_s as in (1.5). Correspondingly we have a decomposition for u:

(1.5')
$$u = v + w,$$

where v is e^{sy} times an element of the range of P, and w is e^{sy} times an element of $H_{s,e}$. According to (1.6'), w belongs to H; therefore v must also belong to H. It follows from (1.7) that $v = e^{sy}v_s$ is also an exponential polynomial; since it belongs to H, i.e., is integrable over $0 \leqq y \leqq \infty$, all of the exponents must have real part less than one[6]. Since the exponents in the representation of v as an exponential polynomial belong to a *finite* set, it follows that all such functions v decay exponentially at a *uniform* rate, i.e.,

(1.8)
$$||v||_n^\infty \leqq \theta^n$$

for n large and for some constant θ less than one which is the same for all v of the form $e^{sy}v_s$, v_s in the range of P.

Since every element u of H can be decomposed as in (1.5'), and since according to (1.6') and (1.8) both components v and w decay exponentially it follows that every element u of H decays exponentially. This completes the proof of Theorem 1.1.

---

[6]It is only at this point that the case $p = \infty$ has to be treated slightly differently.

1.4. In proving Lemma 1.2 we have shown that no element $u(y)$ of an interior compact space can vanish over an interval of the y-axis without, vanishing for all larger values of y. In other words, *elements of an interior compact space have the unique continuation property in the forward direction.*

Naturally, the question arises: do elements of an interior compact space have the property of unique continuation backward? That is, *could an element $u(y)$ of such a space S vanish for all $y > \eta$ without vanishing for all values of y?*

A more general and related question is: *Could an element $u(y)$ of an interior compact space decay faster than any exponential in y without being identically zero?* We shall show in Section 3 that *the answer to this question is "no" if, and only if, the eigenfunctions of the translation operator $T(\eta)$ are complete.* Furthermore, we shall show (Theorem 3.1 of Section 3), on the basis of an important theorem of Beurling [1], that the answer is indeed no if the Banach space B in which the function values lie is finite-dimensional.

Since $T(\eta)$ is completely continuous, one can associate with each point of its spectrum a projection operator whose range is the generalized eigenspace of T at that particular point. As we have shown in subsection 1.3, these generalized eigenfunctions are exponentials in y and powers of y times an exponential. From these projections, we can build a Fourier series for any element u of H:

$$u \sim \sum u_i,$$

where u_i is the projection of u into the i-th eigenspace of T. We think of the eigenvalues λ_i as being arranged in decreasing order of their absolute value. What is of interest, especially in applications, is the convergence of this Fourier series to u or, more generally, its usefulness in representing u.

This question of convergence is dependent on the completeness of eigenfunctions which will be discussed in Section 3. But from the argument given at the end of subsection 1.3 it follows that the Fourier series has at any rate an asymptotic validity.

THEOREM 1.4. *Let H be, as before, an interior compact space of integrable functions; then the Fourier series associated with the translation operators $T(\eta)$ is asymptotic at $y = \infty$, i.e., the difference of u and any section of its Fourier series decays at infinity at the same rate as the first neglected term.*

More precisely, denote by s_n the sum of the first n terms of the Fourier series of u, by r_n the remainder:

$$u = s_n + r_n;$$

then

$$\|r_n\|_y^\infty \leqq \text{const.} \, |\lambda_{n+1}|^y.$$

1.5. We close this section with a few miscellaneous remarks. The first concerns the extension of the notion of interior compactness to linear spaces of functions over the whole real axis. The prototype is the space of harmonic functions in a whole strip, subject to some boundary conditions. Weinstein [16] has proved a theorem concerning this situation which has an abstract extension.[7]

ABSTRACT WEINSTEIN THEOREM. *Let S be an interior compact space of functions $u(y)$ over the real line, invariant with respect to positive and negative translations. Assume every element of S grows at most exponentially, i.e., there exists a constant α such that, for any u in S, $\int_{-\infty}^{\infty} ||u(y)|| e^{-\alpha|y|} dy$ converges. Conclusion: S is finite-dimensional.*

This theorem follows easily from our Phragmén-Lindelöf theorem; one can show namely that the unit sphere in the norm $\int_{-\infty}^{\infty} ||u(y)|| e^{-\alpha|y|} dy$ is compact. We omit the details of the proof. This theorem shows that only those interior compact spaces over the whole axis are of interest which contain elements that grow faster than any exponential.

Another possibly interesting notion is that of an interior compact space over a *finite* interval (a, b), defined as the closure in the norm $||u||_a^b$ of an interior compact translation invariant space over the half-axis (or the whole axis). This would be an abstract notion of the space of solutions of an elliptic equation in a finite cylinder.

Further questions about interior compact spaces of functions over the positive reals arise:

i) Are functions of an interior compact space differentiable over every subinterval (d, ∞), d positive?

ii) Is the linear combination of two interior compact spaces of functions whose values lie in the same Banach space necessarily interior compact?

Instances of pairs of such spaces whose linear combination is also interior compact are the space of solutions of two elliptic equations $L_1 u = 0$ and $L_2 u = 0$, when L_1 and L_2 commute, and the spaces investigated by Laurent Schwartz in [12].

2. Concrete Phragmén-Lindelöf Theorem

In this section we shall show (Theorem 2.1) that the abstract theorem of Section 1 is applicable to the space of solutions of an elliptic equation over a half-cylinder. Furthermore we shall show (Theorem 2.2) that a concrete Phragmén-Lindelöf principle is valid for solutions of elliptic equations over

---

[7]I learned this in a conversation with J. B. Diaz and was led to the formulation of the abstract Weinstein theorem. Weinstein has also considered cases of more general boundary conditions, while Diaz discussed the biharmonic operator.

any unbounded domain, provided that the elliptic operator is *positive* over the underlying domain.

Terminology. By a smooth function defined in a domain of Euclidean space we mean one that has a sufficient number of continuous derivatives. By the m-fold *Dirichlet integral* we mean the integral of the sum of squares of u and all its derivatives up to and including order m. The symbol $||u||$ will now stand for the square root of the Dirichlet integral which is called the *Dirichlet norm*. The index m will not be indicated explicitly; in the present context it always stands for half the order of the elliptic operator under study. The domain over which the integration is extended, when in doubt, will be indicated by a subscript as in $||u||_G$. We shall denote the L_2 scalar product of two functions u, v by the symbol (u, v); the underlying domain, if necessary, will be indicated by a subscript.

Let L be an elliptic operator of order $2m$ with smooth coefficients; solutions of the elliptic equation $Lu = 0$ display an interior compactness of the following kind:

PRINCIPLE OF INTERIOR COMPACTNESS. *Let G be any bounded domain in Euclidean space, G' a subdomain of G with no common boundary points. Assertion: The set of solutions of $Lu = 0$ satisfying $||u||_G \leqq 1$ is precompact in the norm $||u||_{G'}$.*

This result is based on interior estimates (see Friedrichs [3], Gårding [4]) for the square integrals of higher derivatives, and on Rellich's selection principle.

I was able to extend this result to the case when G' and G have boundary points in common, provided that the solutions considered have zero Dirichlet data on a portion of the boundary which includes the common boundary points of G and G'.

GENERALIZED PRINCIPLE OF INTERIOR COMPACTNESS. *Let G be any domain in Euclidean space, L an elliptic operator of order $2m$ with smooth coefficients. Let Σ denote the set of solutions of $Lu = 0$ whose m-fold Dirichlet integral over G does not exceed one and which have zero Dirichlet data on an open, smooth portion γ of the boundary of G. Let G' denote a bounded subdomain of G whose common boundary points with G form a closed subset of γ. Conclusion: The set Σ is compact with respect to the Dirichlet integral over G'.*

The proof of this extension is given in [18]. It can also be deduced from recent results of Louis Nirenberg [10] and F. Browder [2] on the boundary behavior of solutions of elliptic equations.

Let D be the half-cylinder $y > 0$ erected over the base D_0; assume that D_0 is a *bounded* domain with smooth boundary. Let L be an elliptic operator of order $2m$ whose coefficients are smooth functions not depending on the variable y. Denote by S the space of solutions of $Lu = 0$ in D with zero Dirichlet data[8] on the vertical parts of the boundary. Regard these functions u as functions of the variable y alone, the functional values lying in the space of functions over D_0, normed by the square root of the Dirichlet integral over D_0. Clearly, S is translation invariant.

Assertion: S *is an interior compact space in the sense of Section* 1, *with* $p = 2$.

Proof: We have to show that the set $\{\int_a^b \|u\|^2 \, dy \leq 1\}$ is precompact in the norm $\int_{a'}^{b'} \|u\|^2 \, dy$. According to an observation of Nirenberg [10], the Dirichlet integral of a solution u with respect to the full set of variables over any section $\tilde{a} \leq y \leq \tilde{b}$ of the half-cylinder is bounded by a constant times the Dirichlet integral of u with respect to the x variables alone over any larger section (a, b). Apply the generalized principle of interior compactness to the case where G is the section $\tilde{a} < y < \tilde{b}$ of the half-cylinder, G' the section $a' < y < b'$, the interval $(\tilde{a}, \tilde{b})$ being so chosen that it contains (a', b') and is contained in (a, b). This proves our assertion.

We can apply the abstract Phragmén-Lindelöf principle of Theorem 1.1 to obtain

THEOREM 2.1. *Let L be an elliptic operator whose coefficients are independent of y. Let u be a solution of $Lu = 0$ in the half-cylinder D with zero Dirichlet data on the vertical sides, whose Dirichlet integral over D is finite. Assertion: u decays exponentially in y in the sense that the Dirichlet integral of u extended over the section $y \geq N$ of the half-cylinder is less than const. $e^{-\alpha N}$. The value of α is positive and depends only on the operator L and the underlying domain D_0, not on the particular solution u.*

We turn now to the generalization of Theorem 2.1. We shall use the following terminology:

The elliptic operator L is said to be *positive* over a domain D if an inequality

$$(2.1) \qquad\qquad \text{const. } \|u\|^2 \leq (u, Lu)$$

holds for all smooth functions u with compact support in D,[9] the value of the constant independent of u. The completion of the space of smooth functions with finite Dirichlet integral in the sense of the Dirichlet norm will be denoted by H. The operation of multiplying by a factor ϕ, where ϕ is a function with m continuous and bounded derivatives is a *bounded* operation, and hence has a well-defined meaning for elements of H. The completion of the space of smooth functions with compact support in D will be denoted by H^0. An

---

[8]To be precisely defined below.

[9]I.e., which vanish outside a closed, bounded subset of D.

element u of H is said to have *zero Dirichlet data on an open portion B′ of the boundary of D* if ϕu belongs to H^0 whenever the factor ϕ is a smooth function defined over the whole space, whose support is a compact subset of the whole space and includes only those boundary points of D which lie in B'.

It is easy to show that if u has zero Dirichlet data on the whole boundary of D, then in fact u belongs to H^0.

THEOREM 2.2. *Let D be an unbounded domain, L an elliptic operator with smooth coefficients which is positive over D. Let u be a solution[10] of $Lu = 0$ with zero Dirichlet data on that portion of the boundary of D which is included in the half-space $y > 0$, and suppose that u has finite Dirichlet integral over D. Conclusion: u dies down exponentially in the positive y-direction, i.e., the Dirichlet integral of u over that portion of D which is contained in the half-space $y > N$ is less than const. $e^{-\alpha N}$. The value of α depends only on the lower bound of L (the constant occurring in inequality (2.1)) and on bounds for the coefficients of L and a certain number of their derivatives.*

Theorem 2.2 is of a much more elementary nature than Theorem 2.1.

Proof: We introduce the auxiliary functions

$$f_N(y) = \begin{cases} y & \text{for } y < N-1 \\ N & \text{for } N < y \end{cases}$$

which are smoothly joined between $N-1$ and N. Clearly, we can arrange matters so that the first $2m$ derivatives of f_N (but not the functions f_N themselves!) are uniformly bounded by some constant for all y and all N.

If u satisfies the equation $Lu = 0$, the function $u_N = e^{\alpha f_N} u$ will satisfy an equation $L_N u_N = 0$. The operator L_N is related to L by the formula $Lu = e^{-\alpha f_N} L_N e^{\alpha f_N} u$; it differs from L by αM, where M is an operator of order $2m-1$ whose coefficients are polynomials in derivatives of f up to order $2m$ and in α. Since an operator which differs by a small amount from a positive definite one is itself positive definite, and since furthermore L was assumed positive definite, it follows that for α small enough, L_N is positive definite. The choice of α and the magnitude of the lower bound of the quadratic form induced by L_N depend only on the quantities indicated in Theorem 2.2, and not on N.

Next we introduce the auxiliary function

$$c(y) = \begin{cases} 0 & \text{for } y < \frac{1}{3} \\ 1 & \text{for } \frac{2}{3} < y, \end{cases}$$

smooth in the interval $(1/3, 2/3)$. We define v_N as $cu_N = ce^{\alpha f_N} u$. If u has zero

---

[10]By a solution we mean a function with $2m$ continuous derivatives, satisfying the equation. It is well known that, for elliptic equations, definitions which sound more general amount, in fact, to the same thing.

Dirichlet data at all boundary points of D in the half-space of positive y, then clearly v_N belongs to H^0.

We shall exploit this fact plus the positivity of the quadratic form induced by L_N to prove the inequality stated in Theorem 2.2. We have namely by the positivity of L_N

$$(2.2) \qquad (v_N, L_N v_N) \geqq \text{const. } ||v_N||^2.$$

Now if u is a solution of $Lu = 0$, $v_N = cu_N$ is a solution of $L_N v_N = 0$ whenever the factor c is identically zero or one, i.e. outside of the slab $1/3 < y < 2/3$. In this range both the function v_N and the operator L_N are actually independent of the parameter N. Therefore the left side of (2.2) is independent of N. The right side of (2.2) is certainly not less than a constant times the Dirichlet integral of v_N extended over a *subset* of D, namely for $y > N$. On this subset, however, v_N is equal to $e^{\alpha N} u$; so it follows from (2.2) that $e^{\alpha N}$ times the Dirichlet integral of u over the portion of D contained in $y > N$ is uniformly bounded, as asserted in Theorem 2.1.

There is one gap to be filled: in deriving (2.2) we have assumed that the quadratic form induced by L_N is positive for v_N, whereas positivity is defined to mean that $(w, L_N w)$ is positive for *smooth* functions w of compact support. To justify the application to $w = v_N$ a passage to the limit is necessary which we shall carry out.

For the sake of simplicity in writing, we shall drop the subscript N and write v for v_N, L for L_N.

We start by transforming the bilinear form $(w, Lz) = Q(w, z)$, defined for w, z smooth and with compact support, by m-fold integration by parts into an integral over D of a bilinear form q in w and z and their partial derivatives up to order m:

$$(2.3) \qquad Q(w, z) = \int_D q(w, z).$$

This expression for Q exhibits the continuous nature of its dependence on its arguments in the m-norm, and is a representation of the extension of Q to H^0. By continuity, Q is positive definite over H^0. In particular, since $v = v_N$ belongs to H^0, we have

$$(2.4) \qquad Q(v, v) \geqq \text{const. } ||v||^2.$$

Now let $\{v_n\}$ be a sequence of smooth elements with compact support tending to v in the m-norm. By the continuity of Q we have

$$(2.5) \qquad Q(v, v) = \lim_n Q(v_n, v).$$

Since v has square integrable derivatives up to order $2m$ in every compact subset of D, and since the support of v_n is compact, we can integrate the right side of (2.5) by parts to obtain

$$Q(v_n, v) = (v_n, Lv).$$

As remarked before, Lv is zero outside of the interval $(1/3, 2/3)$; therefore if an auxiliary function $d(y)$ is chosen to be 1 over $(1/3, 2/3)$, zero outside $(0, 1)$ and smooth in between, (v_n, Lv) is equal to $(dv_n, Lv) = Q(dv_n, v)$. Letting n tend to infinity we get

(2.6) $$Q(v, v) = Q(dv, v).$$

Using the representation (2.3) for $Q(dv, v)$ we see that its value depends only on the range of values of v in the slab $0 < y < 1$, i.e., is independent of N. This fills the gap in our derivation of inequality (2.2) and the proof of Theorem 2.2 is complete.

The original version of this proof had a somewhat different arrangement; there we showed that the Dirichlet integral of u over that portion of D which is contained in the half-space $y > N$ was less than a constant times the Dirichlet integral over the portion contained in the slab $N-1 < y < N$. The general scheme of such a proof has been suggested by Hans Lewy. The idea of including an exponential factor under the integral sign, incorporated in the present version, has been suggested by Louis Nirenberg.

Positivity of the operator L over D was essential in our proof of Theorem 2.2. Nevertheless the results of Section 1 suggest that more general conclusions are true. Here is

CONJECTURE I. *Let D be a domain with a smooth boundary, contained in a half-cylinder in the $+y$-direction. Let L be an operator uniformly elliptic over D whose coefficients have a sufficient number of continuous bounded derivatives in D. Denote by H the set of solutions of $Lu = 0$ which have a finite Dirichlet integral over D and zero Dirichlet data over those boundary points of D which lie in the half-space $y > 0$. Denote by H_α the subspace of H consisting of those u whose Dirichlet integral over $y > N$ dies down like $e^{-\alpha N}$. Contention: H/H_α is finite-dimensional.*

From a result of this kind one could easily deduce a further concrete generalization of Weinstein's theorem:

CONJECTURE II. *Let D be a domain such that the portion of D outside of some sphere is the union of a finite number of domains of the type considered in Conjecture I. Let L be an elliptic operator over D as in Conjecture I. Let S denote the space of solutions of $Lu = 0$ in D with zero Dirichlet data on the boundary, whose Dirichlet integral over D_R, consisting of those points P of D for which $|P| \leq R$, grows at most like const. e^{kR}. Conclusion: S is finite-dimensional.*

3. Spectral Synthesis

3.1. In this section we shall discuss completeness of the eigenfunctions of the translation operator; we shall work, for the most part, in the abstract framework of Section 1.

Let H be a translation invariant interior compact linear space of vector-valued functions $u(y)$ over the positive reals. Assume that each function $u(y)$ belongs to L_p, p some number ≥ 1, over the positive reals, and denote now the p-norm of u by $||u||$. Take H to be complete under this norm.

According to Theorem 1.2 of Section 1, the translation operators $T(\eta)$, $\eta > 0$, are *completely continuous*. Let $\{u_n\}$ denote the totality of their eigenfunctions (proper and generalized). We shall say that they are *complete in H if the only element u of H whose Fourier coefficients all vanish is the zero element.*

We start with the following observation concerning completeness:

If the Fourier coefficients of u are all zero, then u dies down faster than any exponential in y and conversely.

If a_k is the k-th Fourier coefficient of u corresponding to the eigenvalue λ_k of $T(1)$, the Fourier coefficient of $T(n)u = T(1)^n u$ is, clearly,[11] $\lambda_k^n a_k$. But the Fourier coefficients are bounded linear functionals; therefore, if $T(n)u$ tends to zero faster than λ_k^n, a_k must be zero. If $T(n)u$ decreases in n faster than *any* exponential, all the Fourier coefficients of u have to be zero. Conversely, let $\tilde{H}$ denote the totality of functions u all of whose Fourier coefficients are zero. $\tilde{H}$ is, it is easy to see, closed, translation invariant and, since it is a subspace of H, interior compact. The spectrum of $T(1)$ over $\tilde{H}$ consists of $\lambda = 0$, since all eigenfunctions of T were excluded from $\tilde{H}$. Therefore, by the Gelfand formula for the spectral radius, $\lim ||T(n)||^{1/n} = 0$ over $\tilde{H}$. This is the "faster than exponential" decay asserted for elements of $\tilde{H}$.

In some special cases it is possible to prove directly that no function of the space under consideration decays faster than any exponential. The general case will be discussed at the end of this section.

We shall present now a proof for the completeness of eigenfunctions of interior compact spaces of functions whose values lie in a *finite-dimensional* space, as a corollary of a theorem of Beurling [1].

THEOREM OF BEURLING.[12] *Let H be a translation invariant (not necessarily interior compact) space of scalar-valued functions $u(y)$, each belonging to L_p over*

---

[11]The formula for the coefficients of generalized eigenfunctions is more complicated but leads to the same conclusion.

[12]This is a slightly weaker version of the theorem; Beurling asserts that H contains an exponential function, i.e., that the point spectrum of translations is not empty, but at the expense of taking the closure of H in a weaker topology. In our case, of course, when H is interior compact, the full assertion of Beurling's theorem follows from our formulation.

$0 < y < \infty$. *Assume H closed under the L_p norm. Conclusion: The spectrum of the translation operator $T(\eta)$, $\eta > 0$, contains points besides $\lambda = 0$, except when all functions $u(y)$ in H vanish outside of a fixed interval.*

In view of the Gelfand formula for the spectral radius, this theorem can be stated as follows:

If $T(\eta)$ is quasinilpotent, it is nilpotent.

Applying Beurling's theorem to the space $\tilde{H}$ consisting of functions of an interior compact space whose Fourier coefficients are all zero, we conclude that such functions are identically zero for y sufficiently large. As a matter of fact, we shall show below that such a function is zero for all y.

Proof of Beurling's theorem: Assume that $T(\eta)$ is quasinilpotent; then $||T(\eta)||$ tends to zero as η tends to infinity. Take a to be any number so large that the norm of $T(a)$ is less than one, and take b to be any number greater than a. The space of functions of H considered over the interval $(0, b)$ is *not* dense among all functions in L_p over this interval; e.g. the function

$$v(y) = \begin{cases} 0 & \text{for } 0 < y < a \\ 1 & \text{for } a < y < b \end{cases}$$

cannot be approximated arbitrarily closely; otherwise the norm of $T(a)$ would be one. We can therefore assert, restricting the discussion for the sake of convenience to $p = 2$, that the orthogonal complement of H over $(0, b)$ is not empty, i.e., that there exists a function $h(y)$ in L_2 over $(0, b)$ such that $\int_0^b h(y)u(y)dy$ is zero for all u in H. Take in particular any function $f(y)$ in H and its translates $f(y+\eta)$:

$$\int_0^b h(y)f(y+\eta)dy = 0 \quad \text{for } \eta \text{ positive.}$$

Imagine $f(y)$ defined to be zero for y negative and $h(y)$ likewise for y outside of $(0, b)$. Then the formula

(3.1) $$\int_{-\infty}^\infty h(y)f(y+\eta)dy = g(\eta)$$

defines a function g for all values of η which turns out to be zero for η positive. To bring formula (3.1) into the form of a convolution we switch y to $-y$, and denote $h(-y)$ by $k(y)$:

(3.1') $$\int_{-\infty}^\infty k(y)f(\eta-y)dy = g(\eta).$$

Since $k(y)$ vanishes outside of $(-b, 0)$ and f for negative values of its argument, $g(\eta)$ also vanishes for η less than $-b$, as well as for η positive, as we already know. It also follows easily that g belongs to L_2, and therefore, being zero outside a finite interval, to L_1. The same is true for k.

Denote by $K(z)$, $F(z)$ and $G(z)$ the Fourier transforms of k, f, and g. $K(z)$ and $G(z)$ are analytic and bounded in the half-plane $\mathscr{I}m\ z \leqq 0$, since $k(\eta)$ and $g(\eta)$ are zero for η positive and absolutely integrable for η negative. Since $f(y)$ is zero for y negative, $F(z)$ is an entire function bounded on the real axis. Assume now that T over H is quasinilpotent; then $f(y)$ decreases faster than any exponential for y positive.

Since g is the convolution of k and f, G is the product of K and F:

$$(3.2) \qquad F(z)K(z) = G(z)$$

or

$$(3.3) \qquad F(z) = \frac{G(z)}{K(z)}.$$

Next we use a *division theorem* of function theory, based on the factorization of a function which is analytic and bounded in a half-plane, into a Blaschke product times a bounded function without zeros, and on the representation due to Nevanlinna [9] of a positive harmonic function in a half-plane as the Poisson integral with non-negative mass distribution plus const. y.

Let G and K be functions bounded on the real axis and of exponential growth in the lower half-plane. If their quotient $F = G/K$ is bounded on the real axis and regular in the lower half-plane, then it too is of exponential growth in the lower half-plane.

Having concluded from the divisibility theorem that $F(z)$ is of exponential growth in the lower half-plane, we further conclude from the theorem of Paley-Wiener [11] that $F(z)$ is a finite Fourier transform, i.e., that $f(y)$ vanishes outside a finite interval. This completes the proof of Beurling's theorem.

More precise information about the length of the support of f can be obtained from the *convolution theorem* of Titchmarsh [15], also employed by Kahane in his investigations [7]:

If g is the convolution of k and f, the support of g is the sum of the supports of k and f. In particular, the length of the support of f cannot be greater than that of g. Since the latter is at most b, we conclude: the support of f has at most length b.

The only restriction on b was that for some value a less than b, $\|T(a)\|$ should be less than one. According to Lemma 1.2 of Section 1, for an interior compact space this is the case for any positive a. Hence the support of $f(y)$ is zero. Since f could be any element of $\tilde{H}$, this shows that $\tilde{H}$ contains only the zero function, i.e., we have proved

THEOREM 3.1. *Let H be an interior compact translation invariant linear space of scalar-valued functions $u(y)$ over $0 < y < \infty$ of class L_p, closed in* *the norm. Then the eigenfunctions of the translation operators $T(\eta)$ are complete.*

Of course, the eigenfunctions of $T(\eta)$ are just the exponentials in y and polynomials times exponential functions.

3.2. Beurling's theorem and Theorem 3.1 can be extended easily to functions $u(y)$ whose values lie in a *finite*, say *n-dimensional*, vector space B. We take in Beurling's theorem a subspace H' of H spanned, in the L_2-sense, by all translates of a single function $f(y)$. Operating the same way as before, we consider the *totality* of all functions $k(y)$ orthogonal to $f(y+\eta)$ in the interval $(0, b)$. We deduce the analogue of relation (3.2) which now reads

$$(3.2') \qquad \sum_{j=1}^{n} F_j(z)K_j(z) = G(z),$$

where F_j and K_j are the Fourier transforms of the components of f and k.

Next we shall show that for each z_0 in the lower half-plane the values of $K(z_0)$ span the whole dual space B'. For, assume to the contrary that there is a vector A orthogonal to all $K(z_0)$, i.e., that

$$A \cdot K(z_0) = 0.$$

In terms of the Fourier inverse of K this means

$$\int_0^b A e^{-iz_0 y} h(y) dy = 0$$

for all h orthogonal over $(0, b)$ to all functions $f(y+\eta)$. But, since H' is closed, this implies according to the projection theorem that there is a function $m(y)$ in H' which over the interval $(0, b)$ is equal to $A e^{-iz_0 y}$.

We now show that $m(y)$ is equal to $A e^{-iz_0 y}$ for all positive y. Consider the element

$$n(y) = m(y+\eta) - e^{-iz_0 \eta} m(y)$$

of H, η some small positive number; $n(y)$ is zero over $(0, b-\eta)$ which, for η small enough, includes $(0, a)$. Since the norm of $T(a)$ is less than one, such an element $n(y)$ must be zero for all y. From the relation

$$m(y+\eta) = e^{-iz_0 \eta} m(y), \qquad\qquad y > 0,$$

and from the fact that $m(y)$ equals $A e^{-iz_0 y}$ over $(0, b)$, we conclude at first that $m(y)$ is equal to $A e^{iz_0 y}$ over $(0, b+\eta)$ and, recursively, over the whole positive axis.

But this is a contradiction; for, $m(y)$ is an element of H which is a fixed exponential, whereas all elements of H supposedly die down faster than any exponential. Hence the values of $K(z_0)$ must span the whole space B', which means that n of them, say $K^l(z)$, $l = 1, 2, \cdots, n$, span B'. A relation of the type (3.2') holds for each of these K^l:

$$(3.2'') \qquad\qquad \sum_{j=1}^{n} F_j K_j^l = G^l, \qquad\qquad l = 1, 2, \cdots, n.$$

Solve the equations $(3.2'')$ for the F_j ; the determinant Δ is not zero at $z = z_0$, so we have

$$(3.3') \qquad\qquad F_j = \frac{\Delta_j}{\Delta}.$$

The elements of the determinants Δ_j and Δ are analytic functions bounded in the lower half-plane; therefore, Δ and Δ_j themselves are functions of the same type. Furthermore, Δ is not identically zero. From here on we proceed as before.

3.3. The analogue of Beurling's theorem for functions whose values lie in an infinite-dimensional space is certainly false. Consider a function $f(y)$ with values in Hilbert space, so defined that the n-th component of $f(y)$ is zero for $y \geqq n$. All positive translates of f have this property, and so do all linear combinations of translates. Therefore, the space spanned by the translates of f—in whatever topology—does not contain an exponential. Furthermore, it is easy to give examples of interior compact translation invariant spaces which contain no exponentials. Let $S(t)$, $0 \leq t$, be a strongly continuous one-parameter semigroup of bounded mappings of a Banach space into itself. Associate with this semigroup the set of functions

$$f(y) = S(y)u,$$

u any element of B. Clearly, the set of these functions f is a translation invariant linear space. If the operators S are completely continuous, the associated space of functions is interior compact, and if the operators $S(y)$ are quasinilpotent, so are the translation operators acting on the associated space of functions.

The nullspaces of elliptic operators are nontrivial examples of translation invariant spaces of vector-valued functions for which an adequate spectral synthesis is possible. It is an interesting problem to give an intrinsic characterization of such spaces.

I would like to close with two observations about this problem. One is that one cannot expect a function theoretic proof for completeness in the vector-valued case. For, such a proof shows at the same time that the spectra of the translations consist of points of the form $e^{\mu_n \eta}$, where μ_n are the roots of some analytic function bounded in a half-plane. This, it is well known, places a limitation on the density of the sequence $\{\mu_n\}$ which is not satisfied by the exponents that figure in the separated solutions of partial differential equations.[13]

---

[13] E.g., the characteristic exponents for Laplace's equation in a strip of width π are $\mu_n = -n$.

The second observation is this: If L is of second order and of the form

$$L = \frac{\partial^2}{\partial y^2} + M,$$

where M is self-adjoint in the x-variables, then the translation operators over the nullspace of L are self-adjoint with respect to the L_2 norm over the half-cylinder. Therefore the completeness of their eigenfunctions follows from the theory of self-adjoint operators. Over the nullspaces of higher order operators the translation operators show no obvious tendency toward self-adjointness; indeed the characteristic exponents of the biharmonic operator $\Delta\Delta$ over a half-strip are no longer real.[14] Nevertheless R. C. T. Smith [14] succeeded in proving the completeness of the separated solutions in this case. A different completeness proof has been given by G. Horvay, see [17].

Bibliography

[1] Beurling, A., *A theorem of functions defined on a semi-group*, Math. Scand., Vol. 1, 1953, pp. 127–130.

[2] Browder, F. E., *On the regularity properties of solutions of elliptic differential equations*, Comm. Pure Appl. Math., Vol. 9, 1956, pp. 351–362.

[3] Friedrichs, K. O., *On the differentiability of solutions of linear elliptic differential equations*, Comm. Pure Appl. Math., Vol. 6, 1953, pp. 299–325.

[4] Gårding, L., *Dirichlet's problem for linear elliptic partial differential equations*, Math. Scand., Vol. 1, 1953, pp. 55–72.

[5] Hille, E., *Functional Analysis and Semi-groups*, Amer. Math. Soc. Coll. Publ., New York, 1948

[6] Horvay, G., *Biharmonic eigenvalue problem of the semi-infinite strip*, Quart. Appl. Math., Vol. 15, 1957, pp. 65–81.

[7] Kahane, J., *Sur quelques problèmes d'unicité et de prolongement relatifs aux fonctions approchables par des sommes d'exponentielles*, Ann. Inst. Fourier Grenoble, Vol. 5, 1953–1954, pp. 39–130.

[8] Lax, P. D., *Spectral Theory of Functions on Semi-Groups and the Separation of Variables*, Studies in Eigenvalue Problems, Tech. Rep. 14, Conf. on Partial Diff. Equations, Univ. of Kansas, Summer, 1954.

[9] Nevanlinna, F., and Nevanlinna, R., *Über die Eigenschaften analytischer Funktionen in der Umgebung einer singulären Stelle oder Linie*, Acta Soc. Sci. Fenn. Nova Ser. A., Vol. 50, 1922, 46 pp.

[10] Nirenberg, L., *Remarks on strongly elliptic partial differential equations*, Comm. Pure Appl. Math., Vol. 8, 1955, pp. 649–674.

[11] Paley, R., and Wiener, N., *Fourier Transforms in the Complex Domain*, Amer. Math. Soc. Coll. Publ., New York, 1934.

[12] Schwartz, L., *Étude des sommes d'exponentielles réelles*, Hermann, Paris, 1943.

[13] Schwartz, L., *Théorie générale des fonctions moyenne-périodiques*, Ann. of Math., Vol. 48, 1947, pp. 857–929.

[14] Smith, R. C. T., *The bending of a semi-infinite strip*, Australian J. of Sci. Research. Ser. A., Vol. 5, 1952, pp. 227–237.

---

[14] The imaginary parts are small compared to the real parts.

[15] Titchmarsh, E. C., *The zeros of certain integral functions*, Proc. London Math. Soc., Vol. 25, 1926, pp. 283–302.

[16] Weinstein, A., *Zum Phragmén-Lindelöfschen Ideenkreis*, Abh. Math. Sem. Univ. Hamburg, Vol. 6, 1928, pp. 263–264.

[17] Lax, P. D., *A Phragmén-Lindelöf principle in harmonic analysis, with application to the separation of variables in the theory of elliptic equations*, Lecture Series of Symposium on Partial Differential Equations, Summer 1955, Berkeley, Calif., University of Kansas, 1957.

[18] Halmos, P. R., *Introduction to Hilbert Space and the Theory of Spectral Multiplicity*, Chelsea, New York, 1951.

[19] Treuenfels, P., *Boundary Layer Problems in Elasticity Theory and Exponential Decay of Solutions of Certain Linear Partial Differential Equations*, New York University dissertation, 1956.

Received January, 1957.

TRANSLATION INVARIANT SPACES

To Karl Loewner

BY

PETER D. LAX

New York University [1]

1. Let h denote the space of complex-valued square integrable functions $u(x)$ defined for x real which are zero for x negative. Let H denote the space of functions $U(z)$ which are Fourier transforms of functions in h. The space H is characterized by the one-sided

PALEY–WIENER THEOREM. [2] *Every function U in H can be extended as regular analytic to the upper half-plane, so that*

$$\int_{-\infty}^{\infty} U^*(i\tau+\sigma)\, U(i\tau+\sigma)\, d\sigma \leqslant \text{const},$$

for all τ positive. Conversely, the restriction to the real axis of any such function belongs to H.

For fixed τ, $U(i\tau+\sigma)$ is the Fourier transform of $e^{-x\tau}u(x)$; since $u(x)$ vanishes for negative x, the L_2 norm of $e^{-x\tau}u(x)$ decreases with increasing τ. So by Parseval's formula we have this

COROLLARY. *If U lies in H, its L_2 norm along the line $\mathrm{Im}\, z = \tau$, $\tau \geqslant 0$, decreases with increasing τ.*

The orthogonal complement of h with respect to the space of square integrable functions on the entire real axis is the space of square integrable functions which vanish for x positive. The Fourier transforms of these functions form the orthogonal complement $H^{\perp}$ of H. Functions in $H^{\perp}$ can be continued analytically into the lower halfplane. Also, it is easy to show that $H^{\perp}$ is the conjugate of H:

$$H^{\perp} = H^*$$

---

[1] Work performed at the Atomic Energy Commission Computing and Applied Mathematics Center, Institute of Mathematical Sciences, New York University, under Contract Number AT (30-1)-1480. Reproduction in whole or in part permitted for any purpose of the U.S. Government.

[2] We denote the conjugate of a complex number by *; in section 4 where we deal with matrix valued functions the * denotes the adjoint.

We denote by r any subspace of h which is invariant under right translation. I.e., whenever $g(x)$ belongs to r, we require that $g(x-s)$ should belong to r for all positive s. A subspace will be called left translation invariant and denoted by l if, whenever $g(x)$ belongs to it, the projection of $g(x+s)$ into h (i.e., its restriction to the positive axis) also belongs to l for all positive values of s.

The closure of translation invariant spaces is translation invariant.

The orthogonal complement with respect to h of an r-space is an l-space, and vice versa.

The Fourier transform of an r-space will be denoted by R. Such an R-space can be characterized intrinsically as a subspace of H such that $e^{isz}R$ is contained in R for all positive s.

In this paper we study R-spaces of *vector-valued* functions, i.e., functions whose values lie in a finite-dimensional Hilbert space S over the complex numbers. When we wish to make a distinction, we shall denote the H-space of functions with values lying in S by H_S. Our main result is a unique representation for such spaces:

REPRESENTATION THEOREM. *Every closed R-space is of the form FH_T, where $F(z)$ is an operator-valued function of z mapping a Hilbert space T of possibly lower dimension than S into S. $F(z)$ is regular in the upper half-plane, $\|F(z)\| \leqslant 1$ there, and for z real F is an isometry. This representation of R is unique, save for a multiplication of F on the right by a constant unitary matrix.*

In the scalar case, such a representation of the Fourier transform of an r-space spanned by the translates of a single function has been given by Karhunen in [5]. A similar representation theorem for the Fourier transform of an r-space spanned by the translates of a finite number of functions defined on the positive integers has been given by Beurling in [1]. So in the scalar case my representation theorem is a slight extension of their results.

Beurling and Karhunen use a function-theoretic method, relying on the factorization due to Riesz, Herglotz and Nevanlinna of functions, analytic in the upper half-plane and bounded in a certain integral sense, into an inner and outer factor. The outer factor is the exponential of a Poisson integral of an absolutely continuous measure, the inner factor is the exponential of a Poisson integral with respect to a singular measure times a Blaschke product. My proof employs only Hilbert space methods, specifically the projection of the exponential function into r. The significance of this projection has already been pointed out by Beurling in [1].

In Section 3 we use the representation theorem to reduce problems of division in the ring of bounded analytic functions to problems in the Boolean algebra of r-spaces. In particular we are able to factor functions into inner and outer factors. This decomposition is used to give a new proof of the Titchmarsh convolution theorem.

The proof of the representation theorem in the scalar case is given in Section 2, for the vector-valued case in Section 4.

Many problems of analysis are about translation invariant spaces, such as occur in the theory of approximation by exponentials, in Wiener's theory of Tauberian theorems, and in many others. A representation theorem such as the one given here is often useful in such problems, see e.g. [9]. My own interest in the subject came from the study of solutions of partial differential equations in a half-cylinder which, as explained in [8], can be regarded as an l-space of functions whose values lie in an *infinite-dimensional* space. Whether the theory given in the following pages applies to that situation and just how useful it might be is still to be seen.

2.1. In this section we treat the scalar case for which the representation theorem asserts:

SCALAR REPRESENTATION THEOREM. *Every nonempty closed R-space is of the form FH, where $F(z)$ is a regular analytic function in the upper half-plane, $|F(z)| \leqslant 1$ there. For z real, $|F(x)| = 1$. F is uniquely determined by R, save for multiplication by a complex constant of modulus 1.*

Let l and r be a pair of closed translation invariant subspaces of h which are orthogonal complements of each other with respect to h. Let λ be any complex number in the upper half-plane; the function defined as $e^{i\lambda x}$ for x positive, zero for x negative, belongs then to h. Decompose this function into components by orthogonal projection into l and r:

$$e^{i\lambda x} = a_\lambda(x) + b_\lambda(x), \quad 0 < x, \tag{2.1}$$

a_λ in l, b_λ in r. Take the complex conjugate of (2.1), multiply both sides by $b_\mu(x - s)$, where μ is any complex number in the upper half-plane and s is non-negative, and integrate with respect to x from 0 to ∞:

$$\int_0^\infty b_\mu(x-s) e^{-i\lambda^* x} \, dx = \int_0^\infty b_\mu(x-s) a_\lambda^*(x) \, ds + \int_0^\infty b_\mu(x-s) b_\lambda^*(x) \, dx. \tag{2.2}$$

Denote by $B_\mu(z)$ the Fourier transform of $b_\mu(x)$. The left side of (2.2) is equal to $e^{-i\lambda^* s} B_\mu(-\lambda^*)$. On the right, the first term vanishes since $a_\lambda(x)$ belongs to l while $b_\mu(x-s)$ belongs to r. We transform the second term by Parseval's theorem, [1] using the fact that the Fourier transform of $b_\mu(x-s)$ is $B_\mu(z) e^{izs}$. So we get from (2.2)

---

[1] dz denotes $dz/2\pi$.

$$e^{-i\lambda^* s} B_\mu(-\lambda^*) = \int_{-\infty}^{\infty} B_\lambda^*(z) B_\mu(z) e^{izs} dz, \quad 0 \leqslant s. \tag{2.3}$$

Take the complex conjugate of both sides, interchange the role of λ and μ and replace s by $-s$. We get

$$e^{-i\mu s} B_\lambda^*(-\mu^*) = \int_{-\infty}^{\infty} B_\lambda^*(z) B_\mu(z) e^{izs} dz, \quad s \leqslant 0. \tag{2.3'}$$

Putting $s = 0$ in (2.3), (2.3') shows that $B_\mu(-\lambda^*)$ and $B_\lambda^*(-\mu^*)$ are equal; we denote their common value by $B_{\lambda\mu}$:

$$\int_{-\infty}^{\infty} B_\lambda^*(z) B_\mu(z) dz = B_{\lambda\mu} = B_\lambda^*(-\mu^*) = B_\mu(-\lambda^*). \tag{2.4}$$

Equations (2.3) and (2.3') give the Fourier transform of $B_\lambda^*(z) B_\mu(z)$ in the ranges $s \geqslant 0$ and $s \leqslant 0$ respectively. Therefore, the value of $B_\lambda^*(z) B_\mu(z)$ for z real can be found by Fourier inversion:

$$B_\lambda^*(z) B_\mu(z) = B_{\lambda\mu} \int_{-\infty}^{0} e^{-i\mu s - izs} ds + B_{\lambda\mu} \int_{0}^{\infty} e^{-i\lambda^* s - izs} ds$$

$$= B_{\lambda\mu} \left\{ \frac{i}{\mu + z} - \frac{i}{\lambda^* + z} \right\} = \frac{i B_{\lambda\mu}(\lambda^* - \mu)}{(\mu + z)(\lambda^* + z)}. \tag{2.5}$$

Set $\mu = \lambda$ in (2.4); we get

$$B_{\lambda\lambda} = \int_{-\infty}^{\infty} |B_\lambda(z)|^2 dz$$

which, by Parseval's formula, is equal to

$$\int_{0}^{\infty} |b_\lambda(x)|^2 dx.$$

In particular if $B_{\lambda\lambda} = 0$ it follows that $b_\lambda(x) \equiv 0$ and therefore, in view of (2.1), that $e^{i\lambda x}$ is orthogonal to r. We have assumed that r contains non-zero elements; the Fourier transform of one of these cannot vanish for all z in the upper half-plane; say it does not vanish at $z = \lambda$. Then $B_{\lambda\lambda} \neq 0$. Set μ equal to this λ in (2.5); we get, for real z,

$$B_\lambda^*(z) B_\lambda(z) = \frac{2 \operatorname{Im} \lambda B_{\lambda\lambda}}{|\lambda + z|^2}$$

from which we deduce that

$$B_\lambda(z) = \frac{F(z)}{\lambda + z} G, \tag{2.6}$$

where $G = (2 \operatorname{Im} \lambda B_{\lambda\lambda})^{\frac{1}{2}}$ and $F(z)$ has modulus 1.

481

Equation (2.6) provides an extension of $F(z)$ into the upper half-plane, as a regular analytic function. *We claim that $F(z)$, thus extended, is a bounded function.* For let w be any point in the upper half plane. Then we have (omitting the subscript λ) a Poisson formula for $w\,B(w)$:

$$B(w) = \int_{-\infty}^{\infty} b(x)\, e^{ixw}\, dx = \frac{i}{w} \int_{-\infty}^{\infty} b'(x)\, e^{ixw}\, dx$$

$$= \frac{i}{w} \int_{-\infty}^{\infty} b'(x)\, e^{ix\,\mathrm{Re}\,w - |x|\,\mathrm{Im}\,w}\, dx$$

$$= \frac{i}{w} \int_{-\infty}^{\infty} z\,B(z)\,P(z - \mathrm{Re}\,w)\, dz.$$

The first equality is obtained by integrating by parts, the second by noting that the support of b' is contained in $(0, \infty)$, the third by Parseval's formula, noting that the Fourier transform of b' is $-iz\,B(z)$, and denoting the Fourier transform of $e^{-|z|\,\mathrm{Im}\,w}$ by $P(z)$. Now $P(z) = 2\,\mathrm{Im}\,w/(z^2 + (\mathrm{Im}\,w)^2)$ is positive and its integral is equal to 1. On the other hand, according to (2.6), $z\,B(z)$ is bounded on the real axis. Therefore, from the last integral formula it follows that $w\,B(w)$—and thereby $F(w)$—is bounded in the upper halfplane. Hence by a standard extension of the maximum principle, $|F(z)|$ assumes its maximum on the real axis where it has modulus 1.

Multiply both sides of (2.5) by $B_\lambda(z)$; using formula (2.6) for B_λ, the above formula for $B_{\lambda\mu}$ and the relation $F_\lambda(z)\,F_\lambda^*(z) = 1$, we obtain the following expression for B_μ:

$$B_\mu(z) = i\,\frac{F(z)\,F^*(-\mu^*)}{\mu + z} \tag{2.7}$$

for z real. Since both sides are regular analytic in the upper halp-plane, (2.7) holds for all z in the upper half-plane as well.

This completes the construction of the function F. Formula (2.7) shows as well that $F(z)$ is uniquely determined up to a constant factor of modulus 1. We turn now to showing that the space R is equal to FH.

Denote the space FH by R'. Since $|F(z)| \leqslant 1$ in the upper half-plane, we conclude by the Paley-Wiener theorem that R' is a subspace of H. R' is the Fourier transform of a right translation invariant subspace of h, since $e^{isz}R' = e^{isz}FH = Fe^{isz}H \subset FH = R'$ for s positive. Finally, since $|F(z)| = 1$ on the real axis, R' is closed.

As before, we project $e^{i\mu x}$ into r'.

$$e^{i\mu x} = a'_\mu(x) + b'_\mu(x).$$

Take the Fourier transform of both sides:

$$\frac{i}{\mu + z} = A'_\mu(z) + B'_\mu(z), \tag{2.8}$$

where B'_μ lies in R', A'_μ in L', the orthogonal complement of R' with respect to H.

The spaces H and H^* are orthogonal complements of each other in L_2. Multiplication by F is a unitary mapping of L_2 into itself and therefore complements are preserved. Hence the orthogonal complement in L_2 of $R' = FH$ is FH^*. So A'_μ and B'_μ are of the form

$$A'_\mu = F D, \qquad B'_\mu = F E, \qquad D \in H^*, \qquad E \in H. \tag{2.9}$$

Substitute (2.9) into (2.8) and multiply it by $F^*(z)$. Using the facts that, for z real, $F^*(z) F(z) = 1$ and $z = z^*$ we get

$$\frac{i F^*(z^*)}{\mu + z} = D(z) + E(z). \tag{2.10}$$

We are now in a position to determine D and E explicitly:

$$\left. \begin{aligned} D &= i \frac{F^*(z^*) - F^*(-\mu^*)}{\mu + z}, \\ E &= i \frac{F^*(-\mu^*)}{\mu + z}. \end{aligned} \right\} \tag{2.11}$$

To verify these expressions for D and E, we have to show that they belong to H^* and H respectively. Clearly, $D(z)$ is regular in the lower half-plane and its square integral along any line parallel to the real axis is uniformly bounded. Therefore, by the Paley–Wiener theorem, D belongs to H^*. E, on the other hand, clearly belongs to H.

From (2.10) we get

$$B'_\mu(z) = FE = i \frac{F(z) F^*(-\mu^*)}{\mu + z}.$$

Comparing this with (2.7) we conclude that $B'_\mu(z)$ and $B_\mu(z)$ are identical, i.e., that the projections of $e^{i\mu x}$ into r and r' are identical. Since projections are linear and bounded, it follows that also all linear combinations of exponential functions and their closures have identical projections. Since the set of all functions $e^{i\mu x}$ spans h, it follows that r and r' coincide.

Observe the curious skew symmetry in the dependence of B on μ and z displayed by formula (2.7).

2.2. Denote by d_λ the distance of the normalized exponential function $(2\,\mathrm{Im}\,\lambda)^{\frac{1}{2}}\,e^{i\lambda x}$ from the space l. From (2.4) and (2.7) we have

$$d_\lambda^2 = 2\,\mathrm{Im}\,\lambda \int\limits_0^\infty |b_\lambda(x)|^2\,dx = 2\,\mathrm{Im}\,\lambda \int\limits_{-\infty}^\infty |B_\lambda(z)|^2\,dz = |F(-\lambda^*)|^2. \qquad (2.12)$$

This formula is already contained in Beurling, l.c.; a special case of it goes back to Müntz [10]. Take namely l as the space spanned by the set of exponentials $\{e^{i\lambda_j x}\}$. R, the Fourier transform of its orthogonal complement, has the form FH, where F is the Blaschke product

$$F(z) = \prod \frac{z + \lambda_j^*}{z + \lambda_j}.$$

To show this we note: an element of h is orthogonal to l if and only if it is orthogonal to every exponential function $e^{i\lambda_j x}$, which means that its Fourier transform vanishes at $z = -\lambda_j^*$. So R consists of those elements of H which vanish at $z = -\lambda_j^*$, $j = 1, 2, \ldots$. Clearly, any function of the form FH_1, H_1 in H and F the above Blaschke product, does vanish at $z = -\lambda_j^*$. Conversely, it is well known that any function in H which vanishes at $z = -\lambda_j^*$ can be factorized as FH_1, H_1 in H. Therefore, according to formula (2.12) the distance d of the normalized exponential function $e^{i\lambda x}(2\,\mathrm{Im}\,\lambda)^{\frac{1}{2}}$ to l is

$$|d| = \prod \left| \frac{\lambda_j - \lambda^*}{\lambda_j - \lambda} \right|.$$

For a finite set of exponentials, this formula was derived by Müntz by representing the distance as the ratio of two Gram determinants and evaluating the determinants explicitly. For an infinite set of exponentials the formula was derived by Müntz through a passage to the limit, leading to his celebrated criterion for completeness:

A set of exponentials is complete if and only if the Blaschke product formed of them diverges.

Müntz considered real exponentials only; the analogous treatment of complex exponentials is due to Szász [15].

3.1. In this section F, subscripted possibly by some index, will denote a regular analytic function in the upper half-plane, $|F(z)| \leqslant 1$ there, and $|F(z)| = 1$ for z real. If two such functions differ by a constant multiple, they shall be regarded as equivalent. The functions F form a semigroup; we shall now discuss, with the aid of the representation theorem, division in this semigroup and subsequently in the ring of all bounded analytic functions.

F_1 is divisible by F_2 if $F_1 = F_2 F_3$.

THEOREM 3.1. F_1 *is divisible by* F_2 *if and only if* F_1H *is contained in* F_2H.

Proof. Since $F_3H \subset H$, $F_1H = F_2F_3H \subset F_2H$. Conversely, assume that $F_1H \subset F_2H$, i.e., that to any H_1 in H there exists an H_2 in H such that $F_1H_1 = F_2H_2$. This relation is valid for z real and therefore for z complex and can be expressed as follows: Multiplication by $F_2^{-1}F_1$ maps H into itself. It follows then that multiplication by any power of $F_2^{-1}F_1$ maps H into H. Since the L_2 norm on the real axis is preserved in this multiplication, it follows from the corollary of the Paley–Wiener theorem that it cannot be increased on any line parallel to the real axis. Clearly, this is the case if and only if $F_2^{-1}F$ has modulus $\leqslant 1$, i.e., belongs to the semigroup.

The intersection, linear combination and closure of translation invariant spaces is likewise translation invariant. Given F_1 and F_2 it follows from the representation theorem that the spaces $F_1H \cap F_2H$ and $\overline{F_1H \oplus F_2H}$ are of the form F_3H and F_4H. We shall denote F_3 by $\{F_1, F_2\}$ and F_4 by (F_1, F_2). An immediate consequence of the divisibility criterion in Theorem 3.1 is

THEOREM 3.2. $\{F_1, F_2\}$ *is the least common multiple,* (F_1, F_2) *the greatest common divisor of* F_1 *and* F_2.

If $(F_1, F_2) = 1$, F_1 and F_2 are called relatively prime. We shall show now that the relation between the greatest common divisor and the least common multiple is the usual one:

THEOREM 3.3. $(F_1, F_2)\{F_1, F_2\} = F_1F_2$.

It is easy to show that the above proposition is equivalent with the following one: *If* F_1 *and* F_2 *are relatively prime and if* F_1 *divides* FF_2, *then* F_1 *divides* F. This may be proved as follows:

If F_1 divides F_2F, then according to Theorem 3.1

$$F F_2 H \subset F_1 H.$$

Since $F_1FH \subset F_1H$, we have also

$$F F_2 H \oplus F_1 F H \subset F_1 H. \tag{3.1}$$

The left side is equal to $F(F_2H \oplus F_1H)$; since we have assumed that F_1 and F_2 are relatively prime it follows that $F_2H \oplus F_1H$ is a dense subset of H. Therefore the closure of the space on the left in (3.1) is FH. Since the space on the right is closed, we have $FH \subset F_1H$, i.e., F_1 divides F.

An immediate *corollary* of Theorem 3.3 is that if

$$F_3 F_4 = F_1 F_2,$$

then
$$F_4 = \tilde{F}_1 \tilde{F}_2,$$

where $\tilde{F}_1$ divides F_1, $\tilde{F}_2$ divides F_2.

If F_1 and F_2 are relatively prime, then the linear combination of $R_1 = F_1 H$ and $R_2 = F_2 H$ is dense in H. This is equivalent with the assertion that L_1 and L_2, the orthogonal complement of R_1 and R_2 respectively have only zero in common. The orthogonal complement of $L_1 \oplus L_2$ is $R_1 \cap R_1$ which, according to Theorem 3.3, is $F_1 F_2 H$. Using formula (2.12) for the distance of normalized exponentials from l-spaces we have

THEOREM 3.4. *Let l_1 and l_2 denote two left translation invariant spaces whose intersection is the zero function. Then the distance of $(2 \operatorname{Im} \lambda)^{\frac{1}{2}} e^{i\lambda x}$ from $l_1 \oplus l_2$ is the product of its distances from l_1 and from l_2.*

There seems to be no obvious geometric interpretation of this result.

We turn now to the ring of analytic functions bounded in the upper half-plane. We shall denote elements of this ring by C, possibly subscribed by some index.

It follows from the Paley–Wiener theorem that for any function C, $CH \subset H$. Furthermore since $e^{izs} H \subset H$, also $e^{izs} CH \subset CH$ for $s > 0$, i.e., CH is the Fourier transform of a right invariant subspace of h. Therefore the closure of CH is an R-space, and so can be represented as FH. So by construction $CH \subset FH$, i.e., multiplication by $F^{-1}C$ maps H into H; from this it follows, just as in the proof of Theorem 3.1, that $F^{-1}C = G$ is regular and bounded in the upper half-plane. So we have shown that every bounded analytic function has a unique factorization

$$C = FG.$$

In the terminology of Beurling, F is the inner factor, G the outer factor of C. It follows from our construction that if G is an outer factor, then GH is a *dense* subspace of H.

THEOREM 3.5. *The inner factor of $C_1 C_2$ is the product of the inner factors of C_1 and of C_2.*

Proof. We have to show that $C_1 C_2 H = F_1 F_2 H$. Clearly, $C_1 C_2 H = F_1 F_2 G_1 G_2 H$ is a subspace of $F_1 F_2 H$. Since G_1 and $G_2 H$ are outer factors, multiplication by them—and therefore by their product—maps H into a dense subset of H.

It follows from Theorem 3.5 that divisibility of two bounded analytic functions is equivalent to the divisibility of their inner and outer factors. Concerning divisibility by a n outer factor we have

THEOREM 3.6. *C is divisible by G if and only if CG^{-1} is bounded on the real axis.*

Proof. We have to show that if CG^{-1} is bounded on the real axis, it is bounded in the upper half-plane. First we note that multiplication by CG^{-1} maps GH into H. Secondly, since CG^{-1} is bounded on the real axis, this operation is bounded and so can be extended to the closure of GH. Third, since G is an outer factor, the closure of GH is H. Therefore we have the result: *Multiplication by CG^{-1} maps H into H.* From this the boundedness of CG^{-1} in the upper half-plane can be deduced as in the proof of Theorem 3.1.

3.2. The Convolution Theorem. Let $c_1(x)$ and $c_2(x)$ be a pair of functions which are zero for x negative, in L_1 over the positive reals. Denote their convolution by $c(x)$:

$$c = c_1 * c_2 = \int c_1(y)\, c_2(x-y)\, dy. \tag{3.2}$$

Let d_1, d_2 and d be largest numbers such that the supports of $c_1(x)$, $c_2(x)$, $c(x)$ are contained in $x \geq d_1$, d_2, d respectively. Clearly, $d \geq d_1 + d_2$. What is much less obvious is the

CONVOLUTION THEOREM OF TITCHMARSH:

$$d = d_1 + d_2.$$

Proof. Denote the Fourier transforms of c_1, c_2, c by C_1, C_2, C. Taking the Fourier transform of (3.2) we conclude

$$C = C_1 C_2. \tag{3.3}$$

Since c_1 and c_2 are in L_1 over the positive reals, C_1, C_2, and C are bounded analytic functions in the upper half-plane. Denote their inner factors by F_1, F_2 and F. According to Theorem (3.5) it follows from (3.3) that

$$F = F_1 F_2. \tag{3.4}$$

Since the support of c is contained in $x \geq d$, it follows that e^{idz} divides C. Since e^{idz} is an inner factor, according to Theorem 3.5, it divides the inner factor of C:

$$F = e^{idz} F_3.$$

Combining this with (3.4) we get

$$e^{idz} F_3 = F_1 F_2.$$

According to the corollary of Theorem 3.3, F_4 being taken as e^{idz}, it follows that

$$e^{idz} = \tilde{F}_1 \tilde{F}_2$$

where $\tilde{F}_1$ divides F_1, $\tilde{F}_2$ divides F_2. We use now the

THEOREM: *The only factorization of e^{idz} in the ring of bounded analytic functions is the trivial one*

$$e^{idz} = e^{i\tilde{d}_1 z} e^{i\tilde{d}_2 z}, \qquad \tilde{d}_1 + \tilde{d}_2 = d, \qquad \tilde{d}_1, \tilde{d}_2 \geqslant 0.$$

For the sake of completeness we include a proof of this well-known result. Denote by $h(x,y)$ the function $-\log|\tilde{F}_1(z)|$. Since both $|\tilde{F}_1|$ and $|\tilde{F}_2|$ are $\leqslant 1$, it follows from the above factorization that

$$0 \leqslant h(x,y) \leqslant dy, \quad 0 \leqslant y.$$

I.e., h is a positive harmonic function in the upper half-plane which vanishes at the boundary. We shall show now that the only such functions are constant multiples of y. We continue h into the lower half by reflection, and represent $h(x,y)$ by the Poisson integral along a circle of radius R around the origin:

$$h(x,y) = \int\limits_0^\pi P(x,y,R,\theta)\, h\, d\,\theta;$$

P here is the difference between the values of the Poisson kernel at R,θ and $R, -\theta$. For y positive, P is positive and for large R it is asymptotically equal to $(2y\sin\theta)/R$. Since the integrand in the above representation is positive, it follows that $h(x,y)/h(x',y')$ is asymptotically—and thus actually—equal to y/y'.

We conclude that $\tilde{F}_1 = e^{i\tilde{d}_1 z}$, $\tilde{F}_2 = e^{i\tilde{d}_2 z}$. Since $\tilde{F}_1$ and $\tilde{F}_2$ divide F_1 and F_2, they also divide C_1 and C_2. But then according to the Paley–Wiener theorem (the L_1 variety) it follows that the supports of c_1 and c_2 are contained in $x \geqslant \tilde{d}_1$ and $x \geqslant \tilde{d}_2$. This shows that $d_1 \geqslant \tilde{d}_1$, $d_2 \geqslant \tilde{d}_2$, and so $d_1 + d_2 \geqslant \tilde{d}_1 + \tilde{d}_2 = d$. Combined with the trivial inequality $d \geqslant d_1 + d_2$, this yields the desired result $d = d_1 + d_2$.

Previous proofs of the convolution theorem such as the one by Dufresnoy [3] or Koosis [6], also make use of theorems on positive harmonic functions. For this reason the present proof cannot be called new. Its virtue lies in reducing the convolution theorem to the one about the factorization of e^{idz} swiftly and painlessly.

4. In this section we derive the representation theorem of p. 164 for translation invariant spaces of functions whose values lie in a finite-dimensional Hilbert space S.

The L_2 scalar product of two such functions f and g is defined as

$$\int\limits_0^\infty (f(x), g(x))\, d\,x,$$

where (f,g) denotes the scalar product in S.

The spaces h, l and r and their Fourier transforms are defined analogous to their old definitions.

As before the proof proceeds by projecting $e^{i\lambda x}u$, where u is an arbitrary element of S, into r:

$$e^{i\lambda x}u = a_\lambda(x) + b_\lambda(x), \tag{4.1}$$

a_λ in l, b_λ in r. The value of $b_\lambda(x)$ depends on u, and this dependence is linear; therefore we can write

$$b_\lambda(x) = b_\lambda(x)u,$$

where now $b_\lambda(x)$ denotes an *operator* mapping S into itself. It is easy to show, on account of the finite-dimensionality of S, that $\|b_\lambda(x)\|$, the operator norm of $b_\lambda(x)$, is square integrable.

Making this change also in the meaning of a_λ, (4.1) can be rewritten as

$$e^{i\lambda x}u = a_\lambda(x)u + b_\lambda(x)u. \tag{4.1'}$$

Let v denote any element of S. Take the scalar product of (4.1') with $b_\mu(x-s)v$, s non-negative, and integrate with respect to x. The resulting expression is an analogue of (2.2) and can be transformed, by Parseval's theorem, into

$$e^{-i\lambda^*s}(B_\mu(-\lambda^*)v, u) = \int_{-\infty}^{\infty}(B_\mu(z)v, B_\lambda(z)u)e^{izs}\,dz. \tag{4.3}$$

Take the complex conjugate of (4.3), interchange the role of λ and μ and of u and v and write $-s$ for s. We get the analogue of (2.3); from this and from (4.3) we can determine $(B_\mu(z)v, B_\lambda(z)u)$ for real z by Fourier inversion:

$$(B_\mu(z)v, B_\lambda(z)u) = \frac{i(\lambda^*-\mu)}{(\mu+z)(\lambda^*+z)}(B_{\lambda\mu}v, u), \tag{4.4}$$

where $B_{\lambda\mu}$ abbreviates $B_\lambda^*(-\mu^*)$. The left side of (4.4) can be written as $(B_\lambda^*(z)B_\mu(z)v, u)$; since (4.4) holds for all vectors u and v in S, we conclude that

$$B_\lambda^*(z)B_\mu(z) = \frac{i(\lambda^*-\mu)}{(\mu+z)(\lambda^*+z)}B_{\lambda\mu}. \tag{4.5}$$

Similarly, by setting $s=0$ in (4.3) and transforming the right side by shifting the operator $B_\lambda(z)$ we obtain

$$B_{\lambda\mu} = \int_{-\infty}^{\infty}B_\lambda^*(z)B_\mu(z)\,dz. \tag{4.6}$$

Setting $\lambda=\mu$ we obtain $\qquad B_{\lambda\lambda} = \int_{-\infty}^{\infty}B_\lambda^*(z)B_\lambda(z)\,dz, \tag{4.6'}$

which shows that $B_{\lambda\lambda}$ is a symmetric, non-negative operator.

The *rank* of $B_{\lambda\lambda}$ has some maximum p as λ varies in the upper half-plane. For any value of λ, the nullspace of $B_{\lambda\lambda}$ is then at least $(n-p)$-dimensional, n denoting the dimension of S. Let u be a vector annihilated by $B_{\lambda\lambda}$. According to formula (4.6′),

$$0 = (B_{\lambda\lambda}\, u,\, u) = \int \| B_{\lambda}\,(z)\, u \|^2\, d\,z;$$

by Parseval's relation, we have then

$$\int\limits_0^\infty \| b_\lambda\,(x)\, u \|^2\, d\,x = 0.$$

i.e., $b_\lambda(x)\,u \equiv 0$. Since $b_\lambda u$ is the projection of $e^{i\lambda x}\,u$ into r, we have

LEMMA 4.1. *If u is annihilated by $B_{\lambda\lambda}$, $e^{i\lambda x}\,u$ is orthogonal to r.*

Let $g(x)$ be an arbitrary element of r, $G(z)$ its Fourier transform. According to Lemma 4.1

$$0 = \int\limits_0^\infty (g\,(x),\ e^{i\lambda x}\,u)\, d\,x = (G\,(-\lambda^*),\, u).$$

In other words, the value of any function G in R at $-\lambda^*$ is orthogonal to the nullspace of $B_{\lambda\lambda}$. Since the dimension of this nullspace is at least $n-p$, we have

LEMMA 4.2. *At every point of the upper half-plane, the values of the functions in R lie in a p-dimensional linear subspace of S.*

By a passage to the limit we can deduce the following

COROLLARY. *Let $G_1(z), \ldots, G_k(z)$ be a finite set of functions in R; then for almost all z on the real axis their values lie in a p-dimensional subspace of S.*

Denote by λ a value where the rank of $B_{\lambda\lambda}$ is maximal. Putting $\mu = \lambda$ in (4.5) we get for real z

$$B_\lambda^*\,(z)\, B_\lambda\,(z) = \frac{2\,\mathrm{Im}\,\{\lambda\, B_{\lambda\lambda}\}}{|z + \lambda|^2}.$$

Denote the non-negative square root of $2\,\mathrm{Im}\,\{\lambda\, B_{\lambda\lambda}\}$ by G. Since the nullspace of $B_\lambda(z)$ includes that of G, there exists an operator $F(z)$ such that

$$B_\lambda\,(z) = \frac{F\,(z)\, G}{\lambda + z}. \tag{4.7}$$

$F(z)$ is defined on the range T of G only, and is an isometry there (see e.g. [14], p. 283).

Formula (4.7) serves to extend $F(z)$ as a regular analytic function to the upper half-plane. As in Section 2, it follows that $F(z)$ is bounded there. [1]

Since, for z real, $F(z)$ is an isometry on T, we have

$$F^*(z) F(z) = I. \tag{4.8}$$

Substitute (4.7) into (4.5) and into the expression $B_{\lambda\mu} = B_\lambda^*(-\mu^*)$. We obtain the relation

$$G^* F^*(z) B_\mu(z) = i G^* \frac{F^*(-\mu^*)}{\mu + z}, \tag{4.9}$$

where G is regarded as mapping S into T. Since G is a non-negative hermitean operator $G^* = G$ does not annihilate any element of T and so can be cancelled from both sides of (4.9):

$$F^*(z) B_\mu(z) = i \frac{F^*(-\mu^*)}{\mu + z}. \tag{4.10}$$

Using (4.8), (4.10) can be written as

$$F^*(z) \left\{ B_\mu(z) - i \frac{F(z) F^*(-\mu^*)}{\mu + z} \right\} = 0. \tag{4.11}$$

Let $u_1, u_2, \ldots, u_n$ be a set of n elements spanning S. The functions $B_\lambda(z) u_j, j = 1, \ldots, n$ belong to R and, as formula (4.7) shows, they span the range of $F(z)$. By our choice of λ the range of $F(z)$ has the maximal dimension p. According to the corollary of Lemma 4.2 the range of $B_\mu(z)$ belongs to the range of $F(z)$. But F^* does not annihilate any element on the range of F; therefore the factor $F^*(z)$ can be dropped on the left in (4.11), leaving

$$B_\mu(z) = i \frac{F(z) F^*(-\mu^*)}{\mu + z}. \tag{4.12}$$

There remains to be shown that the space $R' = FH$ is R itself. We shall show this as before by verifying that the orthogonal projection of $e^{i\mu x} u$ into r' is $b_\mu(x) u$, or what is the same, that the projection of $iu/(\mu + z)$ into R' is $B_\mu(z) u$, with B_μ given by formula (4.12). This means that

$$\frac{I - F(z) F^*(-\mu^*)}{\mu + z} u$$

is orthogonal to FH for all u, i.e., that

$$F^*(z) \frac{I - F(z) F^*(-\mu^*)}{\mu + z} u$$

belongs to H_T^*. Using (4.8), this last expression can be rewritten, for z real, as

---

[1] One can show the boundedness of $(F(z) v, u)$ for all vectors u, v in T.

$$\frac{F^*(z^*) - F^*(-\mu^*)}{\mu + z}.$$

This function can be continued as a regular analytic function into the lower half-plane and its square integral along any line parallel to the real axis is uniformly bounded. Therefore, by the Paley–Wiener theorem, it belongs to H^*. This completes the proof of the representation theorem.

Those parts of the division theory developed in Section 3 which do not use commutativity remain valid in the vector-valued case. In particular, every matrix-valued analytic function can be written as the product of an inner and an outer factor, in this order. Even further splitting of inner factors F is possible. Take the Fourier inverse r of $R = FH$ and form its orthogonal complement l. Take the set of all exponential polynomials contained in l and form their orthogonal complement r'. r' is a closed invariant space, and it contains r. Its Fourier transform R' then contains R; according to the representation theorem, R' is of the form BH. According to Theorem 3.1, F is divisible by B on the left:

$$F = BE.$$

In the scalar case, B is a Blaschke product and E the exponential of the Poisson integral with respect to a singular measure. Just how useful this factorization is in the matrix case remains to be seen.

It is already known through the researches of Wiener [17], Wiener and Masani [18], and Helson and Lowdenslager [19], that square matrix valued analytic functions whose determinant does not vanish identically can be written as products of an inner and outer factor.

In [20], Potapov shows that bounded analytic matrix functions with determinant $\neq 0$ can be factored as a Blaschke product times a multiplicative integral of the exponential of the Poisson kernel.

References

[1]. BEURLING, A., On two problems concerning linear transformations in Hilbert space. *Acta Math.*, 81 (1948), 239–255.

[2]. BOAS, R. P., *Entire Functions*. Academic Press, New York, 1954.

[3]. DUFRESNOY, J., Sur le produit de composition de deux fonctions. *C. R. Acad. Sci. Paris*, 225 (1947), 857–859.

[4]. HERGLOTZ, G., Über Potenzreihen mit positivem, reellem Teil im Einheitskreis. *Ber. Verh. Sächs. Akad. Wiss. Leipzig*, 63 (1911), 501–511.

[5]. KARHUNEN, K., Über die Struktur stationärer zufälliger Funktionen. *Ark. Mat.*, 1 (1950), 141–160.

[6]. Koosis, P., Functions mean periodic on a half-line. *Comm. Pure Appl. Math.*, 10 (1957), 133–149.

[7]. ——, Interior compact spaces of functions on a half-line. *Comm. Pure Appl. Math.*, 10 (1957), 583–615.

[8]. Lax, P. D., A Phragmen–Lindelöf theorem in harmonic analysis and its application to some questions in the theory of elliptic equations. *Comm. Pure Appl. Math.*, 10 (1957), 361–389.

[9]. ——, Remarks on the preceding paper. *Comm. Pure Appl. Math.*, 10 (1957), 617–622.

[10]. Müntz, Ch., Über den Approximationssatz von Weierstrass. *H. A. Festschrift*, Berlin, 1914.

[11]. Nevanlinna, R., *Eindeutige analytische Funktionen*, 2nd Edit. Springer, Berlin, 1953.

[12]. Nyman, B., *On the one-dimensional translation group and semigroup in certain function spaces*. Thesis, Uppsala, 1950.

[13]. Paley, R. E. A. C. & Wiener, N., *Fourier Transforms in the Complex Domain*. A.M.S. Colloquium Publications, New York, 1934.

[14]. Riesz, F. & Sz.–Nagy, B., *Leçons d'analyse fonctionnelle*. Akademiai Kiado, Budapest, 1952.

[15]. Szász, O., Über die Approximation stetiger Funktionen durch lineare Aggregate von Potenzen. *Math. Ann.*, 77 (1916), 382–396.

[16]. Titchmarsh, E. C., *Introduction to the Theory of Fourier Integrals*. Clarendon Press, Oxford, 1937.

[17]. Wiener, N., On the factorization of matrices. *Comment. Math. Helv.*, 29 (1955), 97–111.

[18]. Wiener, N. & Masani, P., The prediction theory of multivariate stochastic processes. *Acta Math.*, 98 (1957), 111–150, 99 (1958), 93–137.

[19]. Helson, H. & Lowdenslager, D., Prediction theory and Fourier series in several variables. *Acta Math.*. 99 (1958), 165–202.

[20]. Potapov, V. L., On matrix functions holomorphic and bounded in the unit circle. *Doklady Akad. Nauk SSSR*, 72 (1950), 849–852.

The Time Delay Operator and a Related Trace Formula[†]

PETER D. LAX

Courant Institute of Mathematical Sciences,
New York University,
New York, New York

AND

RALPH S. PHILLIPS

Department of Mathematics
Stanford University
Stanford California

DEDICATED TO MARK G. KREIN ON THE OCCASION OF HIS SEVENTIETH BIRTHDAY

The time delay operator was introduced into quantum mechanics by Eisenbud [2] and Wigner [8]; see also Smith [7] and Jauch and Marchand [3]. In this paper we define the time delay operator for classical wave equations in terms of the Lax–Phillips scattering theory and show its relation to the scattering matrix.

In Section 3 we derive a trace formula which appears to be related to the time delay operator. This formula is an extension to the classical case of one derived by the authors in [5] for the automorphic wave equation.

1. INTRODUCTION

In this note we study one-parameter groups $U(t)$ of unitary operators in a separable Hilbert space $\mathcal{H}$. We think of $U(t)$ as describing the evolution of a physical system in time; that is, if the initial state of the system is f, its state at time t is $U(t)f$. We take the classical rather then the quantum mechanical view, regarding $\|f\|^2$ as the total energy of the state f.

Let $\mathcal{K}$ be some subspace of $\mathcal{H}$ and K the orthogonal projection of $\mathcal{H}$ onto $\mathcal{K}$. The component of f in $\mathcal{K}$ is then Kf, and the energy contained in this component is $\|Kf\|^2$.

The quantity

$$\int_{-\infty}^{\infty} \|K U(t)f\|^2 \, dt \tag{1.1}$$

[†] This work was supported in part by the National Science Foundation under grant numbers MCS76–07039 and MCS76–07289.

197

measures the energy of the $\mathscr{K}$-component of $U(t)f$ throughout its whole history. Since K is an orthogonal projection,

$$K^* = K, \qquad K^2 = K,$$

and since $U(t)$ is a unitary group,

$$U(-t) = U^*(t).$$

Using these facts we deduce that

$$\|KU(t)f\|^2 = (KU(t)f, KU(t)f) = (U(-t)KU(t)f, f).$$

Substituting this above we can rewrite (1.1) as

$$(Tf, f) \tag{1.2}$$

where

$$T = \int_{-\infty}^{\infty} U(-t)KU(t)\,dt; \tag{1.3}$$

the integral is to be taken in the strong sense. T is called the *time delay operator of the subspace* $\mathscr{K}$.

It is convenient to interpret the integral (1.3) in the sense of some summability such as

$$Tf = \lim_{\lambda \to 0} T_\lambda f, \tag{1.4}$$

where

$$T_\lambda = \int_{-\infty}^{\infty} e^{-\lambda|t|}U(-t)KU(t)\,dt. \tag{1.5}$$

The domain of T consists of those f for which the limit in (1.4) exists. If this domain is dense, then T commutes with $U(s)$. For

$$T_\lambda U(s) = \int_{-\infty}^{\infty} e^{-\lambda|t|}U(-t)KU(t+s)\,dt = U(s)T_\lambda',$$

where

$$T_\lambda' = e^{\lambda s}\int_s^{\infty} e^{-\lambda r}U(-r)KU(r)\,ds + e^{-\lambda s}\int_{-\infty}^s e^{\lambda r}U(-r)KU(r)\,ds.$$

If f is in the domain of T, then $T_\lambda'f$ converges with $T_\lambda f$. Using the fact that the integral in (1.1) converges it is easy to prove that $\lim(T_\lambda'f, f) = (Tf, f)$ and by polarization that weak $\lim T_\lambda'f = Tf$. It then follows that

$$U(s)T = TU(s). \tag{1.6}$$

It is helpful to look at T in the spectral representation for U; that is, take $\mathscr{H}$ to be $L_2(\mathbb{R}, \mathscr{N})$, consisting of functions $f(\sigma)$ which take their values in some auxiliary Hilbert space $\mathscr{N}$, where the action of $U(t)$ is multiplication by $e^{i\sigma t}$.

If we suppose that in this representation K can be written as an integral operator with a continuous Hermitian symmetric kernel $q(\sigma, \rho)$,

$$(Kf)(\sigma) = \int q(\sigma, \rho) f(\rho) \, d\rho, \tag{1.7}$$

then

$$(U(-t)KU(t)f)(\sigma) = \int q(\sigma, \rho) e^{i(\rho - \sigma)t} f(\rho) \, d\rho. \tag{1.8}$$

We take f to have bounded support. Multiplying (1.8) by $e^{-\lambda|t|}$ and integrating with respect to t we get

$$T_\lambda f = 2 \int q(\sigma, \rho) \frac{\lambda}{\lambda^2 + (\rho - \sigma)^2} f(\sigma) \, d\sigma.$$

Clearly,

$$\lim_{\lambda \to 0} (T_\lambda f)(\sigma) = 2\pi q(\sigma, \sigma) f(\sigma). \tag{1.9}$$

So, in this representation T is multiplication by $2\pi q(\sigma, \sigma)$ and every f with bounded support belongs to the domain of T. Thus T is defined on a dense domain and since it corresponds to a multiplicative operator in the spectral representation it is essentially self-adjoint.

The choice of the space $\mathscr{K}$ is dictated by physical considerations. In the quantum mechanical case, where $U(t) = \exp(itH)$, H the Hamiltonian, $|f(x)|^2$ is interpreted, for $\|f\| = 1$, as a probability density, and (Kf, f) is the probability of finding the system in $\mathscr{K}$; so (Tf, f) can be interpreted as the expected time that the system spends in the subspace $\mathscr{K}$. Eisenbud [2] and Wigner [8] have chosen $\mathscr{K}$ to consist of functions $f(x)$ whose support lies in some ball $|x| \leqslant R$. In this paper we study the classical analog of the quantum mechanical case. Here the appropriate choice for $\mathscr{K}$ is suggested by the scattering theory developed by the authors [4]. In this theory, which will be outlined below, the Hilbert space is decomposed into three orthogonal subspaces: an incoming subspace $\mathscr{D}_-$, an outgoing subspace $\mathscr{D}_+$, and the interacting subspace $\mathscr{K} = \mathscr{H} \ominus (\mathscr{D}_- \oplus \mathscr{D}_+)$. K is orthogonal projection onto $\mathscr{K}$, and the quantity (1.9) is the total energy of interaction of the wave with the scatterer.

In the Lax–Phillips scattering theory the incoming and outgoing subspaces satisfy the following properties:

$$U(t)\mathscr{D}_- \subset \mathscr{D}_- \qquad \text{for} \quad t \leqslant 0,$$
$$U(t)\mathscr{D}_+ \subset \mathscr{D}_+ \qquad \text{for} \quad t \geqslant 0; \tag{1.10i}$$

$$\wedge U(t)\mathscr{D}_- = 0 = \wedge U(t)\mathscr{D}_+; \tag{1.10ii}$$

$$\overline{\vee U(t)\mathscr{D}_-} = \mathscr{H} = \overline{\vee U(t)\mathscr{D}_+}; \tag{1.10iii}$$

$$\mathscr{D}_- \perp \mathscr{D}_+. \tag{1.10iv}$$

Associated with $\mathscr{D}_-$ and $\mathscr{D}_+$ are two unitary translation representations T_- and T_+ of U onto $L_2(\mathbb{R}, \mathscr{N})$,

$$T_\pm : f \in \mathscr{H} \to k_\pm(s) \in L_2(\mathbb{R}, \mathscr{N}), \tag{1.11}$$

taking $\mathscr{D}_-, \mathscr{D}_+$ onto $L_2(\mathbb{R}_-, \mathscr{N})$, $L_2(\mathbb{R}_+, \mathscr{N})$, respectively, with the action of $U(t)$ going into right translation:

$$U_t : k(s) \to k(s - t) \tag{1.12}$$

(here $\mathscr{N}$ is an auxiliary Hilbert space). The incoming and outgoing spectral representations of U are obtained by Fourier transforming the translation representations:

$$\tilde{k}_\pm(\sigma) = (1/\sqrt{2\pi}) \int e^{i\sigma s} k_\pm(s)\, ds. \tag{1.13}$$

The scattering operator S is defined as the mapping

$$S : k_- = T_- f \to k_+ = T_+ f. \tag{1.14}$$

It is easy to show that S is unitary on $L_2(\mathbb{R}, \mathscr{N})$, commutes with translation U_t, and, in the presence of (iv), is causal, that is,

$$SL_2(\mathbb{R}_-, \mathscr{N}) \subset L_2(\mathbb{R}_-, \mathscr{N}). \tag{1.15}$$

Thus S can be realized as a distribution-valued convolution operator with support in $\mathbb{R}_-$. The spectral representer of S, called the scattering matrix and denoted by $\mathscr{S}$, is a multiplicative operator, that is,

$$\mathscr{S} : \tilde{k}_-(\sigma) \to \tilde{k}_+(\sigma) = \mathscr{S}(\sigma)\tilde{k}_-(\sigma). \tag{1.16}$$

Moreover, $\mathscr{S}(\sigma)$ is unitary on $\mathscr{N}$ for each σ in $\mathbb{R}$ and is the boundary value for an operator-valued function $\mathscr{S}(z)$, holomorphic in the lower half-plane, $\mathrm{Im}\, z < 0$, where $\mathscr{S}(z)$ is a contraction operator on $\mathscr{N}$.

We denote the orthogonal projections of $\mathscr{H}$ onto $\mathscr{D}_-^\perp$ and $\mathscr{D}_+^\perp$ by P_- and P_+, respectively. The operators

$$Z(t) = P_+ U(t) P_-, \qquad t \geqslant 0, \tag{1.17}$$

annihilate $\mathscr{D}_- \oplus \mathscr{D}_+$ and form a semigroup of contraction operators on $\mathscr{H}$. This semigroup of operators plays an important role in the theory. For instance, there is a one-to-one correspondence between the points λ in the spectrum of the infinitesimal generator B of Z and the points z in the lower half-plane at which $\mathscr{S}(z)$ is not invertible; this correspondence is $\lambda = -iz$.

The time delay operator T for $K = P_- P_+$ and the operators Z, being built of the same ingredients, are related. One of these relations can be seen by taking f to belong to $\mathscr{K}$; then

$$f = P_- f = P_+ f.$$

Furthermore, by (1.10i) $U(t)$ maps $\mathscr{D}_-^\perp$ into itself for $t \geqslant 0$ and $\mathscr{D}_+^\perp$ into itself for $t \leqslant 0$. Hence, for f in $\mathscr{K}$

$$(Tf, f) = \int_{-\infty}^{\infty} \|P_- P_+ U(t)f\|^2 \, dt = \int_{-\infty}^{0} [\|P_- U(t)f\|^2 + \int_0^{\infty} \|P_+ U(t)f\|^2] \, dt$$

$$= \int_0^{\infty} \|Z^*(t)f\|^2 \, dt + \int_0^{\infty} \|Z(t)f\|^2 \, dt. \tag{1.18}$$

Another relation is exhibited by some explicit formulas for T and Z. In Section 2 we shall show that in the outgoing spectral representation T is represented by

$$T \sim -i\mathscr{S}(\sigma)\,\partial_\sigma \mathscr{S}^*(\sigma). \tag{1.19}$$

In Section 3 we shall show that for any ψ of class $C_0^\infty(\mathbb{R}_+)$ for which

$$C = \int_0^{\infty} \psi(t)Z(t) \, dt \tag{1.20}$$

is of trace class,

$$\operatorname{tr} C = (-i/\sqrt{2\pi}) \int \Psi(\sigma) \operatorname{tr}[\mathscr{S}(\sigma)\partial_\sigma \mathscr{S}^*(\sigma)] \, d\sigma; \tag{1.21}$$

here Ψ is the Fourier transform of ψ. Formula (1.21) is reminiscent of the well-known trace formula for self-adjoint operators due to Birman and Krein [1].

In Section 3 we also show how to apply this theory to the wave equation in the exterior of a star-shaped obstacle with Dirichlet boundary conditions.

2. THE TIME DELAY OPERATOR

We assume—and this is satisfied in all cases of physical interest—that the scattering matrix $\mathscr{S}(\sigma)$ introduced in (1.16) has the following additional properties:

(i) $\mathscr{S}(\sigma)$ is analytic at every point of the real axis.

(ii) $|\mathscr{S}(z)| \leqslant \text{const}/|\operatorname{Im} z|$ for z in the lower half-plane.   (2.1)

With $\mathscr{D}_-$ and $\mathscr{D}_+$ defined as in Section 1, we set

$$\mathscr{K} = \mathscr{H} \ominus (\mathscr{D}_- \ominus \mathscr{D}_+).$$

Since by (1.10iv) the subspaces $\mathscr{D}_-$ and $\mathscr{D}_+$ are orthogonal, the projection K onto $\mathscr{K}$ is

$$K = P_- P_+, \tag{2.2}$$

where P_- and P_+ are the orthogonal projections onto $\mathscr{D}_-^\perp$ and $\mathscr{D}_+^\perp$.

We shall determine the time-delay operator T associated with $\mathcal{K}$. To be able to apply formula (1.9) we shall represent K as an integral operator in the, say, outgoing spectral representation. Let f be any element of $\mathcal{H}$, $f(\sigma)$ its outgoing spectral representer:

$$f \underset{+}{\leftrightarrow} f(\sigma).$$

In the outgoing translation representation P_+ acts as multiplication by h_+:

$$h_+(s) = \begin{cases} 1 & \text{for} \quad s < 0 \\ 0 & \text{for} \quad s > 0. \end{cases} \tag{2.3}_+$$

Therefore, in the outgoing spectral representation P_+ acts as convolution with H_+, the Fourier transform of h_+:

$$H_+(\sigma) = (-i/2\pi\sigma) + \tfrac{1}{2}\delta(\sigma), \tag{2.4}_+$$

$$P_+ f \underset{+}{\leftrightarrow} \int H_+(\rho - \sigma) f(\sigma) \, d\sigma.$$

To determine the action of P_- we first have to switch to the incoming representation; this is accomplished by multiplying with $\mathscr{S}^*(\rho)$:

$$P_+ f \leftrightarrow \mathscr{S}^*(\rho) \int H_+(\rho - \sigma) f(\sigma) \, d\sigma.$$

The action of P_- in this representation is convolution with H_-, the Fourier transform of h_-:

$$h_-(s) = \begin{cases} 0 & \text{for} \quad s < 0 \\ 1 & \text{for} \quad s > 0 \end{cases} \tag{2.3}_-$$

$$H_-(\sigma) = (i/2\pi\sigma) + \tfrac{1}{2}\delta(\sigma), \tag{2.4}_-$$

$$P_- P_+ f \leftrightarrow \int H_-(\tau - \rho) \mathscr{S}^*(\rho) \int H_+(\rho - \sigma) f(\sigma) \, d\sigma \, d\rho.$$

Finally, we switch back to the outgoing representation through multiplication by $\mathscr{S}(\tau)$:

$$P_- P_+ f \leftrightarrow \mathscr{S}(\tau) \iint H_-(\tau - \rho) \mathscr{S}^*(\rho) H_+(\rho - \sigma) f(\sigma) \, d\sigma \, d\rho. \tag{2.5}$$

(2.5) represents $P_- P_+$ as an integral operator, with kernel

$$q(\tau, \sigma) = \mathscr{S}(\tau) \int H_-(\tau - \rho) H_+(\rho - \sigma) \mathscr{S}^*(\rho) \, d\rho. \tag{2.6}$$

Using formulas (2.4)$_\pm$ for $H_\pm$ we get

$$q(\tau, \sigma) = \mathscr{S}(\tau) \int \left[\frac{i}{2\pi(\tau - \rho)} + \tfrac{1}{2}\delta(\tau - \rho) \right] \left[\frac{-i}{2\pi(\rho - \sigma)} + \tfrac{1}{2}\delta(\rho - \sigma) \right] \mathscr{S}^*(\rho) \, d\rho. \tag{2.7}$$

We take $\tau \neq \sigma$; then $\delta(\tau - \rho)\delta(\rho - \sigma) = 0$. Using this and the partial fraction decomposition

$$\frac{1}{(\tau - \rho)}\frac{1}{(\rho - \sigma)} = \frac{1}{\tau - \sigma}\left(\frac{1}{\tau - \rho} + \frac{1}{\rho - \sigma}\right) \tag{2.8}$$

in (2.7) we get

$$q(\tau, \sigma) = \mathscr{S}(\tau)\int\left[\frac{1}{(2\pi)^2(\tau - \sigma)}\left(\frac{1}{\tau - \rho} + \frac{1}{\rho - \sigma}\right)\right.$$

$$\left. + \frac{i}{4\pi}\frac{\delta(\rho - \sigma)}{\tau - \rho} - \frac{i}{4\pi}\frac{\delta(\rho - \tau)}{\rho - \sigma}\right]\mathscr{S}^*(\rho)\,d\rho. \tag{2.9}$$

Next we use the following consequence of Cauchy's formula:

Let $g_+(z)$ be analytic in the upper half-plane, and

$$|g_+(z)| \leqslant \frac{\text{const}}{\text{Im } z}, \qquad \text{Im } z > 0;$$

then

$$\int\frac{g_+(\rho)}{\rho - \sigma}\,d\rho = \pi i g_+(\sigma). \tag{2.10$_+$}$$

Similarly, for $g_-(z)$ analytic in the lower half-plane and satisfying there

$$|g_-(z)| \leqslant \frac{\text{const}}{|\text{Im } z|}, \qquad \text{Im } z < 0,$$

$$\int\frac{g_-(\rho)}{\rho - \sigma}\,d\rho = -\pi i g_-(\sigma). \tag{2.10$_-$}$$

Using (2.10)$_+$ in (2.9) with $g_+(\rho) = \mathscr{S}^*(\rho)$ we get

$$q(\tau, \sigma) = \frac{i\mathscr{S}(\tau)}{4\pi}\left[\frac{1}{\tau - \sigma}(\mathscr{S}^*(\sigma) - \mathscr{S}^*(\tau)) + \frac{\mathscr{S}^*(\sigma)}{\tau - \sigma} - \frac{\mathscr{S}^*(\tau)}{\tau - \sigma}\right]$$

$$= \frac{-i}{2\pi}\mathscr{S}(\tau)\frac{\mathscr{S}^*(\tau) - \mathscr{S}^*(\sigma)}{\tau - \sigma}. \tag{2.11}$$

We claim that the limiting form of (2.11) as σ tends to τ is valid on the diagonal; in other words, that

$$q(\sigma, \sigma) = (-i/2\pi)\,\mathscr{S}(\sigma)\,\partial_\sigma\mathscr{S}^*(\sigma). \tag{2.12}$$

To prove this we shall verify that the integral operator whose kernel g is given by (2.11) and (2.12) is indeed the projection onto $\mathscr{K}$. We have to verify that

$$\int q(\tau, \sigma) f(\sigma) d\sigma = 0$$

when f represents an element of $\mathscr{D}_+$ or $\mathscr{D}_-$, and that

$$\int q(\tau, \sigma) f(\sigma) d\sigma = f(\tau)$$

when f represents an element of $\mathscr{K}$.

From (2.11),

$$\int q(\tau, \sigma) f(\sigma) d\sigma = \frac{i}{2\pi} \mathscr{S}(\tau) \int \frac{\mathscr{S}^*(\sigma) - \mathscr{S}^*(\tau)}{\tau - \sigma} f(\sigma) d\sigma. \qquad (2.13)$$

An $f(\sigma)$ representing an element of $\mathscr{D}_+$ is analytic in the upper half-plane. The integrand in (2.13) is an analytic function of σ in the upper half-plane, so by the Cauchy integral theorem we can shift the line of integration to $z = \sigma + i\mu$, $\mu > 0$. Estimating the resulting integral by Schwarz's inequality shows that the integral (2.13) is zero.

Because of the unitarity of $\mathscr{S}$, we can rewrite (2.13) in the form

$$\int q(\tau, \sigma) f(\sigma) d\sigma = \frac{i}{2\pi} \int \frac{\mathscr{S}(\tau) - \mathscr{S}(\sigma)}{\tau - \sigma} \mathscr{S}^*(\sigma) f(\sigma) d\sigma. \qquad (2.13')$$

Now if $f(\sigma)$ represents an element of $\mathscr{D}_-$, $\mathscr{S}^*(\sigma) f(\sigma)$ is analytic in the lower half-plane. In this case the integrand is analytic in the lower half-plane, so again the same argument shows that the integral is zero.

We rewrite (2.13) as

$$\int q(\tau, \sigma) f(\sigma) d\sigma = \frac{i}{2\pi} \mathscr{S}(\tau) \int \frac{\mathscr{S}^*(\sigma) f(\sigma) d\sigma}{\tau - \sigma} - \frac{i}{2\pi} \int \frac{f(\sigma) d\sigma}{\tau - \sigma}. \qquad (2.13'')$$

Now when $f(\sigma)$ represents an element of $\mathscr{K}$, then $f(\sigma)$ is analytic in the lower half-plane and $\mathscr{S}^*(\sigma) f(\sigma)$ is analytic in the upper half-plane. Applying $(2.10)_+$ to the first term in (2.13'') with $g_+ = \mathscr{S}^* f$ and $(2.10)_-$ to the second term with $g_- = f$ we get that

$$\int q(\tau, \sigma) f(\sigma) d\sigma = \tfrac{1}{2} \mathscr{S}(\tau) \mathscr{S}^*(\tau) f(\tau) + \tfrac{1}{2} f(\tau) = f(\tau).$$

This completes the verification that the integral operator whose kernel q is given by formulas (2.11) and (2.12) is indeed the projection onto $\mathscr{K}$.

We now apply formula (1.9) to our operator K to conclude

THEOREM 2.1. *Suppose $\mathscr{S}(\sigma)$ satisfies conditions (2.1). Then the time-delay operator T as defined by (1.5) is essentially self-adjoint and has in the* *outgoing spectral representation the form*

$$-i\mathscr{S}(\sigma)\,\partial_\sigma\mathscr{S}^*(\sigma). \tag{2.14}$$

From its definition, the operator T is nonnegative; therefore, the operator (2.14) is nonnegative for all σ. This is easy to deduce directly from the fact that $\mathscr{S}(z)$ is unitary for z real and a contraction for $\operatorname{Im} z \leqslant 0$; for then

$$|\mathscr{S}(\sigma - i\mu)f|^2 \leqslant |\mathscr{S}(\sigma)f|^2$$

for $\mu < 0$, which implies that

$$\frac{d}{d\mu}|\mathscr{S}(\sigma - i\mu)f|^2 \leqslant 0.$$

From this the nonnegativity of (2.14) is easily deduced.

The operator T represented as multiplication by (2.14) is a bounded operator if and only if $\partial_\sigma\mathscr{S}^*(\sigma)$ is uniformly bounded on the whole σ axis. Whether this is the case or not depends on the location of the poles of $\mathscr{S}(z)$ in the upper half-plane. When these poles approach the real axis, $\partial_\sigma\mathscr{S}(\sigma)$ cannot be uniformly bounded; on the other hand, if $\mathscr{S}(\sigma)$ is holomorphic and bounded in a strip $\operatorname{Im} z < \beta, \beta > 0$, then $\partial_\sigma\mathscr{S}(\sigma)$ is bounded on the whole real axis, and so the time delay operator T is bounded.

It is worth remarking that in the latter case the boundedness of T can be deduced directly from definition (1.5) or (1.3), without appealing to formula (2.14). We have shown in [4, p. 84] that $\mathscr{S}(z)$ is holomorphic and bounded in a strip above the real axis if and only if $Z(t)$ decays exponentially:

$$\|Z(t)\| \leqslant \operatorname{const} e^{-\alpha t}, \tag{2.15}$$

α any positive number. To deduce the boundedness of T from (2.15), we decompose $\mathscr{D}_-$ and $\mathscr{D}_+$ as follows:

$$\mathscr{D}_- = \bigcup_{j<0} \mathscr{D}_-^{(j)}, \qquad \mathscr{D}_+ = \bigcup_{j>0} \mathscr{D}_+^{(j)},$$

where

$$\mathscr{D}_+^{(j)} = U(j-1)\mathscr{D}_+ \ominus U(j)\mathscr{D}_+, \qquad j = 1, 2, \ldots,$$

$$D_-^{(j)} = U(1+j)\mathscr{D}_- \ominus U(j)\mathscr{D}_-, \qquad j = -1, -2, \ldots.$$

Every f in $\mathscr{H}$ can be decomposed as

$$f = \sum_{-\infty}^{\infty} f_j, \qquad f_0 \in \mathscr{K}, \qquad f_j \in \begin{cases} \mathscr{D}_+^{(j)} & \text{for } j > 0 \\ \mathscr{D}_-^{(j)} & \text{for } j < 0. \end{cases}$$

Since the decomposition is orthogonal,

$$\|f\|^2 = \sum \|f_j\|^2. \tag{2.16}$$

Since for $j < 0$, $f_j \in U(1 + j)\mathcal{D}_-$, it follows from (1.10i) that for $t < -j - 1$

$$U(t)f_j \in \mathcal{D}_-.$$

Hence,

$$P_- U(t)f_j = 0 \qquad \text{for} \quad t < -j - 1. \tag{2.17}$$

On the other hand, $U(-j)f_j$ is orthogonal to $\mathcal{D}_-$; so,

$$\begin{aligned} P_- U(-j)f_j &= U(-j)f_j, \\ U(t)f_j &= U(t + j)P_- U(-j)f_j \qquad \text{for} \quad t > -j. \end{aligned} \tag{2.17'}$$

Finally, $U(t)f_j \in \mathcal{D}_+$ for $t > 0$ and $j > 0$, so

$$P_+ U(t)f_j = 0 \qquad \text{for} \quad j > 0, \quad t > 0. \tag{2.17''}$$

Combining (2.17), (2.17'), and (2.17'') we deduce the following for $t > 0$:

$$\begin{aligned} P_+ P_- U(t)f &= \sum_{-\infty}^{\infty} P_+ P_- U(t)f_j \\ &= P_+ P_- U(t)f_{-[t]} + \sum_{-[t] < j \leq 0} Z(t + j)U(-j)f_j; \end{aligned} \tag{2.18}$$

here $[t]$ denotes the integer part of t. By the triangle inequality and using estimate (2.15) we get

$$\|P_+ P_- U(t)f\| \leq \|f_{-[t]}\| + \text{const} \sum_{-[t] < j \leq 0} e^{-\alpha(t + j)}\|f_j\|.$$

Using the Schwarz inequality yields

$$\|P_+ P_- U(t)f\|^2 \leq k\left(\|f_{-[t]}\|^2 + \text{const} \sum_{-[t] < j \leq 0} e^{-\alpha(t + j)}\|f_j\|^2\right), \tag{2.19}$$

where

$$k = (1 + \text{const} \sum_{[-t] < j} e^{-\alpha(t + j)});$$

so,

$$\int_0^\infty \|P_+ P_- U(t)f\|^2 \, dt \leq k \sum_{j < 0} \|f_j\|^2 + \frac{\text{const}}{\alpha} \sum_{j \leq 0} \|f_j\|^2 \leq k'\|f\|^2.$$

The integral from $-\infty$ to 0 is estimated similarly, using Z^* in place of Z. This shows that the quadratic form

$$(Tf, f) = \int_{-\infty}^\infty \|P_+ P_- U(t)f\|^2 \, dt$$

is bounded, and, hence, that T is a bounded operator.

3. A Related Trace Formula

We turn now to the proof of the trace formula (1.21) for operators of the type

$$C = \int_0^\infty \psi(t)Z(t)\,dt, \tag{3.1}$$

where ψ lies in $C_0^\infty(\mathbb{R}_+)$. We shall denote the trace class norm of an operator by the subscript 1 as in $\|\cdot\|_1$.

THEOREM 3.1.   *Suppose that the scattering matrix $\mathscr{S}$ is of the form*

$$\mathscr{S}(\sigma) = e^{-i\sigma\rho}\mathscr{S}_0(\sigma), \tag{3.2}$$

where

$$\mathscr{S}_0(\sigma) = I + K(\sigma) \tag{3.3}$$

and $\partial_\sigma K(\sigma)$ is an operator of trace class on $\mathscr{N}$ with $\|\partial_\sigma K(\sigma)\|_1$ being of polynomial growth. If, in addition, C is of trace class, then

$$\operatorname{tr} C = (-i/\sqrt{2\pi}) \int_{-\infty}^\infty \Psi(\sigma) \operatorname{tr}[\mathscr{S}_0(\sigma)\partial_\sigma\mathscr{S}_0^*(\sigma)]\,d\sigma. \tag{3.4}$$

Proof.   We first compute (Cf, f) in the outgoing translation representation. It will be convenient to work with all f in $\mathscr{D}_- \oplus \mathscr{K} = \mathscr{D}_+^{\perp}$ rather than in $\mathscr{K}$ alone; in other words, we consider all f with $T_+ f = k$ having support in $\mathbb{R}_-$. For such f and $t \geqslant 0$,

$$Z(t)f = P_+ U(t)P_- f.$$

As in the proof of Theorem 2.1, P_+ can be characterized in the outgoing translation representation as multiplication by h_+ and P_- as Sh_-S^*. We can therefore write

$$T_+ Z(t)f = h_+ U_t Sh_- S^* T_+ f, \tag{3.5}$$

where U_t denotes right translation by t units. Setting $k = T_+ f$ we therefore have

$$(Z(t)f, f) = (h_+ U_t Sh_- S^* k, k)_{L_2} = (h_- S^* k, S^* U_{-t} h_+ k)_{L_2}.$$

Since we are considering only k with support in $\mathbb{R}_-$, we can write this, at least formally, as

$$(Z(t)f, f) = \int_0^\infty du\left(\int_{-\infty}^\infty ds\, S^*(u-s)k(s), \int_{-\infty}^\infty dr\, S^*(u-r)k(r+t)\right)_{\mathscr{N}}; \tag{3.6}$$

the action of h_- is taken care of by the u range of integration. Finally, if we include ψ and the t integration this becomes

$$(Cf, f) = \int_0^\infty dt \int_0^\infty du \int_{-\infty}^\infty ds \int_{-\infty}^\infty dr \left(S^*(u - s)k(s), \overline{\psi}(t)S^*(u + t - r)k(r) \right)_{\mathcal{N}}.$$
(3.7)

The expression (3.7) can be rigorized if we replace the distribution-valued S^* by a mollified version. In this respect we treat the above two S^*'s differently. The one on the right in (3.7) can be thought of as already mollified by $\overline{\psi}(-t)$ and the resulting operator is

$$S_\psi^*(s) = \int_{-\infty}^\infty e^{-i\sigma s} \overline{\Psi(\sigma)} \mathscr{S}^*(\sigma) \, d\sigma.$$
(3.8)

Note that this requires an interchange in the order of the t and r integrations, but since ψ has compact support this is easily justified. We obtain a mollification for the S^* on the left by replacing C in the outgoing translation representation by

$$C_\epsilon k = C(\varphi_\epsilon * k),$$
(3.9)

where

$$\varphi_\epsilon(s) = (1/\epsilon)\varphi(s/\epsilon)$$

and φ in C_0^∞ is $\geqslant 0$ with $\int \varphi \, ds = 1$. Since we are treating C as an operator on $\mathbb{R}_-$ we should really write C, as $C(h_+(\varphi_\epsilon * k))$. However, this h_+ is superfluous since its action is nullified by that of P_+ on the left in $Z(t) = P_+ U(t)P_-$. Hence, convolution with φ_ϵ in (3.9) can be shifted over from k to the left S^* in (3.7). Denoting the so-mollified S^* by $S_\epsilon^*(s)$ and the Fourier transform of φ_ϵ by Φ_ϵ, we have

$$S_\epsilon^*(s) = \int e^{-i\sigma s} \Phi_\epsilon(\sigma) \mathscr{S}^*(\sigma) \, d\sigma.$$
(3.10)

We can now write

$$(C_\epsilon f, f) = \int_0^\infty du \int_{-\infty}^\infty ds \int_{-\infty}^\infty dr \left(S_\epsilon^*(u - s)k(s), S_\psi^*(u - r)k(r) \right)_{\mathcal{N}}.$$
(3.11)

Because of causality (1.15), the support of $S^*(s)$ is contained in $\mathbb{R}_+$. By choosing φ so that its support also lies in $\mathbb{R}_+$, the mollified $S_\epsilon^*(s)$ will inherit this property.

We note that

$$\text{is } S_\epsilon^*(s) = \int e^{-i\sigma s} \partial_\sigma(\Phi_\epsilon(\sigma) \mathscr{S}^*(\sigma)) \, d\sigma.$$

For any integer n we have the estimate

$$|\Phi_\epsilon(\sigma)|, |\partial_\sigma \Phi_\epsilon(\sigma)| \leqslant \text{const} (1 + \sigma^2)^{-n},$$

and since we have assumed that $\|\partial_\sigma \mathscr{S}^*(\sigma)\|_{\mathscr{N}}$ is of polynomial growth, it follows that $\|sS_\epsilon^*(s)\|_{\mathscr{N}}$ is bounded and square integrable. The same is trivially true of $\|S_\epsilon^*(s)\|_{\mathscr{N}}$. It follows from this that $\|S_\epsilon^*(s)\|_{\mathscr{N}}$ is integrable, and the same is true of $\|S_\psi^*(s)\|_{\mathscr{N}}$. The integrand in (3.11) is therefore absolutely integrable.

We note that C_ϵ is the product of a trace class operator with the bounded operator defined as convolution with φ_ϵ and, hence, C_ϵ is itself of trace class. Moreover, since convolution with φ_ϵ converges strongly to the identity it follows that

$$\lim_{\epsilon \to 0} \|C - C_\epsilon\|_1 = 0 \tag{3.12}$$

and, hence, that

$$\operatorname{tr} C = \lim_{\epsilon \to 0} \operatorname{tr} C_\epsilon. \tag{3.13}$$

Next we choose an orthonormal basis $\{n_k\}$ for $\mathscr{N}$ and denote by P_k the orthogonal projection of $L_2(\mathbb{R}_-, \mathscr{N})$ onto $L_2(\mathbb{R}_-)n_k$. Then

$$\operatorname{tr} C = \sum_k \operatorname{tr}[P_k C P_k]. \tag{3.14}$$

The analogs of (3.12) and (3.13) hold for $P_k C P_k$, so that

$$\operatorname{tr}[P_k C P_k] = \lim_{\epsilon \to 0} \operatorname{tr}[P_k C_\epsilon P_k].$$

It follows that

$$\operatorname{tr} C = \sum_k \lim_{\epsilon \to 0} \operatorname{tr}[P_k C_\epsilon P_k]. \tag{3.15}$$

It is clear from (3.11) that $P_k C_\epsilon P_k$ is an integral operator on $L_2(\mathbb{R}_-)$ with kernel

$$\int_0^\infty du \, (S_\epsilon^*(u - s)n_k, \, S_\psi^*(u - r)n_k)_{\mathscr{N}}. \tag{3.16}$$

This is obviously continuous and integrable along the diagonal. The trace of $P_k C_\epsilon P_k$ can, therefore, be evaluated as the integral of this kernel along the diagonal:

$$\operatorname{tr}[P_k C_\epsilon P_k] = \int_{-\infty}^0 ds \int_0^\infty du \, (S_\epsilon^*(u - s)n_k, \, S_\psi^*(u - s)n_k)_{\mathscr{N}}.$$

Making the substitution

$$v = u - s \quad \text{and} \quad w = u + s$$

and performing the w integration, this becomes

$$\operatorname{tr}[P_k C_\epsilon P_k] = \int_0^\infty dv \, (vS_\epsilon^*(v)n_k, \, S_\psi^*(v)n_k)_{\mathscr{N}}. \tag{3.17}$$

Because of causality the lower limit of integration can be replaced by $-\infty$. Since the Fourier transform of $ivS_\epsilon{}^*(v)$ is $\partial_\sigma \mathscr{S}_\epsilon{}^*(\sigma)$, the spectral version of (3.17) is

$$\text{tr}[P_k C_\epsilon P_k] = -i \int_{-\infty}^{\infty} (\partial_\sigma \mathscr{S}_\epsilon{}^*(\sigma)n_k, \mathscr{S}_\psi{}^*(\sigma)n_k)_{\mathscr{N}} \, d\sigma$$

$$= -i \int_{-\infty}^{\infty} (\partial_\sigma(\Phi_\epsilon(\sigma)\mathscr{S}^*(\sigma)n_k), \overline{\Psi(\sigma)}\mathscr{S}^*(\sigma)n_k)_{\mathscr{N}} \, d\sigma. \quad (3.18)$$

Passing to the limit as $\epsilon \to 0$, it is easy to see that

$$\lim_{\epsilon \to 0} \text{tr}[P_k C_\epsilon P_k] = \frac{-i}{\sqrt{2\pi}} \int_{-\infty}^{\infty} \Psi(\sigma)(\partial_\sigma \mathscr{S}^*(\sigma)n_k, \mathscr{S}^*(\sigma)n_k)_{\mathscr{N}} \, d\sigma. \quad (3.19)$$

We now make use of the relations (3.2) and (3.3):

$$\mathscr{S}(\sigma) = e^{-i\sigma\rho}\mathscr{S}_0(\sigma), \qquad \mathscr{S}_0(\sigma) = I + K(\sigma)$$

(by assumption, $\partial_\sigma \mathscr{S}_0(\sigma) = \partial_\sigma K(\sigma)$ is of trace class). The integrand in (3.19) can be written in terms of $\mathscr{S}_0$ as

$$i\rho\Psi(\sigma) + \Psi(\sigma)(\partial_\sigma \mathscr{S}_0{}^*(\sigma)n_k, \mathscr{S}_0{}^*(\sigma)n_k)_{\mathscr{N}}.$$

Using this together with the fact that

$$\int \Psi(\sigma) \, d\sigma = (2\pi)^{1/2}\psi(0) = 0$$

and combining the resulting form of (3.19) with (3.15), we get

$$\text{tr } C = (-i/\sqrt{2\pi}) \sum \int_{-\infty}^{\infty} \Psi(\sigma)(\partial\mathscr{S}_0{}^*(\sigma)n_k, \mathscr{S}_0{}^*(\sigma)n_k)_{\mathscr{N}} \, d\sigma. \quad (3.20)$$

By assumption, $\|\mathscr{S}_0(\sigma)\partial_\sigma \mathscr{S}_0{}^*(\sigma)\|_1$ is of polynomial growth. Hence,

$$\sum |\Psi(\sigma)| \, |(\partial_\sigma \mathscr{S}_0{}^*(\sigma)n_k, \mathscr{S}_0{}^*(\sigma)n_k)_{\mathscr{N}}|$$

is integrable and we can interchange the order of integration and summation in (3.20). Since

$$\text{tr}[\mathscr{S}_0(\sigma)\partial_\sigma \mathscr{S}_0{}^*(\sigma)] = \sum (\partial_\sigma \mathscr{S}_0{}^*(\sigma)n_k, \mathscr{S}_0{}^*(\sigma)n_k)_{\mathscr{N}}$$

this gives us the desired relation (3.4) and completes the proof of Theorem 3.1.

We conclude this section by showing that Theorem 3.1 is applicable to the scattering problem for the wave equation with Dirichlet boundary conditions in an exterior domain G in $\mathbb{R}^3$:

$$\begin{aligned} u_{tt} &= \Delta u & \text{in} \quad G, \\ u &= 0 & \text{on } \partial G. \\ u(x, 0) &= f_1, & u_t(x, 0) = f_2. \end{aligned} \quad (3.21)$$

In this case $\mathscr{H}$ consists of all data $f = \{f_1, f_2\}$ obtained by completion from $C_0(G)$ data with respect to the energy norm:

$$\|f\|^2 = \int_G (|\nabla f_1|^2 + |f_2|^2)\, dx. \tag{3.22}$$

The solution operator $U(t)$ taking initial data f into $\{u(\cdot, t), u_t(\cdot, t)\}$ is generated by

$$A = \begin{bmatrix} 0 & I \\ \varDelta & 0 \end{bmatrix};$$

here $\varDelta$ is defined in the weak sense on functions vanishing on ∂G. We shall suppose that the scattering object is contained in a sphere of radius ρ about the origin. For further details on how this problem is set up see [4, Chapter 5].

For φ in $C_0^\infty(\mathbb{R}_+)$ we define

$$Z(\varphi) = \int_0^\infty \varphi(t) Z(t)\, dt. \tag{3.23}$$

It is easy to prove that

$$Z(\varphi * \theta) = Z(\varphi) Z(\theta). \tag{3.24}$$

LEMMA 3.2.   *If ∂G is of class C^3 and φ in $C_0^\infty(\mathbb{R})$ has support in $(2\rho, \infty)$, then $Z(\varphi)$ is Hilbert–Schmidt.*

Since the product of two Hilbert–Schmidt operators is of trace class, it follows from Lemma 3.2 and the relation (3.24) that

COROLLARY 3.3.   *If φ and θ belong to $C_0^\infty(\mathbb{R})$ and both have their support in $(2\rho, \infty)$, then $Z(\varphi * \theta)$ is of trace class.*

Proof of Lemma 3.2.   Let $U_0(t)$ denote the solution operator for the wave equation in free space and set

$$M = U(2\rho) - U_0(2\rho). \tag{3.25}$$

If we treat data f in $\mathscr{H}$ as extended to be zero in the complement of G, then it makes sense to consider the action of M on $\mathscr{H}$. As shown in [4, pp. 152–153], M maps $\mathscr{H}$ into $\mathscr{H}$ with the following properties:

(a) $\|M\| \leqslant 2$,
(b) Mf depends only on the behavior of f in the sphere $|x| < 5\rho$,
(c) $Z(2\rho) = P_+ M P_-$.

For φ in $C_0^\infty(\mathbb{R}_+)$ with support in $(2\rho, \infty)$, $\varphi_0(t) = \varphi(t + 2\rho)$ lies in $C_0^\infty(\mathbb{R}_+)$ and

$$Z(\varphi) = Z(2\rho) Z(\varphi_0) = P_+ M P_- P_+ U(\varphi_0) P_- = P_+ M U(\varphi_0) P_-; \tag{3.26}$$

the P_-P_+ in the next-to-last member is redundant because (1) $U(\varphi_0)\mathcal{D}_-^\perp \subset \mathcal{D}_-^\perp$ and (2) both $U(2\rho)$ and $U_0(2\rho)$ take $\mathcal{D}_+$ into itself and therefore the $\mathcal{D}_+$ component of $U(\varphi_0)P_-$ will be annihilated by the P_+ on the left. We now choose ζ in $C_0^\infty(\mathbb{R}^3)$ so that

$$\zeta(x) = \begin{cases} 1 & \text{for} \quad |x| < 5\rho \\ 0 & \text{for} \quad |x| > 6\rho. \end{cases}$$

Then by property (b) above

$$MU(\varphi_0) = M\zeta U(\varphi_0). \tag{3.27}$$

In view of (3.26) and (3.27) it suffices to prove that $\zeta U(\varphi_0)$ is Hilbert–Schmidt. Now for any g in $\mathscr{H}$, $\{u_1, u_2\} = U(\varphi_0)g$ belongs to the domain of all powers of the generator A of U. Hence if ∂G is of class C^3 it follows from elliptic theory and the Sobolev lemma that

$$|u_1(x)|, |\Delta u_1(x)|, |u_2(x)| \leq \text{const} \, \|g\| \tag{3.28}$$

for all x in G. By the Riesz representation theorem there exist functions $k_{ij}(x, y)$ $(i, j = 1, 2)$ with

$$\{k_{i1}(x, \cdot), k_{i2}(x, \cdot)\} \quad \text{in} \quad \mathscr{H},$$

of norm uniformly bounded for all x in G, and such that

$$u_i(x) = (\{k_{i1}(x, \cdot), k_{i2}(x, \cdot)\}, g).$$

It follows that $\zeta U(\varphi_0)$ is an integral operator with kernel matrix

$$(\zeta(x)k_{ij}(x, y))$$

and by (3.28) that

$$\iint_{G \times G} (|\nabla_x \nabla_y \zeta k_{11}|^2 + |\nabla_x \zeta k_{12}|^2 + |\nabla_y \zeta k_{21}|^2 + |\zeta k_{22}|^2) \, dx \, dy \leq \text{const}.$$

This proves that $\zeta U(\varphi_0)$ is Hilbert–Schmidt and hence, by (3.26) and (3.27), so is $Z(\varphi)$. This completes the proof of Lemma 3.2.

LEMMA 3.4. *If*

$$\|[i\sigma - B]^{-1}\| \leq \text{const}$$

for all real σ, then for the wave equation (3.21) in an exterior domain, $\partial_\sigma \mathscr{S}_0(\sigma)$ is of trace class and

$$\|\partial_\sigma \mathscr{S}_0(\sigma)\|_1 \leq \text{const} \, \sigma^2. \tag{3.29}$$

Remark. The hypothesis of Lemma 3.4 is satisfied for the wave equation with Dirichlet boundary conditions when the scattering object is star shaped (see [6]).

Proof. The operator $K(\sigma)$ in (3.3) can be constructed as follows (see [4, pp. 167–171]): Let v denote the outgoing scattered wave; that is, v is the solution of the differential equation

$$\Delta v + \sigma^2 v = 0 \tag{3.30}$$

satisfying for a given unit vector ω the boundary condition

$$v = -e^{-i\sigma x \cdot \omega} \quad \text{on} \quad \partial G \tag{3.31}$$

and having asymptotic behavior

$$v(x) \sim s(\theta, \omega; \sigma)e^{i\sigma r}/r \quad \text{for large } r, \tag{3.32}$$

where $x = r\theta$ and $\theta = x/|x|$. The kernel of $K(\sigma)$ is given by

$$k(\omega, \theta; \sigma) = -\overline{s(-\theta, \omega; \sigma)}. \tag{3.33}$$

Next we choose $\zeta \in C^\infty(\mathbb{R}^3)$ so that

$$\zeta(x) = \begin{cases} 1 & \text{on the scattering object} \\ 0 & \text{for} \quad |x| \geqslant \rho \end{cases}$$

and set

$$w = v + \zeta e^{-i\sigma \cdot \omega}. \tag{3.34}$$

Then w satisfies the differential equation

$$\Delta w + \sigma^2 w = 2\nabla\zeta \cdot \nabla e^{-i\sigma x \cdot \omega} + (\Delta\zeta)e^{-i\sigma x \cdot \omega} \equiv h, \qquad w = 0 \quad \text{on} \quad \partial G. \tag{3.35}$$

We note that

$$\|\{0, h\}\|, \|\{0, \partial_\sigma h\}\| = O(|\sigma|) \tag{3.36}$$

uniformly in ω. Rewriting (3.35) in vector form we get

$$[i\sigma - A]\begin{bmatrix} w \\ i\sigma w \end{bmatrix} = \begin{bmatrix} 0 \\ -h \end{bmatrix}. \tag{3.35'}$$

It can be shown (see [4, pp. 162–163]) that for $|x| < \rho$,

$$[i\sigma - B]\begin{bmatrix} w \\ i\sigma w \end{bmatrix} = \begin{bmatrix} 0 \\ -h \end{bmatrix}. \tag{3.37}$$

It now follows from our assumption on the resolvent of B and the estimate (3.36) that

$$\int_{G_\rho} (|\nabla w|^2 + \sigma^2|w|^2)\,dx = O(\sigma^2), \tag{3.38}$$

where $G_\rho = G \cap \{|x| < \rho\}$. Recall that

$$\partial_\sigma [i\sigma - B]^{-1} = -i[i\sigma - B]^{-2}.$$

Hence, inverting (3.37) and differentiating with respect to σ gives

$$\begin{bmatrix} \partial_\sigma w \\ iw + i\sigma\,\partial_\sigma w \end{bmatrix} = -i[i\sigma - B]^{-2} \begin{bmatrix} 0 \\ -h \end{bmatrix} + [i\sigma - B]^{-1} \begin{bmatrix} 0 \\ -\partial_\sigma h \end{bmatrix}.$$

Again invoking our assumption on the resolvent of B and using the estimates (3.36) and (3.38), we obtain

$$\int_{G_\rho} (|\nabla \partial_\sigma w|^2 + \sigma^2 |\partial_\sigma w|^2)\, dx = O(\sigma^2). \tag{3.39}$$

We now choose η in $C^\infty(\mathbb{R}^3)$ so that

$$\eta(x) = \begin{cases} 0 & \text{on the support of } \zeta \\ 1 & \text{for } |x| > \rho. \end{cases}$$

Then ηw is smooth over all of $\mathbb{R}^3$ where it satisfies the differential equation

$$\Delta(\eta w) + \sigma^2(\eta w) = 2\nabla\eta \cdot \nabla w + (\Delta\eta)w \equiv g, \tag{3.40}$$

and it follows from (3.38) and (3.39) that

$$\int |g|^2\, dx, \ \int |\partial_\sigma g|^2\, dx = O(\sigma^2). \tag{3.41}$$

The free space solution of (3.40) with the required asymptotic behavior is given by

$$\eta w = \int \gamma(x - y; \sigma)g(y; \omega, \sigma)\, dy, \tag{3.42}$$

where

$$\gamma(x; \sigma) = -e^{i\sigma|x|}/4\pi|x|.$$

It is clear from (3.34) that $\eta w = v$ for $|x| > \rho$. From the relation (3.42) one easily shows that

$$k(\omega, \theta; \sigma) = \overline{s(-\theta, \omega; \sigma)} = -(i\sigma/8\pi^2) \int e^{i\sigma y \cdot \theta} \overline{g(y; \omega, \sigma)}\, dy. \tag{3.43}$$

Finally the kernel for the integral operator $\partial_\sigma \mathscr{S}_0 = \partial_\sigma K$ is

$$\partial_\sigma k(\omega, \theta; \sigma) = (\sigma/8\pi^2) \int e^{i\sigma y \cdot \theta} y \cdot \theta \overline{g(y; \omega, \sigma)}\, dy$$

$$- (i\sigma/8\pi^2) \int e^{i\sigma y \cdot \theta} \partial_\sigma \overline{g(y; \omega, \sigma)}\, dy - (i/8\pi^2) \int e^{i\sigma y \cdot \theta} \overline{g(y; \omega, \sigma)}\, dy.$$

$$\tag{3.44}$$

We treat each term on the right side of (3.44) as the kernel of the composition of two operators: the g factor being the kernel of an operator on $L_2(S_2)$ into $L_2(G_\rho)$, and the exponential being the kernel of an operator on $L_2(G_\rho)$ into $L_2(S_2)$ (here S_2 denotes the unit sphere in $\mathbb{R}^3$). Each of these factors correspond to Hilbert–Schmidt kernels; the Hilbert–Schmidt norm of the g factors are by (3.41) of order $|\sigma|$, as are those of the second factors. It follows that $\partial_\sigma \mathscr{S}_0$ is of trace class with trace norm of order σ^2. This concludes the proof of Lemma 3.4.

Lemmas 3.2 and 3.4 show that the trace formula (3.4) applies to the wave equation (3.21) in the exterior of a star-shaped obstacle for functions $\psi = \varphi * \theta$ where φ, θ belong to $C_0^\infty(R)$ with support in $(2\rho, \infty)$.

REFERENCES

1. M. Sh. Birman and M. G. Krein, On the theory of wave operators and scattering operators, *Dokl. Akad. Nauk. SSSR* **144** (1962), 475–478.
2. L. Eisenbud, Ph.D. Thesis, Princeton Univ., 1948.
3. J. M. Jauch and J. P. Marchand, The time delay operator for simple scattering systems, *Helv. Phys. Acta* **40** (1967), 217–229.
4. P. D. Lax and R. S. Phillips, "Scattering Theory," Academic Press, New York, 1967.
5. P. D. Lax and R. S. Phillips, "Scattering Theory for Automorphic Functions," Annals of Mathematical Studies, Princeton Univ. Press, Princeton, New Jersey, 1976.
6. P. D. Lax, C. S. Morawetz, and R. S. Phillips, Exponential decay of solutions of the wave equation in the exterior of a star-shaped obstacle, *Commun. Pure and Appl. Math.* **16** (1963), 427–455.
7. F. T. Smith, Lifetime matrix in collision theory, *Phys. Rev.* **118** (1960), 349–356.
8. E. P. Wigner, Lower limit for the energy derivative of the scattering phase shift, *Phys. Rev.* **98** (1955), 145–147.

Integral Equations and Operator Theory, Vol. 4/3, 1981
©Birkhäuser Verlag, CH—4010 Basel (Switzerland), 1981

THE TRANSLATION REPRESENTATION THEOREM

Peter D. Lax and Ralph S. Phillips[*]

Short of a new theorem on semigroups of operators, a
new proof of an old theorem on this subject is a suitable
offering to Einar Hille on his 85th birthday.

TRANSLATION REPRESENTATION THEOREM

Let $\mathcal{K}$ be a Hilbert space, $U(t)$ a strongly continuous
one parameter group of unitary mappings. A closed subspace
$\mathcal{D}$ of $\mathcal{K}$ is called an <u>incoming subspace</u> with respect to the
group U if

(1)
  i) $U(t)\mathcal{D} \subset \mathcal{D}$ for $t \leq 0$
  ii) $\cap\, U(t)\mathcal{D} = 0$
  iii) $\cup\, U(t)\mathcal{D} = \mathcal{K}$.

CONCLUSION: $\mathcal{K}$ can be represented as $L_2(\mathbb{R},N)$, N some
auxiliary Hilbert space, so that $U(t)$ is right translation
by t, and $\mathcal{D} = L_2(\mathbb{R}_-, N)$.

Remark: Outgoing subspaces can be defined similarly;
a similar representation holds for them. Note that if $\mathcal{D}$
is an outgoing subspace, its complement $\mathcal{D}^{\perp}$ is incoming,
and conversely.

The Translation Representation Theorem is due to
Sinai [5], who deduces it (and more) from v. Newmann's
theorem on the Heisenberg commutation relation.

In Section 7, Chapter I of our book we showed that
v. Neumann's theorem can be easily deduced from the Translation Representation Theorem; thus the two theorems are
equivalent. For further information concerning this relation,
as well as generalizations of it, where the group of translations

---

[*]
The work of both authors was supported in part by the
National Science Foundation, the first author under
Grant No. MCS-76-07039 and the second author under
Grant No. MCS-77-04908 A 01.

of R is replaced by a non-commutative group, we refer to George Mackey, see e.g. [3] and [4] .

Our new proof of the Translation Representation Theorem is based on a representation theorem for semigroups, which can be considered as the dual of an important early theorem of JLB Cooper [1] .

We start our proof by introducing P, the orthogonal projection onto D , and defining

(2) $S(t) = PU(t)$

LEMMA 2:   S(t), mapping D into D , is a strongly continuous one-parameter semigroup of contractions for $t \geq 0$ which tends to zero strongly as $t \to \infty$.

PROOF: S(t), being the product of a projection and of a unitary operator, is a contraction. Since U(t) is strongly continuous, so is S(t). To prove the semigroup property we write for $r,t \geq 0$,

$$S(r)S(t)d = PU(r)\ PU(s)d =$$

$$= PU(r)[U(s)d+c] = PU(r+s)d + PU(r)c = S(r+s)d + PU(r)c,$$

where   c   is some vector orthogonal to D .

We claim that also U(r)c is orthogonal to D , for according 1)i $U^*(r) = U^{-1}(r) = U(-r)$ maps D into itself and hence its adjoint U(r) maps $D^{\perp}$ into itself. It follows then that $PU(r)c = 0$, which by the above proves the semigroup property.

Note that for every d in D , and $r > 0$

(3) $S(r)\ U(-r)d = d.$

To prove the last part of the Lemma we show $U(-s)D^{\perp}$, $s > 0$ is dense in $\mathcal{H}$. For suppose not; then there is an $h \in \mathcal{H}$ orthogonal to $U(-s)D^{\perp}$ for all $s > 0$. Since U is an isometry,

it follows that U(s)h is orthogonal to $D^{\perp}$ for all $s > 0$, which means that $U(s)h \in D$ for all $s > 0$. But this contradicts property 1)ii.

Using the result just proved we assign to any   d   in D and any $\varepsilon > 0$, a vector $c \perp D$ and $s > 0$ such that

$$|d - U(-s)c| < \varepsilon$$

Since   U   is unitary, we conclude that

$$|U(s)d-c| < \varepsilon$$

Since $c \mathrel{L} D$, it follows that

$$|S(s)d| = |PU(s)d| < \varepsilon$$

This completes the proof of Lemma 2.

THEOREM 3: $S(t)$, $t > 0$ is a strongly continuous one-parameter semigroup of contractions defined on a Hilbert space K . We assume that $S(t)$ tends to 0 strongly as $t \to \infty$:

$$\lim_{t \to \infty} |S(t)k| = 0 \qquad \text{for every k in } K .$$

Then K can be represented isometrically as a subspace of $L_2(\mathbb{R}_-, N), N$ some auxiliary Hilbert space, so that the action of $S(t)$ is translation to the right followed by restriction to $\mathbb{R}_-$:

(4) $S(t) \leftrightarrow RT(t)$

Note: This is Theorem 2.1 in Chapter III or our book; for sake of completeness we include the proof:

Let k be any vector in $D(B)$, the domain of the generator B of S in K ; we shall represent k by the following function f on $\mathbb{R}_-$:

(5) $f(t) = S(-t)k, \quad t < 0.$

As defined above f is a function with values in $D(B)$. We shall define a new norm in $(D(B))$, denoted as $|\ |_N$, so that the $L_2(\mathbb{R}_-, N)$ norm of f equals the norm of k in K . Using (5), we want

(6) $|k|^2 = \int_{-\infty}^{0} |f(t)|_N^2 \, dt = \int_{0}^{\infty} |S(t)k|_N^2 \, dt .$

We apply this to $S(r)k$ in place of k:

$(6)_r \quad |S(r)k|^2 = \int_{0}^{\infty} |S(t+r)k|_N^2 \, dt = \int_{r}^{\infty} |S(t)k|_N^2 \, dt$

We differentiate both sides of $(6)_r$ and set $r = 0$:

(7) $2 \operatorname{Re}(Bk,k) = -|k|_N^2 .$

Since S is a contraction, the left side of (7) is nonpositive; this shows that (7) can be used to define $|k|_N$ as a nonnegative Hilbert seminorm.

We define now N as the completion of $D(B)$ in the N-norm, modulo the space of vectors with $|k|_N = 0$. We claim that with this definition of N and the N-norm, (6) holds for all k in D(B). For then S(t)k also belongs to D(B), and by definition

$$|f-(t)|_N^2 = |S(t)k|_N^2 = -2 \text{ Re}(BS(t)k, S(t)k) =$$
$$= -\frac{d}{dt} |S(t)k|^2 .$$

Integrate this from 0 to r and let $r \to \infty$. Using the hypothesis that $|S(r)k| \to 0$ as $r \to \infty$ we deduce the isometry of the representation for all k in D(B). Since D(B) is a dense subset of K the proof of Theorem 3 is complete.

LEMMA 4: If k belongs to D(B), the function f representing k is continuous in the N-norm.

PROOF: By definition (7),

(8) $|f(t) - f(r)|_N^2 = -2 \text{ Re}(B[S(t) - S(r)]k, [S(t) - S(r)]k)$

$\leq 2|[S(t) - S(r)] Bk| |[S(t) - S(r)]k|$

by the Schwarz inequality. The result follows from the strong continuity of S.

We apply now Theorem 3 to the situation described in Lemma 2: we take K to be $\mathcal{D}$, and S the semigroup defined by (2). The content of Lemma 2 is that the hypotheses of Theorem 3 are satisfied.

Theorem 3 gives a representation of S as translation followed by truncation. We shall show now that it furnishes a translation representation for U(t).

LEMMA 5: Let d be a vector in $\mathcal{D}$; denote by f the function representing d in the representation constructed in Theorem 3.

Let r > 0; then by 1)i, U(-r)d belongs to $\mathcal{D}$. We claim that $d_r = U(-r)d$ is represented by f_r defined as

(9) $f_r(t) = \begin{cases} f(t+r) & , \quad t \leq -r \\ \\ 0 & , \quad -r < t \leq 0 \end{cases}$

PROOF: According to 3), $S(r)d_r = d$. It follows then from (5) that (9) holds for $t \leq -r$. This shows that

$$|d_r|^2 = \int_{-\infty}^{0} |f_r(t)|_N^2 \, dt \geq \int_{-\infty}^{-r} |f_r(t)|_N^2 \, dt$$

$$= \int_{-\infty}^{0} |f(t)|_N^2 \, dt = |d|^2 \, ,$$

with equality holding iff $f_r(t) = 0$ for $-r < t < 0$. But
equality does holds since $d_r = U(-r)d$ and $U(-r)$ is isometric,
this completes the proof of (9).

Preparatory to extending the representation constructed in
Theorem 3 we extend the functions $f(t)$ representing vectors in
$\mathcal{D}$ by setting $f(t) = 0$ for $t > 0$. Having done this we define
the representer of $U(r)d$, $r > 0$, d in $\mathcal{D}$, to be $f(t-r)$. It
follows from Lemma 5 that if $U(r)d$ belongs to $\mathcal{D}$, then this
definition agrees with the original one. Since $U(r)$ is an
isometry, the representation thus extended remains isometric;
clearly, it is a translation representation for U.

LEMMA 6: The completion of the translation representation
defined above maps $\mathcal{K}$ onto $L_2(\mathbb{R}, N)$.

PROOF: The translation representation was defined above
for all vectors in $\cup U(r)\mathcal{D}$; since by 1)iii these are dense
in $\mathcal{K}$ we see that we can, by completion, define a function in
$L_2(\mathbb{R}, N)$ representing any vector in $\mathcal{K}$.

If we compare the definition (2) of $S(t) = PU(t)$ with
relation (4) describing the action of $S(t)$ in our representation
as $RT(t)$, we conclude that action of the projection P in our
representation corresponds to multiplication by the characteristic function of $\mathbb{R}_-$.

To show that every function in $L_2(\mathbb{R}, N)$ is the representer
of some vector in $\mathcal{K}$, we take any d in $\mathcal{D}$; we denote by n
the vector in N which corresponds to d , and by f the
function f representing d by formula (4). It has the
following properties

a) $f(0) = n$

b) $f(t) = 0$ for $t > 0$

c) f is continuous.

Let δ be a small positive number; define the vector $h = h(\delta, n)$
in $\mathcal{K}$ to be

$$h = (I-P) \, U(\delta)d$$

According to the foregoing h is represented by the function

517

that is equal to $f(t-\delta)$ for $t > 0$, and $= 0$ for $t < 0$. It
follows from the properties of f listed above that this
function differs little from n in the interval $(0,\delta)$, and is
zero elsewhere. Clearly, the linear span of such functions is
dense in $L_2(\mathbb{R},N)$ This concludes the proof of the Translation
Representation Theorem.

BIBLIOGRAPHY

1. Cooper, J.L.B.: One parameter semigroups of isometric
    operators in Hilbert space, Ann. Math. Ser 2, Vol. 48,
    1947, p. 827-842.

2. Lax, P.D. and Phillips, R.S.: Scattering Theory, Academic
    Press, New York, 1967.

3. Mackey, G.: In Functional Analysis and Related Fields,
    F. Browder, Ed., Springer-Verlag, New York, 1970.

4. Mackey, G.: Unitary Group Representation in Physics,
    Probability and Number Theory, Benjamin, 1978.

5. Sinai, Ja.G.: Dynamical Systems with Countable Lebesgue
    Spectrum. I, Izv. Akad. Nauk. USSR 25, 1961, p. 899-924.

P.D. Lax,   R.S. Phillips,
Courant Institute of   Stanford University,
Mathematical Sciences,   Stanford,
New York University,   California 94305,
New York,   U.S.A.
N.Y. 10012,
U.S.A.

Trace Formulas for the Schroedinger Operator

PETER D. LAX

Courant Institute

To Henry McKean, with affection and admiration.

In this note we give an exceedingly simple, although not quite rigorous, derivation of the well-known trace formula expressing a periodic potential in a one-dimensional Schroedinger operator in terms of its periodic, antiperiodic, and Dirichlet respectively Neumann spectrum. The items listed in the bibliography deal with results of this kind. Then we use our method to suggest a trace formula for the potential in a two-dimensional Schroedinger operator.

The idea is to deform the potential. Thus we deal with one-parameter families $L(t)$ of selfadjoint operators that have a common domain, and which depend differentiably on t. We take the case that $L(t)$ has a discrete spectrum $\{\lambda_n\}$, tending to ∞:

$$(1) \qquad Lw_n = \lambda_n w_n , \qquad \lim_{n \to \infty} \lambda_n = \infty .$$

The eigenvectors w_n form an orthonormal basis:

$$(1)' \qquad (w_n, w_m) = \delta_{nm} .$$

Standard perturbation theory gives

$$L\dot{w} + \dot{L}w = \lambda\dot{w} + \dot{\lambda}w , \qquad \dot{} = \frac{d}{dt}$$

from which we get

$$(2) \qquad (w, \dot{L}w) = \dot{\lambda} .$$

Differentiating $(1)'$ gives

$$(3) \qquad (\dot{w}_n, w_m) + (w_n, \dot{w}_m) = 0 .$$

Take $L(t)$ to be of the form

$$(4) \qquad L = -\partial_x^2 + q , \quad q = q(x; t) , \quad q(x + 2\pi; t) = q(x; t) ,$$

under boundary conditions to be specified later at the endpoints of the interval $[0, 2\pi]$; the scalar product is $L^2[0, 2\pi]$.

Communications on Pure and Applied Mathematics, Vol. XLVII, 503–512 (1994)
© 1994 John Wiley & Sons, Inc.   CCC 0010–3640/94/040503–10

The perturbation formula (2) for the operator (4) is

$$(5) \qquad \int \dot{q} w_n^2 \, dx = \dot{\lambda}_n \, .$$

Integrating this with respect to t from, say, 0 to 1 gives

$$(6) \qquad \int_0^1 \int \dot{q} w_n^2 \, dx \, dt = \lambda_n - \lambda_n(0) \, .$$

We specify that during the deformation $\int q \, dx$ remains constant; then

$$(7) \qquad \int \dot{q} \, dx = 0 \, .$$

We may then rewrite (6) as

$$(6)' \qquad \int_0^1 \int \dot{q}(w_n^2 - M) \, dx \, dt = \lambda_n - \lambda_n(0) \, ,$$

M any constant independent of x. In view of the normalization (1)′, we choose $M = 1/2\pi$.

Now comes the step for which we offer presently no justification: sum (6)′ over all n:

$$(8) \qquad \int \int \dot{q} \sum (w_n^2 - M) \, dx \, dt = \sum \lambda_n - \lambda_n(0) \, .$$

We claim:

$$(9) \qquad S = \sum (w_n^2 - M)$$

is independent of t.

Proof: Differentiate (9):

$$(10) \qquad \dot{S} = \frac{d}{dt} \sum (w_n^2 - M) = 2 \sum w_n \dot{w}_n \, .$$

Expand $\dot{w}_n$ as

$$\dot{w}_n = \sum a_{nm} w_m \, , \qquad a_{nm} = (\dot{w}_n, w_m) \, .$$

Setting this into (10) gives

$$(10)' \qquad \dot{S} = 2 \sum a_{nm} w_n w_m \, .$$

Since according to (3) the matrix a_{nm} is antisymmetric, the right side of (10)′ is zero; this proves that $\dot{S} = 0$.

Since S is independent of t, the t integration on the left of (8) can be carried out, leading to

$$(11) \qquad \int [q - q(0)] S \, dx = \sum \lambda_n - \lambda_n(0) \, .$$

We consider now four boundary conditions assuming q to be periodic in x with period 2π.

(P) Periodic: $\qquad\qquad w(x + 2\pi) = w(x) \, .$
(A) Antiperiodic: $\qquad\;\; w(x + 2\pi) = -w(x) \, .$
(D) Dirichlet: $\qquad\qquad w(0) = 0 = w(2\pi) \, .$
(N) Neumann: $\qquad\qquad w_x(0) = 0 = w_x(2\pi) \, .$

Given any potential $q(x) = q(x, 1)$ we deform it into a constant $q(x; 0) = c$ so that condition (7) is satisfied; integrating (7) shows that the constant is

$$(12) \qquad \int q(x) \, dx = 2\pi c \, .$$

We compute now $\lambda_n(0)$ and S in all four cases; for simplicity we take $c = 0$, i.e., $q \equiv 0$ at $t = 0$:

$$(P) \qquad w_0^P = \frac{1}{\sqrt{2\pi}} \, , \qquad w_{2n-1}^P = \frac{1}{\sqrt{\pi}} \sin nx \, , \qquad w_{2n} = \frac{1}{\sqrt{\pi}} \cos nx \, .$$

$$(13)_P \qquad \lambda_0^P(0) = 0 \, , \quad \lambda_{2n-1}^P(0) = \lambda_{2n}^P(0) = n^2 \, , \qquad n = 1, 2, \ldots \, ,$$

$$(14)_P \qquad S^P = \sum \left(w_k^2 - \frac{1}{2\pi} \right) = \sum \left(\frac{\sin^2 nx}{\pi} + \frac{\cos^2 nx}{\pi} - \frac{2}{2\pi} \right) = 0 \, .$$

$$(A) \qquad w_{2n-1}^A = \frac{1}{\sqrt{\pi}} \sin(n - 1/2)x, \; w_{2n}^A = \frac{1}{\sqrt{\pi}} \cos(n - 1/2)x \, .$$

$$(13)_A \qquad \lambda_{2n-1}^A(0) = \lambda_{2n}^A(0) = \left(n - \frac{1}{2} \right)^2 \, , \qquad n = 1, 2, \ldots \, ,$$

$$(14)_A \qquad S^A = 0 \, .$$

$$(D) \; w_m^D = \frac{1}{\sqrt{\pi}} \sin \frac{m}{2} x \, , \qquad \lambda_m^D = \left(\frac{m}{2} \right)^2 \, ; \quad \text{setting } m = 2n - 1, 2n \text{ gives}$$

$$(13)_D \qquad \lambda_{2n-1}^D(0) = \left(n - \frac{1}{2} \right)^2 \, , \qquad \lambda_{2n}^D(0) = n^2 \, .$$

Using the identity $\sin^2 \alpha = \dfrac{1 - \cos 2\alpha}{2}$ we get

$(14)_D \qquad S^D = \sum_1^\infty \left(\dfrac{1}{\pi} \sin^2 \dfrac{m}{2} x - \dfrac{1}{2\pi} \right) = -\dfrac{1}{2\pi} \sum_1^\infty \cos mx = -\dfrac{1}{2} \delta(x) + \dfrac{1}{4\pi} .$

$(N) \ w_0^N = \dfrac{1}{\sqrt{2\pi}} , \quad w_m^N = \dfrac{1}{\sqrt{\pi}} \cos \dfrac{m}{2} x , \quad \lambda_m^N = \left(\dfrac{m}{2} \right)^2 ; \quad \text{setting } m = 2n - 1, 2n$

gives

$(13)_N \qquad \lambda_0^N(0) = 0 , \quad \lambda_{2n-1}^N(0) = \left(n - \dfrac{1}{2} \right)^2 , \quad \lambda_{2n}^N(0) = n^2 .$

Using the identity $\cos^2 \dfrac{\alpha}{2} = \dfrac{1 + \cos 2\alpha}{2}$ we get

$(14)_N \qquad S^N = \sum_1^\infty \left(\dfrac{1}{\pi} \cos^2 \dfrac{mn}{2} x - \dfrac{1}{2\pi} \right) = \dfrac{1}{2\pi} \sum_1^\infty \cos mx = \dfrac{1}{2} \delta(x) - \dfrac{1}{4\pi} .$

We set now these values into formula (11) for each of the boundary conditions; in the periodic and antiperiodic case we get, using formulas (13) and (14),

$(15)_P \qquad 0 = \lambda_0^P + \sum \left[\lambda_{2n-1}^P + \lambda_{2n}^P - 2n^2 \right] .$

$(15)_A \qquad 0 = \sum \left[\lambda_{2n-1}^A + \lambda_{2n}^A - 2 \left(n - \dfrac{1}{2} \right)^2 \right] .$

Since $q \equiv 0$ at $t = 0$ and $\int q = 0$, we get in the Dirichlet and Neumann case

$(15)_D \qquad -\dfrac{1}{2} q(0) = \sum \left[\lambda_{2n-1}^D + \lambda_{2n}^D - \left(n - \dfrac{1}{2} \right)^2 - n^2 \right] .$

$(15)_N \qquad \dfrac{1}{2} q(0) = \lambda_0^N + \sum \left[\lambda_{2n-1}^N + \lambda_{2n}^N - \left(n - \dfrac{1}{2} \right)^2 - n^2 \right] .$

Now add $(15)_P$ and $(15)_A$ and subtract twice $(15)_D$:

$(16)_D \qquad q(0) = \lambda_0^P + \sum_{(2)} \left[\lambda_k^P + \lambda_k^A - 2\lambda_k^D \right] ,$

where $\sum_{(2)}$ means that the even and odd terms have to be bracketed.

Similarly, replacing the Dirichlet by the Neumann case leads to the formula

$$(16)_N \qquad -q(0) = \lambda_0^P - 2\lambda_0^N + \sum_{(2)} \left[\lambda_k^P + \lambda_k^A - 2\lambda_k^N \right].$$

The difference in sign in $(16)_D$ and $(16)_N$ is quite typical for formulas involving the Dirichlet and Neumann problems.

In deriving formulas $(16)_D$ and $(16)_N$ we have assumed that the mean value of q is zero. Replacing q by $q + c$, c a constant, increases each eigenvalue of L by c. Therefore the right side of $(16)_D$ and $(16)_N$ increases, respectively decreases by c. This shows that $(16)_D$, $(16)_N$ are valid for arbitrary periodic potentials q.

To obtain formulas expressing q at any point z other than 0, we use the Dirichlet or Neumann eigenvalues at the points z and $z + 2\pi$. As remarked in the Introduction, formula $(16)_D$ is very well known, and formula $(16)_N$ is known.

We turn now to the multidimensional case. Let L denote the operator

$$(17) \qquad\qquad L = -\Delta + q ,$$

$q(x, y; t)$ of period 2π in x and y. We impose, in addition to periodic and antiperiodic boundary conditions also mixed conditions:

(AP)   Antiperiodic in x, periodic in y:

$$w(x + 2\pi, y) = -w(x, y) , \qquad w(x, y) + 2\pi) = w(x, y) .$$

(PA)   Periodic in x, antiperiodic in y:

$$w(x + 2\pi, y) = w(x, y) , \qquad w(x, y + 2\pi) = -w(x, y) .$$

These will be compared to the Dirichlet and Neumann spectrum.

We compute now the eigenfunctions and eigenvalues of the Laplacian $-\Delta$ with zero potential on the square $2\pi \times 2\pi$ under all six boundary conditions:

(P)

$$(18)_0 \qquad\qquad w_0^P \equiv \frac{1}{2\pi} , \qquad \lambda_0^P(0) = 0 ,$$

$$w_n^P = \frac{1}{\sqrt{2\pi}} \frac{\sin}{\cos} nx ,$$

$$(18)_n \qquad\qquad \lambda_n^P(0) = n^2 \text{ of multiplicity 2 .}$$

$$w_{,k}^P = \frac{1}{\sqrt{2\pi}} \frac{\sin}{\cos} ky ,$$

$$(18)_k \qquad\qquad \lambda_{,k}^P(0) = k^2 \text{ of multiplicity 2 ,}$$

$$w_{n,k}^P = \frac{1}{\pi} \frac{\sin}{\cos} nx \frac{\sin}{\cos} ky \,,$$

$(18)_{n,k}$ $\qquad\qquad \lambda_{nk}^P(0) = n^2 + k^2$ of multiplicity 4 .

(AP)

$$w_n^{AP} = \frac{1}{\sqrt{2\pi}} \frac{\sin}{\cos} \left(n - \frac{1}{2}\right) x \,,$$

$(19)_n$ $\qquad\qquad \lambda_n^{AP}(0) = \left(n - \frac{1}{2}\right)^2$ of multiplicity 2 .

$$w_{n,k}^{AP} = \frac{1}{\pi} \frac{\sin}{\cos} \left(n - \frac{1}{2}\right) x \frac{\sin}{\cos} ky \,,$$

$(19)_{n,k}$ $\qquad\qquad \lambda_{n,k}^{AP}(0) = \left(n - \frac{1}{2}\right)^2 + k^2$ of multiplicity 4 .

(PA)

$$w_{,k}^{PA} = \frac{1}{\sqrt{2\pi}} \frac{\sin}{\cos} \left(k - \frac{1}{2}\right) y \,,$$

$(20)_k$ $\qquad\qquad \lambda_k^{PA}(0) = \left(k - \frac{1}{2}\right)^2$ of multiplicity 2 .

$$w_{n,k}^{PA} = \frac{1}{\pi} \frac{\sin}{\cos} nx \frac{\sin}{\cos} \left(k - \frac{1}{2}\right) y \,,$$

$(20)_{n,k}$ $\qquad\qquad \lambda_{n,k}^{PA}(0) = n^2 + \left(k - \frac{1}{2}\right)^2$ of multiplicity 4 .

(A)

$$w_{n,k}^A = \frac{1}{\pi} \frac{\sin}{\cos} \left(n - \frac{1}{2}\right) x \frac{\sin}{\cos} \left(k - \frac{1}{2}\right) y \,,$$

(21) $\qquad\qquad \lambda_{n,k}^A(0) = \left(n - \frac{1}{2}\right)^2 + \left(k - \frac{1}{2}\right)^2$ of multiplicity 4 .

(D)

$$w_{m,\ell}^D = \frac{1}{\pi} \sin \frac{m}{2} x \sin \frac{\ell}{2} y \,;$$

set $m = 2n - 1, 2n$ and $\ell = 2k - 1, 2k$:

$(22)_{n,k}$ $\qquad \lambda_{n,k}^P(0) = \left(n - \frac{1}{2}\right)^2 + \left(k - \frac{1}{2}\right)^2 \,,$

$$\left(n - \frac{1}{2}\right)^2 + k^2 \,, \; n^2 + \left(k - \frac{1}{2}\right)^2 \,, \; n^2 + k^2 \,.$$

(N)

$(23)_0$
$$w_0^N \equiv \frac{1}{2\pi}, \qquad \lambda_0^N(0) = 0.$$

$$w_n^N = \frac{1}{\sqrt{2\pi}} \cos \frac{m}{2} x;$$

set $m = 2n - 1, 2n$:

$(23)_n$
$$\lambda_n^N(0) = \left(n - \frac{1}{2}\right)^2, n^2.$$

$$w_{,k}^N = \frac{1}{\sqrt{2\pi}} \cos \frac{\ell}{2} y;$$

set $\ell = 2k - 1, 2k$:

$(23)_k$
$$\lambda_k^N(0) = \left(k - \frac{1}{2}\right)^2, k^2.$$

$$w_{m,\ell}^N = \frac{1}{\pi} \cos \frac{m}{2} x \cos \frac{\ell}{2} y;$$

set $m = 2n - 1, 2n$ and $\ell = 2k - 1, 2k$:

$(23)_{n,k}$
$$\lambda_{n,k}^N = \left(n - \frac{1}{2}\right)^2 + \left(k - \frac{1}{2}\right)^2, \; \left(n - \frac{1}{2}\right)^2 + k^2,$$
$$n^2 + \left(k - \frac{1}{2}\right)^2, \; n^2 + k^2.$$

In all these formulas, m, ℓ, n, and k parametrizing the eigenvalues are positive integers.

We compute now the sum S defined by (9) with $M = 1/4\pi^2$ in all six cases. Just as in the one-dimensional case

(24)
$$S^P = S^{AP} = S^{PA} = S^A = 0.$$

In the Dirichlet and Neumann cases the series (9) diverge even in the sense of distributions; but put together the combined sum converges:

(25)
$$S^P + S^N = \sum_{m,\ell} \left[\left(w_{m,\ell}^D\right)^2 + \left(w_{m,\ell}^N\right)^2 - \frac{2}{4\pi^2} \right]$$
$$+ \sum_n \left[\left(w_n^N\right)^2 - \frac{1}{4\pi^2} \right] + \sum_k \left[\left(w_{,k}^N\right)^2 - \frac{1}{4\pi^2} \right].$$

We evaluate now the above sums, using the formulas for w^D and w^N given above and the trigonometric identities used before:

$(25)_{\ell,m}$

$$\sum_{m,\ell} = \frac{1}{4\pi^2} \sum_{m,\ell} \left[4\sin^2 \frac{m}{2}x \sin^2 \frac{\ell}{2}y + 4\cos^2 \frac{m}{2}x \cos^2 \frac{\ell}{2}y - 2 \right]$$

$$= \frac{1}{4\pi^2} \sum_{m,\ell} \left[(1 - \cos mx)(1 - \cos \ell y) \right.$$

$$\left. + (1 + \cos mx)(1 + \cos \ell y) - 2 \right]$$

$$= \frac{1}{4\pi^2} \sum 2\cos mx \cos \ell y = \frac{1}{2}\left(\delta(x) - \frac{1}{2\pi}\right)\left(\delta(y) - \frac{1}{2\pi}\right).$$

Similarly

$(25)_n$
$$\sum_n = \frac{1}{4\pi^2} \sum_n \left[2\cos^2 \frac{m}{2}x - 1 \right] = \frac{1}{4\pi^2} \sum \cos mx = \frac{1}{4\pi}\delta(x) - \frac{1}{8\pi^2} ,$$

and

$(25)_k$
$$\sum_k = \frac{1}{4\pi^2} \sum_k \left[2\cos^2 \frac{\ell}{2}y - 1 \right] = \frac{1}{4\pi^2} \sum_k \cos \ell y = \frac{1}{4\pi}\delta(y) - \frac{1}{8\pi^2} .$$

Setting $(25)_{m,\ell}$, $(25)_n$, and $(25)_k$ into (25) gives

(26)
$$S^D + S^N = \frac{1}{2}\delta(x)\,\delta(y) - \frac{1}{8\pi^2} .$$

We apply now the basic formula (11) to the four cases P, A, AP and PA; we get in view of (18), (19), (20), (21), and (24) that

$(27)_P$
$$0 = \lambda_0^P + \sum_n \left[\sum_2 \lambda_n^P - 2n^2 \right]$$

$$+ \sum_k \left[\sum_2 \lambda_k^P - 2k^2 \right] + \sum_{n,k} \left[\sum_4 \lambda_{n,k}^P - 4(n^2 + k^2) \right].$$

$(27)_{AP}$
$$0 = \sum_n \left[\sum_2 \lambda_n^{AP} - 2\left(n - \frac{1}{2}\right)^2 \right]$$

$$+ \sum_{n,k} \left[\sum_4 \lambda_{n,k}^{AP} - 4\left(\left(n - \frac{1}{2}\right)^2 + k^2\right) \right].$$

$(27)_{PA}$
$$0 = \sum_k \left[\sum_2 \lambda_k^{PA} - 2\left(k - \frac{1}{2}\right)^2 \right]$$

$$+ \sum_{n,k} \left[\sum_4 \lambda_{n,k}^{PA} - 4\left(n^2 + \left(k - \frac{1}{2}\right)^2\right) \right].$$

$(27)_A$
$$0 = \sum_{n,k} \left[\sum_4 \lambda_{n,k}^A - 4\left(\left(n - \frac{1}{2}\right)^2 + \left(k - \frac{1}{2}\right)^2\right)\right].$$

Since $q \equiv 0$ at $t = 0$, we get in the combined Neumann-Dirichlet case

$(27)_{D+N}$
$$\iint q(S^D + S^N)\,dxdy$$
$$= \lambda_0^N + \sum_n \left[\sum_2 \lambda_n^N - \left(n - \frac{1}{2}\right) - n^2\right]$$
$$+ \sum_k \left[\sum_2 \lambda_{,k}^N - \left(k - \frac{1}{2}\right)^2 - k^2\right]$$
$$+ \sum_{n,k} \left[\sum_4 \left(\lambda_{n,k}^D + \lambda_{n,k}^N\right)\right.$$
$$\left. - 4\left(\left(n - \frac{1}{2}\right)^2 + n^2 + \left(k - \frac{1}{2}\right)^2 + k^2\right)\right].$$

Use formula (26) for $S^D + S^N$; since we have assumed that $\iint q = 0$, we can express the left side of $(27)_{D+N}$ as $\frac{1}{2}q(0).$, Now add $(27)_P$, $(27)_{AP}$, $(27)_{PA}$, $(27)_A$ and subtract twice $(27)_{D+N}$. The result is

(28)
$$-q(0) = \lambda_0^P - 2\lambda_0^N$$
$$+ \sum_{(4)} \left[\lambda_j^P + \lambda_j^{AP} + \lambda_j^{PA} + \lambda_j^A - 2\lambda_j^D - 2\lambda_j^N\right].$$

The symbol $\sum_{(4)}$ means that in the sum one should bracket together all four terms that emerge from the eigenvalues of multiplicity four of $-\Delta$ with the corresponding four eigenvalues in the Dirichlet and Neumann case. This is a tall order because the four eigenvalues $(22)_{n,k}$ are in general not consecutive; e.g., for $n = 5$, $k = 10,$

$$\left(n - \frac{1}{2}\right)^2 + \left(k - \frac{1}{2}\right)^2 = 110.5, \qquad n^2 + k^2 = 125,$$

and between them comes $6^2 + 9^2 = 117$!

There are two additional complications: some eigenvalues have multiplicity greater than four, because some integers can be represented as a sum of two squares in more than one way. One can avoid this by considering instead of a square a rectangle whose sides have irrational ratio. The second complication is that during deformation the order of eigenvalues may change; the principle of avoidance of crossing may save us from that.

Acknowledgments. I thank Fritz Gesztesy for several enlightening electronic conversations.

This work is supported by the Applied Mathematical Sciences Program of the United States Department of Energy under contract DE-FG02-88ER25053.

Bibliography

[1] Flaschka, H., *On the inverse problem for Hill's operator*, Arch. Rat. Mech. Anal. 59, 1975, pp. 293–309.

[2] Gesztesy, F., Holden, H., Simon, B., and Zhao, Z., *Trace formulae and inverse spectral theory for Schroedinger operators,* Bull. AMS 29, 1993, pp. 250–255.

[3] McKean, H. P., and van Moerbeke, P., *The spectrum of Hill's equations*, Invent. Math. 30, 1975, pp. 217–274.

[4] Trubowitz, E., *The inverse problem for periodic potentials,* Comm. Pure Appl. Math. 30, 1977, pp. 321–337.

Received July 1993.

COMMENTARY ON PART VII

14

Let D be a nonpositive self-adjoint operator in a Hilbert space H from the spectral representations of solutions of $du/dt = Du$. We conclude that no solution of such an equation, except one that is identically zero, can tend to zero faster than some exponential as t tends to infinity. Here we show that the same is true for all functions $u(t)$ that merely satisfy an inequality of the form $\|du/dt - Du\| < k\|u\|$, provided that the spectrum of D contains infinitely many gaps of size $d > 2k$.

The conditions on D can be relaxed; it need not be selfadjoint. We merely require that there be a sequence λ_n tending to $-\infty$ such that the norm of the resolvent $(D - \lambda I)^{-1}$ is less than or equal to c on the lines Re $\lambda = \lambda_n$, where c is any number less than $1/k$.

The first result is applied to show that solutions of elliptic equations of the form $\Delta + P$, where P is a first-order operator, cannot have a zero of infinite order, unless identically zero.

It seems that both theorems should have other applications.

P.D. Lax

17

Let X denote some Banach space over $\mathbb{C}^1$, S a linear space of locally integrable functions $u(y)$ defined on the positive axis $y > 0$, whose values lie in X. Assume that S has these properties:

(i) S is translation invariant; that is, when $u(y)$ belongs to S, so does $u(y + t)$ for every positive t.

(ii) ii) For any interval I on the positive reals define the norm $|u|_I$ to be $\int_I |u(y)|dy$. We require that the unit ball $|u|_I \leq 1$ in S be precompact in the norm $|u|_J$ whenever J is contained in the interior of I.

(iii) A space with these properties is called *interior compact*, and has the following property: There is a positive number c such that all functions u in S for which

$$\int_0^\infty |u(y)|dt < \infty$$

29

satisfy

$$\int_0^\infty |u(y)| e^{cy} dy < \infty.$$

The proof is rather tricky but amusing, and would be indignantly rejected by anyone with the slightest leaning to constructivity. This theorem is used to prove the exponential decay of solutions of elliptic equations defined in a half-cylinder, subject to coercive boundary conditions on the mantle.

P.D. Lax

22

This paper contains two results. One is an extension of a theorem of Beurling, which says that a closed subspace of the space H of square-integrable analytic functions in the unit disk that is invariant under multiplication by bounded analytic functions can be expressed as fH, where f is a bounded analytic function in the unit disk whose absolute value is 1 on the unit circle. The extension is to functions whose values lie in a finite-dimensional space. An extension to functions whose values lies in Hilbert space has been given by Paul Halmos.

The second result uses Beurling's theorem, transplanted to the upper half- plane, to show that if f and g are bounded analytic functions in the upper half-plane, and if e^{iaz} and e^{ibz} are the highest powers of e^{iz} that divide f and g respectively within the algebra of bounded analytic functions, then the highest power of e raised to the power iz that divides the product fg is $e^{i(a+b)}$. This is equivalent to the Titchmarsh convolution theorem. The argument shows that Beurling's theorem is a kind of principal ideal theorem; algebraists please note.

References

[1] Halmos, P. Shifts on Hilbert spaces. *Crelle's J.* **208** (1961), 102–112.

P.D. Lax

132

In this elegant note on trace formulas, Peter Lax suggests a potential deformation method to derive several classical trace formulas for periodic Schrödinger operators in one dimension and indicates how to obtain extensions to periodic Schrödinger operators in the plane.

After briefly recalling his principal one-dimensional results, we will put them into a historical perspective and point to their connection with the inverse periodic spectral problem and

30

especially the periodic KdV initial value problem. The latter area owes much to Peter Lax's extraordinary insights; he influenced its rapid development in the most profound manner. Finally, we comment on his two-dimensional results and some subsequent multidimensional generalizations.

Peter Lax considers a potential deformation $q = q(x, t)$, $t \in [0, 1]$, where q is real-valued, continuous, and of period one with respect to x, such that $\int_0^1 q(x, t)dx$ is conserved, that is,

$$\int_0^1 q_t(x, t)dx = 0. \tag{1}$$

(One may think of $q(x, t) = tq(x)$ with $\int_0^1 q(x)dx = 0$.) Subsequently, he derives the identity

$$\int_0^1 [q(x - 1) - q(x, 0)]S(x)dx = \sum_{n \in I}[\lambda_n(1) - \lambda_n(0)], \tag{2}$$

modulo a technical detail. Here λ_n denote the eigenvalues of the one-dimensional Schrödinger operator $L(t) = -\partial_x^2 + q(x, t)$ in $L^2([0, 1]; dx)$ with fixed self-adjoint boundary conditions at $x = 0, 1$; $I = \mathbb{N}$ or $I = \mathbb{N}_0 = \mathbb{N} \cup \{0\}$ is an appropriate index set; and $S(x)$ denotes the distribution

$$S(x) = \sum_{n \in I}[w_n(x, t)^2 - 1], \tag{3}$$

where $w_n(x, t)$ are the eigenfunctions of $L(t)$ corresponding to $\lambda_n(t)$. The distribution S turns out to be t-independent because of (1). Since the sum in (3) converges in the sense of distributions but not pointwise, the derivation of (2) becomes a delicate matter on which we will comment a bit later.

Taking $t = 0$ and $q(x, 0) = 0$, Peter Lax explicitly computes S in the case of periodic (P), antiperiodic (A), Dirichlet (D), and Neumann (N) boundary conditions and obtains

$$S^P = S^A = 0, \quad S^D = -\left(\frac{1}{2}\right)[\delta(x) - 1] = -S^N. \tag{4}$$

Thus, (2) implies the set of trace formulas

$$0 = \lambda_0^P + \sum_{n \in N}[\lambda_{2n-1}^P + \lambda_{2n}^P - 8\pi^2 n^2], \tag{5}$$

$$0 = \sum_{n \in N}[\lambda_{2n-1}^A + \lambda_{2n}^A - 8\pi^2(n - (1/2))^2], \tag{6}$$

$$-q(0)/2 = \sum_{n \in N}[\lambda_{2n-1}^D + \lambda_{2n}^D - 4\pi^2(n - (1/2))^2 - 4\pi^2 n^2], \tag{7}$$

$$q(0)/2 = \lambda_0^N + \sum_{n \in N}[\lambda_{2n-1}^N + \lambda_{2n}^N - 4\pi^2(n - (1/2))^2 - 4\pi^2 n^2], \tag{8}$$

31

and

$$q(0) = E_0 + \sum_{n \in \mathbb{N}} [E_{2n-1} + E_{2n} - 2\lambda_n^D], \tag{9}$$

$$-q(0) = E_0 - 2\lambda_0^N + \sum_{n \in \mathbb{N}} [E_{2n-1} + E_{2n} - 2\lambda_n^N], \tag{10}$$

where, for simplicity, we wrote $q(0) = q(0,1)$, $\lambda^{P,A,D,N} = \lambda^{P,A,D,N}(1)$, and

$$E_0 = \lambda_0^P, \; E_{4n-1} = \lambda_{2n-1}^P, \; E_{4n} = \lambda_{2n}^P, \; E_{4n-3} = \lambda_{2n-1}^A, \; E_{4n-2} = \lambda_{2n}^A, \; n \in \mathbb{N}.$$

Trace formulas of the type (5)–(10) have a prominent history. Those in (7) and (8) originated by Gelfand and Levitan [6] in 1953 and subsequently generated a lot of work (we refer, e.g., to [2], [5], [10], [14, Chs. 8, 9], [15, Sect. 1.14], [16, Sect. 5.5], [17], [18, Ch. VI], [23], and the references therein). Trace formulas of the type (9) go back to a paper by Hochstadt [11] in 1965. Together with the Dubrovin equations for Dirichlet eigenvalues, they played a fundamental role in solving the inverse spectral problem for periodic Schrödinger operators with finitely many bands in their spectra as well as the corresponding periodic algebra-geometric initial value problem for the Korteweg–de Vries (KdV) equation as discussed by Dubrovin [3], Flaschka [4], Its and Matveev [12], and McKean and van Moerbeke [20] in 1975. (These considerations extend from the periodic to the quasi-periodic algebra-geometric case.) A discussion of the general periodic inverse spectral problem based on the trace formula (9) and the analogue of Dubrovin's equations for Dirichlet eigenvalues (see also [21]) was given by Trubowitz in 1977. This circle of ideas keeps its fascination, as the more recent literature on the subject attests (cf., e.g., [1], [7], [13], [14, Chs. 8–12], [19], [24], and the references therein).

The step involving S in the derivation of the trace formulas (5)–(10) suggests the use of a regularization procedure. Indeed, heat kernel and resolvent regularizations come to mind (cf. [8]). Concentrating on the former, one notes the asymptotic expansion

$$\mathrm{tr} \left[e^{-sL^P} + e^{-sL^A} - 2e^{-sL^D} \right] \underset{s \downarrow 0}{=} 1 - sq(0) + o(s), \tag{11}$$

where L^P, L^A, and L^D denote the Schrödinger operator $-\partial_x^2 + q$ in $L^2([0,1]; dx)$ with periodic, antiperiodic, and Dirichlet boundary conditions at $x = 0$, 1, respectively. Assuming $q \in C^1([0,1])$, which implies $\sum_{n \in \mathbb{N}} |E_{2n} - E_{2n-1}| < \infty$, and noticing that $E_{2n-1} \le \lambda_n^D \le E_{2n}$, the asymptotic expansion (11) proves the trace formula (9). An analogous argument works for the case of Neumann eigenvalues in (10) and more generally for any separated self-adjoint boundary conditions (cf.[8]). As shown in [7] for the Dirichlet case and subsequently in [8] for Neumann and other self-adjoint boundary conditions, the trace formulas (9) and (10) permit vast extensions to general (nonperiodic) potentials q by studying the asymptotic expansion

$$2 \, \mathrm{tr} \left[e^{-sL} - e^{-sL^{D_y}} \right] \underset{s \downarrow 0}{=} 1 - sq(y) + o(s), \tag{12}$$

as long as $y \in \mathbb{R}$ is a point of Lebesgue continuity of q, and $q \in L^1_{\mathrm{loc}}(\mathbb{R})$ and bounded from below (the latter condition can be removed; see [22]). Here L denotes the Schrödinger operator $-\partial_x^2 + q$ in $L^2(\mathbb{R}; dx)$, and L_y^D denotes the corresponding operator in $L^2((-\infty, y]; dx) \oplus$

32

$L^2([y, \infty); dx)$ with a Dirichlet boundary condition at the point y. The asymptotic relation (12) then yields the general trace formula (involving an Abelian regularization)

$$q(x) = E_0 + \lim_{\varepsilon \downarrow 0} \int_{E_0}^{\infty} e^{-\varepsilon \lambda}[1 - \xi(x, \lambda)]d\lambda, \tag{13}$$

where $E_0 = \inf(\mathrm{spec}(L))$ and $\xi(x, \lambda)$ denotes the spectral shift function associated with the pair (L_x^D, L) (cf. [7]). If $q \in C^1(\mathbb{R})$ and periodic, (13) is easily seen to reduce to (9) because of the step function behavior of $|\xi(x, \lambda)|$. Again (13) extends to more general self-adjoint boundary conditions at the point x. Moreover, (12) implies

$$\mathrm{tr}\left[e^{-sL_y^D} - e^{-sL_y^N}\right] \underset{s \downarrow 0}{=} 1 - sq(y) + o(s), \tag{14}$$

and this asymptotic expansion naturally extends to higher dimensions (cf. (16)). We also note that higher-order heat kernel expansions lead to the corresponding trace formulas for higher-order KdV invariants.

In the two-dimensional case Peter Lax considers $q = q(x, y)$ to be real-valued on the square with periods one with respect to x and y. In this case he also considers mixed boundary conditions AP and PA, which denote the cases of boundary conditions antiperiodic in x and periodic in y and periodic in x and antiperiodic in y, respectively. Using the analogous deformation method he obtains

$$-q(0, 0) = \lambda_0^P - 2\lambda_0^N - 2\lambda_0^N + \sum_{(4)} [\lambda_j^P + \lambda_j^{AP} + \lambda_j^{PA} - 2\lambda_j^D - 2\lambda_j^N], \tag{15}$$

where the symbol $\sum_{(4)}$ means that all four terms that emerge from the eigenvalues of $-\Delta$ of multiplicity four (as t tends from 0 to 1) are taken together with the corresponding four Dirichlet and Neumann eigenvalues. The latter requirement is problematic, since the eigenvalues in the periodic case do not necessarily occur in consecutive order. Moreover, some eigenvalues may have multiplicity greater than four (this can be avoided by considering a rectangle whose sides have an irrational ratio), and during the deformation the order of eigenvalues may change.

Inspired by Peter Lax's attempt to derive a two-dimensional trace formula, the following multidimensional asymptotic expansion (an extension of (14)) was derived in [9],

$$\mathrm{tr}\left(\sum_{A \subseteq \{1,\dots,\nu\}} (-1)^{|A|} e^{-s(-\Delta_A + q)}\right) \underset{s \downarrow 0}{=} 1 - s\langle q \rangle + o(s), \tag{16}$$

where $|A|$ denotes the number of points in A, $\langle q \rangle$ the average of q at the 2^ν corners of $[0, 1]^\nu$, and $-\Delta_A, A \subseteq \{1, \dots, \nu\}$, denotes the Laplace operator in $L^2([0, 1]^\nu; d^\nu x)$ with mixed Dirichlet–Neumann boundary conditions of the type

$$w(x) = 0, \quad x_j = 0 \text{ or } x_j = 1, \, j \in A,$$
$$\partial_{x_j} w(x) = 0, \quad x_j = 0 \text{ or } x_j = 1, \, j \in \{1, \dots, \nu\} \backslash A.$$

33

In this manner one can also derive an Abelianized version of Lax's two-dimensional trace formula (15) of the form

$$\text{tr} \left(e^{-sL^P} + e^{-sL^A} + e^{-sL^{PA}} + e^{-sL^{AP}} - 2e^{-sH^N} - 2e^{-sH^D} \right) \underset{s\downarrow 0}{=} -1 + sq(0,0) + o(s).$$

References

[1] Craig, W. The trace formula for Schrödinger operators on the line, *Commun. Math. Phys.* **126**, 379–407 (1989).

[2] Dikii, L.A. Trace formulas for Sturm–Liouville differential operators, *Amer. Math. Soc. Transl. Ser.* (2), **18**, 81–115 (1961).

[3] Dubrovin, B.A. Periodic problems for the Korteweg–de Vries equation in the class of finite band potentials, *Funct. Anal. Appl.* **9**, 215–223 (1975).

[4] Flaschka, H. On the inverse problem for Hill's operator, *Arch. Rat. Mech. Anal.* **59**, 293–309 (1975).

[5] Gelfand, I.M. On identities for the eigenvalues of a second-order differential operator, *Uspehi Mat. Nauk*, **11:1**, 191–198 (1956) (Russian). English translation in Izrail M. Gelfand, Collected Papers Vol. I (S.G. Gindikin, V.W. Guillemin, A.A. Kirillov, B. Kostant, S. Sternberg, eds.), Springer, Berlin, 1987, pp. 510–517.

[6] Gelfand, I.M.; Levitan, B.M. On a simple identity for the eigenvalues of a second-order differential operator, *Dokl. Akad. Nauk SSSR*, **88**, 593–596 (1953) (Russian). English translation in Izrail M. Gelfand, Collected Papers Vol. I (S.G. Gindikin, V.W. Guillemin, A.A. Kirillov, B. Kostant, S. Sternberg, eds.), Springer, Berlin, 1987, pp. 457–461.

[7] Gesztesy, F.; Simon, B. The xi function, *Acta Math.* **176**, 49–71 (1996).

[8] Gesztesy, F.; Holden, H.; Simon, B.; Zhao, Z. Higher order trace relations for Schrödinger operators, *Rev. Math. Phys.* **7**, 893–922 (1995).

[9] Gesztesy, F.; Holden, H.; Simon, B.; Zhao, Z. A trace formula for multidimensional Schrödinger operators, *J. Funct. Anal.* **141** 449–465 (1996).

[10] Halberg, C.J.A.; Kramer, V.A. A generalization of the trace concept, *Duke Math. J.* **27** (1960), 607–617.

[11] Hochstadt, H. On the determination of a Hill's equation from its spectrum, *Arch. Rat. Mech. Anal.* **19**, 353–362 (1965).

[12] Its, A.R.; Matveev, V.B. Schrödinger operators with finite-gap spectrum and N-soliton solutions of the Korteweg–de Vries equation, *Theoret. Math. Phys.* **23**, 343–355 (1975).

34

[13] Kotani, S.; Krishna, M. Almost periodicity of some random potentials, *J. Funct. Anal.* **78**, 390–405 (1988).

[14] Levitan, B.M. Inverse Sturm–Liouville Problems, VNU Science Press, Utrecht, 1987.

[15] Levitan, B.M.; Sargsjan, I.S. *Introduction to Spectral Theory*, Amer. Math. Soc., Providence, R. I., 1975.

[16] Levitan, B.M.; Sargsjan, I.S. *Sturm–Liouville and Dirac Operators*, Kluwer, Dordrecht, 1991.

[17] Lyubishkin, V.A. On the trace formulas of Gel'fand–Levitan and Kreĭn, *Math. USSR Sbornik* **74**, 531–540 (1993).

[18] Magnus, W.; Winkler, S. *Hill's Equation*, Dover, New York, 1979.

[19] Marchenko, V.A. The Cauchy problem for the KdV equation with non-decreasing initial data, in *What is Integrability?*, V.E. Zakharov (ed.), Springer, New York, 1991, pp. 273–318.

[20] McKean, H.P.; van Moerbeke, P. The spectrum of Hill's equation, *Invent. Math.* **30**, 217–274 (1975).

[21] McKean, H.P.; Trubowitz, E. Hill's operator and hyperelliptic function theory in the presence of infinitely many branch points, *Commun. Pure Appl. Math.* **29**, 143–226 (1976).

[22] Rybkin, A. On the trace approach to the inverse scattering problem in dimension one, *SIAM J. Math. Anal.* **32**, 1248–1264 (2001).

[23] Sadovnichiĭ, V.A.; Lyubishkin, V.A. Trace formulas and perturbation theory, *Sov. Math. Dol.* **37**, 789–791 (1988).

[24] Sodin, M.; Yuditskii, P. Almost periodic Sturm–Liouville operators with Cantor homogeneous spectrum, *Comment. Math. Helvetici* **70**, 639–658 (1995).

[25] Trubowitz, E. The inverse problem for periodic potentials, *Commun. Pure Appl. Math.* **30**, 321–337 (1977).

F. Gesztesy

35

PART VIII

ANALYSIS

COMMUNICATIONS ON PURE AND APPLIED MATHEMATICS, VOL. XXIV, 133–135 (1971)

Approximation of Measure Preserving Transformations*

PETER D. LAX

The investigations described in this note were instigated by some recent numerical studies of the structure of the orbits of certain measure preserving transformations by Laslett [3], Hénon [2], and others. Since these calculations were performed on computers with a finite word length, the transformations actually carried out were only approximations to the transformations which these authors set out to study. These approximate transformations were only approximately measure preserving; unlike the original transformations they were not even invertible.

The purpose of this note is to prove that every continuous measure preserving transformation can be approximated uniformly by measure preserving transformations which can be carried out on finite word length machines. Our method leads in principle, but hardly in practice, to explicit construction of such approximations. It is not clear if the approximations constructed here are useful in studying the orbits of the originally given transformation, either theoretically or experimentally. We describe now more precisely the context of this discussion.

Let T be a measure preserving mapping which is one-to-one and continuous. For the sake of simplicity we take the domain of T to be the n-dimensional unit torus Ω. Divide each of the n generators of the torus into $N = 10^p$ equal subintervals; in this way the torus itself is divided into 10^{np} cubes, each of side length 10^{-p}. An approximate transformation T_N which can be evaluated on a machine employing words p-digits long maps each cube into some other cube of this subdivision. T_N *is measure preserving if and only if this mapping of the set of cubes into itself is a permutation.*

To make T_N an approximation to T, we demand that *the image of any cube Q under T_N should have a point in common with the image of Q under T.* This means that in defining T_N we may choose as image of a cube Q any of the cubes which have a point in common with $T(Q)$.

Let $Q_1, \cdots, Q_k$ be a collection of any k cubes. Consider all cubes which are eligible as images for some $Q_j, j = 1, \cdots, k$. Since an ineligible cube has no point in common with the image of any Q_j under T, it follows that the eligible

---

* This research represents results obtained at the Courant Institute, New York University, under the sponsorship of the Atomic Energy Commission, contract AT(30-1)-1480. Reproduction in whole or in part is permitted for any purpose of the United States Government.

133

cubes cover the image of $Q_1 \cup Q_2 \cup \cdots \cup Q_k$ under T. Since T is an invertible measure preserving transformation, the volume of this image is the same as the volume of $Q_1 \cup \cdots \cup Q_k$, i.e., k times the volume of one cube. Since we saw that the image is covered by the eligible cubes, we conclude that *for any collection of k cubes there are at least k cubes eligible*. According to the celebrated *Marriage theorem* (see [4]) it is then possible to assign in a one-to-one way to each cube Q another cube $T_N Q$ which is eligible as approximate image.

How good is this approximation? Denote by $K(s)$ the modulus of continuity of T, i.e.,

$$(1) \qquad\qquad\qquad |Tx, Ty| \leqq K(|x, y|) \; ;$$

we claim that, for every x,

$$(2) \qquad\qquad\qquad |Tx, T_N x| \leqq K(d) + d \, ,$$

where d is the diagonal of a cube. For, let Q be the cube to which x belongs, and let x_0 be that point in the cube whose image under T falls inside $T_N Q$. Then $T_N x$ and Tx_0 fall into the same cube; therefore,

$$(3) \qquad\qquad\qquad |T_N x, T x_0| \leqq d \, .$$

Similarly, since x and x_0 fall inside the same cube,

$$(4) \qquad\qquad\qquad |x, x_0| \leqq d \, .$$

Using the triangle inequality and (1), (3) and (4), we get

$$|Tx, T_N x| \leqq |Tx, Tx_0| + Tx_0, T_N x| \leqq K(d) + d \; ;$$

this proves (2).

We summarize our result:

THEOREM. *Let T be a continuous invertible measure preserving transformation mapping the n-dimensional torus Ω onto itself. Divide each generator of Ω into N equal parts, thus dividing Ω into N^n congruent n-dimensional cubes. Then there exists a transformation T_N which permutes these cubes, and approximates T in the sense that*

$$|Tx, T_N x| \leqq K(d) + d \; ;$$

here $d = \sqrt{n}/N$ is the diameter of one of the cubes, $K(s)$ is the modulus of continuity of T, and $|x, y|$ is the shortest distance between two points x and y of Ω.

Our theorem is related to an earlier theorem of Halmos [1]; he deals with measure preserving transformations which are invertible but not necessarily continuous, and shows that these can be approximated by permutations T_N *in the weak topology*, i.e., so that the symmetric difference of sets under T and T_N is small for N large. Neither theorem implies the other.

Halmos shows that T_N can be chosen to be a *cyclic* permutation. In the continuous case it may be worthwhile to pursue the following possibility: Is there any relation between the number of fixed points of $T^j, j = 1, 2, \cdots$, and the number of cycles of order j contained in the permutation T_N, not for all T_N but at least for "almost all" T_N, which approximates T?

The approximation theorem of this paper can be thought of as a cruder version of a theorem of Oxtoby and Ulam (Theorem 12, p. 919 of [6]); they prove that every measure preserving automorphism of a unit cube can be approximated arbitrarily closely in the maximum norm by another one T^* which is continuous and which permutes rigidly a denumerable set of cubes whose total measure is 1.

Bibliography

[1] Halmos, P., *Lectures on Ergodic Theory*, The Mathematical Society of Japan, Tokyo, 1956.

[2] Hénon, *Numerical Study of Quadratic Area-Preserving Mappings*, to appear.

[3] Laslett, L. J., McMillan, E. M., and Moser, J., *Long-Term Stability for Particle Orbits*, AEC Research and Devel. Report, NYO-1480-101, 1968.

[4] Ore, Øystein, *Theory of Graphs*, Amer. Math. Soc., 1962.

[6] Oxtoby, J. C. and Ulam, S., *Measure preserving homeomorphisms and metrical transitivity*, Annals of Math., Vol. 42, 1941, pp. 874–920.

Received June, 1970.

On the Factorization of Matrix-Valued Functions

PETER D. LAX

A Toeplitz operator T_S associated with the doubly infinite sequence $S = \{s_k\}$, $k = 0, \pm 1, \cdots$, maps simply infinite sequences $\{u_k\}$, $k = 0, 1, \cdots$, into sequences $\{v_k\}$, $k = 0, 1, \cdots$, according to the following rule:

$$(1) \qquad (Tu)_k = v_k = \sum_0^\infty s_{k-j} u_j, \qquad k = 0, 1, \cdots.$$

Operators of this kind arise in many problems of analysis, most recently in the theory of difference approximations to partial differential equations, see [5] and [6].

When the entries u_k, and s_k are scalars, T_S is called a scalar Toeplitz operator; when u_k is a vector, the s_k are matrices and T_S is called a matrix Toeplitz operator.

An alternative representation of T_S is obtained by taking the Fourier transform of (1). Introducing

$$(2) \qquad \sum_0^\infty u_k e^{ik\theta} = U(\theta), \qquad \sum_0^\infty v_k e^{ik\theta} = V(\theta),$$

$$(3) \qquad \sum_{-\infty}^\infty s_k e^{ik\theta} = S(\theta),$$

one can rewrite (1) as follows:

$$(4) \qquad V(\theta) = PS(\theta)U(\theta),$$

where the operator P removes all Fourier components of negative index.

If the sequence $\{s_k\}$ is such that $S(\theta)$ is a bounded function on the unit circle, then the operator $U \to V$ defined by (4) is continuous in the L_2 norm. If the sequence $|k|^n |s_k|$ is bounded for any exponent n, then $S(\theta)$ is a C^∞ function and (4) maps U into V continuously in the C^∞ topology. It follows from (2) that both U and V are functions on the unit circle which are boundary values of functions analytic in the unit disk.

683

The theory of scalar Toeplitz operators is classical. The main results, stated below, are valid in both the L_2 and C^∞ topologies:

THEOREM 1. *Suppose that the complex-valued function $S(\theta) \neq 0$ anywhere on the unit circle. Then the operator $T_S : U \to V$ defined by (4) is a Fredholm operator; its index, i.e., the difference of the dimension of its null space and the codimension of its range, equals $-W_S$, where W_S is the winding number of $S(\theta)$ around the origin.*

THEOREM 2. *Suppose $S(\theta) \neq 0$ for any θ on the unit circle. Then, if $W_S \geqq 0$, the operator T_S is one-to-one and its range has codimension W_S; if $W_S \leqq 0$, T_S is onto and has null space of dimension $-W_S$.*

Clearly, Theorem 2 implies Theorem 1. The general case of Theorem 2 is easily reduced to the case when $W_S = 0$. The proof in this case can be given by factoring $S(\theta)$ as a product

$$(5) \qquad S(\theta) = S_-(\theta) S_+(\theta),$$

where $S_-(\theta)$ is the boundary value of a function which is analytic outside the unit circle and free of zeros there, while S_+ is the bounding value of a function analytic inside the unit circle and free of zeros there. Such a factorization of S is easily constructed from an additive decomposition of $\log S$:

$$(6) \qquad \log S(\theta) = R_-(\theta) + R_+(\theta),$$

where R_- is analytic outside, R_+ inside the unit circle. Such a decomposition is possible since it follows from the assumption $W_S = 0$ that $\log S(\theta)$ is a continuous function. Exponentiating (6) yields (5), with $S_- = \exp R_-$, $S_+ = \exp R_+$.

With the help of the factorization (5) the inverse of the operator T_S can be written down explicitly:

$$(7) \qquad T_S^{-1} = S_+^{-1} P S_-^{-1}.$$

That (7) is indeed the inverse of T_S can be verified easily.

The basic theory of matrix Toeplitz operators is contained in [1]; Theorem 1 has a straightforward generalization.

THEOREM 3. *Let $S(\theta)$ be a matrix-valued function on the unit circle, such that $\det S(\theta) \neq 0$ anywhere. Then T_S defined by (4) is a Fredholm operator, and its index is $-W$, where W is the winding number of $\det S(\theta)$ around the origin.*

The matrix version of Theorem 2 is false, as may be seen from the simple example

$$(8) \qquad S(\theta) = \begin{pmatrix} e^{i\theta} & 0 \\ 0 & e^{-i\theta} \end{pmatrix},$$

since $\det S(\theta) = 1$, $W = 0$. But clearly, T_S has a one-dimensional null space and its range has codimension 1.

It is still true that if S has a factorization of the form (5), then T_S is invertible, its inverse being given by formula (7). The converse of this proposition is also true: if T_S is invertible, S has a factorization of the form (5). To see this, denote by E_j the j-th unit vector, and by F_j the solution of

$$T_S F = E_j.$$

These vector equations can be combined into a single matrix equation

$$T_S F = I,$$

where F is the matrix whose j-th column is F_j. Using the definition of T_S we can write, by (4), this equation as

$$PSF = I;$$

this can be rewritten as

$$(9) \qquad SF = S_-,$$

where S_- is a function whose Fourier coefficients of positive order are zero, and whose zero-th coefficient is I. Taking the determinant of (9), we see that

$$(10) \qquad \det S \det F = \det S_-.$$

It follows from the invertibility of T_S that $\det S(\theta) \neq 0$ and

$$(11) \qquad W(\det S) = 0.$$

It can be shown that $\det F(\theta) \neq 0$; so we conclude from (10) that

$$(12) \qquad W(\det S) + W(\det F) = W(\det S_-).$$

Since F is analytic inside, and S_- outside the unit circle, it follows that

$$(13) \qquad W(\det F) \geqq 0, \qquad W(\det S_-) \leqq 0.$$

Combining the inequalities (13) with (11) and (12), we conclude that in both relations in (13) the sign of equality holds. That implies that both F and S_- are invertible. So we deduce from (9) that

$$(14) \qquad S = S_- F^{-1} = S_- S_+,$$

the asserted factorization of S.

What fails in the matrix case is the deduction of a factorization of the form (5) from condition (11). Since the logarithmic function fails to satisfy the usual functional equation for a noncommuting matrix argument, it is not possible to pass from an additive decomposition of the form (6) to a multiplicative decomposition of the form (5).

The purpose of this note is to point out how to obtain a factoring of the form (5) by solving a Dirichlet problem for a system of nonlinear partial differential equations.

Suppose a factorization of the form (14) is possible; we denote by $A(z)$ the analytic matrix-valued function in $|z| \leq 1$ whose boundary value is S_+, and by $B(z)$ the antianalytic matrix-valued function in $|z| < 1$ whose boundary value is S_-. Then, denoting differentiation with respect to z and $\bar{z}$ by subscripts, we have

$$(15) \qquad A_{\bar{z}} = 0, \qquad B_z = 0.$$

The relation

$$(16) \qquad S(z) = B(z)A(z)$$

constitutes an extension of S to $|z| \leq 1$.

Differentiating (16) with respect to z and $\bar{z}$, respectively, we get, using (15),

$$(17) \qquad S_z = BA_z, \qquad S_{\bar{z}} = B_{\bar{z}}A.$$

One more differentiation gives

$$S_{z\bar{z}} = B_{\bar{z}}A_z.$$

Expressing A_z, $B_{\bar{z}}$ from (17) we can rewrite this as

$$(18) \qquad S_{z\bar{z}} = S_{\bar{z}}A^{-1}B^{-1}S_z = S_{\bar{z}}S^{-1}S_z,$$

where (16) was used in the last step.

Conversely, suppose that S satisfies (18); this implies that $S^{-1}S_z$ is analytic. We then solve the ordinary differential equation

$$(19) \qquad AS^{-1}S_z = A_z$$

subject to the initial condition, say, $A(0) = I$. The solution A is an analytic function. We claim that the solution is invertible for every z. To see this we note that (19) implies that $\det A(z) = a(z)$ satisfies

$$a_z = \operatorname{tr}(S^{-1}S_z)a ;$$

it follows from this that $a(z) \neq 0$, which in turn implies the contention.

Having found A, we set

$$(20) \qquad B = SA^{-1} .$$

It follows from (17) and (18) that B is antianalytic; thus (20) yields the desired factorization of S.

To factor a given function $S(\theta)$ on the unit circle, we have to find a solution S of (18) in $|z| \leq 1$ whose value on $|z| = 1$ is given; i.e., we have to solve a Dirichlet problem for equation (18). Equation (18) is a system of the form

$$(21) \qquad \Delta s = Q(s_z, s_{\bar{z}}, s) ,$$

where Q is quadratic in s_z, $s_{\bar{z}}$. Such systems of equations arise in differential geometry; their theory has been investigated by E. Heinz in [2], and subsequently by S. Hildebrandt in [3]. These authors show that the Dirichlet problem for (20) can be solved provided that the Q_m satisfy an inequality of the form

$$|Q|^2 \leq a(|s_z|^2 + |s_{\bar{z}}|^2) + b ,$$

with a small enough.

Our system (18) does not satisfy such an inequality, and our analysis shows that indeed the Dirichlet problem for (18) can be solved if and only if $S(\theta)$ has a factorization of the form (5). Now example (8) shows that not every function $S(\theta)$ has such a factorization. On the other hand, M. Krein has shown that among all C^∞ matrix functions $S(\theta)$ satisfying the condition

$$W(\det S) = 0 ,$$

the ones which can be factored form an open and dense set. Thus (18) is an example of a nonlinear elliptic system for which the Dirichlet problem can be solved for all but a nowhere dense set of boundary data.

Some of the results described in this note are contained in [4].

Bibliography

[1] Gohberg, I. C., and Krein, M. G., *Systems of integral equations on a half-line with kernels depending on the difference of arguments*, Amer. Math. Soc. Transl., Vol. 14, 1960, pp. 217–288.

[2] Heniz, E., *On certain nonlinear elliptic differential equations and univalent maps*, J. d'Analyse Math., Jerusalem, Vol. 5, 1956/57, pp. 197–272.

[3] Hildebrandt, S., *Maximum principle for minimal surfaces of constant mean curvature*, Math. Z. 128, 1972, pp. 253–269.

[4] Lax, P. D., *Toeplitz Operators, Lectures on Differential Equations*, ed. Aziz, Vol. 2, Van Nostrand, 1969, pp. 257–282.

[5] Osher, S., *Systems of difference equations with general homogeneous boundary conditions*, Trans. Am. Math. Soc. 137, 1969, pp. 177–201.

[6] Strang, G., *Wiener–Hopf difference equations*, J. Math. and Mech., Vol. 13, 1964, pp. 85–96.

Received June, 1976.

A Short Path to the Shortest Path

Peter D. Lax

This note contains a demonstration of the isoperimetric inequality. Our proof is somewhat simpler and more straightforward than the usual ones; it is eminently suitable for presentation in an honors calculus course.

1. *The Isoperimetric Inequality* says that a closed plane curve of length 2π encloses an area $\leq \pi$. Equality holds only for a circle.

Let $x(s), y(s)$ be the parametric presentation of the curve, s arclength, $0 \leq s \leq 2\pi$. Suppose that we have so positioned the curve that the points $x(0), y(0)$ and $x(\pi), y(\pi)$ lie on the x-axis, i.e.

$$y(0) = 0 = y(\pi). \tag{1}$$

The area enclosed by the curve is given by the formula

$$A = \int_0^{2\pi} y\dot{x} \, ds, \tag{2}$$

where the dot ˙ denotes differentiation with respect to s. We write this integral as the sum $A_1 + A_2$ of an integral from 0 to π and from π to 2π, and show that each is $\leq \frac{\pi}{2}$.

According to a basic inequality,

$$ab \leq \frac{a^2 + b^2}{2};$$

equality holds only when $a = b$. Applying this to $y = a$, $\dot{x} = b$, we get

$$A_1 = \int_0^{\pi} y\dot{x} \, ds \leq \frac{1}{2}\int_0^{\pi} (y^2 + \dot{x}^2) \, ds. \tag{3}$$

Since s is arclength, $\dot{x}^2 + \dot{y}^2 = 1$; so we can rewrite (3) as

$$A_1 \leq \frac{1}{2}\int_0^{\pi} (y^2 + 1 - \dot{y}^2) \, ds. \tag{3'}$$

Since $y = 0$ at $s = 0$ and π, we can factor y as

$$y(s) = u(s)\sin s, \tag{4}$$

u bounded and differentiable. Differentiate (4):

$$\dot{y} = \dot{u}\sin s + u\cos s.$$

Setting this into (3') gives

$$A_1 \leq \frac{1}{2}\int_0^{\pi} \left[u^2(\sin^2 s - \cos^2 s) - 2u\dot{u}\sin s \cos s - \dot{u}^2 \sin^2 s + 1 \right] ds. \tag{5}$$

The product $2u\dot{u}$ is the derivative of u^2; integrating by parts changes (5) into

$$A_1 \leq \frac{1}{2}\int_0^{\pi} (1 - \dot{u}^2 \sin^2 s) \, ds,$$

clearly $\leq \pi/2$. Equality holds only if $\dot{u} \equiv 0$, which makes $y(s) \equiv$ constant $\sin s$. Since equality in (3) holds only if $y = \dot{x} = \sqrt{1 - \dot{y}^2}$, $y(s) \equiv \pm\sin s$, $x(s) \equiv \mp\cos s$ + constant. This is a semicircle.   Q.e.d.

Courant Institute of Mathematical Sciences
New York University
251 Mercer Street
New York, NY 10012

Change of Variables in Multiple Integrals

Peter D. Lax

*Dedicated to the memory of Professor Clyde Klipple, who taught me real variables
by the R. L. Moore method at Texas A & M in 1944.*

1. Let $y = \varphi(x)$ be a differentiable mapping of the interval $S = [c, d]$. Denote by T the interval $[a, b]$ with $\varphi(c) = a$, $\varphi(d) = b$. Let f be a continuous function of y. The change of variable formula says that

$$\int_S f(\varphi(x)) \frac{d\varphi}{dx} \, dx = \int_T f(y) \, dy. \tag{1.1}$$

The usual proof uses the fundamental theorem of calculus. Denote by g an anti-derivative of f:

$$f = \frac{d}{dy} g. \tag{1.2}$$

According to the fundamental theorem of calculus,

$$\int_T f(y) \, dy = g(b) - g(a), \tag{1.3}$$

where a and b are the endpoints of the interval T. On the other hand, by the chain rule the derivative of the composite $g \circ \varphi$ is given by

$$\frac{d}{dx} g(\varphi(x)) = \frac{dg}{dy} \frac{d\varphi}{dx}.$$

Using (1.2) we see that the x derivative of $g \circ \varphi$ is the integrand on the left in (1.1); therefore by the fundamental theorem of calculus,

$$\int_S f(\varphi(x)) \frac{d\varphi}{dx} \, dx = g(\varphi(d)) - g(\varphi(c)), \tag{1.4}$$

where c and d are the endpoints of the interval S. Since $\varphi(c) = a$ and $\varphi(d) = b$, the right sides of (1.3) and (1.4) are the same; this completes the proof of (1.1).

The usual proof of the change of variable formula in several dimensions uses the approximation of integrals by finite sums; see for instance [7]. The purpose of this note is to show how to use the fundamental theorem of calculus to prove the change of variable formula for functions of any number of variables. Then, as a surprising byproduct, we obtain a proof of the Brouwer fixed point theorem. In the last section we compare our proof with other known analytic proofs of the fixed point theorem.

I thank Daniel Velleman for suggesting a substantial simplification of the argument.

2. In this section we study mappings $\varphi(x) = y$ of n-dimensional x space into n-dimensional y space. We impose two assumptions:

i) φ is once differentiable.

ii) φ is the identity outside some sphere, say the unit sphere:

$$\varphi(x) = x \quad \text{for } |x| \geq 1.$$

Change of variable theorem:. *Let f be a continuous function of compact support. Then*

$$\int f(\varphi(x))J(x)\,dx = \int f(y)\,dy, \qquad (2.1)$$

where J is the Jacobian determinant of the mapping φ:

$$J(x) = \det\frac{\partial\varphi_j}{\partial x_i}; \qquad (2.2)$$

here φ_j is the jth component of φ.

We prove this for functions f that are once differentiable and for mappings φ that are twice differentiable; since functions and mappings can be approximated by differentiable ones, this suffices. The approximation can be accomplished by mollification, that is, by convolving each component of φ with a smooth, nonnegative, spherically symmetric function m with small support whose integral equals 1. As the support of m shrinks to zero, $m * \varphi$ and its first derivatives tend to those of φ. If φ is the identity, so is $m * \varphi$.

Define

$$g(y_1, y_2, \ldots, y_n) = \int_{-\infty}^{y_1} f(z, y_2, \ldots, y_n)\,dz. \qquad (2.3)$$

Clearly, $\frac{\partial g}{\partial y_1} = f$. Since f is once differentiable, so is g. Since f has compact support, we can choose c so large that f is zero outside the c-cube

$$|y_i| \le c, \quad i = 1, 2, \ldots, n.$$

It follows from (2.3) that $g(y_1, \ldots, y_n) = 0$ when $|y_j| \ge c$ for any $j \ne 1$, and when $y_1 \le -c$.

Take $c \ge 1$; then the c-cube contains the unit ball. Since φ is the identity outside the unit ball, $f(\varphi(x))$ is zero outside the c-cube in x-space. So in the integrals in (2.1) we may restrict integration to the c-cube.

In the left side of (2.1), express f as the partial derivative of g:

$$\int \frac{\partial g}{\partial y_1}(\varphi(x))J(x)\,dx. \qquad (2.4)$$

We denote by D the gradient with respect to x; the columns of the Jacobian matrix $\partial\varphi/\partial x$ are $D\varphi_1, \ldots, D\varphi_n$.

Observation:. *The integrand in (2.4) can be written as the following determinant:*

$$\det(Dg(\varphi), D\varphi_2, \ldots, D\varphi_n). \qquad (2.5)$$

Proof: By the chain rule

$$Dg(\varphi) = \sum_{j=1}^{n} (\partial_{y_j}g)D\varphi_j. \qquad (2.6)$$

We set this into the first column in (2.5). Formula (2.6) expresses $Dg(\varphi)$ as a linear combination of the vectors $D\varphi_1, D\varphi_2, \ldots, D\varphi_n$; the last $n - 1$ of these vectors are the last $n - 1$ columns of the matrix in (2.5), and therefore these can be subtracted from $Dg(\varphi)$ without altering the value of the determinant (2.5). This leaves us with

$\det((\partial_{y_1} g(\varphi)) D\varphi_1, D\varphi_2, \ldots, D\varphi_n)$; factoring out the scalar $(\partial_{y_1}) g(\varphi)$ gives $(\partial_{y_1}) g(\varphi) J$, the integrand in (2.4). ∎

The next step is to expand the determinant (2.5) according to the first column; we obtain

$$M_1 \partial_{x_1} g(\varphi) + \cdots + M_n \partial_{x_n} g(\varphi), \tag{2.7}$$

where $M_1, \ldots, M_n$ are the cofactors of the first column of the Jacobian matrix. Setting (2.7) into the integrand in (2.4) we get

$$\int \left(M_1 \partial_{x_1} g(\varphi) + \cdots + M_n \partial_{x_n} g(\varphi) \right) dx. \tag{2.8}$$

Since φ is twice differentiable, we can integrate each term by parts over the c-cube and obtain

$$-\int g(\varphi) \left(\partial_{x_1} M_1 + \cdots + \partial_{x_n} M_n \right) dx + \textit{boundary terms}. \tag{2.9}$$

We use now the following classical identity:

$$\partial_{x_1} M_1 + \cdots + \partial_{x_n} M_n \equiv 0. \tag{2.10}$$

We sketch a proof: We can write the left side of (2.10) symbolically as

$$\det(D, D\varphi_2, \ldots, D\varphi_n). \tag{2.11}$$

For $n = 2$ we have

$$\det(D, D\varphi_2) = \partial_1 \partial_2 \varphi_2 - \partial_2 \partial_1 \varphi_2 = 0.$$

For $n > 2$ we note that the cofactors M_j are multilinear functions of the φ_j. Using the product rule of differentiation, we write (2.11), again symbolically, as

$$\sum_{2 \leq k \leq n} \det(D, D\varphi_2, \ldots, D\varphi_n)_k, \tag{2.12}$$

where the subscript k means that the differential operator D in the first column acts only on the kth column. We leave it to the reader to verify that each of the determinants in the sum (2.12) is zero.

The identity (2.10) shows that the n-fold integral in (2.9) is zero.

We turn now to the boundary term in (2.9). Since $g(\varphi(x)) = g(x)$ on the boundary of the c-cube, the only nonzero boundary term is from the side $x_1 = c$; since $M_1 = 1$ when $\varphi(x) \equiv x$, that boundary term is

$$\int g(c, x_2, \ldots, x_n) \, dx_2 \cdots dx_n. \tag{2.13}$$

Using the definition (2.3) of g in (2.13) gives

$$\iint_0^c f(z, x_2, \ldots, x_n) \, dz \, dx_2 \cdots dx_n,$$

which is the right side of equation (2.1). This completes the proof of the change of variables formula.

3. In our proof of the change of variables formula, we assumed neither that φ is one-to-one, nor that it is onto. We claim:

A mapping φ having properties i) and ii) of the change of variables theorem maps $\mathbb{R}^n$ onto $\mathbb{R}^n$.

Suppose some point y_0 were not the image of any x. Since φ is the identity outside the unit ball, y_0 would lie inside the unit ball. Since φ maps $|x| \leq 1$ into a closed set, it would follow that some ball B_0 centered at y_0 would be free of images of φ. Now take any function f supported in the ball B_0, whose integral is nonzero:

$$\int f \, dy \neq 0. \tag{3.1}$$

By the change of variable formula

$$\int f(\varphi(x)) J \, dx = \int f \, dy \neq 0. \tag{3.2}$$

Since the range of φ avoids B_0, and since the support of f lies in B_0, the integrand on the left in (3.2) is identically zero; then so is the integral. This contradicts (3.1), and so the claim is established.

Intermediate Value Theorem:. *Let φ be a continuous map of the unit ball in $\mathbb{R}^n$ into $\mathbb{R}^n$ that is the identity on the boundary*:

$$\varphi(x) = x \quad \text{for } |x| = 1.$$

Then the image of φ covers every point in the unit ball.

Proof: Extend φ to be the identity outside the unit ball. Then approximate the extended map by differentiable maps, each the identity outside the unit ball. According to our claim, each of these maps covers the unit ball. By compactness, so does their limit. ∎

The following well-known argument shows how to deduce the Brouwer fixed point theorem from the intermediate value theorem.

Let ψ be a continuous mapping of the unit ball into the unit ball; we claim that it leaves a point fixed. If not then for every x there is a ray from $\psi(x)$ through x. This ray pierces the unit ball at a point that we denote by $\varphi(x)$. Clearly, φ is a continuous mapping; it is the identity for x on the unit sphere and maps the unit ball into the unit sphere. This contradicts the intermediate value theorem. ∎

4. The Brouwer fixed point theorem has many analytical proofs. How do they compare with the present one? Hadamard [3] employed the identity (2.10) about the Jacobian matrix; so did Dunford-Schwartz [2, pp. 467–470].

Samelson [6] used Stokes' theorem to give an extremely short proof of the Brouwer fixed point theorem. This proof was rediscovered by Kannai [5]. According to Laurent Schwartz, as related by Haim Brézis, such a proof was current in Paris in the thirties.

Báez-Duarte [1] proved formula (2.1) using exterior forms and Stokes' theorem and deduced from it the intermediate value theorem. My deduction is the same as Báez-Duarte's.

The integration of exterior forms over chains presupposes the change of variable formula for multiple integrals. It is amusing that the change of variables formula alone implies Brouwer's theorem.

In conclusion we call attention to Erhardt Heinz's beautiful analytic treatment of the Brouwer degree of a mapping.

REFERENCES

1. L. Báez-Duarte, Brouwer's Fixed-Point Theorem and a Generalization of the Formula for Change of Variables in Multiple Integrals, *J. Math. Anal. Appl.* **177** (1993) 412–414.
2. N. Dunford and J. Schwartz, *Linear Operators, Part I*, Wiley-Interscience, New York, 1958.
3. J. Hadamard, Sur quelques applications de l'indice de Kronecker, pp. 437–477 in J. Tannery, *Introduction à la théorie des fonctions d'une variable*, vol. 2, Paris, 1910.
4. E. Heinz, An Elementary Analytic Theory of the Degree of Mapping n-Dimensional Space, *J. Math. Mech.* **8** (1959) 231–248.
5. Y. Kannai, An elementary proof of the no retraction theorem, *Amer. Math. Monthly* **88** (1981) 264–268.
6. H. Samelson, On the Brouwer fixed point theorem, *Portugal. Math.* **22** (1963) 189–191.
7. J. Schwartz, The formula for change in variables in a multiple integral, *Amer. Math. Monthly* **61** (1954) 81–85.

PETER LAX was born in Hungary in 1926; he came to the U.S. in December, 1941 on the last boat. He is a fixture at the Courant Institute of New York University; his mathematical interests are too numerous to mention. He has always liked to teach at all levels, hence this paper.
Courant Institute, NYU, 251 Mercer St., New York, NY 10012
lax@cims.nyu.edu

Change of Variables in Multiple Integrals II

Peter D. Lax

Dedicated to the memory of Professor E.C. Klipple, who taught me real variables by the R.L. Moore method at Texas A&M in 1944.

In a paper of the same title [1], published in this MONTHLY in the summer of 1999, I gave a simple, algebraic derivation of the change of variables formula for multiple integrals. This is the result proved there:

Theorem 1. *Let φ be a once continuously differentiable mapping of $\mathbb{R}^n$ into $\mathbb{R}^n$ that is the identity outside the unit ball:*

$$\varphi(x) = x \quad \text{for} \quad \|x\| \geq 1.$$

Let f be a continuous function of compact support; then

$$\int f(y)dy = \int f(\varphi(x))J(x)dx. \tag{1}$$

Here $J(x)$ is the Jacobian determinant of the mapping φ: $J(x) = \det \frac{\partial \varphi_j}{\partial x_i}$, where $\varphi = (\varphi_1, \ldots, \varphi_2)$.

Several correspondents have pointed out that the mappings φ that come up in real life are *not* the identity outside the unit, or any other, ball. The purpose of this note is to sketch a way to extract the garden variety change of variable formula from Theorem 1.

Theorem 2. *Let D be a domain in $\mathbb{R}^n$, and let ψ be a mapping of D into $\mathbb{R}^n$ that is once continuously differentiable on the closure of D. We impose the following conditions on ψ:*

 i) ψ is a 1-to-1 mapping on the closure of D.

 ii) The Jacobian determinant J of ψ is positive on the closure of D.

Let f be a continuous function whose support lies in the image of D under ψ. Then

$$\int f(y)dy = \int_D f(\psi(x))J(x)dx. \tag{2}$$

Since we have required the support of f to be contained in $\varphi(D)$, the integrand on the right in (2) is zero on the boundary of D, and so it can be extended to be zero outside D, and remain continuous. Therefore the x integral in (2) can be evaluated as an integral over any box containing D.

We deduce Theorem 2 from Theorem 1 in two steps:

Step 1: Take a partition of unity in y-space.

$$\Sigma p_j(y) \equiv 1,$$

where each p_j is a continuous function whose support lies in some ball of radius ε. Multiply this relation by f to get a decomposition of f as

$$\Sigma f_j = f, \qquad f_j = p_j f.$$

Since p_j is a factor of f_j, each f_j is supported in a ball of radius ε. If we can prove (2) for each f_j, adding these formulas gives (2) for f. Thus it suffices to prove (2) for functions f whose support is contained in a ball B of radius ε.

Step 2: To prove (2) when the support of f is contained in an ε-ball B we construct a continuous differentiable mapping φ of $\mathbb{R}^n$ into $\mathbb{R}^n$ with the following properties:

    a) $\varphi(x) = \psi(x)$ at all points x in $\psi^{-1}(B)$, the set of points x that are mapped into B by ψ.

    b) $\varphi^{-1}(B) = \psi^{-1}(B)$.

    c) φ is the identity outside some ball: $\varphi(x) = x$ for $\|x\| > r$.

Before we carry out Step 2 we show how to use it to deduce Theorem 2. It follows from a) that $f(\varphi(x)) = f(\psi(x))$ for all x in $\psi^{-1}(B)$, and that the Jacobian determinants of ψ and φ are equal at all these points. By definition of B, $f(y) = 0$ for y not in B. Therefore $f(\psi(x)) = 0$ when x is not in $\psi^{-1}(B)$. According to b) such an x is not in $\varphi^{-1}(B)$ either, so that $f(\varphi(x)) = 0$.

Combining these two observations, we conclude that the integrands on the right in (1) and (2) are equal everywhere, and so therefore are their integrals.

It follows from c) that Theorem 1 applies to φ, and therefore (1) is true; but then so is (2). ∎

The construction of φ is carried out in four stages. We start with the observation that since ψ is continuously differentiable and 1-to-1, and $J \neq 0$, the inverse of ψ is continuous. The preimage $\psi^{-1}(B)$ of every ε-ball is therefore contained in a ball of radius δ, where $\delta = \delta(\varepsilon)$ tends to zero when ε tends to zero. Since we can replace the mapping $\psi(x)$ by $\psi(x - a) - b$, we may assume that the center of the ball B is the origin 0 in y-space, and its preimage $\psi^{-1}(0)$ is the origin 0 in x-space.

The size of ε and the parameter d appearing in the construction below will be specified later.

Stage 1: Define, for $\|x\| \leq d$,

$$\varphi(x) = \psi(x). \tag{3}$$

Stage 2: Denote by M the Jacobian matrix of ψ at $x = 0$. Then by Taylor's formula

$$\psi(x) = Mx + N(x), \tag{4}$$

$$\|N(x)\| \leq o(x)\|x\|, \tag{5}$$

where $o(x)$ denotes a function that tends to zero when $\|x\|$ tends to zero. Choose any function $s(t)$ that goes smoothly and monotonically from 1 to 0 as t goes from 0 to 1. Define, for $d \leq \|x\| \leq 2d$,

$$\varphi(x) = Mx + s(\|x\|/d - 1)N(x). \tag{6}$$

© THE MATHEMATICAL ASSOCIATION OF AMERICA [Monthly 108

Stage 3: We need the following basic result of linear algebra:

Lemma. *A real matrix M whose determinant is positive can be deformed smoothly into the identity matrix I, so that the determinant of the deformed matrices lies between* $\det M$ *and* 1.

Proof. In case the first column c_1 of M is not proportional to the first unit vector $e_1 = (1, 0, \ldots, 0)^t$, let $R(t)$ be a one-parameter family of rotations in the plane spanned by c_1 and e_1. Here $R(0) = I$ and $R(1)$ is rotation of c_1 into a multiple of e_1. Then $R(t)M$ is a smooth deformation of M into a matrix M_1 whose first column is ke_1, k positive. Next deform the first row of M_1 into $(k, 0, \ldots, 0)$. Finally, deform the first column ke_1 into e_1, and at the same time deform the second column c_2 into kc_2. During these deformations the determinant remains constant. The final product of these deformations is a matrix whose first row is e_1', its first column e_1. Now we are poised for an inductive proof of the lemma with respect to the order of the matrix. ∎

Denote the deformation of M into I by $M(t)$; $M(0) = M$ and $M(1) = I$. Define, for $2d \leq \|x\| \leq 3d$,

$$\varphi(x) = M(\|x\|/d - 2)x. \tag{7}$$

Stage 4: For $\|x\| > 3d$, define

$$\varphi(x) = x. \tag{8}$$

We now verify that φ as defined in stages 1-4 has all three required properties stipulated in Step 2, provided that d and ε are small enough.

a) requires $\varphi(x)$ to be equal to $\psi(x)$ for all x in $\psi^{-1}(B)$. Since $\psi^{-1}(B)$ is contained in the ball $\|x\| < \delta$, it follows from (3) that this condition is fulfilled if $\delta(\varepsilon)$ is less than d.

b) requires every x in $\psi^{-1}(B)$ to be in $\varphi^{-1}(B)$, and conversely. By (3) this is true for all x in the ball $\|x\| \leq d$. Since $\psi^{-1}(B)$ is contained in that ball, no point x with $\|x\| > d$ is in $\psi^{-1}(B)$. We now show that such an x is not in $\varphi^{-1}(B)$ either; this is the same as saying that for $\|x\| > d$, $\varphi(x)$ is not in B, which would follow from $\|\varphi(x)\| > \varepsilon$.

For $\|x\|$ between d and $2d$, $\varphi(x)$ is defined by (6). Since M is invertible at every point x and depends continuously on x, $\|Mx\| \geq m\|x\|$ for all x, where m is some positive number. We choose d so small that for x in the ball $\|x\| \leq 2d$, $o(x) \leq m/2$. Then by (5), $\|N(x)\| \leq (m/2)\|x\|$. Setting this in (6) and using the fact that $|s(t)| \leq 1$ we conclude that for $d \leq \|x\| \leq 2d$

$$\|\varphi(x)\| \geq \|Mx\| - \|N(x)\| \geq m\|x\| - \frac{m}{2}\|x\| \geq \frac{m}{2}d,$$

which is greater than ε if $\varepsilon < md/2$.

The argument for $\|x\|$ between $2d$ and $3d$ is very similar. Here $\varphi(x)$ is defined by (7). The one-parameter family of matrices $M(t)$ is uniformly bounded, and its determinant is uniformly bounded from below. Therefore $M(t)$ is invertible for each x and t, and it follows from Cramer's rule that there exists a positive constant ℓ such that $\|M(t)x\| \geq \ell\|x\|$ for all x and all t. Setting this in (7) we see that for $\|x\| \geq 2d$,

$$\|\varphi(x)\| \geq \ell\|x\| \geq 2\ell d,$$

which is greater than ε if $\varepsilon < 2\ell d$.

c) For $\|x\| \geq 3d$, $\varphi(x)$ is defined in (8) to be x.

We verify now that all the conditions imposed on d and ε can be satisfied:

i) For $\|x\| \leq d$, $\quad o(x) < m/2$.

ii) $\delta(\varepsilon) < d, \varepsilon < md/2, \quad \varepsilon < 2\ell d$.

Since $o(x)$ tends to zero as $\|x\|$ tends to zero, i) is satisfied for d small enough. Since $\delta(\varepsilon)$ tends to zero as ε tends to zero, the first condition in ii) is satisfied when ε is small enough; so are the other two.

In Theorem 2 we have required that the mapping ψ be 1-to-1 and its that Jacobian determinant be positive. This can be relaxed to requiring only that the *degree* of the mapping ψ be one at each point of the support of f. Recall that the degree of a mapping ψ of $\overline{D}$ into $\mathbb{R}^n$ at some point y in $\mathbb{R}^n$ that is not the image of a boundary point of D is defined as the algebraic sum of the number of times y is covered by ψ; the covering $y = \psi(x)$ is counted as positive or negative depending on the sign of $J(x)$. At points $y = \psi(x)$ where $J(x) = 0$, this definition does not apply. Call the set of such points y the *critical set* of the mapping ψ; according to Sard's lemma, the critical set has measure zero.

Theorem 3. *Let D be a domain in $\mathbb{R}^n$, and let ψ be a mapping of D into $\mathbb{R}^n$ that is once continuously differentiable on the closure of D. Suppose that the mapping ψ has degree 1 on an open set O in the image space. Let f be a continuous function whose support is contained in the set O. Then the change of variables formula (2) holds.*

Sketch of Proof. According to Sard's lemma, there is an open set C covering the critical set of ψ whose volume $V(C)$ is less that any prescribed positive number η. We furthermore require that each point of C have distance less that a from the critical set, a a small positive number to be specified later. Let y be any point in the support of f that lies outside the set C. Since the critical set is contained in C, y has a finite number of preimages $x_1, \ldots, x_k$, $\psi(x_j) = y$, $J(x_j) \neq 0$. Let B be a ball centered at y of radius ε, so small that

i) *Each point in B has distance greater than $a/2$ from the critical set.*

ii) *$\psi^{-1}(B)$ consists of k connected sets, and $J(x)$ has the same sign on each of these sets.*

Additional restrictions on ε are specified later.

Since the support of f is a compact set, so is its intersection with the complement of C. It follows that a finite number of ε-balls B_i, $i = 1, \ldots, N$ cover it, and $\cup B_i \cup C$ covers the support of f. Let $\sum_1^{N+1} p_i(y) \equiv 1$ be a partition of unity where the support of p_i is contained in B_i, $i = 1, \ldots, N$, and the support of p_{N+1} is contained in C. Decompose f as

$$f = \sum_1^{N+1} f_i, \quad f_i = p_i f.$$

We claim that the change of variable formula (2) holds for each $f_i, i = 1, \ldots, N$. To see this we observe that the inverse image of the support of such an $f = f_i$ is contained in the inverse image of $B = B_i$, and so by property ii) is contained in a finite number of disjoint connected sets $X_1, \ldots, X_k$. We claim that for all j,

$$\int_{X_j} f(x) J(x) dx = \sigma_j \int f(y) dy, \tag{9}$$

where σ_j is the signature of $J(x)$ on X_j. Where $\sigma_j = 1$, this is Theorem 2; where $\sigma_j = -1$, compose the mapping ψ with the reflection $y_1 \rightarrow -y_i, y_i \rightarrow y_i, i = 2, \ldots, n$, and apply Theorem 2.

We have assumed that the mapping ψ has degree one at all points of the support of f; this means that $\sum_1^k \sigma_j = 1$. Summing (9) gives therefore the change of variables formula (2) for $f = f_i, i = 1, \ldots, N$.

What about f_{N+1}? Since the support of f_{N+1} is contained in C,

$$\left| \int f_{N+1}(y) dy \right| \leq \max |f| \, \text{Vol} \, (C) \leq \max |f| \, \eta. \tag{10}$$

Each point of C has distance less than a from the critical set. It follows that $|J(x)| < \delta$ at each point of $\psi^{-1}(C)$, where δ tends to zero as a tends to zero:

$$\left| \int f_{N+1}(\psi(x)) J(x) dx \right| \leq V \max |f| \delta, \tag{11}$$

where V is the volume of $\psi^{-1}(C)$.

Clearly, both (10) and (11) tend to zero as η and a tend to zero. ∎

The requirement in Theorem 1 that φ be the identity outside the unit ball guarantees that the degree of the mapping φ is one everywhere.

In a 1983 article [2], Leinfelder and Simader prove the change of variable formula, and the Brouwer fixed point theorem, by deforming the mapping to the identity. A key point in their argument is the identity (2.10) in [1]. The same identity was used by Dunford and Schwartz, by Hadamard (see [1]), and by Kronecker, who presumably discovered it. Michael Taylor has shown how to obtain this identity in a natural fashion by using the language of the calculus for exterior forms. Another proof can be found in [3, Lemma 1.1, p. 8].

REFERENCES

1. P. D. Lax, Change of variables in multiple integrals, *Amer. Math. Monthly* **106** (1999) 497–501.
2. H. Leinfelder and C.G. Simader, The Brouwer fixed point theorem and the transformation rule for multiple integrals via homotopy arguments, *Exp. Math.* **4** (1983) 349–355.
3. C. B. Morrey, Multiple integral problems in the calculus of variations and related topics, University of California Press, Berkeley, 1943.
4. M. Taylor, Differential forms and the change of variable formula for multiple integrals, to appear.

PETER LAX was born in Hungary in 1926; he came to the U.S. in December, 1941 on the last boat. He is a fixture at the Courant Institute of New York University; his mathematical interests are too numerous to mention. He has always liked to teach at all levels, hence this paper.
Courant Institute, NYU, 251 Mercer St., New York, NY 10012
lax@cims.nyu.edu

COMMENTARY ON PART VIII

[59]

Let f be a measure-preserving transformation of the unit n-cube onto itself. Subdivide the cube into 2^{Kn} small cubes of edge length 2^{-K}. One can think of every numerical approximation to f as a mapping of small cubes into small cubes. To make the approximation measure-preserving, the approximate map should be a permutation. In this paper, the marriage theorem is used to show that such approximations exist.

Jürgen Moser has made an ingenious use of the above approximation theorem to prove that every volume-preserving transformation of the n-cube can be approximated in the Koopman topology by infinitely differentiable diffeomorphisms.

Steve Alpern has given a new proof and new applications of the approximation theorem.

References

[1] Jürgen Moser, 1928–1999. *Ergodic Theory & Dynamical Systems*, **22** (2002), 1337–1342.

[2] Alpern, S.R.; Prasad, V.S. Typical dynamics of volume preserving homeomorphisms, *Cambridge Tracts in Mathematics* #139, Cambridge Press, 2000.

P.D. Lax

[76]

If $S(\theta)$ is a complex-valued function on the unit circle with Fourier coefficients s_k, then the semi-infinite matrix $T_S = (s_{j-k})_{j,k\geq 0}$ is called the *Toeplitz matrix with symbol S*. When S is bounded, it represents a bounded operator on $\ell^2(\mathbf{Z}^+)$; if $S \in C^\infty$, then it also acts continuously on the space of rapidly decreasing sequences. The paper is not concerned with the weakest conditions on S. It is assumed that $S \in C^\infty$, and the space on which T_S acts can be either of those mentioned.

The basic fact about invertibility of T_S is that the following three conditions are equivalent:

(i) T_S is invertible;

(ii) $S(\theta) \neq 0$ and there is a factorization $S(\theta) = S_-(\theta)S_+(\theta)$, where S_- is the boundary value of a function analytic and nonzero outside the unit circle (including ∞) and S_+ is the boundary value of a function analytic and nonzero inside the unit circle;

(iii) $S(\theta) \neq 0$, and $w(S)$, the winding number of $S(\theta)$ about the origin, is equal to zero.

36

In general, if $S(\theta) \neq 0$, then T_S is a Fredholm operator with index $-w(S)$, and (iii) gives a simple test for invertibility.

What if $S(\theta)$ is a matrix-valued function? A partial answer is that if $S(\theta)$ is invertible, then T_S may be a Fredholm operator with index $-w(\det S)$. But in distinction to the scalar case, T_S may be Fredholm of index zero yet not be invertible. The functions S for which T_S is invertible form a proper open and dense subset of those for which it is Fredholm of index zero.

On the other hand, there is still the equivalence of

(i) T_S is invertible;

(ii) $S(\theta)$ is invertible and there is a factorization $S(\theta) = S_-(\theta)S_+(\theta)$, where S_- is the boundary value of a function analytic and invertible outside the unit circle and S_+ is the boundary value of a function analytic and invertible inside the unit circle.

Again in distinction to the scalar case, there is no simple way to determine whether there is a factorization as in (ii). There are results in special cases, but nothing as general as in the scalar case.[†]

In the paper under review Lax gives yet another equivalent for invertibility. He first observes that the existence of a factorization as in (ii) is equivalent to S having an invertible extension $S(z)$ to the closed unit disk that has a factorization

$$S(z) = B(z)A(z),$$

where A is analytic in the interior and B antianalytic. He deduces from this that S must satisfy

$$S_{z\bar{z}} = S_{\bar{z}}S^{-1}S_z,$$

and he shows that the existence of an extension satisfying this equation is also a sufficient condition for there to be a factorization as above.

Since S must have the boundary function $S(\theta)$, Lax has produced a simple example of a nonlinear elliptic system of the form

$$\nabla s = Q(s_z, \ s_{\bar{z}}, \ s),$$

where Q is quadratic in s_z and $s_{\bar{z}}$, for which the Dirichlet problem can be solved for all but a dense set of boundary data.

Harold Widom

[†]When $S(\theta)$ is invertible there is a factorization $S(\theta) = S_-(\theta)D(\theta)S_+(\theta)$, where $D(\theta)$ is a diagonal matrix $\mathrm{diag}(e^{ik_j\theta})$. The k_j are uniquely determined integers called the *partial indices*. The index of T_S equals $-\sum k_j$, and the condition for invertibility is that all $k_j = 0$. These results were established by I.C. Gohberg and M.G. Krein, *Systems of integral equations on a half-line with kernels depending on the difference of arguments*, Uspehi Mat. Nauk **13** (1958); English transl., *Trans. Math. Monographs* (2) **14** (1960), Amer. Math. Soc. There is no general way to compute the partial indices.

37

Being required from time to time to teach elementary calculus, I've become convinced in my jaded old age that there is nothing one can say about this subject, Anno Domini 2001, that hasn't been said a thousand times before. I was shocked and delighted, therefore, to learn two years ago of a new proof of one of the basic theorems of multivariate calculus, the change of variables formula, which is *much* simpler than any of the known proofs. Since these comments are written for grown-ups I will couch this proof (which I will give in toto below) in the language of differential forms following a presentation of Lax's result by Mike Taylor in [1].

Let $\varphi : \mathbb{R}^n \to \mathbb{R}^n$ be a once-differentiable mapping that is equal to the identity outside a large ball. Lax proves the following theorem:

Theorem. *If* $\omega = f(x_1, \cdots, x_n)dx_1 \wedge \cdots \wedge dx_n$ *is a compactly supported n-form, then*

$$\int \varphi^* \omega = \int \omega.$$

Here is the proof: For $a \in \mathbb{R}$, $a > 0$, let

$$\omega_a = f(x_1 + a, x_2, \cdots, x_n)dx_1 \wedge \cdots \wedge dx_n.$$

Lemma. *There exists a compactly supported* $(n-1)$*-form* μ *for which* $\omega_a - \omega = d\mu$.

Proof. Let

$$g(x_1, \ldots, x_n) = f(x_1 + a, x_2, \ldots, x_n) - f(x_1, \ldots, x_n)$$

and let

$$h(x_1, \ldots, x_n) = \int_{-\infty}^{x_1} g(t, x_2, \ldots, x_n)dt.$$

Then $\mu = h\, dx_2 \wedge \cdots \wedge dx_n$ does the trick. $\qquad\square$

Now we're done, because

$$\int \omega_a = \int \omega + \int d\mu = \int \omega$$

and

$$\int \varphi^* \omega_a = \int \varphi^* \omega + \int d\varphi^* \mu = \int \varphi^* \omega,$$

and for a large, $\varphi^* \omega_a = \omega_a$.

38

One can easily eliminate the "form" formalism from this proof, as Lax does in the first of the papers above, to make it comprehensible to a calculus student. This makes the proof a bit longer (but not much).

The one objection one might have to this proof is that what's proved is *not* the usual change of variables formula, which has to do with two bounded domains in $\mathbb{R}^n$ and a map between them. However, in the second of the two papers above Lax shows that the usual change of variables formula can be extracted from the formula above by a simple partition of unity argument.

One can but hope that the next generation of calculus texts will see this proof replacing the classic "chopping into small rectangles and approximating φ on each rectangle by a linear mapping" proof.

References

[1] Taylor, M. Differential forms and the change of variable formula for multiple integrals, to appear, 2005.

V. Guillemin

39

PART IX

ALGEBRA

ON MATRICES WHOSE REAL LINEAR COMBINATIONS ARE NONSINGULAR

J. F. ADAMS, PETER D. LAX[1] AND RALPH S. PHILLIPS[2]

Let Λ be either the real field R, or the complex field C, or the skew field Q of quaternions. Let $A_1, A_2, \cdots, A_k$ be $n \times n$ matrices with entries from Λ. Consider a typical linear combination $\sum_{j=1}^{n} \lambda_j A_j$ with real coefficients λ_j; we shall say that the set $\{A_j\}$ "has the property P" if such a linear combination is nonsingular (invertible) except when all the coefficients λ_j are zero.

We shall write $\Lambda(n)$ for the maximum number of such matrices which form a set with the property P. We shall write $\Lambda_H(n)$ for the maximum number of Hermitian matrices which form a set with the property P. (Here, if $\Lambda = R$, the word "Hermitian" merely means "symmetric"; if $\Lambda = Q$ it is defined using the usual conjugation in Q.) Our aim is to determine the numbers $\Lambda(n)$, $\Lambda_H(n)$.

Of course, it is possible to word the problem more invariantly. Let W be a set of matrices which is a vector space of dimension k over R; we will say that W "has the property P" if every nonzero w in W is nonsingular (invertible). We now ask for the maximum possible dimension of such a space.

In [1], the first named author has proved that $R(n)$ equals the so-called Radon-Hurwitz function, defined below. In this note we determine $R_H(n)$, $C(n)$, $C_H(n)$, $Q(n)$ and $Q_H(n)$ by deriving inequalities between them and $R(n)$. The elementary constructions needed to prove these inequalities can also be used to give a simplified description of the Radon-Hurwitz matrices.

The study of sets of real symmetric matrices $\{A_j\}$ with the property P may be motivated as follows. For such a set, the system of partial differential equations

$$u_t = \sum_j A_j u_{x_j}$$

is a symmetric hyperbolic system in which the sound speeds are nonzero in every direction. For such systems the solution energy is propagated to infinity and a scattering theory can be developed.

To give our results, we require the Radon-Hurwitz numbers [2],

Received by the editors September 10, 1963.

[1] Sloan Fellow.

[2] Sponsored by the National Science Foundation, contract NSF-G 16434.

318

[3]. We set $n = (2a+1)2^b$ and $b = c+4d$, where a, b, c, d are integers with $0 \leq c < 4$; then we define

$$\rho(n) = 2^c + 8d.$$

THEOREM 1. *We have*

$$R(n) = \rho(n), \qquad R_H(n) = \rho(\tfrac{1}{2}n) + 1,$$
$$C(n) = 2b + 2, \qquad C_H(n) = 2b + 1,$$
$$Q(n) = 2b + 4, \qquad Q_H(n) = 2b + 1.$$

The results for $\Lambda = Q$ are included so that topologists may avoid jumping to the conclusion that the subject is directly related to the Bott periodicity theorems. If this were so then it would be surprising to see the case $\Lambda = Q$ behaving like the case $\Lambda = C$.

The proof of Theorem 1 will be based on a number of simple constructions, which we record as lemmas.

LEMMA 1. $R_H(n) \leq C_H(n) \leq Q_H(n)$.

This is clear, since a matrix with entries from R may be regarded as a matrix with entries from C, and similarly for C and Q.

LEMMA 2. (a) $C(n) \leq R(2n)$, (b) $Q(n) \leq C(2n)$.

PROOF. We may regard our matrices as Λ-linear transformations of coordinate n-space Λ^n. Now by forgetting part of the structure of C^n it becomes a real vector space of dimension $2n$ over R, i.e., an R^{2n}. Thus any C-linear transformation of C^n gives an R-linear transformation of R^{2n}. Similarly for C and Q.

LEMMA 3. $\Lambda(n) + 1 \leq \Lambda_H(2n)$.

PROOF. Let W be a k-dimensional space of $n \times n$ matrices with entries from Λ which has the property P. For each $A \in W$ and $\lambda \in R$, consider the following linear transformation from $\Lambda^n \oplus \Lambda^n$ to itself.

$$B(x, y) = (Ay + \lambda x, A^*x - \lambda y).$$

It is clear that its matrix is Hermitian, and that such B form a $(k+1)$-dimensional space. We claim that this set $\{B\}$ has property P. For suppose that some B is singular; then there exist x, y not both zero such that

$$Ay + \lambda x = 0, \qquad A^*x - \lambda y = 0.$$

Evaluating x^*Ay in two ways, we find

$$\lambda(x^*x + y^*y) = 0.$$

Hence $\lambda = 0$. Thus either A or A^* is singular; so A is singular and $A = 0$. This proves the lemma.

LEMMA 4. (a) $C_H(n) + 1 \leq C(n)$, (b) $Q_H(n) + 3 \leq Q(n)$.

PROOF. Let W be a k-dimensional space of $n \times n$ Hermitian matrices with entries from Λ which has the property P. Consider the matrices

$$A + \mu I,$$

where A runs over W and μ runs over the pure imaginary elements of Λ. We claim that they form a space with the property P and of dimension $k+1$ if $\Lambda = C$ or $k+3$ if $\Lambda = Q$. In fact, suppose that such a matrix is singular; then there is a nonzero x such that

$$Ax = -\mu x;$$

arguing as is usual for the complex case, we find

$$-\mu x^*x = x^*Ax = (-\mu x)^*x = \mu x^*x.$$

So μ is zero, A is singular, and thus A is zero. This proves the lemma.

LEMMA 5. $R_H(n) + 7 \leq R(8n)$.

PROOF. Let W be a k-dimensional space of real symmetric matrices with the property P. We require also the Cayley numbers K, which form an 8-dimensional algebra over R. We can thus form the real vector space

$$R^n \otimes_R K$$

of dimension $8n$. For each $A \in W$ and each pure imaginary $\mu \in K$ we consider the following linear transformation from $R^n \otimes_R K$ to itself:

$$B(x \otimes y) = Ax \otimes y + x \otimes \mu y.$$

We claim that the $(k+7)$-dimensional space formed by such B has property P. For suppose that some B is singular, and suppose, to begin with, that μ is nonzero. Then the elements $1, \mu$ form an R-base for a sub-algebra of K which we may identify with C. Now every two elements of K generate an associative sub-algebra; in particular, K is a left vector space over C. Choose a C-base of K; this splits $R^n \otimes_R K$ as the direct sum of 4 copies of $R^n \otimes_R C$. Since B acts on each summand, it must be singular on at least one. That is, the real symmetric matrix A has a nonzero complex eigenvalue which is purely imaginary—a contradiction. Hence μ must be zero and $B = A \otimes 1$. Now choose an R-base of K; this splits $R^n \otimes_R K$ as the direct sum of 8 copies of R^n. Since B acts on each summand, it must be singular on at least one. That is, A must be singular; hence $A = 0$. This completes the proof.

PROOF OF THEOREM 1. First we consider $R_H(n)$. If we use the fact that $R(n) = \rho(n)$, Lemmas 3 and 5 give

$$\rho(\tfrac{1}{2}n) + 1 \leq R_H(n) \leq \rho(8n) - 7.$$

But using the explicit definition of ρ, we have

$$\rho(8n) - 7 = \rho(\tfrac{1}{2}n) + 1.$$

This disposes of $R_H(n)$.

It follows from this argument that if we have a set of $\rho(n)$ $n \times n$ matrices with the property P, then by applying successively the constructions given in the proofs of Lemma 3 (taking $\Lambda = R$) and Lemma 5, we obtain a set of $\rho(2^4 n)$ $2^4 n \times 2^4 n$ matrices with the property P. Now the set of 1, 2, 4, and 8 matrices which express the respective actions of R, C, Q, and K on R^m, C^m, Q^m, and K^m for $m = 2a + 1$ can be used to start the induction for the different cases $b \equiv 0$, 1, 2, and 3 (mod 4). This gives a slight variation of the construction of Hurwitz and Radon [2], [3]; the iterative procedures used by these authors require more steps and do not involve the Cayley numbers explicitly.

Next we consider $C(n)$. Lemmas 3 and 4(a) give

$$C(n) + 2 \leq C(2n).$$

Now induction shows that

$$C(n) \geq 2b + 2,$$

which gives us our inequality one way. Applying Lemma 2(a) directly only gives a good inequality for certain values of b, so we proceed as follows. Choose $e \geq 0$ such that $b + e \equiv 0$ or 1 (mod 4). Then by induction we have

$$C(n) + 2e \leq C(2^e n);$$

by Lemma 2(a) we have

$$C(2^e n) \leq \rho(2^{e+1} n),$$

and by our choice of e we have

$$\rho(2^{e+1}n) = 2b + 2e + 2.$$

This gives

$$C(n) \leqq 2b + 2,$$

and proves the assertion made about $C(n)$.

Lemmas 3 and 4(a) now show that

$$C_H(n) = 2b + 1.$$

Again, Lemmas 1, 4(b) and 2(b) show that

$$Q_H(n) - C_H(n) \geqq 0,$$
$$Q(n) - Q_H(n) \geqq 3,$$
$$C(2n) - Q(n) \geqq 0.$$

But $C(2n) - C_H(n) = 3$, so all these inequalities are equalities. This completes the proof of Theorem 1.

REFERENCES

1. J. F. Adams, *Vector fields on spheres*, Ann. of Math. (2) **75** (1962), 603–632.

2. A. Hurwitz, *Über die Komposition der quadratischen Formen*, Math. Ann. **88** (1923), 1–25.

3. J. Radon, *Lineare Scharen orthogonalen Matrizen*, Abh. Math. Sem. Univ. Hamburg **1** (1922), 1–14.

MANCHESTER UNIVERSITY, MANCHESTER, ENGLAND,
    NEW YORK UNIVERSITY, AND
        STANFORD UNIVERSITY

CORRECTION TO "ON MATRICES WHOSE REAL LINEAR COMBINATIONS ARE NONSINGULAR"

J. F. ADAMS, PETER D. LAX AND RALPH S. PHILLIPS

We are grateful to Professor B. Eckmann for pointing out an error in the proof of Lemma 4(b) of our paper [1]. This error invalidates Lemma 4(b) and that part of Theorem 1 which states the values of $Q(n)$, $Q_H(n)$. The error occurs immediately after the words "arguing as is usual for the complex case, we find"; it consists in manipulating as if the ground field Λ were commutative.

The proof of Lemma 4(b) can be repaired, as will be shown below, but it leads to a different conclusion from that given. Our paper should therefore be corrected as follows.

(i) In Theorem 1, the values of $Q(n)$ and $Q_H(n)$ should read

$$"Q(n) = \rho(\tfrac{1}{2}n) + 4, \qquad Q_H(n) = \rho(\tfrac{1}{4}n) + 5."$$

The two sentence paragraph following Theorem 1 should be deleted. It remains interesting to ask what topological phenomena (if any) can be related to our algebraic results.

(ii) In Lemma 4, part (b) should read

$$"Q_H(n) + 3 \leq R(4n)."$$

The proof is as follows.

Let W be a k-dimensional space of $n \times n$ Hermitian matrices with entries from Q which has the property P. The space Q^n is a real vector space of dimension $4n$. For each $A \in W$ and each pure imaginary $\mu \in Q$ we consider the following real-linear transformation from Q^n to itself:

Received by the editors December 20, 1965.

$$B(x) = Ax + x\mu.$$

We claim that the $(k+3)$-dimensional space formed by such B has the property P. For suppose that such a B is singular; then there is a nonzero x such that

$$Ax = -x\mu;$$

then we have

$$x^*(Ax) = -x^*x\mu,$$
$$(x^*A)x = (-x\mu)^*x = \mu x^*x.$$

Since x^*x is real and nonzero, we have $\mu = 0$; hence A is singular and $A = 0$. This completes the proof.

(iii) In Lemma 5, there should be added a second part, reading

$$\text{“(b)} \quad R_H(n) + 3 \leqq Q(n).\text{”}$$

PROOF. Let W be a k-dimensional space of $n \times n$ real symmetric matrices which has the property P. Consider the matrices

$$A + \mu I,$$

where A runs over W and μ runs over the pure imaginary elements of Q. We claim that they form a space of dimension $k+3$ with the property P. In fact, suppose that such a matrix is singular; and suppose to begin with, that μ is nonzero. Then the elements 1, μ form an R-base for a subalgebra of Q which we may identify with C. Choose a C-base of Q; this splits Q^n as the direct sum of two copies of C^n. Since the matrix $A + \mu I$ acts on each summand, it must be singular on at least one. That is, the real symmetric matrix A has a nonzero complex eigenvalue which is purely imaginary, a contradiction. Hence μ must be zero and $B = A$. Now choose an R-base of Q; this splits Q^n as the direct sum of 4 copies of R^n. Sinc A acts on each summand, it must be singular on at least one. That is, A must be singular; hence $A = 0$. This completes the proof.

(iv) The final paragraph of the paper should be deleted, and replaced by the following proof.

"Finally, Lemmas 5(b), 3 and 4(b) show that

$$Q(n) - R_H(n) \geqq 3,$$
$$Q_H(2n) - Q(n) \geqq 1,$$
$$R(8n) - Q_H(2n) \geqq 3.$$

But we have already shown that

$$R(8n) - R_H(n) = 7,$$

so all these inequalities are equalities. This completes the proof of Theorem 1."

We note that this method provides an alternative proof of Lemma 5 $(R(8n) - R_H(n) \geqq 7)$, without using the Cayley numbers.

BIBLIOGRAPHY

1.   J. F. Adams, Peter D. Lax and Ralph S. Phillips, *On matrices whose real linear combinations are nonsingular,* Proc. Amer. Math. Soc. 16 (1965), 318–322.

MANCHESTER UNIVERSITY, MANCHESTER, ENGLAND
   NEW YORK UNIVERSITY, AND
      STANFORD UNIVERSITY

On Sums of Squares

To Olga Taussky-Todd, with affection and admiration

Anneli Lax and Peter D. Lax
Courant Institute of Mathematical Sciences
New York University
New York, New York 10012

Submitted by Richard Varga

ABSTRACT

We show that the fourth order form in five variables,

$$\sum_{i=1}^{5} \prod_{j \neq i} (x_i - x_j),$$

is nonnegative, but cannot be written as a sum of squares of quadratic forms.

In the 1971 International Mathematical Olympiad for high school students (see [4]), the following problem was posed:

Let n be a positive integer, $x_1, x_2, \ldots, x_n$, n real numbers. Define the form $A_n(x)$ to be

$$A_n(x) = \sum_{i=1}^{n} \prod_{j \neq i} (x_i - x_j). \tag{1}$$

For which integers n is the form A_n nonnegative?

The answer is: only for $n = 3$ and $n = 5$. Clearly, for n even, A_n is of odd degree and so changes sign. Suppose n is odd and $\geqslant 7$; set

$$x_1 = x_2 = x_3 = 0, \qquad x_4 = 1, \qquad x_5 = x_6 = \cdots = x_n = 2.$$

All terms but the 4th in the sum (1) are zero, so

$$A_n(x) = (-1)^{n-4} = -1 < 0.$$

LINEAR ALGEBRA AND ITS APPLICATIONS 20, 71–75 (1978)   71

© Elsevier North-Holland, Inc., 1978   0024-3795/78/0020-0071/$1.25

For $n = 3$ one can write

$$A_3(x) = \tfrac{1}{2}\left[(x_1 - x_2)^2 + (x_1 - x_3)^2 + (x_2 - x_3)^2 \right],$$

clearly nonnegative. For $n = 5$ one makes use of the invariance of $A_n(x)$ under any permutation of the variables x_i and assumes that the x_j are nondecreasing with the index j. Under this condition the sum of the first two terms in (1), which can be written as

$$(x_1 - x_2)\left[(x_1 - x_3)(x_1 - x_4)(x_1 - x_5) - (x_2 - x_3)(x_2 - x_4)(x_2 - x_5) \right],$$

is clearly $\geqslant 0$. Similarly the sum of the last two terms is $\geqslant 0$. The middle term, being the product of two nonnegative and two nonpositive numbers, also is $\geqslant 0$. This shows that $A_5(x) \geqslant 0$ for all x.

The proof for the case $n = 3$ differs from the proof for $n = 5$ inasmuch as A_3 was exhibited as a sum of squares, while A_5 wasn't. The purpose of this note is to point out the

THEOREM. A_5 cannot be written as a sum of squares of quadratic forms.

That higher order nonnegative forms cannot always be written as sums of squares was first pointed out by Hilbert in [5]; he credits Minkowski for conjecturing this. A review of this subject is given by Olga Taussky in [7]. The present example is of degree 4, ostensibly in 5 variables but in reality only in 4, since $A_5(x)$ depends only on the differences of the variables x_j. This is the smallest possible number of variables, since, as shown by Hilbert in [5], every nonnegative quartic form in 3 variables can be written as a sum of squares.

Hilbert conjectured in his famous collection of problems, [6], that every nonnegative form can be written as a sum of squares of *ratios of forms*. This was proved to be true by Artin in [1]. It may be interesting to determine the smallest number of squares of ratios of forms into which A_5 can be decomposed.

Proof. Suppose A_5 could be represented as

$$A_5 = \sum Q_j^2, \tag{2}$$

each Q_j a quadratic form. Clearly $Q_j(x) = 0$ whenever $A_5(x) = 0$. Now $A_5(x) = 0$ whenever each x_j equals some other x_k; this is the case for instance when

$$x_1 = x_2 \quad \text{and} \quad x_3 = x_4 = x_5, \tag{3}$$

and what (3) becomes when the indices are permuted.

LEMMA. *A quadratic form Q which is zero whenever condition (3), or one of its permutations, is satisfied is identically zero.*

This lemma, combined with the above observation, shows that each Q_j in the representation (2) is identically zero. This proves that (2) cannot hold and proves the theorem. ∎

Proof of Lemma. Write

$$Q(x) = \sum c_{jk} x_j x_k, \qquad c_{jk} = c_{kj}. \tag{4}$$

By hypothesis $Q(x) = 0$ when $x_1 = x_2 = y$, $x_3 = x_4 = x_5 = z$:

$$0 = Q(y, y, z, z, z) = (c_{11} + 2c_{12} + c_{22}) y^2$$

$$+ 2(c_{13} + c_{14} + c_{15} + c_{23} + c_{24} + c_{25}) yz$$

$$+ (c_{33} + c_{44} + c_{55} + 2c_{34} + 2c_{35} + 2c_{45}) z^2.$$

Since this holds for all y, z, it follows that

$$c_{11} + 2c_{12} + c_{22} = 0, \tag{5}$$

$$c_{13} + c_{14} + c_{15} + c_{23} + c_{24} + c_{25} = 0, \tag{6}$$

$$c_{33} + c_{44} + c_{55} + 2c_{34} + 2c_{35} + 2c_{45} = 0. \tag{7}$$

Since the relations obtained by permuting the indices above are also true, we deduce from (5), using the permutation $12345 \rightarrow 34125$, that

$$c_{33} + 2c_{34} + c_{44} = 0.$$

Subtracting this from (7), we get

$$c_{55} + 2c_{35} + 2c_{45} = 0. \tag{8}$$

Consider all permutations of the indices which keep 5 fixed; the relations resulting from (8) remain true, so we deduce that

$$c_{j5} + c_{k5} = -\tfrac{1}{2} c_{55}$$

for all $j \neq k$ and $j, k \neq 5$. This implies that

$$c_{15} = c_{25} = c_{35} = c_{45}. \tag{9}$$

Transposing 1 and 5, and then 2 and 5, we deduce from (9) that

$$c_{51} = c_{21} = c_{31} = c_{41},$$

$$c_{12} = c_{52} = c_{32} = c_{42}.$$

Using the symmetry of c_{jk}, we deduce from these relations that

$$c_{1j} = c_{2j} = c_{12} \qquad \text{for} \quad j = 3, 4, 5.$$

Substituting this into (6), we conclude that $6c_{12} = 0$. Since the pair $1, 2$ can be replaced by any other, we conclude that

$$c_{jk} = 0 \qquad \text{for} \quad j \neq k.$$

Substituting this into (8), we deduce that $c_{55} = 0$. Since 5 can be replaced by any other index, we conclude that

$$c_{jj} = 0.$$

This completes the proof of the lemma.   ∎

Other interesting examples have appeared in a paper of M. D. Choi [2]. M. D. Choi and T. Y. Lam [3] have announced that the very simple quartic form $x^2 y^2 + y^2 z^2 + z^2 x^2 + w^4 - 4xyzw$ is positive semidefinite but not a sum of squares.

REFERENCES

1 E. Artin, Über die Zerlegung definiter Funktionen in Quadrate, Abh. Math. Sem. Univ. Hamb. 5 (1926) 100–115.
2 M. D. Choi, Positive semi-definite biquadratic forms, Linear Algebra Appl. 12 (1975), 95–100.
3 M. D. Choi and T. Y. Lam, Positive semi-definite homogeneous polynomials, Am. Math. Soc. Not. 23 (1976), A-353.
4 S. L. Greitzer, International Mathematical Olympiad, NML, Vol. 27, MAA, 1977.
5 D. Hilbert, Über die Darstellung definiter Formen als Summen von Formenquadraten, Math. Ann. 32 (1888), 342–350.

6 D. Hilbert, Mathematical developments arising from Hilbert problems, *Proceedings of Symposia of Pure Math*, Vol. 28, 1976, AMS.
7 O. Taussky, Sums of squares, *Am. Math. Mon.* 77 (1970), 805–830.

Received 22 November 1976

BULLETIN (New Series) OF THE
AMERICAN MATHEMATICAL SOCIETY
Volume 6, Number 2, March 1982

THE MULTIPLICITY OF EIGENVALUES

BY PETER D. LAX

There are many examples of first order $n \times n$ systems of partial differential equations in 2 space variables with real coefficients which are strictly hyperbolic; that is, they have simple characteristics. In this note we show that in 3 space variables there are no strictly hyperbolic systems if $n \equiv 2(4)$. Multiple characteristics of course influence the propagation of singularities. For a different context see Appendix 10 of [2].

M denotes the set of all real $n \times n$ matrices with real eigenvalues. We call such a matrix *nondegenerate* if it has n distinct real eigenvalues.

THEOREM. *Let A, B, C be three matrices such that all linear combinations*

(1) $$\alpha A + \beta B + \gamma C,$$

α, β, γ *real, belong to* M . *If $n \equiv 2 \pmod 4$, then there exists α, β, γ real, $\alpha^2 + \beta^2 + \gamma^2 \neq 0$ such that (1) is degenerate.*

REMARK 1. The theorem applies in particular to A, B, C real symmetric.

REMARK 2. The theorem shows that first order hyperbolic systems in three space variables of the indicated order always have some multiple characteristics.

PROOF. Denote by N the set of nondegenerate matrices in M . The normalized eigenvectors u of N is N ,

$$Nu_j = \lambda_j u_j, \quad |u_j| = 1, \quad j = 1, \dots, n,$$

are determined up to a factor ± 1.

Let $N(\theta)$, $0 \leqslant \theta \leqslant 2\pi$, be a closed curve in N. If we fix $u_j(0)$, then $u_j(\theta)$ can be determined uniquely by requiring continuous dependence on θ. Since $N(2\pi) = N(\theta)$,

(2) $$u_j(2\pi) = \tau_j u_j(0), \quad \tau_j = \pm 1.$$

Clearly

(i) Each τ_j is a homotopy invariant of the closed curve.

(ii) Each $\tau_j = 1$ when $N(\theta)$ is constant.

Suppose now that the theorem is false; then

(3) $$N(\theta) = \cos \theta A + \sin \theta B$$

Received by the editors June 4, 1981.

1980 *Mathematics Subject Classification.* Primary 15A18, 57R99, 35L40.

213

is a closed curve in N. Note that $N(\pi) = -N(0)$; this shows that

(4)
$$\lambda_j(\pi) = -\lambda_{n-j+1}(0) \quad \text{and}$$
$$u_j(\pi) = \rho_j u_{n-j+1}(0), \quad \rho_j = \pm 1.$$

Since the ordered basis $\{u_1(\theta), \ldots, u_n(\theta)\}$ is deformed continuously, it retains its orientation. Thus the ordered bases

$$\{u_1(0), \ldots, u_n(0)\} \quad \text{and} \quad \{\rho_1 u_n(0), \ldots, \rho_n u_1(0)\}$$

have the same orientation. For $n \equiv 2 \pmod 4$, reversing the order reverses the orientation of an ordered base; this proves that

$$\prod_1^n \rho_j = -1.$$

This implies that there is a value of k for which

(5)
$$\rho_k \rho_{n-k+1} = -1.$$

Next we observe that $N(\theta + \pi) = -N(\theta)$; it follows from this that $\lambda_j(\theta + \pi) = -\lambda_{n-j+1}(\theta)$ and by (4) that

$$u_j(2\pi) = \rho_{n-j+1} u_{n-j+1}(\pi).$$

Combining this with (4) we get that $\tau_j = \rho_j \rho_{n-j+1}$. By (5), $\tau_k = -1$; this shows that the curve (3) is not homotopic to a point.

Suppose that all matrices of form (1), $\alpha^2 + \beta^2 + \gamma^2 = 1$, belonged to N. Then since the sphere is simply connected the curve (4) could be contracted to a point, contradicting $\tau_k = -1$.

See [1] for related matters.

ADDED IN PROOF. S. Friedland, J. Robbin and J. Sylvester have proved the theorem for all $n \equiv \pm2, \pm3, \pm4 \pmod 8$, and have shown it false for $n = 0$, $\pm 1 \pmod 8$. They have further results involving linear combinations of more than 3 matrices.

REFERENCES

1. F. John, *Restriction on the coefficients of hyperbolic differential equations*, Proc. Nat. Acad. Sci. U. S. A. 74 (1977), 4150–4151.
2. V. L Arnold, *Math. methods in classical mechanics*, Lecture Notes in Math., vol. 60, Springer-Verlag, Berlin and New York, 1978.

COURANT INSTITUTE OF MATHEMATICAL SCIENCES, NEW YORK UNIVERSITY, NEW YORK, NEW YORK 10012

On the Discriminant of Real Symmetric Matrices

PETER D. LAX

Courant Institute

Let S be a real symmetric matrix of order $n \times n$. We define its discriminant $d(S)$ in the usual fashion as

$$(1) \qquad d(S) = \prod_{i<j} (\lambda_i - \lambda_j)^2,$$

where the λ_i are the eigenvalues of S. One can write $d(S)$ as the discriminant of the characteristic polynomial; this shows that $d(S)$ is a homogeneous polynomial of degree $n(n-1)$ in the entries of the matrix S. Since the eigenvalues of a real symmetric matrix are real, it follows from (1) that $d(S)$ is nonnegative. It is zero if and only if S is *degenerate*, i.e., if S has a multiple eigenvalue.

It was Hilbert and Minkowski who observed that, unlike quadratic forms, nonnegative forms of degree greater than 2 and in more than two variables cannot always be written as a sum of squares of real polynomials. Hilbert gave the first such example (see [2]); another example can be found in [4]. The question of which nonnegative polynomials can be written as a sum of squares of polynomials is left open.[1] In this note we show that the discriminant of real symmetric matrices can be so written.

THEOREM 1 *The discriminant of a real symmetric matrix is a sum of squares of polynomials.*

1 Suppose the discriminant can be written as a sum of squares:

$$(2) \qquad d(S) = \sum r_j^2(S).$$

Since the discriminant of an $n \times n$ matrix is a homogeneous polynomial of degree $n(n-1)$, each r_j is likewise a homogeneous polynomial of degree $n(n-1)/2$ in the entries of the matrix S. The null set of the discriminant is the set of degenerate matrices, that is, those with multiple eigenvalues. Clearly, each r_j in (2) must vanish on the null set.

It will now be shown that, conversely, if there is a nonzero polynomial r of degree $n(n-1)/2$ that vanishes when S is degenerate, then we can construct a

---

[1] In the seventeenth problem of his Paris address, Hilbert posed the question if every nonnegative form can be written as a quotient of sums of squares of polynomials. E. Artin [1] has shown in a brilliant fashion that this can always be done.

Communications on Pure and Applied Mathematics, Vol. LI, 1387–1396 (1998)
© 1998 John Wiley & Sons, Inc.   CCC 0010–3640/98/111387-10

representation formula (2). This is done as follows: Define the polynomial q by the formula

$$(3) \qquad q(S) = \int r^2 \left(OSO^\mathsf{T} \right) dO \,.$$

The integration is with respect to the invariant measure over the set of all special orthogonal matrices. Since r is not identically zero, neither is q; since r is of degree $n(n-1)/2$, q is of degree $n(n-1)$.

Claim.

(i) q is nonnegative.

(ii) $q(S) = 0$ when S is degenerate.

(iii) q is invariant, that is, $q(RSR^\mathsf{T}) = q(S)$ for any special orthogonal matrix R.

PROOF: (i) follows since the integrand is nonnegative. For (ii), we note that if S is degenerate, so is OSO^T. (iii) follows because the integration is with respect to the invariant measure.   ∎

We claim further that the only polynomial q with these properties is a constant times the discriminant. For, take S to be a diagonal matrix D with diagonal entries $d_1, d_2, \ldots, d_n$. Then $q(D)$ is a polynomial of degree $n(n-1)$ in the variables $d_1, d_2, \ldots, d_n$. By property (ii), $q(D)$ is zero when $(d_i - d_j)$ is zero. Therefore q is divisible by each of these and hence by their product

$$(4) \qquad q(D) = \prod_{i<j}(d_i - d_j)p \,,$$

where p is a polynomial of degree $n(n-1)/2$. But p is also zero when $(d_i - d_j)$ is zero; otherwise, q would change sign when $(d_i - d_j)$ changes sign. This shows that p, too, is divisible by the product of these factors; since p is of degree $n(n-1)/2$, it must be of the form

$$(5) \qquad p(D) = \text{const} \prod_{i<j}(d_i - d_j) \,,$$

and so

$$(6) \qquad q(D) = \text{const} \prod_{i<j}(d_i - d_j)^2 \,.$$

Any symmetric matrix can be diagonalized by an orthogonal matrix

$$RSR^\mathsf{T} = D,$$

the diagonal entries of D being the eigenvalues of S. Since q is rotation-invariant, it follows from (6) that for any S with eigenvalues λ_i,

$$q(S) = \text{const} \prod_{i<j}(\lambda_i - \lambda_j)^2 = \text{const} \cdot d(S),$$

as claimed.

Formula (3) thus represents the discriminant as an integral of squares; this can easily be turned into a finite sum of squares. Polynomials of degree $n(n-1)/2$ form a finite-dimensional linear space; let $r_1, r_2, \ldots, r_k$ be a basis for this space. Then $r(OSO^\mathsf{T})$ can be expressed uniquely as a linear combination of the r_j:

$$r(OSO^\mathsf{T}) = \sum_1^k c_j(O)r_j(S).$$

Substituting this into (3) and identifying q as d gives

(7) $$d(S) = \int \left(\sum_1^k c_j(O)r_j(S) \right)^2 dO = \sum u_{ij}r_i(S)r_j(S),$$

where

$$u_{ij} = \int c_i(O)c_j(O)dO.$$

Using this integral representation of u_{ij}, we can write the associated quadratic form as an integral:

$$\sum u_{ij}\xi_i\xi_j = \int \sum c_i(O)c_j(O)\xi_i\xi_j \, dO = \int \left(\sum c_i(O)\xi_i \right)^2 dO.$$

This proves that the quadratic form on the right in equation (7) is nonnegative; therefore it can be written as a sum of squares of linear combinations of the r_j.

To complete the proof of Theorem 1, all that remains is to construct a non-trivial polynomial r of degree $n(n-1)/2$ that is zero whenever S is degenerate.

2 We need the following characterization of degenerate real symmetric matrices:

THEOREM 2 *A real symmetric matrix S is degenerate if and only if it commutes with some nonzero real antisymmetric matrix A:*

(8) $$AS - SA = 0, \quad A^\mathsf{T} = -A.$$

PROOF: Relation (8) is invariant under orthogonal transformation, so it is sufficient to consider diagonal matrices $S = D$. Obviously $AD - DA = 0$ can be satisfied by an antisymmetric A if and only if D is degenerate.

We can regard the commutation relation (8) as a system of homogeneous linear equations for the entries of A. The number of equations is $n(n+1)/2$, since the commutator of a symmetric and an antisymmetric matrix is symmetric; the number of unknowns, the entries of A, is $n(n-1)/2$. Out of this overdetermined system we select the $n(n-1)/2$ equations that come from the off-diagonal entries in $AS - SA$. The resulting determined system of equations can be written in matrix form as

$$(9) \qquad\qquad\qquad Ma = 0,$$

where M is an $\frac{n(n-1)}{2} \times \frac{n(n-1)}{2}$ matrix, and a is a vector whose components are the entries of A. The entries of M are linear functions of S.

According to Theorem 2, if S is degenerate, the system of homogeneous equations (9) has a nontrivial solution. This can be only if the determinant of M is zero. Since $\det M$ is a polynomial of degree $\frac{n(n-1)}{2}$ in the entries of S, it will serve as the polynomial $r(S)$ we are looking for; all that remains is to verify that $\det M$ is not identically zero. This is not hard to accomplish.

Take $i < j$; the ij^{th} entry of AS is a sum that contains the term $a_{ij}s_{jj}$; the ij^{th} entry of SA is a sum that contains the term $s_{ii}a_{ij}$. Therefore the ij^{th} entry of $AS - SA$ contains the term $(s_{jj} - s_{ii})a_{ij}$. So the diagonal entries of the matrix M in (9) are $(s_{ii} - s_{jj})$; no other entry of M contains any diagonal entries of S. When $\det M$ is expressed as a sum of products, one term is the product

$$\prod_{i<j}(s_{ii} - s_{jj})$$

of all the diagonal entries of M; all other products contain fewer than $n(n-1)/2$ diagonal entries of S. This proves that $\det M \not\equiv 0$ and completes the proof of Theorem 1.   ∎

To illustrate, we write down explicitly the polynomials constructed above for the lowest two values of n. For $n = 2$,

$$r_2(S) = s_{11} - s_{22}.$$

For $n = 3$,

$$r_3(S) = (s_{11} - s_{22})(s_{22} - s_{33})(s_{33} - s_{11}) + (s_{11} - s_{22})s_{12}^2$$
$$+ (s_{22} - s_{33})s_{23}^2 + (s_{33} - s_{11})s_{12}^2.$$

3 The argument of the first two sections gives no reasonable estimate of the number of squares needed to represent the discriminant. For $n = 2$,

$$S = \begin{bmatrix} a & c \\ c & b \end{bmatrix};$$

a brief calculation gives

$$d(S) = (a - b)^2 + 4c^2,$$

a sum of two squares.

How about $n > 2$? According to a classical result of von Neumann and Wigner, the degenerate, real symmetric matrices form a variety of codimension 2. For, degenerate matrices S_d can be represented as

(10) $$S_d = OD_dO^\mathsf{T},$$

where D_d is a degenerate diagonal matrix; that is, two of its diagonal elements are equal. The degenerate diagonal $n \times n$ matrices are the union of a finite number of linear spaces of dimension $n - 1$. The orthogonal matrices form a variety of dimension $n(n-1)/2$, but a degenerate D_d is left-invariant by a one-parameter group of rotations. The orthogonal group modulo this subgroup is of dimension one less, so (10) is an $\frac{n(n-1)}{2} - 1 + n - 1 = \frac{n(n+1)}{2} - 2$ dimensional variety.

In light of the von Neumann–Wigner result, one would expect $d(S)$ to be represented as a sum of two squares; we show now that this is not possible for $n > 2$.

Suppose $d(S)$ could be represented as a sum of two squares,

(11) $$d(S) = p^2(S) + r^2(S).$$

Let c and s be a pair of real numbers, $c^2 + s^2 = 1$. Define

(12) $$p' = cp + sr, \qquad r' = -sp + cr.$$

Clearly

(11') $$d(S) = p'^2(S) + r'^2(S).$$

We call (11) and (11') equivalent representations of $d(S)$.

The polynomial $d(S)$ has only a finite number of inequivalent representations; for, (11) implies the factorization

$$d = (p + ir)(p - ir),$$

and a polynomial has only a finite number of inequivalent factorizations.

Replace S in (11) by OSO^T; since $d(S)$ is invariant under this replacement, we get

$$(11_0) \qquad\qquad d(S) = p_0^2(S) + r_0^2(S)\,,$$

where

$$(13) \qquad p_0(S) = p(OSO^T)\,, \qquad r_0(S) = r(OSO^T)\,.$$

The polynomials p_0 and r_0 depend continuously on O; since the inequivalent representations form a discrete set, the representations $(11)_0$ are equivalent with (11). So by (12)

$$(14) \qquad p_0 = c(O)p + s(O)r\,, \qquad r_0 = -s(O)p + c(O)r\,.$$

The matrix relating (p, r) to (p_0, r_0),

$$R(O) = \begin{bmatrix} c(O) & s(O) \\ -s(O) & c(O) \end{bmatrix},$$

is a rotation of the plane. It follows from (13) and (14) that

$$R(O_1 O_2) = R(O_1)R(O_2)\,,$$

i.e., that $R(O)$ is a representation of the n-dimensional orthogonal group by rotations of the plane. For $n > 2$, there is no such representation except the trivial one $R(O) \equiv I$; therefore $p_0 \equiv p$ and $r_0 \equiv r$. Since the real symmetric matrix S can be diagonalized,

$$(15) \qquad\qquad r(S) = r_0(S) = r(D)\,,$$

D a diagonal matrix whose entries are the eigenvalues λ_j of S. It follows from (11) that $r(S) = 0$ when S is degenerate; therefore we deduce from (15) that

$$(16) \qquad\qquad r(S) = \text{const} \prod_{i<j} (\lambda_i - \lambda_j)\,.$$

But the right side of (16) is not a polynomial in the entries of S. This is a contradiction into which we were led by the mistaken assumption that the discriminant of S can be represented by a sum of two squares.

We show next that for $n = 3$ we need at least six squares; this can be seen as follows: Take for S a 3×3 symmetric matrix whose diagonal elements are zero:

$$S_0 = \begin{bmatrix} 0 & d & e \\ d & 0 & f \\ e & f & 0 \end{bmatrix}.$$

The characteristic polynomial of S_0 is

(17) $$\lambda^3 - (d^2 + e^2 + f^2)\lambda - 2def.$$

The discriminant of the cubic $\lambda^3 + u\lambda + v$ is

(18) $$-27v^2 - 4u^3.$$

Setting the coefficients in (17) into formula (18) gives

(19) $$\frac{1}{4}d(S_0) = (d^2 + e^2 + f^2)^3 - 27d^2e^2f^2.$$

This formula shows that the nonnegativity of the discriminant of matrices S_0 expresses the classic inequality between the arithmetic and geometric means of three positive numbers.

Expanding (19) gives

$$d(S_0) = 4d^6 + 4e^6 + 4f^6$$
$$+ 12d^4e^2 + 12d^4f^2 + 12e^4d^2 + 12e^4f^2 + 12f^4d^2 + 12f^4e^2$$
$$- 84d^2e^2f^2.$$

A brief calculation shows that this can be written as the sum of squares

$$d(S_0) = (d(2d^2 - e^2 - f^2))^2 + 15(d(e^2 - f^2))^2$$
$$+ (e(2e^2 - f^2 - d^2))^2 + 15(e(f^2 - d^2))^2$$
$$+ (f(2f^2 - d^2 - e^2))^2 + 15(f(d^2 - e^2))^2.$$

I believe that six is the smallest number of squares needed to express $d(S_0)$; so, at least six squares would be needed to express $d(S)$ for arbitrary symmetric 3×3 matrices S.

I have no further estimate on the number of squares needed to represent the discriminant except for the following observation:

THEOREM 3   *The number of squares needed to represent the discriminant of $n \times n$ real symmetric matrices is an increasing function of n.*

PROOF: We have to show that, if the discriminant of $n \times n$ real symmetric matrices can be represented as a sum of N squares, then the discriminant of real symmetric matrices of order $(n-1) \times (n-1)$ can be represented as a sum of not more than N squares. To show this, we turn an $(n-1) \times (n-1)$ real symmetric matrix T into an $n \times n$ matrix S by adding to T a row and a column of zeros:

$$(20) \qquad S = \begin{bmatrix} T & & 0 \\ & & \vdots \\ 0 & \cdots & 0 \end{bmatrix}.$$

The eigenvalues of S are the eigenvalues of T and 0. So

$$(21) \qquad d(S) = d(T)[\det T]^2 .$$

Suppose that $d(S)$ is represented as a sum of N squares:

$$(22) \qquad d(S) = \sum_1^N r_j^2(S) .$$

Each r_j is zero when S is degenerate. If we take S to be of form (20), the r_j are functions of the matrix T. We shall denote these functions by $r_j(T)$; some may be identically zero. Since $r_j(S) = 0$ when S is degenerate, it follows from (21) that $r_j(T) = 0$ when $\det T = 0$. ∎

LEMMA 1   *A polynomial $r(T)$ that is zero when $\det T = 0$ is divisible by $\det T$ in the ring of polynomials.*

PROOF: We introduce, in the space of real symmetric matrices, a basis of positive definite matrices T_k. Any T can be expressed uniquely as a linear combination of T_k:

$$T = \sum t_k T_k .$$

Both $r(T)$ and $\det T$ are polynomials in t_k. Now fix all t_k but one, call it t_j, and regard $\det T$ as function of t_j alone. We claim that all zeros of $\det T$ as a function of the complex variable t_j are real. For, if $\det T = 0$, T annihilates some possibly complex vector v:

$$(23) \qquad Tv = 0 .$$

We can write T as $T = T^{(j)} + t_j T_j$, where

$$T^{(j)} = \sum_{k \neq j} t_k T_k \,.$$

Taking the hermitian scalar product of (23) with v gives

$$(Tv, v) = (T^{(j)}v, v) + t_j(T_j v, v) = 0 \,.$$

The first term is real and the factor $(T_j v, v)$ in the second term is real and positive; therefore, t_j is real.

We now consider the quotient $q(t) = r(t)/\det T$; since by assumption $r(t) = 0$ when $\det T = 0$, it follows that, as function of t_j, $q(t)$ is analytic in the whole complex t_j-plane provided that the zeros of $\det T$ are simple. A quotient of polynomials that has no pole is a polynomial in the variable t_j. Therefore, the partial derivative of q with respect to t_j of high enough order is zero. Since this holds for all variables t_j, only a finite number of mixed partial derivatives of q are nonzero. This proves that q is a polynomial in all variables.

To satisfy the condition that the zeros of $\det T$ in each variable t_j are simple, all we have to do is to perturb T by a small amount. In this way we can show that q is a polynomial in all variables in some open set; but then q is a polynomial for all values of its arguments. This completes the proof of the lemma.   ∎

Now we combine (21) and (22) and use the lemma:

$$d(T) = \frac{d(S)}{[\det T]^2} = \sum_1^N \frac{r_j^2(T)}{[\det T]^2} = \sum_1^N q_j^2(T)$$

represents $d(T)$ as the sum of no more than N squares of polynomials, as asserted in Theorem 3.

After completing this paper I came across Ilyushechkin's article [3], where he gives a beautiful, explicit representation of the discriminant as a sum of $n!$ squares.

Bibliography

[1] Artin, E., *Über die Zerlegung der definiten Functionen in Quadrate*, Abhandlungen aus dem mathematischen Seminar der Hamburgschen Universität 5, 1926, pp. 102–115.

[2] Hilbert, D., *Über die Darstellung definiter Formen als Summen von Quadraten*, Math. Annalen 32, 1888, pp. 342–350.

[3] Ilyushechkin, N. V., *The discriminant of the characteristic polynomial of a normal matrix*, Mat. Zametki, 51, 1992, pp. 16–23; English translation in Math. Notes 51, 1992, pp. 230–235.

[4] Lax, A., and Lax, P. D., *On Sums of Squares*, Linear Algebra Appl. 20, 1978, pp. 71–75.

PETER D. LAX
Courant Institute
251 Mercer Street
New York, NY 10012
E-mail: lax@cims.nyu.edu

Received November 1997.
Revised January 1998.

COMMENTARY ON PART IX

These papers on algebra, except [85], deal with the algebra of matrices, especially symmetric matrices. They all grew out of problems concerning solutions of hyperbolic partial differential equations.

Not included here is a paper that appeared in 1958, where I conjectured that every homogeneous hyperbolic polynomial p in three variables can be written as a determinant:

$$p(\tau, \xi, \eta) = \det(\tau I + \xi A + \eta B). \tag{1}$$

Hyperbolic means that for any real choice of ξ and η the roots τ of $p = 0$ are real. Here I is the identity matrix and A, B are real, symmetric matrices; the coefficient of τ^n in p is 1.

When the degree of p is 2, we can take $p = \tau^2 - \xi^2 - \eta^2$, which can be written as

$$\det \begin{pmatrix} \tau - \xi & \eta \\ \eta & \tau + \xi \end{pmatrix}.$$

For arbitrary n we can apply an orthogonal transformation that diagonalizes the matrix A. So the number of parameters on the right side of (1) is $n + n(n+1)/2$; this equals the number of parameters in p, normalized as above.

Recently, A.S. Lewis et al. succeeded in deducing this conjecture from an observation of Helton and Vinnikov, based on a deep result of Vinnikov.

References

[1] Helton, J.W.; Vinnikov, V. Linear matrix inequality representation of sets, *Technical Report*, Math. Dept., UCSD, 2003.

[2] Lax, P.D. Differential equations, difference equations, and matrix theory, *CPAM* **16** (1958),175–194.

[3] Lewis, A.S.; Parillo, P.A.; Raman, M.V. The Lax conjecture is true, Preprint, 2003.

[4] Vinnikov, V. Selfadjoint determinental representation of real plane curves, *Math. Ann.* **296** (1993), 453–479.

P.D. Lax

40

36

In our work on scattering in Euclidean space, Ralph Phillips and I were studying symmetric hyperbolic systems of the form

$$u_t = \sum_1^d A_j u_{x_j},$$

where each A_j is a real symmetric matrix, and where all signals are propagated to ∞. This requires that all nontrivial linear combination of the matrices A_j with real coefficients be invertible. It is natural to ask what is the maximum dimension d for which such a collection of symmetric matrices exists.

In the 1920s, Radon and Hurwitz, independently, constructed linear families of orthogonal matrices; the existence of such families has been related to the existence of continuous vector fields on spheres; this deep problem in topology was solved by J.F. Adams. Using Adams's result we were able to solve our problem, and we submitted a manuscript to the Proceedings. We received a detailed referee's report, which showed that the referee understood much more about the problem than we did; so we wrote to the editor offering to withdraw our paper in favor of the referee. The referee was J.F. Adams, and he proposed to make it a joint publication. He even extended the result to other fields of coefficients, including quaternions. This gave rise to a glitch, which he was able to correct readily.

In our correspondence we received a course in topology in exchange for an abbreviated tutorial in scattering theory.

P.D. Lax

85

Most examples of positive forms that cannot be written as a sum of squares are artificial. The one described here arose naturally.

I have always been fascinated by the problem of expressing nonegative functions as a sum of squares, see paper #143 below. This problem also comes up in proving the stability of difference schemes, see P. D. Lax, On the Stability of Difference Schemes for Hyperbolic Equations with Variable Coefficients, *CPAM* 14 (1961), 497–520.

P.D. Lax

99

Let M_n be the set of $n \times n$ real matrices. A matrix $A \in M_n$ is called *nondegenerate* if it has n distinct eigenvalues. Let $\mathcal{U}, \mathcal{H} \subset M_n$ be subspaces. Then $\mathcal{U}$ is called *nondegenerate* if

41

any nonzero matrix in $\mathcal{U}$ is nondegenerate. Otherwise, $\mathcal{U}$ is called *degenerate*. The subspace $\mathcal{H}$ is called *hyperbolic* if any $A \in \mathcal{H}$ has only real eigenvalues. The subspace $S_n \subset M_n$ of symmetric matrices is hyperbolic. In what follows we assume that $\mathcal{H}$ is hyperbolic with a basis $A_1, \ldots, A_m$. Then the system $\frac{\partial u}{\partial t} = \sum_{i=1}^{m} A_i \frac{\partial u}{\partial x_i}$ is called a first-order $n \times n$ hyperbolic system in m space variables. If $\mathcal{H}$ is nondegenerate, then the corresponding PDE system is strictly hyperbolic. It is easy to prove the existence of two-dimensional nondegenerate hyperbolic subspaces for any $n \geq 2$. (For example, A_1 is any irreducible tridiagonal matrix and A_2 a diagonal matrix with distinct diagonal entries.) Strictly hyperbolic systems have simple characteristics and have nice numerical schemes. In his seminal paper Lax showed that any hyperbolic subspace of M_n of dimension $m \geq 3$ is degenerate for $n \equiv 2 \pmod 4$.

Friedland, Robbin, and Sylvester [5] characterized the degenerate hyperbolic subspaces for any $n \geq 2$ as follows. Write $n = (2a + 1)2^{c+4d}$ where a, c, d are nonnegative integers with $c \in \{0, 1, 2, 3\}$. Let $\rho(n) = 2^c + 8d$ be the Radon–Hurwitz number. Let $\sigma(n) = 2$ for $n \not\equiv 0, \pm 1 \pmod 8$ and $\sigma(n) = \rho(4b)$ for $n = 8b, 8b \pm 1$. Then any hyperbolic subspace $\mathcal{H} \subset M_n$ of dimension $\sigma(n) + 1$ is degenerate. Moreover, there exists a nondegenerate hyperbolic subspace $\mathcal{H} \subset S_n$ of dimension $\sigma(n)$. Similar results are shown in [5] for real degenerate subspaces of $n \times n$ complex-valued matrices.

The main idea of the proof of these results is as follows. The n eigenspaces of a nondegenerate A induce a decomposition of $\mathbf{C}^n$ to a direct sum of n lines. If in addition, A has real eigenvalues, then one has the corresponding decomposition of $\mathbf{R}^n$. Let $S^k \subset \mathbf{R}^{k+1}$ be the standard unit sphere. Then one has the continuous map $\Phi : S^{m-1} \to M_n$ given by $\mathbf{x} = (x_1, \ldots, x_m) \mapsto \sum_{i=1}^{m} x_i A_i$. Assume that $\mathcal{H}$ is nondegenerate. Then one has a trivial $\mathbf{R}^n$ bundle on S^{m-1} that is a sum of n line bundles. Using the identity $\Phi(-\mathbf{x}) = -\Phi(\mathbf{x})$ one shows that the existence of p copies of the canonical line bundle on the real projective space of dimension $m - 1$, whose Whitney sum is trivial, for some $p \in \{\lfloor \frac{n}{2} \rfloor, \lceil \frac{n}{2} \rceil\}$. I.e., there exists a continuous odd map $\Psi : S^{m-1} \to \mathrm{GL}(p, \mathbf{R})$, $\Psi(-\mathbf{x}) = -\Psi(\mathbf{x})$. The fundamental theorem of Adams [1] yields the results in [5].

Let $A \in \mathcal{H}$ and arrange the eigenvalues of A in decreasing order $\lambda_1(A) \geq \cdots \geq \lambda_n(A)$. If A is degenerate, then $\lambda_i(A) = \lambda_{i+1}(A)$ for some i, and we say that $\lambda_i(A)$ is a multiple eigenvalue. Assume that $\mathcal{H}$ is hyperbolic of dimension $\sigma(n) + 1$ at least. Then $\mathcal{H}$ contains a nonzero A with a multiple $\lambda_i(A)$. Which i should one expect? From the proof of the Corollary in §5 of [5] it follows that for an even n any $(\rho(n) + 1)$ -dimensional hyperbolic subspace $\mathcal{H} \subset M_n$ contains a nonzero matrix A such that $\lambda_{\frac{n}{2}}(A) = \lambda_{\frac{n}{2}+1}(A)$. If n is even and $\mathcal{H} \subset S_n$, then there exists $0 \neq A \in \mathcal{H}$ with $\lambda_{\frac{n}{2}}(A) = \lambda_{\frac{n}{2}+1}(A)$ if $\dim \mathcal{H} \geq \rho(\frac{n}{2}) + 2$. The proof of this result uses a nonlinear generalization of a theorem of Adams–Lax–Philips [2]. Recall that for $n > 2$ any n-dimensional subspace of S_n contains a nonzero matrix with a multiple first eigenvalue $(i = 1)$, and this result is sharp [4].

We conclude with another generalization of Lax's theorem due to Falikman, Friedland, and Loewy [3]. Let $q \in [2, n]$ be an integer and $n \equiv \pm q \pmod{2^{\lceil \log_2 2q \rceil}}$. Then any $\binom{q+1}{2}$-dimensional subspace of S_n contains a nonzero matrix A with $\lambda_i(A) = \cdots = \lambda_{i+q-1}(A)$ for some $i \in [1, n - q + 1]$, and this result is sharp. The case $q = 2$ is Lax's theorem.

42

References

[1] Adams, J.F. Vector fields on spheres, *Annals Math.* 75 (1962), 603–632.

[2] Adams, J.F.; Lax, P.D.; Philips, R.S. On matrices whose real linear combinations are nonsingular, *Proc. Amer. Math. Soc.* 16 (1965), 318–322; Correction *Proc. Amer. Math. Soc.* 17 (1966), 945–947.

[3] Falikman, D.; Friedland, S.; Loewy, R. On spaces of matrices containing a nonzero matrix of bounded rank, *Pacific J. Math.* 207 (2002), 157–176.

[4] Friedland, S.; Loewy, R. Subspaces of symmetric matrices containing matrices with a multiple first eigenvalue, *Pacific J. Math.* 62 (1976), 389–399.

[5] Friedland, S.; Robbin, J.; Sylvester, J. On the crossing rule, *Comm. Pure Appl. Math.* 37 (1984), 19–37.

S. Friedland

[143]

The year before the above article appeared, Wiley published PDL's treasure-trove graduate-level textbook *Linear Algebra*. He chose a cover that illustrates the intriguing and little-known phenomenon (outside the physics community) called avoidance of crossing. If A and B belong to Sym (n), the space of $n$by$n$ real symmetric matrices, then in general, for *all* real t the matrix $A + tB$ will have n distinct eigenvalues $\lambda_1(t) < \lambda_2(t) < \cdots < \lambda_n(t)$. In other words, the subset, or variety, $\mathcal{D} \subset$ Sym (n) whose elements have at least one multiple eigenvalue is missed by almost all lines in Sym (n). Physicists call such matrices degenerate. On the other hand, well-known matrices such as O, I, and the discretized Laplacian on a square or cube with suitable boundary conditions all belong to $\mathcal{D}$.

During the development of quantum mechanics, Wigner and von Neumann explained the phenomenon ($\mathcal{D}$ has codimension 2) and clearly, this strange set is described by the equation $d(S) = 0, S \in$ Sym (n). For $n > 1, d(S) := \prod_{i<j}(\lambda_i - \lambda_j)^2$ is the discriminant of S and is a polynomial in the entries that is homogeneous of degree $n(n-1)$ and is manifestly nonnegative. Fo $n > 2, d(S)$ is not a quadratic form, and following an observation of Hilbert, there is a nasty possibility that $d(S)$ might not be expressible as a sum of squares of polynomials of degree $(n\,2) := n(n-1)/2$.

The main purpose of the paper is to allay our fears. The proof is elegant (of course) and indirect and has three parts. First PDL characterizes $d(S)$, to within a constant multiple, as the only nonnegative homogeneous polynomial that vanishes only on $\mathcal{D}$ and is invariant under the special orthogonal group, $S \to QSQ^T$, $\det(Q) = 1$. Second, he constructs one polynomial $r(S)$ of the correct degree that vanishes on $\mathcal{D}$ by showing that $S \in \mathcal{D}$ if and only if

43

S commutes with a nonzero skew-symmetric matrix K, $K + K^T = O$. Now $KS - SK = O$ is a linear homogeneous system of $(n + 12)$ equations in $(n\,2)$ unknowns (the entries of K). The square subsystem corresponding to the off-diagonal terms in the commutator may be written in standard form as $Mk = O$, $k = \text{vec}\,(K)$, and each nonzero entry in M is either a diagonal difference $S(i, i) - S(j, j)$ or a multiple of an off-diagonal entry. Thus $r(S); = \det(M)$ is the desired polynomial. Third, PDL exhibits $d(S)$ as an integral,

$$d(S) = \int r^2(QSQ^T)dQ,$$

over the special orthogonal group. The integral may be turned into a sum because the space of (homogeneous) polynomials of degree $(n\,2)$ in $(n + 12)$ variables has finite dimension, but one that grows quickly with n.

The rest of the paper consists of a few observations on the number of terms in that sum. As the paper went to press, PDL learned that six years earlier, in 1992, the Russian mathematician Ilyushechkin had expressed $d(S)$, $S \in \text{Sym}\,(n)$, as a sum of at most $((n + 12)n)$ terms.

Perhaps the most important fact is that for $n > 2$, only one term includes all $(n\,2)$ off-diagonal entries and all $(n\,2)$ diagonal differences in S.

For $n = 2$, $d(S)$ is a quadratic form and is the sum of two squares. For $n = 3$, $\mathcal{D}$ is composed of exactly ten cubic subvarieties. Bernd Sturmfels has extended the results of Ilyushechkin and Lax to the context of algebraic geometry. See [S]. Beresford Parlett exhibited several square matrices depending on S whose determinants equal $d(S)$ in [P].

Surprise, surprise. One group of physicists in Germany and one group on image compression in California have expressed interest in detecting matrices in $\mathcal{D}$ for low values of n when S is a function of a few parameters.

B. Parlett

References

[P] Parlett, B. The matrix discriminant as a determinant, *J. Lin. Alg. Appls.* vol. 355 (2002) 85–101.

[S] Sturmfels, B. *Solving Systems of Polynomial Equations*, CBMS Lecture Notes 97, AMS, Providence, RI, 2002, Section 7.5.

44